1st International Conference on Environmental Sciences (ICES2018)

Disaster Mitigation, Environmental and Sustainable Development

IOP Conference Series: Earth and Environmental Science
Volume 314

Padang, Indonesia
15 - 16 November 2018

Editors:

Rusnardi Rahmat Putra
Hamdi
Nasfryzal Carlo
Abdul Razak
Siti Fatimah
Iswandi Umar
Heldi
Dedi Hermon
Yudi Antomi
Ramli
Alizar
Erianjoni
Yohandri
Skunda Diliarosta

ISBN: 978-1-5108-9343-6
ISSN: 1755-1307

Printed from e-media with permission by:

Curran Associates, Inc.
57 Morehouse Lane
Red Hook, NY 12571

CURRAN ASSOCIATES INC. proceedings.com

Some format issues inherent in the e-media version may also appear in this print version.

Institute of Physics
Dirac House, Temple Back
Bristol BS1 6BE UK

Phone: 44 1 17 929 7481
Fax: 44 1 17 920 0979

techtracking@iop.org

Additional copies of this publication are available from:

Curran Associates, Inc.
57 Morehouse Lane
Red Hook, NY 12571 USA
Phone: 845-758-0400
Fax: 845-758-2633
Email: curran@proceedings.com
Web: www.proceedings.com

TABLE OF CONTENTS

ENVIRONMENTAL CHEMISTRY PHYSICS

ENVIRONMENTAL MAPPING TECHNOLOGY

CROSS-CUTTING ENVIRONMENTAL ISSUES

PREFACE

Welcome to the 1st International Conference on Environmental Sciences (ICES), a joint effort between the Postgraduate Program in Environmental Sciences of Universitas Negeri Padang, the Indonesian Centre for Environmental Studies Cooperation Agency (BKPSL) and the Indonesian Environmental Sciences Study Program Association (PEPSILI). This international conference is organized by the Postgraduate Program in Environmental Sciences of Universitas Negeri Padang which aims to accommodate the use of innovations and trends in the fields of environment, science, education and technology to overcome global challenges.

The 1st ICES was held on 15-16 November 2018 in the city of Padang, West Sumatera, Indonesia with the theme "Disaster Mitigation, Environmental and Sustainable Development" and sub-themes: (1) Physical Environmental Chemistry, (2) Education, Socio-cultural Economy, Local Wisdom, and Ecotourism, (3) Environmental Mapping Technology, (4) Cross-Environmental Problems. It is an honor for us to have more than one hundred national and international experts, practitioners and observers to explain the results of their research and discuss it through this conference and also to accommodate the collaboration among researchers.

Through a strict peer-review process by the board across disciplines, over 100 selected manuscripts had been presented during the conference from authors and were also qualified to be published into the present conference proceeding.

We would like to express our highest gratitude to eight keynote speakers in the conference: Dr. Ir. Siti Nurbaya Bakar, M.Sc. (Minister of Environment and Forestry), Simone Maynard, PhD (Lead of the Ecosystem Services Thematic Group for IUCN's Commision on Ecosystem Management), Dr. Indrajit PAL (Assistant Professor & Chair Disaster Preparedness, Mitigation and Management Asian Institute of Technology, Thailand), Dr. Jose M. Regunay (University of the Philippines, Diliman, Quezon City), Prof. Dr. Syamsul Maarif (Universitas Pertahanan), Agus Rahardjo. ST. M.Sc. Mgt. (Chairman of Corruption Eradication Commission/KPK), Prof. Dr. Eri Barlian, M.S. (Universitas Negeri Padang) and Dr. Indang Dewata, M.Si (Chairman of BKPSL) for giving some insights and valuable information from their disciplines.

We would like to thank the organizing committee, the member of reviewer and the editors for the kind assistance, precious time and patience to read and revise the manuscripts in this proceeding, as well as to IOP Publishing for their helpful service in publishing the output of this conference.

Thank you very much and we are looking forward for your next participation on next ICES.

General Chairman of ICES2018

Prof. Dr. Eri Barlian, M.S.

IOP Conf. Series: Earth and Environmental Science **314** (2019) 011002 doi:10.1088/1755-1315/314/1/011002

EDITORS

Rusnardi Rahmat Putra, S.T., M.T., Ph.D.En.
Department of Civil Engineering, Faculty of Engineering,
Universitas Negeri Padang,
Jl. Prof. Dr. Hamka, Air Tawar, Padang 25231, Indonesia
rusnardi.rahmat@gmail.com

Dr. Hamdi, M.Si
Department of Physics, Faculty of Mathematics and Natural Sciences,
Universitas Negeri Padang,
Jl. Prof. Dr. Hamka, Air Tawar, Padang 25231, Indonesia
rifai.hamdi@gmail.com

Prof. Dr. Nasfryzal Carlo, M.Sc.
Faculty of Civil Engineering and Planning
Universitas Bung Hatta
Jl. Sumatera Ulakkarang Padang, 25133, Indonesia
carlo@bunghatta.ac.id

Dr. Abdul Razak, M.Si.
Department of Environmental Science, Post Graduate Program,
Universitas Negeri Padang,
Jl. Prof. Dr. Hamka, Air Tawar, Padang 25231, Indonesia
ar710322@gmail.com

Dr. Siti Fatimah, M.Pd, M.Hum
Department of Environmental Science, Post Graduate Program,
Universitas Negeri Padang,
Jl. Prof. Dr. Hamka, Air Tawar, Padang 25231, Indonesia
siti_fatimahunp@yahoo.com

Dr. Iswandi Umar, M.Si
Department of Geography, Faculty of Social Science
Universitas Negeri Padang,
Jl. Prof. Dr. Hamka, Air Tawar, Padang 25231, Indonesia
iswandi_u@yahoo.com

Ir. Heldi, M.Si., Ph.D
Department of Environmental Science, Post Graduate Program,
Universitas Negeri Padang,
Jl. Prof. Dr. Hamka, Air Tawar, Padang 25231, Indonesia
heldiensten@yahoo.co.id

Dr. Dedi Hermon, M.P.
Department of Environmental Science, Post Graduate Program,
Universitas Negeri Padang,
Jl. Prof. Dr. Hamka, Air Tawar, Padang 25231, Indonesia
dihermon006@gmail.com

IOP Conf. Series: Earth and Environmental Science **314** (2019) 011002 doi:10.1088/1755-1315/314/1/011002

Dr. Yudi Antomi, M.Si.
Department of Geography, Faculty of Social Science
Universitas Negeri Padang,
Jl. Prof. Dr. Hamka, Air Tawar, Padang 25231, Indonesia
tmy_bima@yahoo.com

Dr. Ramli
Department of Physics, Faculty of Mathematics and Natural Sciences,
Universitas Negeri Padang,
Jl. Prof. Dr. Hamka, Air Tawar, Padang 25231, Indonesia
ramli@fmipa.unp.ac.id

Alizar, Ph.D
Kampus FMIPA UNP, Universitas Negeri Padang,
Jl. Prof. Dr. Hamka, Air Tawar, Padang, West Sumatera, Indonesia
alizarulianas@yahoo.com

Dr. Erianjoni, M.Si.
Department of Sociology, Faculty of Social Science
Universitas Negeri Padang,
Jl. Prof. Dr. Hamka, Air Tawar, Padang 25231, Indonesia
erian_joni@yahoo.com

Yohandri, Ph.D
Department of Physics, Faculty of Mathematics and Natural Sciences,
Universitas Negeri Padang,
Jl. Prof. Dr. Hamka, Air Tawar, Padang 25231, Indonesia
yohandri@fmipa.unp.ac.id

Dr. Skunda Diliarosta, M.Pd.
Department of Environmental Science, Post Graduate Program,
Universitas Negeri Padang,
Jl. Prof. Dr. Hamka, Air Tawar, Padang 25231, Indonesia
skunda2isberd@gmail.com

ORGANIZING COMMITTEE

Steering Committees
Prof. Ganefri Ph.D., Universitas Negeri Padang, Indonesia
Prof. Dr. Yunia Wardi, M.Si., Universitas Negeri Padang, Indonesia
Syahril, ST, M.T, Ph.D., Universitas Negeri Padang, Indonesia
Dr. Dedi Hermon, M.P., Universitas Negeri Padang, Indonesia
Prof. Dr. Rusdinal, M.Pd., Universitas Negeri Padang, Indonesia
Prof. Yenni Rozimela, M.Ed., Ph.D., Universitas Negeri Padang, Indonesia
Prof. Dr. Atmazaki, M.Pd., Universitas Negeri Padang, Indonesia
Dr. Khairudin, M.Kes.AIFO. Universitas Negeri Padang, Indonesia

General Chair
Prof. Dr. Eri Barlian, M.S., Universitas Negeri Padang, Indonesia

Co-Chair
Dr. Indang Dewata, M.Si.
Zainal Abdul Haris, M.Si.

Secretary
Mira Gusniwati, A.Md.
Alizar, M.Sc., Ph.D.
Agus Teguh Prihartono, SP., M.Si.

Treasurer
Hesti Palupi, M.Kom.
Novita Eka Dewi, S.E.
Linda Handayuni, S.K.M., M.Si.

Secretariat
Serly Mutia Sari, S.T.
Sanny Edinov, M.Si
Osronita, M.Pd.
Hestilia Anggraini, S.Hut
Syafril, A.Md.
Drs. Bambang Supriadi
Rudi Mulya, M.Kom.
Diki Atmarizon, M.Pd

Keynote Speaker Coordinator

Prof. Ir. Yonariza, M.Sc., Ph.D.

Journal Coordinator
Dr. Ramli

IT Coordinator
Bayu Ramadhani Fajri, M.Ds.

Seminar Paralel Coordinator
Dr. Nurhasan Syah, M.Pd.

Seminar Parallel Team
Dr. Zul Amri, M.Ed.
Dr. Azwir Anhar, M.Si.
Yohandri, M.Si.
Dr. Siti Fatimah, M.Si.
Dr. Erianjoni, M.Si.
Dr. Rahadian Z., M.Si.

Reviewer Coordinator
Dr. Abdul Razak, M.Si.

Event
Alexander Syam, M.Pd.
Ansosry, M.T.
Haris Satria

Guest and Accommodation
Drs. H. Dasrizal, M.P.
Aidil Onasis, M.Kes.
Nella, S.Pd.
Arum Pramesti, S.Pd.
Olivia Oktorie, M.Pd.

Equipment and Place
Drs. Helfia Edial, M.P.
Dedy Kurniawan, M.Kom.
Didi Diandi

Documentation, Finance and Transportation

Drs. Ikhwan, M.Si.
Yunhendri Danhas, M.Si.
Ir. Haryani, M.T.
Mulya Gusman, M.T.
Admizal Nazki, M.Si.
Auwilla Putri, S.T., M.Si.
Faizal Muchtar
Hendri Kurnia

Sustenance
Berta Dewina, S.Hut.
Rahmadhani, S.Pd.
Yusfiandrita, S.Hut.
Elfitri Oktavia, S.T.
Aulia Rinadi, S.Si.
Evi Handayani, S.I.P.

Public Relations
Nofri Elfida, M.Pd.
Bakri
Agusmardi

Conference Photographs

The 1st International Conference on Environmental Sciences (ICES2018)

November 15-16, 2018, Universitas Negeri Padang, Padang, Indonesia

Chairman of ICES 2018, Prof. Dr. Eri Barlian, M.S., delivered the opening speech on November 15th 2019 at Auditorium of Universitas Negeri Padang

Participant of ICES 2018 at Auditorium of Universitas Negeri Padang on November 15th 2018

IOP Conf. Series: Earth and Environmental Science **314** (2019) 011004 doi:10.1088/1755-1315/314/1/011004

Group Photo with Keynote Speaker of ICES

Welcome Party for Participant of ICES 2018, on Wednesday November 14th 2018

IOP Conf. Series: Earth and Environmental Science **314** (2019) 011004 doi:10.1088/1755-1315/314/1/011004

Partisipant of ICES 2018, Room PPS01114

Partisipant of ICES 2018, Room PPS2103

Partisipant of ICES 2018, Room PPS2102

Partisipant of ICES 2018, Room PPS2101

Partisipant of ICES 2018, Room PPS01113

Partisipant of ICES 2018, Room 2104

Participant of ICES, Room PPS01106

One of the participants was explaining the results of his research, in the present of keynote speakers; Simone Maynard, PhD (Lead of the Ecosystem Services Thematic Group for IUCN's Commision on Ecosystem Management) and Dr. Indrajit PAL (Assistant Professor & Chair Disaster Preparedness, Mitigation and Management Asian Institute of Technology, Thailand)

Peer review statement

All papers published in this volume of *IOP Conference Series: Earth and Environmental Science* have been peer reviewed through processes administered by the proceedings Editors. Reviews were conducted by expert referees to the professional and scientific standards expected of a proceedings journal published by IOP Publishing.

Strategy 3M plus to reduce incidence disease dengue haemorrhagic fever in Public Health Centre (PHC) *Tigo Baleh Bukittinggi* West Sumatra-Indonesia

S Alhamda* and E Barlian

Department of Environmental Science, Postgraduate Program, Universitas Negeri Padang, Prof. Dr. Hamka Street, West Air Tawar, Padang - West Sumatra 25131 Indonesia

*syukra909604@gmail.com

Abstract. Disease of Dengue Haemorrhagic Fever (DHF)) is contagion which because of virus dengue and contagious through larva mosquito of Aedes aegypti. This disease can attack everybody and death. In Bukittinggi represent non-area of endemic DHF, but almost every year the progress of occurrence DHF, District which at most case of DHF is Aur Birugo Tigo Baleh (ABTB), besides wide of smallest area also there is environmental factor which related case of DHF. This Research aim to know related of existence larva mosquitoes Aedes aegypti and Implementation of 3M Plus with improvement incidence disease Dengue Haemorrhagic Fever in PHC Tigo Baleh Bukittinggi. Research method is analytic survey with study design of case control use test of chi square. Research location in District of ABTB, its research responder is 78 with technique of totally sampling. The research show existence snap fingers at house snapping fingers at and suffering DHF 61,5%, do not snap fingers at and suffer DHF 38,5%. Value of House Index (HI) 43,6%, Container Index (CI) 17,9%. Implementation of 3M Plus unfavorable and suffer DHF 69,2%, while implementation of 3M Plus good and suffer DHF 30,8%. From result of research p_{value}=0,003 ($p<0,05$), there are relation existence larva mosquitoes Aedes aegypti with Improvement incidence disease Dengue Haemorrhagic Fever and p_{value}=0,040 ($p<0,05$), there are relation of 3M Plus with Improvement incidence disease Dengue Haemorrhagic Fever. Suggested, Public Health Service in Bukittinggi and PHC Tigo Baleh conduct socialization and enableness of society to join in to conduct eradication of mosquito Aedes aegypti regularly so that can degrade number incidence of disease DHF.

1. Background

The progress of national development for decades in Indonesia has had an impact on the health sector despite the recent economic crisis. The future of Indonesian society to be achieved through health development is the society, nation and state characterized by its people living in a healthy environment, have the ability to maintain, and have the highest health level throughout the territory of the Republic of Indonesia [1].

The optimal level of health for the community, health efforts are carried out with maintenance approach, health promotion (promotive), preventive prevention, curative disease, health restoration (rehabilitative), and traditional health service efforts are implemented thoroughly, integrated and Sustainable success [2].

To achieve optimal public health status, the Eradication Program focuses on preventing disease, reducing morbidity and mortality and reducing the adverse effects of infectious and non-communicable diseases. Infectious diseases are still a priority issue in the development of public health in Indonesia. In the list of Infectious Disease Surveillance, (SK Menkes No.145710 October 2003) a number of infectious diseases are listed as issues that must be a priority by the region. The problem of infectious diseases is still apprehensive, some types of diseases even show a tendency to increase and not successfully overcome such as pulmonary, malaria, and dengue haemorrhagic fever [3].

Dengue Haemorrhagic Fever (DHF) is an infectious disease caused by dengue virus and is transmitted through the bite of Aedes aegypti mosquito. This disease can attack all people and can lead to death[3]. Generally, affects children under 15 years of age, but can also attack adults, and often cause extraordinary events (outbreaks) or outbreaks [4].

In Indonesia, DHF emerges as an outbreak for the first time found in Surabaya in 1968. Data from around the world shows Asia ranks first in the number of DHF patients each year. Meanwhile, from 1968 to 2009, the World Health Organization (WHO) noted Indonesia as the country with the highest dengue fever case in Southeast Asia and the second highest in the world after Thailand [5].

In the last 5 years the number of cases and infected areas has been increasing and widespread. It is estimated that every year there are 300 million cases in Indonesia, 500,000 cases of dengue that require hospitalization and at least 12,000 of them die, especially children [5].

DHF prevention and eradication program has lasted for approximately 43 years and succeeded in reducing mortality from 41.3% to 0.87% in 2010, but has not succeeded in reducing morbidity. The number tends to increase, spread more broadly, attacking not only children but also older age group[3]. In 2011 through August 24,362 cases were recorded with 196 deaths [6].

In 2012 the number of DHF patients in Indonesia reported as many as 90,245 cases with the number of deaths 816 people. In line with the increase in number / morbidity, the number of districts / cities affected by dengue fever in 2012 also increased from 374 (75.25%) to 417 districts / cities (83.9%) [6].

Ministry of Health mentions in 2013 Indonesia is still a hotbed of dengue fever or already endemic. Until the middle of this year, dengue fever cases occurred in 31 provinces with 48,905 people, 376 of whom died. The number of dengue fever patients in the first half of this year shows an increase compared to last year [7].

To overcome dengue problems in Indonesia since 2004 the Ministry of Health has worked closely with Provincial Health Office and District Health Offices to implement a national program of dengue haemorragic fever prevention. The program includes epidemiological surveillance / early alert system and outbreak response, counseling, vector eradication for adult mosquitoes by spraying, focusing and periodic larva examination, larvalidation and vector surveys. In addition, cross-program cooperation through Pokjanal DHF and 3M movement period, treatment / case management including doctor training and procurement of suggestions for buffer stock [3].

Based on West Sumatera Provincial Health Profile in 2012, there are several regencies / cities in West Sumatera that belong to dengue endemic areas of Padang City with 1,626 cases per 100,000 population (868 cases and 758 cases), Pesisir Selatan Regency Number of cases 634 per 100,000 population (Male 305 cases and Women 329 cases), Kota Bukittinggi with number of cases 107 per 100,000 population (Male 54 cases and Women 53 cases).

In Bukittinggi, during 2010-2013 has been recorded cases of dengue from year to year has increased. In 2010 dengue cases 83 cases, decreased in 2011 to 68 cases, and in 2012 increased again 107 cases, so that in 2013 has not decreased 142 cases, reported from each PHC found in Bukittinggi. PHC that have the highest cases of DHF by 39 cases are PHC Tigo Baleh. In this case, the geographical condition of Bukittinggi, which consists of 3 subdistricts, among others, is the smallest subdistrict of Aur Birugo Tigo Baleh which has an area of 6.25 km^2, consisting of 8 Kelurahan which is the work area of PHC Tigo Baleh. During the year 2013 cases of dengue fever recorded at PHC Tigo Baleh are spread in 8 *Kelurahan*.

PHC Tigo Baleh supported by the local government made various efforts in overcoming the emergence of cases of dengue fever by doing a variety of counseling about the dangers, prevention and control of mosquito Aedes aegypti through the program of Mosquito Nest. The most important effort, easy and cheap is emphasized to the community is doing the mosquito nest in a physical way that the environment management with 3M plus. Regular larva monitoring activities are also routinely conducted through trained jumantik cadres, as well as selective abuse activities ie abate powder in schools, public places and houses with positive larva [8].

Until now, efforts to eradicate the DHF vector that has been done and It has not shown optimal results so that cases of DHF are still high and increasing every year in the work area of PHC Tigo Baleh. The results of interviews and observations in the community show that they rarely do mutual cooperation to clean up the environment, there are still visible places where used pieces of cans into puddles and become a hotbed of mosquitoes.

The most appropriate way to eradicate Aedes aegypti mosquitoes is to eradicate the mosquito larva in the breeding place. Eradication of mosquito nest DHF can be done with 3M method and abatesasi technique. The 3M program consists of: draining the bathtub once a week, closing the water reservoir both inside and outside the house and bury secondhand items that can accommodate the rain water and allow the water inundated inside [9].

From the results of Sulina [9] research in Medan city 2012 obtained the result that there is a relationship presence of larva Aedes aegypti and implementation of 3M Plus with the incidence of DHF.

2. Methods

Subjects in this study were all people who tested positive for dengue and listed as a patient in PHC Tigo Baleh, in the period January to December 2016 and all those who declared free of dengue residing in PHC Tigo Baleh Bukittinggi and do not live with case as many as 39 peoples are free of dengue. The research method used design of *case control and analytic approach*, data collection as well at a time when the independent variables and the dependent variable in carefully while at the same time. In this case we have seen that the existence of Aedes aegypti mosquito larvae and implementation of 3M plus as the independent variables and the incidence of DHF as the dependent variable in PHC Tigo Baleh Bukittinggi.

3. Result and Discussion

3.1 Univariate Analysis

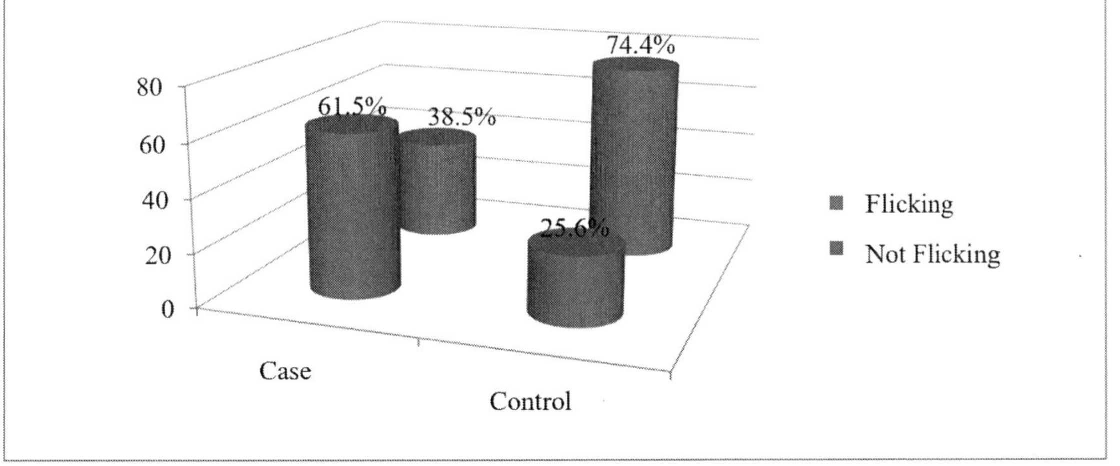

Figure 1. Frequency Distribution of Aedes aegypti Mosquito larva at PHC Tigo Baleh

Based on figure 1 can be seen that the presence of larvae that flicked and suffered dengue is 24 respondents (61.5%), and the presence of larvae that do not flick and suffer from DHF is 15 respondents (38.5%).

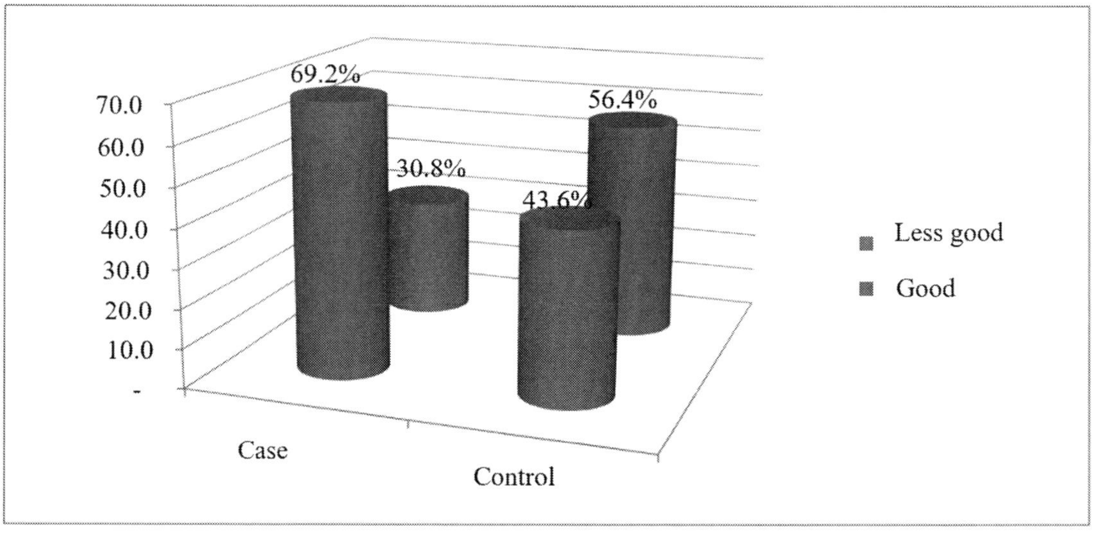

Figure 2. 3M Plus Implementation Frequency Distribution at PHC Tigo Baleh

Based on figure 2 can be seen that the implementation of 3M plus less good and suffering from DHF are 27 respondents (69.2%) and the implementation of 3M plus good and suffering from DHF is 12 respondents (30.8%).

3.2 Bivariate Analysis

Bivariate analysis was conducted to see the relationship between independent variables i.e. the presence of mosquito larvae and 3M implementation plus the dependent variable is as follows:

Table 1. Relationship of Existence of Aedes aegypti Mosquito larva with Increased Dengue Disease incidence at PHC Tigo Baleh

The presence of larvae	DHF				*P* value	Odds Ratio (OR)	CI 95%
	Cases		Control				
	f	%	*f*	%			
Flicking	24	61,5	10	25,6	0,003	4,640	1,766 – 12,189
No Flicking	15	38,5	29	74,4			
Amount	39	100	39	100			

From Table 1 it can be seen that the presence of larvae that flick and suffer dengue is 24 respondents (61.5%), and the presence of larvae that do not flick and suffer from DHF that is 15 respondents (38.5%). Larval larvae and 10 dengue fever (25.6%), and larvae with no larvae and no dengue fever which was 29 respondents (74.4%). From the analysis of the presence relationship of Aedes aegypti mosquito larva with chi-square test obtained p = 0,003 (p <0,05), meaning Ha accepted that there is correlation between existence of larva of Aedes aegypti mosquito with increasing incidence of dengue disease.

Table 2. 3M Plus Implementation Relationship with Increased Dengue Haemorrhagic Fever at PHC Tigo Baleh

Implementation of 3M Plus	DHF				*P* value	Odds Ratio (OR)	CI 95%
	Cases		Control				
	f	%	*f*	%			
Not Good	27	69,2	17	43,6	0,040	2,912	1,150 – 7,272
Good	12	30,8	22	56,4			
Amount	39	100	39	100			

From Table 2 it can be seen that 3M plus less good and suffering dengue is 27 respondents (69.2%) and 3M plus good implementation and suffering from DHF that is 12 respondents (30,8%), while implementation of 3M plus less good and not Suffering from DHF that is 17 respondents (43,6%), 3M plus good implementation and not suffering from DHF that is 22 respondents (56,4%). From the analysis result of 3M implementation relationship plus with chi-square test obtained p value = 0,040 (p <0,05), meaning Ha accepted that there is relation between 3M implementation plus with increasing incidence of DHF.

Table 3. Density Figure, House Index, and Container Index

DF	HI	CI
1	1-3	1-2
2	4-7	3-5
3	8-17	6-9
4	18-28	10-14
5	29-37	15-20
6	38-49	21-37
7	50-59	28-31
8	60-76	32-40
9	77+	41+

Source: WHO

From the results of this research obtained HI value of 43.6% and CI of 17.9% which we compare with the Grouping of Density Figure (DF) obtained from the combination of Hi and CI values on a scale of 1-9 with the provisions of DF 1 = low density, DF 2 = medium density and DF 3 = high density as shown in the table above, it can be conclusion that HI is in category DF 6 = high density and CI 5 = medium density. This is supported by the incident rate indicating as shown in the table below.

Table 4. The Number of Incidence Rate of DHF

PHC	Cases of DHF	Death	CFR
PHC Tigo Baleh	39	1	3.1%

At this public health centre had been seen incident rate of 39 with 1 case of death with number CFR 3.1%. this indicates that the incidence rate of DHF in this PHC areas needs further assessment.

Based on the research that has been done can be seen that the respondent cases that flicked 24 (61.5%) while the control respondent who flicked 10 (25.6%). So the value of pvalue obtained by 0,003 (p <0,05), mean existence of Aedes aegypti mosquito larvae have correlation with increasing incidence of dengue disease and Odds Ratio (OR) = 4,64 mean home that found mosquito larva at risk 4.64 times suffer disease DHF compared to homes that are not found mosquito larvae. Based on the observations it can be concluded that Bukittinggi is predicted to be not a dengue endemic area because geographically the altitude is more than 900 m above sea level, where at a height the Aedes aegypti mosquito cannot breed. However, in reality the District of ABTB environment field is still found larva existence therefore the District of ABTB is not classified as Free Number of larvae so that the positive house of larva becomes the driving factor of increasing the incidence of DHF in Bukittinggi.

This shows that there are still people in the District of ABTB who have not taken action in the form of prevention of DHF, which is to prevent dengue fever, its transmitting mosquito must be eradicated, because the vaccine to prevent it does not yet exist. The right way to eradicate Aedes aegypti mosquitoes is to eradicate larva in breeding grounds [1].

Based on the research that has been done by Sulina [9] in line with the results of research by researchers using Exact Fisher test obtained p value of 0.002 (p value <0.005) which means that the presence of larvae has a significant relationship with the occurrence of dengue disease in Environment XVIII Kelurahan Binjai.

According to the researcher's assumption that the relation of presence of Aedes aegypti mosquito larvae with increasing incidence of DHF, due to dengue fever vector mosquito disease, and more and more is found existence of mosquito larva have big chance for dengue disease in environment District of ABTB.

Based on the research that has been done with the result shows the respondents of the less good case of 3M Plus implementation is 27 (69,2%), while the control respondent is not good implementation of 3M Plus equal to 17 (43,6%) so that p value of 0,04 (p <0.05), it means that 3M Plus implementation is related to increasing incidence of DHF and Odds Ratio (OR) = 2,912 means that people who have poor implementation 3M Plus will be at risk 2,912 times suffering from DHF compared to people who are good implementation of 3M Plus.

To prevent dengue disease the right way to combat Aedes aegypti mosquitoes are: 1) with Insecticide, 2) without insecticide is the most important way in vector control is environmental management with a view to prevent or reduce the development of vector and human contact, vector, pathogen. Eradication of larva of Aedes aegypti known as chemistry, biology and physical / 3M Plus Movement, each family must carry out regular at least once [1].

The results of this study are in line with the results of research that has been done by Sulina [9] which states there is a relationship of 3M Plus implementation with the incidence of dengue disease in XVIII Neighbourhood Kelurahan Binjai, which the analysis results using Exact Fisher test obtained p value of 0.047 (p value <0.005).

According to the researcher, most of 3M plus implementation included in good category, but still there are 27 respondents which still in less good category. This could have an impact on the increasing number of water reservoirs that could potentially become a breeding ground so as to increase the occurrence of dengue cases. The implementation of 3M Plus is still less good indicates that there is still a lack of public awareness of the importance.

Maintain cleanliness of the house and the surrounding environment to prevent the occurrence of DHF. Therefore, it is necessary to increase the motivation to make the community more effective in implementing 3M Plus through providing mutual assistance to clean up the environment given by local government officials, such as the Head of RT and the extension of the local PHC. In addition to increased motivation, coaching also needs to be done on community groups, such as youth, adolescent and youth mosques in order to increase preventive measures in the form of regular implementation of 3M Plus.

4. Conclusion

Based on the results of research conducted at PHC Tigo Baleh against 39 respondents' cases and 39 respondents' control so that amounted to 78 respondents, it can be concluded:

1. The presence of mosquito larva is more common in cases compared to control PHC Tigo Baleh.
2. Implementation of 3M Plus is less good in cases than controls at PHC Tigo Baleh.
3. The incidence of DHF in cases comparable to controls at PHC Tigo Baleh.
4. There is a relationship presence of Aedes aegypti mosquito larva with the incidence of dengue disease at PHC Tigo Baleh.

5. There is a relationship of 3M Plus implementation with the incidence of dengue disease at PHC Tigo Baleh.

After doing this research, it is suggested that the community should do more efforts to prevent the occurrence of DHF through mosquito nets eradication, such as the implementation of 3M Plus and to carry out environmental sanitation efforts such as cleaning home ban and doing environmental sanitation Around regularly, and pay attention to any place that has the potential to become a breeding ground for Aedes aegypti mosquitoes. This research can be used as consideration to add curriculum in the promotion of health. PHC are expected to conduct more intensive and continuous extension activities and conduct health promotion on dengue fever both individually and in groups when the community visits the PHC as an effort to increase the behaviour and increase periodic larva monitoring activity for once a week.

Acknowledgement
Special thank to The Director of Polytechnic of Health Padang which has given the author the opportunity to conduct this research, thanks also to the Head of the Bukittinggi City Health Office and to the Head of the Public Health Center Tigo Baleh. The highest appreciation to fellow lecturers who have provided full support to carry out this research, and all parties who have helped that cannot be mentioned individually.

References

[1] Ministry of Health 2008 *Indikator Indonesia Sehat 2010* (Jakarta: Ministry of Health, Republic of Indonesia)
[2] Undang-undang Republik Indonesia Nomor 36 Tahun 2009 tentang Kesehatan *(Law of the Republic of Indonesia No. 36 of 2009 concerning Health)*
[3] Ministry of Health 2010 *Buletin DBD Volume 2, Agustus 2010* (Jakarta: Center of Data and Epidemiological Surveillance, Ministry of Health, Republic of Indonesia)
[4] Ministry of Health 2011 *Profil Kesehatan Indonesia 2010* (Jakarta: Ministry of Health, Republic of Indonesia)
[5] Firdaus 2013 *Pengantar Epidemiologi Penyakit Menular Untuk Mahasiswa Kesehatan Masyarakat* (Jakarta: Trans Info Media)
[6] "Kemenkes: Indonesia Masih Endemis Demam Berdarah." Retrieved from https://nasional.tempo.co/read/500085/kemenkes-indonesia-masih-endemis-demam-berdarah/full&view=ok on June 2017
[7] *Profil Puskesmas Tigo Baleh 2013* (Bukittinggi, West Sumatera, Indonesia: Puskesmas Tigo Baleh)
[8] Sitio Anton 2008 Hubungan perilaku tentang pemberantasan sarang nyamuk dan kebiasaan keluarga dengan kejadian demam beradarah dengue di Kecamatan Medan Perjuangan Kota Medan *Tesis* (Semarang: Program Pasca Sarjana Universitas Diponegoro)
[9] Sulina P S 2012 Hubungan keberadaan jentik Aedes aegypti dan pelaksanaan 3M Plus dengan kejadian penyakit DBD di Lingkungan XVIII Kelurahan Binjai Kota Medan Tahun 2012 *Skripsi* (Medan: Fakultas Kesehatan Masyarakat, Universitas Sumatera Utara)

IOP Conf. Series: Earth and Environmental Science **314** (2019) 012002 doi:10.1088/1755-1315/314/1/012002

The potential and contamination of metals Pb and Zn on the soil around Tamangapa Antang landfill

Andi Artiningsih[1,*], Hazairin Zubair[2], A M Imran[1] and Sri Widodo[3]

[1] Geological Engineering Department, Faculty of Engineering, Hasanuddin University 90245, Indonesia

[2] Agricultural Sciences, Faculty of Agricultural Science, Hasanuddin University 90245, Indonesia

[3] Mining Engineering Department, Faculty of Engineering, Hasanuddin University 90245, Indonesia

*andi.artiningsih@umi.ac.id

Abstract. A Study on the potential and contamination of Pb and Zn metal on the soil around Tamangapa Antang Landfill has been conducted. The study aims to analyze the potential and contamination of heavy metals of Pb and Zn on the soil around the landfill. The sample points are divided into 4 lines namely, line A, line B, line C and Line D, at the point of 5 m (laterally) with the depth of 0 to 10 cm, 10 to 20 cm and 20 to 30 cm (vertically). The total soil samples whose heavy metal elements analyzed using *Atomic Absorption Spectroscopy* method are twelve samples. The concentrations of Pb and Zn metals in Tamangapa Antang Landfill soil have spread horizontally on the soil surface (top soil). The concentration of Pb for Line A has exceeded the threshold (R_2 = 0.10714) in every depth, while in Line B, Line C and Line D, the concentration is still below the threshold. Zn concentrations for Line A, Line B, Line C and Line D are above the threshold, except Line B at the depth of 20 to 30 cm is still below the threshold (67 ppm) which is allowed by the Government of Indonesia and The General of Drug and Food Control as a contaminant in the soil. Soils pH varies from 4.82 to 6 (acid) and the average abundance of heavy metal content in the soil samples at Tamangapa Antang Landfill site is Zn> Pb. Pollution of this heavy metal has been felt by local residents from a radius of 5 to 50 m. Where the Tamangapa Antang Landfill is in the middle of a settlement, so that it can affect the health of the population around the landfill, requiring residents who are still using dug wells. Request assistance from the local government to take immediate action to avoid problems with humans and the local environment.

1. Introduction

Landfill for Makassar citizens is an essential requirement to create a comfortable environment in a big city. When the waste in Makassar City is not transferred to the landfill in one day, then the city will be full of waste so it disturbs the public convenience. Tamangapa Antang Landfill in its operation uses Open Dumping in which the waste is dumped and left open daily with the soil [1]. Since its opening, Tamangapa Antang Landfill is estimated to have 1,240,000 tons of organic waste disposed [2]. If the landfill does not have proper management, it will produce solid waste which contains many toxic compounds such as heavy metals [3]. Uncontrolled Landfill is very harmful to health and can damage the environment [4,5]. Landfill is one of the main sources of the increased levels of metal in the soil.

The movement of contaminants from the landfill to the ecosystem is a complex process which involves a variety of geochemical activities [6].

Heavy metals are very harmful to the environment and organisms. This can be enriched through the food chain. Soil which is contaminated with heavy metals is difficult to managed [7]. In the past, soil contamination was not considered as important as the air and water pollution since soil contamination is more difficult to control than the air and water pollution [7]. But in the last years, soil pollution in developed countries such as in Indonesia, especially in South Sulawesi becomes serious. Thus, more attention is given and it also becomes a hot topic for environmental protection worldwide. The presence of high concentrations of metal elements in the soil is caused by natural and anthropogenic factors [6].

Soil surface is the recipient of various pollutants, especially heavy metals such as copper, nickel and zinc which can be used as indicators to determine the environmental quality [8]. In their activities, humans always interact with the soil. Soil is formed by the decomposition of rock and organic materials for years. The characteristics of Soil vary from one place to another place with different composition of bedrock, climate and other factors. Certain chemical elements occur naturally in soil as mineral components, but may be toxic at some concentration [9]. Most of the people around Tamangapa Antang Landfill Makassar City, using dug wells as a source of drinking water. If the soil is contaminated with heavy metals Pb and Zn then it can contaminate the well water. Consuming water contaminated with heavy metals, will settle in the body. Heavy metals such as Pb and Zn, if already absorbed into the human body will not be destroyed and can cause health effects for humans themselves depending on which part of the heavy metal can be bound in the body. If heavy metals are already bound in the human body will be poison [5].

Soil which is contaminated by heavy metals is potentially polluting the environment. Plants grown on soil contaminated with heavy metals such as Pb and Zn greatly disrupt human health. Soil which is contaminated by heavy metals is colorless and odorless, thus making it difficult to control. It inexplicitly damages the environment in a short time, but when it exceeds the environmental tolerance, the heavy metals in the soil will be active and cause ecological damage [7].

The location of Tamangapa Antang Landfill located in Bangkala Sub-District, Manggala District, Makassar City, South Sulawesi Province (see Figure 1), has been used since 1995 until now with the area of 18.8 hectare. According to the planing, Tamangapa Antang Landfill that was originally designed for the needs of 10 years, but in fact, until now the Landfill is still in use, it means that the age of the Landfill is 21 years, and cannot accommodate the volume of waste that is in Makassar City is reaching 800 tons or about 4,000 cubic per day. Based on the records of the Department of Hygiene and Environment, Makassar with an approximate total population of 1.3 million people, is producing about 3,800 m^3 or equal to 300 tons of public waste every day. Whereas the maximum capacity of Tamangapa Antang Landfill is only about 2,800 m^3 for accommodate public waste every day. Additional landfill is needed for the disposal of 1000 m^3 of the over waste. About 87% of waste in Makassar City is organic waste and about 13% is inorganic waste, such as plastic and paper.

Considering these facts, it can be assumed that there is pollution in Tamangapa Antang Landfill of Makassar, such as environmental pollution, soil contamination and groundwater contamination which can affect or risk the environmental sanitation and cause air pollution. The movement of pollutants has spread to the northwest of the Southeast Sea of which the pollution is around 300 tom 450m from the waste of Tamangapa Antang Landfill [10]. The direction of the groundwater flow in Tamangapa Antang Landfill is in the same direction to the slope of rock layer that is Northwest-Southeast [11]. This greatly affects the shallow groundwater quality. The pollution of shallow groundwater caused by liquid waste infusion (Leachate) from the landfill's waste which can also contaminate the wells of the surrounding residences. This pollution has been felt by the residents around the landfill, especially for those who use the groundwater as a source of drinking water [11].

This research is triggered by the amount of pressure received by the soil due to the piles of waste from domestic activities. When a hazardous substance has polluted the soil surface, it can evaporate, swept away by the rainwater into the soil then settled as a toxic chemical on the ground. This study aims to analyze the potential spread of heavy metals of Pb and Zn on the soil around Tamangapa Antang Landfill. The research is focused on how the impact of heavy metal metals can be spread on the soil around Tamangapa Antang Landfill of Makassar City to the life of the people around the landfill area, at the point of 5 m (laterally) with the depth of 0 to 10 cm, 10 to 20 cm [12] and 20 to 30 cm (vertically).

2. Research Methodology

2.1. Research Area

The location of this study is administratively included in the Bangkala Sub-District, Manggala District, and Makassar City, South Sulawesi Province, Indonesia. Geographically, Tamangapa Antang landfill is situated 119°29'10" to 119°29'40" East Longitude and 5°10'20" to 5°10'40" South Latitude. The location of the study area is shown in Figure 1.

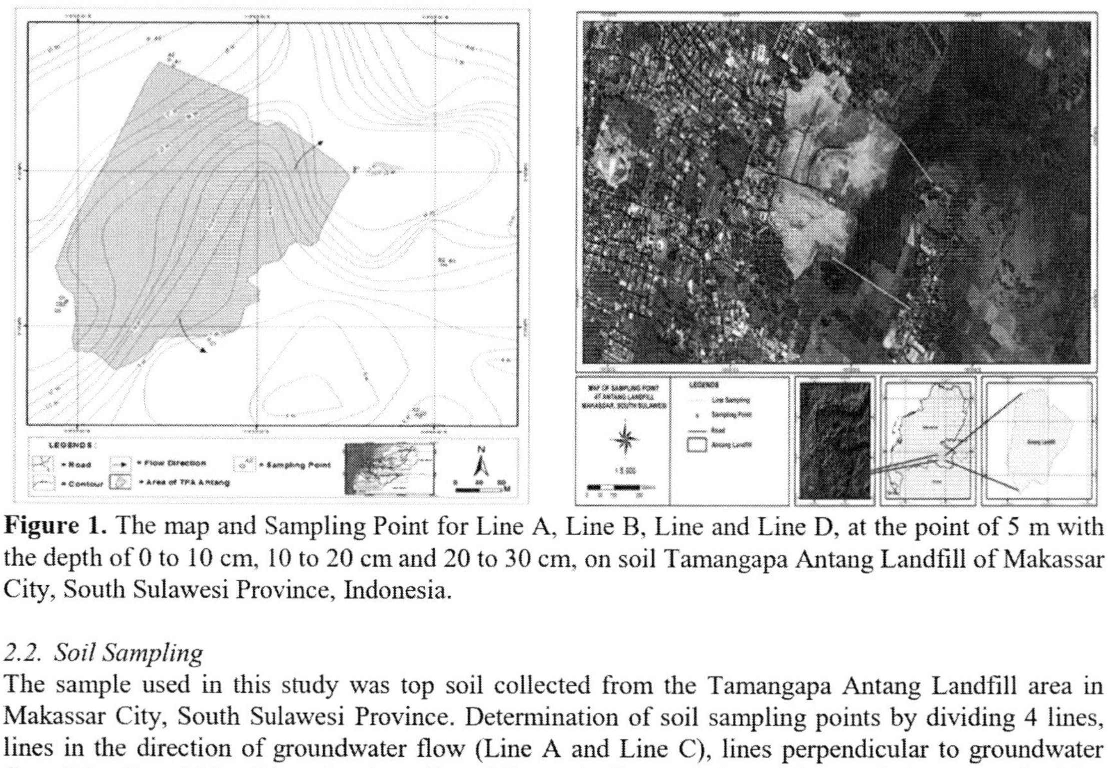

Figure 1. The map and Sampling Point for Line A, Line B, Line and Line D, at the point of 5 m with the depth of 0 to 10 cm, 10 to 20 cm and 20 to 30 cm, on soil Tamangapa Antang Landfill of Makassar City, South Sulawesi Province, Indonesia.

2.2. Soil Sampling

The sample used in this study was top soil collected from the Tamangapa Antang Landfill area in Makassar City, South Sulawesi Province. Determination of soil sampling points by dividing 4 lines, lines in the direction of groundwater flow (Line A and Line C), lines perpendicular to groundwater flow (Line B and Line D), at 5 m laterally, while vertically consisting of dept variations, namely, Line A with a dept of 0 to 10 cm, 10 to 20 cm and 20 to 30 cm, Line B with a dept of 0 to 10 cm, 10 to 20 cm and 20 to 30 cm, Line C with a dept of 0 to 10 cm, 10 to 20 cm and 20 to 30 cm and line C and Line D with a dept of 0 to 10 cm, 10 to 20 cm and 20 to 30 cm [12]. The landfill are located about 10 m from the residential area (especially in the western and northern part of landfill area, see Figure 1). The total soil samples to be analyzed for heavy metal elements using Atomic Absorption Spectroscopy Method. Soil samples on every sampling points were taken 0.5 kg for analysis. Measurement of temperature is carried out directly on the field (landfill), while pH measurements are carried out in the laboratory. Sampling poits are shown in Figure 1 (marked by green points in line A, B, C, and D).

2.3. Soil Samples Testing

Before being tested, the soil was saved in the oven for 48 hours at 106°C [6,13,14] first to ease the filtering process so that the grains are not attached to each other. It was then crushed with wooden or ceramic hammer [6], sieved using stainless steel of 2 mm sieves [6,15,12] and stored in the desiccator before being analyzed [14]. The soil sample that has been prepared was then brought to the laboratory for analysis.

3. Results

The measurement results of the total concentrations of metals Pb and Zn using *Atomic Absorption Spectroscopy* method for Line A, Line B, Line C and Line D at point 5 m (laterally) with the depth of 0 to 10 cm, 10 to 20 cm [12] and 20 to 30 cm (vertically). The analysis result of the heavy metals of Pb and Zn for Line A, at point 5 m with the depth of 0 to 10 cm, 10 to 20 cm, and 20 to 30 cm can be seen in (Figure 2).

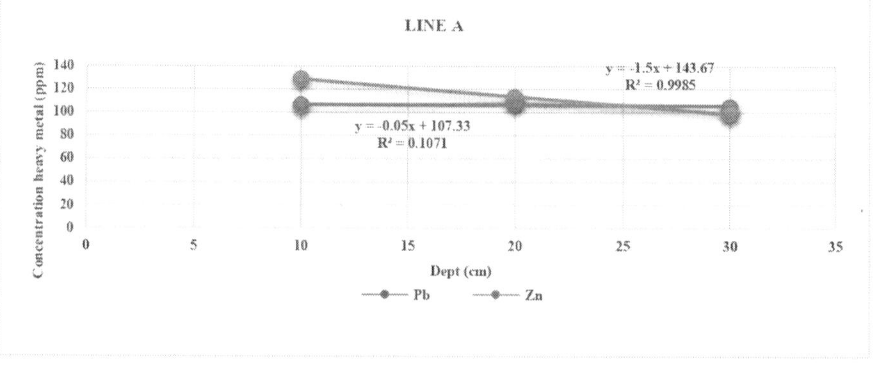

Figure 2. The analysis result of heavy metals Pb and Zn for Line A, at point 5 m with the depth of 0 to 10 cm, 10 to 20 cm and 20 to 30 cm on soil Tamangapa Antang Landfill of Makassar City, South Sulawesi Province, Indonesia.

For Line A, the concentration of metal Pb at the depth of 0 to 10 cm ranges from 106 ppm, at the depth of 10 to 20 cm, it increases to 108 ppm and at the depth of 20 to 30 cm the concentration decreases to 105 ppm. Zn metal concentration at the depth of 0 to 10 cm is the highest level of 129 ppm, at the depth of 10 to 20 cm and 20 to 30 cm the concentration decreases to 113 ppm and 99 ppm respectively. The results of the analysis of the heavy metals Pb and Zn for Line B, at point 5 m with the depth of 0 to 10 cm, 10 to 20 cm and 20 to 30 cm can be seen in Figure 3.

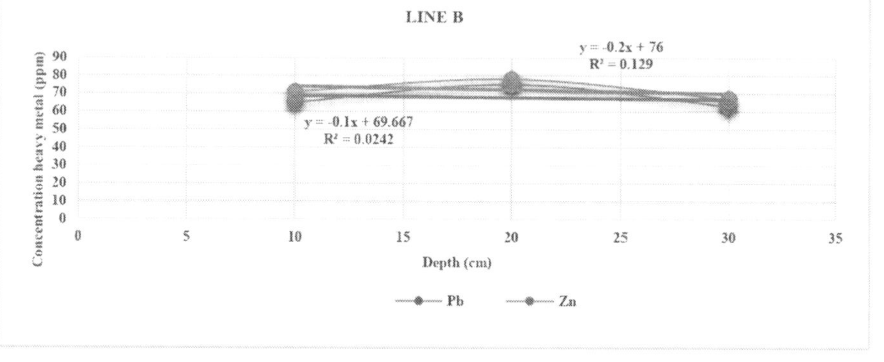

Figure 3. The analysis result of heavy metals Pb and Zn for Line B at point 5 m with the depth of 0 to 10 cm, 10 to 20 cm and 20 to 30 cm on soil Tamangapa Antang Landfill of Makassar City, South Sulawesi Province, Indonesia.

For Line B, the concentration of metal Pb at the depth of 0 to 10 cm is about 65 ppm, at the depth of 10 to 20 cm, the concentration increases around 75 ppm, while at the depth of 20-30 cm the concentration decreases to around 63 ppm. The concentration of Zn metal at the depth of 0 to 10 cm is about 71 ppm, at the depth of 10 to 20 cm, the concentration increases to around 78 ppm, while at the depth of 20 to 30 cm, the concentration decreases to 67 ppm. The analysis result of heavy metals of Pb and Zn for Line C, at point 5 m with the depth of 0 to 10 cm, 10 to 20 cm and 20 to 30 cm can be seen in Figure 4.

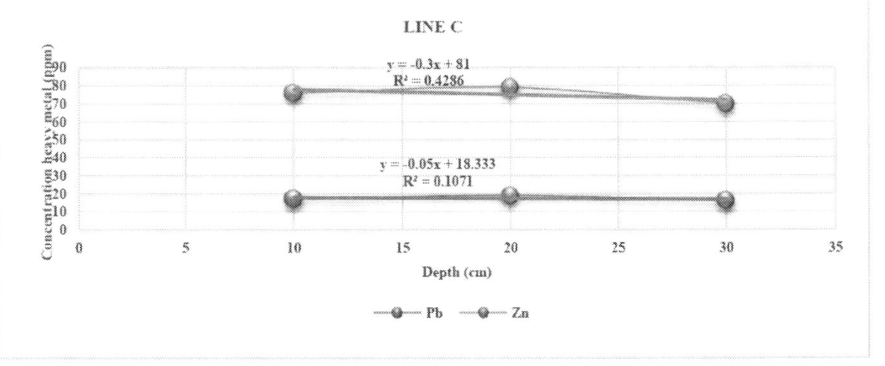

Figure 4. The analysis result of heavy metals Pb and Zn for Line C, at point 5 m with the depth of 0 to 10 cm, 10 to 20 cm and 20 to 30 cm on soil Tamangapa Antang Landfill of Makassar City, South Sulawesi Province, Indonesia.

For Line C, Pb metal concentration at the depth of 0 to 10 cm ranges 17 ppm, at the depth of 10 to 20 cm, the concentration increases to 19 ppm, whereas at the depth of 20 to 30cm, the concentration decreases to 16 ppm. Zn concentration at the depth of 0 to 10 cm concentration ranges from 76 ppm, at the depth of 10 to 20 cm, the concentration increases to 79 ppm, while at the depth of 20 to 30 cm, the concentration decreases to 70 ppm. The analysis result of heavy metals of Pb and Zn for Line C, at point 5 m with the depth of 0 to 10 cm, 10 to 20 cm and 20 to 30 cm can be seen in Figure 5.

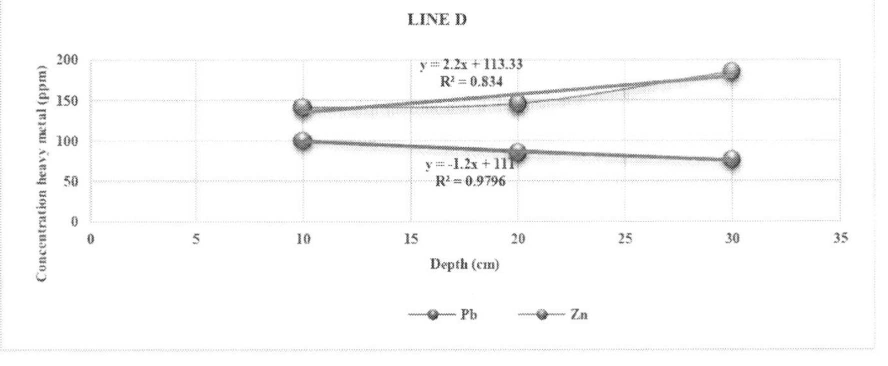

Figure 5. The analysis result of heavy metals Pb and Zn for Line D, at point 5 m with the depth of 0 to 10 cm, 10 to 20 cm and 20 to 30 cm on soil Tamangapa Antang Landfill of Makassar City, South Sulawesi Province, Indonesia.

For Line D, Pb metal concentration at the depth of 0 to 10 cm is the highest level which is around 100 ppm, while at the depth of 10 to 20 cm and 20 to 30 cm, the concentration decreases exponentially which is around 85 ppm and 76 ppm. The concentration of Zn metal at the depth of 0 to 10 cm is about 141 ppm, at the depth of 10 to 20 cm, the concentration increases to 146 ppm, while at the depth of 20 to 30 cm, the concentration increases exponentially to 185 ppm.

4. Discussion

The concentration of Pb and Zn heavy metals in the location of the Tamangapa Antang Landfill in Makassar City, South Sulawesi Province has spread both laterally and vertically. The concentration of the two metal elements are varied in each line and depth on the soil surface (top soil), with the level of enrichment already exceeding the expected threshold value by the soil environment. This condition can cause cancer (carcinogen) especially for the comunity who live around the Tamangapa Antang Landfill, Makassar City, South Sulawesi Province, where the Tamangapa Antang Landfill is situated around the residential area.

The mean concentration of Pb metal in the soil of Tamangapa Antang landfill at Line A and Line D ranges from 76 to 108 ppm which are above the average threshold of each depth unless Line D at the depth of 10 to 20 cm and 20 to 30 cm which are still below the average threshold. Meanwhile, in Line B and Line C, the average concentration ranges between 16 to 75 ppm which is still below the average threshold. Pb metal in the soil is as the oxidant of Pb^{2+}. When the pH increases, the ion of Pb $^{2+}$ ions are less soluble as the oxidant. At higher pH, it will form complexes in which the organic material will precipitate as carbonate, hydroxide and phosphate. The increase of concentration of lead (Pb) in Line A can be attributed to industrial, chemical and anthropogenic waste disposal and lead is positively correlated with other metals [6]. Pb metal is one of heavy metals which is very harmful to the living things because such metal is carcinogenic which can cause mutations and decomposes in a long time and its toxicity does not change [16].

The average concentration of Zn metal in the soil of Tamangapa Antang Landfill is between 67 to 185 ppm. For Line A, Line B, Line C and Line D, the concentration has exceeded the threshold for all depth. Except Line B which at the depth of 20 to 30 cm is about 67 ppm, thus it is still below the average threshold. The average pH ranges from 4.82 to 6. The oxidant of Zn metal in the soil is about Zn^{2+} which is under the acid oxidizing conditions. Zn^{2+} is one of the most heavily soluble and moving heavy metals in clays as well as hydrocarbon oxides and organic materials in which the solubility in the soil is very potential at acidic pH [6].

The average abundance of heavy metal content in soil samples at the Tamangapa Antang Landfill site is Zn>Pb. The maximum concentration of Zn metal is 67 to 185 ppm which is higher than the result obtained by [17] (80 ppm), but the result is lower than [18] (600 ppm). The result of Metal Pb (16 to 108 ppm) obtained is higher than [19] (5.67 ppm), but lower than [18] (280 ppm) and [20] (29 to 138 ppm). The enrichment of heavy metal elements of Pb and Zn on top soil surface around Tamangapa Antang landfill soil is still high accumulated. This is due to anthropogenic or human activity, Leachate and soil characteristics [21,22,23,24], supported by a very acidic pH ranging from 4.82 to 6. At heavy acidity, the mobility of heavy metals' pH in the soil is relatively high [25]. If there is a very high and continuous rainfall, it can result to permeation to the bottom of landfill of which the water seepage will come out into the river. The heavy metals are under Leachate and will settle in the soil, so heavy metals such as Pb, and Zn will continue to accumulate in the soil [7]. The heavy metals Pb and Zn will affect the quality of the soil and groundwater around the landfill, this greatly affects the lives of people around the Tamangapa Antang Landfill, because most people around the Tamangapa Antang Landfill still use dug wells as a source of drinking water. Pollution of Pb and Zn heavy metals has been felt by residents in Antang Tamangapa Landfill from a radius of 5 to 50 m.

The soil of Tamangapa Antang Landfill includes Utisol soil (a reddish soil contains a lot of clay), thus showing plasticity, encouraging flooding and surface water pollution [12]. The color of the soil is due to the metal content, especially the oxidized iron and aluminum (weathered soil). The soil pH analyzed at Tamangapa Antang Landfill of Makassar City ranges from 4.82 to 6 which is acidic with an average value of 4.84. The temperature of soil ranges from 25 to 31°C. pH plays an important role in the bioavailability of heavy metals and toxicity of the soils to the surrounding area [26][25]. pH affects the mobility of metals in the soil [12,26]. The low pH or acid pH will assist in the availability, mobility and redistribution of heavy metals of Pb and Zn in various fractions due to the increased solubility of ions in acidic soil environments [26].

The relatively acid soils have increased solubility and mobility of the micronutrients so that the concentrations of heavy metals in the soil increases [8,26]. If the soil solution is too acidic, the plant cannot take the advantage of N, P, K and other nutrients they need. In acid soil (low pH), the soil is dominated by Al, Fe and Mn ions. These ions will bind the nutrients need by plants especially P, K, S,

Mg so that the plants cannot absorb food well although the nutrient content in the soil is a lot. The heavy metals of Pb and Zn are very harmful to the health of the population [7], especially the residents who are living around the landfill area, because they can cause cancer and lung damage and the children will also have low IQ or low level of intelligence. Therefore, it becomes the threat for human beings because it can cause death [7,27,9,28,29].

5. Conclusion

The potential of metal content of Pb and Zn has spread horizontally above the soil surface (top soil). The metal concentration of Pb for Line A has exceeded the threshold (R_2 = 0.10714) in every depth, whereas in Line B, Line C and Line D the concentration is still below the threshold. The potential of metal content of Zn for Line A, Line B, Line C and Line D is above the threshold, while for Line B at the depth of 20 to 30 cm, it is still below the threshold (67 ppm) which is still allowed by the Government of the Republic of Indonesia and the General of Drugs and Foods Control as a contaminant in the soil. The soil pH varies from 4.82 to 6 which is acidic.

The metal concentration of Pb and Zn at Tamangapa Antang Landfill of Makassar City, South Sulawesi Province is influenced by the content of organic matter, temperature, texture and soils pH. The heavy metals of Pb and Zn still have high accumulation. If there is a continuous high rainfall, it can result to the permeation to the bottom of the landfill site, of which residents around Tamangapa Antang landfill still use dug wells. Therefore, the water leachate will affect the soil quality around the landfill and it is damage for the residents because the landfill is located in the resediential area and has been operating for 21 years.

Authorized government must immediately take appropriate action by means of bioremediation, namely by purifying well water, to remove heavy metals Pb and Zn, using zeolite activated charcoal and chitosan.

References

[1] Arbain, N K Mardana dan I B Sudana 2011 Pengaruh air lindi Tempat Pembuangan Akhir Sampah Suwung terhadap kualitas air tanah dangkal di sekitarnya di Kelurahan Pedungan Kota Denpasar *Ecotrophic* **3**(2) 55-60

[2] Peraturan daerah Kota Makassar Nomor 6 Tahun 2006 Tentang Rencana Tata Ruang Wilayah Kota makassar 2005-2015

[3] Hogland W, Matiasson L, Mårtensson L and Manandhar D R 2007 Analysis of pollutants from city dump/landfills in Kathmandu Valley Nepal with emphasis on heavy metals and persistent organics. SWE-2002-055

[4] Manimekalai M B and P Vijayalakshmi 2012 Analysis of leachate contamination potential of a municipal landfill using Leachate Pollution Index *IOSR Journal of Environmental Science, Toxicology and Food Technology (IOSR-JESTFT)* **2**(1) 16-39

[5] Palar H 2004 *Pencemaran dan Toksikologi Logam Berat, Cetakan kedua* (Jakarta: Penerbit Rineka Cipta)

[6] Parth V, N N Murthy and Raj Saxena P 2011 Assessment of heavy metal contamination in soil around hazardous waste disposal sites in Hyderabad city (India) *E. J. Environ. Res. Manage.* **2**(2) 027-034

[7] Su C, Jiang L and Zhang W 2014 A review on heavy metal contamination in the soil worldwide: Situation, impact and remediation techniques *Environmental Skeptics and Critics* **3**(2) 24-38

[8] Lu S, Wang H and Guo J 2010 Magnetic response of heavy metals pollution in urban soil: Magnetic proxy parameters as an indicator of heavy metals pollution *19th World Congress of Soil Science, Soil Solutions for a Changing World 1 – 6 August 2010, Brisbane, Australia*

[9] Gangadhar Z S 2014 Environmental impact assessment on soil pollution issue about human health *Int. Research J. Environ. Sci.* **3**(11) 78-81

[10] Imran A M 2008 Perkembangan Polutan dalam air tanah pada Tempat Pembuangan Akhir Sampah Antang Kota Makassar Provinsi Sulawesi Selatan; *Jurnal Penelitian Enjiniring Fakultas Teknik Unhas;* ISSN 1411-6243; Volume 12, No. 2 Mei-Agustus 2008, 187-410.
[11] Arifin M F, Imran A M, Ramli M dan Mukhsan P H 2014 Optimalisasi sains dan aplikasinya dalam peningkatan daya saing Bangsa Makassar. Tinjauan geohidrologi lokasi TPA sampah Kota Makassar, Provinsi Sulawesi Selatan (Studi Kasus TPA Sampah Tamangapa Antang) *Prosiding Seminar Nasional Geofisika*)
[12] Amos-Tautua B M W, Onigbinde Adebayo O and Dyapreye E 2014 Assessment of some heavy metals and physicochemical properties in surface soils of municipal open waste dumpsite in Yenagoa, Nigeria *Academic Journal. Org/ Afr. J. Environ. Sci. Technol.* **8**(1) 41-47
[13] Oluseyi T, Adetunde O and Amadi E 2014 Impact assessment of dumpsites on quality of near-by soil and undeground water: A case study of an abandoned and a functional dumpsite in Lagos, Nigeria *Int. J. Sci. Environ. Technol.* **3** (3) 1004-1015
[14] Abiodum O M, Akinlolu A F and Lawal O 2011 Potential soil contamination with toxic metal in the vicinity of active and abandoned dumpsite *Agric. Biol. J. N. Am.* **2**(5) 785-790
[15] Mtunzi F M, Dikoi E D and Moja S J 2015 Evaluation of heavy metal pollution on soil in Vaderbijl Park South Africa *International Journal of Environmental Monitoring and Analysis 2015;* **3**(2) 44-49; ISSN: 2328-7659 (Print): ISSN: 2328-7667 (Online).
[16] Brass G and M Strauss W 1981 *Air Pollution Control Part IV* (New York: John Willey & sons)
[17] Andras P, Turisova I, Lackova E, Kharbish S, Krnac J and Cmielova L 2013 Environmental risk due to heavy metal contamination caused by old copper mining activity at Lubietova Deposit, Slovakia *Energy and Environment Research* **Vol. 3** No. 2 (2013)
[18] Kouame I K, Dibi B, Koffi K, Savane I and Sandu I 2010 Statistical approach of assessing horizontal mobility of heavy metals in the soil of Akouedo Landfill nearby Ebrie Lagoon (Abidjan-Cote D'ivoire) *Int. J. Conserv. Sci.* **1**(3) 149-160
[19] Onder S, Dursun S, Gezgin S and Demirbas A 2007 Determination of heavy metal pollution in grass and soil of city centre green areas (Konya, Turkey) *Polish J. of Environ. Stud.* **16**(1) 145-154
[20] Habib H R, Awadh S M and Muslim M Z 2012 Toxic heavy metal in soil and some plants in Baghdad, Iraq *Journal of Al-Nahrain University* **15**(2) June 2012, 1-6
[21] Bityukova L, Shogenova A and Birke M 2000 Urban Geochemistry: A study of element distribition in the soil of Tallin (Estonia) *Environ. Geochemistry and Health* **22**(2) 173-193
[22] Al-Jibury D A and Essa S K 2016 Using pollution indexes to assessment of heavy metals pollution in highway-side soil around Baghdad City *Journal of Environment and Earth Science* **Vol. 6** No. 10 (2016)
[23] Soran N Sadeq and Mohammad H A 2016 Impact of leachate on soil and groundwater quality in vinicity of landfill sites of Kirkuk City (Kirkuk-Iraq) *Int. J. Curr. Res. Aca.* **4**(8) 1-21
[24] Sayadi M H, Sayyed M R G and Shabani N 2009 Quantification of heavy metal pollutants in the surface soils of Chitgar industrial area Tehran, Iran with spatial references to their spatial pattern *Pollution Research* **28**(3) 345-351
[25] Schmitt H S and Sticker H 1991 *Heavy metal compounds in the soil.* Merian, E. (Ed.), *Metals and their compounds in the Environment* (Weinheim VCH) pp 311-326
[26] Chimuka L, Mugwedi R, Moboladisoro B H and Odiyo O J 2005 Metals in environmental media: A study of trace and platinum group metals in Thohoyandou, South Africa *Water SA* **31**(4) 581-588
[27] Yabe J, Ishizuka M and Umemura T 2010 Current levels of heavy metal pollution in Africa *J. Vet. Med. Sci.* **72**(10) 1257-63
[28] Tijhuis L, Brattli B and Saether O M 2002 A Geochemical survey of top soil in the city of Olso, Norway *Environmental Geochemistry and Health* **24** 67-94
[29] Nagajyoti P C, Lee K D and Sreekanth T V M 2010 Heavy metal, occurrence and toxicity for plants: a review *Environ. Chem. Lett.* **8** 199-216

IOP Conf. Series: Earth and Environmental Science **314** (2019) 012003 doi:10.1088/1755-1315/314/1/012003

Pollution load capacity of Batang Kuranji River, Padang City

A Azhar[1,2], E Barlian[2], I Dewata[2], A Amran[2], A Teguh[1,2] and Y Danhas[1,2]

[1] Badan Penelitian dan Pengembangan Provinsi Sumatera Barat, Sudirman Street No. 51 Padang, West Sumatera, Indonesia

[2] Graduate Program of Universitas Negeri Padang, Prof. Dr. Hamka Street, West Air Tawar, Padang-West Sumatera 25131 Indonesia

*aulia.azhar@yahoo.com

Abstract. This study aimed to determine the pollution load capacity (DTBP) of Batang Kuranji River against the parameters of Total Suspended Solid (TSS), Biochemical Oxygen Demand (BOD) and Chemical Oxygen Demand (COD). In general, pollutant sources are caused by developments and activities of urban communities such as domestic activities, industry, agriculture and other activities. This research is descriptive quantitative by using Qual2KW method with the sampling location point on the representative of land use located along the kuranji stem river as many as 24 sampling locations, which are divided into 24 segments, spatially from upstream to downstream. The results showed that the parameters of Total Suspended Solid (TSS) were seen from upstream to downstream in the area of sections 8, 10, 14 must be reduced pollution load of 648 kg / day (section 8), 15,552 Kg / day (section 10) and 22,032 Kg / day (section 14), this is due to (1) economic activities of the community in the agricultural sector, in the form of rice fields directly adjacent to the Batang Kuranji river which can cause erosion, (2) livestock sector activities in the form of chicken farming, where the community directly discarding the remnants of chicken feed is discharged directly into the river so that it can increase the pollution load of TSS parameters. Parameters of BOD in Section 6 and Section 16 must be reduced by a pollution load of 453.60 Kg / day (Section 6), 1,308.96 Kg / day (Section 16), This is due to this high pollutant source such as: (1) domestic / domestic waste activities, (2) rice field agriculture. Whereas the COD parameters in areas 3, 7, 14, 15, and 16 must be reduced by 3,862 kg / day (section 3), 16,783.20 kg / day (section 7), 3,149.28 Kg / day (segment 14), 9,629.28 Kg / day (section 15) and 22,589.28 Kg / day (section 16), this is because there are still many agricultural, plantation and livestock areas so that many sources of pollutants enter the Batang Kuranji river from chemicals organic so that the river water in Batang Kuranji can still accept pollution loads, for the COD parameter the approach is more inorganic chemistry and there are many inorganic chemical compounds in the disposal of liquid waste in the Company / Industry. With the quality conditions of the current Batang Kuranji River in the Padang City segment, control efforts are needed for the pollution. Related to the results of the study, and from the conclusions obtained, the efforts made for optimal control are the application of environmentally sound spatial principles.

1. Introduction

Padang city is the capital of West Sumatra Province. In Padang city, there is a river called the river of Batang Kuranji. The river flows in four (4) districts, namely (1) Kuranji Subdistrict, Subdistrict Pauh (2), (3) Sub Nanggalo, and (4) of North Padang sub district. Pauh, there are Sub on 8 (eight) village.

But the village irrigated River Rod Kuranji only four (4) wards, namely (1) the village Hull Hill, (2) Wards Quarts Tangah, (3) subdivisions of Kapalo Koto and (4) the village of Kampung Binuang inside. At Kuranji Subdistrict 9 (nine) village. But not all the keluarahan irrigated by Sungai Batang Kuranji. There are 5 (five) wards are irrigated by Sungai Batang Kuranji, namely (1) the village Korong, Gadang, Kalumbuk Village (2) and (3) the village Bottom Leech Kuranji, (4) and (5) Pasa Ambacang.

In Nanggalo there is 6 (six) village. In contrast to that Kecamatan Kuranji Pauh and not all neighborhood and flowed the river Rods Kuranji, in district Nanggalo to 6 wards, all of which is irrigated area the river Rods Kuranji. The subdistricts are (1) the village of Kampung Chi, (2) Desert Laweh Village, (3) the village Surau, Gadang, and (4) Wards Kurao Pagang, Lapai Village (5) and (6) Neighborhood Tabing Banda Gadang. Meanwhile, Padang Utara subdistrict has 7 (seven) the village, but the village irrigated River Rod Kuranji only three villages only. Each village was (1) the village of freshwater West, (2) the village of freshwater East, and (3) the village North Reef Ulak. Thus, it can be concluded that of the 4 existing sub-districts in the city of Padang, there were 18 (eighteen) villages which flowed the river Rods Kuranji. In other words, a Batang River Kuranji, Padang city segments are at 18 areas villages spread over four (4) districts.

In fact, a Batang Kuranji River is one of the rivers that pressure gets high enough. This is caused by the development and activity of the society of the city such as the activities of domestic, industrial, agricultural or other activities. Any existing community activities, especially those that reside on each village irrigated by Sungai Batang Kuranji is impacting on the quality of the river Rods Kuranji. In terms of demographics, Padang city from year to year has increased the number of inhabitants in quantity. Reported by BPS Padang in numbers (2013) that the population in the city of Padang has reached number 1,261,000 souls. Rate of population growth of the city of Padang which amounted to 2.23%, will project a large enough population.

Population growth will have implications for Padang City against adding space to a residential area. Neighborhoods will potentially against the pollution and destruction of the environment, because in the region that the occurrence of various activities of the society. These circumstances, of course, will have an impact on power support (carrying capacity) environments, especially in the environment of the river Rods Kuranji in terms of capacity against the burden of pollution caused by population growth.

To maintain the water quality of the river Rods Kuranji so as not to experience a decrease in the future, required a study of water quality of the River in the existing conditions. Research on water quality of the River in question is the determination of capacity of existing tamping. It is useful to know the potential or ability of the River in receiving waste is corrupted it. With the known capacity of the capacity of the River, a Batang Kuranji will be determined step forward other strategic measures related to control measures of the quality of the river water Rods Kuranji. Given the function of river water as one of the water resources in the form of a body of water, then it is very important that the determination of this capacity. In addition, the research capacity is considered urgent because of other variables that are in the system of the city, such as the rate of population growth, community activity pattern of the city, as well as the development of land use in city space is extremely complex, so It's hard to be examined in order to control the quality of the river water. On the contrary, through the determination of the power this pollution burden of tamping, then the results can be used as a foundation for controlling variables in such a complex above.

The formulation of the problem in this paper is: (a) How much is the pollution load capacity of Padang City Batang Kuranji River from upstream to downstream to TSS parameters; (b) How much is the pollution load capacity of Padang City Batang Kuranji River from upstream to downstream towards BOD parameters; (c) How much is the pollution load capacity of Padang City Batang Kuranji River from upstream to downstream towards COD parameters. The objectives of this study are: (a) Analyzing the pollution load capacity of Padang City Batang Kuranji River towards TSS parameters; (b) Analyzing the pollution load capacity of the Batang Kuranji River against BOD parameters; (c)

Analyzing the pollution load capacity of the Padang City Batang Kuranji River against the COD parameters.

Water quality management strategies with QUAL2Kw in the Bagmati River, Nepal, during the critical period (drought) to maintain targeted water quality where DO is a minimum of 4 mg / L or more; Maximum BOD 3 mg / L; Total N 2.5 mg / L; Total P 0.1 mg / L; water temperature is small or equal to 20◦C; pH in the range 6.5–8.5, taking into account modification of pollutant load, addition of discharge and local oxygenation. The results of the study indicate that local oxygenation is effective in maintaining minimum DO content in the river. The combination of modified wastewater, addition of flow and local oxygenation is suitable for maintaining permissible water quality limits [1]

1.1 Pollution Load Capacity

According to the decision of the Minister of the environment No. 110 Year 2003 that the pollution load capacity is the ability of the water at a water source to receive the input of the load without resulting in contamination of the water are being polluted. While the burden of pollution is the amount of a contaminant elements contained in water or waste water which is expressed in units of mass per time. [2]

Pollution of waters caused by loading various human activities that go beyond the capabilities of the amount recovered from the waters of self or other activities. Pollutants can be either solid chemical compounds (organic and inorganic) and pathogen bacteria all of which resulted in a decline in water quality and value to the waters as a source of raw water for various activities. Capacity is the ability of the environment to absorb the energy and matter, or other components that enter or put into it. [3]

1.2 Determination of the Capacity load of Pollution by Qual2KW Method

One method that can be used to determine the load capacity of the pollution of the river is by the method Qual2KW. This method is the development of Qual2 by the United States Environmental Protection Agency (US EPA). Development of the method of Qual2 is the final revision and Qual2E Qual2KW. Through this method can simulate or predict changes in the quality of the River from upstream to downstream if the flow of liquid waste is reduced or added. And thus, be aware of the load capacity of the pollution of the River in accordance with the criteria of quality raw environment that has been established.[4]

2. Methods

Research conducted is a quantitative descriptive, in determining the location of a sample of researchers using the applications of GIS (Geographical Information System) as well as using GPS (Geographical Portable System), the application can determine the distribution of land use and the coordinate points of sampling to be taken before taken directly spaciousness (location of sampling point from upstream to downstream is presented in Figure 1).

Figure 1. Location of Sampling Point River Batang Kuranji Padang City.

Determination of water quality parameters of river Rod Kuranji refers to government regulation Number 82 Year 2001 as indicated in table 1.

Table 1. Water quality parameter based on the Government Regulation Number 82/2001

No	Parameters	Unit	Tool	Method	SNI
1.	TSS	mg/l	Analytical balance	Gravimetric	06-6989.3-2004
2.	BOD	mg/l	Buret	Titrimetric	6989.72:2009
3.	COD	mg/l	Spectrophotometer	Spectrophotometry	6989.2:2009

The selected sampling technique is the grab sampling, where sampling is carried out as many as 1 (one) point for each sampling time. Provision of distance sampling is + 1/3 and 2/3 the width of the River at a depth of 1/2 meters + from the surface of the water stream by using an ordinary bottle or a plastic bucket.[5]

3. Results and discussion

3.1 Parameters TSS

Table 2. Pollution Load Capacity Batang Kuranji River Padang City to parameters TSS

No	Location	Range Site		Pollution Load	Capacity	Pollution Load Capacity	Description
		Up (km)	Down (km)	(kg/day)	(kg/day)	(kg/day)	
1	Location 1	12,80	12,30	129,60	9.072,00	-8.942,40	pollution load may be added
2	Location 2	12,29	11,70	3.888,00	6.480,00	-2.592,00	pollution load may be added
3	Location 3	11,69	11,20	6.480,00	6.480,00	0,00	pollution load may be added
4	Location 4	11,19	10,60	-3.888,00	6.480,00	-10.368,00	pollution load may be added
5	Location 5	10,59	10,00	1.296,00	6.480,00	-5.184,00	pollution load may be added
6	Location 6	9,99	9,48	1.296,00	6.480,00	-5.184,00	pollution load may be added
7	Location 7	9,47	8,95	1.296,00	6.480,00	-5.184,00	pollution load may be added
8	Location 8	8,94	8,40	7.128,00	6.480,00	648,00	pollution load must be reduced
9	Location 9	8,39	7,84	-5.184,00	6.480,00	-11.664,00	pollution load may be added
10	Location 10	7,83	7,25	22.032,00	6.480,00	15.552,00	pollution load must be reduced
11	Location 11	7,24	6,71	-2.592,00	6.480,00	-9.072,00	pollution load may be added
12	Location 12	6,70	6,17	-3.240,00	6.480,00	-9.720,00	pollution load may be added
13	Location 13	6,16	5,63	-6.480,00	6.480,00	-12.960,00	pollution load may be added
14	Location 14	5,62	5,10	28.512,00	6.480,00	22.032,00	pollution load must be reduced
15	Location 15	5,09	4,53	-25.920,00	6.480,00	-32.400,00	pollution load may be added
16	Location 16	4,52	3,98	-11.664,00	6.480,00	-18.144,00	pollution load may be added

17	Location 17	3,97	3,43	-3.888,00	6.480,00	-10.368,00	pollution load may be added
18	Location 18	3,42	2,89				
19	Location 19	2,88	2,35				
20	Location 20	2,34	1,81				
21	Location 21	1,80	1,27				
22	Location 22	1,26	0,72				
23	Location 23	0,71	0,35				
24	Location 24	0,34	0,00				
	Total			24.753,60	99.792,00	-75.038,40	

Parameter TSS on sections 8, 10 and 14 of the load should be reduced pollution of 648 kg/day (section 8), 15,552 Kg/day (section 10) and 22.032 Kg/day (section 14); This is caused due to (1) the economic activities of the community in the agricultural sector, in the form of Rice that is directly adjacent to the river Rods Kuranji that could lead to the occurrence of erosion, (2) the activity of the farm sector in the form of a chicken farm, where the community immediately dispose of the remains of chicken feed were disposed of directly to the river so that it can increase the burden of the pollution parameters of TSS.

From the results of this study when compared to the results of relevant research results can be seen monitoring parameter TSS in the Metro the city of Malang, showed increased from 1 station (Metro/hulu Sungai Karang Besuki Village) to station 2 (River span Metro/Village Pisangcandi) and 3 (Metro/River downstream of the village Bandungrejosari). TSS value at station 1 of 34.0 mg/l and then rising at 2 stations of 50.4 mg/l and increases at 3 stations of 62.6 mg/l. This is due to an increase in TSS values of river water on Metro stations 2 and 3, due to the large number of rather the function of the land into a residential/woke up around Metro basin, causing the soil solids-solids entering the river flow through run off increasing [6]

3.2 Parameter BOD (Biochemical Oxygen Demand)

Table 3. Pollution Load Capacity Batang Kuranji River Padang City to parameters BOD

No	Location	Range Site		Pollution Load	Capacity	Pollution Load Capacity	Description
		Up (km)	Down (km)	(kg/day)	(kg/day)	(kg/day)	
1	Location 1	12,80	12,30	152,93	583,20	-430,27	pollution load may be added
2	Location 2	12,29	11,70	141,26	259,20	-117,94	pollution load may be added
3	Location 3	11,69	11,20	145,15	375,84	-230,69	pollution load may be added
4	Location 4	11,19	10,60	145,15	440,64	-295,49	pollution load may be added
5	Location 5	10,59	10,00	149,04	427,68	-278,64	pollution load may be added
6	Location 6	9,99	9,48	907,20	453,60	453,60	pollution load must be reduced
7	Location 7	9,47	8,95	0,65	518,40	-517,75	pollution load may be added
8	Location 8	8,94	8,40	0,65	453,60	-452,95	pollution load may be added

9	Location 9	8,39	7,84	0,65	388,80	-388,15	pollution load may be added
10	Location 10	7,83	7,25	142,56	453,60	-311,04	pollution load may be added
11	Location 11	7,24	6,71	142,56	388,80	-246,24	pollution load may be added
12	Location 12	6,70	6,17	142,56	453,60	-311,04	pollution load may be added
13	Location 13	6,16	5,63	142,56	440,64	-298,08	pollution load may be added
14	Location 14	5,62	5,10	142,56	557,28	-414,72	pollution load may be added
15	Location 15	5,09	4,53	142,56	635,04	-492,48	pollution load may be added
16	Location 16	4,52	3,98	1.944,00	635,04	1.308,96	pollution load must be reduced
17	Location 17	3,97	3,43	0,00	635,04	-635,04	pollution load may be added
18	Location 18	3,42	2,89	0,00	635,04	-635,04	pollution load may be added
19	Location 19	2,88	2,35	0,00	0,00	0,00	
20	Location 20	2,34	1,81	0,00	0,00	0,00	
21	Location 21	1,80	1,27	0,00	0,00	0,00	
22	Location 22	1,26	0,72	0,00	0,00	0,00	
23	Location 23	0,71	0,35	0,00	0,00	0,00	
24	Location 24	0,34	0,00	0,00	0,00	0,00	
Total				2.498,04	6.829,92	-4.331,88	

Parameters BOD on sections 6 and 16 must be reduced load pollution of 453.60 Kg/day (section 6), 1,308.96 Kg/day (section 16). This is because in this segment very high polluters source such as: (1) the activity of the household/domestic, industrial waste (2) and (3) rice paddy agriculture.

From the results of this study when compared to the results of relevant research Results can be seen monitoring parameters BOD on the Metro the city of Malang, showed increased from station 1 to station 3. The value of BOD at station 1 for 4.7 mg/l, the 2 stations of 6.1 mg/l and the 3 of 6.25 mg/l, soaring numbers BOD on River Metro can originate from organic materials derived from domestic waste and agricultural. [6]

3.3 Parameter COD (Chemical Oxygen Demand)

Table 4. Pollution Load Capacity Batang Kuranji River Padang City to parameters COD

No	Location	Range Site		Pollution Load	Capacity	Pollution Load Capacity	Description
		Up (km)	Down (km)	(kg/day)	(kg/day)	(kg/day)	
1	Location 1	10,57	10,10	1.425,60	4.406,40	-2.980,80	pollution load may be added
2	Location 2	10,00	9,67	777,60	3.304,80	-2.527,20	pollution load may be added
3	Location 3	9,66	9,17	7.128,00	3.265,92	3.862,08	pollution load must be reduced
4	Location 4	9,16	8,12	-2.073,60	3.304,80	-5.378,40	pollution load may be added
5	Location 5	8,11	7,62	-1.036,80	3.304,80	-4.341,60	pollution load may be added
6	Location 6	7,61	6,67	-648,00	3.304,80	-3.952,80	pollution load may be added

7	Location 7	6,66	6,04	20.088,00	3.304,80	16.783,20	pollution load must be reduced
8	Location 8	6,03	5,54	-12.960,00	3.304,80	-16.264,80	pollution load may be added
9	Location 9	5,53	5,14	-6.480,00	3.304,80	-9.784,80	pollution load may be added
10	Location 10	5,13	4,64	-1.296,00	3.304,80	-4.600,80	pollution load may be added
11	Location 11	4,63	3,99	1.555,20	3.317,76	-1.762,56	pollution load may be added
12	Location 12	3,98	3,22	1.101,60	3.330,72	-2.229,12	pollution load may be added
13	Location 13	3,21	2,07	1.036,80	3.330,72	-2.293,92	pollution load may be added
14	Location 14	2,06	1,65	6.480,00	3.330,72	3.149,28	pollution load must be reduced
15	Location 15	1,64	0,80	12.960,00	3.330,72	9.629,28	pollution load must be reduced
16	Location 16	1,64	0,80	25.920,00	3.330,72	22.589,28	pollution load must be reduced
17	Location 17	1,64	0,80	-41.472,00	3.330,72	-44.802,72	pollution load may be added
18	Location 18	1,64	0,80	0,00	0,00	0,00	
19	Location 19	1,64	0,80	0,00	0,00	0,00	
20	Location 20	1,64	0,80	0,00	0,00	0,00	
21	Location 21	1,64	0,80	0,00	0,00	0,00	
22	Location 22	1,64	0,80	0,00	0,00	0,00	
23	Location 23	1,64	0,80	0,00	0,00	0,00	
24	Location 24	1,64	0,80	0,00	0,00	0,00	
Total				28.058,40	50.751,36	-22.692,96	

Parameters COD in Area 3, section 7, 14, 15 dan section 16 must be reduced load its pollution of 3,862 Kg/day (section 3), 16,783.20 Kg/day (section 7), 3,149.28 Kg/day (section 14), the 9,629.28 Kg/day (section 15) and 22,589.28 Kg/day (section 16);

This is because there are still many areas of agriculture, forestry and animal husbandry so that the source of the many polluters went into the Batang Kuranji River comes from organic chemistry so that river water Batang Kuranji can still accept the burden of pollution, to his approach is rather to the COD parameter inorganic chemistry and inorganic chemical compounds of this many on the disposal of liquid waste on a company/industry.

When compared with the results of relevant research can be seen that the results of the monitoring of the parameters COD in River Metro Malang denotes increased from station 1 to station 3. The value of COD at station 1 for 11.11 mg/l, at 2 stations of 15.97 mg/l and the 3 of 17.56 mg/l. high COD Numbers, indicating the greater the level of pollution that occurs, increasing the value of COD Metro caused by river water waste disposal were sourced from the area woke up/residential and farming areas around the river [6]

4. Conclusion

For the Parameter Total Suspended Solid (TSS) viewed from upstream to downstream in sections 8, 10, 14 it pollution burden must be reduced of 648 kg/day (section 8), 15,552 Kg/day (section 10) and 22.032 Kg/day (section 14), the parameters of the BOD on the area roads 6 and segment 16 it pollution burden must be reduced of 453.60 Kg/day (section 6), 1,308.96 Kg/day (section 16), to the COD Parameter on the sections 3, 7, 14, 15, and 16 roads must be reduced load it pollution of 3,862 Kg/day (section 3), 16,783.20 Kg/day (section 7), 3,149.28 Kg/day (section 14), the 9,629.28 Kg/day (section 15) and 22,589.28 Kg/day (section 16).

With the quality conditions of the River at Batang Kuranji existing segment of the city of Padang at the moment, it takes the efforts of control of pollution. Related to the research results, and of conclusions, efforts are being made to the optimal control is the application of spatial environmental environmentally. The arrangement of space that will be able to control the pollution of Sungai Batang Kuranji is resting on two aspects, namely (1) the restructuring of the space, and (2) of regulation pattern space. To the two aspects of this subject matter, focused towards the area that has the area of DAS Batang Kuranji in Padang city.

Reference

[1] Kannel P R, Lee S, Lee Y S, Kanel S R and Pelletier G J 2007 Application of Automated QUAL2Kw for Water Quality Modeling and Management *Ecological Modelling* **202**(2007) 503–517
[2] Keputusan Menteri Lingkungan Hidup Republik Indonesia Nomor 110 Tahun 2003 tentang Pedoman Penetapan Daya Tampung Beban Pencemaran Air Pada Sumber Air
[3] Soemarwoto O 1997 *Ekologi, Lingkungan hidup dan Pembangunan* (Jakarta: Djambatan)
[4] Ginting S 2009 *Daya Tampung Sungai Kampar* (Pekanbaru: KNLH-PPLH Regional Sumatera)
[5] John Syahrul 2006 *Teknik Pengambilan Sampel dan Penanganan Sampel* (Riau: KLH PPLH Regional Sumatera)
[6] Azwar Ali 2013 Kajian Kualitas Air dan Status Mutu Air Sungai Metro di Kecamatan Sukun Kota Malang *Jurnal Bumi Lestari* **13**(2) 265-274

Evaluation of land use to critican Batang Kuranji watershed in Padang City

Helfia Edial[1,2], Bustari Muchtar[3] and Indang Dewata[3]

[1]Doctoral Programe of Environmental Sciences, Universitas Negeri Padang, Padang, Indonesia
[2]Lecturers of Geography Department, Social Sciences Faculty, Universitas Negeri Padang, Padang, Indonesia
[3]Post graduate Lecturers in Environmental Sciences, Universitas Negeri Padang, Padang, Indonesia

*helfia.edial@fis.unp.ac.id

Abstract. This study aims to 1). find out the land use index 2). analyze the suitability of land use based on land capability, 3). How is the spread and extent of critical watersheds 4). calculate the level of erosion hazard in the Padang City Kuranji Watershed. This research uses quantitative descriptive method. Quantitative analysis is done by classifying, scoring, and weighting with existing scales and criteria. Samples are taken based on land units obtained from overlapping maps of landform units, slopes, land use, land and geology. The results showed 1). there are three classes of land use index, namely good class 11,375 km2 (97.23%), medium class 37.50 km2 (43.91%) and bad class 214.22km2 (16.71%). 2). suitability of land use consists of two classes, namely: Appropriate class 19,099 km2 (85.97%) and class not appropriate 3.13 km2 (14.3%). 3) there are three classes to watershed criterion namely uncritical class (70.94%), critical potential class (24.54%) and critical class (4.53). 4). erosion danger level there are five classes: Very light (68.64%), Lightweight class (9.75%), Moderate class (14.68%), Weight class (4.79%) and Very Heavy class (2, 14%).

1. Introduction

Land criticality has occurred in many regions, this is one of the consequences of population growth that requires land for their livelihood, both for settlements as well as for business. The lands that were formerly used as buffer or conservation land, are now replaced by built land, as a result this change often brings problems for the many communities around them, namely the occurrence of groundwater and river droughts, floods and landslides. Likewise with increasing the air temperature in a watershed

The watershed management program (DAS) in Indonesia has been started since Pelita I of the new order period of the 70s which was implemented in the form of reforestation and reforestation projects. The first activity began in 1973 in the form of the Upper Watershed Management and Upland Development Project in the Bengawan Solo Watershed with the assistance of FAO / UNDP. Whereas lately a watershed management project carried out by the government under the Ministry of Forestry is the National Movement for Forest and Land Rehabilitation (GNRHL / GERHAN) since 2003. The watershed management activities have sought to maintain and improve the quality of watersheds in

Indonesia so they can be used sustainably. But the reality on the ground shows that watershed conditions in Indonesia are getting worse and the problems are increasingly complex [1].

Watershed problems in Indonesia often do not guide the direction of land area functions which eventually cause various natural disasters. The description of the condition of the watershed in Indonesia which is increasingly damaged can be observed based on the number of priority watersheds that are increasing from year to year. In 1984 of the 458 watersheds in Indonesia there were 20 super priority watersheds (Priority I) and 37 in 1992. In 1999, the number of priority I watersheds increased to 60 watersheds. 2006 increased to 108. [1]

In the city of Padang there are 4 watersheds that often bring various disasters to the community, namely, Air Dingin Watershed, Kuranji watershed and Arau watershed and Kalumpang watershed. According to PP No. 17 of 1980, the area of Padang City was 694.96 Km2 The effective area including the river was 205,007 km2, while the area of the hilly area including the river was 486,209 km2. Padang City consists of 11 Districts and 103 Villages with the widest sub-district is Koto Tangah which reaches 232.25 Km2. Of the total area of Padang City, most or 52.52% is in the form of forests protected by the government, in the form of buildings and yards covering 9.01% or 62.63 km2, while those used for paddy fields cover 7.52% or 52.25 km2 [2].

Land use in Padang City can be grouped into 3 categories, namely forest land, rice fields and non-rice fields. More than 51% of Padang City's land is in the form of forests, but there are some places in urban areas that are not properly used by the community, so often causing landslides, especially in the Kuranji watershed such as land clearing, agricultural activities, development of residential areas, trade or services. which is not in accordance with environmental concepts. Efforts to utilize protected areas into cultivated areas are increasing. In addition, there is still a tendency for people to use shifting cultivation systems as well as land clearing for agricultural purposes and built-up areas [3], which has the potential for floods and landslides as had happened several years earlier.

In the use of space for various activities in the city of Padang, there are various problems such as the conversion of forest land into residential land and gardens, development that violates established boundaries and development that is not in accordance with land use, land use in the area along the watershed can interfere with the ecological function of the watershed so that the impact on residential areas downstream. Batang Kuranji watershed has an area of 22,544 ha [4]. Throughout 2008-2009, showed runoff coefficient value (c) and the River Regime Coefficient (KRS) was high, so that it could be said that the watershed conditions were bad. Therefore the watershed includes the Super Priority Watershed. The last Banjir Bandang on July 24, 2012, caused high runoff, large erosion and sedimentation. Based on these problems, it is necessary to do research with the title "Evaluation of Land Use Against the Criticality of Batang Kuranji Das in Padang City"

2. Research Methods

2.1. The materials used in this study are as follows:

Basic map for research: Padang Sheet Topographic Map (0715-32) Scale 1: 50,000 in 1984, Solok Sheet Topography Map (0815-11) Scale 1: 50,000 in 1984, Map of the Land Type of Padang City Scale 1: 100,000 in 2008, Padang City Land Use Map Scale 1: 100,000 in 2013, Geological Map of Padang City Scale 1: 100,000 in 2008, 2012 Padang Quickbird Image g) DEM (Digital Elevation Model)

2.2. Research Sample

Samples were taken based on the land unit map found in the Padang City Kuranji Watershed. Maps of watershed land units are obtained from overlapping maps of landform units, slopes, land use, types of soil and geology. Determination of samples drawn by purposive sampling is withdrawal based on calculations, namely samples taken as much as one sample per unit of land. The land unit sample in the study area was 26 land units. The type of data needed is; Slope, soil texture, effective depth of soil, drainage, crop management taken from the everlay map and primary data in the field.

To determine watershed criticality, Landsat TM +8 satellite imagery, thematic maps and topographic maps were corrected and analyzed, both visually and digitally for the Kuranji sub-watershed. After identifying the location of the observation from the field data taken then do a scoring for each parameter. After that, do the overlay process of the maps with the Geographic Information System so that the critical watershed level can be obtained in the Kuranji watershed division.

Critical watershed analysis all these parameters are scored. The critical level of watershed in terms of land use is based on the total score of the parameters used with the following formula or model:

$$\sum = IPL + KPL + TBE + PL \quad (1)$$

Where:

IPL : Index of Land Closure by Vegetation
KPL : Suitability of Land Use
TBE : Danger Level of Erosion
PL : Land Management

To determine the critical level of watershed in the Kuranji watershed, the scoring and weighting method is used where the score of each critical land parameter is multiplied by the weight of each critical land parameter.

Table 1. Total Score on watershed criticality level

No	watershed criticality	Score
1	Critical	4 - 9,3
2	Critical Potential	>9,3 - 14,7
3	Not Critical	>14,7 - 20

3. Research Results and Discussion

3.1. Land Use Index in the Kuranji Watershed of Padang City

Land use in the Kuranji watershed is mostly used for forests by 59.87% and only 7.78% is used for settlements, for details, see table 2.

Table 2. Use of watersheds. Kuranji

Land Use	Area (km^2)	%
Forest	133,63	59,87
Settlement	17,36	7,78
Garden	19,98	8,95
Rice Fields	52,24	23,40
Quantities	223,22	100

Permanently restricted watershed area is obtained from the land use map of 1: 150,000 scale in 2013. The permanent vegetation referred to is annual crops and forests that function as protection or conservation. The land use index is obtained from a permanent vegetation map with a watershed map. The next stage is to calculate the area of permanent vegetation and sub-watersheds divided by the area of the sub-watershed. The classification of land use index in the Kuranji watershed can be seen in the sub-watershed table 3.

Table 3. Land Use Index (IPL) of the Kuranji Sub-watershed

No	Sub Watershed	LVP (Ha)	Area Sub Watershed (Ha)	Percentage (%)	IPL
1	Watershed Kuranji	2250,58	5501,11	40,91	Medium
2	Sub Watershed Belimbing	524,15	892,73	58,71	Medium
3	Sub Watershed Limau Manis	3040,9	3129,33	97,45	Good
4	Sub Watershed Lubuk Gajah	976,00	3039,63	32,11	Medium
5	Sub Watershed Padang Janiah	2120,07	2126,07	99,72	Good
6	Sub Watershed Padang Karuah	4733,00	4738,93	99,88	Good
7	Sub Watershed Sapih	214,22	1282,21	16,71	Bad
8	Sikabu Gadang	1481,36	1612,82	91,86	Good
	Jumlah	15340,28	22322,83	68,72	**Medium**

LVP: Permanent Vegetation Areas

Land use in the Kuranji watershed can be divided into 3 classes of indices, namely: "Good" index class where permanent vegetation in a large watershed of 75% is spread in the Padang Karuah Sub-watershed, Sikabu Gadang Watershed (91.86%) Sub Lake Limau Manis watershed (97%) and Padang Janiah sub-watershed, with an average index class of 97.23% and all land area of 11.375 km2.

"Medium" Index where permanent vegetation in the watershed 30% -75% is spread in the Belimbing Sub-watershed, Lubuk Gajah Sub-watershed, and Kuranji watershed with an average index class of 43.91% covering 3.750 km2, while index-class land use " Ugly "scattered in the Sapih Sub-watershed with index class 16.71%, the number of permanent vegetation is less than 30%, the dominant land use in this area is settlements and rice fields. Overall class of overall land use index of watershed. Bt. Kuranji is "Medium".

3.2. Land Use Suitability (KPL)

The suitability of land use in the Kuranji watershed consists of 2 classes, namely: "Appropriate" class (85.97%) which is scattered in frictional shoals, alluvial and volcanic plains, where this area is dominated by forest land cover, rice fields and plantations. Class "not suitable" (14.03%) spread in alluvial plains and alluvial fans with residential land use, plantations and rice fields with large erosion. The suitability of land use in the Kuranji watershed can be seen in the table 4.

Table 4. Suitability Level of Use of Kuranji Watershed

No	Suitability Level of Use	Area Amount Amount (km^2)	%
1	accordingly appropriate	191,99	85,97
2	not appropriate	31,32	14,03
	Amount	223,31	100

Source: Primary Data Processing 2018

3.3. Distribution and Extent of Critical Watersheds

The criticality level of the Kuranji watershed consists of 3 classes: "uncritical" class (70.94%) spread in the form of volcanic land, critical potential class (24.54%) spread in the central part of the Kuranji watershed with the use of rice fields and plantations, and Critical class (4.53%) is spread in alluvial fan land area with residential land use, plantations and rice fields. For more details, see Table 5 and map critical watershed levels.

Table 5. Watershed criticality level

No	Watershed criticality level	Area (ha)	%
1	Critical	1011,06	4,53
2	Critical Potensial	5478,78	24,54
3	No Critical	15840,00	70,94
	Amount	22329,84	100

Land at a critical level is generally characterized by turbid river flow caused by the amount of sedimentation due to lack of permanent vegetation and the amount of erosion in the area around the river, also due to inappropriate land use, both based on the damage and geology.

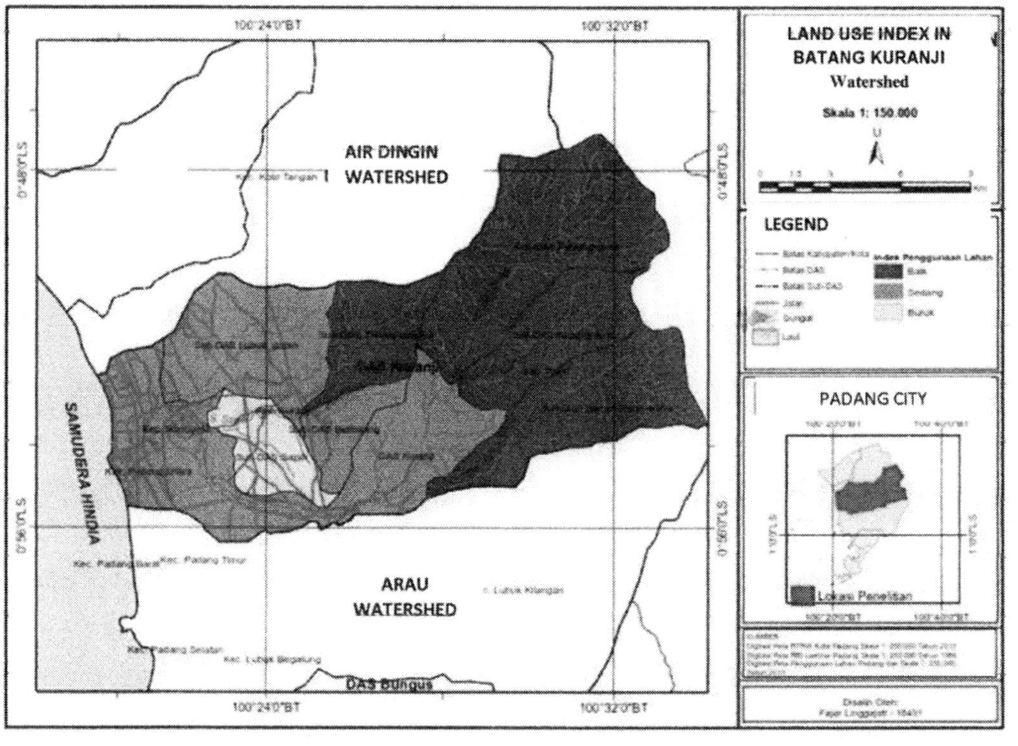

Figure 1. Map of Land Use Index in Batang Kuranji Watershed, Padang City

3.4. Danger Levels of Erosion

There are 5 classes of erosion hazard in the Kuranji watershed: Very light (68.64%) spread in the middle part of the watershed with the use of plantation land and rice fields with flat slope, Lightweight (9.75%) scattered in the alluvial fan area with use of plantation land and rice fields, Medium class (14.68%) spread in the middle area of watershed and volcanic central slope with forest land use, Weight class (4.79%) spread in alluvial fan areas with forest land use and very steep slope, Very Heavy class (2.14%) spread in alluvial fan areas with residential land use, plantations and rice fields. More details can be seen in table 6 and the map on the next page.

Table 6. Danger Level of Erosion

No	Danger Level of Erosion	Area(ha)	%
1	Very Mild	15328,24	68,64
2	Mild	2178,21	9,75
3	moderate	3179,37	14,68
4	Heavy	1068,95	4,79
5	Very Heavy	477,36	2,14
	Amount	22332,13	100

The high erosion in the study area was caused by the shallow soil solubility in the upper part of the Kuranji watershed and the degree of slope. To reduce erosion, the forest in this area must be maintained and the community is prohibited from clearing land for other purposes. Erosion hazard land distribution can be seen in figure 2.

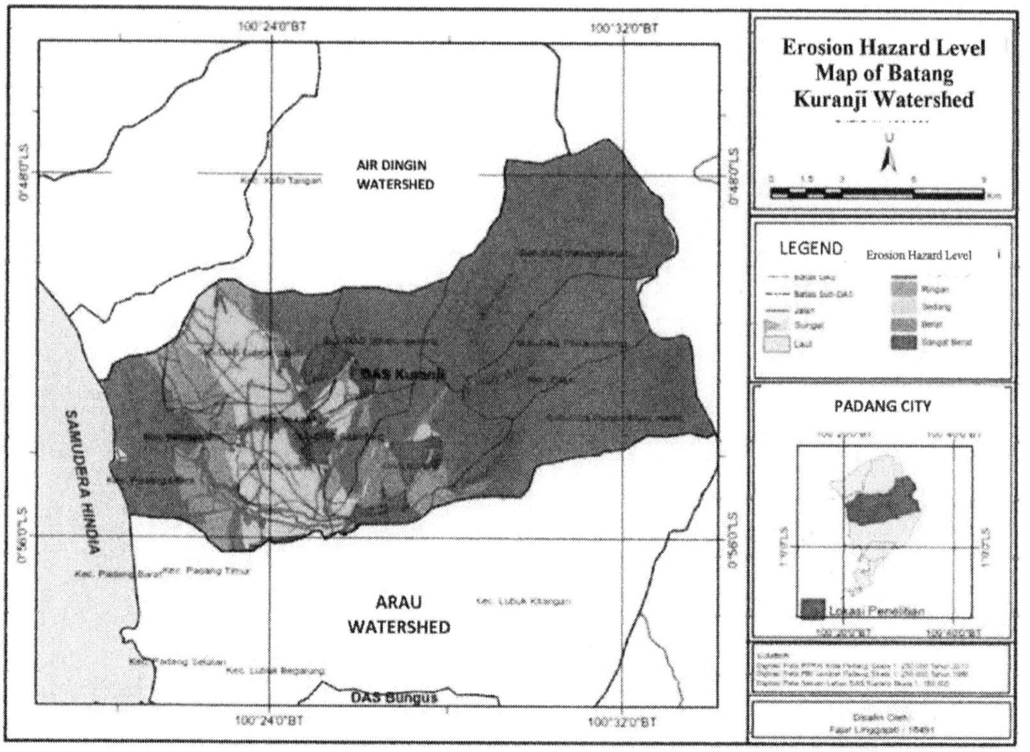

Figure 2. Erosion Hazard Level Map of Batang Kuranji Watershed in Padang City

4. Conclusion

From the results of research that has been obtained in the Padang City Kuranji Watershed, conclusions can be drawn as follows:

a. There are 3 classes of land cover index in the Kuranji watershed namely good class 11,375 km2 or 97,23%, medium class 3750,73 km2 or 43,91% and bad class (214,22 km2) or 16,71%
b. The suitability of land use in the Kuranji watershed in many downstream areas that do not match what, residential land use, rice fields and plantations are found in the land suitability

class of class V-VIII. There are 3132.41 ha or 14.3%. land that is not suitable for the Kuranji watershed.

c. There are three classes of watershed criticality in the Kuranji watershed namely 70.94% non-critical class, 24.54% critical potential class and 4.53% critical class.
d. The erosion hazard level of the Kuranji watershed is classified as 4.79% by weight and very heavy by 2.14%. The greater erosion that occurs in an area will result in the loss of topsoil, covered soil pores, high run-off and the amount of sedimentation that adversely affects the watershed.

5. Suggestions

a. At the location of the high watershed critical level it is recommended to make green open spaces and plant trees in the home garden.
b. Land use by the community is as good as being able to guide land use maps to reduce the risk of landslides, drought and damage to soil structure.
c. At the upstream part of the river, it should be carried out roboisasi and still be conserved in order to reduce damage to the watershed.
d. In high erosion class locations, land clearing and existing protective vegetation should not be damaged and the community is prohibited from settling there for the safety of the risk of lonsor disaster.

References

[1] Asdak C 2014 *Hydrology and Management of Watersheds* (Yogyakarta : Gajah Mada University Press) p 535 – 537
[2] Arsyad 2015 *Soil and Water Conservation* (Bogor : Department of Soil Science and Land Resources IPB Press) P 56
[3] Linggajati 2017 The criticality of the Kuranji watershed in terms of land use *Skripsi* (Padang: FIS UNP) p 88 – 97
[4] Govers G, Roel M, Bas VW and Kristof VO 2017 Soil Conservation in the 21st Century: Why We Need Smart Agricultural Intensification *Soil Journal* **3** 45-59

IOP Conf. Series: Earth and Environmental Science **314** (2019) 012005 doi:10.1088/1755-1315/314/1/012005

The impact of ultrasonic power and time for the removal of Total Petroleum Hydrocarbon from low permeability contaminated soils

Agus Jatnika Effendi [1,*] and Marita Wulandari[2]

[1] Environmental Engineering Department Faculty of Civil and Environmental Engineering, Institut Teknologi Bandung, St. Ganeca No 10 Bandung, West Java 40132 Indonesia

[2] Environmental Engineering Department, Institut Teknologi Kalimantan, St. Soekarno Hatta Km 15, East Kalimantan 76127 Indonesia

*jatnika@indo.net.id

Abstract. Ultrasonic was widely studied in soil or sediment with several types of contaminants, both heavy metals and organic matter. Ultrasonic may enhance the desorption or leaching of contaminants from the soil and play an important role in the formation of radicals (•OH) that involved in the oxidation of hydrocarbon contaminants in the soil. This research has consisted of a tube reactor made from stainless steel in 21 cm × 21 cm × 18 cm of volume. According to previous research, ultrasonic power not only increases the desorption process or leaching contaminants from the soil but also plays an important role in the formation of radicals (•OH) which are oxidizers involved in the oxidation process. The main reactor has a dimension of 21 cm × 21 cm × 18 cm. The system used was a bath system, where the transducer was attached under the bath (reactor) so that it indirectly produces sonication. Contaminated soil has low permeability and obtained from conventional petroleum mining in Grobogan Regency, Central Java. Initial concentrations of TPH was 334100 mg / kg. Based on the optimum time experiment, it can be seen that the TPH removal was significant at the initial 15 minutes of sonication. At optimum conditions, TPH efficiency removal at 160 watts of power, 48 kHz of frequency, 1: 9 (gr/ml) of soil / liquid ratio was 61.03%.

1. Introduction

Petroleum hydrocarbons contaminated soil have serious problems with the environment, soil ecology, human health, and requires effective soil remediation technology [1]. The contaminants are processed and reduced to the concentration of Total Petroleum Hydrocarbons (TPH) having a final value requirement of less than 1% in accordance with the Minister of Environment Decree Number 128 of 2003 concerning Biological Procedures and Technical Requirements for Petroleum Waste and Contaminated Soil. Alternative techniques for processing contaminated soil are physical and chemical remediation, one of them is soil washing. However, one of the disadvantages of this technology is the use of surfactants which have an impact on toxicity, environmental balance, and biodegradability [1]. Therefore, the other alternative technology is ultrasonic irradiation.

Ultrasonic has been shown to be an effective method for removing adsorbed material in soil particles, separating soil from water in high suspension concentrations, decreasing the stability of the

IOP Conf. Series: Earth and Environmental Science **314** (2019) 012005 doi:10.1088/1755-1315/314/1/012005

water/oil emulsion [1]. The decay of cavitation bubbles and shock waves (shockwaves) by ultrasonic is known to break down aggregate soil particles and support the release of contaminants that adhere to the soil [1,2]. The success rate of this ultrasonic method can be influenced by several factors such as soil type, soil liquid ratio, temperature, wave frequency, and energy/power used [3].

Soil clay (clay) has a low permeability value, making it difficult to treat. The desorption of chemical compounds on finer soil such as clay is more difficult than with coarser soil particles [4]. Song also stated that the desorption of contaminants in clay or dust type soil takes longer than the soil with a larger fraction such as sand. However, according to some studies, ultrasonic has been proven to be able to process the soil with a fine texture if the required ultrasonic energy is available. Some studies reveal the ultrasonic effects on contaminated soil. However, only a few studies have discussed the comparison of the effects of ultrasonic time and power on desorption of TPH contaminant in low-probability soil. In this study, experiment was carried out by comparing the efficiency of contaminant reduction in several treatments. They are ultrasonic irradiation of various power and ultrasonic time.

2. Methods

2.1 Soil Preparation

Contaminated soil sampling using a disturbed grab sampling method with a depth of 0-15 cm. Samples of petroleum contaminated soil come from around the petroleum mining site that has been contaminated for decades (aged-oil contaminated soil) located in Grobogan Regency with location coordinates that are 7°11'36,28 " S and 111°11'42, 26 " E. Then, contaminated soil was prepared before the experiment. Soil preparation begins with silencing the soil for several days (7-14 days) and contacting it with open air without being exposed to direct sunlight so that air-dried soil was obtained. After drying, the soil was filtered on a sieve measuring 2 mm / 10 mesh. The soil that does not pass through the sieve was crushed again using the pestle and filtered again. After that, the soil was homogenized by the cone & quartering technique.

2.2 Ultrasonic Reactor

The main reactor has a dimension of 21 cm × 21 cm × 18 cm. In the ultrasonic bath system, the ultrasonic transducer is attached under the bath (reactor) so that it indirectly produces sonication. Bath itself can be used as a reaction container to receive ultrasonic waves. Figure 1 shows the scheme of an ultrasonic reactor.

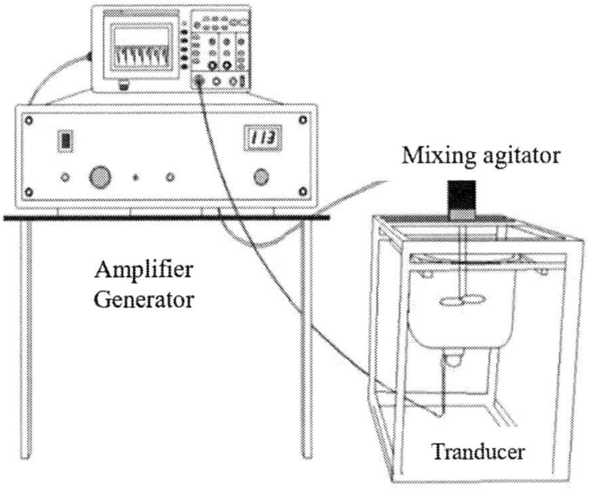

Figure 1. The Scheme of Ultrasonic Reactor [5]

2.3 Soil Characteristics

The quality of physical and chemical characteristics such as water content, pH, a conductivity of several other parameters, are the factors that significantly reduce the petroleum contaminants [6,7]. The physical parameters really affect the effectivity of the ultrasonic remediation. The soil coefficient of permeability value is low. The petroleum contaminated soil used in this study is categorized as sandy silt loam. The physical and chemical characteristics of the soil can be seen in Table 1.

Table 1. Physical and chemical characteristics of the contaminated soil

Parameter	Values
Soil Properties	
Gravel	5,2 %
Sand	15,81 %
Silt	45,46 %
Clay	33,53 %
Permeability	$2{,}81\times 10^{-5}$ cm/s
pH	9,30
N Total	0,130 %
C Organic	4,64 %
Oil & grease	8,77 %
TPH	3,41 %

The surface area of soil particles is a significant factor that affects the physical and chemical kinetics on the processing of the contaminated soil. Along with the increased surface area, the adsorption capacity and biological activity. The sand and silt particles have a relatively smaller surface area compared to clay fraction. The silt fraction, or commonly referred to dust or loam, is a fine fraction with a size range of 0.05-0.02 mm. The clay fraction is the finest size fraction with a size less than 0.002 mm. Soil texture is one of the important parameters in the soil washing process because the larger the size of the soil particles the easier it is to be drained by water [8]. Compared to sand, the clay structure tends to be more complex and a lot of layers and micropores between soil matrices, which means there are more contaminants that adsorbed and trapped [4].

2.4 Ultrasonic Process and Analysis

The sample of the contaminated soil that has been dried and filtered is mixed with a washing solution (deionized water) with soil ratio: 1: 3 and 1: 12 of liquid for both mechanical shaking only and mechanical shaking + ultrasonic irradiation. After the suspension looks homogeneous, a reactor was placed in ultrasonic with the desired frequency and power for a certain period of time. The ultrasonic test was done by three times of replications in each combination. The sample will be stirred with a stirrer for 10 minutes with a rotation speed of 113 rpm, and samples are collected in every 1 minute for TPH and oil & grease measurements on the soil and water, then a TPH elimination plot was made per time unit of the process. After the soil samples have reached the equilibrium conditions, the liquid phase is separated for 20 minutes by centrifugation at 3000 rpm, then filtered (filtration) to separate the soil (solids) and liquid. After that, the soil was drying by air dried for ±3 days. Then analyzed by liquid-liquid extraction to determine the balanced concentration.

2.5 Hydrocarbon Compounds Analysis with GC-MS

The separation of soil with liquid is done by filtration, and then extracted and analyzed by TPH compounds on the soil with GC-MS. The principle of GC is to separate organic compounds by passing the gas flow through the stationary phase. The mooring time (retention time) on the GC shows the identity of the compound that can give qualitative information, which is the presence or absence of certain compounds and qualitatively can show the number of each compound in a mixture. The GC device usage results in a separation of compounds in hydrocarbon fraction so that degradation can be detected by looking at the growth or loss of various peaks which are the identities of hydrocarbon compounds. The injector temperature is 250 ° C, detector temperature is 300° C and column temperature is 40°/ 2 minutes, 10° C / minute, 200° C/ 1 minute, and 290 ° C/ 10 minutes.

3. Results and Discussion

3.1 Effects of ultrasonic time on TPH

Based on the optimum time experiment, it can be seen that the degradation of TPH was significant at the initial 15 minutes of sonication. As the ultrasonic process continued, more oil was transferred to the solvent (deionized water) until they reached the maximum release capacity. Referring to these results, it is likely that at that time a desorption and re-adsorption process of contaminants into soil particles has occurred, or it can be referred to reaching the equilibrium phase. However, it still needed to be proven by calculating the desorption capacity through desorption isotherm. The balance between oil re-adsorption and the injection of oil into the solution/emulsion occurred very quickly [9].

The initial TPH value was 32437 mg/kg then degraded to 22296 mg/kg at 13 minutes. This means that there was 45.48 % of degradation. The degradation of TPH can be seen in Figure 2. The test results have shown that time gives a positive impact on the desorption of petroleum hydrocarbon, and it is confirmed by the Wilcoxon signed-rank test.

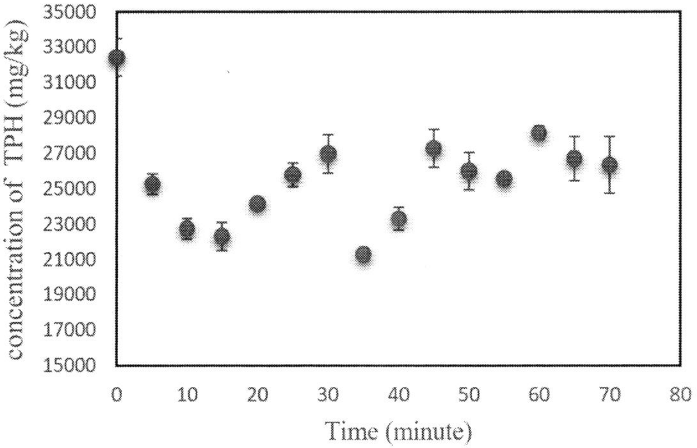

Figure 2. Graphic of the degradation of TPH (70 minutes)

Based on statistical testing with the Wilcoxon signed-rank test, the obtained P value was 0.000061 or can be written as p value < 0.05. This means that H_0 is rejected by the difference in the average TPH value before the ultrasonic process and after the ultrasonic process. Based on the optimum time test results, the optimum time for ultrasonic remediation used in the next test in this study is 10-15 minutes. This result was also supported by Li et al., (2013). The research by the optimum time for the ultrasonic remediation was 15 minutes, where the desorption process was able to reach the equilibrium point at that time.

The reduction of TPH increases at 1 to 5 minutes [10]. However, the ultrasonic remediation at more than 5 minutes, no significant change occurred, and finally the reduction of TPH degraded at 8 minutes. So, it can be concluded that the processed sample of soil with longer ultrasonic, does not provide a greater reduction of the contaminants compared to the shorter processed samples. During processing (sonication), oil was desorbed into the slurry phase and resulting in a mixture of solution/emulsion between water and oil [11]. The efficiency of the ultrasonic period is important to do because it will affect the economic factor (costs required) related to the energy used.

IOP Conf. Series: Earth and Environmental Science **314** (2019) 012005 doi:10.1088/1755-1315/314/1/012005

3.2 Effect of power and intensity on TPH removal

Power setting in ultrasonic irradiation can regulate how strong the vibration occurs, so the process occurred in the form of how strong the collisions occurred between particles. Power is closely related to ultrasonic intensity. Intensity is directly related to ultrasonic wave amplitude. In general, increased intensity can increase the sonochemical effect in a liquid medium, therefore, the greater the power used, then the cavitation effect will become better [11]. The intensity produced by the transducer is equal to the electrical power divided by the transducer surface area. That is, the higher the power, the higher the intensity [12]. Ultrasonic intensity values based on power values can be seen in table 2.

Table 2. Intensity values based on powers variation

No	Power (watt)	Intensity (Watt/cm^2)
1	70	0,79
2	100	1,13
3	130	1,47
4	160	1,81

Degradation of TPH concentration based on the time during the occurred irradiation process can be seen in Figure 3.

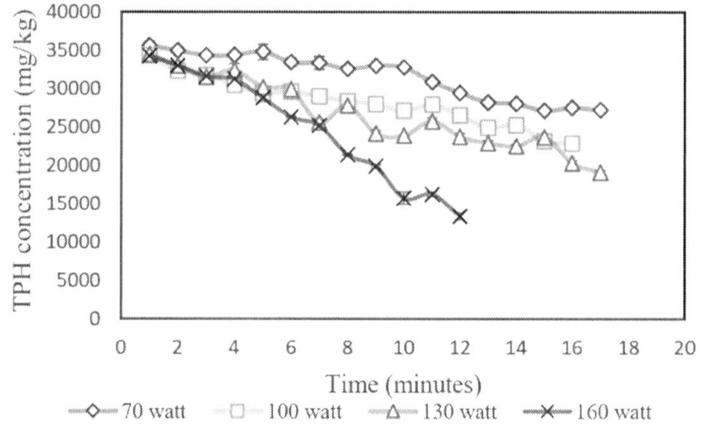

Figure 3. The degradation of TPH concentration at various powers

According to Figure 3, it can be seen that the concentration of TPH undergoes degradation during the ultrasonic irradiation process that occurred up to a point. The removal efficiency at the power of 70 watts, 100 watts, 130 watts, and 160 watts are respectively 30.80%, 33.44%, 44.53%, and 61.03%. Power and intensity are closely related to the desorption process and degradation of the contaminants in the slurry phase when ultrasonic is applied [1,10].

If the frequency used is in a high range, the intensity needed will also be greater. This means that the greater the power, then the possibility in contaminant removal will also be greater [11]. Based on the results of this study, 70 watts of power was actually sufficient as a condition to produce cavitation effects in a liquid medium for 48 kHz of frequency.

ANOVA test results for power variables towards TPH contaminant, obtained calculated F value of 4.319, while the F table value is 2.75. Therefore, the calculated F > F table, then H_0 is rejected and H_1 is accepted. Then it can be concluded that the difference in power gives a significant effect on the degradation of TPH.

The degradation in the value of the contaminant can be explained by the main mechanism that plays the most role, i.e. desorption. The results indicate that desorption is an important mechanism that has the potential to reduce TPH from the soil. The available energy must be sufficiently large and strong enough so that cavitation can occur then the adsorbed contaminants can be desorbed into the solution. The size of soil particles affects the energy needed. Because sand has a smaller surface area, the energy needed is also lower, and on the opposite, loam or clay type soil, has a larger surface area, and so the energy needed to remove contaminant is greater. Ultrasonic processing can be successfully applied to treat coarse as well as fine soil when sufficient energy is available. However, fine soil has a low hydraulic conductivity which is like to inhibit the transfer and desorption of contaminant values [13]. Figure 4 shows the removal efficiency of oil & grease and TPH at various powers during certain times.

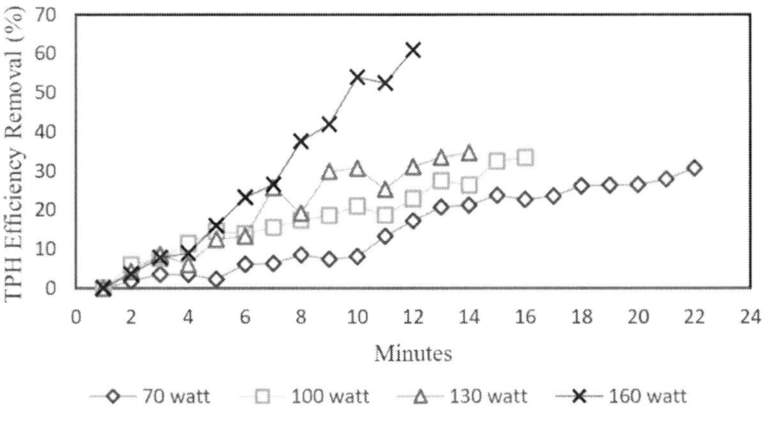

Figure 4. TPH Efficiency Removal at various powers

Based on Figure 4 it can be seen that the highest contaminant removal happens at the ultrasonic process with 160 watts of power. When viewed, there are differences in the equilibrium time between the four powers. Where the lowest power (70 watts) has an equilibrium time that tends to be longer than ultrasonic irradiation at other power. The minimum power needed for organic decontamination processes including hydrocarbon petroleum is 100-600 watts [14]. As previously explained, power will affect intensity. In order to occur a reaction that is cavitation formed, the intensity is 0.35 watts/cm^2. This means that even with 70 watts of power, cavitation can occur. Removing oil contaminants occur at 100 watts of power usage [3]. In this study, it means that the best intensity used so that the remediation process runs optimally is 1.81 watts/cm^2.

3.3 Mixing Effect

Mixing can help the dissolution process because it is capable to distribute soil to all sections of the solution so that larger the area of contact between the solid and liquid phase. Mass transfer and oxidation of hydrocarbon compounds only occurred in the solution or slurry phase in order to achieve the dispersion process of the contaminant into the solvent. The dispersion effect is investigated by comparing the remediation process with mechanical stirring alone to ultrasonic process that equipped with mechanical stirring. Ultrasonic waves are able to produce a droplet with a very small size, and that creates emulsion between two phases [15]. After the sonication process, the two phases can be separated easily, due to its density differences. Figure 5 shows the oil & grease concentration and TPH in two different processes (mechanical stirring alone and ultrasonic process equipped with mechanical stirring).

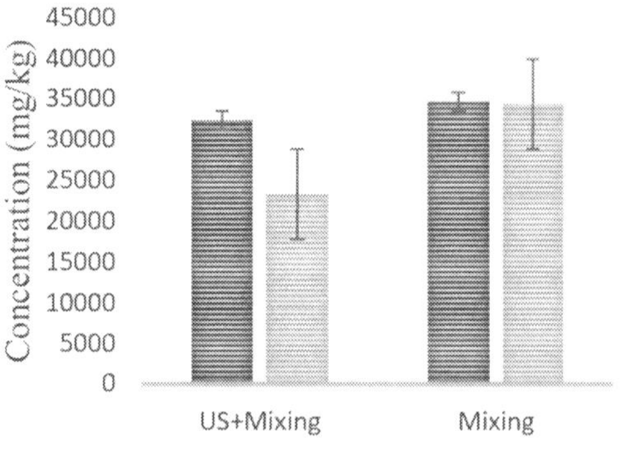

Figure 5. The differences in the degradation of TPH concentration between two processes (us+ mixing and mixing only)

An experiment to prove the effect of stirring was done in one hour with the mixing speed 113 rpm both with mechanical mixing alone and mechanical mixing equipped with ultrasonic. Based on Figure 5, it can be seen that the removal of the contaminant's concentration occurred in the ultrasonic process equipped with mechanical mixing where the removal is 50.81% for oil & grease and 28.12% for TPH. Meanwhile, the percentage of contaminants in the mechanical stirring only process is 0.87% for oil & grease and 0.96% for TPH. TPH removal percentage on ultrasonic + mechanical stirring tends to be small. This is because the contaminants removal efficiency was occurred at 10-15 minutes sonication, as a result after one hour of sonication, the degradation of contaminants concentrations is not significant and even tends to increase due to the re-adsorption process.

The mechanical mixing process can create better turbulence and mixes, thus contributes to making physical contact between the contaminated soil particles and washing solution [16]. As a result, the pollutants that bind weakly with soil particles, are able to be released from the soil surface. Meanwhile, the pollutants that bind strongly with soil pore matrices will be more difficult to desorb if using mechanical mixing alone. Ultrasonic/mechanical mixing produces higher removal efficiency due to the desorption of both macroscale and microscale which not only remove the contaminants from the soil surface but also the pollutants that trapped in the pores of soil particles [17]. This argument is also supported by the research conducted by [16] whose results have shown that the removal efficiency of heavy metals with HCl washing solution was higher in ultrasonic/mechanical mixing processes compared to mechanical mixing alone or ultrasonic alone.

3.4 Analysis of the contaminants with GC MS

GCMS test (gas chromatography- mass spectrophotometry) is used to find out the organic compounds contained in the soil sample. GCMS test is used to confirm the occurred breakups of the organic compounds through the ultrasonic remediation process. Another goal is to find out the characteristics of the contaminants contained in the soil sample. Figure 6 (a) and (b) show hydrocarbon compounds before and after ultrasonic remediation.

IOP Conf. Series: Earth and Environmental Science **314** (2019) 012005 doi:10.1088/1755-1315/314/1/012005

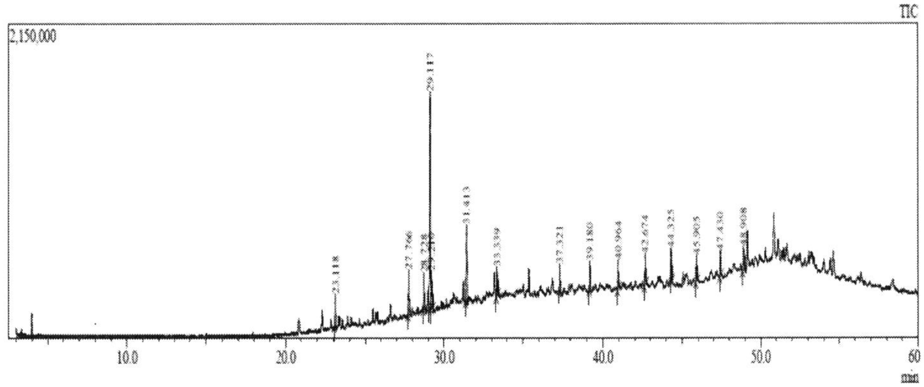

(a)

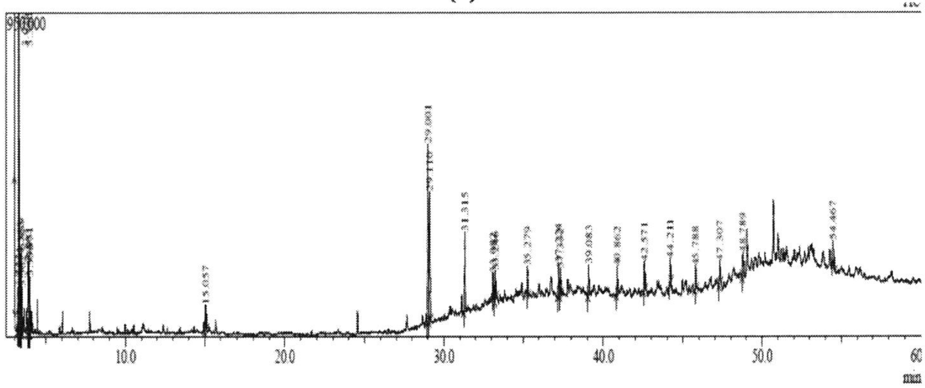

(b)

Figure 6. GC MS results on contaminated soil (a) before ultrasonic remediation (b) after ultrasonic remediation

The compounds in contaminated soil are classified as aged petroleum hydrocarbon because the soil has been contaminated for a long range of time. Most of the compounds detected are alkane compounds or also known as a saturated hydrocarbon. Based on the literature, ultrasonic cavitation has the potential of increasing the destruction and degradation of compounds, which when viewed from the carbon chains, are partially stable. Changes of hydrocarbon compounds can be seen in Figure 6. Compared to the initial GC MS graphs, there is a slight movement of peak up to retention time that reaches 60 minutes. The results of the degradation of petroleum by ultrasonic might strengthen the degradation of some chains of organic compounds into a relatively simple form and tend to be volatile. GC MS profile before and after ultrasonic remediation changes with a peak that tends to decrease in the final GC results. TPH component undergoes a change due to the pyrolysis of long-chain organic molecules, turned into intermediate compounds that have a lighter molecular weight [18]. In his research, it was seen that all peaks of n-alkanes degraded. This indicates that ultrasonic processing is not selective in degrading TPH.

In this study, after the remediation process, short-chain and volatile compounds like C_2, C_8, C_{10} were found, which were not previously found at the initial condition before remediation. Most of the compounds were still aliphatic, but the number of aromatic compounds were increasing. It is likely that the degraded aliphatic compounds that had enough energy to achieve stability formed the aromatic groups.

Some of the organic compounds (long-chain hydrocarbon) that destructed can occur due to the production of hydroxyl radicals [9]. This usually occurs at 200- 400 kHz of middle frequency.

Whereas in this study, the used frequency was low at 48 kHz. Hence, the destruction change of hydrocarbon compounds is not significant. However, the energy released from ultrasonic cavitation under the condition of 48 kHz of frequency is sufficient to achieve affinity of hydrocarbon molecules and increasing the desorption.

4. Conclusion

The results of this study and statistical analysis showed that the time and power of ultrasonic irradiation had a significant effect on the contaminants removal (oil & grease and TPH) in petroleum-contaminated clay. According to the optimum time test, the degradation of contaminants occurred at the initial 15 minutes while the rest is relatively stable and even had an escalation that can be happened due to the re-adsorption process. The highest contaminations removal occurred at 160 watts of power with an intensity of 1.89 watts/cm^2 with a removal efficiency of 63.15%

Acknowledgments

The authors acknowledge the financial support for this study from the Environmental Engineering Department, ITB.

References

[1] Li J, Song X, Hu G and Thring R W 2013 *Journal of Environmental Science and Health, Part A: Toxic/Hazardous Substances and Environmental Engineering* **48**(11) 1378-1389
[2] Mason T J, Collings A and Sumel A 2004 *Ultrasonic Sonochemistry* **11**(3-4) 205-210
[3] Kim Y U and Wang M C 2003 *Ultrasonics* **41**(7) 539-542
[4] Song X 2007 *Remediation of salt and petroleum hydrocarbon contaminated ultrasound and soil washing* (Canada: Thesis University of Northern British Columbia)
[5] Wulandari and Effendi A J 2018 Effect of frequency and ratio solid-liquid on ultrasonic remediation of petroleum contaminated soil *AIP Conference Proceedings 2014, 020120 (2018);* https://doi.org/10.1063/1.5054524
[6] Pathak H, Bhatnagar K and Jaroli D P 2011 *IJ FALS* **1**(3) 84-89
[7] Onojake M C and Osuji L C 2012 *Arch. Appl. Sci. Res.* **4**(1) 48-58
[8] Bagarello V, Sferlazza S and Sgroi A 2009 *Geoderma* **154**(1-2) 52-58
[9] Feng D and Aldrich 2000 *Advances Environmental. Res.* **4**(2) 103–112
[10] Zhang J, Li J, Thring R and Liu L 2013 Application of ultrasound and fenton reaction process for the treatment of oily sludge *Procedia Environmental Sciences* **18** (2013) 686 – 693
[11] Mason T J and Lorimer J P 2002 *Applied sonochemistry: the uses of power ultrasound in chemistry and processing* (Newyork: Wiley-VCH)
[12] Mason T J (ed.) 1990 *Sonochemistry: the uses of ultrasound in chemistry* (Cambridge, Eng. : Royal Society of Chemistry)
[13] Collings A F, Farmer A D, Gwan P B, Pintos A P S and Leo C J 2006 *Minerals Engineering* **19**(5) 450-453
[14] Pham T D 2014 ISBN 978-952-265-644-5 Lappeenranta University of Technology University Press: Finlandia
[15] Deshpande A, Bassi A and Prakash A 2005 *Energy Fuels* **19**(1) 28-34; doi: 10.1021/ef0340965
[16] Park B and Son Y 2017 *Ultrasonic Sonochemistry* **35**(Part B) 640-645
[17] Son Y, Cha J, Lim M, Ashokkumar M and Khim J 2011 *Ind. Eng. Chem. Res.* **50**(4) 2400-2407
[18] Zhang W, Li J, Han Y and Ong W 2010 Experimental study on the remediation of oil contaminated soil using ultrasound *2010 4th International Conference on Bioinformatics and Biomedical Engineering, iCBBE 2010.* 1-4. doi: 10.1109/ICBBE.2010.5516737

A dynamics condition of coastal environment in Padang City-Indonesia

Febriandi[1,2,*], Dasman Lanin[2], Dedi Hermon[1], Siti Fatimah[2], Triyatno[1,2] and Aprizon Putra[1]

[1] Department of Geography - Padang State University, Indonesia
[2] Postgraduate - Padang State University, Indonesia

*febriandigeografi@gmail.com

Abstract. Padang City is located on the West coast of Sumatra Island that borders directly with the Indian Ocean. The coastal region of the city stretches from Pasia Jambak beach, Muara Anai in the Northto Sungai Pisang in the South. The coast of South section is different from the coast of Northsection, because the Southern coastline is heterogeneous because there are many small bay, small islands and narrow beaches directly adjacent to the volcanic hills. This research aims to determine the dynamics condition of a coastal environment based on the characteristic (length, velocity, energy, and height), sediment transport value, accretion/abrasion factor value, and coastal equilibrium in Padang City.The method used in this study is the calculation of wavelength (L_0), wave velocity (C), wave energy (E), run-up height (H_b), and sediment transport (Q) to determine accretion/abrasion factor and equilibrium (G_0). Furthermore, it is done by *Geographic Information System* (GIS) approach to spatial information of research location. Research result shows that coastal region(beach)in Padang City tends to experience accretion along the beach of Pasia Sabalah, Panjalinan North, Panjalinan South, Parkit, Purus South, Purus North, Muaro Padang, Air Manis North and Pasia Putih. While the abrasion conditions occur inPasia Jambak, Parupuk Tabing, Bung Hatta University beach, Air Manis South and Sungai Beremas.

1. Introduction

The coastal region is very dynamic since the energy from ocean and land meet. The changes around coastal region occur slowly or fast, the depends on the equilibrium between topographic, rock structure and its behavior towards the wave, tidal, wind, and biological factor [1]. The process of equilibrium between the sediment source and the force acting on the shore will tend to cause the shoreline to change in the form of abrasion and accretion. [2] explain that the abrasion will cause the coastline to retreat to the land, while the advance coastline towards the sea is known as accretion. Abrasion disasters are often heard because they are more detrimental than accretion that only interfere with the management of ports and shipping lanes. The development of abrasion disasters is an illustration of the negligence of coastal utilization that is only for the benefit of the economy without understanding the natural processes of the coast. This problem is almost encountered in all the beaches in the world [3,4].

[5] explain if the shoreline experience abrasion and accretion that occur in a place, and it is difficult to overcome because the process usually works with high frequency for a long time, where the

development of coastal areas should consider the suitability of land between planned with their characteristics.The equilibrium dynamics are the common problems faced along Padang City coastal region, in the form of accretion and abrasion of coastal, caused by the wave and current movement. These events induced properties damages and loss of economic activities. [6,[7] add the coastal hazards become the social and economic problems that are difficult to completely overcome by the local government. Material detriments in the future become bigger along with climate changes, which affect ocean wave behavior. Moreover, the wave behavior influences the dynamics of the coastal region [8]. To reduce the negative effect, well planning and management of coastal region must be applied. This research aims to determine the dynamics of a coastal environment based on the characteristic (length, velocity, energy, and height), sediment transport value, accretion/abrasion factor value, and coastal equilibrium in Padang City.

2. Research Methods

The research was conducted along Padang City coastal region, started from the beach of Pasia Jambak, Muara Anai in North part to Sungai Pisang in the South part. Padang City has about ± 65 km of shoreline. It located in Sumatera west section coastal region between 0°44'00''-1°08'35''S and 100°05'05'' -100°34'09''E (Figure 1).

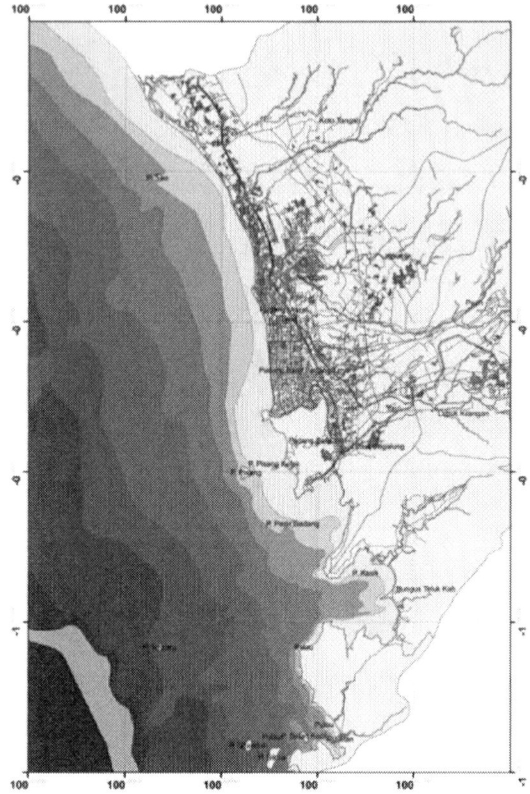

Figure 1. Map of research location.

The value derived from series of measurements, wave height (H_0), wave period (T), seashore slope ($Tg\alpha$) and median of sediment size/percentile (D_{50}), was substituted into data analysis formula in order to find wave parameter that meets the research purpose. Data analysis employed in this research show by following formulas:

Wavelength

$$L_0 = 1.56\, T^2 \quad (1)$$

where:

L_0 = Wavelength (m)
T = Wave period (sec) [9].

Wave Velocity

$$C = L/T \quad (2)$$

where:
C = Wave velocity (m/sec)
L_0 = Wave length (m)
T = Wave period (sec) [9].

Wave Energy

$$E = 1/8\, \rho\, gH_0^2 \quad (3)$$

where:
E = Wave energy (Kg/sec^2)
ρ = Sea water specific gravity (1.025 kg/m^3)
g = Earth gravity (9.81 m/sec^2)
H_0 = Wave height (m) [9].

Run-up Height

$$H_b = 0.39 \text{ x } g^{1/5}(T \text{ x } H_0^2)^{2/5} \quad (4)$$

where:
H_b = Run-up height (m)
T = Wave period (sec)
G = Gravity constants (9.8 met/sec^2)
H_0 = Wave height (m) (Triatmodjo, 1999).

Sediment Transport

$$Q = 1.646 \text{ x } 106\, H_b^2 \quad (5)$$

where:
H_b = Run-up height (m)
Q = Sediment volume (m^3/day) [10].

Factor of Accretion and Abrasion

$$G_o = (H_o/L_o).(Tg\ Q)^{0.27}(D_{50}/L_0)^{-0.67} \quad (6)$$

where:
H_0 = Wave height (m)
L_0 = Wave length (m)
D_{50} = Median of sediment size value (mm)
$Tg\,\alpha$ = Seashore slope (^{0}) [10].

Coastal region can be categorized as abrasion, accretion or equilibrium after we found the G_0 value, which then matches the following criteria:
< 0.0556 = abrasion
> 0.111 = accretion
0.556 - 0.111 = equilibrium [10].

3. Results and Discussion

The results of field measurement show found the value of wave height (*H*), seashore slope (α), the median of sediment size value(D_{50}), and wave direction (λ) at the research location as listed in Table 1 below.

Table 1. Value of *H, T,* α and D_{50}.

Location	*H (m)*	*T (sec)*	α %	α (°)	D_{50}	λ (°U)
Pasia Jambak	1.63	17	10	2.8 = 2°51'0'	0.221	216
Pasia Sabalah	1.29	14	22	8.2 = 8°7'0'	1.632	221
Panjalinan North	1.18	13	23	11.15 = 11°13'0'	1.132	213
Panjalinan South	1.71	23	35	17.1 = 17°2'0'	1.053	237
Parupuak Tabing	1.09	13	17	8.3 = 8°9'0'	0.341	239
Parkit	1.79	17	22	11.1 = 11°13'0'	1.567	242
Bung Hatta beach	1.15	14	20	8.2 = 8°7'0'	0.342	222
Purus North	1.50	14	23	8.2 = 11°13'0'	1.652	267
Purus South	1.38	14	24	8.2 = 10°19'0'	1.032	266
Muaro Padang	1.37	15	26	10.2 = 10°17'0'	1.609	249
Air Manis North	1.19	12	19	9.0 = 9°2'0'	1.591	243
Air ManisSouth	1.47	16	16	4.6 = 4°33'0'	0.386	208
Sungai Baremas	1.02	15	27	11.1 = 11°11'0'	0.912	195
Pasia Putih	1.10	14	22	9.47 = 9°29'0'	1.691	197

Source: Data Analysis, 2017

3.1 Wavelength

The results of wavelength values were can be seen in Table 2 below.

Table 2. Wavelength value

Location	($L_0 = 1.56T^2$)
Pasia Jambak	450.84
Pasia Sabalah	305.76
PanjalinanNorth	263.64
PanjalinanSouth	825.24
Parupuak Tabing	263.64
Parkit	450.84
Bung Hatta beach	305.76
Purus North	305.76
Purus South	305.76
Muaro Padang	351.00
Air Manis North	224.64
Air ManisSouth	399.36
Sungai Baremas	351.00
Pasia Putih	305.76

Source: Data Analysis, 2017.

The based on formula mentioned before, that wave velocity proportionally related to wavelength. the rising of wavelength will be followed by the rising of wave velocity value. The highest wave velocity (35.88 m/sec) took place at PanjalinanSouth beach, while the slowest (18.72 m/sec) took place at Air Manis North. Wave velocity average along Padang City around 23.51 m/sec.

3.2 Wave Velocity

The results of wave velocity values were can be seen in Table 3 below.

Table 3. Wave velocity value

Location	$(C = L/T)$
Pasia Jambak	26.52
Pasia Sabalah	21.84
PanjalinanNorth	20.28
PanjalinanSouth	35.88
Parupuak Tabing	20.28
Parkit	26.52
Bung Hatta beach	21.84
Purus North	21.84
Purus South	21.84
Muaro Padang	23.04
Air Manis North	18.72
Air ManisSouth	24.96
Sungai Baremas	23.04
Pasia Putih	21.84

Source: Data Analysis, 2017.

The based on formula mentioned before, that wave velocity proportionally related to wavelength. the rising of wavelength will be followed by the rising of wave velocity value. The highest wave velocity (35.88 m/sec) took place at PanjalinanSouth beach, while the slowest (18.72 m/sec) took place at Air Manis North. Wave velocity average along Padang City around 23.51 m/sec.

3.3 Wave Energy
The results of wave energy values were can be seen in Table 4 below.

Tabel 4. Wave energy value

Location	$E = 1/8\ \rho\ gH_0^2$
Pasia Jambak	3.34
Pasia Sabalah	2.09
PanjalinanNorth	1.75
PanjalinanSouth	3.68
Parupuak Tabing	1.49
Parkit	4.03
Bung Hatta beach	1.66
Purus North	2.83
Purus South	2.39
Muaro Padang	2.36
Air Manis North	1.78
Air ManisSouth	2.72
Sungai Baremas	1.31
Pasia Putih	1.52

Source: Data Analysis, 2017

Amount of wave energy is determined by wave height ρ and g value are the constants of sea water specific gravity (ρ = 1.025 kg/m^3, g = 9.81 m/sec^2). The higher the wave, the bigger the energy of the wave produces. Parkit beach had the biggest value of wave energy (4.03 Kg/s^2) and the smallest value in Sungai Beremas beach (1.31 Kg/s^2). The wave energy average of all the coastal region in Padang City around 2.35 Kg/s^2.

3.4 Run-up Height

The results of run-up height values were can be seen in Table 5 below.

Table 5. Run-up height value

Location	$H_b=0.39xg^{1/5}\ (T\,x\,H_0^2)^{2/5}$
Pasia Jambak	2,343.99
Pasia Sabalah	599.30
PanjalinanNorth	356.39
SouthPanjalinan	4,029.74
Parupuak Tabing	239.69
Parkit	3,743.54
Bung Hatta beach	337.43
Purus North	1,273.95
Purus South	839.64
Muaro Padang	867.49
Air Manis North	343.15
Air ManisSouth	1,316.06
Sungai Baremas	198.45
Pasia Putih	270.18

Source: Data Analysis, 2017.

The run-up height is determined by the period value and the wave height. If the period and the higher the wave, the run-up height also greater. After observing close we conclude that wave period value not much affected in determining run-up height value. The based on field survey result, the biggest run-up height value took place at Panjalinan North beach about 4,029.74 m and the smallest at Sungai Beremas beach about 198.45 m. The run-up height average of all the coastal region in Padang City around 1,197.07 m.

3.5 Sediment Transport

The results of sediment transport values were can be seen in Table 6 below.

Table 6. Sediment transport value

Location	$Q = 1.646 \times 106\ H_b^2$
Pasia Jambak	958,619
Pasia Sabalah	763.3
PanjalinanNorth	22,160
PanjalinanSouth	2,356
Parupuak Tabing	10,024
Parkit	2,830
Bung Hatta beach	19,866
Purus North	1,395
Purus South	123,004
Muaro Padang	971
Air Manis North	20,545
Air ManisSouth	1,287
Sungai Baremas	6,872
Pasia Putih	404

Source: Data Analysis, 2017.

Sediment transport is the parameter that related to run-up height, where the high run-up height values will result in high sediment transport value. [11] explain run-up the high also related with wave height and energy, high wave brings the high amount of wave energy which produces the high value of run-up height and then triggers high amount of sediment transport. The highest sediment transport appeared at Pasia Jambak beach about 958,619 m^3/day, and the smallest amount at Pasia Putih beach about 404 m^3/day with the average of sediment transport value around 82,935m^3/day.

3.6 The factor of Accretion and Abrasion

The results of accretion and abrasion factors (G_0) values for all calculate parameter results were can be seen in Table 7.

Table 7. G_0 Value

Location	$G_0 = (H_0/L_0).(Tg\,\alpha\,)^{0,27}(D_{50}/L_0)^{-0.67}$	Dynamic
Pasia Jambak	0.001	abrasion
Pasia Sabalah	14.413	accretion
PanjalinanNorth	10.855	accretion
PanjalinanSouth	6.728	accretion
Parupuak Tabing	0.003	abrasion
Parkit	17.601	accretion
Bung Hatta beach	0.002	abrasion
Purus North	18.769	accretion
Purus South	10.201	accretion
Muaro Padang	14.518	accretion
Air Manis North	17.139	accretion
Air ManisSouth	0.003	abrasion
Sungai Baremas	0.005	abrasion
Pasia Putih	11.899	accretion

Source: Data Analysis, 2017

G_0 is describing the dynamic condition that occurs in the coastal region. Based on the data calculation analysis shows G_0 value less than 0.00556, it means coastal region tend to recede (abrasion), G_0 value between 0.0556 and 0.111 means the coastal region is in the equilibrium condition, and if G_0 value above 0.111 expresses the accretion (augmented). G_0 is the calculation results of H_0 (wave height), L_0 (wavelength), $Tg\,\alpha$ (seashore slope), and D_{50} (median of sediment size value/percentile).

The based on G_0 value in Table 7, there are 9 locations in the coastal region of Padang City that the accretion, ie in Pasia Sabalah beach, Panjalinan North beach, Panjalinan South beach, Parkit beach, Purus North beach, Purus South beach, Muaro Padang beach, Air Manis South beach and Pasia Putih beach. Meanwhile, the 5 locations are in abrasion condition, ie in Pasia Jambak beach, Parupuk Tabing beach, Bung Hatta University beach, Air Manis South beach and Sungai Beremas beach.

The dynamic form of coastal environment can be observed by wave parametric measurement and calculation that define as G_0 value. Although wave period, velocity, energy, run-up height, and sediment transport values were high they were not always bringing negative effects to the coastal region. [12] add positive or negative effects depend on G_0 value at the end of the calculation. The high value of wave period, velocity, energy, run-up height, and sediment transport at Pasia Sabalah beach, Panjalinan South beach, Parkit beach and Purus South beach cannot be straightly assumed as abrasion area. In fact, based on field survey, those beaches lead to accretion.

The existence of beach building also contributes in forming dynamic of the coastal environment. Pasia Jambak beach, Panjalinan South beach, Parkit, Purus North beach and Air Manis South beach had the high value of wave parameters (except G_0 value). The existence of beach protection building has provided resilience/protection to the disaster that hit in the coastal region of Padang City.

[13] explain the collaboration of wave energy, sediment supply from the estuary and the existence of the beach building, took part in forming dynamic of the coastal environment. This makes the link between G_0 parameters, H_0 as wave high, L_0 as wavelength, $Tg\,\alpha$ as seashore slope, and D_{50} as median sediment size are difficult to correlate. Whereas G_0 is the value that determines the formation of dynamics of the coastal environment. It seemed there was "*X*" variable that cannot be ignored, which influenced G_0 and it was the collaboration of coastal-estuary adjacency and the existence of beach building. The coastal condition also one of the determinants of dynamic of the coastal environment. [14] add beaches protected by bay safer than those that adjacent to the open ocean. We assume that bays might be the natural protection of the coastal region from abrasion, which initiates by wave dynamics energy.

By identifying the relation between G_0 value, another parameter, existence of the beach building, and coastal physical condition (estuary and bay) along in the coastal region of Padang City, can be concluded that during a dry season/during research activities (July 2017), condition of the coastal region in a state of abrasion and accretion.

3.7 The condition of Coastal Environment

Jambak beach, Pasia Sabalah beach, and North Panjalinan beach stretch from Panjalinan estuary to Batang Anai estuary in the Northern part of Padang City. In the estuary (North and South) of Panjalinan estuary, the government has built a *Jetty* type beach building. *Jetty* is a beach building that is made perpendicular to the beach that is placed on both sides of the estuary of the river that serves to withstand sediment/sand that travels along the coast and settles in the mouth of the river [15]. With this *Jetty* building in the research location causing sediment transport from the South coast to the North, coast tends to become obstructed. Due to the energy value and run-up height relatively small and also due to the existence of *Jetty*, resulting in sediment stuck and sediment combing towards the North coast. This indicates a wave of energy-depleting sediment transporting through the *Jetty*.

South Panjalinan beach, Parupuak Tabing beach, and Parkit beach stretch from the Kuranji estuary in the South to the mouth of the Panjalinan river in the North. Based on the condition of the field, the condition of Parkit beach until Patenggangan beach has been built *groin* and *revetment*. [16] explain *groin* is a coastal protective building that is usually made perpendicular to the coastline and serves to capture and retain sediment transport along the coast so that the presence of *groin* will be able to reduce/stop coastal erosion. While the *revetment* is a coastal building built with a parallel shoreline that serves as a protective coastal wall against erosion and ground wave overtopping. [7] add the *revetment* causes the shore to be protected from the erosion process and the presence of *groin* leads to the sediment being caught which in turn causes the sediment to enter larger than the exit volume so that the beach is accretion.

Groin and *revetment* conditions built at the site of research have begun to accumulate sand. Indicates there has been a high sediment buildup. When associated with coastal conditions, the sedimentation originating from the Kuranji watershed has been carried by wave power from the Kuranji estuary to the area along the North of Parkit beach. Large sediment supply and large wave energy have caused sediment to transport to the North of this coast. This is what causes the high accumulation of sediment in the area along the North coast of Parkit beach to Patenggangan beach.

The beach on the campus of Bung Hatta University and North Purus beach stretch from the mouth of the Banjir Kanal estuary in the South to the Kuranji estuary in the North. Muara Banjir Kanal almost the same with the Panjalinan estuary. Where both sides of the river estuary have been built *Jetty*. With the construction of *groin* and *revetment* around the campus beach of Bung Hatta University, means the existence of coral reef ecosystems and coastal buildings has caused sediments from the South of deposited. With the existence of beach buildings on the South side of the coast, conjecture strong even in the storm season, Bung Hatta Beach tends to be vulnerable to abrasion.

South Purus beach and Muaro Padang beach stretch from Batang Arau estuary in the South to the mouth of the Banjir Kanal estuary in the North. Beach Muaro Padang is precisely in the park area has experienced accretion. With the low sediment supply from the South, causing the coast to lack of sediment. But with the existence of *groin* and *revered*in the North, the beach will not undergo much change. Low sediment supply and wave energy will also not be able to erode the beach due to the presence of beach buildings. Means in this area tend to be safe from the danger of abrasion in both dry and storm season.

North Air Manis beach and South Air Manis beach stretch from Batang Arau estuary in the North up to the legend Stone Malin Kundang in the South. The beach in this area experienced abrasion in the South and accretion in the North. conjecture strong with not yet unbuilt beach buildings, waves during hurricane seasons has caused abrasion on the South side of the coast.

Sungai Beremas beach and Pasia Putih beach are directly adjacent to the volcanic hills and are located on the edge of the bay. This coastal area has several bays such as Bayur bay and Bungus bay which is the largest and most important bay on the western coast of Sumatra island [17]. Pasia Putih beach located on Bungus bay. If associated with a coastal location within the bay, then the existence of the bay causes the sediment to not get out and move away from the shore. [18,19,20] explain the existence of bays that surround the coast, causing relatively small values of wave parameters to be measured, and resulting in relatively protected coast from coastal abrasion hazards. But in these storm seasons, the beach is abrasion.

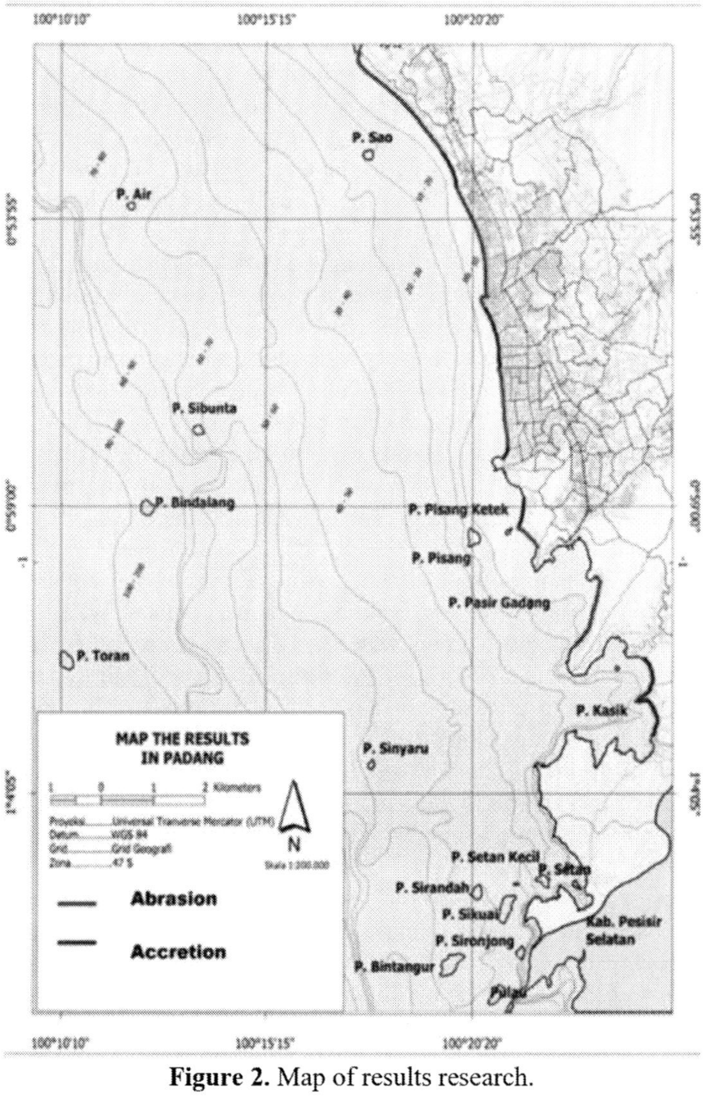

Figure 2. Map of results research.

4. Conclusion

The wave energy value, run-up Height and the sediment transport value are strongly influenced by the wave height value in the form of a proportional relationship (the higher the wave value, the wave energy value, the high wave value and the sediment transport value are also high, and vice versa). The existence of beach building and the natural physical state of the beaches in Padang City (river estuary and bay), contributes to the coastal dynamics that occur. The average run-up height during the dry season (end of July 2017) was 2.35 kg/s^2 and the average sediment transport along the coast of Padang City during the dry season (end of July 2017) was 82,935.6 m^3/day. Most of the beaches in Padang City during the dry season (late July 2017) tend to experience accretion ie in Pasia Sabalah beach, Panjalinan North beach, Panjalinan South beach, Parkit beach, Purus North beach, Purus South beach, Muaro Padang beach, Air Manis South beach and Pasia Putih beach. Meanwhile in abrasion condition ie in Pasia Jambak beach, Parupuk Tabing beach, Bung Hatta University beach, Air Manis South beach and Sungai Beremas beach. Along the Beach of Padang City during the dry season (end of July 2017) there is no equilibrium of beaches found.

Acknowledgments

Acknowledgments are given to Chairman of Geography Department - Padang State University, who has provided support for us in carrying out this research. We also express our gratitude to all those who have helped with the administration process as well as those who helped us in the field.

References

[1] Hu J and Wang X H 2016 Progress on upwelling studies in the China seas *Review of Geophysics* **54**: 653-673

[2] Putra A, Wisha U J and Kusumah G 2017 Spatial Analysis of River Line and Land Cover Changes in Kampar River Estuary: The Influence of Tidal Bore Bono Phenomenon *Forum Geografi.* **31**(2): 220-231

[3] Hapke C J 2004 *The Measurement and Interpretation of Coastal Cliff and Bluff Retreat* (U.S. Geological Survey *Professional Paper* 1693) 39-50

[4] Putra A 2017 Evaluasi Kesesuaian Pemanfaatan Ruang Pada Kawasan Pesisir Teluk Bungus Kota Padang *Tesis* (Padang: Pascasarjana Universitas Andalas)

[5] Prasita V D 2015 Determination of Shoreline Changes from 2002 to 2014 in the Mangrove Conservation Areas of Pamurbaya Using GIS *Procedia Earth and Planetary Science* **14**: 25-32

[6] Bishop M J, Mayer-Pinto M, Airoldi L, Firth L B, Morris R L, Loke L H L, Hawkins SJ, Naylor L A, Coleman R A, Chee S Y and Dafforn K A 2017 Effects of ocean sprawl on ecological connectivity: impacts and solutions *J. Experimental Marine Biology and Ecology* **492**: 7-30

[7] Hermon D, Putra A and Oktorie O 2018 Suitability Evaluation of Space Utilization Based on Enviromental Sustainability at The Coastal Area of Bungus Bay in Padang City *International Journal* **14**(41): 193-202

[8] Honesti L, Majid M Z A, Muchlian M and Djali N 2014 Assessing Building Vulnerability to Tsunami Hazard in Padang *J. Teknologi (Sciences & Engineering)* **69**(6): 5-9

[9] Pethick J 1984 *An Introduction to coastal Geomorphology* (London: Edward Arnold a Division of Holder and Stougthon)

[10] Triatmodjo B 1999 *Coastal Engineering* (Yogyakarta: Beta Offset)

[11] Frid C, Andonegi E, Depestele J, Judd A, Rihan D, Rogers S I and Kenchington E 2011 The environmental interactions of tidal and wave energy generation devices *Environmental Impact Assessment Review* **05732**: 1-7

[12] Fuad M A Z and Fais M D A 2017 Automatic Detection of Decadal Shoreline Change onNorthern Coastal of Gresik, East Java-Indonesia. *Conference Series: Earth and Environmental Science* **98**: 012001

[13] Crowell M, Leatherman S P and Buckley K 1991 Historical Shoreline Change: Error Analysis and Mapping Accuracy *J. Coastal Research.* **7**(3): 87-105
[14] Black K P, Kurian N P, Mathew J and Baba M 2004 Open Coast Monsoonal Beach Dynamics *J. Coastal Research.* **24**(1): 1-12
[15] Flor G, Flor-Blanco G and Rey J 2015 Dynamics and morpho-sedimentary interactions in the lowermesotidal estuary of Villaviciosa (NW Spain): A managementproposal. *Geologica Acta.* **13**(2): 107-121
[16] Dean G B and Dalymple R A 2004 Coastal Processes With Engineering Applications (New York: Cambridge University Press)
[17] Kunzmann A and Samsuardi 2017 A Century of Change in an Indonesian Coral Reef:Sluiter's Brandewijnsbaai (1890) Revisited *Annual Research & Review in Biology* **13**(3): 1-7
[18] Rio L D and Gracia F J 2009 Erosion risk assessment of active coastal cliffs in temperate environments *Geomorphology* **112**(1–2): 82-95
[19] Putra A and Mutmainah H 2016 The Mapping of Temporary Evacuation Site (TES) and Tsunami Evacuation Route in North Pagai Island, Mentawai Islands Regency-Indonesia *IOP Conference Series: Earth and Environmental Science* **47**(1): 012020
[20] Febriandi 2017 The Estimation of Mangroves Area and Density of Mangroves Changes Use the Remote Sensing Data at Northheast Province of Aceh, Indonesia *Sumatra Journal of Disaster, Geography and Geography Education* **1**(1): 105-110

The vegetable oil in the production of polymers and plastics; an effort of creating green products

Flora Elvistia Firdaus

Jayabaya University, Jl. Pulomas Selatan Kav. 23 Jakarta Timur 13210 Indonesia

*flora_elvistia@yahoo.com

Abstract. Manufacturing of polymers and plastics derived from petroleum feedstocks are called synthetic products. The existing of synthetics throughout the plastic series of process is detrimental to the environment. Polyurethane is classified as a family of plastics. Vegetable oil is the potential of producing polyurethane in the presence of unsaturated chain in triglyceride, is being considered intermediates of polymers product. An effort of replacing petroleum with green resources is urgently required of its cheap price, and abundantly available. The vegetable now is not just a cooking oil but turned into various polymeric products. This research is aimed to identify the vegetable oil content in polyurethane's formula with comprehensive low production cost and potentially resulted in the best polyurethane product properties.

1. Introduction

Plastics are polymeric materials, the global production and consumption of plastic have grown rapidly to a commercialized desire product from household to industrial scale. The plastic littering has become big issues of worldwide which essentially fulfil the modern life style. The impact of its non-biodegradable properties is worried, has the burden of exceedingly of earth carrying capacity without knowing when is destroyed. Plastics are actually a shortened form of thermoplastic, the materials can be shaped and reshaped by heating. The production of synthetic polymers globally has dominated by polyolefins, polyethylene, and polypropylene.

According to Jambeck *et. al*, the estimated plastic world consumption is approximately 335 million metric tonnes (MMT) in 2016. It confirmed and upward trend over the past years. Indonesia is the world second-ranked mass of mismanaged plastic waste after China [1].

Plastics are polymeric materials is made from petroleum feedstocks. Reducing the plastics amount in a series of processes or replaced to green materials are possibly step to save the environment. The economic losses and crisis of petroleum shortages have created serious innovative findings which have to fuel the growing interest of worldwide researcher on using renewable resources [2]. Despite the lack of limitations to suitably fabricate plastics, the proper method can solve and established a sustainable environment [3][4][5].

Triglyceride for Polyurethane
Polyurethane is one class of plastics. Green resources are renewables, the availability if served as the main ingredient in plastic fabrication is much promising besides it saved the environment. It should be part of upstream into downstream of the process. Triglycerides in vegetable are typically contained 10 or more different fatty acids, it can either be saturated or unsaturated. The triglyceride has the answered for the world awareness to the environment, the functional groups at least one unsaturated fatty acid to be introduced to a desired of soy-polyol. The name and structure of fatty acid are illustrated in Table 1.

Table 1. Fatty acid of Vegetable Oils

Name	Structure
Myristic acid	$CH_3(CH_2)_{12}COOH$
Palmitic acid	$CH_3(CH_2)_{14}COOH$
Palmitolec acid	$CH_3(CH_2)_5CH{=}CH(CH_2)_7COOH$
Stearic acid	$CH_3(CH_2)_{16}COOH$
Oleic acid	$CH_3(CH_2)_7CH{=}CH\text{-}CH_2\text{-}CH{=}CH(CH_2)_7COOH$
Linolenic acid	$CH_3CH_2CH{=}CHCH_2CH{=}CHCH2CH{=}CH(CH_2)_7COOH$
α-Eleostearic acid	$CH_3(CH_2)_3CH{=}CHCH{=}CHCH{=}CH(CH_2)_7COOH$
Ricinoleic acid	$CH_3(CH_2)_4CHCHCH_2CH{=}CH(CH_2)_7OHCOOH$

The soybean oil contains carbon-carbon double bonds with no hydroxyl groups, the epoxidation reaction will form reactive group oxirane rings is an intermediate to fabricated fatty acid derivatives [6]. Soybean oil has in the majority the unsaturated fatty acid chains which are approximately 85%, the reactivity of fatty acid depend on the amount of carbon double bond, the structure of a triglyceride, in general, is depicted in Figure 1.

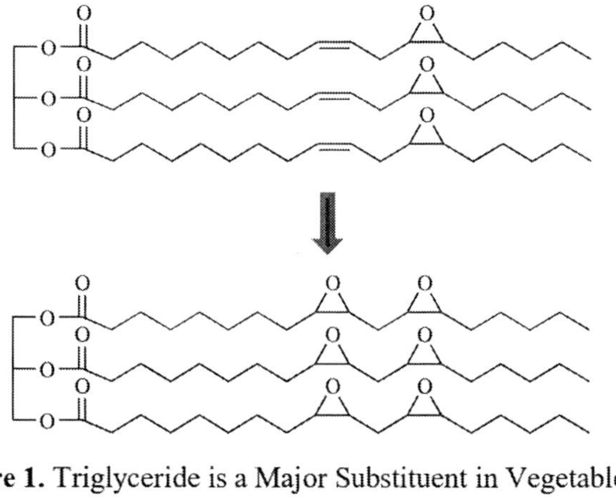

Figure 1. Triglyceride is a Major Substituent in Vegetable Oils

The polymerization by epoxidation reaction of three carbon-carbon π-bond of linolenic acids, linoleic (18:2) and oleic (18:1) [7][8]. The soy epoxidized is one of the raw material, the ring opening has made the epoxides transformed into polyols which is the meat of foam formulation, the ingredients bind to forming the molecular chain of polyurethane [9].

Methylene diisocyanate (MDI) is the material reacts with polyol forming chemical reaction and builds the complex shape of polyurethane flexible foam. The hydroxyl functionality of polyol is used to increase isocyanates reactivity. The carbon chain before and after the hydroxyl acted as a plasticizer to improved properties of polyurethane and reduce rigidity [10]. High hydroxyl values can be caused by high production costs because of the high isocyanates requirement. The number of reactive hydroxyl groups per polyol molecule ultimately control the degree of cross-linking between molecules, this effected to mechanical properties of polyurethane. The maximum content of vegetable oils for petroleum replacement is approximately 15-20% of the polyol content. The mechanism reaction of polyurethane can be seen in Figure 2.

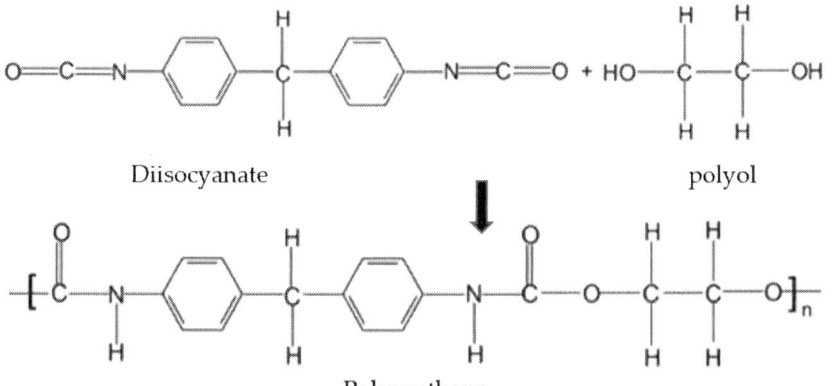

Figure 2. Mechanism Reaction of Polyurethane from Polyol and Diisocyanate

2. Methods

Polyurethane is manufactured by spontaneously cell opening. The molding method can catalyze to the perfect blowing of foam. The properties possessed are from flexible to rigid. In the series of the process, the petrochemicals are partially reduced.

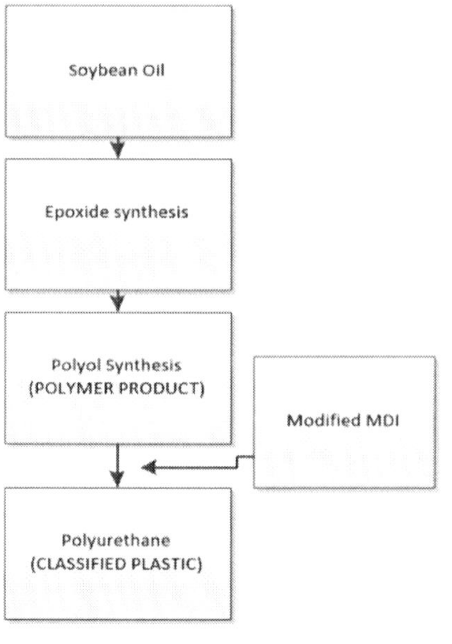

Figure 3. Fabrication of Polymers and Classified Plastic from Soybean Oil

The material used is soybean oil from Variatama with the specification of iodine value of 53.89 gram Iod/100 gr sample, viscosity 65.5278 cps, and acid value 0.024 mgr KOH/sample.

Environmental set-up
The vegetable used in this research is soybean oil, it contains triglycerides of long chain saturated and unsaturated carboxylic acids, the triglycerides are charged to peracetic acid using a sulfur acid catalyst. The concentration of catalyst is designed 1%, 2%, and 3% (v/v), temperature is maintained 60± 0.50C. The flow diagram of the research is depicted in Figure 3. Polyols are synthesized by blending of soy epoxide to ethylene glycol in stoichiometric calculation into 1% (v/v) acid catalyst at 1170C. The producing of polyurethane is using the molding method.

3. Results and discussion

The stepwise of the research is meant to find the optimized conditions. The oxiranes values are measuring the unsaturated fatty acid. The best oxirane number occurred is by 1% (v/v) sulfur acid was 6.7% and 7.3%; while 2% (v/v) of the acid catalyst the oxirane number was 3.9% and 4.6% and 3% (v/v) of sulfur acid was 3.3% and 4.5%. The optimized result is the catalyst 1% (v/v) as can be seen in Figure 4, this condition can fasten the reaction with a low concentration of catalyst where it considered reduce the production cost.

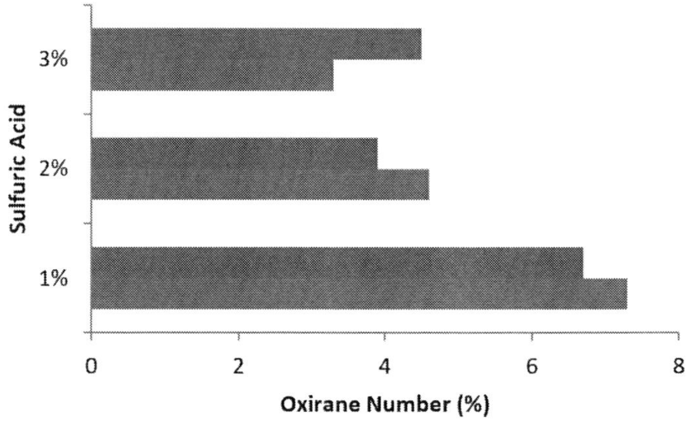

Figure 4. Optimization of the Epoxidation catalyst

The time reaction is other factors to be considered because it related to cost-effectively of polyol product. The polyol creates hydroxyl sites to be linked into MDI which resulted in polyurethane. The steady increase of hydroxyl number may be attributed to the continued breaking of the ester bond. The high hydroxyl number is achieved from partially of epoxidized and ethylene glycol, the high hydroxyl number will require high MDI.

The 1 hour of reaction and 1:3 (mo/mol) the lowest concentration of epoxide/EG in the designated formula as illustrated in Figure 5, is fairly produce polyurethane in the range of flexible properties. The flexibility of polyurethane is mainly the focus of the work as if it uses a blending of synthetic polyol and soy polyol 4:1 (v/v) as the optimized the concentration which resulted in elasticity 86.66% and resilience 13.34 %, while using 100% soy polyol resulted in elasticity 71.42% and resilience 28.58%. The polyurethane from soy-based also has good compressive strength with 3.3 x 10^{-3} MPa in 10% deformation.

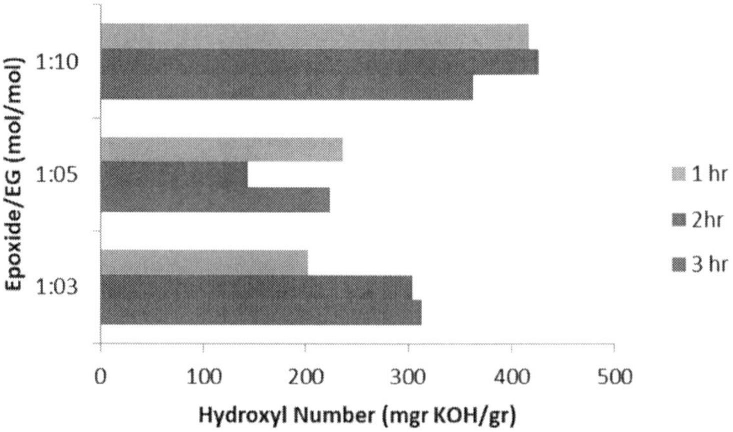

Figure 5. The Effect of Time reaction to Hydroxyl Number of Soy Polyol-Synthetic Ethylene Glycol (EG)

The appearance of polyurethane can be seen in Figure 6.

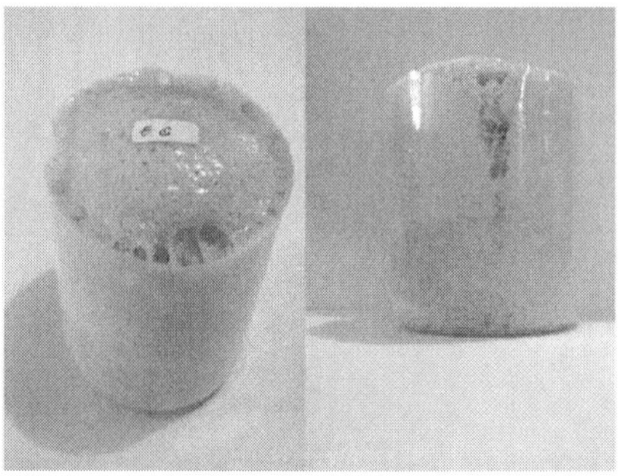

Figure 6. The polyurethane of Soybean oil based

4. Conclusions

The shortages of petroleum fuel feedstock have rising focuses of world researches on using vegetable oil as renewable resources for a variety of polymer products. The unsaturated chain in triglyceride is a functional site of vegetable oil conducted to resulted polymers and its derived in the form of classified plastic. The polyol is taking part in the process, is the only candidate from green resources in polyurethane formula for petroleum replacement. Polyol soybean-based are represented as one amongst vegetable oils, are viable, and has most an economic point of view.

Acknowledgments

The author gratefully acknowledges the support of the Indonesia Ministry of Research, Technology and Higher Education as funding the research in the scheme of National Grant Competence.

References

[1] T. R. S. Jenna R. Jambeck, Roland Geyer, Chris Wilcox, "Plastic Waste Inputs from Land into the ocean," *Clim. Chang. 2014 Impacts, Adapt. Vulnerability Part B Reg. Asp. Work. Gr. II Contrib. to Fifth Assess. Rep. Intergov. Panel Clim. Chang.*, no. January, pp. 1655–1734, 2015.

[2] L. Maisonneuve, G. Chollet, E. Grau, and H. Cramail, "Vegetable oils: a source of polyols for polyurethane materials," *Ocl*, vol. 23, no. 5, pp. 1–10, 2016.
[3] J. O. Akindoyo, M. D. H. Beg, S. Ghazali, M. R. Islam, N. Jeyaratnam, and A. R. Yuvaraj, "Polyurethane types, synthesis and applications-a review," *RSC Adv.*, vol. 6, pp. 114453–114482, 2016.
[4] M. R. Islam, M. D. H. Beg, and S. S. Jamari, "Development of vegetable-oil-based polymers," *J. Appl. Polym. Sci.*, vol. 131, no. 18, pp. 9016–9028, 2014.
[5] P. Saithai, V. Tanrattanakul, W. Chinpa, K. Kaewtathip, and E. Dubreucq, "Synthesis and Characterization of Triglyceride-Based Copolymer from Soybean Oil," *Mater. Sci. Forum*, vol. 695, pp. 320–323, 2011.
[6] F. E. Firdaus, "Chain extender on property relationships of polyurethane derived from soybean oil," *World Acad. Sci. Eng. Technol.*, vol. 81, 2011.
[7] K. F. Adekunle, "A Review of Vegetable Oil-Based Polymers : Synthesis and Applications," *Open J. Polym. Chem.*, vol. 5, no. August, pp. 34–40, 2015.
[8] A. E. Kale, D. G. Goswami, P. S. Zade, and M. B. Mandake, "Recent Advances in Epoxidation of Vegetable Oils," *J. Emerg. Technol. Innov. Res.*, vol. 4, no. 04, pp. 1–5, 2017.
[9] F. E. Firdaus, "Optimization of soy epoxide hydroxylation to properties of prepolymer polyurethane," *World Acad. Sci. Eng. Technol.*, vol. 81, 2011.
[10] F. E. Firdaus, "The selection reaction of homogeneous catalyst in soy-epoxide hydroxylation," in *Journal of Physics: Conference Series*, 2014, vol. 495, no. 1.

IOP Conf. Series: Earth and Environmental Science **314** (2019) 012008 doi:10.1088/1755-1315/314/1/012008

Biohydrogen as a renewable energy and its potential production from the conversion of palm oil mill effluent by anaerobic processes

A Gumilar*, M Syafila, M Handajani and M I Angga

Department of Environmental Engineering, Institut Teknologi Bandung, Ganesha Street No 10 Bandung, Indonesia

*kang.andrie@gmail.com

Abstract. Biohydrogen is one of the promising alternatives to be used as a source of energy in the future. Several studies related to the production of biohydrogen through the process of conversion of organic wastewater by anaerobic process have been done with different reactor conditions. However, biohydrogen is not the main products produced in the process, even biohydrogen is produced in relatively few amounts compared to other acidogenesis products such as volatile acids. The aims of the study are to determine the optimum conditions of biohydrogen production through the control of pH are 5.5; 6.5; 7.5 and 8.5. This factor affects to the metabolism pathway. The study was conducted using the wastewater from the palm oil mill effluent as a substrate, pure culture and mixed culture bacteria as a biomass and performed in the laboratory using an anaerobic batch reactor which will be operated for 72 hours. Sampling was done every 6 hours and then analysed the product of acidogenesis produced. The results show that the optimum conditions for biohydrogen production occurs in the reactor with the pH 5.5-6.5 are 15.27% ± 0.404% to 15.40% ± 0.265%.

1. Introduction

The utilization of renewable fuels is the main topic of discussion to overcome the problem of fossil fuels. This is due to the limited number of fossil fuels from fossils while the world's energy needs continue to increase. At present, the average energy use per capita in the world is 2.5 kW. The population in the world in 2050 is predicted to be 9.6 billion people, then the energy needed to meet all human needs is 24 TW. The use of energy per capita in the world tends to increase along with the standard of living in 2050 to 6.5 kW, so the energy needs in the world in 2050 are also predicted to increase to 50 TW [1].

Bioenergy such as biogas, bioethanol, methanol, biodiesel, and biohydrogen are high-quality renewable fuels that can be used to replace fossil fuels [2]. Biohydrogen is one of the feasible alternatives to be developed and used as an alternative to fossil fuels [3]. The choice of biohydrogen compared to other biogas such as methane is based on several considerations, including hydrogen having a greater heating value of 142 MJ kg^{-1}compared to methane (55 MJ kg^{-1}) and combustion from hydrogen in the form of O_2 and H_2O which is very environmentally friendly [4].

The availability of cheap raw materials which are not classified into food is one of the obstacles in the bioenergy industry in Indonesia [5]. So far, the raw materials used for the production of biohydrogen is cane sugar, corn and cassava [6]. However, these materials are economic commodities that are economical and classified as food commodities, so it is necessary to strive to use non-food raw materials to support the realization of the domestic biofuel industry [7]. To overcome this problem, it can be done using wastewater containing high concentration organic compounds from agro industrial activities such as palm oil, rubber, cocoa, sugar and coffee wastewater which has a chemical oxygen demand (COD) content of more than 4000 mg L^{-1} through the process anaerobes [8]. The utilization of palm oil industry wastewater as a substrate for bioenergy formation has also been carried out, which produces ethanol of 2267.29 mgCOD L^{-1} [9].

The production of biohydrogen by anaerobic microorganisms is distinguished by pathway conversion of pyruvic acid to acetaldehyde which is then reduced to hydrogen and ethanol by alcohol dehydrogenase [10,11,12]. The first pathway is the conversion of pyruvic acid to acetyl-CoA and CO_2 and then reduced to acetaldehyde. This pathway is used by mixed anaerobic culture bacteria which can utilize high concentration organic wastewater as a substrate [13]. The second pathway is the direct conversion of pyruvate acid to acetaldehyde with the help of pyruvate decarboxylase enzyme used by *Zymomonas mobilis*, *Sarcina ventriculi*, *Erwinia amylovora*, acetic acid bacteria and *Saccharomyces cerevisiae* [14].

The metabolism pathway and growth rate of microorganisms are defined by the pH of a reactor during the anaerobic processes. Any change in hydrogen ion concentration results in a change in pH and eventually leads to a change in redox potential [15].

2. Materials and Method

2.1. Raw material and biomass

The fresh raw palm oil mill effluent (POME) was collected from the influent basin of PT. Condong, Garut, Indonesia. Biomass as mixed culture bacteria was taken from the sludge of palm oil mill effluent mixed by cow rumen with the ratio of 50:50 and has been acclimated to the POME. The acclimated biomass concentration used for the anaerobic batch reactor in terms of mixed liquor suspended solids (MLSS) and mixed liquor volatile suspended solids (MLVSS), were about 4.81 g SS/L and 3.72 g VSS/L respectively.

2.2. Circulating bed reactor

Circulating bed reactor (CBR) with the working volume of 5L were operated with the flushing N_2 1L/min for 24 h and continued operates for 72h by internal biogas (Figure 1). The reactor filled with a mixture of consisted of 20% (v/v) mixed culture bacteria and 80% (v/v) of POME as a substrate. The pH of the reactor was set at 5.5; 6.5; 7.5 and 8.5 in the room temperature. Sample was collected every 6h and analyze for biogas (biohydrogen), volatile fatty acid and volatile suspended solid.

2.3. Analytical method

Biological oxygen demand (BOD), chemical oxygen demand (COD), TSS, oil and grease, total nitrogen and pH were analyzed according to standard methods. Gas composition was analyzed using Gas Chromatography (Shimadzu GC 8A) equipped with column Porapak P 8A serial D-4167 and TCD detector types, operating condition at 100°C, injector and detector temperature are 130°C, He pressure at 60 kPa with sample injection of 1 μL. Volume of biogas produced during the anaerobic process measured by displacement water method and the pressure measurement using manometer Lutron PM9100.

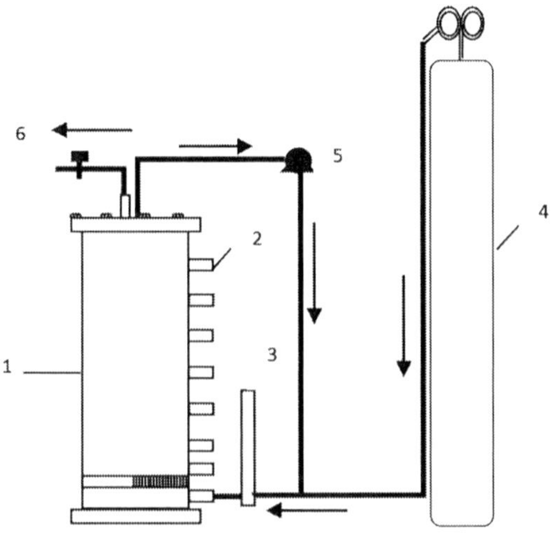

Figure 1. Circulating bed reactor (CBR), (1) CBR; (2) sampling port; (3) flowmeter; (4) nitrogen; (5) circulation pump; (6) gas outlet

3. Results and Discussion

3.1. Characteristic of palm oil mill effluent

The characteristics of palm oil mill effluent (POME) are given in Table 1**Table**. POME contains complex carbohydrate polymers such as hemicelluloses and lignocelluloses which resulted to high COD concentration about 15,000-100,000 mg/L [16], and it can be used as a substrate for biohydrogen production by anaerobic processes [16-21]. Anaerobic wastewater treatment is one of the main processes that can produce energy. In general, the processes that occur in anaerobic wastewater treatment process divided up into four stages, namely hydrolysis, acidogenesis, acetogenesis and methanogenesis [22] as represented in Figure 2.

Table 1. Characteristic of palm oil mill effluent (POME)

Parameter	Unit	Results
BOD	mg/L	16,810
Soluble COD	mg/L	19,475
Total COD	mg/L	23,270
TSS	mg/L	9,400
Oil and Grease	mg/L	245
Nitrogen Total	mg/L N	186
Ethanol	mg/L	23.8
Volatile fatty acids	mg/L	1024.8
pH	-	5.63

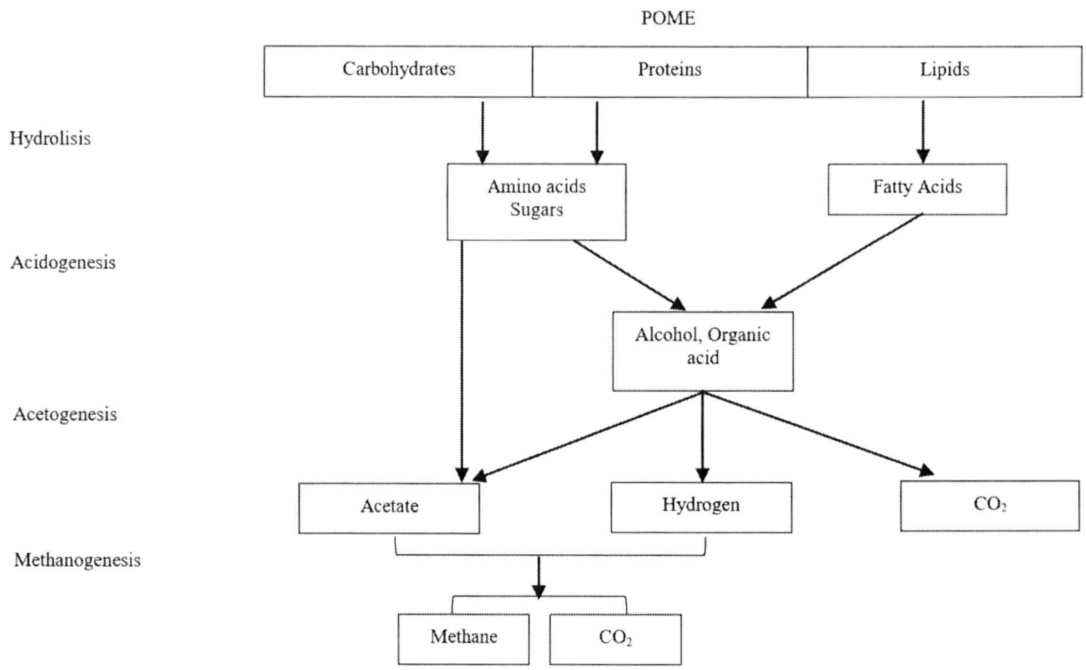

Figure 2. Metabolism pathway in anaerobic processes [23].

3.2. Effect of pH on biohydrogen production

In Figure 3 shown the effect of pH on biohydrogen production. The maximum biohydrogen production obtained in the different condition pH of reactors for 5.5; 6.5; 7.5 and 8.5 are 15.40% ± 0.265%; 15.27% ± 0.404%; 8.77% ± 0.129% and 8.10% ± 0.176%. The results show a trend of pH 5.5 > 6.5 > 7.5 > 8.5. These results indicate that the optimum condition for biohydrogen production by anaerobic processes occurs at the pH value about 5.5-6.5. The similar results have been reported by Liu et al. (2011), the pH 5.5-6.8 are the optimum range for biohydrogen production and pH 4.5 seems to inhibit hydrogen production [15,24]. If the rate of hydrolysis phase faster than acidogenesis and acetogenesis phase, it will cause volatile fatty acid to accumulate and reduce the pH which can inhibit hydrogen production [25]. Another result showed that pH 6.0 is an optimum pH and the maximum hydrogen yield of 28.3 mL g^{-1} COD was obtained [26]. Metabolism pathway and the growth of microorganism defined by the condition of pH. Any changes of the hydrogen concentration also influence to the pH and eventually leads to the changes of redox potential [24].

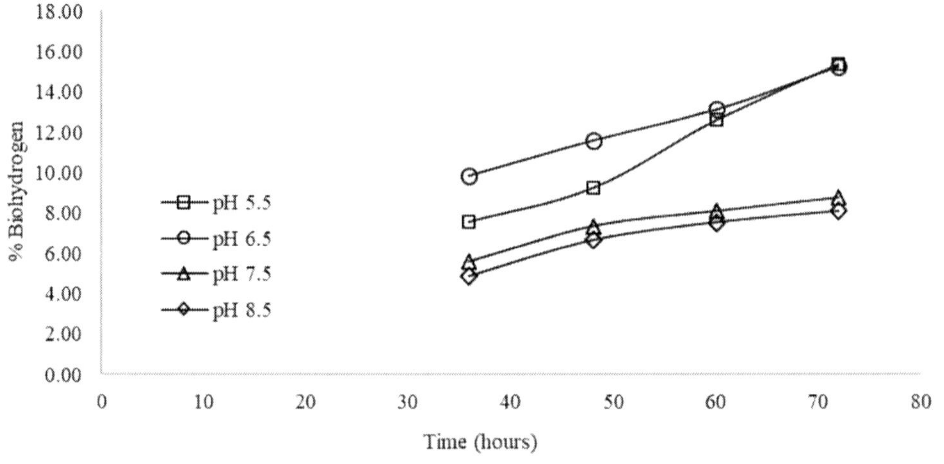

Figure 3. Biohydrogen production in the different pH condition

3.3. Effect of the control pH on anaerobic product production

Volatile fatty acids (VFA's) such as acetate, propionate, butyrate, valerate and biogas (CO_2 and CH_4) was produced during the experiments. Acetate and propionate are the main acidogenic product in the anaerobic processes. In Figure 4 shown the volatile fatty acid in every reactor. Maximum acetate and propionate concentration in the reactor by control are 3318.4; 2864.7; 2163.5 and 2691.5 mg/L for acetate and 9953.2; 8792.2; 8229.1 and 6726.4 mg/L for propionate respectively. Acidogenic product formation shown in Table

***Table*2.**

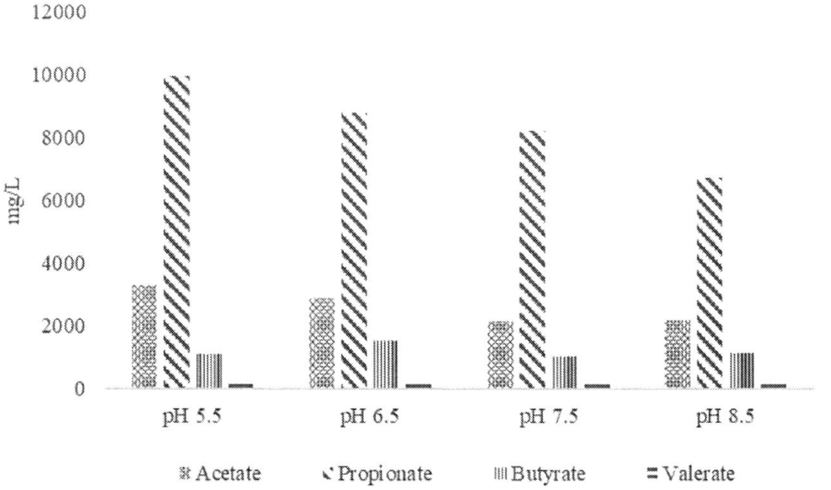

Figure 4. Volatile fatty acid production in the different pH condition

Table 2. Acidogenic product formation in anaerobic processes

Parameter	Unit	pH			
		5.5	6.5	7.5	8.5
Biohydrogen	%	15.4	15.27	8.8	8.10
Acetate	mg/L	3318.4	2864.7	2163.5	2188.4
Propionate	mg/L	9953.2	8792.2	8229.1	6726.4
Butyrate	mg/L	1107.2	1527.4	1044.2	1126.2
Valeric acid	mg/L	172.3	208.4	248.8	267.5

4. Conclusion

Controlling the reactor pH can be influence to the biohydrogen production and another product such as volatile fatty acid (acetate, propionate, butyrate, valeric acid) in anaerobic processes. The optimum biohydrogen production occurs in the reactor with the pH 5.5 was about 15.40% ± 0.265%.

Acknowledgments

This research was funded by *Program Penelitian, Pengabdian kepada Masyarakat dan Inovasi (P3MI) tahun 2018.*

References

[1] Dincer I and Zamfirescu C 2016 *Sustainable hydrogen production* (Elsevier, Oxford) pp 65-66
[2] Mshandete A M and Parawira W 2009 Biogas technology research in selected sub-Saharan African countries–A review *African Journal of Biotechnology* **8** 116-125
[3] Rafieenia R, Girotto F, Peng W, Cossu R, Pivato A, Raga R and Lavagnolo M C 2017 Effect of aerobic pre-treatment on hydrogen and methane production in a two-stage anaerobic digestion process using food waste with different compositions *Waste Management* **59** 194-199
[4] Guwy A, Dinsdale R, Kim J, Nicolau J and Premier G 2011 Fermentative biohydrogen production systems integration *Bioresource Technology* **102** 8534-8542
[5] Murdiyatmo U 2006 *Pengembangan Industri Etanol: Prospek, Kendala dan Tantangan* (Workshop Nasional Bisnis Biodiesel dan Bioetanol di Indonesia) Jakarta pp 80-86
[6] Dillon H S and Laan T 2008 *Biofuels – at What Cost?: Government Support for Ethanol and Biodiesel in Indonesia. One of a series of reports addressing subsidies for biofuels in developing countries.* Prepared for: The Global Subsidies initiative (GSI) of the International Institute for Sustainable Development (IISD) Geneva, Switzerland pp 28-35
[7] Sari D A and Hadiyanto 2013 Proses produksi bioenergi berbasiskan bioteknologi: review *Jurnal Aplikasi Teknologi Pangan* **2**(3) 108-113
[8] Grady L C P Jr, Daigger G T and Lim H C 1999 *Biological Wastewater Treatment, 2nd Edition* (Revised dan Expanded, Marcel Dekker, Inc.) p 32
[9] Andrio D 2014 *Mekanisme dan Strategi Kontrol Pembentukan Etanol Sebagai Usaha Pemanfaatan Kembali Limbah Cair yang Mengandung Senyawa Organik Konsentrasi Tinggi* (Disertasi Program Doktor, Institut Teknologi Bandung) pp 130-134
[10] Algapani D E, Qiao W, Su M, Pumpo F, Wandera S M, Adani F and Dong R 2016 Bio-hydrolysis and bio-hydrogen production from food waste by thermophilic and hyperthermophilic anaerobic process *Bioresource Technology* **216** 768-777
[11] Rafieenia R, Girotto F, Peng W, Cossu R, Pivato A, Raga R and Lavagnolo M C 2017 Effect of aerobic pre-treatment on hydrogen and methane production in a two-stage anaerobic digestion process using food waste with different compositions *Waste Management* **59** 194-199

[12] Rojas M del P A, Zaiat M and Wever H De 2016 Improvement of hydrogen production via ethanol-type fermentation in an anaerobic down-flow structured bed reactor *Bioresource Technology* **202** 42-49

[13] Syafila M, Handajani M and Prayascitra A 2010 The effect of nitrogen gas flushing on intermediate products formation in acidogenic stage of anaerobic process of cocoa sweatings *Institut Teknologi Bandung Journal of Engineering Science* **42**(2) 129-136

[14] Kim B H and Gadd G M 2008 *Bacterial Physiology and Metabolism* (Cambridge, UK: Cambridge University Press) pp 252-289

[15] Liu I C, Whang L M, Ren W J and Lin P Y 2011 The effect of pH on the production of biohydrogen by clostridia: thermodynamic and metabolic considerations *Int. J. Hydrogen Energy* **36** 439-449

[16] Chong M L, Rahman N A, Rahim R A, Aziz S A, Shirai Y and Hassan M 2009 Optimization of biohydrogen production by Clostridium butyricum EB6 from palm oil mill effluent using response surface methodology *International Journal of Hydrogen Energy* **34** 7475-7482

[17] Chong M L, Rahman N A, Yee P L, Aziz S A, Rahim R A and Shirai Y 2009 Effects of pH, glucose and iron sulfate concentration on the yield of biohydrogen by Clostridium butyricum EB6 *International Journal of Hydrogen Energy* **34** 8859-8865

[18] Ismail I, Hassan M, Rahman N A and Soon C S 2010 Thermophilic biohydrogen production from palm oil mill effluent (POME) using suspended mixed culture *Biomass and Bioenergy* **34**(1) 42-47

[19] Jamil Z, Annuar M, Ibrahim S and Vikineswary S 2009 Optimization of phototrophic hydrogen production by Rhodopseudomonas palustris PBUM001 via statistical experimental design *International Journal of Hydrogen Energy* **34** 7502-7512

[20] Prasertsan P, Thong S and Birkeland N K 2009 Optimization and microbial community analysis for production of biohydrogen from palm oil mill effluent by thermophilic fermentative process *International Journal of Hydrogen Energy* **34** 7448-7459

[21] Yusoff M, Hassan M, Aziz S and Rahman N 2009 Start-Up of biohydrogen production from palm oil mill effluent under non-sterile condition in 50 L continuous stirred tank reactor *International Journal of Agricultural Research* **4** 163-168

[22] Demirel B and Scherer P 2008 The roles of acetotrophic and hydrogenotrophic methanogens during anaerobic conversion of biomass to methane: a review *Rev. Environ. Sci. Biotechnol.* **7** 173-190

[23] Garritano A N, Faber M O, Livian R V de Sa and Ferreira-Leitao V S 2018 Palm oil mill effluent (POME) as raw material for biohydrogen and methane production via dark fermentation *Renewable and Sustainable Energy Reviews* **92** 676-684

[24] Ruggeri B, Tommasi T and Sanfilippo S 2015 *BioH$_2$ & BioCH$_4$ through Anaerobic Digestion* (London: Springer-Verlag)

[25] Khan M A, Ngo H H, Guo W S, Liu Y, Nghiem L D, Hai F I, Deng L J, Wang J and Wu Y 2016 Optimization of process parameters for production of volatile fatty acid, biohydrogen and methane from anaerobic digestion *Bioresour. Technol.* **219** 738-748

[26] Pisutpaisal N, Tanikkul P and Phoochinda W 2014 Improvement of mesophilic biohydrogen production from palm oil mill effluent using ozonation process, The International Conference on Technologies and Materials for Renewable Energy (Environment and Sustainability TMREES14) *Energy Procedia* **50**(2014) 723-728

Study of coastal abrasion disasters and their causes in Pariaman City

Haryani[1,*], Agus Irianto[2] and Nurhasan Syah[2]

[1]Doctor Programe of Environmental Science, Padang State University and Faculty of Civil Engineering and Planning, Bung Hatta University, Padang City, Indonesia
[2]Environmental Science Padang State University, Padang, Indonesia

*irharyanimtp@yahoo.co.id

Abstract. In 2003 until 2016 on the coast of West Sumatra Province there have been disasters of abrasion and accretion in 32 points in 6 Regencies and Cities covering an area of 732.69 Ha of abrasion and accretion of 55.4 Ha. Pariaman City is one of the coastal cities in West Sumatra Province. The purpose of this study is to examine the characteristics of coastal abrasion in the last 15 years (2003 to 2018) in Pariaman City and its causal factors. The analytical method used is the 2003 coastline comparison method with the coastline in 2018. Congratulations on 15 years of observation, there was coastal abrasion in the coastal area of Pariaman City covering an area of 197.65 ha and accretion of 285.38 ha. This proves that coastal abrasion causes a substantial reduction in land in Pariaman City, which is an average of 13.18 ha / year, while land addition is 19.03 ha / year. Physical factors that most influence the abrasion and accretion of coastal Pariaman City are current factors, shoreline shape, coastal typology and vegetation cover while low wave factors are quite influential.

1. Introduction

Indonesia's reputation as a disaster-prone country is noted in the report of The Atlas of the Human Planet 2017 [1]. The data shows that the threat of natural disasters has doubled in the last 40 years in line with the increasing population.

Historically, West Sumatra is an area that has experienced a lot of disasters from 1943 to 2011. Disasters that have occurred in West Sumatra based on the highest percentage sequence include 43% floods, 18% landslides, 7% fires, floods and land 7% landslide, 6% earthquake, 3% tidal wave / abrasion, and another 7% disaster [2]. Climatologically, summer and rain in West Sumatra Province experience drastic changes and extreme, resulting in vulnerability to landslides, floods and tidal waves and abrasion. To why 3.00% of the tidal and abrasion nature, the coastal areas of West Sumatra are at a high-risk level, 1.71% are at medium risk and 5, 41% are at a low risk level.

From 2003 to 2016 on the coast of West Sumatra Province there have been abrasion and accretion disasters in 32 points in 6 Regencies and Cities covering an area of 732.69 Ha of abrasion and accretion of 55.4 Ha [3, 4]. Abrasion disasters cause a substantial reduction in land in West Sumatra Province, which is an average of 56.3 Ha / year, while the addition of land is only 4.26 Ha / year. This figure proves insignificant between the extent of coastal abrasion and coastal accretion in West Sumatra.

Pariaman City is one of the coastal cities in West Sumatra Province. Geographically, Pariaman City is located between 0 ° 33'00 " - 0 ° 40'43 " LS and 100 ° 04 '46' '- 100 ° 10'55' 'BT. The land area is 79.22 km2 and the sea area is 282.69 km² with 6 small islands namely Bando Island, Gosong Island, Ujung Island, Tangah Island, Angso Island and Kasiak Island with a beach length of approximately 12.7 km. Pariaman City consists of 4 (four) sub-districts namely North Pariaman District, Central Pariaman, East Pariaman and South Pariaman with a total of 16 villages and 55 villages belonging to 12 (twelve) villages [5].

The purpose of this study is to examine the characteristics of coastal abrasion in the last 15 years (2003 to 2018) in Pariaman City and its causal factors.

2. Study of literatur

Almost all regions of Indonesia have a disaster risk both in the hilly, lowland, and coastal areas. Indonesia is one of the countries with high numbers of natural disaster vulnerability. The National Disaster Management Agency [6] noted that 2,175 natural disasters occurred throughout 2017. Although Indonesia is a disaster-prone country, there is a downward trend in the number of disasters that occurred in 2017 when compared to 2016. In 2017 there were 2,271 incidents a disaster that resulted in 372 deaths, 3.45 million displaced, 44,539 housing units damaged and 1,999 public facilities damaged.

Disasters can be caused by natural disasters (natural disaster) or by human activities (man-made disaster). Factors that can cause disasters include natural hazards and human-caused dangers. According to the United Nations International Strategy for Disaster Reduction (UN-ISDR) [7] can be grouped into geological hazards (geological hazards), hydrometeorological hazards (biological hazards), biological hazards, technological hazards (technological hazards) and environmental degradation (environmental degradation), high vulnerability of the community, infrastructure and elements in the city / region at risk of disaster and low capacity of various components in the community.

More than 90% of disaster events are hydrometeorological events [6]. Likewise, with the disaster of coastal abrasion that occurred in Indonesia which increased from time to time. In 2011 there were only 17 abrasions recorded, but in 2012 it increased to 29 times, as did the following year, in 2013 there were 36 incidents. But in 2014 the incidence of abrasion decreased by only 20 times. In 2016 the incidence of coastal abrasion in Indonesia increased by 300% more than the previous year (in 2015), which was only 7 times to 23 incidents. Coastal abrasion in Indonesia occurs in an area of 1,888,085 Ha with a total of 4,917,327 people exposed. Physical losses caused by coastal abrasion are Rp. 22,042,350 (M) and economic losses of Rp. 1,290,842 (M) and cause environmental damage of 460,252 Ha [6].

The coastal area is one of the areas threatened by tsunami, coastal abrasion, sea water intrusion, storms, floods, extreme waves / tidal waves. For this reason, actions that can be carried out on disaster management include prevention, mitigation, preparedness and emergency response [8]. Threat / danger is a natural or artificial phenomenon that has the potential to threaten human life, property losses and environmental damage.

Beach erosion, which is also called abrasion, has recently tended to increase in various regions. Abrasion is erosion or reduction of land (beach) due to wave, current and tidal activity. In this connection land compaction caused the land surface to fall and be flooded by sea water so the coastline changed. In this connection land compaction caused the land surface to fall and be flooded by sea water so the coastline changed. The beach is said to experience abrasion when sediment transport occurs to a point greater than the amount of sediment transported out of that point [9].

The main parameters needed to understand coastal abrasion as one of the problems in the coastal zone are:

- Coastal geomorphology, which is the type of coastline and the sensitivity of coastal occurrence
- Wind is the main factor that causes waves

- Waves are the most important force causing coastal abrasion
- Tide is an influential force on coastal morpho dynamics
- Coastal vegetation is a component that maintains coastal slope stability and shoreline protection
- Human activities along the coast affect coastline stability; onshore and offshore activities, sand mining and coral reefs, and estuary dredging can increase coastal abrasion (shore abrasion).

In addition to the influence of oceanographic hydrocoography, currents and tides, human intervention in the form of exploitation of coastal resources and rock conversion and mangrove forests helps accelerate the rate of coastal degradation of Ambon Bay [10]. Strong currents and waves, as well as the destruction of coral reef ecosystems along the waters of Padang Beach, have caused worsening of the abrasion, which is compounded by the damage to natural and artificial beach protectors which have a negative impact on residential areas along the coast of Padang [11].

Factors that cause abrasion besides being influenced by the ocean and waves are also caused by wind. Longshore currents due to deep ocean waves and lifted by wind. The abrasion process occurs in the concentrated area of coastal waves with lithology composed of non-compact rock formations. Coastal and topographic geomorphology also affects coastal changes, as well as lithology of coastal sediments consisting of alluvium material with the dominance of gravel and sand and triggered by sediment transport from rivers [12].

Abrasion and accretion occur at different levels which indicate that ecosystems play an important role in controlling abrasion, especially in areas where mangrove ecosystems are located. Abrasion is more due to natural factors than human activity. Therefore, integrated management in the form of synchronization programs with relevant agencies, initiated and developed a series of schemes that become regional development priorities [13].

The cause of abrasion is based on the Engineering Details of the Abrasion and Rob Handling of Demak Regency [14] namely; a) Land Subsidence, b) Mangrove Forest Damage, c) Damage due to wave hydrodynamic forces, d) Damage caused global climate change and e) Damage due to other human activities.

3. Research methods

3.1. Research Approach Method

The study approach method used is descriptive quantitative, visual and triangulation. Quantitative descriptive approach, namely data and information obtained, processed and presented using forms of frequency tables and percentages. The 2003 Pariaman City coastline map with 2018 Pariaman City coastline map was overlaid. The results obtained are any beaches in Pariaman City that experience abrasion and / or accretion.

Visual approach, namely direct depiction of objects and the analytical tool used in the form of maps and the latest object images.

Triangulation approach, namely the examination of the validity of the results of the initial analysis with the results of interviews on the factors that cause abrasion and accretion. Triangulation is carried out with different techniques and various data sources, namely interviews, observations and useful documents to further strengthen the results of the study.

3.2. Method of Data Collection

Data collection methods used are primary survey methods and secondary surveys. Primary survey is a method of conducting direct observations into the field, namely observing the progress of abrasion or accretion in the coastal area of Pariaman City in order to find out the existing (actual) conditions.

Secondary surveys are methods by collecting data from various agencies and literature studies related to the substance of study objects in Pariaman City in the form of thematic maps, books and articles in relevant scientific journals that are processed into one data and information. In detail the maps (secondary data) and sources can be seen in the following table.

Table 1. Types of Secondary Data and Resources

Data	Data Type	Analitical Method	Data Source
Landsat ETM image data for 2003 2018 Landsat OLI image data of Pariaman City	Secondary data are: Satellite imagery from Landsat obtained from USGS *(U.S. Geological Survey)*	Using the Band-Ratio method and the Single Band Method to obtain delineation of land and water boundaries (pixel value limits) whose purpose is to obtain the coastline in each good Landsat ETM image in 2003 and Landsat OLI 2018.	USGS *U.S.* or the United States Department of Domestic Geological Survey

3.3. Analysis Method
The analytical method used is the 2003 coastline comparison method with the coastline in 2018. The analysis method can be seen in Table 2.

Table 2. Analysis Method

The first stage	Second Stage	Third phase	Fourth Stage
2003 Landsat ETM image data and 2018 Landsat OLI or acquisition year adjacent to the main criteria of each selected dataset is cloud-free, the Landsat dataset requirements for analysis and download are cloud content data $\leq$ 10% and have not experienced defects on Scan Line Corrector (SLC)	To facilitate the analysis, the multi-spectral band types of both datasets (Landsat-7 ETM and Landsat-8 OLI) are made in the form of stacking layer bands (merged) in each dataset. Because the scope of the research area (all of the West Sumatera Coastal Areas) requires 3 scenes (paths / rows), this step produces 6 stacking files for the needs of coastline analysis.	Delineation Land and water boundaries to get the coastline by using the Band-Ratio method to obtain a more informative pixel value limit. In the Band-Ratio method, the NIR band ratio with the Green band (b4 / b2 on Landsat-7; b5 / b3 on Landsat-8) will produce a land-water boundary on the coastal area covered by vegetation. Land areas that are not vegetated are included in water (sea) pixels. In contrast to the SWIR-1 band ratio with the Green band (b5 / b2 on Landsat-7; b6 / b3 on Landsat-8) the coastline is obtained from areas covered by sand and soil.	To help extract the land-sea boundary information that will be a shoreline feature, a composite band technique or a combination of false colours is used to display the limit of each object observed. raster to vector conversion results. This polyline vector file is the same vertex size as the spatial resolution of the original dataset (Landsat ETM and Landsat OLI) which is 30 meters. So as to refine and edit it, improvements are made. Improvements are made using line smooth tools in ArcGIS

4. Results and discussion
4.1. Beach Abration of Pariaman City
Kota Pariaman is one of the coastal cities in West Sumatra Province which borders the Indian Ocean. As a coastal city, it has various disaster threats, one of which is coastal abrasion.

The following is a profile of coastal abrasion that occurred several years back in Pariaman City and actions and efforts made by the Pariaman City Government and the local community.

IOP Conf. Series: Earth and Environmental Science **314** (2019) 012009 doi:10.1088/1755-1315/314/1/012009

Table 3. Beach Abrasion of Pariaman City 2013-2017

10/10/2017	• abrasion on the Nareh coast of North Pariaman District • abrasion along the coast of Padang Biriak-Biriak Village, North Pariaman Subdistrict, residents are quite dense. • "Pasia Maelo"; land ownership rights or beaches that become land are threatened with loss • some of the land around the shoreline is certified, if the abrasion continues to expand then the ownership rights are automatically lost • 12 houses were abrasion • has eroded up to 80 meters inland • only two breakwater stones have been installed, at least need seven.	 **Figure 1.** Abration October 2017 • APBD Amendments and central government assistance in installing breakwater stones at some points. • the local community was also asked to work together to build a temporary embankment
11/1/ 2017	• abrasion of Gandoriah Beach, Anas Malik Beach, Angso Duo Island, Kasiak Island and Nareh Beach in North Pariaman District • Gandoriah Beach to Anas Malik Beach which is located in the District of Central Pariaman is approximately affected by abrasion around 35 to 40 meters • tourism facilities and infrastructure that have been built are also threatened (Development of Anas Malik Park area, Republic of Indonesia Navy Monument, ASEAN Monument, and a number of other tourist parks) • beach abrasion that occurs is predicted to be purely due to natural phenomena due to extreme weather • Around 500 meters of Nareh Beach's lips have not been protected	 **Figure 2.** Abration January 2017 breakwater stones need Rp. 30 billion
2014	• coast of Nareh Beach, North Pariaman District	BNPB provided assistance of Rp 30 billion for the installation of breakwater stones
2013	• coast of Nareh Beach, North Pariaman District	BNPB provided assistance of Rp 40 billion for the installation of breakwater stones

4.2. Analysis of The Development of Abrasion and Accretion

4.2.1. Pariaman City Coastline in 2003 and 2018. The coastline in 2003 of Pariaman City was taken from Landsat ETM (Enhanced Thematic Mapper) Image Data in 2003 or acquisitions that were close to the main criteria of each selected dataset which were cloud-free. The requirements of the Landsat dataset used for analysis and download are cloud content data $\leq 10\%$ and have not experienced defects in Scan Line Corrector (SLC).

The coastline of 2003 Pariaman City was taken from the 2018 Landsat OLI Image Data or the acquisition which is adjacent to the main criteria of each selected dataset is cloud-free. The requirements of the Landsat dataset used for analysis and download are cloud content data ≤ 10% and have not experienced defects in Scan Line Corrector (SLC).

4.2.2 Abrasion and Accretion of Pariaman City in 2003 and 2018. The results of the analysis of the study of the development of coastal abrasion and accretion in the City of Pariaman are by overlapping the 2003 map with the 2018 map of the coastline so that the coastlines experience abrasion (shoreline retreat) or accretion (beach addition). Map analysis is done by delineating land and water boundaries to get a coastline covered by vegetation. Land areas that are not vegetated are included in water (sea) pixels. In contrast to the SWIR-1 band ratio with the Green band (b5 / b2 on Landsat-7; b6 / b3 on Landsat-8) the coastline is obtained from areas covered by sand and soil.

Beach abrasion and accretion of Pariaman City in 2003 to 2018 can be seen in the following tables and maps.

Table 4. Coastal abrasion and accretion of Pariaman City in 2003-2018

No	Coordinate		Long Change (m)	Change	Location
	X	Y			
1	623781	9932734	22.18	Abrasion	Pauh
2	623695	9932913	30.25	Abrasion	Pauh
3	623916	9932448	29.8	Abrasion	Pauh
4	624010	9932246	33.35	Abrasion	Pauh
5	624095	9932028	16.05	Abrasion	Pauh
6	624151	9931862	8.62	Abrasion	Pauh
7	624152	9931766	25.14	Accretion	Pauh
8	624177	9931552	22.36	Accretion	Pauh
9	624076	9930827	21.56	Abrasion	Pasir
10	624134	9930601	22.05	Accretion	Pasir
11	624388	9930184	8.54	Accretion	Pasir
12	624599	9929941	4.53	Abrasion	Kel.Lohong
13	624827	9929640	25.03	Accretion	Kel.Lohong
14	624891	9929547	39.35	Accretion	Karan Aur
15	625708	9928331	56.45	Abrasion	Karan Aur
16	627383	9926841	39.15	Abrasion	Marunggi
17	623020	9934204	31.89	Accretion	Apar
18	622809	9934573	11.99	Accretion	Manggung
19	622022	9935799	25.88	Accretion	Naras 1
20	622270	9935399	61.43	Abrasion	Naras Hilir
21	622452	9935115	41.06	Accretion	Naras Hilir
22	626160	9928069	109.24	Abrasion	Taluak
23	621432	9936983	38.05	Abrasion	Padang Birik 2
24	621658	9936445	32.09	Accretion	Balai Naras

Source: Results of analysis of Arc Gis Image Map, 2018

There were 24 points in 12 villages observed abrasion and accretion along the coast of Pariaman City from 2003 to 2018 (15 years of observation). There are 5 locations that experience abrasion and accretion at the same time, namely Pauh Village, Pasir Village, Lohong Village, Karan Aur Village, Naras Hilir Village. This shows normality where one location experiences abrasion while experiencing accretion. For more information on abrasion and accretion locations in Pariaman City, see the following table.

Table 5. Sub-districts that experience abrasion and accretion in Pariaman City in 2003-2018

No	Sub Districts	Abrasion	Accretion
1	Pauh	v	v
2	Pasir	v	v
3	Lohong	v	v
4	Karan Aua	v	v
5	Marunggi	v	-
6	Apar	-	v
7	Manggung	-	v
8	Naras 1	-	v
9	Naras Hilir	v	v
10	Taluak	v	-
11	Padang Birik-birik	v	-
12	Balai Naras	-	v
	Jumlah	**8**	**9**

Source: Data processing results, 2018

There were 13 points experiencing abrasion and 11 accretion points scattered in 12 villages in Pariaman City. During the last 15 years there has been abrasion (land loss) as far as the range of 4.51 m to 109.24 m while the accretion (addition of land) occurs as far as 8.54 m to 41.06 m.

4.3 Area Analysis of Coastal Pariaman Abrasion and Accretion in 2003-2018

With the same method, the results of the extensive abrasion and accretion analysis that occurred on the coast of Pariaman City were by overlapping the map in 2003 with a map of 2018 obtained by shoreline experiencing abrasion (shoreline retreat) or accretion (addition of beaches) and the length of the area experienced change. By knowing the length and changes that occur to the shoreline, the area of the coast / coast that experiences abrasion or beach accretion is obtained.

There were 24 points that experienced changes in both abrasion and accretion during this period spread along the coast of Pariaman City. Beach abrasion in Pariaman City is found in 8 sub-district (13 points / locations) where the smallest area is 1.57 ha in Padang Birik-Birik and the widest coastal abrasion is at point 22, which is 33.42 ha in the Village / Taluak Village. The widest beach accretion (beach addition) occurs at point 14 of the sub-district / Karan Aur Village, namely selus 45.12 ha and the smallest accretion at point 19 of the Village / Village Naras 1, which is 4.09 ha. The results of the analysis on the extent of coastal abrasion and accretion in Pariaman City period 2003 - 2018 can be seen in Table 6.

Table 6. Area of abrasion and accretion of Pariaman City in 2003-2018

No	Coordinate		Length (m)	Area (Ha)	Change	Location (Sub-districts)
	X	Y				
1	623781	9932734	22.18	15.65	Abrasion	Pauh
2	623695	9932913	30.25	20.66	Abrasion	Pauh
3	623916	9932448	29.8	19.98	Abrasion	Pauh
4	624010	9932246	33.35	9.87	Abrasion	Pauh
5	624095	9932028	16.05	9.47	Abrasion	Pauh
6	624151	9931862	8.62	21.62	Abrasion	Pauh

7	624152	9931766	25.14	25.39	Accretion	Pauh
8	624177	9931552	22.36	10.28	Accretion	Pauh
9	624076	9930827	21.56	6.26	Abrasion	Pasir
10	624134	9930601	22.05	9.86	Accretion	Pasir
11	624388	9930184	8.54	21.78	Accretion	Pasir
12	624599	9929941	4.53	8.70	Abrasion	Lohong
13	624827	9929640	25.03	21.74	Accretion	Lohong
14	624891	9929547	39.35	45.12	Accretion	Karan Aur
15	625708	9928331	56.45	25.79	Abrasion	Karan Aur
16	627383	9926841	39.15	16.89	Abrasion	Marunggi
17	623020	9934204	31.89	28.97	Accretion	Apar
18	622809	9934573	11.99	13.58	Accretion	Manggung
19	622022	9935799	25.88	4.09	Accretion	Naras 1
20	622270	9935399	61.43	7.77	Abrasion	Naras Hilir
21	622452	9935115	41.06	7.79	Accretion	Naras Hilir
22	626160	9928069	109.24	33.42	Abrasion	Taluak
23	621432	9936983	38.05	1.57	Abrasion	Padang Birik2
24	621658	9936445	32.09	27.06	Accretion	Balai Naras

Source: Results of analysis, 2018

Over the past 15 years there has been a significant loss of land in Pariaman City due to coastal abrasion, fearing that the day is getting bigger. In Pariaman City there were 13 abrasion locations with abrasion area of 197.65 ha and 11 beach accretion points with an accretion area of 285.38 ha. This is different from the beaches in the region of West Sumatera Province in general where on the coast of Pariaman there are additional beaches of 285.38 ha or 144.39% greater than the area of beach loss which is only 197.65 ha. More can be seen in the following table 7.

Table 7. Extent of abrasion and accretion in Pariaman City in 2003-2018

No	Sub-Districts	Abrasion (hectare)	Accretion (hectare)
1	Pauh	97,25	35,67
2	Pasir	6,26	31,64
3	Lohong	8,7	21.74
4	Karan Aua	25,79	45.12
5	Marunggi	16,89	-
6	Apar	-	28.97
7	Manggung	-	13.58
8	Naras 1	-	4.09
9	Naras Hilir	7,77	7.79
10	Taluak	33,42	-
11	Padang Birik-birik	1,57	-
12	Balai Naras	-	27.06
	Total	**8**	**9**

Source: Data processing results, 2018

Table 8. Extensive annual abrasion and accretion in Pariaman City

Total Point	Area(ha)	Abrasion/Accretion	Average hectare / year
13	197,65	Abrasi	13,18
11	285,38	Akresi	19,03

Source: Results of analysis, 2018

Theoretically, if an area occurs abrasion, there will be other areas of coastal accretion. The city of Pariaman also experienced coastal abrasion and beach accretion even though in a less balanced comparison.

The results of the analysis of the image map, it can be concluded that the annual average of land reduction in Pariaman City is 13.18 ha / year. This land loss rate is quite high and very dangerous considering Pariaman City is a coastal city that is quite densely populated and as a center of government and a trade center. The addition of the coastal land of Pariaman City is 144.39% equivalent to 19.03 ha / year greater than the land loss which is only 13.18 ha / year.

This condition if no coastal mitigation efforts are carried out, will threaten the balance of the coastal environment, where the loss of the coast continues to threaten the City of Pariaman. Meanwhile accretion that occurs will also raise new problems related to land ownership status due to beach accretion (growing land).

4.4 Analysis of Factors Affecting Beach Abrasion and Accretion

There has been abrasion and accretion over the last 15 years along the coast of the City of Paraiman. The factors analyzed by map processing with the GIS Arc method include wave factors, coastal typologies, shoreline shapes, vegetation and currents. The results of the study can be seen that the main factor influencing coastal abrasion in Pariaman City is the current factor. This can be seen from the high currents of the Pariaman City waters which are 0.513 - 0.7 (m / s). With high current velocities, the high threat of abrasion disasters occurs.

The second factor is the land cover / vegetation factor. In general, the areas in the North Pariaman and South Pariaman areas have a thicker vegetation cover that is 30% - 65% compared to Central Pariaman with a density of only 30-32%. Then the southern and northern regions of Pariaman City have a Medium index, while the Central Pariaman District in the High index is based on the classification of BNPB regulation No. 2 of 2012 [15]. This means that the coast of Pariaman City lacks coastal vegetation as a mitigation thus triggering an increasingly high abrasion disaster. Lack of land cover by vegetation or beach structure adds to the high threat of abrasion that will occur on the coast of Pariaman City.

The third factor that influences the coastal abrasion of Pariaman City is due to the shoreline form factor. The form of the beach in Pariaman City is generally a straight and straight beach form with a straight length of 12.7 Km. Looking at the data in the form of a straight beach is included in the high index, which is mostly located in the District of North Pariaman, while the form of a straight beach is almost located on the majority of the coast of Pariaman City which belongs to the medium category. This means that using a straight or straight coastline, the higher the abrasion. The fourth factor that influences the coastal abrasion in Pariaman City is the low-grade wave factor. The waters of North Pariaman Subdistrict, Central Pariaman and South Pariaman have a wave height of 0.69 to 0.92 m, where this condition is included in the Low category.

The fifth factor is caused by coastal typology factors in Pariaman City, namely sandy and muddy. Typology of sandy beaches is included in the medium index and typology of muddy beaches including high index categories. The results of the analysis showed that the coastal typology in Pariaman City was included in the medium to high typology which affected the occurrence of coastal abrasion. The factors that influence the abrasion and accretion on the coast of Pariaman City can be seen in the following table.

Table 9. Indicators that affect coastal abrasion and accretion

No	Indicators	Index Class			Total weight
		Low	Medium	High	
1	Wave height	< 1 m	1 – 2,5 m	> 2,5	30 %
2	Current	< 0,2	0,2 – 0,4	> 0,4	30 %
3	Land cover/vegetation	> 80 %	40 – 80 %	< 40 %	15 %
4	Beach shape	Hug	Straight hug	Straight	15 %
5	Beach typology	Rocky	Sandy	Muddy	10 %

Source: Perda BNPB No 2 Year 2012

Table 10. Factors influencing the abrasion and accretion of Pariaman City

Wave category	Coastal typology index	Shoreline shape index	Vegetation index	Current index
0,78 - 0,83 m	Muddy & sandy	Straight & straight-headed	30% - 65%	0,513 - 0,7 (m/s)
Low	Medium-high	Medium-high	Medium-high	Low

Source: Processing Analysis of Arc Gis, 2018

5. Conclusion

From 2003 to 2018 (15 years of observation) on the coast of Pariaman City there has been beach abrasion and accretion in 24 points or 12 villages. The number of abrasion points occurred in 13 points or 8 villages with an abrasion area of 197.65 ha and 11 points of beach accretion or 9 villages with an accretion area of 285.38 ha. Congratulations on 15 years of observation, there was a coastal abrasion on the coast of Pariaman City covering an area of 197.65 ha and accretion of 285.38 ha. This proves that coastal abrasion causes a substantial reduction in land in Pariaman City, which is an average of 13.18 ha / year, while land addition is 19.03 ha / year. This figure proves quite significant between the area of land / beach loss and the addition of beaches in Pariaman City.

During the last 15 years there has been an abrasion as far as the range 4.53 m to 109.24 m while the accretion occurs as far as the range 8.54 m to 41.06 m. The farthest abrasion is in point 22 (Taluak), which is as far as 109.24 m or an average of 7.28 m / year and closest to point 12 (Lohong) as far as 4.53 m. Whereas the accretion in which the furthest coastline advances is found at point 21 (Naras Hilir) which is 41.06 m or an average of 2.74 m / year. Almost all physical factors affect the coastal abrasion and accretion of Pariaman City, namely current factors, shoreline shape, coastal typology and vegetation cover except low wave factors.

References

[1] The Atlas of the Human Planet 2017
[2] Data dan Informasi sumber Bencana Indonesia (DIBI)
[3] Haryani, Agus I and Nurhasan S 2018 Coastal Abrasion and Accretion Studies of West Sumatera Province in Period 2003-2016 *Journal of Environmental Science and Engineering A* 7 **7** (1) 22-29
[4] Haryani 2012 Model Mitigasi Bencana di Wilayah Pesisir dengan Pemberdayaan Masyarakat *Jurnal Nasional Tataloka* **14**(3) ISSN 0852-7458
[5] BAPPEDA (Badan Perencanaan dan Pembangunan Daerah) Kota Pariaman 2017 Revisi RTRW Pariaman 2010-2030
[6] Badan Nasional Penanggulangan Bencana 2016 *Penurunaan Indeks Resiko Bencana di Indonesia* (Direktorat PRB, BNPB)
[7] United Nations International Strategy for Disaster Reduction (UN-ISDR)
[8] Badan Koordinasi Nasional Penanggulangan Bencana 2007 *Panduan Pengenalan Karakteristik Bencana dan Upaya Mitigasinya di Indonesia* ed Triutomo, Sugeng, Widjaja, B. Wisnu ,

Amri, M.Robi (Jakarta)

[9] Kurnia Damaywanti Dampak Abrasi Pantai terhadap Lingkungan Sosial (Studi Kasus di Desa Bedono, Sayung Demak) *Prosiding Seminar Nasional Pengelolaan Sumberdaya Alam dan Lingkungan 2013*

[10] T J Kakisina 2015 Analysis of The Impact of Land Use on The Degradation Of Coastal Areas at Ambon Bay-Mollucas Province Indonesia *International Journal of Procedia Environmental Sciences* **23** 266 – 273

[11] Ferli F, Rifardi, Afrizal T 2013 Integrating Local Knowledge and Perception for Assessing Vulnerability to Climate Change in Economically Dynamic Coastal Areas: The Case of Natural Protected Area Aiguamolls de l'Empordà Spain *International Journal of Ocean & Coastal Management* **85** 90-102

[12] Tirza J K, Sutrisno A, Agus H, Suripin 2016 NEMOS (Nearshore Modelling of Shoreline Change) Model for Abrasion Mitigation at the Northern Coast of Ambon Bay *Aquatic Procedia* **7** 242 – 246

[13] Rostika R 2016 The Managing Plan for Abrasion in Coastal Area of Garut Regency *International Journal of Procedia Environmental Sciences* **33** 512 – 519

[14] Kimpraswil 2006 Detail Engineering Penanganan Abrasi dan Rob Kabupaten Demak

[15] Peraturan Kepala Badan Nasional Penanggulangan Bencana Nomor 02 Tahun 2012 Pedoman Umum Pengkajian Risiko Bencana

Characteristics of melanic epipedon based on biosequence in the physiography of Marapi - Singgalang, West Sumatra

Dedi Hermon[1,*], Ganefri[2], Aprizon Putra[1] and Olivia Oktorie[2]

[1]Department of Geography, Padang State University, Indonesia
[2]Postgraduate - Padang State University, Indonesia

*dihermon006@gmail.com

Abstract. This research aim to express descriptively melanic epipedon based on biosequence middle bevel physiography of Marapi-Singgalang mount representing Upper watershed area from Batang Anai watershed. The sample determined by *Stratified Random Sampling* at each *biosequence*. From result of research obtained by difference of melanic epipedon of Marapi mount characteristic with melanic epipedon of Singgalang mount characteristic, so that for the farm in the middle bevel of Singgalang need the existence of land conservation action, passing reboisasi program, utilize to improve again land that function as buffer zone in Batang Anai watershed.

1. Introduction

The soil is the result of natural weathering of rocks or organic (vegetation and animals). The characteristic of the soil comes from the climate and rock material in certain topography condition for a certain period of time. Bedrock, topography and time are the passive factors, while climate and organism are the active factors of the soil forming [1,2,3. 4,5] adds soil that forms on the surface of the earth directly or indirectly develops from rock minerals. Through the weathering process, both physically and chemically and aided by the influence of the atmosphere (rainfall and temperature), the rock disintegrates produce loose parent materials, and subsequently under the effect of pedogenic processes develop into the soil. Soil that forms in the tropics with conditions of rainfall and relatively high temperatures each year would make the process faster. The effect of climate, especially rainfall and temperature, on the development of soil profiles is very large. Climate affects weathering and soil development processes. Indirectly rainfall affects the chemical characteristics of the soil [6,7,8].

According to [9] the effect of vegetation, soil organism, animals and human also very large for soil development. the role of human (farmer), especially in the rural areas are very diverse vertically characterized by the formation of the horizons. The soil horizon is the soil layer due to the process of soil formation. This is the difference of the soil horizon i.e: 1) O-Horizon, is a horizon on the surface of the soil due to the weathering of organic material; 2) A-Horizon is the horizon formed below horizon O which is characterized by the percentage of sand is more dominant than the percentage of dust and clay; 3) B-Horizon, is the horizon below the A-Horizon which is characterized by the percentage of clay is more dominant than the percentage of sand and dust; and 4) C-Horizon, is the lowest horizon and above the R-Horizon (bedrock) which is characterized by the loose rock. While according to [9,10], the entirety of A-Horizon and half of B-horizon will form epipedon. Epipedon is an identifier for the soil classification which formed in topsoil. In the wet tropic areas, epipedon formed on Andisols soil, that is soil formed from volcanic ash, is melanic epipedon. Epipedon is characterized by relatively high C-organic and N-content, base saturation <50%, value and chroma 2 or smaller, melanic index <1.70, and thickness 30 - 40 cm from ground level.

Furthermore, [11,12] explain that the characteristic of melanic epipedon can change or having degradation due to rainfall, vegetation, microorganism, animals and human. The relatively high rainfall every year causes the leaching process of A-Horizon because of infiltration, so the elements K^{+}, Na^{+}, Ca^{2+}, Mg^{2+}, organics, clay and another characteristic of melanic epipedon disappear and accumulate on B-horizon. Other than that, the effect of vegetation, microorganism and animals also very large to the characteristic of melanic epipedon. The vegetation, microorganisms, and animals that live in it (biosequence) will give different effects on the characteristic of melanic epipedon and other biosequences. Forest biosequence and bush which hasn't been disturbed by human activities in Andisols soils tend to form melanic epipedon. According to [13,14,15,16,17], this epipedon identified by the high organic material in the soil. This epipedon will have the degradation process and developing into Umbria epipedon, plan epipedon, or anthropic if the forest changed into agriculture area. As a result, the soil will have a low resistance due to run off.

The physiographic region of the middle slope of a young volcano in the western part of Marapi and the physiographic region of the middle slope of the volcanic mountain east of Singgalang has relatively the same biosequence. This area is located in Tanah Datar Regency, which is the upper watershed of the Batang Anai watershed which is very vulnerable to land crises. Identification of melanic epipedon and differences in melanic epipedon in biosequence of forest and bush on Mount of Marapi and Singgalang will provide recommendations for land reclamation and conservation as a result. So, the damage to land in the upper Batang Anai watershed can be minimal and the capacity of the land can be sustainably. Therefore, the characteristics of melanic epipedon will not be degraded and in accordance with the specified criteria, so that the function of the area as a buffer zone in the Batang Anai watershed will be optimal.

2. Research Methods

This research was using descriptive survey method, based on biosequence. Determination and sampling are done by *Stratified Random Sampling* and biosequence are considered to be the strata. In the research location Marapi mount, 2 observation points were determined, on natural forest land (Ht) and scrubland (sm), as well as the research location on Mount Singgalang. Soil analysis (observed parameters) and the methods used are: 1) analysis of soil C-organic with the Walkley and Black method; (2) analysis of N-total soil by the Kjeldahl method; 3) base cations can be exchanged (Caddy), Mgdd, Kdd, Nadd¬) with extraction method NH4Oac 1N pH 7; 4) KTK total using NH4Oac 1N washing method pH 7; 5) P2O5 extracted by 1% citric acid solution method; 6) melanic index by Honna method; 7) P retention with the Blackmore method; 8) pH (H2O and KCl) with glass electrode method; 9) NaF pH with combination electrode method; 10) soil texture analysis (3 fractions) with pipette method; and 11) BV using the graphimetry method.

3. Results and Discussion

The results of observations on soil morphology in the form of observations of colours, genetic horizons, thickness, structure and consistency soil layers can be seen in Table 1.

Table 1. Nature of morphology of soil research location

Physiography	Biosequence		Genetic horizon	Thick (cm)	Soil coulor	Soil structure	Consistency	Horizon limit
					10 YR 2/2	crumb	free	average
			Ap	0-24	10 YR 3/4	crumb	loose	gradually
			AB	24-55	7,5 YR	glob	loose	gradually
	forest	I	B1	55-67	4/4	glob	loose	average
			BW	67-88	10 YR 5/6	glob	somewhat	----
			2Bl	>88	7,5 YR		firm	
					4/4			
1					10 YR 2/2	crumb	free	average
			Oa	0-2	7,5 YR	crumb	loose	gradually
			A2	2-52	4/4	glob	somewhat	flat
	forest bush	II	B1	52-85	7,5 YR	glob	firm	average
			B2	85-98	5/4	grain	somewhat	average
			BC	>98	7,5 YR		firm	----
					4/4		free	
					10 YR 5/8			
						crumb	loose	average
			Ap	0-36	10 YR 2/2	crumb	loose	gradually
			AB	36-69	10 YR 3/4	glob	somewhat	gradually
	forest	III	B1	69-98	10 YR 4/4	glob	firm	average
			B2	98-120	10 YR 4/6	grain	somewhat	----
			BC	>120	7,5 YR		firm	
					4/4		somewhat	
2							firm	
					10 YR 2/1	crumb	free	average
			Oa	0-3	10 YR 2/2	crumb	loose	average
			A1	3-24	7,5 YR	glob	loose	average
	bush	IV	B1	24-79	3/4	glob	somewhat	gradually
			B2	79-103	7,5 YR	glob	firm	----
			BC	>103	4/4		firm	
					7,5 YR			
					3/4			

Note: 1. Physiography of Singgalang's Old Middle Volcanic Slope
2. Physiography of the Western Volcanic Middle Slope of Western Marapi
KP. Profile Code

According to Table 1 above morphology characteristics on biosequence of forest and bush both at the Marapi-Singgalang mount still characterize andisols soil properties, the properties of soil that characterize Andisols soil, the soil colour is generally black (2-5 and chroma 1-6), crumb soil structure and the consistency of the soil is generally loose. These also characterize the characteristics of melanic epipedon. The results of the analysis of the physical characteristics of the soil for supporting the identification of melanic epipedon can be seen in Table 2.

Table 2. The result of the analysis of the physical aspect in research location

Physiography	Biosequence		Genetic horizon	Thick (cm)	Particle Size Distribution(%) Sand	Dust	Clay	Texture Class	BV g/cm³
1	forest	I	Ap	0-24	49,09	12,48	38,77	Sandy sand	0,74
			AB	24-55	31,11	37,55	30,25	Clay	0,77
			B1	55-67	23,45	41,07	34,98	Clay	0,73
			BW	67-88	74,56	25,12	0,39	Sandstone	0,69
			2Bl	>88	21,92	48,13	28,87	Clay	0,66
	bush	II	Oa	0-2	29,42	46,66	23,05	Clay	-----
			A2	2-52	49,99	48,05	3,52	Sandy Clay	0,79
			B1	52-85	33,21	22,92	43,12	Clay	0,81
			B2	85-98	49,05	17,72	33,21	Sandy Clay	0,83
			BC	>98	31,07	36,88	31,28	Clay	0,61
2	forest	III	Ap	0-36	29,98	52,64	15,51	Dusty Clay	0,70
			AB	36-69	23,51	56,09	18,36	Dusty Clay	0,78
			B1	69-98	19,44	48,72	31,02	Dusty Clay	0,80
			B2	98-120	25,92	42,08	31,12	Clay	0,82
			BC	>120	37,71	25,22	36,99	Clay	0,75
	bush	IV	Oa	0-3	33,88	47,94	16,70	Clay	-----
			A1	3-24	27,07	60,91	10,23	Dusty Clay	0,59
			B1	24-79	25,84	58,07	14,18	Dusty Clay	0,68
			B2	79-103	21,61	50,15	26,01	Clay	0,17
			BC	>103	29,34	41,05	28,59	Clay	0,58

Note: 1. Physiography of Singgalang's Old Middle Volcanic Slope
2. Physiography of the Western Volcanic Middle Slope of Western Marapi
KP. Profile Code

Based on Table 2, the observed soil profile has BV values that meet the requirements of the soil properties andik (Andisols soil characteristics), which has BV <0.85 g/cm³. One of the condition of melanic epipedon is part of the A-Horizon and B-Horizon, as well as the overall A-Horizon and B-Horizon, or the entire horizon has Andik soil properties [17,18,19]. BV values ranging from 0.58-0.83 g/cm³ also specify the Melanic epipedon. In profiles I and II on Singgalang physiography (Table 2), the sand fraction is more dominant than the dust fraction and clay fraction, this indicates that the land has undergone an intensive horizontal and vertical washing process. In profiles III and IV in Marapi physiography (Table 2), the dust fraction was more dominant than the sand and clay fractions, so the characteristics of the melanic epipedon in Marapi mount were thicker than the epipedon characteristics of melanic epipedon on Singgalang physiography. In addition to the physical properties and identification of soil morphology, analysis of soil chemical properties is also needed for identification of melanic epipedon characteristics at the research location. The results of the analysis of soil chemical properties can be seen in Table 3 below.

Table 3. Results of analysis of chemical characteristics of soil characteristics of melanic epipedon

Biosequence		Genetic Horizon	Thick (cm)	pH Soil H₂O	KCl	NaF	*C-Organik* %	N-total	P_2O_5 ppm	*Kation Basa dd* Ka	Ca		Mg	Kej-Basa %	KTK me/100g
										Me/100g					
Forest	I	Ap	0-24	5,71	5,16	11,34	5,67	0,61	260,3	1,23	0,18	9,06	1,60	47,8	25,2
		AB	24-55	5,47	4,82	11,61	4,68	0,48	113,7	1,16	0,14	5,39	1,22	39,7	21,0
		B1	55-67	5,89	5,34	11,48	2,09	0,33	134,5	0,20	0,01	7,71	1,30	52,6	17,5
		BW	67-88	5,82	5,14	11,43	1,88	0,34	94,2	0,11	0,02	4,85	0,18	31,1	16,7
		2Bl	>88	5,75	5,21	11,48	3,85	0,42	129,8	0,43	0,12	5,98	0,10	38,3	17,3
Bush	II	Oa	0-2	5,73	5,09	11,06	11,58	0,77	378,9	1,53	0,55	11,01	1,64	32,1	29,4
		A2	2-52	5,25	4,97	10,87	6,68	0,58	224,7	0,48	0,42	8,96	0,69	47,0	22,4
		B1	52-85	5,72	5,33	11,72	3,59	0,30	94,2	0,60	0,43	6,68	1,35	51,1	17,7

		B2	85-98	5,63	4,85	11,53	5,64	0,44	113,7	0,22	0,28	9,90	0,33	50,6	21,2
		BC	>98	5,76	5,11	10,53	1,14	0,08	70,4	0,33	0,87	1,58	0,63	21,0	16,2
		Ap	0-36	6,25	5,56	10,67	6,61	0,65	177,2	1,81	0,81	5,83	0,67	30,1	30,2
		AB	36-69	5,51	4,72	11,24	5,37	0,54	100,1	0,39	0,34	3,08	0,20	14,1	28,4
Forest	III	B1	69-98	5,37	4,40	10,94	4,07	0,31	123,8	0,42	0,10	3,40	0,36	16,1	26,5
		B2	98-120	5,26	5,01	10,88	2,17	0,26	Tu	0,10	0,22	1,79	0,63	8,8	31,0
		BC	>120	5,37	4,91	10,73	1,76	0,14	Tu	0,09	0,18	1,54	0,71	9,1	27,7
		Oa	0-3	5,92	5,33	10,90	13,84	0,81	212,8	0,97	0,38	8,01	2,05	33,6	33,9
		A1	24-3	6,01	5,42	11,11	10,58	0,63	123,8	0,22	0,17	6,65	1,81	32,4	27,3
bush	IV	B1	24-79	5,88	5,17	11,24	4,27	0,34	100,1	0,56	0,16	3,11	0,15	13,1	30,3
		B2	79-103	5,74	5,28	11,25	4,48	0,36	94,2	0,45	0,09	5,80	0,38	20,9	32,1
		BC	>103	5,72	4,99	10,84	2,17	0,26	64,5	0,08	0,09	0,94	0,12	4,8	25,3

Note: 1. Physiography of Singgalang's Old Middle Volcanic Slope
2. Physiography of the Western Volcanic Middle Slope of Western Marapi
KP. Profile Code and tu. not measurable

pH value of NaF (Table 3) for 60-minute measurements show a range of 10.53-11.72. This NaF pH value indicates that the soil profile observed also has an inherent characteristic. According to [20], that NaF pH>10 for as large a horizon shows indications of reliable ingredients. In profile I (Table 3), melanic epipedon characteristics are found at 24 cm thickness, this is characterized by the value of chroma in moist 2, organic C-5.67% (<6%), base saturation 47.89% (<50%), the index melanic 1.39 (<1.7) and retention P >85% (89.23%).

In profile II (Table 3), melanic epipedon has a thickness of 50 cm (upper limit of organic layer 2 cm thick), organic C-6.68% (>6%), base saturation 47.01% (<50%), index Melanic 1.66 (<1.7), and retention P>85% (90.14%). The value of chroma and value in epipedon thickness has undergone changes, i.e value and chroma 4 (>2), but other supporting characteristics show more characteristic of Melanic epipedon at a thickness of 2-52 cm. In profile III (Table 3), Melanic epipedon has a thickness of 36 cm (at a depth of 0-36 cm), organic C-6.61% (>6%), base saturation 30.19% (<50%), Melanic index 1.45% (<1.7), and P retention >85% (91.14%). The characteristics of melanic epipedon in profile IV (Table 3) have a thickness of 21 cm (3-24 cm from ground level), value and chroma 2, C-organic 10.58% (>6%), base saturation 32.42% (>50%), index melanic 1.23 (<1.7), and retention P 94.75% (>85%).

In the [17] it was explained that the characteristics of epipedon in volcanic ash soil is melanic epipedon, which has a thickness of 30 cm or more in total thickness of 40 cm, C-organic 6% or more, value and chorma 2 or smaller, melanic index <1.70, and at epipedon thickness has Andik soil characteristics. Then [18,19] say that the melanic epipedon criteria, in the epipedon, have a base saturation of <50%. Based on these results, the soil formed in the eastern volcanic slope of the eastern Singgalang slope was categorized as having melanic epipedon as the surface soil characterization horizon. Profile I with natural forest (Ht) has a thickness of melanic epipedon 24 cm which is classified as having changed from the required thickness (30 cm or more). In forest management, it does not refer to soil conservation techniques, so that parts of the forest are burnt due to land clearing for agriculture, so that soil erosion because of run-off. Profile II under the shrub (Sm) has a thickness of melanic epipedon 50 cm and has undergone changes in value and chroma value. The thickness of melanic epipedon in profile II is formed through the addition of organic matter due to the intensive weathering of plant debris.

The characteristics of epipedon formed on the physiography of the middle slope of the young volcanic west part of Marapi mount were also categorized in melanic epipedon. Profile III with which is the same as profile I (forest vegetation) has epipedon thickness 36 cm, C-organic >6%, N-total >0.3%, base saturation <50%, and melanic index <1.7. Profile IV which also has the same profile II (shrub vegetation) also has melanic epipedon characteristics with epipedon thickness of 21 cm. Thus, profiles III and IV have sufficient requirements to characterize melanic epipedon. The characteristics of melanic epipedon in the study location based on biosequence can be seen in Table 4.

Table 4. The characteristics of melanic epipedon in research location based on biosequence

Criteria	Natural Forest Biosequence (Ht)		Shrub Biosequence (Sm)	
	Singgalang Profil I (Ht)	Marapi Profil III (Ht)	Singgalang Profil II (Sm)	Marapi Profil IV (Sm)
Thickness ((cm)	24	36	50	21
Value*	2	2	4	2
Chroma*	2	2	4	2
Melanic index	1,39	1,45	1,66	1,23
Retention P	89,23	91,41	90,14	94,75
Soil Structure	CR	CR	CR	CR
Consistency	LO	FR	FR	FR
Texture Class	SC	SiL	SL	SiL
BV (g/cm^3)*	0,74	0,70	0,79	0,59
pH H_2O	5,71	6,25	5,25	6,01
pH KCl	5,16	5,56	4,97	5,42
pH NaF*	11,34	10,67	10,87	11,11
C-Organik (%)*	5,67	6,61	6,68	10,58
BO (%)	9,74	11,36	11,50	18,19
N-Total (%)	0,61	0,65	0,58	0,63
P_2O_5 (ppm)	260,33	177,27	224,74	123,87
K_{dd} (me/100g)	1,23	1,81	0,58	0,22
Na_{dd} (me/100g)	0,18	0,81	0,42	0,17
Ca_{dd} (me/100g)	9,06	5,83	8,96	6,65
Mg_{dd} (me/100g)	1,60	0,67	0,69	1,81
Kejenuhan Basa (%)*	47,89	30,19	47,01	32,42
KTK (me/100g)	25,2	30,2	22,4	27,3

Note: Cr, crumbs; LO, release; FR, loose; SC, sandy clay; SL, sandy clay; SiL, dusty clay; * Melanic epipedon condition

4. Conclusion

The characteristics of melanic epipedon found both at the mount Singgalang and at the mount Marapi based on biosequence are still in the stage of completing the conditions for melanic epipedon. The characteristics of melanic epipedon in forest (Ht) and shrub (Sm) in the Marapi location are still haven't any degradation of epipedon properties, whereas in forests (Ht) and shrubs (Sm) in Singgalang locations have undergone degradation of C-organic thickness and content. This is due to the role of humans (farmers) who's not referring to soil and water conservation techniques in land management, so the process of soil erosion by run-off. Thus, the role of mount Marapi as Upper Batang Anai watershed area is still capable of being a buffer zone for its downstream area and mount Singgalang locations need land reclamation and land conservation through reforestation programs, to increase soil resistance to erosion or erosion.

Acknowledgments

The author to express our gratitude to all those who have helped with the administration process as well as those who helped us in the field.

References

[1] Hermon D 2009 *Geografi Tanah* (Padang: UNP Press)
[2] Buol S W Southard R J Graham R C and Mc Daniel P A 2011 *Soil genesis and classification* (John Wiley & Sons)
[3] Coleman D C Callaham M A and Crossley Jr D A 2017 *Fundamentals of soil ecology* (Academic press)

[4] Hansen J and Lebedeff S 1987 Global trends of measured surface air temperature *J. geophysical research: Atmospheres* **92**(D11) 13345-13372

[5] Sparks D L 2018 *Kinetics and mechanisms of chemical reactions at the soil mineral/water interface* In *Soil Physical Chemistry* (Second Edition CRC press)

[6] Hermon D 2017 *Climate Change Mitigation* (Jakarta: Rajawali Pers - Radjagrafindo)

[7] Dixit R Malaviya D Pandiyan K Singh U Sahu A Shukla R and Paul D 2015 Bioremediation of heavy metals from soil and aquatic environment: an overview of principles and criteria of fundamental processes *Sustainability* **7**(2) 2189-2212

[8] Alloway B J 2013 *Sources of heavy metals and metalloids in soils* (Springer)

[9] Putra A Triyatno Syarief A and Hermon, D. (2018). Penilaian Erosi Berdasarkan Metode USLE dan Arahan Konservasi Pada DAS Air Dingin Bagian Hulu Kota Padang-Sumatera Barat *Jurnal Geografi* **10**(1): 1-13

[10] Palacio A G 2005 *Caracterización y manejo de los suelos de la Península de Yucatán: implicaciones agropecuarias, forestales y ambientales* (Instituto Nacional de Ecología)

[11] Isbell R 2016 *The Australian soil classification* (CSIRO publishing)

[12] Osman K T 2012 *Soils: principles, properties and management* (Springer Science & Business Media)

[13] Kairis O Karavitis C Salvati L Kounalaki A and Kosmas K 2015 Exploring the impact of overgrazing on soil erosion and land degradation in a dry Mediterranean agro-forest landscape (Crete, Greece) *Arid land research and management* **29**(3) 360-374

[14] Hermon D Putra A and Oktorie O 2018 Suitability Evaluation of Space Utilization Based on Environmental Sustainability at The Coastal Area of Bungus Bay in Padang City *International Journal of Geomate* **14** (41) 193-202

[15] Bautista F Duarte Y A and Pereira E D 2011 Soils as natural reactors for swine wastewater treatment. Tropical and subtropical agroecosystems. **13**(2) 199-210

[16] Ponziani F Pandolfo C Stelluti M Berni N Brocca L and Moramarco T 2012 Assessment of rainfall thresholds and soil moisture modeling for operational hydrogeological risk prevention in the Umbria region (central Italy) *Landslides* **9**(2) 229-237

[17] Jury W A and Stolzy L H 2018 Soil physics (*Handbook of Soils and Climate in Agriculture* CRC Press)

[18] Soil Survey Staff 2003 *Keys to soil taxonomy* (Department of Agriculture: Natural Resources Conservation Service)

[19] Michéli E, Owens P R, Láng V, Fuchs M and Hempel J 2014 *Organic Carbon as a major differentiation criterion in soil classification systems* (Springer)

[20] Shoji S and Ono T (1978) Physical and chemical properties and clay mineralogy of Andosols from Kitakami Japan *Soil Science* **126**(5) 297-312

Analysis of JHA and JSA at KIP 16 Bangka Ocean Mining Units PT Timah (Persero) Tbk Bangka Belitung Islands Province

Riri Rahmahwati Joni[1,*], Rusli HAR[2] and Indang Dewata[1]

[1] Department of Environmental Science, Postgraduate Program, Universitas Negeri Padang, Prof. Dr. Hamka Street, West Air Tawar, Padang-West Sumatra 25131 Indonesia

[2] Department of Mining Engineering, Faculty of Engineering, Universitas Negeri Padang, Prof. Dr. Hamka Street, West Air Tawar, Padang-West Sumatra 25131 Indonesia

*ririrahmahwatijoni29495@gmail.com

Abstract. This research discusses about safety and health occupational control in Bangka Ocean KIP Timah 16 Units using some risk management methods. The methods are Job Hazard Analysis (JHA) and Job Safety Analysis (JSA). Purpose of the research is to explain safety and health occupational in KIP Timah 16 Unit it is for get the improvement of productivity work, production, welfare and comfortability workers and achievement of zero accident. The type of the research is descriptive. There are two data had collected. They are primary and secondary data. Primary data get from interview with the employees and secondary data get from K3 documents in the enter company of PT. Timah UPLB and KIP 16. Based on result of research can get some conclusions. First, to use method JHA and JSA for identification and control occupational accident.

1. Introduction

PT Timah is a company that was developed on August 2, 1976 and it is a State-Owned Enterprise (BUMN) that participate in tin mining and has been listed in the Indonesia Stock Exchange since 1995.

In carrying out mining operations, not all process activities will always function easily. This is because there will be interactions between humans and mining equipment and material with the environment. This interaction is very risky and can cause work accident. According to Nugraha [1], the occurrence of work accidents is caused by 4 processes including the interrelated elements of production, it was people, equipment, material, environment (PEME) that interact to produce a product and service.

According to ILO (1989), work accidents can be caused by 3 (three) factors it was human factors, their work and the environment in the workplace. Human factors can be age, level of education and work experience. Job factors such as the form of shifts and types of work. Meanwhile, for environmental factors such as physical environment (lighting and noise), chemical environment and biological environmental factors.

Accident prevention function is to minimize accidents and losses. Accident prevention work is carried out after determining the cause of accidents in the production process so that made steps how to handle workplace accidents correctly. Control of workplace accidents can be done by several approaches including hazard approach, human approach, technical approach, administrative approach and management approach.

Based on data from workplace accidents in 2005 to 2017, there were 132 cases of accidents that occur in the PT. Timah (Persero) Tbk and this is high. The case consists of small accidents such as tripping, cuts and scratches, typically only causing minor injuries, so workers can keep working. Lost time accident is an accident of work requiring workers to be taken to the hospital and took a temporary break, resulting in loss of working time. some of the causes of accidents lost time accident is twisted, hit and pinched (to drop). While the accident is fatality accidents that can result in death. Details of the incident that occurred in work accidents PT. Timah (Persero) Tbk is 26.5% small accidents; 37.1% lost accident control and 36.4% died.

From the details of the accident data, it is proof of the lack of control and management of Occupational Safety and Health at PT. Timah (Persero) Tbk. So, the researchers were interested in analyzing JHA and JSA in Bangka Ocean KIP Timah 16 Units. Before, we have to know what is hazard, hazard is the potential for harm. In practical terms, a hazard often is associated with a condition or activity that, if left uncontrolled, can result in an injury or illness. To controlled, it is need safety [2]. Next, OSHA [2] explained too, job hazard analysis is a technique that focuses on job tasks as a way to identify hazards before they occur. It focuses on the relationship between the worker, the task, the tools, and the work environment. Besides, Mohammmadi [3] explained that job safety analysis is a technique that focuses on job tasks as a way to identify hazards before they occur. They two have connection each other. The purpose of the study was to analyze Job Hazard Analysis (JHA) and Job Safety Analysis (JSA) in Bangka Ocean KIP Timah 16 Units.

2. Research Method

This research uses a descriptive method. The source of information is gotten from primary data and secondary data. Primary data is gotten from direct observation / field observation and interviews, while secondary data is gotten from library materials, articles, journals, documentation, company internal data and other supporting documents such as JHA and JSA.

3. Result

Based on the results of interviews conducted at Tin 16 KIP, the causes of accidents that often occur in KIP are as follows: careless / inaccurate at work, the amount of oil fluid that is scattered everywhere, do not use APD during work, there is no protective fence on rotating machines, place APD around rotating machines, the number of holes in the floor, tools that are scattered (irregular), don't understand how the machine works, not following the procedure, workplace is not safe, works not in their fields, lack of coordination, nonstandard tools.

3.1. Job Hazard Analysis (JHA) in KIP Timah 16

The JHA is an analysis and improvement process that can literally transform workplace safety. The JHA is a structured process that can discover the causes for the vast majority of workplace injuries and illnesses. Before carry out an analysis of the occupational hazards found in KIP 16, the first work that must be done is to make the steps for drafting JHA, including the following:

1. Form a JHA team, they are safety professionals, engineers, supervisors and workers.
2. Determine JHA priorities, it is occupations with the highest disease rates, jobs that have a potential level of serious injury, work where one simple human error can cause injury, work that is complex enough to have written instructions, jobs that significantly have technological change process or procedure.
3. involving workers.
4. Re-discuss the incident that happened before.

5. doing to review job.
6. there is a checklist, scale and priority setting for hazardous work.
7. Describe steps or tasks.
8. Involve workers who have an understanding about work, help minimize negligence so that the analysis is quality and people who get direct benefits.

After preparing the JHA steps, the next step is to identify the type of work, specific work, the danger of work and proper control. For more details, see Table 1:

Table 1. JHA in KIP 16

Type of work	Detail work	Danger	Control
digging tin ore using ladder	digging tin ore using a ladder which is controlled from the command room	- Intermittent wire split - Ladder caught in a working hole - Ladder demage	Monitor ship rotation speed regularly when excavating tin ore
the process of washing tin ore using a rotary filter, primary jig, secondary jig and sluice box	The work of separating tin ore from debasement minerals based on grain size and specific gravity using rotary filter, primary jig, secondary jig and sluice box	The swivel filter is located on the 3rd floor and the potential hazards found in the swivel filter environment: - fall -Stumble - knock - Fall from a height The primary jig is located on the 2nd floor and the potential hazards found in the primary jig environment: - fall - knock -Stumble The secondary jig is located on the 1st floor, so the surface of the first floor is often inundated by seawater and potential hazards in the secondary jig environment: - fall - slipping - slip -Stumble - knock - Fall into the sea -Pinched - hit	Using APD such as: safety shoes, safety helmet, buoys, general work gloves, body hardness, work clothes (coverall) and be careful in working

		The sluice box is located on the 1st floor close to the secondary jig, the potential danger in the sluice box environment: - fall - slipping - slip -Stumble - knock - Fall into the sea	
Welding work	Work uses a power source (welding transformer)	- Affected by bursts of sparks from the welding machin -Bending -Rum -Hot -Burnt skin -Noisy -Damage to the eyes -Fatigue due to hot temperatures	Workers have to observe some sources of strains that are supplied to the welding machine and must use complete PPE such as: -Safety shoes -Glass glasses - Welding mask - Welding gloves - Welding work apron -Earplug -Fill -Working clothes (coverall)
work on monitoring and controlling machinery, such as ground pump machines, hydraulic machines and rudder propeller machines	control engine performance directly from the source	Potential hazards found in the ground pump engine environment: - Fast heat - Noisy -Pinched - knock - Hot exposure Potential hazards found in a hydraulic engine environment: - spinning - twisted - cut off -Pinched - knock - Noisy Potential hazards found in a rudder propeller engine environment: -Pinched - cut off - High heat pressure - twisted - knock - Noisy - Hot exposure	-Give safety on a planetary object - Install warning signs and so on -Use PPE such as: safety shoes, safety helmets, buoys, general work gloves, earplugs, dust masks, gas masks, work clothes (coverall) and be careful in working

Work in limited space	Work carried out in narrow spaces such as pontoon repairs	Potential hazards found in a confined space environment: - Lack of oxygen and excess oxygen - Stuck on the way out - Flammable and explosive material - Toxic materials (gas, fume, steam) - Solid or liquid substances stored in it - Mechanical energy, electricity and heat	Use of special PPE such as: - General work gloves - Safety shoes - Respiratory equipment - Work clothes (coverall) - Earplug - Safety helmet - Google - Chemical gloves - Electric gloves - Fire resistant clothes

3.2. Job Safety Analysis (JSA) in KIP Timah 16

NSC (2009, p. 240d) in Glenn [4] explained that JSA is an analytical process of developing safer job procedures and to the document that is developed as a result of the analysis. The methods used in JSA techniques include: observation (observation) method, discussion method (consultation), and review method. Before doing out an analysis to the work of the JSA, the steps that need to be taken are:

1. Choose the job to be analyzed.
2. Divide the work, which describes the sequence of work procedures.
3. Identify hazards in work steps, and identify various possibilities that have the potential for accidents.
4. Provide control to avoid accidents that have been identified in each step.

After preparing the steps of the JSA, the next step is to identify the type of work, potential hazards and recommended actions. For more details, it can be seen in Table 2.

Table 2. JSA in KIP 16

Type of work	Potential hazard	Recommended actions or procedures
digging tin ore using ladder	Personal: near miss-injury-fatallity Property: interrupted wire breaks and ladder broken Environmental: sea water is cloudy, this can cause disruption of the activities of marine life so that it affects the difficulty of fishermen to find fish	- All workers must use PT standard safety equipment such as: safety shoes, safety helmet, buoys, general work gloves, body hardness, work clothes - Perform induction safety 1 (one) time in a month - Perform a meeting tool box 1) times a week - Conduct PJSM (Pre Job Safety Meeting) before work - Do safety talk every day - Training to improve skills in safety
the process of washing tin ore using a rotary filter, primary jig,	Personal: near miss and personal injury Property: damage to jig	Perform daily maintenance

secondary jig and sluice box	components such as: rooster, bed, wire screen, afsluiter underwater, balance, eccentric, membrane and spigot handlebar Environmental: sea water due to tailings from the washing process	
Welding work	Personal: personal injury Property: damage to the welding machine Environmental: the environment around the welding area gets hot	Using APD completely like: - Safety helmet - Safety shoes - Welding glasses - Welding mask - Welding gloves - Welding work apron - Earplug - Buoy - Work clothes (coverall)
work on monitoring and controlling machinery, such as ground pump machines, hydraulic machines and rudder propeller machines	Personal: near miss-injury-fatallity Property: damage to the engine due to improperly controlled Environmental: o	- Using PPE equipment - Installation of warning signs - Installing rotating object safety
Work in limited space	Personal: fatallity Property: leak on the pontoon Environmental: fire	Use special safety equipment such as: - General work gloves - Safety shoes - Respiratory equipment - Work clothes (coverall) - Earplug - Safety helmet - Google - Chemical gloves - Electric gloves - Fire resistant clothes

The results of the JHA and JSA analysis obtained will be used as material references to develop the KIP Timah standard operational procedure (SOP). besides the results of this analysis should also show how the condition of JHA and JSA in KIP Timah.

4. Conclusion

JHA is one of the risk management methods that identify all activities prior to the occurrence of hazards. The types of work in the KIP 16 was identified was the work of excavation, the washing process, the welding job, a job in charge of controlling the machinery and work in confined spaces. After identification, the control performed in accordance with the kinds of work. Additionally, hazard identification methods in KIP Timah 16 JSA is not yet realized.

JSA is one of the risk management methods that identify hazards at each work contained in the KIP Timah 16 JSA identified components include: personal, property and environmental based work processes. During this time the condition of the JSA in KIP Lead 16 is not yet realized, so that it impacts on the frequent occurrence of minor accident (minor accidents).

Reference

[1] Nugraha, OMC. 2015. "Gambaran Penerapan Sistem Manajemen Keselamatan Dan Kesehatan Kerja (SMK3) berdasarkan OHSAS18001:2007". Semarang: Universitas Negeri Semarang.
[2] U. S. Departement of Labor. 2002. "Job Hazard Analysis". Occupatioanal Safety and Heald Administration 3071. United States of America.
[3] Mohammadi, Mohsen. 2017. Job Safety Analysis, Universal Journal Of Management, 5(7):355-363.
[4] Glenn, D David. 2011. Job Safety Analysis, Risk Management.Chicago.www.asse.org.

IOP Conf. Series: Earth and Environmental Science **314** (2019) 012012 doi:10.1088/1755-1315/314/1/012012

Institutional capacity building for local disaster prevention board of West Pasaman Regency

K Khairul[1,*], H Purnaweni[2] and S Suwitri[1]

[1] Public Administration Department, Faculty of Social Scienes and Politics, Diponegoro University, Semarang, Indonesia
[2] Environmental Science, School of Postgraduate Studies, Diponegoro University, Semarang, Indonesia

*65khairul65@gmail.com

Abstract. This study aimed at analysing institutional capacity building of BPBD of West Pasaman in local-based disaster prevention and management. The study applied a qualitative methodology with phenomenological approach. The disaster prevention and management institutional capacity building was performed by a coordination. The coordination did not restrict itself with formal institutions, but also local community institutions and organisations, such as Kogami, HSNI, and three local community elements (tungku tigo sajarangan). This study found that socialisation in disaster mitigation and prevention was still inadequate because of lacking local community involvement. Furthermore, it also found lacking clarity of authority and responsiblity between central, provincial, and regency governments. West Pasaman Regency needed support from the provincial government in terms of post-disaster equipments. This study had a theoretical contribution of extending the knowledge in public management and practical contribution of helping the local government implement the institutional capacity building of BPBD in the disaster prevention and management

1. Introduction

West Sumatera Province has been applying 24-hour disaster preparedness for years due to the fact that it has a high disaster prone. Natural disasters may occur anytime, anywhere [1]. A similar fact is also gathered from the Center for Volcanology and Geological Disaster Mitigation (PVMBG), that West Sumatera has 144 districts with a moderate-to-high risk of natural disaster in terms of mass movement, which is responsible for causing landslide and flood. Such disaster prone can be found in sixteen regencies/municipalities [2,3,4].

West Pasaman is one of regencies under the administrative territory of West Sumatera, which is highly prone to disasters. The Local Disaster Prevention Board (BPBD) of West Pasaman Regency reports detailed the geographical stature of the regency as the reason for being exposed by frequent disasters. The topographical profile of West Pasaman, being either hilly, mountaneous, or coastal, contribute to the threats of vary natural disasters, from landslide, to flood, to abrasion, to tsunami. Such condition requires awareness and preparedness around the clock [1]. Several districts, including Talamau, Gunung Tuleh, Sungai Aua, Ranah Batahan, and Parit Kota Balingka, are siuated in mountaneous, landslide prone area. The regency also has rivers that potentially cause flood, i.e. Batang Pasaman and Batang Sikabau, which are all drought at Sasak Estuary [1]. High level of preparedness from the local government of West Pasaman are crucial to deal with the highly potential disaster threats [5].

Analysis of institution and institutional capacity building is necessary to help the disaster institutions perform its duties effectively and efficiently. Community involvement, especially those who received the immediate impact of the disaster, is crucial. In the context of West Pasaman, local-borne communities, which play important roles in dealing with the disasters, e.g. Komunitas Siaga Tsunami (Kogami), HSNI, and three elements of local leadership (*tungku tigo sajarangan*) are considered necessary to take part into disaster prevention and management [6,7,8].

2. Method

This research performed a qualitative methology with a phenomenological approach. Phenomenology influences attitudes and viewpoint, becoming the only way to know other's experience by asking them questions through interviews. The phenomenon of the institutional capacity building of BPBD of West Pasaman can be studied by the phenomenological approach. This method aims at revealing reality from the viewpoint of the individuals who have direct experience. It relates to the natures of human experience and the meaning adjacent to it.

3. Theoretical Reviews

3.1. Natural Disaster

Disaster is an extreme functional disturbance of a particular community, which causes social, material, and environmental lossess, extending and surpassing the ability of the affected community and their own resources. Earthquakes, floods, typhoons, and tsunamis are considered disasters once they have caused serious problems and affected human life, living, and property or belongings. A harmful event (e.g., earthquake, flood, typhoon, or tsunami) becomes a disaster when it causes injury, death, and loss of properties or living, as well as damage of infrastructures [9,10].

Law 24/2007 defines disaster as "an event or a series of events that threaten and disturb life and living of community due to either natural or human factors as such that these events cause death, environmental damage, loss of properties, and psychological disorders" [11].

Natural disaster is a consequence of a combination between earth activities, either physical (e.g. volcanic erruption, earthquake, landslide) or human activities. Human powerlessness due to lacking preparedness and emergency management is responsible for causing financial and structural losses, not to mention life. Natural symptoms are those very natural that occur overtime as long as the Earth stands; however they may become the disasters when causing deaths and physical losses [12]. A natural disaster occurs quickly and spontaneously with huge impacts on the environment and socio-economic systems [13]. The disaster impact is enormous over time, as the United Nations reports that the annual toll of the disaster victims have reached 200 people. Such figure is seven times bigger than war victims. In just a few years different natural disasters, i.e. earthquake, tsunami, typhoon, and flood, have caused severe damages across the globe [14].

In addition to being natural-borne, natural disasters are, in part, also caused by human, such as technological failure, which triggers people, community, individual, and environmental responses to rendering a far-reaching enthusiasm, in the forms of protests, petitions, or causes, that oppose to the further application of the tehnology [15].

3.2. Institutional capacity building

Major focuses of capacity building are investing in people, investing in organisation, and investing in network. Investing in people deals with th notion that human as an actor must be capable of doing something beneficial to life and future [15]. Investing in organisation goes with organisational consideration as a tool or a medium used by human in attaining mutual objective in efficient and effective manners. Capacity building comprises three dimensions, i.e. human resource development, organisational empowerment, and institutional reform. The focus of institutional reform and macrostructure system deals with the following types of activities: economic and political rule of the game, change in policy and regulation, and constitutional reform [16].

3.3. Previous researches

Buchari et al. (2017), in their research on institutional capacity building in Tarokong Kaler District, Garut Regency, West Java Province, conclude that disaster resilient village capacity building is defined as efforts to improve capabilities in dealing with disaster management from pre-response to post-disaster phases [17].

Caymaz et al. (2013) in their research in Turkey concerning problems of effective and efficient public institution management on disasters find that in the National Earthquake Strategy and Action Plan 2012-2023, the focus on earthquake did not apply effectively and efficiently in practice [18].

Cheema et al. (2015) promote the importance of a historial case study approach to disaster management and mitigation. Their empirical research presents approaches to natural disasters in Pakistan. It gives priority to the serious awareness of planning and management during the disasters in developing countries at different political scales, from local to national, in particular when resources are limited [19].

Hagelsteen and Burke (2016) contend that capacity builing in disaster risk reduction (DRR) becomes one of leading methods towards the loss reduction due to the disaster. The research reminds us to challenges faced by organisations to materialise the guidance for capacity building to be put in pratice. Among the challenges is limited resource in project management [20].

A research that discusses climate change phenomenon in urban area in Asia with a case study of Hat Yai city, Thailand, reveals key factors towards successful pubic policy in disaster management. It finds three important factors, i.e. balance between self- and individual interests, private sector participation, and effective implementation of policies in response to obstacles [21].

Siriwardana et al. (2017) in their research on efficiency and effectiveness of disaster management frameworks in Sri Lanka find that the SLNDMP, the national institution responsible for disaster issues, is not capable of implementing the Post Disaster Assessment effectively; PDA is an important feature in this country to reduce the disaster risk. The SLNDMP is also unable to produce an extensive and inclusive national disaster profile, so that many of the disaster prone areas are overlooked, therefore, causing more losses [22].

Khairul et al. (2016) in their research on institutional empowerment of disaster management in Padang, West Sumatera conclude that the BPBD of Padang Municipality as an independent leading sector must have a stronger authority by being posited under immediate order of the local head, i.e. Mayor of Padang. The Board is also supposed to cooperate and coordinate better with Police Department, the Army, and disaster experts [6].

4. Results

4.1. Natural Disasters in West Pasaman Regency

Flood that occurred during October 2018 in West Pasaman Regency took the tolls of eleven different districts. A report from BPBD of West Pasaman Regency explained that the flood occurred in Wonisari Kinali, Simpang Tiga, Jambak Luhak Nan Duo, Kapa, Sasak, Batang Saman, Pasaman, Gunung Tuleh, Sungai Aur, Koto Sawah Ujung Gading, Ranah Batahan, Air Bangis, and Parit. Furthermore, the flood was also followed by another disaster, landslide. [23] Flood disaster caused damages of two important bridges located in two districts, i.e. Pasaman and Ranah Batahan, isolating as many as 875 households [24].

Flood did not only take human victim, but also natural resources. In some districts, i.e. Sasak, Pasaman Barat, Ranah Batahan, Lembah Malintang, and Sungai Beremas, the flood caused rice field and plantation damage. Accordingly, the local farmers sufferred from harvest loss [25].

Flood disaster in West Pasaman did not only occur in the current year. The previous years showed that River Batang Batahan became the major cause of flood in local settlements of Kampung Baru Nagari Batahan [26]. In the past five years, West Pasaman has been suffering from vary disasters, i.e.

earthquake, flood, landslide, and abrasion. Table 1 illustrates the natural disasters that took place in the regency from 2013 in many districts.

Table 1. Natural disasters in West Pasaman Regency, 2013-2018

No.	Year	Disaster	Location
1	2013	Abrasion	Sasak
2	2014	Flood; Landslide	Simpang Ampek; Kinali; Gunung Tuleh; Lembah Melintah
3	2015	Earthquake	Simpang Ampek; Kinali
4	2016	Landslide	Talamau
5	2017	Flood	Simpang Ampek; Talamau
6	2018	Flood	Kinali; Simpang Tigo; Ujung Gading; Ranah Batahan; Air Bangis; Parit; Jambak Luhak Nan Duo; Pasaman; Gunung Tuleh; Sungai Aur; Koto Sawah Ujung Gading

Sources: [3,26,27]

4.2. Role of BPBD of West Pasaman in disaster prevention and management

4.2.1. Pre-disaster efforts. BPBD of West Pasaman has a function of formulating and making policies on disaster prevention and management, as well as refugee coordination, by taking timely and on-target actions, effectively and efficiently. In order to perform all activities related to the disaster prevention and management BPBD of West Pasaman already has a Standard Operation Procedure (SOP) in reference to the SOP issued by BPBD of West Sumatera Province. The Board, in addition, is also responsible for coordinating activities dealing with the disaster management and prevention in planned, integrated, and inclusive manners. For a while, BPBD of West Pasaman has already equipped itself with Standard Operational Procedure (SOP) in dealing with the disasters. It directly involves on the disaster site and coordinates with BPBDs from other regencies/municipalities, Indonesia Red Cross (PMI), Quick Reaction Team, and Search and Rescue (SAR) Team. However, the mechanism is under the coordination of BPBD of West Pasaman [1].

The threat of landslide became an importance reminder for BPBD of West Pasaman to inform to the public at large during the 2018 *Idul Fitri* holidays. The Board gave a warning to commuters to stay awake of the landslide potential danger during their road trip. The Regency Government of West Pasaman, through its Public Works and Spatial Management Office, had been preparing heavy duties and equipments needed for dealing with the landslide. A 24-hour preparedness was applied supported by any necessary equipments and aides, such as operational vehicles, boats, and diving equipments [1][28].

BPBD of West Pasaman has issued a 2016-2021 Strategic Plan to give guidance for planning and implementation of programs and activities, policy direction and strategies for empowerment and implementation of disaster prevention in the regency. This Strategic Plan includes objective of accountability system of the BPBD as one of key institutions in Pasaman Regency that deal with disaster towards efficient and effective management of the existing resources.

4.2.2. Post-disaster efforts. Flood and landslide in November 2014 were responded by BPBD of West Pasaman by constructing the data obtained directly from the location. The results were then endorsed to the Provincial Government of West Sumatera. The Board also expected helps from externalities to relieve the post-disaster impacts [26].

During the 2016 landslide BPBD of West Pasaman, in collaboration with Public Work Office and Sector Police (*Polsek*) of Talamau District, made a joint effort to resolve traffic jam. The landslide disconnected the transportation from West Pasaman Regency to Pasaman Regency. When the landslide occurred, no adequate equipments were available to clean the debris. The Board, the Police, and PMI coordinated and organised the situation. Talamau is a highly disaster prone regency [4].

The 2017 flood was responded by BPBD of West Pasaman by a coordination with River Area Center (BWS) Sumatera II in River Batahan Normalisation to prevent the flood. However, in practice, the coordination was hampered, therefore, allowing the flood to occur [4].

5. Discussion

5.1. Institutional capacity building of BPBD of West Pasaman

The pre-, during, and post-disaster phases were marked by the Board's limited coordination with official governmental institutions. Community institutions and groups, such as Kogami, HSNI, *tungku tigo sajarangan,* were not given opportunities to take part.

Should the focus of institutional capacity building be investing in organisation, in which the organisation becomes a medium towards mutual objective, the practical finding during the research on the disaster prone in West Pasaman offerred a phenomenon. Historically, the regency had been "familiar" with natural disasters, but the BPBD still found classical difficulties in performing its duties to deal with the disasters. No further efforts were taken towards investing in organisation by developing the institutional capacity in preventing exceeded losses due to the disasters.

The ongoing coordination was ineffective. It proved a possibility of overlapping duties. The disaster preparedness, as shown in the historically lanslide prone areas, proved the unprepared manner of the institution.

Furthermore, no mutual cooperation was found in terms of active participation between authorities and local community groups, such as NGOs, social groups, and traditional entities (e.g. the *Nagari*). The disaster management by local community using inadequate equipments and knowledge proved that the institution did not capable of transmitting knowledge and preparedness to the affected community.

Inadequate institutional capacity building did not only cost human losses, as what happend in flood disasters, but also environmental problems. The negative impact of the natural disasters on the environment was the loss of ecological equilibrium, threatening sustainable environment. The local institutions' capacity in taking the "correct" and timely actions during the disasters will decrease dependency on other externalities in terms of post-disaster aids, financially or materially.

In the capacity building the coordination must be performed with institutions that concern with the disaster issues, such as Agency of Meteorology, Climatology, and Geophysics (Badan Meteorologi, Klimatologi, dan Geofisika, or BMKG) and those dealing with volcanology. The purpose of such coordination is to share initial, pre-disaster knowledge amongst the affected citizens, including that of related to disaster symptoms.

The existing disaster prevention and management performed by BPBD West Pasaman is lacking socialization of disaster mitigation. The socialization can be manifested in mitigation, simulation, and training with the purpose of to enable the disaster management, to make the affected people more sensitive and more responsive to the disaster, and to minimize the negative impacts of the disaster. This socialization can be performed in collaboration with local community groups, such as Kogami, HSNI, and three local leadership elements (*tungku tigo sajarangan*).

The normalization of River Batahan proposed by BPBD West Pasaman has not been properly responded by the central government because its authority is significant. This lacking central government response has prolonged flood threats on the local community. There must be a clarity of authority and responsibility between central government, provincial government, and regency/municipal government to avoid irresponsible manner, for example, the central government claims that it does not have responsibility for what is happening in the provincial or regency territory during the disaster. Furthermore, The Talamau landslide has become a serious problem for the affected people due to lacking equipment. The provincial government must help the regency government provide the adequate equipment by coordinating with concerned institutions and bodies.

6. Conclusion

West Pasaman had proven itself as the highly prone to natural disaster regency. Flood, landslide, earthquake, abrasion, and tsunami had been continuing to pose serious threats. This research applied a phenomenological approach because, on one hand, West Pasaman had been for generations identically with natural disasters, and, on the other hand, the authorized institutions did not seem to have knowledge, experience and capacity in dealing with the disasters.

The institutional capacity building of BPBD of West Pasaman is a mandatory option towards prime public service in terms of preferrable disaster prevention and management.

7. Recommendations

This research recommended the following points:

a. Institutional capacity building of BPBD of West Pasaman based on investing in organisation and investing in networks;
b. Active and direct involvement of disaster affected parties, such as NGOs, local leaders, community leaders, and *Nagari* government. Further effort can be taken by giving knowledge and extended information about tsunami and abrasion to the community living in the coastal areas; and
c. The institution will perform well if it has a strong network. Therefore, investing in networks is also considered necessary to be taken into account.

References

[1] "Landslide Alert for Commuters in West Pasaman During Holidays." Retrieved from [https://sumbar.antaranews.com/berita/227223/pemudik-yang-melintas-di-pasaman-barat-agar-waspada-daerah-longsor] on 26 October 2018. [in Indonesian]
[2] "National Preparedness Capacity is the Key to Disaster Free Development." Retrieved from [https:www.bnpb.go.id/kapasitas-dan-kesiapsiagaan-nasional-kunci-menuju-pembangunan-aman-bencana] on 26 October 2018. [in Indonesian]
[3] "Lanslides Hit West Pasaman Again." Retrieved from [https://harianhaluan.com/news/detail/62656/pasbar-kembali-dihantam-longsor] on 25 October 2018. [in Indonesian]
[4] "144 Districts Have Medium-to-High Potential of Landslides." Retrieved from [https://www.bnpb.go.id/144-kecamatan-sumbar-berpotensi-gerakan-tanah-menengah-hingga-tinggi] on 26 October 2018. [in Indonesian]
[5] "West Pasaman BPBD Alerts Citizens of Disaster Threats." Retrieved from [https://sumbar.antaranews.com/berita/111901/bpbd-pasaman-barat-ingatkan-warga-wasapadai-bencana] on 26 October 2018. [in Indonesian]
[6] Khairul K, Purnaweni H and Suwitri S 2016 *Proceedings of the 2016 International Conference on Public Management (ICPM 2016)*. DOI: dx.doi.org/10.2991/icpm-16.2016.31.
[7] Rivera C, Tehler H and Wamsler 2015 *International Journal of Disaster Risk Reduction* **14** 445
[8] Couto N, Latour B 2013 *Bruno Latour and ANT (Actor Network Theory)*. Retrieved from [http://dc352.4shared.com/doc] on 10 January 2016
[9] Ulum M C 2014 *Manajemen Bencana: Suatu Pengantar Pendekatan Proaktif* (Malang, Indonesia: Universitas Brawijaya Press)
[10] Asian Disaster Reduction Center 2003 *Glossary on Natural disasters*. Retrieved from [http://adrc.asia] on 23 October 2018.
[11] Law 24/2007 on Disaster Prevention. [in Indonesian]
[12] Khambali I 2017 *Disaster Prevention Management* (Yogyakarta, Indonesia: Andi) [in Indonesian]
[13] Alexander D E 1999 *Natural Disasters* (Dordrecht, Netherlands: Kluwer Academic Publishers)
[14] Mauch C and Pfister C (eds.) 2009 *Natural Disasters, Cultural Responses: Case Studies toward a Global Environmental Histor* (Plymouth, United Kingdom: Lexington Books)
[15] Eade D 1997 *Capacity Building: An Approach to People-centered Development* (Oxford, United

Kingdom: Oxfham)
[16] Grindle M S (ed.) 1997 *Getting Good Government: Capacity Building in the Public Sectors of Developing Countries* (Boston, United States: Harvard Institute for International Development) pp 1-28
[17] Buchari A, Santoso M B and Marlina N 2017 *Jurnal Analisis dan Kebijakan Publik* **3**(1) 49-62
[18] Caymaz E, Akyon F V and Erenel F 2013 *Procedia - Social and Behavioral Sciences* **99** 189-195
[19] Cheema A R, Mehmood A and Imran M 2015 *Disaster Prevention and Management* **25**(4) 449-463
[20] Hagelsteen M and Burke J 2016 *International Journal of Disaster Risk Reduction* **16** 43-52
[21] Siriporananon S and Visuthismajarn P 2018 *Kasetsart Journal of Social Sciences* **39** 269-276
[22] Siriwardana C S A, Jayasiri G P and Hettiarachchi S S L 2018 *Procedia Engineering* **212** 1091-1098
[23] "11 District in West Pasaman Suffered Flood." Retrieved from [https://cnnindonesia.com/nasional/20181012092446-337893/banjir-pasaman-barat-rendam-11-kecamatan] on 26 October 2018. [in Indonesian]
[24] " Flood in West Pasaman: Two Bridges Broken, 875 Households Isolated." Retrieved from [https://www.inews.id/daerah/sumbar/2-jembatan-di-pasaman-barat-terputus-akibat-banjir-875-kk-terisolasi/278741] on 26 October 2018. [in Indonesia]
[25] "West Pasaman Flood Destroyed Hundreds of Hectare Plantation." Retrieved from [https://www.viva.co.id/berita/nasional/1084526-ratusan-hektare-perkebunan-di-pasaman-barat-rusak-parah-akibat-banjir] on 24 October 2018. [in Indonesian]
[26] "Two Casualties Due to West Pasaman Flood Disaster." Retrieved from [https://republika.co.id/berita/nasional/daerah/14/12/06/ng53ns-dua-meninggal-akibat-banjir-di-pasaman-barat] on 24 October 2018. [in Indonesian]
[27] "A 6.0 RS Earthquake Shook West Pasaman This Evening." Retrieved from [https://wartakota.tribunnews.com/2015/03/03/gempa-kekuatan-60-sr-guncang-pasaman-tadi-sore] on 27 October 2018. [in Indonesian]
[28] "West Sumatera Has High Potential of Natural Disasters." Retrieved from [https://sumbar.antaranews.com/berita/226100/sumbar-miliki-potensi-bencana-alam-yang besar] on 25 Oktober 2018. [in Indonesian]

Stand performance of revegetation of post coal mining

Kissinger[1,2,*], Hamdani[1,3] and Rina Muhayah Noor Pitri[2]

[1]Center for Environmental Research University of Lambung Mangkurat, Indonesia
[2]Faculty of Forestry University of Lambung Mangkurat, Indonesia
[3]Faculty of Agriculture University of Lambung Mangkurat, Indonesia

*durror2ali@yahoo.com

Abstract. The success of revegetation can be monitored by an assessment of the performance of the revegetation stand. This study aims to analyze and comparing the performance of post-coal revegetation stands conventional cropping patterns and hydroseeding planting patterns. The research location is revegetation area of PT. Adaro Site Tanjung Tabalong. The research took place for 2 months from data collection, analysis of data and reporting article. Data is collected using nested plot, the main plot size used is 10 m x 40 m. The main plot is divided into 4 measuring plots measuring 10m x 10m. The main plot measuring 10m x 40m was purposively placed on 2 locations of conventional cropping revegetation stands and 4 locations for revegetation stands of hydroseeding planting patterns. Data collected are stand dimensions data in the form of individual numbers, diameter and height of trees. The composition of plant species was analyzed by tabulation matrix, the percentage of plant growth was calculated based on number of plant growth/ha, stand diversity was assessed by the Shanon-Wiener index (H '), stand volume was analyzed by volume formula, plant health was analyzed qualitatively from observations of visualization of revegetation stand. The composition of plant species used in revegetation is 6 species. The percentage of planting success for conventional planting patterns is 93.5%, while the hydroseeding planting pattern is can't be determined by the percentage of growth. The number of revegetated plants compared to the two cropping patterns has exceeded the established revegetation success rules (625 individuals / ha). Conventional plants have fewer plants/ ha than the hydroseeding pattern, but have a higher average diameter, height, stand volume and diversity index value. Based on visual observations, there was no difference in plant health between the two revegetation plots. The revegetation stand performance from the results of this study can be categorized as good.

1. Introduction

The Various types of vegetation cover change as a result of mining activities. Coal mining has changed the existing vegetation formation into open areas. Land clearing that has occurred must be following with reclamation and revegetation activities.

Based on [1,2], the area of after mining is required to carry out reclamation activities that aim to restore the land condition as according to its designation. The importance of reclamation activities in mining operation makes the techniques in reclamation activities must be planned so that reclamation activities can reach the target. Reclamation is an activity carried out throughout the mining business stage to organize, restore, and improve the quality of the environment and ecosystem so that it can function again according to its designation [1]. Revegetation is an effort to repair and restore damaged vegetation through planting and maintenance activities on ex-forest land use [3]. Revegetation is a plant development activity in order to cover open land with certain types of plants so that it is expected to reduce land degradation that will occur.

Government Regulation Number 60/2009 concerning guidelines for evaluating the success of forest reclamation explains that the criteria for success in forest reclamation is indicated by the activities of land management, erosion control and sedimentation, and the existence of revegetation or tree planting [4]. Revegetation is part of land reclamation activities. Revegetation is an effort to restore land cover in the ecosystem of various types of plants that are adaptive to land after coal mining. The success of reclamation activities is an important stage in the improvement of the degraded environment.

The purpose of revegetation can include the re-establishment of indigenous plant communities, reducing erosion or surface flow, improving biodiversity and restoring landscape aesthetics. Landscape recovery is directly beneficial to the environment through improved wildlife habitat, biodiversity, soil productivity and water quality. There are several methods to do revegetation, such as planting adaptive seeds conventionally using planting holes and applying hydrosedding techniques to after coal mining land. Enrichment planting on land that has reached a certain age is done to increase crop diversity

The implementation of revegetation based on several research results has experienced many problems. These problems can be indicated by the bad performance of revegetation plants. Various characteristics that can be identified from the inhibition of plant growth can be seen from stagnation of growth, yellowing leaves, low growth percentages and some other plant health symptoms.

Monitoring of the success of revegetation is very important as a reference to identify problems in implementing revegetation. Stand performance is a measure that can be used to assess the success of revegetation. Stand performance was measured by approach: species composition, percentage of success of revegetation plant growth, plant diversity, pontential of stand and plant health. This study aims to analyze and compare the performance of the revegetation stand between conventional and hydroseeding cropping patterns.

2. Methods

2.1. *Object and tools of research*

The monitoring object is in the form of plants in the reclamation and revegetation area of PT. Adaro Site Tanjung Tabalong, South Kalimantan. The main plant species is dominated by Acacia (*Acacia mangium*) and Sengon (*Paraserianthes falcataria*) and Johar (*Cassia siamea*). Enrichment plants are planted under the main vegetation stand. Some species planted are types of Gaharu (*Aquilaria malaccensis*), Jabon (*Anthocepalus cadamba*) and Pulai (*Alstonia scholaris*).

The tools used in the study are: diameter bands to measure diameter, Christen meters to measure tree height, ropes to make measuring plots, GPS to determine the coordinates of the measuring plot, Tally sheet to record measurement results, Computer hardware and software for data analysis, and documentation equipment.

2.2. *Time and location of research*

The implementation of the study lasted for ± 2 months from the stage of preparation, data collection, reference review, data analysis and preparation of reports. The time of data collection was conducted in March 2018. Data collection in the field lasted 4 days covering 1 day for the planning and preparation stages and 3 days of effective observation in the field.

The planting technique in revegetation activities used by the company is conventional planting using a spacing of 2 m x 5 m and planting using hydroseeding techniques. The study was conducted at 6 revegetation plant locations. 2 locations in the form of revegetation plants with conventional techniques and 4 locations with hydroseeding techniques

2.3. *Procedure of data collection*

Vegetation data collection is carried out using a double plot, where the main plot size used is 10 m x 40 m. The main plots are placed in several different revegetation plant locations. The main plot is divided into 4 plots measuring 10 m x 10 m. Data collected in the form of the number of individuals, number of types and circumference / diameter.

Measurements are made on all types of revegetation plants contained in the measurement plot. Each tree in the plot is measured in height Measurement of tree diameter using phi band and measuring diameter at breast height (DBH). There are 7 locations for revegetation plots with different locations of revegetated land that are monitored. Descriptive observation through visualization is done on tree health

2.4. Analysis of data

Some analytical approaches that indicate the performance of the revegetation stand are as follows:

1. Type composition is analyzed by species list tabulation matrix,
2. The percentage of revegetation success is calculated by number of growing plants per ha. Comparative standard for success of revegetation refers to [3] that has exceeded the amount specified as a revegetation success indicator = 625 stems/ ha
3. The potential of stand is calculated by using the tree volume per hectare. The volume formula approach is:

   $$v = 1/4\ \pi^2 dt \quad (1)$$

   Where:
   v: tree volume
   d: tree diameter
   t: tree height
4. Stand diversity is calculated using the Shannon-Wiener Diversity Index [5]. The grouping of plants for the calculation of diversity index is all plants that have a diameter of $\geq$ 1 cm. Shanon-Wiener diversity is calculated by the formula:

   $$H = - \sum Pi \ln Pi \quad (2)$$

   with:
   H = Shannon Diversity Index
   Pi = proportion of the number of species i to the total number of species

   The diversity index value category used is high for H '=> 3, while for H' = 2-3, and low for H '= <2 [6]
5. Plant health is detected from appearance or visualization of the leaves, skin and stems of plants and the presence of disturbing plants

3. Result and Discussion

3.1. Revegetation stand of conventional planting patterns

The conventional planting plot is a revegetation area that was planted on 2013. Conventional cropping patterns use planting holes and spacing of 2m x 5m. The main plot for conventional planting is plot 1 and plot 2. Plot 1 is located at the coordinates of 9758769.78 mN and 333857.88 mE, and plot 2 is located at coordinates 9757674.26 mN 333723.85 mE. The results of the calculation of plant species composition, stand volume, tree density / ha and tree diversity in the conventional cropping pattern plots are listed in Table 1.

Table 1. Ecological characteristics of revegetation plot of conventional planting pattern

No	Plant Species	mean diameter (cm)	mean Tinggi (cm)	Volume (m^3)	N/ha	H'
1	Akasia (*Acacia mangium*)	11,78	5,59	3,68	50	0,17
2	Gaharu (*Aquilaria malaccensis*)	2,83	4,00	0,13	50	0,16
3	Johar (*Cassia siamea*)	10,63	6,82	18,75	280	0,37
4	Jabon (*Anthocepalus cadamba*)	5,20	5,00	0,11	10	0,06
5	Pulai (*Alstonia scholaris*)	3,50	3,32	0,68	220	0,34
6	Sengon (*Paraserianthes falcataria*)	12,49	6,93	41,90	325	0,35
		7,74	**5,28**	**65,25**	**935**	**1,44**

There are 6 species of plants in this plot. 3 species are fast-growing plants and 2 species have medium growth species (gaharu and pulai). Early revegetation plants were *A.mangium, C.siamea* and *P.falcataria*. Enrichment activities were carried out by planting species of Pulai (*Alstonia malaccensis*) and gaharu (*Aquilaria malaccensis*). Plant species with a high tree diameter size among other species are *A.mangium, P.falcataria* and *C.siamea*. While the species that have the highest size are *P.falcataria* and *C.siamea*.

The growth of revegetation plants in conventional pattern plots is relatively good, although the value of the diversity index is still relatively low (H '= 1.44). The number of individual plants in this location ranges from 935 stems / ha. Based on the cropping pattern used, the percentage of plant growth exceeds 93,5%. The number of plants in this plot refer to [3] has exceeded the amount set as a revegetation success indicator = 625 stems / ha (Table 2). Percentage of revegetated plant growth of after coal mine land in PT Kitadin Embalut site in East Kalimantan's Kutai Kartanegara Regency for Sengon buto, trembesi and Johar plants ranged from 88% for 7-year-old plants, 77.7% for 6-year-old plants and 83,6 % for 5-year-old revegetated plants [7]. The results of the analysis of plant health, land conditions and the percentage of plant growth in plot 1 are listed in Table 2.

Table 2. Plant health and the percentage of plant growth in conventional planting plots

Plant Health	Percentage of Growth
The health of the plants from the appearance of the stems, leaves, roots and canopy of the plant is considered healthy. Number of plants / ha: 935 stems	Based on spacing 2m x 5m, the percentage of growth is 93,5% Based on RI Minister of Forestry Regulation No: P. 4 / Menhut-II / 2011 concerning Forest Reclamation Guidelines, the number of plants exceeded the target of 625 stems / ha or equivalent planting distance of 4m x 4m

The results of health analyze of plants are classified as healthy, some plants are disturbed by weeds. Weed disorders are relatively low and do not kill plants. Revegetation stand performance with conventional cropping patterns can be categorized as good.

3.2. Revegetation stand of hydroseeding pattern

The main plot of observation for revegetation of the hydroseeding pattern consists of 4 observation plots (plots 3,4,5, and plots 6). Plot 3 is located at the coordinates: 9757917.10 mN; 334266.67 mE. Plot 4 is located at the coordinate point: 9757867.23 mN; 334134.93 mE. Plot 5 is located at the coordinate point: 9757938.16 mN; 333616.04 mE, and plot 6 is located at coordinates: 9757674.26 mN; 333723.85 mE. The results of the calculation of plant species composition, stand volume, tree density / ha and tree diversity in the conventional cropping pattern plots are listed in Table 3.

Table 3. Ecological characteristics of revegetation plot of hydroseeding pattern

No	Plant Species	mean diameter (cm)	mean high (m)	Volume (m3)	N/ha	H'
1	Akasia (*Acacia mangium*)	7,21	5,13	25,70	1055	0,29
2	Gaharu (*Aquilaria malaccensis*)	2,61	2,99	0,11	73	0,13
3	Jabon (*Anthoce palus cadamba*)	2,42	3,38	0,08	58	0,11
4	Johar (*Cassia siamea*)	5,52	4,68	1,36	13	0,19
5	Pulai (*Alstonia scholaris*)	2,48	2,97	1,18	125	0,19
6	Sengon (*Paraserianthes falcataria*)	8,38	5,28	8,53	315	0,28
		4,77	**4,07**	**36,95**	**1639**	**1,19**

A.mangium, C.siamea and *P.falcataria* are species with the highest dimensions of trees among other species (indicated by diameter and height). *A. mangium* is the type with the highest number of individuals in the location of plot revegetation. The plot of revegetation stand is good, although the value of the diversity index is still relatively low (H '= 1.19).

Percent growth of revegetation plot of hydroseeding pattern cannot be measured at the beginning of this monitoring because there is no definite spacing of the hydroseeding system. The number of plants in plot 2 if referring to [3] far exceeds the amount determined as a revegetation success indicator of 625 stems / ha (Table 4). The results of the analysis of plant health, land conditions and the percentage of plant growth in hydroseeding plots are listed in Table 4.

Table 4. Plant health and percentage of growing plots of plants planting hydroseeding patterns

Plant Health	Percentage of Growth
The health of plants from the appearance of stems, leaves, roots and plant roots is classified as healthy. Number of plants / ha: 1639 stems	The percentage of growth cannot be calculated because of planting using the hydro seeding method Based on RI Minister of Forestry Regulation No: P. 4 / Menhut-II / 2011 concerning Forest Reclamation Guidelines, the number of plants exceeded the target of 625 stems / ha or equivalent planting distance of 4m x 4m

The results of the health analysis of plants are classified as healthy. The advantages of the hydroseeding system are that soil cover is more dense than conventional systems, so that from the aspect of soil protection against erosion it is relatively better. Disorders of plant weeds also occur in hydroseeding planting plots, but are still within tolerable limits. The revegetation stand performance with a hydroseeding planting pattern can be categorized as good.

3.3. The performance comparison the revegetation plots between conventional patterns and hydroseeding patterns

Comparison of performance between conventional pattern and hydroseeding revegetation patterns is shown in Table 5. The dimensions of stand dimensions used were diameter, height, tree volume, stand density (N / ha) and diversity index (H ').

Table 5. The Comparison of Ecological Characteristics between Plot of Conventional Pattern and Hydroseeding Pattern

Planting Pattern of Revegetation Stand	mean diameter (cm)	mean high (m)	Volume (m3)	N/ha	H'
Conventional Planting Pattern	7,74	5,28	65,25	935	1,44
Hydroseeding Pattern	4,77	4,07	36,95	1639	1,19

The crown closure at this location is relatively tight. The advantages of the hydroseeding system are that soil cover is more dense than conventional systems, so that from the aspect of soil protection against erosion it is relatively better. The lack of a hydroseeding cropping pattern is that the amount of stand biomass is relatively smaller per unit area compared to conventional cropping patterns. Based on visual observations, there was no difference in plant health between the plot of revegetation stands.

Revegetation activities can be carried out using conventional techniques using planting holes and using hydroseeding techniques. Conventional planting is generally carried out on relatively flat, easily accessible and light slopes. Planting a hydroseeding system is carried out on relatively extreme land, which is difficult to access manually.

There are differences in biomass (volume) and stand density between conventional revegetation systems using planting holes and hydroseeding systems. The selection of revegetation techniques can be adjusted to the needs, field conditions, character of the habitat and funding.

4. Conclusion

There are six types of plants used in revegetation activities in the observation plot. The percentage of planting success can only be measured on hydroseeding plots with a value of 93.5%. Based on the number of stands / ha, both observational plots of both revegetation with conventional and hydroseeding planting patterns exceed the established revegetation success rules (625 individuals / ha). There is a difference in the size of the comparison between the revegetation stand of conventional planting patterns and the hydroseeding pattern. Conventional plants have fewer plants than the hydroseeding pattern, but have a higher average diameter, height, stand volume and diversity index value. Based on visual observations, there was no difference in plant health between the two revegetation plots. The revegetation stand performance from the results of this study can be categorized as good.

Acknowledgement

Thank you to the Center for Environmental Research, Lambung Mangkurat University, Banjarbaru and PT. Adaro Indonesia Tanjung Tabalong South Kalimantan Site which has facilitated research.

References

[1] Peraturan Menteri Energi dan Sumber Daya Mineral Republik Indonesia Nomor 07 Tahun 2014 tentang Pelaksanaa Reklamasi dan Pascatambang pada Kegiatan Usaha Pertambangan Mineral dan Batubara (Jakarta)

[2] Peraturan Pemerintah Republik Indonesia Nomor 78 Tahun 2010 tentang Reklamasi dan Pascatambang (Jakarta)

[3] Peraturan Menteri Kehutanan Republik Indonesia Nomor. P.4/Menhut-II/2011 tentang Pedoman Reklamasi Hutan (Jakarta)

[4] Peraturan Pemerintah Republik Indonesia Nomor 60 Tahun 2009 tentang Perubahan Peraturan Pemerintah Nomor 45 Tahun 2004 tentang Perlindungan Hutan (Jakarta)

[5] Ludwig J A and Reynold J F. 1989. *Statistical ecology: A primer on methods on computing* (John Willey and Son)

[6] Magurran A E. 1983. *Ecological diversity and Its Measurement*. University College of North Wales. ISBN 07099 35390

[7] Budiana I G E, Jumani and Biantary M P 2017 Evaluation of the Level of Success of Revegetation of Former Coal Mining Land at PT. Kitadin East Kalimantan Regency Kutai Kartanegara Embalut Site *AGRIFOR Journal* **XVI** (2) ISSN P: 1412-6885. ISSN O: 2503-4960. https://media.neliti.com/media/publications/217443-none.pdf

[8] Istomo and Aji D A N 2016 The Species Structure and Composition Stand with Root System of Plants in Karst Area in Bantimurung-Bulusaraung National Park, Resort Pattunuang-Karaenta *Tropical Silviculture Journal* **7**(1) 58-67 ISSN: 2086-8227. https://jurnal.ipb.ac.id/index.php/jsilvik/article/download/13298/9990

[9] Ismaini L, Lailati M, Rustandi and Sunandar D. 2015. Composition and plant diversity analysis on Mount Dempo, South Sumatera. *Proceedings of the National Seminar on the Indonesian Biodiversity Society* **1**(6) ISSN:2407-8050 http://biodiversitas.mipa.uns.ac.id/M/M0106/M010623.pdf

[10] Setiadi and Adinda 2013 Evaluation of Growth in Post-Mining Revegetation Land PT Vale Indonesia Tbk. Sorowako, South Sulawesi *Tropical Silviculture Journal* **4**(1) 19 – 22 ISSN: 2086-8227 https://media.neliti.com/media/publications/126529-ID-none.pdf

[11] Setyowati, RrDN., Amala, NA. and Aini, NNU 2017 Study of Selecting Revegetation Plants for Main Used Land Reclamation Success. *Al-Ard Journal of Environmental Technology.* https://jurnalsaintek.uinsby.ac.id/index.php/alard/article/view/256, https://www.researchgate.net/publication/328429548

The characteristics of fecal disposal system in Penyengat Island, Kepulauan Riau

Novi Asti Lalasati[1,*] and M Pramono Hadi[2]

[1]Department of Environmental Geography, Faculty of Geography, Universitas Gadjah Mada, Yogyakarta, Indonesia
[2]Center for Environmental Studies, Universitas Gadjah Mada, Yogyakarta, Indonesia

*novi.asti.l@mail.ugm.ac.id

Abstract. Indonesia is the largest archipelagic country in the world, there are thousands of small islands that have to be managed. One of the main managements in small island is related with sanitation issue, especially the fecal (black water). This research is aimed to identify the characteristics of fecal disposal system in a small island and analyse the related factors affecting it. The research was conducted in Penyengat Island, Kepulauan Riau. Descriptive and Inferential Statistics were applied by using SPSS. The characteristics of fecal disposal system is determined by latrine and sewage disposal type. Meanwhile, the related factors included are settlement type, education, income, and ethnicity. The result shows that fecal disposal system of Penyengat is dominated with modern latrine (71.4% Gooseneck Latrine) and conventional sewage disposal (53.5% simple pit sewage). Preferency of fecal disposal type is related with settlement type, ethnicity, and highest education level of household members. Besides, the income level is not related with the fecal disposal type at all. The result of this research is expected as a reference of sanitation programs in small island.

1. Introduction

The number of islands in Indonesia is about 16,056 [1] consisting of hundreds of large islands and thousands small islands. Small islands are island with an area smaller than or equal to 2000 km^2 and its ecosystem according to Law Number 27/2007 on the Management of Coastal Areas and Small Islands). Indonesia's status as an archipelagic country certainly indicates that the development of the region is not only centered in the large islands such as Sumatra, Java, Kalimantan, Sulawesi, and Papua but also the small islands. Regional development is related to the growth and population density that encourages the new settlement area not only in terrestrial area and mainland categories but also in the small islands. Population growth and widespread settlement expansion in various regions, especially small islands, of course bring various adverse effects. One of the impacts is environmental pollution due to the disposal of liquid waste, especially fecal. Increasing population activity has an impact to increasing wastewater volume too. This is often not supported by the provision of environmental sanitation infrastructure.

Kepulauan Riau (Kepri) is a region with high economic growth that is 6.33% which exceeds the national economic growth is 5.12% [2]. Economic growth in a region encourages infrastructure development and population growth [3]. Population growth caused the widespread of settlement land. Two types of settlements are commonly found in small islands, which are land settlements and coastal settlements. Land settlements are located in the central part of the island and the settlements built on land, while coastal settlements are located jutting into the sea and some of it built on the sea. Coastal settlements in Kepri are referred to as "Pelantar" as shown in Figure 1. The pelantar settlements are commonly found in Tanjungpinang City, Kepri. Tanjungpinang City as the capital of Kepri has 18 sub-districts, two of which are separated from Bintan Island (mainland). One of them is Penyengat Sub-district located in Penyengat Island.

Penyengat island has an area of 4.6 km2 with the population amount of 2,577 people consisting of 775 households and population density of 1,073 people per km2 [4]. Penyengat is the second highest density sub-district in Tanjungpinang Kota District. However, in reality most of the land in Penyengat Island cannot be used for settlements because it consists of forest and cultural heritage sites. Penyengat Island is a historical part of the Riau Lingga (Malay) Kingdom with various cultural heritage sites that are still preserved. The Malay historical in this island made the state of society not only depend on the wealth, education level, and other aspect that common in some region but also the royal lineage or the ethnicity. It is the value proposition of Penyengat Island as the research location than the other small islands in Kepri. The existence of the Malay heritage site causes people that have royal lineage to settle in Penyengat Island for generations. The impact is increasing needs against settlement land due to limited land area. Increased land area due to economic and social sectors led to emergence of various problems. One of the problems that need to be addressed is the potential for environmental pollution due to domestic wastewater.

Figure 1. 'Pelantar' settlement in Penyengat Island
Source: Researcher Documentation, 2015

Domestic wastewater management, both latrines and non-latrines, has been regulated through the Minister of Public Works and Public Housing Regulation Number 04/2017 on the Implementation of Domestic Wastewater Management System. However, the regulation does not further explain the system of managing domestic wastewater (blackwater, fecal) in a particular ecosystem that causes no differences management in continental area (large islands) and small islands. In fact, the carrying capacity of small island is significantly different with large island. The level of vulnerability for small island groundwater to pollute is high because of the very limited water resources. In addition, the disposal of fecal into the sea also causes threats to marine life and indirectly to the sustainability of community's economy [5].

Fecal disposal is intended to isolate the fecal so that infectious seedlings in it cannot reach the new host. The methodology chosen for different areas will depend on several factors including geological and hydrogeological conditions, culture and customs of community, availability of local materials and both short-term and long-term costs [6]. Therefore, the study of fecal disposal system characteristics in

Penyengat Island, Kepri becomes important as an illustration of wastewater management in referring to other small islands. The study of fecal disposal system characteristics can provide information about latrine and sewage disposal types and its distribution including factors related to the selection of a disposal system. This can give an idea of how the real condition of domestic wastewater management, especially fecal disposal in a small island ecosystem, that is in Penyengat Island, Kepri.

2. Methods of the study

2.1. Research design

This research is a quantitative descriptive research with two segments of research. The research segment is differentiated by population and the method used. The first segment is intended to identify the characteristic and distribution of fecal disposal system in the research area. The second segment is aimed to analyze the related factors of fecal disposal system selection in the research area. The population which are selected in the second segment is based on the results of the first research segment.

2.1.1. Segment 1. The first research objective will be answered in segment 1. Identification of the fecal disposal system characteristics and the distribution is done by mapping in the research area using the census of 301 units of houses as the research population. In the segment 1, the information that have to collect are latrine type and sewage disposal type. There are three types of latrine and sewage disposal might be found in this research area such as shown in Table 1.

Table 1. Latrine and Sewage Disposal Type

No.	Latrine Type	Sewage Disposal Type
1.	Pit latrine (*jamban cubluk*)	Simple pit sewage
2.	Gooseneck trap/latrine (*jamban leher angsa*)	Individual septic tank
3.	Simple pour-flush latrine (*jamban pelengsengan*)	Communal septic tank

The latrine type and sewage disposal type of the research population would make new categories population into a smaller group known as strata to simplify the research of segment 2.

2.1.2. Segment 2. This segment is aimed to analyze the factors related to fecal disposal system characteristics in the research area or the second research objective. The related factors information is collected based on the sampling of sewage disposal system and latrine type findings that is mentioned in segment 1 explanation. The sample is a part of the population which are collected randomly by strata based on the result of segment 1 which means the segment 2 used stratified random sampling as a data collection method.

Stratified random sampling is a technique used if the population has a member or elements which are heterogenic. It is method of sampling that involves the division of a population into smaller groups or subgroups known as strata. In stratified random sampling or stratification, the strata are formed based on members' shared attributes or characteristics either qualitative or quantitative data. Stratified random sampling is also called proportional random sampling or quota random sampling.

Stratified random sampling is a probabilistic sampling option. The first step in stratified random sampling is to split the population into strata, i.e. sections or segments. The strata are chosen to divide a population into important categories in relevance to the research interest [7], for this research there are nine strata that might be found based on the latrine and sewage disposal type, i.e. 1) pit latrine-simple pit sewage, 2) pit latrine-individual septic tank, 3) pit latrine-communal septic tank, 4) gooseneck latrine-simple pit sewage, 5) gooseneck latrine-individual septic tank, 6) gooseneck latrine-communal septic tank, 7) simple pour-flush latrine-simple pit sewage, 8) simple pour-flush latrine-individual septic tank, and 9) simple pour-flush latrine-communal septic tank. The number of samples is accounted for according to Slovin method [8] the formula is shown as follows.

$$n = \frac{N}{1+N\left(d^2\right)} \quad (1)$$

Definition :
n : Number of sample
N : Number of population
d : Error level

The number of samples which are used in segment 2 of the research is based on the results of segment 1. The error level of this research is 5% or 0.05. The less error level, the more accurately the sample describes the population and lead to the higher number of samples. Actually, the n value defines the new population (segment 2) for each strata and sample size. The next step is selecting the sample by using random sampling method. Each element of a random sample is chosen entirely by chance and has an equal probability of being selected. Random sampling can be done by using dice in a conventional way but this research used *Microsoft Excel (part of Microsoft Office)* as a software that supports picking random data by using formula 'randbetween'. All of the data is named by numbers that represent the latrine type, sewage disposal type, house number and coordinate. The numbers of houses will be used in random sampling process with randomize feature in *Microsoft Excel.*

This research has four independent variables to fulfill the second objective which is to analyze the related factor of fecal disposal system in Penyengat Island. First, settlement type is categorized according to the foundation of houses, which is land or sea. Second, education level that affect attitude and knowledge of the resident in many aspects of sanitation. Education level of resident could not be only represented by householder's education level, so the highest education level of the resident's member is used because that it might influence the decision making in the household. Third, income level as a variable to determine economic condition in the household that is represented by householder's income and the total income of the resident's member to minimalize bias of household with retired or unemployed householder. Last, the ethnicity as an also important variable to use in this research. Penyengat island is part of Malay kingdom historical, most of community is from Malay ethnic and some of them have royal linkage. Profile book of Penyengat sub-district 2017 mentioned that 86% community ethnic is Malay, it means the ethnicity may be affected the ways of community to agree and connect them on fecal disposal system selection.

2.2. Data Collection
The data collection is conducted with field survey and interview using questionnaire as the tool. The data that needs to be collected in this research is included as follow.

Table 2. Data Collection

No.	Data	Source	Method
1.	Settlement map of Penyengat Island	Secondary	Digital processing of Google Earth Imagery using Geography Information System (GIS)
2.	Absolute location of houses, black water disposal system type, and laterine type.	Primary	Information collection using GPS plotting (survey) and Census method checklist
3.	Socio-economic condition of residents	Primary	Information collection using questionnaire with Random Sampling Method

2.3. Data Analysis
The data analysis which is used to identify the characteristic of fecal disposal system in Penyengat Island is descriptive and inferential statistical analysis. There are several differences and aims between those type of statistical analysis as mentioned below that are appropriate to this research's segment and objectives.

2.3.1. Segment 1 (Descriptive statistic analysis). Descriptive statistic is a statistic used to analyze data with describing and picturing the collected data as it is without the intention to conclude for generalization. A few things that can be done is data presentation using table, graphs, pie charts, pictogram, modus calculation, median, mean (central tendency measurement), decile, percentile, data distribution calculation using average and standard deviation calculation, and percentage calculation. Descriptive statistic tests that is used in this research is frequency table and cross tabulation. The input data of this analysis are number of latrine and sewage disposal type in Penyengat Island. Frequency table can show the domination of latrine and sewage disposal type, meanwhile the cross tabulation of it with the settlement type would explain about the tendency correlation of those variables. The segment 1 data analysis is a manner to identify the fecal disposal characteristics in Penyengat Island as first research objectives.

2.3.2. Segment 2 (Inferential statistis). Inferential statistic is a statistic used to analyze the data sample and the results is implemented for the population. This statistic is suitable to use if the sample is taken randomly from a clear population. This method is also often known as inductive statistic or probabilistic statistic because the conclusions which is implemented to the population based on the data sample and the truth is a probability that is marked by a significance level and level of confidence. Commonly, inferential statistics are started by doing normality test to identify the distribution of data. The assumption of that normality test would be the indicator for the next steps in inferential statistics. It means that if the data are not distributed normally, the statistical analysis could not be conducted. SPSS provides various type of normality test such as Shapiro-wilk, Kolmogorov-smirnov, and Liliefors, and Skewness-Kurtosis [9].

Inferential statistic test in this research is Chi Square Test. Chi square is used to test the relationship between two nominal variable and measure the significance relationship between those variables with coefficient of contingency [10]. The significance level determination or X^2 on chi square is formulated as follow.

$$x^2 = \sum \frac{(O - E)^2}{E} \tag{2}$$

Definition:
O : Observation result frequency
E : Expected Frequency
Number of E : Number of row * number of column/number of data (n)

H_0 or hypothesis is always declared as no relationship between the variable, if the value of Asym Sig or Sig. X^2 (Hα) is more than the significance level stated before, then the hypothesis is accepted and vice versa. Chi Square obligated only nominal data processed so that some variables in the other data scale has to be changed into a nominal data. Although, the significance level determination requires a prior calculation of the available data, but based on the population number of 524 with four independent variable indicating that there are more than 2x2 cells, it could be possible to conduct a Chi Square using 95% as a significance level or the value of $\alpha = 0{,}05$ and the relationship is said to be significant if the value of $p \leq 0.05$ and there is no significant relationship if the value of $p > 0{,}05$ as other studies in general [11].

3. Results

3.1. The characteristics of fecal disposal system in penyengat island

Data collection of the latrine and sewage disposal type in Penyengat Island involves 377 houses. Based on the census, there are 440 houses but 73 of it are houses without toilets and some are used as storage warehouses. The data collected into the population in characteristics of fecal disposal system in Penyengat Island research is only 301 houses because 76 others can be said to be unidentified data (see the following reasons in Table 3). 68.4% of the unidentified data are abandoned houses, generally the reason for the buildings not being used is that the homeowners or families of the residents have migrated out of the island. The reason for the migration is to get a job or access to health and other livelihoods.

Table 3. Excuse of Unidentified Data Frequency

Excuse of Unidentified Data	Frequency	Percent	Valid Percent	Cumulative Percent
Uninhabitable/Damaged House	17	22.4	22.4	22.4
Alternative House (Investment)	5	6.6	6.6	28.9
Abandoned House	52	68.4	68.4	97.4
Unwilling to respond	2	2.6	2.6	100.0
Total	76	100.0	100.0	

As a consequent of unidentified data, the population in this study is only 301 houses even though there are 377 houses. The data in segment 1 of this study refers to the questionnaire containing the information about the type of settlement, latrine, sewage disposal, and fecal disposal system in Penyengat Island. Based on the result of the census, settlement type in Penyengat is dominated by coastal settlement as shown in Table 4, even though the difference between the number of settlement type is not significant. This condition is linier with the previous research by [12] that stated the houses type in small island are dominated by stilt house, which is it part of coastal settlement.

Most of the community in small island prefer to live in coastal settlement cause of several excuses such as the feasibility to have their own port in their 'backyard', to easier predict the weather for their mobility, and is part of the local wisdom that they believe land is the sacral area. This statement is also supported by settlement geography study, a line settlement group are generally built linier with road or coastline. The type of settlement would affect the community's preference of their household sanitation facilities, that are the latrine and sewage disposal type.

Table 4. Settlement Type Frequency

Settlement Type	Freq	%	Valid %	Cum. %
Land Settlement	145	48.2	48.2	48.2
Coastal Settlement	156	51.8	51.8	100.0
Total	301	100.0	100.0	

Gooseneck latrine is considerable latrine type in Penyengat, it more than 70% of the population as shown in Table 5. This latrine type is semi-modern but with affordable price than the pour-flush. Unfortunately, most of sewage disposal that was found in Penyengat is simple pit sewage. It means the community already cared about their toilets interior facility but they did not realize that the sewage has potential to contaminate the soil and sea water. Even though 44,5% of the population used individual septic tank it still led the potential soil contamination. There are only six houses connected to two communal septic tanks. The sewage disposal information can be identified in Table 6.

Table 5. Latrine Type Frequency

Latrine Type	Freq.	%	Valid %	Cum. %
Pit Latrine	82	27.2	27.2	27.2
Gooseneck Latrine	215	71.4	71.4	98.7
Simple Pour-flush Latrine	4	1.3	1.3	100.0
Total	301	100.0	100.0	

Table 6. Sewage Disposal Type Frequency

Sewage Disposal Type	Freq.	%	Valid %	Cum. %
Simple Pit Sewage	161	53.5	53.5	53.5
Individual Septic Tank	134	44.5	44.5	98.0
Communal Septic Tank	6	2.0	2.0	100.0
Total	301	100.0	100.0	

Based on the census data included, the latrine and sewage disposal type characteristics' of fecal disposal in Penyengat can be stratified into six categories or strata that are (1) pit latrine - simple pit sewage, (2) pit latrine - individual septic tank, (3) gooseneck latrine - simple pit sewage, (4) gooseneck latrine - individual septic tank, (5) gooseneck latrine - communal septic tank, and (6) simple pour-flush latrine - simple pit sewage. The strata can be name as fecal disposal type (FDT) 1, 2, 3, 4, 5, 6 and will be mentioned throughout this research. The result of strata number shown that the hypothesis to find nine strata of fecal disposal is rejected. There are no houses used pit latrine are not connected into communal septic tank. Most of FDT 1 houses are part of coastal settlement. The one and only latrine type that used various sewage disposal type is gooseneck with the domination as FDT 4. This fecal disposal type is generally found in land area. See Figure 2 that represent the portrait of latrine type domination. The detail information about each FDT number in Penyengat is shown in Table 7.

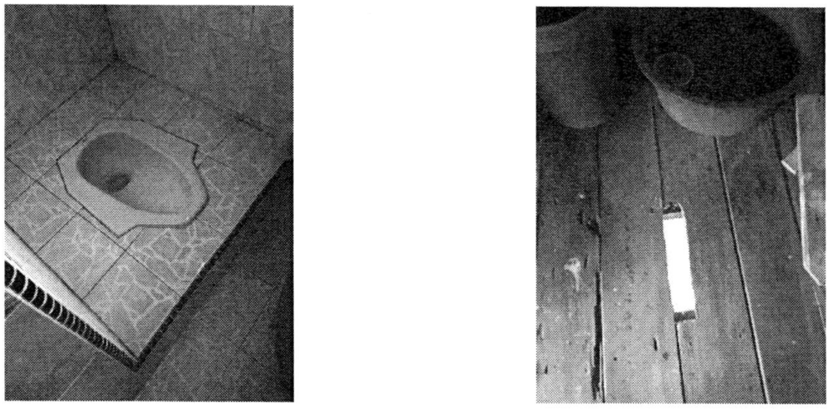

Figure 2. Portrait of FDT 4 and FDT 1 as the dominated FDT in Penyengat
Source: Researcher Documentation, 2018

Table 7. Frequency of Fecal Disposal Type

FDT	Frequency
1	80
2	2
3	78
4	131
5	6
6	4
Total	301

The domination of FDT that is found in Penyengat is FDT 4 (43%) which means the community used the gooseneck latrine and individual septic tank. It led to the assumption that community were already prefer to use modern latrine and aware to good sanitation as they used septic tank. But we could not ignore another fact that most of them used simple pit sewage (FDT 1, FDT 3, and FDT 6) even with three types of latrine. If we calculated the total of simple pit sewage percentage in FDT strata, we can conclude that, most of community dispose their fecal to the sea because they might not have any option to dispose the fecal to the land. Land is the sacral area in Penyengat cause of spiritual-historical value. The characteristics of fecal disposal system in Penyengat Island is shown by the latrine and sewage disposal type. The highest number of latrine type that is used is gooseneck latrine (71.4%) and the highest number of sewage disposal type is simple pit sewage (53.5%). However, by the strata of fecal disposal system, the highest type is FDT 4 (43%). This result is not totally wrong because the second highest number of latrine type (44.5%) is individual septic tank. Nevertheless, we could not ignore the fact that Penyengat island has high potential to pollute the sea environment by disposing the fecal directly to the sea through simple pit sewage. Then, we also need to be concern about the second highest fecal disposal type which is FDT 1.

There are 27% of houses in this small island that still use pit latrine and simple pit sewage. It represented the main problem of our national obstacle in the specific area like small island as discussed in *Buku Penuntun Opsi Sanitasi yang Terjangkau untuk Daerah Spesifik* or Guidebook of Affordable Sanitation Option for Specific Area [13]. This book explained that the community in small islands were using conventional fecal disposal system because of their financial and the geography factors. The income level and transport cost of the sanitation facilities market in small island will influence the willingness to pay of the community. The related factors of fecal disposal system characteristics in each small island is different. In order to identify the related factors of fecal disposal system in Penyengat Island, this research is continued by analyzing several factors as explained in segment 2.

The distribution of fecal disposal type in Penyengat Island is literally depending on the latrine type, sewage disposal type, and settlement type. The houses that are located in the coastline or part of settlement type have simple pit sewage and pit latrine. Besides, the houses that located in land used sewage disposal, that is septic tank. Most of the septic tank type is individual (44.5%) and only 2% of it is communal septic tank. The houses in land area also have various latrine type than houses in coastline. Meanwhile, disposing the fecal directly to the sea should not be recommend. As you could see in the figures below, the distribution of fecal disposal in Penyengat is dominated in north of the island.

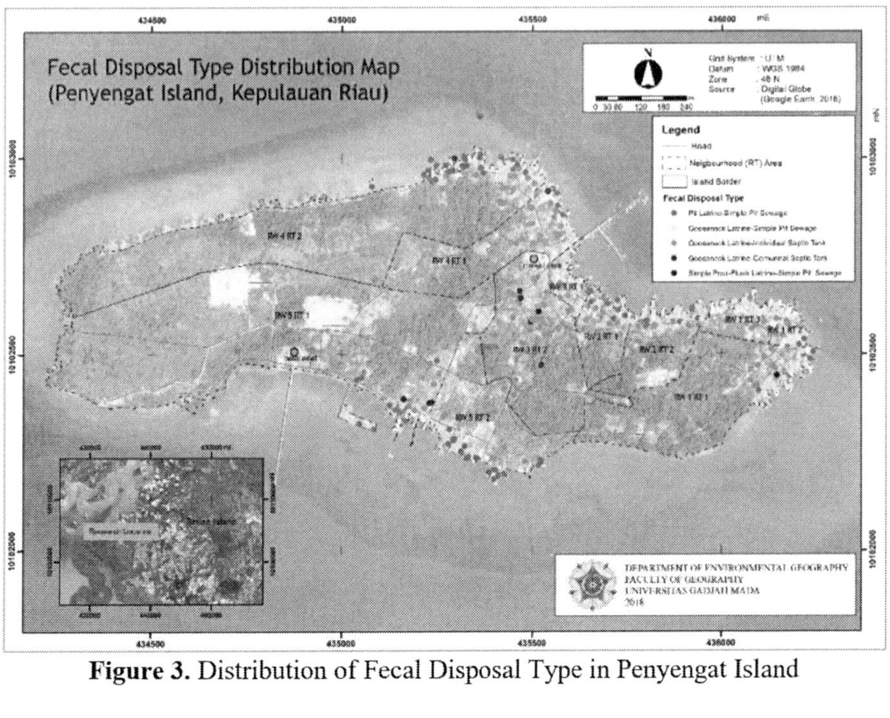

Figure 3. Distribution of Fecal Disposal Type in Penyengat Island

3.2. The related factors with fecal disposal system in Penyengat Island

The inferential statistics used is Chi Square Test with 95% significance level or Hα value is 0.05. This test is used for analyze that both of variables (dependent and independent) are related or not. The dependent variable is fecal disposal type and the independent variables are settlement type, education level (householder and the highest), monthly income (householder and total), and ethnicity. The result of chi square test of those variables is shown in Table 8.

Table 8. Chi Square Test Result

No	Variables	Hα
1	Settlement Type	0.000
2	Education Level of Householder	0.087
3	Highest Education Level of Household Members	0.016
4	Monthly Income Level of Householder	0.916
5	Monthly Income Level of Household Members (Total)	0.990
6	Ethnicity	0.003

3.2.1. Relation of settlement type with fecal disposal system in Penyengat Island.. Settlement type in Penyengat Island is divided into two type that are land settlement and coastal settlement. Coastal settlement is more dominant than land settlement, 54% houses are coastal settlement. Besides, 46% houses are land settlement (displayed in Figure 4). With an area only 4.6 km^2, it is normal if the community prefer to build the house along the coastal. Settlement type are the first variable that we assumed have relation with fecal disposal characteristics in small island like [12] mentioned in their previous research.

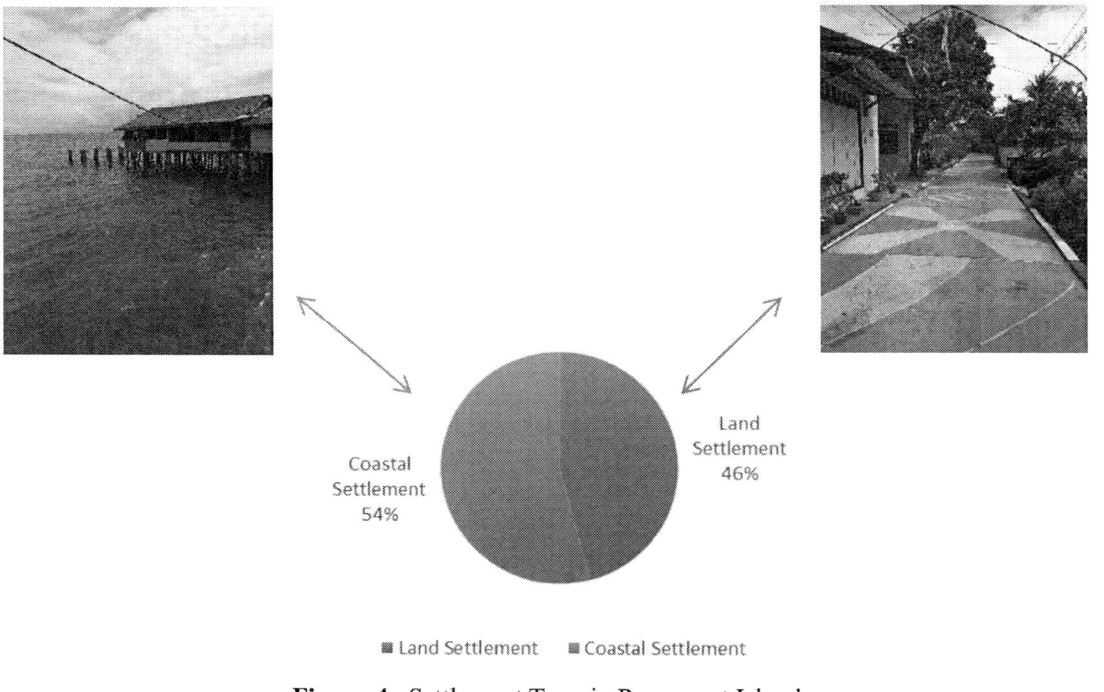

Figure 4. Settlement Type in Penyengat Island

Based on chi square test in this research, **Hα value of FDT with settlement type is 0.000** (see Table 8). It means that the hypothesis is rejected, furthermore we know that settlement type has relation with FDT. This statement was proved by the linier data of settlement type and sewage disposal type. 54% of houses are part of settlement type and 53.5% of houses are used simple pit sewage.

3.2.2. Relation of education level with fecal disposal system in Penyengat Island. Education is something that really influence us in making decision. It is considered an investment in human capital which in turn contributes to knowledge of economy and development of a nation. Education is a diverse and dynamic sector and has been ever changing as per the needs of society. Education is a form of learning in which the knowledge, skills, values, beliefs, and habits of a group of people are transferred from one generation to the next through teaching, training, research, or simply through auto didacticism [14]. Socio-economic status (SES) emerges as one of the most significant factors influencing the choices of the individuals regarding their education. Socio-economic status is defined as a relative standing in society based on an individual's income, power, occupation, education, and prestige [15].

There is an indicator to identity the education variables, commonly used the education level. The education level represented the community status by which level of school they had graduated. In this research, education level is not reviewed only by householder, besides we take the information of the highest education level of their household members. It aimed to control the fact that might be the decision making in one household are a result of discussions. So, the members have the same possibility and rights to influence their decision. The decision in this context is about their preferences of latrine and sewage disposal type. Based on chi square test for education level variables, we can see that householder education level's is not related with FDT is shown by its $H\alpha$ value is 0.087. On the other side $H\alpha$ value of FDT with highest education level of household members is 0,016. This is an interesting point that the highest education level in one household has affect to their decision making of FDT more than the only the householder education background.

3.2.3. Relation of income level witth fecal disposal system in Penyengat Island. As mentioned before by [15] socio-economic is defined based on an individual's income, power, occupation, education, and prestige. Many important economic decisions such as labor supply, residential location, buying insurance, buying new goods, investing in stocks and bonds or in children's education also the willingness to pay sanitation facilities in house often made by households rather than by individuals [16]. Most financial education and counseling took place at the individual level, whereas financial decisions took place at household and intrahousehold levels. However, all family members in the household including spouses (partners) and children influence financial decisions [17].

This research collected two indicators of income level. First, the monthly income level of householder and second is the monthly income level of all members in the household which means the total income of working members. Chi square test of this variable shown that income level either the householder or total members are not related with FDT. The $H\alpha$ value of monthly income level of householder and all members (total income) is bigger than 0.05, respectively 0.916 and 0.990. This fact shown that income level or economic status do not influence their sanitation facilities. It is linier with the data of latrine and sewage disposal type number.

As we discussed before, 71% of the household can buy the modern latrine meanwhile more than 50% were still using simple pit sewage. Latrine is part of sanitation facilities that is built inside the house or especially in the toilets. People will make efforts to fulfill their comfort, without exception for their toilets. So, the willingness to pay for the modern latrine is high but it is contradictive with the sewage disposal. Sewage disposal are part of sanitation facilities that is built outside of the house. If people are not concerned and aware about their environmental issues, the good sewage disposal are not particularly important to pay for them. Matt Damon as actor and founder of Water.org quotations is related with this phenomenon, he said in National Geographic Magazine Edition November, 2017.

"2.4 billion people in the world have bad sanitation facilities, but most of them have smartphones rather than the toilets"

3.2.4. Relation of ethnicity with fecal disposal system in Penyengat Island. Ethnicity is part of socio-culture society, its social identity of individuals. In Indonesia, the ethnicity has important roles to give special characteristics of each individuals and regions. Ethnicity is also part of prestige and in other ways as power. While, the power and prestige of the ethnicity is locally. Penyengat Island has strong connection with Malay Kingdom (Johor-Riau Lingga), which means the Malay ethnic has more power and prestige than the others there. The Malay ethnic is also divided into two types i.e. royalist and non-royalist. The royalist linkage has special name given as their first name e.g. Raja, Datok, Sultan, Daeng, and Engku. The royal family linkage would act or might be giving punishment to whoever break the rules of their local wisdom. This statement is supported by [18], they mentioned about culture in Penyengat Island and almost of others Malay (Melaka – Riau Lingga – Johor) Kingdom Island that is distributed in Malaysia, Singapore, and Indonesia (Kepri), the community in those islands have to protect their island from any "foul and messy" activities in direct or indirect meanings.

The stakeholders (hamlet and neighbourhood) of the Penyengat Island commonly is someone who has royal linkage. The community have local wisdom that really appreciate the royalist. People in Penyengat who are not Malay will be considered as an immigrant the whole time. The ethnicity as culture aspect also influence the decision making. Assumption of this variable is the royalist Malay might be having sense of belonging for their island more than the non-royalist Malay and non-Malay. Cultural aspect as a variable that influence sanitation systems in community is already explained by [19] they stated the local people's in Pakistan have some primary criterion to select sanitation model i.e. prestige, comfort, privacy and had little motivation for environmental. Every household in their research location wanted water in the toilet or latrine for anal cleansing because of their culture as Muslims. It is proved that sanitation systems including latrine, sewage disposal, and fecal disposal might be affect by the culture or in this context is the ethnicity.

The result of chi square test of fecal disposal type and ethnicity shown that both variables are related. The Hα value is 0.003, it means the ethnicity in Penyengat Island is influencing the community to choose their fecal disposal system. This result indicates that community have prestige and social status by their ethnic, not their occupation or economical aspect. Even though, most of previous researches stated that economic aspect divided coastal and small island communities' layers as mentioned by [12] that most of coastal communities in Indonesia have vertical classes by their socio-economic condition. But this research proved that for some special region with pilgrim typical, socio-culture has more influence rather than the wealth and assets.

4. Conclusion

The characteristic of fecal disposal system in Penyengat Island in term of latrine and sewage disposal type are dominated by 43% gooseneck latrine-individual septic tank (FDT 4), 27% pit latrine-simple pit sewage (FDT 1), and 26% gooseneck latrine- simple pit sewage (FDT 3). Meanwhile, 71% used the gooseneck latrine and 53% used the simple pit sewage. The result preferred to have modern latrine but old or traditional sewage disposal.

The characteristics of fecal disposal system in Penyengat Island is related to settlement type, the highest-level education of household's members, and ethnicity. Education level of householder is not related to FDT. Besides, the income level as a representative of economical aspect, either the householder income nor the total income is not related with the FDT preference.

Acknowledgment

The authors would like thank to the Center for Enviromental Studies (PSLH) Universitas Gadjah Mada, Faculty of Geography Universitas Gadjah Mada, Government Official at Kepulauan Riau, Tanjungpinang, and Penyengat, also all parties who have helped the activities from the initial preparation to the implementation of this research.

References

[1] Kompas 2017 *Indonesia Daftarkan 16.056 Pulau ke PBB* Bogor: Edisi 21 Agustus 2017
[2] BI [Bank Indonesia] 2015 *Laporan Keuangan Triwulan II 2014 Kepulauan Riau* (Jakarta: BI)
[3] Rachman H 2010 Kajian Pola Spasial Pertumbuhan Kawasan Perumahan dan Permukiman di Kecamatan Limboto Kabupaten Gorontalo *Tesis* (Semarang: Program Pascasarjana

Pembangunan Wilayah dan Kota, Universitas Diponegoro)
[4] Monograph of Penyengat 2017 *Profil Kelurahan Penyengat 2017* (Tanjungpinang: Penyengat Sub-district Government Official)
[5] Marfai M A, Sarastika T, Trihatmoko E, Rahantan R, Sarihati R dan Suriadi 2018 *Kajian Daya Dukung dan Ekosistem Pulau Kecil Studi Kasus Pulau Pari* (Yogyakarta: Gadjah Mada University Press)
[6] WHO [World Health Organization] 1992 *A Guide to the Development of On-Site Sanitation* (Jenewa: WHO)
[7] Agresti A and Finlay B 2008 *Statistical Options for the Social Sciences, 4th edition* (New Jersey: Prentice Hall)
[8] Notoatmodjo 2007 *Promosi Kesehatan dan Ilmu Perilaku* (Jakarta: Rineka Cipta)
[9] Myoung P H 2008 Univariate Analysis and Normality Test using SAS, Stata, and SPSS. *Working Paper* (Indiana: University Information Technology Services (UITS) Center for Statistical and Mathematical Computing)
[10] Usman H dan Purnomo S A 2000 *Pengantar Statistika* (Jakarta: Bumi Aksara)
[11] Hastono S P 2001 *Modul Analisis Data* (Jakarta: Fakultas Kesehatan Masyarakat Universitas Indonesia)
[12] Hidayat A R T, Wijayanti W P, Kumiati S and Maha E A 2016 Challenge and Opportunity of Settlement and Infrastructure Development in the Small Island (Case Study: Pulau Laut Kelautan District, Kotabaru Regency) *Procedia-Social Behavioral Science* **227** 286-293
[13] WSP [Water and Sanitation Program] 2011 *Opsi Sanitasi yang Terjangkau untuk Daerah Spesifik* (Jakarta: World Bank Office Jakarta)
[14] Jabeen S S and Rafiuddin A 2015 Factors Influencing the Education Decision Making Process. *Second 21st Century Academic Forum Harvard* **5**(1) 208-220
[15] Gati I, Landman S, Davidovitch S, Asulin-Peretz L and Gadassi R 2010 From Career Decision-making Styles to Career Decision-making Profiles: A multidimensional Approach *Journal of Vocational Behavior* **76** 277-291
[16] Carlsson F, Martinsson P, Qin P and Sutter M 2009 Household Decision Making and the Influence of Spouses' Income, Education, and Communist Party Membership: A Field Experiment in Rural China. *IZA Discussion Paper Series 4139.*
[17] Kim J, Gutter M S and Spangler T 2017 Review of Family Financial Decision Making: Suggestions for Future Research and Implications for Financial Education *Journal of Financial Counseling and Planning* **28**(2) 253–267
[18] Hutchinson F E and Chongm T 2016 The SIJORI Cross-Border Region: Transnational Politics, Economics, and Culture (Singapore: ISEAS Publishing)
[19] Nawab B, Nyborg I, Esser K B and Jenssen P D 2006 Cultural Preferences in Designing Ecological Sanitation Systems in North West Frontier Province, Pakistan *Journal of Environmental Psychology* **26**(3) 236-246

IOP Conf. Series: Earth and Environmental Science **314** (2019) 012015 doi:10.1088/1755-1315/314/1/012015

The relationship between work period and use of personal protective equipment with respiratory disorder complaints in brick craftsman in Sintuk Toboh Gadang District Padang Pariaman Regency 2017

Pirna Lastri*, Indang Dewata and Mila Sari

Department of Environmental Science, Post Graduate Program, Universitas Negeri Padang, Prof. Dr. Hamka Street, West Air Tawar, Padang-West Sumatera 25131 Indonesia

*pirnalastri7070@gmail.com

Abstract. Respiratory disorders are not only caused by high concentration of dust, but also affected by employee's habit and characteristics. Based on the data from Sintuk Health Center, there are 145 cases of ISPA in this region. Over all, the brick craftsman smoke and don't use the personal protective equipment. This study aims to determine the relationship between work periods and use of personal protective equipment with respiratory disorders on brick craftsman in Sintuk Toboh Gadang, Padang Pariaman Regency 2017. The research is *descriptive analytic* research with *cross sectional* approach which was conducted on 11-21 August 2017. This study used *Cluster Sampling* with 73 people of total sampling which then using *Accidental Sampling* and questionnaire to collect the data. The data analysed by using *Chi-Square* test. The result shows that among 73 respondents, there were 47 people (64,4%) infected respiratory disorders, 41 (56,2%) of times work, 39 (53,4%) work periods, and 55 people (75,3%) do not use personal protective equipment while working. Statistical test revealed that the variables which have relationship with the habit of respiratory disorders were times work (p=0,003, OR=5,304), work periods (p=0,000, OR=7,857), and use of personal protective equipment variables (p=0,004, OR=5,857). From the research, it was found that times work, work periods, and use of personal protective equipment have a strong impact to respiratory disorders. It is suggested that brick craftsman need to use the personal protective equipment in producing craftsman. Therefore, long work periods do not give impact to the respiratory disorders.

1. Introduction

Air is an environmental component which has a very important role for human life. Air pollution is the presence of one or more physical, chemical or biological substances in the atmosphere that can endanger the health of human, animal and plant, disrupt aesthetics and comfort, and damage the property [1].

The health of workers is influenced by several factors, one of which is the workplace environment. According to Law No. 1 of 1970, the work environment is a place where there are many dangerous sources, one of which is air contamination which is generally known as air pollution [2].

Based on WHO data (*World Health Organization*) in 2010, among all the occupational diseases, 30% to 50% are silicosis and other pneumoconiosis. *The International Labor Organization* (ILO) revealed that the causes of deaths workers are caused by 34% of cancer, 25% of accidents, 21% of respiratory diseases, 15% of cardiovascular diseases, and 5% caused by other factors. These diseases are caused by the exposure of dust in the workplace occurring throughout the world every year [3].

Acute respiratory infection (ARI) is a major health problem as evidenced by the prevalence of ARI in Indonesia reaches 25.5% (range: 17.5% -41.4%) in 16 provinces including prevalence above the national rate and pneumonia 2.1% (range: 0.8% - 5.6%) [3].

Acute respiratory infection (ARI) is an acute infection of the upper and lower respiratory tract caused by microorganisms or bacteria, viruses, or rickets with or without pulmonary parenchyma inflammation. Several risk factors that affect the prevalence of ARI are environmental factors, individual characteristics, and worker behavior. Environmental factors include air pollution (cigarette smoke, air pollution due to industrial products, and high concentration smoke from burning fuel for cooking). Individual factors such as age, gender and education level can also influence the risk of susceptibility to ARI. Behavior of workers includes smoking and the use of masks [4].

Brick making is the fastest-growing industrial sector in many countries (such as China, India, Bangladesh and Pakistan) and between three sectors, together with vehicle smoke and road dust resuspension, contribute to air pollution and health problems in Dhaka (Bangladesh) [5].

The brick making factory is a business that produces bricks. Brick workers work every day with an average working time of ± 8 hours per day. In a day one worker is able to produce ± 2000 bricks. So the risk of respiratory tract disorders in brick workers is very high [6].

The research conducted by Asfawi (2015) showed that there was a relationship between years of work with subjective complaints of breathing in female poultry traders at Penggaron RPU, Semarang City. It was proved by using the *Pearson Product Moment test* which showed a significance value of 0.01 [7].

Based on the study conducted by Josephus (2015), it was known that there was a relationship between the use of personal protective equipment *respirator* with lung vital capacity based on statistical test value of *P values* 0.00 <0.05 [8].

Sintuk Toboh Gadang is one of the Subdistricts in Padang Pariaman Regency which is famous for its abundant brick production. The majority of the population in this region works as brick workers. The brick business has been around for decades. The workers are mostly men but infrequently women are also involved in the manufacturing process, for example in printing bricks. The process of making brick takes a long time up to 2 weeks to be ready for the market. The process starts from softening clay, printing, drying, burning, and cooling bricks. The combustion process uses firewood and lasts for 3 days 2 nights. However, in processing and burning bricks themselves, air quality measurements have never been measured.

Based on the results of the annual disease report from all puskesmas in 2016, there were 24,577 cases of ISPA or 31.8% cases. ARI is in the first rank of the top ten diseases most often suffered by the community and the highest number of the cases is in the working area of Sintuk Community Health Center [9].

Based on the data of the top ten diseases in 2016 from the Sintuk Health Center, ARI is the highest disease suffered by the resident with 145 cases. Then the data obtained from the local Midwives in January to April 2017, there were 30 workers from Jorong Palembayan and 42 workers from Jorong Batang Tapakis who were diagnosed with ARI. [10,11].

Based on the preliminary observation, there were 3 from 5 workers had respiratory complaints during the burning process such as shortness of breath, cough and flu due to smoke from the burning bricks. The workers commonly have bad smoking habit and do not use personal protective equipment (PPE).

In the brick industry, the amount of wood burned is quite large which is around 6-8 tons for each burning process. The combustion process uses high temperatures then lead the thick smoke appears and it certainly cause bad effects on health. In addition, the distance between one brick factory and the other is close together, the burning of bricks area is very close to residential areas and public roads.

Since the working period and personal protective equipment are closely related to respiratory problems in brick workers, the researchers need to examine the relationship between them.

2. Research Methods

This study was *descriptive analytic research*. It used a*cross-sectional* design where all independent and dependent variables were measured at the same time when the study took place. The research was conducted on 11-21 August 2017 in Jorong Palembayan and Jorong Batang Tapakis Sintuk Toboh Gadang District Padang Pariaman Regency. The sampling used was *Cluster Sampling* with 73 people of total sampling which then using *Accidental Sampling* and questionnaire to collect the data. The data was analysed by using *Chi-Square test* [12].

3. Results and Discussion

Based on the study conducted on the Relationship between Duration of Work Period and Use of Personal Protective Equipment with Respiratory Disorders Complaints in Brick craftsman in Sintuk Toboh Gadang, Padang Pariaman Regency in 2017, the results were as follows:

3.1. Univariate Results

3.1.1. Frequency distribution of working period. Based on the table 1 above, it clearly shows that there are 39 respondents (53.4%) who have worked more than six years.

Table 1. Frequency Distribution of Respondents' Working Period

No	Working Period of Respondents	F	(%)
1	Length (> 6 years)	39	53.4
2	New (≤ 6 years)	34	46.6
	Total	73	100

The result of this study is in line with the research conducted by Hariyono, et al (2016) which found that there are 67 respondents from 82 workers have the longest work period for more than ten years working [13]. Working period is the length of time a person works in a particular field of work. Duration of work also affects the occurrence of work accidents. It because the length of time a person works will affect his work experience [14].

3.1.2. Frequency distribution of the use of ppe. Table 2 above shows that there are 55 workers from 73 respondents do not use protective equipment.

Table 2. Frequency Distribution of the Use of PPE Respondents

No	Use of PPE	F	(%)
1	No Use	55	75.3
2	Use	18	24.7
	Total	73	100

However, there is only 18 respondents use the protective equipment. It means that the workers who didn't use protective equipment are greater than the workers who use protective equipment. This study has the same result with the study conducted by Fahmi (2012) related to the use of personal protective equipment (PPE). The result showed that there were 27 workers from 45 respondents (60%) used PPE. In gold producers in the Tatelu community mining area, dimembe district was found that 48.8% workers did not use masks and 51.2% workers used masks [15]. Personal protective equipment (PPE) is a tool that must be used while working in order to maintain the safety of the workers themselves and the people around them [16].

3.1.3. Frequency distribution of complaints of respiratory disorders. Based on the table 3 above, it can be seen that there are 47 (64.4%) workers from 73 respondents have respiratory tract disorders complaints.

Table 3. Frequency Distribution of Respiratory Tracts complaints on Respondents

No	Complaints of Respiratory Disorders	f	(%)
1	There are Complaints	47	64.4
2	No Complaints	26	35.6
	Total	73	100

The amount is greater than the workers who have no complaints which are only 26 workers (35.6%). The result of this study is in line with the research conducted by Hariyono, et al (2016) which found that there were 50 workers (61.0%) from 82 respondents who have respiratory complaints. Furthermore, this study is also in line with the research conducted by Sandra (2013) which found that from 42 respondents, there were 40 workers (95%) have respiratory complaints [13].

In the course of the disease, the symptoms may be more severe. If it more severe, it can be fall into respiratory failure and causes death. When respiratory failure occurs, more complicated management is needed, however mortality is still high. Thus, it is necessary to do quick actions related to the disease so that it is not fall into respiratory failure [17].

3.2. Bivariate Results

3.2.1. Relationship work period with complaints disorders respiratory on respondents. The results of the analysis of table 4 explains about the relationship between the work period and respiratory disorders. It is found that there are 33 workers (84.6%) who have respiratory problems from 39 respondents with long working period (> 6 years). While from 34 respondents who have a new service period (≤ 6 years), there are 14 respondents (41.2%) who have respiratory complaints.

Table 4. Relationship of Work Period and Respiratory Disorder Complaints on Respondents

Work Period	Complaints of Respiratory Tract Disorders				Total		P *value*	OR CI 95%
	There are Complaints		No Complaints					
	N	%	N	%	n	%		
Old	33	84.6	6	15.4	39	100	0,000	7.857 (2.600 to 23.745)
New	14	41.2	20	58.8	34	100		
Total	47	64.4	26	35.6	73	100		

The results of the statistical test shows that *Pvalue* = 0,000, where *Pvalue*<0.05. It can be concluded that there is a significant relationship between work period and complaints of respiratory disorders on the respondents. Based on the analysis, it also finds that OR = 7.857. It means that those who have long working period (> 6 years) have an opportunity to have respiratory problems 8 times greater than those who have new working period (≤ 6 years). This result is supported by the research of Pangemanan (2015) on the relationship between work period and the incidence of pneumoconiosis with p value 0.002. The significant value of the relationship between work period and the incidence of pneumoconiosis is smaller than 0.05. Thus, it can be concluded that there is a significant relationship between the work period and the incidence of pneumoconiosis in cement collecting workers in the cement packing unit of PT. Tonasa Line Bitung City. The analysis result of the relationship between work period and the incidence of pneumoconiosis shows that the OR value is 11,333. It can be assumed that the workers who have long working period (≥ 10 years) have the possibility to get pneumoconiosis 11,333 times than those who have a new working period (<10 years) [18].

The study above is also supported by Suma'mur's statement which says that work period determine how long the workers' have a risk on dust exposure. the greater the dust exposure of a person, the greater the risk of lung disease. Working period is the length of time a person works in a particular field of work. Duration of work also affects the occurrence of work accidents. Therefore, work period will affect someone's work experience [14].

3.2.2. Relationship between the use of ppe and complaints of respiratory disorders in respondents. The table above clearly shows that there are 41 respondents (74.5%) who did not use PPE have respiratory complaints, whereas from 18 workers using PPE, only 6 of them (33.3%) have respiratory complaints.

Table 5. Relationship the Use of PPE and Respiratory Disorder Complaints on Respondents

Using PPE	Complaints of Respiratory Disorders				Total		P *value*	OR CI 95%
	There are Complaints		No Complaints					
	n	%	N	%	n	%		
Not Use	41							74.5 14
Wear	6	33.3	12	66.7	18	100		25.5 55
Total	47	64.4	26	35.6	73	100		100 0.004 5.857 (

Based on the statistical test results, it is known that P *value is* 0.004 which is less than 0.05 (<0.05). It can be concluded that there is a significant relationship between the use of PPE and respiratory complaints of the respondents. Based on the analysis, it also found that the value of OR is 5.857, which means that respondents who do not use PPE have a possibility of having respiratory problems 6 times greater than those using PPE. This research is in line with the research conducted by Sudarmaji (2015), the Chi Square test results shows a significant relationship between the use of masks and respiratory complaints (p-value = 0.014). Personal protective equipment (breathing) must be used by all workers in the industry which produce dust in the production process. The types of personal protective equipment that are widely used by Unit Y workers are masks made of cloth / t-shirts and disposable masks [19].

The use of these type of masks do not guarantee that they can protect workers from the possibility of breathing dust and small-sized particles. Types of respiratory protective equipment that can be used by wood industry workers are such as catheter respirators, and canisters. The obligation to use masks is one of the company's efforts to protect workers from dust exposure. The type of mask used must be suitable for the potential hazards and risk factors in the work environment since the level of mask protection is influenced by the type of dust, mask, and the ability of the mask to filter dust [20].

4. Conclusion

From the study it was found that more than 50% of brick workers worked more than 8 hours per days for more than 6 years and did not use complete PPE so there are many complaints of respiratory problems. It shows that working period and the use of PPE have strong relationship with respiratory complaints. It is recommended that brick workers can reduce their time working in a day and use PPE at work so that the long working period does not affect the respiratory tract.

References

[1] Government Regulation of the Indonesia Republic Numb.41 of 1999, *About Air Pollution Control*
[2] Suma'mur, 2013. *Corporate Hygiene and Work Safety (Hiperkes).* (Jakarta: CV Sagung Seto)
[3] Basic Health Research, 2013. *National Report on Riskesdas 2013.* Health Research and Development Agency of the Indonesian Ministry of Health. Jakarta
[4] WHO Interin Guidelines, 2007. *Prevention and Control of ARI.* Trust Indonesia
[5] Skinder et al 2014 Brick Kiln Emissions and Its Environmental Impact: A Review *J. Ecol. Nat. Environ.* **6**(1) 1-11

[6] Basri S and S Erniatin 2015 *Knowledge Relationship and Occupational Health Attitudes with Disease Caused by Work in Brick Workers*. Wiralodra University. Journal

[7] Asfawi Supriyon 2015 *Factors Associated with Respiratory Subjective Complaints on Female Poultry Traders Due to Air Condition in Rp. 4 Penggaron in Semarang City in 2015* (Semarang: Dian Nuswantoro University)

[8] Manoppo A, Kandou G D dan Josephus J 2015 The Relationship Between the Period of Service and Use of Personal Protective Equipment (Respirator) with Lung Vital Capacity at Firefighters at the Manado City Fire Service *PHARMACON: Jurnal Ilmiah Farmasi-UNSRAT* **Vol. 4** No. 4 November 2015

[9] Annual Report of the Public Health Service Padang Pariaman District, 2016

[10] Annual Report of the Public Health Center Sintuk Toboh Gadang District, Padang Pariaman Regency, 2016

[11] Annual Report of PTT Midwife Jorong Palembayan And Batang Tapakis Padang Pariaman Regency, 2016

[12] Notoatmodjo S 2010 *Health Research Methodology, Revised Edition* (Jakarta: PT Rineka Cipta)

[13] Hariyono, Widodo, et al., 2016. *Relationship of Work Period and Use of Personal Protective Equipment with Complaints of Respiratory Disorders in Employees at PT. Madubaru, Bantul Regency*. IENACO National Seminar - 2016

[14] Sucipto, Cecep Dani, 2014. *Occupational Health and Safety*. (Yogyakarta: Gosyen Publishing)

[15] Fahmi T 2012 Relationship between Work Period and Use of PPE with Lung Function Capacity in Textile Workers in Frame Spinning I Ring Parts in Pt. Pekalongan Regency *Journal of Public Health* **Volume 1** Number 2 Year 2012

[16] Buntarto, 2015. *The Simple Guide to Occupational Health & Safety for Industry* (Yogyakarta: New Library Press)

[17] Irianto, Koes, 2012. *Anatomy and Physiology*. (Bandung: Alfabeta)

[18] Pangemanan, et al. 2015. The Relationship Between Dust Levels, Working Period, Use of Masks and Smoking With Pneumokoniosis Events in Cement Collecting Workers in the Cement Packing Unit of PT. Tonasa Line Bitung City. *JIKMU*, **Vol. 5**, No. 2b April 2015

[19] Sudarmaji and Anindya Mar'atus, 2015. Relationship between Characteristics of Workers and Total Dust Levels with Respiratory Complaints in Wood X Industry Workers in Lumajang Regency. *Journal of Environmental Health Perspective* **Vol. 1**, No. 1 January 2015: 1–12

[20] Minister of Manpower and Transmigration Regulation Number 8 of 2010 concerning Personal Protective Equipment

The content of mercury in sediments around Artisanal Small-scale Gold Mining (ASGM) Bumela district, Gorontalo Regency, Gorontalo Province, Indonesia

Fitryane Lihawa[1,*] and Marike Mahmud[2]

[1] Center of Environmental Study, the State University of Gorontalo, Indonesia
[2] Civil Department, Faculty of Engineering, State University of Gorontalo, Indonesia

*fitryane.lihawa@ung.ac.id

Abstract. Artisanal Small-scale Gold Mining (ASGM) in Gorontalo spread across several regions. One is located in the village Bumela, District Bilato, Gorontalo regency. The processing of gold at the mine site Bumela do with the amalgamation method. The purpose of this study is to assess the content of mercury in sediments around the ASGM Bumela. Sediment sampling sites in Totopo River, Motebo River and ASGM Bumela tailings. The number of sampling point sediment in the river are 15 points. The number of sampling point sediment in tailings location is 2 points. Sampling was done by varying the depth of 0-20 cm and 20-40 cm. To determine the concentration of mercury in the sediment used Atomic Absorption Spectroscopy (AAS) without a flame in LPPMHP Gorontalo Province. The quality standard levels of mercury in the sediment used European Safety Standard. The average content of mercury in the sediment was 71.36 mg/kg. The average content of mercury in the tailings is 31.95 mg/kg. The content of mercury in Totopo River, Motebo River and the tailings are exceeding the quality standard of European Safety Standard.

1. Introduction

In general, gold mining in Indonesia and Gorontalo Province uses mercury [1,2]. Mercury is a metal that is liquid at room temperature. Mercury is widely available in nature, and therefore the mercury often pollutes the environment. Mercury has the ability to dissolve the metal to form a component called amalgam. Mercury in nature is in the form of inorganic mercury and organic mercury (organomercury) [3,4]. Research on environmental pollution caused by gold mining activities have been carried out [1,2,5,6,7,8].

Pollution of river water and sediment caused by the activity of Artisanal Small-scale Gold Mining (ASGM) has occurred. The results of research at ASGM Buru Island showed that mercury concentrations varied. The concentration of mercury at the end of the treatment process is 652 mg/kg. The concentration of mercury in the tailings pond was 10.7 mg/kg and in the drain toward the river at 7.31 mg/kg [5]. Results of research on location ASGM Buladu, North Gorontalo District showed that mercury concentrations in sediments along the river Wubudu and Orchid has exceeded the quality standards set by the World Health Organization (WHO). The level of mercury in fish snapper was 0.5 ug/g. In addition, symptoms of health problems that occur are a tremor, bluish gums, Babinsky reflex and labial reflex [1].

Mercury goes into the sediment will accumulate and be absorbed by plants and fish. This will cause the plants and animals of water will be polluted by mercury [6,7]. Therefore it is necessary to investigate mercury pollution in the sediment at the site of ASGM Bumela, Gorontalo Regency.

2. Research methods

2.1. Research Locations

The study was conducted in Artisanal Small-scale Gold Mining (ASGM) Bumela Village, District Bilato, Gorontalo regency. Distance ASGM Bumela to County Government Center Gorontalo is 54 km. Access to the location can be reached by car or motorcycle with a travel time ± 1 hour. Distance from the village center to the mine site Bumela ± 8 km and can be reached by foot or by motorcycle.

Sediment samples were taken from the river Totopo (station S1 - S13) and the River Motebo (station S14 and S15). Sediment samples from the tailings totaled 2 samples (stations ST1 and ST2) (Figure 1). Sampling sediment using a hand drill. At each point, sediment samples were taken at a depth of 0-20 cm and a depth of 20-40 cm. The equipment used should be sufficient to guarantee the security of the sample. The distance of each sampling point of processing site is shown in Table 1.

Table 1. The distance of the sampling point from the the gold processing site

Sampling Point	Name Rivers	Distance (km)
S1	Totopo	7.5
S2	Totopo	5.47
S3	Totopo	2.14
S4	Totopo	4.16
S5	Totopo	3.50
S6	Totopo	5.16
S7	Totopo	1.85
S8	Totopo	1.45
S9	Totopo	1.26
S10	Totopo	1.14
S11	Totopo	908
S12	Totopo	746
S13	Totopo	467
S14	Left Motebo	28
S15	Right Motebo	39

Source: Results of the interpretation of the map, 2018

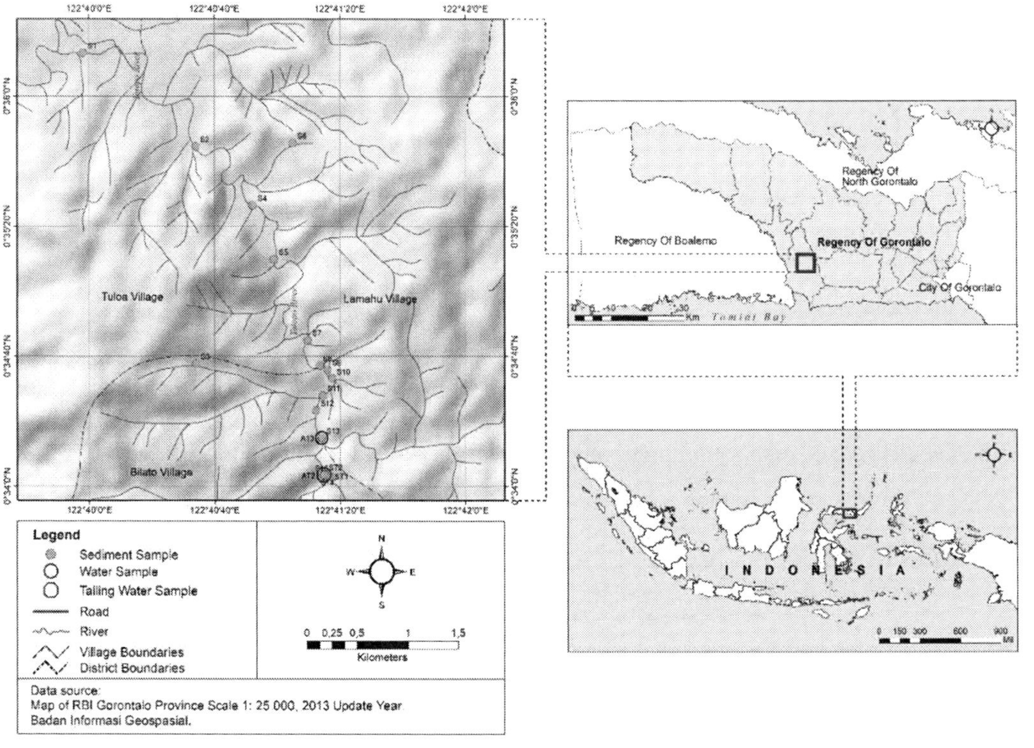

Figure 1. Map of research sites

2.2 Analysis procedures

To determine the concentration of mercury in the water used Atomic Absorption Spectroscopy (AAS) without flame in the Laboratory of PPMHP, Gorontalo Province. Without flame atomization performed by passing electrical energy in a carbon rod (CRA - Carbon Rod Atomizer) or carbon tubes (GTA - Graphite Tube Atomizer) having two electrodes. Samples were inserted into the CRA or GTA. Electric current is applied so that the rod or tube becomes hot (temperature rose to high) and elements analyzed will be atomized. The temperature can be set up to 3000 °C. Heating the sample solution through three stages: 1) drying to evaporate the solvent; 2) ashing, the furnace temperature was raised gradually until decomposition and evaporation of the organic compounds present in a sample in order to obtain salt or metal oxide; 3) atomization. The standard material used is a Certified Raw Material (CRM). The concentration of mercury in the sediment compared with the European Safety Standard. Safety limit value content of mercury in the sediment is 2 mg/kg.

3. Results and discussion

3.1 The general overview of the study site

Artisanal Small-scale gold mining (ASGM) Bumela is one mine site located in the district of Gorontalo. ASGM Bumela known as the "Pertambangan Tangga 2000". ASGM Bumela has been operating since several decades ago. Miners in this location from the local area (district of Gorontalo and vicinity) and from outside the area Gorontalo. The number of miners varies depending on the result of the mine. There are no data on the number of miners. If mining results increases, the number of miners will increase. Conversely, if the mining declined, the number of miners will be reduced. At the time of the survey, the number of miners is ± 150 person. Total operating drum when it is 12 units. In each unit consists of 10 pieces of the drum.

The processing method in the ASGM Bumela is the amalgamation method. Amalgamation is the gold extraction process by mixing gold ore with mercury (Hg). This method uses a round drum so that the material will be crushed and mercury mines will bind gold compounds (Figure 2).

(a) (b)

Figure 2. a) Drum for gold processing; b) Waste gold mining

The interview with one of the miners explained that one drum will grind 5 kg of rock. In one drum will be mixed mercury (Hg) as much as 3 ounces. Processing results will then be squeezed with a cloth. This process aims to separate the mercury from amalgam. Mercury obtained can be reused for the next process. In each processing, the mercury will be reduced 0.1 ounces. End process is burning to vaporize the mercury, so that left only a gold alloy. The average gold produced in the ASGM Bumela is 2.5 grams per drum [9].

3.2 Mercury in the sediment
Sediment sampling carried out at the same location as the location of water sampling. Sediment sampling by variations in depth that is at a depth of 1-20 cm and 20-40 cm. The result of the content of mercury in the sediment analysis are shown in Table 2.

Table 2. The content of mercury in river sediments Totopo, Gorontalo Regency

Sampling Point	The content of Mercury (mg/kg)			Remarks
	Mercury	The average per-point sampling	*European Safety Standard*	
S1.1	30.99	33.545	2	Exceeds
S1.2	36.10			Exceeds
S2.1	67.71	60.59	2	Exceeds
S2.2	53.47			Exceeds
S3.1	31.32	71.005	2	Exceeds
S3.2	110.69			Exceeds
S4.1	68.27	70.74	2	Exceeds
S4.2	73.21			Exceeds
S5.1	104.82	107.295	2	Exceeds
S5.2	109.77			Exceeds
S6.1	117.73	117.195	2	Exceeds
S6.2	116.66			Exceeds

S7.1	104.56	103.905	2	Exceeds
S7.2	103.25			Exceeds
S8.1	107.12	107.425	2	Exceeds
S8.2	107.73			Exceeds
S9.1	34.56	34.535	2	Exceeds
S9.2	34.51			Exceeds
S10.1	58.72	58.615	2	Exceeds
S10.2	58.51			Exceeds
S11.1	42.40	82.9	2	Exceeds
S11.2	40.50			Exceeds
S12.1	79.45	76.295	2	Exceeds
S12.2	73.14			Exceeds
S13.1	77.51	78.005	2	Exceeds
S13.2	78.50			Exceeds
S14.1	53.77	53.585	2	Exceeds
S14.2	53.40			Exceeds
S15.1	16.5	16.6	2	Exceeds
S15.2	16.7			Exceeds
mean	71.36			
SD	31.39			
Max	117.3			
Min	16.5			

Source: results of laboratory analysis, 2018

The results of the study show that mercury content varies with depth. At the sampling point S1 S3, S4, S5, S13, mercury content at a depth 20 – 40 cm was higher than the mercury content at a depth of 10 – 20 cm. The other sampling point (S2, S6, S7, S8, S10, S11, S12), mercury content at a depth of 20 – 40 cm is lower than the mercury content at a depth 10 – 20 cm. At the sampling point S9, S14, S15, mercury content is relatively the same in both depths. This shows that in some locations there has been an accumulation of mercury in sediments.

Comparison charts levels of mercury in the river sediment quality standards European Safety Standard is shown in Figure 3.

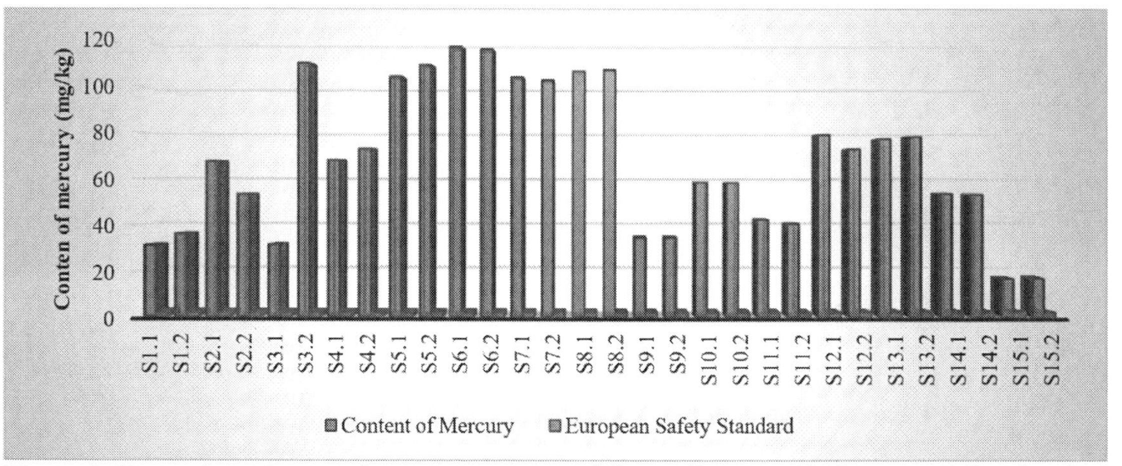

Figure 3. Comparison charts on the mercury content of river sediments with European Safety Standard

Table 3. The content of mercury in the sediment of tailings at the site Bumela ASGM, Gorontalo Regency

Sampling Point	The content of Mercury (mg/kg)			Remarks
	Mercury	The average per-point sampling	*European Safety Standard*	
ST1.1	32.40	26.915	2	Exceeds
ST1.2	21.43			Exceeds
ST2.1	22.77	36.985	2	Exceeds
ST2.2	51.20			Exceeds
mean	31.95			
SD	11.89			
Max	51.20			
Min	21.43			

Source: results of laboratory analysis, 2018

Comparison charts level of mercury content in tailings sediment with European Safety Standard shown in Figure 4.

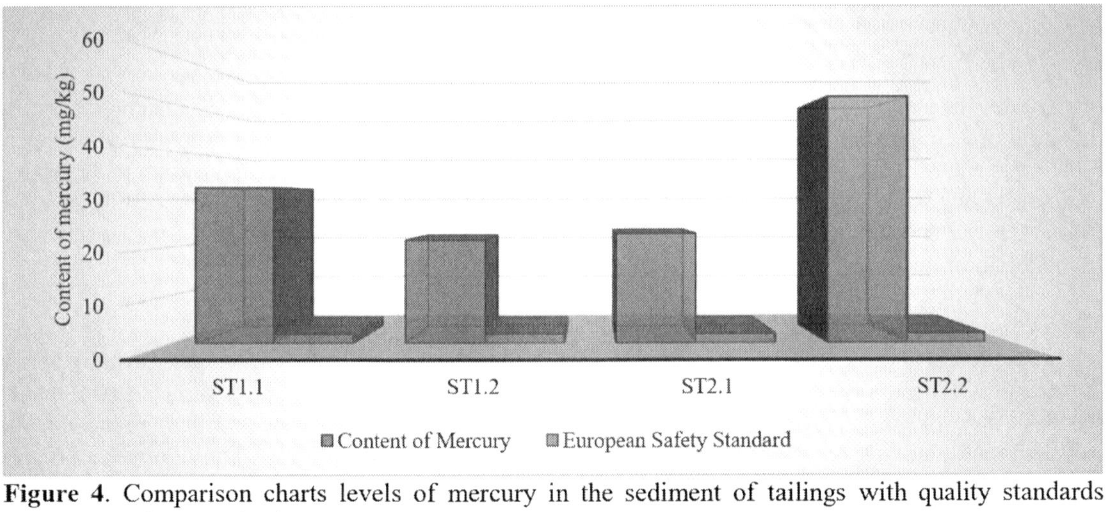

Figure 4. Comparison charts levels of mercury in the sediment of tailings with quality standards European Safety Standard

The results of content analysis of mercury in river sediments Totopo showed that sediment load has exceeded the European Safety Standard. The average content of mercury in river sediments 71.36 mg/kg is higher than the European Safety Standard. Similarly, the content of mercury in ASGM Bumela tailings. The average content of mercury in the tailings is 31.95 mg/kg higher than the European Safety Standard (2 mg/kg). This is due to the process of gold panning in the river Totopo.

3.3 The concentration of mercury in sediments based on the distance from the mine site

The distance of each sampling point of the gold processing site varies (Figure 5).

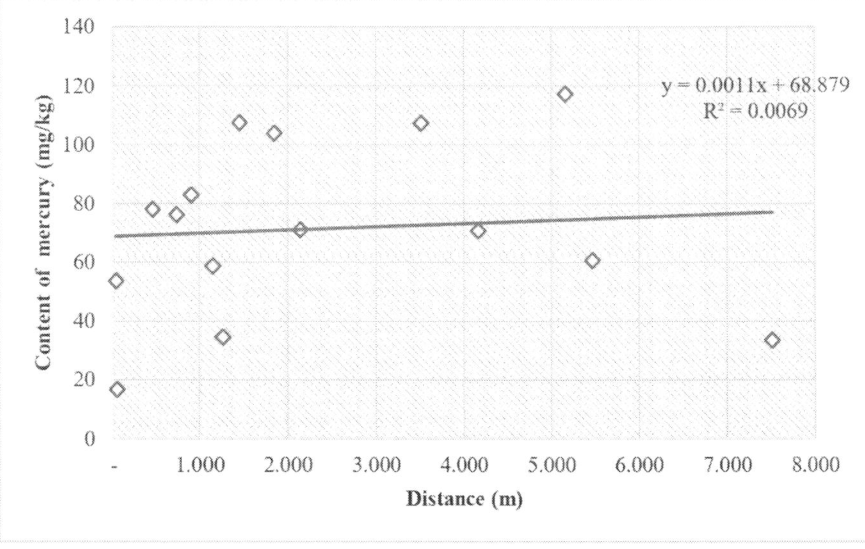

Figure 5. Charts the relationship between the content of mercury in the sediments at a distance from the location of the gold processing

The results showed that the content of mercury in sediment was not affected by distance from the location processing. The only determinant coefficient of 0.0069. Figure 5 shows the highest mercury content is at a distance of 5.159 km of the location processing. This is due to the location to be around buildup tailing will be recycled by the public. On the other hand, the content of mercury in the river Motebo lower than the levels of mercury in the river Totopo (Table 2). The Motebo river is the first river to be the location for the disposal of gold processing waste.

3.4 Discussion

Processing of gold mines in Gorontalo Province in general use mercury. Mercury miners illegally obtain a price of IDR 750,000 - IDR 1,500,000 per kilogram. Results of interviews with miners found that to produce 5 kg of gold from rocks used 85 grams of mercury. After the process, the mercury will be reduced by 2.8 grams. To produce 1 gram of gold, used 20 grams of mercury. After the amalgamation process, 19 grams of mercury released into the tailings and 1 gram evaporate into the air [10]. Mercury in the tailings will be washed away by the water and into a river. Finally, will settle to the bottom of the river. This causes the mercury in the sediment to be high. Several studies on the content of mercury in the sediments indicate that the content of mercury in the sediment has exceeded the quality standard [1,2,5].

The research results obtained ASGM Buru Island in the mercury content of the sediment in the channel leading to the river is 7.31 mg/kg [5]. The content of mercury in the sediments at the ASGM Buladu is 13-17 mg/kg [1]. Mercury in water will be consumed by fish, so it will happen accumulation and biomagnification in fish. Fish are eaten by humans and finally, Mercury bioaccumulation will occur in the human body.

If the terms of the processing locations within the mine with a mercury content of mercury showed that the distribution pattern is not dependent on the distance of the source of the pollutants. Mercury distribution pattern is influenced by several variables, including variable geochemical and hydrological systems. Deposition of mercury in the river affected by the water flow characteristics and the physical condition of the riverbed [11].

The solution to overcoming the problem of the use of mercury in gold mining must be done comprehensively. Some of the strategies that can be done are to educate miners. Educational materials in the form of environmental health, the alternative method in mining processing. Some techniques reduced mercury releases including retorts, mill leaching, vat-leaching, and others replaced mercury from the process such as magnets, direct smelting, sluices, and borax [3,12,13]. Clean Tech Mine has been applied in Artisanal and Small-scale Gold Mining (ASGM) by the natives of Mozambique. Application of Clean Tech Mine is an innovative and inexpensive practice in mining activities [14]. Phytoremediation is one method that can be done to overcome the problem of mercury pollution in the environment [15,16].

4. Conclusion

Activity Artisanal Small-scale Gold Mining (ASGM) in the village of Bumela, District Totopo, Gorontalo Regency has resulted in environmental pollution. One indicator is the mercury content in the river sediment. The average content of mercury in the sediment was 71.36 mg/kg. The average content of mercury in the tailings is 31.95 mg/kg. The content of mercury in river sediments and tailings Totopo is above the quality standard of European Safety Standard.

References

[1] Arifin Y I, Sakakibara M and Sera K 2015 Impacts of artisanal and small-scale gold mining (ASGM) on environment and human health of North Gorontalo Regency, Gorontalo Province, Indonesia *Geosciences* **5**(2)160-176

[2] Mahmud M, Bull B, Desei F, Lihawa F and Saleh Y 2016 Evaluation of mercury concentration in water and sediment at Buladu Artisanal Gold Mine in Gorontalo Province, Indonesia. *International Journal of Innovative Research and Development* **V**(9) 363-371 Retrieved from www.ijird.com

[3] Robles I, Bustos E and Lakatos J 2016 Adsorption study of mercury on lignite in the presence of different anions *Sustainable Environment Research* **26**(3) 136-141

[4] Mallongi A 2017 *The impact of wastewater from institutional and industrial activity* (Yogyakarta: Goshen Publishing)

[5] Male Y T, Reichelt-Brushett A J, Pocock M and Nanlohy A 2013 Recent mercury contamination from artisanal gold mining on Buru Island, Indonesia - Potential future risks to environmental health and food safety *Marine Pollution Bulletin* **77**(1-2) 428-433

[6] Purnawan S, Sikanna R and Prismawiyanti 2013 Distribusi logam merkuri pada sedimen laut di sekitar muara sungai Paboya *Online Journal of Natural Science* **2**(1) 18-24

[7] Mortazavi A, Hatamikia M, Bahmani M and Hassanzadazar H 2016 Heavy metals (mercury, lead and cadmium) determination in 17 species of fish marketed in Khorramabad City, West of Iran *Journal of Chemical Health Risks* **6**(1) 41-48

[8] Tomiyasu T, Kono Y, Kodomatani H, Hidayati N and Rahajoe J S 2013 The distribution of mercury around the small-scale gold mining area along the Cikaniki river, Bogor, Indonesia *Environmental Research* **125** 12-19

[9] JAWE (2018, April 17) *Gold Mining Processing Buladu* (F Lihawa, Interviewer)

[10] Telmer K and Stapper D 2007 Evaluating and monitoring small scale gold mining and mercury use: building a knowledge-base with satellite imagery and field work. UNDP / GEF / UNIDO Project EG / GLO / 01 / G34. Final Report to the United Nations Industrial Development

[11] Moreno - Brush M, Rydberg J, Gamboa N, Storch I and Biester H 2016 Is mercury from small-scale gold mining prevalent in the southeastern Peruvian Amazon? *Environ. Pollut.* **218** 150-159

[12] Zolnikov T R and Ortiz D R 2018 A systematic review on the management and treatment of mercury in artisanal gold mining *Science of The Total Environment* **633**(2018) 816-824

[13] Veiga M M, Angeloci - Santos G and Meech J A 2014 A review of barriers to reduce mercury use in artisanal gold mining *The Extractives Industries and Society* **1**(2014) 351-361

[14] Drace K, Kiefer A M, Veiga M M, Williams M K, Ascari B, Knapper K A, Logan K M, Breslin V M, Skidmore A, Bolt D A, Geist G, Reidy L and Cizdziel J V 2012 Mercury-free, small-scale artisanal gold mining in Mozambique: utilization of magnets to isolate gold mine at clean tech. *Journal of Cleaner Production* **32** 88-95

[15] Zulkoni A, Rahyuni D and Nasirudin 2017 Pengaruh pemangkasan akar jati dan inokulasi jamur Mikoriza Arbuskula terhadap fitoremediasi tanah tercemar merkuri di Kokap Kulonprogo Yogyakarta *Jurnal Manusia dan Lingkungan* **24**(1) 17-22
[16] Liu Z, Wang Li-ao, Xu J, Ding S, Feng X and Xiao H 2017 Effects of different concentrations of mercury on accumulation of mercury by five plant species *Ecological Engineering* **106** 273-278

IOP Conf. Series: Earth and Environmental Science **314** (2019) 012017 doi:10.1088/1755-1315/314/1/012017

Community-based solid waste management planning in the Administrative Village of Surau Gadang, Padang City

Anggrieka Maharani[1,*], Yommi Dewilda[2], Yeggi Darnas[2] and Indang Dewata[1]

[1] Department of Environmental Science, Postgraduate Program, Universitas Negeri Padang, Prof. Dr. Hamka Street, Air Tawar Padang, West Sumatra, Indonesia 25131

[2] Environmental Engineering Department, Faculty of Engineering, Andalas University Limau Manis, Padang City, West Sumatra, Indonesia 25163

*anggarieka10@gmail.com

Abstract. The community-based solid waste management system in Administrative Village of Surau Gadang is planned based on SNI-20-3242-2008, with a design period of 8 years (2011-2018). The service level is planned to reach 100%. The total solid waste generation is 64.18 m^3/d (organic 60.81 m^3/d, inorganic is 0.75 m^3/d, while other waste is 2.61 m^3/d). The sorting process is according to the type of waste. Residential waste collection uses motorized trash tricycles to TPST (integrated waste treatment plant) and non-residential uses pickup cars to TPS (temporary disposal site). Processing carried out at the source is composting the *Takakura* method, while in TPST is composting the *Bokhasi* method, sorting other waste for sale, recycling and making handicrafts. The residue from the remaining processing which was disposed to TPS was 18.43%. The institution formed is a representative of all RW (citizens association). Regulation based on Law No.18 of 2008. The investment in making TPST and its supplementary facilities for eight years is Rp.1,322,000,000, with operational and maintenance costs of Rp.26,678,067/month. The profit obtained is Rp.102,117,284 /month. Community participation can be seen from involvement in technical and non-technical aspects.

1. Introduction

Solid waste can cause problems for life and environmental health, especially human life. The habit of throwing trash carelessly, such as throwing solid waste at the roadside can cause uncomfortable feelings because of the aesthetic disturbance and throwing solid waste in the body of water continuously can bring vector disease and even cause flooding. As population increases rapidly, the amount of waste produced also increases. The increasing amount of waste produced will cause complex problems if it is not followed by improvements in waste management facilities and infrastructure. Increasing the amount of waste produced if it is not followed by improvements in waste management facilities and infrastructure will cause problems to be complex. Therefore, to minimize the impacts that occur, proper and comprehensive management is needed.

Household waste production also increases every day due to an increasing number of products and consumption patterns of the people. So, the thing that must be done to overcome the increase in waste volume is by extracting the volume of waste from its source through community empowerment [1]. Community-based solid waste management is a waste handling system that is planned, organized, operated, managed and owned by the community [2]. The principle of community-based solid waste management technology is fairly simple, easy to operate, inexpensive, accepted by the community and not high technology [3].

Administrative Village of Surau Gadang is one of the potential areas for the implementation of community-based residential waste management systems, because this area is included in the area prioritized in implementing a regional scale waste management system with a new paradigm (community-based solid waste management using the 3R method) in Padang City. This area is also included in the development area I (WP I) according to the regional spatial plan (RTRW) of Padang City in 2008-2028. At this time the solid waste is still collected and stacked at the temporary disposal site (TPS) to be transported to the final disposal site (TPA). Solid waste management in this area is still not fully served by municipal waste institution, namely Dinas Kebersihan dan Pertamanan (DKP) Kota Padang and the local private business entity. Therefore, a community-based solid waste management plan was made in Administrative Village of Surau Gadang, covering a management system that includes operational technical aspects (separation and sorting, storage, collection, processing and transportation of residues) and non-technical aspects (institutional / organizational and management, legal and regulations, benefits, and community participation).

2. Literature Review

Regional waste management is a communal waste management activity to serve part or all of the waste resources in the area where management is located [4]. Whereas community-based solid waste management is a waste handling system that is planned, organized, operated, managed and owned by the community. The goal is the independence of the community in maintaining environmental cleanliness through environmentally friendly waste management [2].

Based on SNI 20-3242-2008 waste management is carried out using the 3 R method (reduce, reuse, and recycle). The waste management system is basically various components that are connected to each other. The policy of managing urban waste in Indonesia defines that waste management, outlined in 5 (five) subsystem components, which consists of operational engineering aspects, institutional/ management and organizational aspects, regulatory/legal aspects, financing aspects and aspects of community participation [4]. Management operates operational techniques, management and financing can refer to SNI 19-2454-2002, while for review Regulations and Law can refer to Law No.18 of 2008 about Waste Management in Indonesia.

3. Methods

The research began with a literature study on solid waste management. After that, secondary data collection was carried out regarding the profile of Administrative Village of Surau Gadang, namely population demographics, social, economic, cultural aspects and other supporting facilities and infrastructure. In addition, data is also needed regarding the generation and composition of waste, storage, collection, transportation and final disposal. Primary data collection with a questionnaire aims to determine the existing conditions in managing waste and the willingness of the community to participate in community-based solid waste management. The type of house in the sampling area is classified based on SNI 20-3242-2008. The general design of a community-based solid waste management system contains:

- The design period is in accordance with Padang City Phase I RTRW (2008-2018);
- Population projection;
- Waste generation projection;
- System determination.

Detail designs of operational and non-technical technical aspects include:

- Service area and service level;
- The operational technical aspects include sorting and separation systems, residue collection, processing and transportation;

- Non-technical aspects that include institutional/management organizations, laws and regulations, financing and community participation.

Technical specifications contain the types of materials used in community-based solid waste management systems and work steps in their implementation. These specifications are tailored to the needs and conditions of the field, as well as aspects of costs and operational convenience. While the calculation for the budget plan, covering the investment, operational and maintenance costs of all the facilities and infrastructure needed, as well as the need for implementation in community-based TPST planning is made based on the basic unit price. While the budget plan, covering the investment, operational and maintenance costs of all the facilities and infrastructure needed, as well as the need for implementation in community-based TPST planning. The calculation is made based on the basic unit price. The area to be served and the management system made, is shown in Figure 1 below.

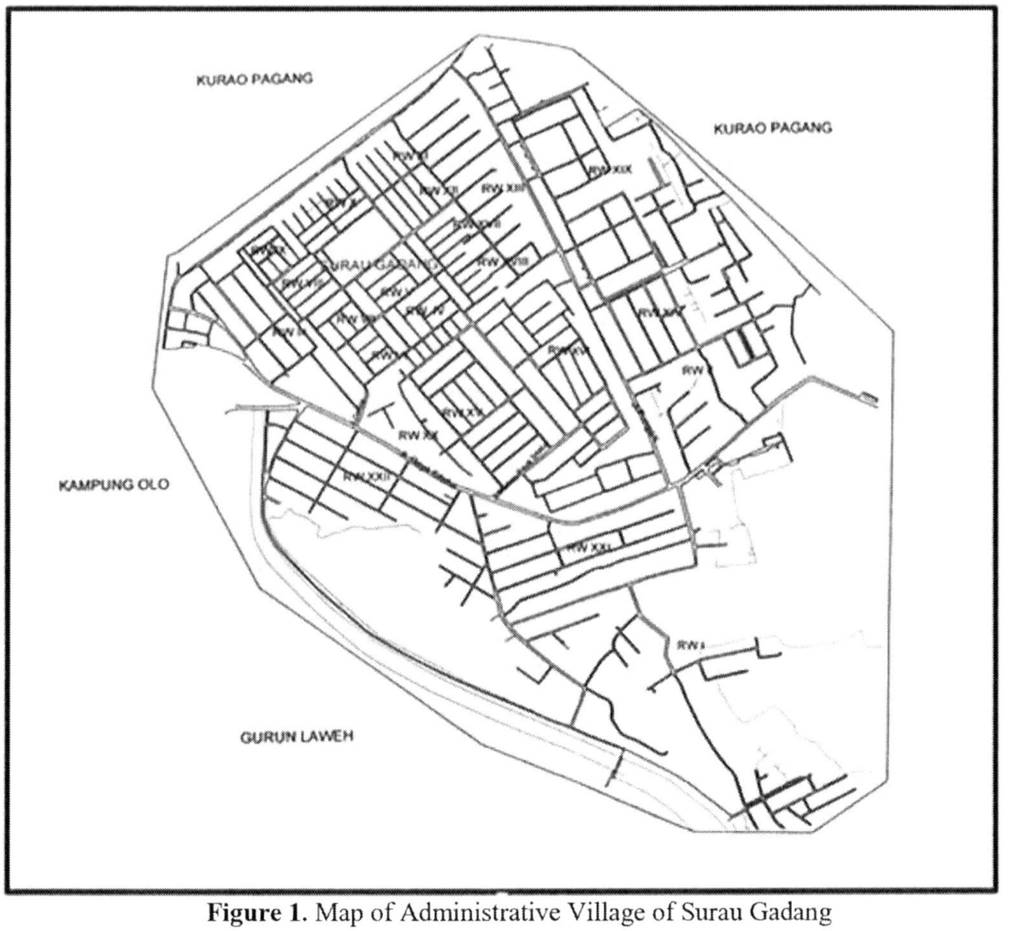

Figure 1. Map of Administrative Village of Surau Gadang

4. Results

Based on the results of the study and evaluation, it was found that waste management in Surau Gadang still did not fulfill the concept of waste management with a new paradigm (SNI-20-3242-2008 concerning waste management in settlements using the 3 R concept and involving the community). Where community involvement in managing waste is still minimal, even though Surau Gadang is one of the priority areas for implementing regional scale waste management systems and included in the development area I, according to the regional spatial plan of Padang City in 2008-2028. Temporary waste management is carried out by related institutions namely the Department of Sanitation and Landscaping of Padang City and third parties for transportation services to temporary dumps.

Supporting facilities for managing waste with this new paradigm are not yet available. The community still uses the remaining packaging bags to store mixed garbage, and it is still rare to find people who have the awareness to sort out their own garbage. The garbage container used is also not as it should be. Recapitulation of solid waste management evaluation in this area can be seen in Table 1 below.

Table 1. Recapitulation of Solid Waste Management Evaluation in Administrative Village of Surau Gadang

Aspect	Existing	Regulation	Description
Regulation	Not available	SNI 20-3242-2008	Not comply
Institutional/organizational management	Not available	SNI 20-3242-2008	Not comply
Community participation	Not available	SNI 20-3242-2008	Not comply
Operational Technical			
- Area and service level	Available	Permen PU no.21/PRT/M/2006	Not comply
- Storage	Available	SNI 19-2454-2002	Not comply
- Collection	Available	SNI 19-2454-2002	Not comply
- Transportation	Available	SNI 19-2454-2002	Not comply

In order that waste management in the Surau Gadang can be better, then waste management planning is made based on the new paradigm with reference to the applicable regulations in Indonesia. Management that is planned can be seen in the following description.

4.1. General Design

The stages of planning for community-based solid waste management systems in Administrative Village of Surau Gadang are adjusted to the Phase I RTRW of Padang City (2009-2018). Where this aims to improve the quality of facilities and infrastructure services and prepare the development of service centers. Before the planning is carried out, the population projection is calculated first. Projection of population is taken based on the chosen method, namely arithmetic method. The population of Administrative Village of Surau Gadang was obtained at the end of the planning (in 2018), which was 20,704 people. Determination of the amount of waste generation and the composition of the household of Surau Gadang refers to previous studies [5]. More details can be seen in the following Tables 2 and 3.

Table 2. Household Waste Generation in Administrative Village of Surau Gadang

Classification	Percentage (%)	Generation Volume (l/o/h)	Generation Weight (kg/o/h)
High Income	20	2.90	0.54
Medium Income	60	2.68	0.44
Low Income	20	2.92	0.57
Average		2.83	0.52

The generation projection at the end of the year (qn) is 3.10 l/p/d. The management system area plan includes all RW in the area. The planned service level is 100% of the total population. In the area of garbage housing is managed by the community. Whereas for non-residential areas spread throughout the Surau Gadang area, management will be carried out by a private entity that already exists in the region, namely the Badan Usaha Ikhlas. The management meant here is management up to polling stations.

Table 3. Composition of Waste in Administrative Village of Surau Gadang

Description	Composition (%)	Composition (l/d)
Organic Waste		
Food waste	75.14	43110.7
Garbage	6.37	3657.30
Paper	4.43	2543.90
Plastic	8.41	4824.22
Rubber	0.40	229.63
Amount of Organic Waste	94.76	54365.52
Inorganic Waste		
Metal	0.56	319.48
Glass	0.61	349.44
Others	4.07	2336.23
Amount of Inorganic Waste	5.24	3005.15
Total Generation of waste	100	57370.67

Source: [5]

4.1.1 Operational Technical Plan. Operational technical plans can be seen in Table 4 below. The source of garbage served is divided into two parts. This division is based on SNI 20-3242-2008. Can be divided as follows:

- Residential / household areas in the form of kitchen waste, yard waste and another household waste.
- Non-residential area
  - Commercial areas in the form of garbage from restaurants, markets and shops / kiosks;
  - Institutional areas in the form of garbage from offices, health centers / *posyandu*, schools and educational institutions as well as mosques and *musholla* (praying room).

Table 4. Recapitulation of Operational Technical Aspects Selected Design

Description	Source	
	Residential	Non-residential
Sorting	- At source: Sorting according to type a. Organic waste is easy to rot (green). b. Paper and plastic waste (yellow). c. Inorganic & other waste (red) - At TPST (area): sorted by type of processing.	Based on the type of waste, the container is adjusted to the type of garbage.
Storage	Individually: 10-60 L sacks / plastic bags and plastic / plastic bin covered in volume from 10 to 60 L.	- Individual: 10-60 L sacks / plastic bag, closed plastic bin volume 10-60 L and plastic bin covered 200-400 L - Communal: For market waste, used insulated containers with a cover measuring 6 m3.
Collection	Individually indirect by using a motorized trash tricycles with a capacity of 1.5 m^3.	Individuals do not go directly to the polling station using the pickup truck (owned by the Badan Usaha Ikhlas)
Treatment	- Household scale: Takakura composting - Regional scale (TPST) a. Bokashi Method b. Sales of trash worth selling can be recycled and made into handicrafts.	
Residue	TPS: container 6 m^3.	

4.1.2 Non-technical Plan

- Institutional / Organizational and Management Aspects. The development plan in this aspect focuses on management and personnel. All officials and field personnel are expected to be from Surau Gadang residents and have representation from all RW.
- Legal and Regulatory Aspects. Law No. 18 of 2008 concerning waste includes material needed in national waste management.
- Financing Aspect. To manage the system, investment costs from the government are needed, while the operational and maintenance costs are obtained from waste levies and the sale of waste processing results from TPST.
- Aspects of Community Participation. Various forms of community participation in Administrative Village of Surau Gadang in managing the expected management include the following:
  - Community participation in the implementation of the planned system;
  - Community participation in the construction or supervision of the construction of processing facilities;
  - Participating in institutions;
  - Participating in financing (payment of levies);
  - Participate in maintaining and cleaning roads and channels
  - Participating in the separation, provision of garbage containers, collection and processing;
  - Community participation in reducing overall waste generation

4.2. Detail Design

Details of community-based system development in Surau Gadang using the 3R method created using predetermined criteria and criteria in general. Detail design cover technical aspects and non-technical aspects (institutional, management, law and regulation). Management ends at TPS (SNI 20-3242-2008). Table 5 Potential in Surau Gadang area, concrete based on the type of waste.

Table 5. Potential of Waste in Administrative Village of Surau Gadang

Composition	Percentage (%)	Waste generation (m^3/d)
Waste generation projection		64.18
Organic		
Food waste	75.14	48.23
Paper	4.43	2.84
Plastic	8.41	5.40
Rubber	0.40	0.26
Yard junk	6.37	4.09
Total Organic	94.76	60.81
Inorganic		
Glass	0.61	0.39
Metal	0.56	0.36
Total Inorganic	1.17	0.75
Others	4.07	2.61

Regional plans in all RW in Surau Gadang area with a service level of 100% of the total population. For non-residential areas scattered in the Surau Gadang area is managed (done) by the Badan Usaha Ikhlas, which already exists there. The total is 22 RW (RW I to RW XXII). The amount of waste served can be seen in Table 6.

Table 6. Amount of Waste in Administrative Village of surau gadang

Description	Unit	Amount
Total population	people	20704
Solid waste generation unit	l/p/d	3.10
Solid waste generation	m^3/d	64.18

4.2.1 Technical Aspects of Operations. The Party that is categorized based on the source of waste and the source of waste is classified by type of waste. More details can be seen in the following Tables 7 and 8.

Table 7. Waste Management

Description	Classification
Residential	Society
Non-residential	Badan Usaha Ikhlas

Separation and Sorting, through the following stages:

a. Residential area
   - Community / source of waste separates its waste into 3 types of separation, namely organic waste, plastic, paper, and other inorganic waste;
   - Waste is placed in each container;
   - Separate waste is collected individually not directly to TPST.

b. Non-residential area
   - Provide garbage containers in each source of waste according to its type;
   - Separate waste is collected individually indirectly;
   - Demolition of garbage from pickups owned by the Business Entity at the nearest polling station.

The storage system is planned to provide 3 different types of containers, i.e.:

- Green containers for wet / decaying organic waste.
- Plastic and paper yellow containers
- Metal, glass, textile, rubber and other red containers.

Table 8. Source of Waste

Source	Classification	Type
Residential	Kitchen and yard	Food waste, paper, plastic, metal, glass, rubber, textile, wood leaves, and others.
Non-residential		
Commercial	Restaurant, market dan shop/kiosk	Food waste, fruit and vegetable skins, paper, plastic, metal, glass, rubber, textiles, etc
Institution	Office, school, public health centre mosque & *musholla*, others	Food waste, paper, plastic, metal, glass, rubber, textiles, leaves, wood, medical, and others.

The planned storage system can be seen in Table 9. Placement of TPST is planned in RW I (based on consideration of accessibility, land availability and collection costs incurred). Land is land owned by the local government.

Table 9. Plan for the Storage Systems

Source	Storage pattern	Type of storage	Placement	Containers	Provided by
Residential	Individual	Varied size plastic bags, plastic baskets (10-60 L), closed plastic bins (10-60 L).	On the front yard or kitchen	4141 unit	Community participation
Non-residential					
Commercial	Individual	Sacks / plastic bags 10-60 L, plastic bin covered 10-60 L, except for market waste used containers 6 m3.	In the shop / kiosk yard, kitchen / restaurant yard, in a strategic place at the market location	276 unit	Community participation and government
Institution	Individual	Varied size sacks / plastic bags, 10-60 L closed plastic bin and 200-240 L closed plastic bin.	The main road or other strategic location	201 unit	Community participation and non-governmental
TPST (residue)	Communal	Container 6 m^3 with lid.	Residue area (TPS)	2 unit	Government or administrator

The pattern of collection from waste sources is done individually indirectly, that is:

- Door to door housing waste is scheduled to start at 06.00 in the morning using 9 units of motorized trash tricycles with a capacity of 1.5 m^3. The collection gathering is 4 rites per day.
- Non-residential waste will be collected using pickup trucks belonging to the Badan Usaha Ikhlas. The collection path will be planned by the business entity itself. Waste collected will be taken to the polling station.

After knowing the amount of solid waste based on the composition, the waste processing potential of Surau Gadang area is obtained as follows.

Table 10. Potential of Waste Management in Administrative Village of Surau Gadang

Type of waste	Weight (kg/d)	Potency	Processed (kg/d)	Residue (kg/d)
Food (TPST)	6405.75	84.01	5381.37	1024.28
Food (Household)	1601.44	84.01	1345.37	256.07
Yard (TPST)	543.05	94.73	514.43	28.62
Yard (Household)	135.76	94.73	128.61	7.15
Plastic	896.20	92.4	830.24	65.96
Paper	472.08	84.20	397.49	74.59
Metal	59.68	87.80	52.40	7.28
Glass	65.00	64.23	41.75	23.25
Rubber	42.63	0.00	0.00	42.63
Others	433.71	0.00	0.00	433.71
Total	10655.29		8691.75	1963.54

The potential level of reduction in the amount of waste that might be achieved is 81.57% with residues to be disposed to TPS as much as 18.43% (11.83 m^3 / d). The transport of residues from polling stations is adjusted to the garbage transportation schedule by DKP, which is carried out periodically. The type of TPST used is type III TPST (SNI 20-3242-2008). Plans for facilities and infrastructure at the TPST include collection and transportation facilities; can be seen in Table 11.

Table 11. Plan for Facilities and Infrastructure

Infrastructure	Capacity	Amount
Collector / transporter: motorized insulated trash tricycles	1.5 m^3	9 unit
a. TPST Land area	480 m^2	
b. TPST Building area	254.6 m^2	
c. Organic chopper machine	1500 kg/hour	1 unit
d. Compost sieving machine	2000 kg/hour	1 unit
e. Composter	550 liters	308 unit
f. Residue container/TPS	6 m^3	2 unit
g. Shovel and hoe		10 /unit
h. Sacks		360 unit
i. Masks and gloves		30/unit
j. Stationery, tables and chairs		1 package

After calculation, the TPST area is 254.6 m2 which is divided into 9 areas, namely:

- Composting area;
- Compost enumeration and sieving area;
- Sorting area;
- Office and toilet;
- Warehouse equipment and garage motorized trash tricycles;
- Warehouse resulting from compost processing;
- Warehouse as a result of processing inorganic waste;
- Storage area for residue.

4.2.2 Non-Technical Aspects. The area's waste management agency is proposed to have a Work Letter (SK) from the Administrative Village of Surau Gadang officials or the regional government to have clearer laws. Plans for increasing community participation are carried out in stages (for 1 year). The form of the managerial organization structure can be seen in Figure 2 below.

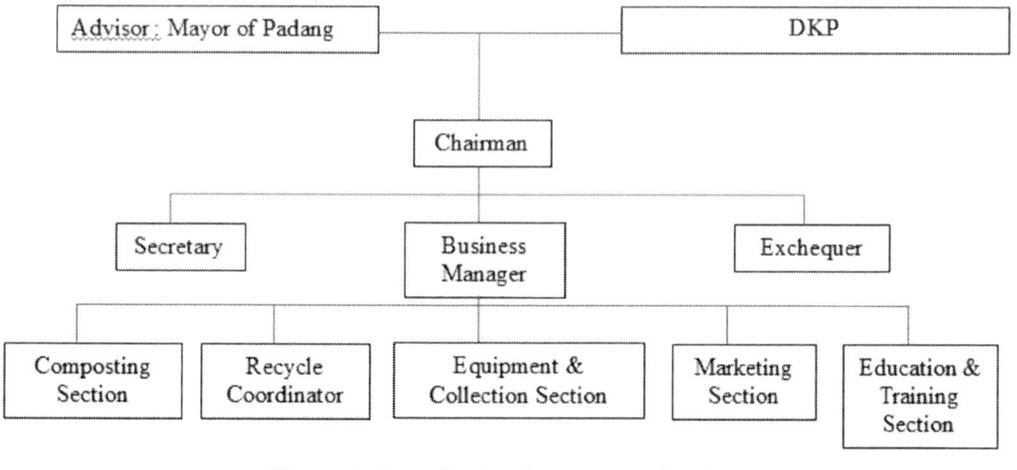

Figure 2. Organizational Structure Planning

Legal aspects and very necessary, with the existence of various regulations will result in the implementation of the system having clear and firm boundaries. Regulations need to be implemented by all parties so that waste management activities can run as optimally as possible. Law No. 18 of 2018 concerning waste includes material needed in national waste management.

Investment costs are expected to come from financing for the implementation of the 3R waste management program from provincial service. The types of investment costs required are as follows:

- Procurement of land and procurement of buildings intended for TPST and facilities as well as procurement of equipment.
- Procurement of operational facilities which include communal refinement, collection and transportation (motorized trash tricycles) and facilities for processing.
- Operational and maintenance costs are carried out independently. Funds are obtained from sales of processing and withdrawal fees by the city government. This fee is grouped into:
  - Labour costs / employee wages (basic salary, allowances and work clothes);
  - The cost of fuel and oil;
  - Maintenance costs for equipment, such as motorized trash tricycles, chopper machines and so on.

Community participation is very necessary in supporting the waste management system. Type of role has been described previously. The program for raising awareness and education is designed to increase public understanding and awareness of waste management. One method is to conduct education and training. Training is a process of debriefing the choice of technology used, a means of socialization and advocacy of citizens' concerns about the program. Submission of material can work with community organizations such as PKK (organization for family welfare development), Youth Organization, community leaders, universities and the school. Training material is general and technical in nature. Other efforts are also carried out, such as:

- Carry out a campaign to reduce and recycle waste;
- Distribution of media kits (brochures, leaflets, posters, guides / banners);
- The process of socialization can be through social gathering forums, community meetings, holding related competitions, and other activities.

4.3. Technical Specifications

Provision of needed equipment needs to be planned because it deals with operational and maintenance costs. In the selection and determination of the equipment used it is adjusted to the needs of technical and economic aspects. The types of equipment and supporting facilities proposed include:

- Sacks and plastic bags are used sacks and plastic bags made of 10-60 liters of polystyrene.
- Waste basket is a three-color basket made of plastic, with a size of 10-60 liters. In the use of parts in baskets coated with used plastic.
- Plastic bin used is a bin covered with three colors, made of plastic fiberglass, with a size of 10-60liter and 200-240 liters.
- The container used is made of metal, with an insulation size of 6 m^3.
- The motorized trash tricycles used are made of iron with a size of 1.5 m^3, with insulation. Trash tricycles are equipped with mechanical levers.
- Takakura composter with a capacity of 60 liters for household scale needs.
- Composter 550 liters for regional scale in TPST.
- Organic waste counting machine with a capacity of 1500 kg / hour. Fuel needs 1.47 liters / day and operates for 3.93 hours.
- Compost sieving machine with a capacity of 2000 kg / hour, the required fuel is 0.37 liters with usage for 0.98 hours / day.
- TPST has an area of 254.6 m^2 intended for composting areas; enumeration and sieving; sorting area; office and toilet; equipment warehouse and motorized trash tricycles, compost processing warehouse; warehouse for processing inorganic waste; and the area for residue. The building is built using river stone foundation, zinc roof, light steel frame and walls.

4.4. Budget Plan
The budget plan includes the calculation of the cost of implementing a waste management system. Determination of the amount of the budget plan is based on the performance unit. The cost component consists of investment costs, operational and maintenance costs and administrative and operational personnel costs.

- Calculation of investment costs includes the cost of building TPST worth Rp. 744,000,000 and the purchase of machinery and equipment worth Rp. 578,000,000. The machines needed are organic counting machines, compost sieving machines, and motorized trash tricycles. The equipment needed by composter, container, shovel, hoe, mask, and other supporting equipment.
- Operational costs include costs for labor, material and fuel costs. Costs that will be needed are as much as Rp. 311,716,799.
- Maintenance costs consist of maintenance costs for TPST, machinery and collecting equipment, composter, and maintenance of containers. The costs to be incurred are Rp. 8,420,000.
- Total expenditure is the sum of operational costs and maintenance, amounting to Rp. 26,678,066 / month.
- Analysis of unit costs with the amount of processed waste 9.18 tons / day is 95.30 / kg.
- The estimated source of income comes from levies drawn by the government and sales of waste processing results. The income from retribution withdrawal is Rp. 10,352,000 / month and from processing and selling sellable waste for Rp. 118,443,350 / month.
- Then the estimated profit earned per month is Rp. 102,117,284.
- Break-even point is obtained in the second month of operation.

5. Discussion
Until now, waste management with the 3 R concept is an alternative effort for waste management in settlements. With the integration among the community, the government and the private sector are expected to manage waste with this paradigm. It can run more effectively and efficiently. Considering that until now there are still limitations on the part of the government (management) in covering all service areas, as well as the increase in waste generation which is increasing, resulting in increasingly limited capacity in the landfill.

With the existence of waste management with a new paradigm, it is hoped that it can provide a win-win solution for all three parties; community, government and private parties in this area. The community can be more independent in managing their waste, and can generate financial benefits from the processing of waste. Public awareness is also expected to increase with the extension and availability of facilities. With the availability of TPST it is also a place for the community to channel creativity, establish ties and enhance the spirit of mutual cooperation. Government burden is also helped, with the smaller service area and the amount of waste that must be transported to the landfill, with other benefits of operational costs and landfill investment can be minimized. The private sector in this area also participated in the transportation of waste to polling stations. Counseling and increasing public awareness must always be improved, so that the waste management using the 3R method becomes a lifestyle and something that cannot be separated in daily life, and does not prioritize mere financial benefits. The government also constantly monitors the sustainability of TPST operations by providing training, allocating funds, and providing needed assistance.

Based on the results of the questionnaire, it was also found that the community wanted to participate and play an active role in waste management with the 3R method, only a small proportion of the people who submitted their objections. This is due to the unavailability of facilities, and a minimal understanding of waste management using the 3R and community-based methods. It is hoped that with the prioritization of the Surau Gadang urban village to become one of the regions that applies the regional scale waste management system, the attention and support from the community and the government can be more significant.

6. Conclusions

Based on the planning carried out, the following conclusions are obtained:

a. The community-based solid waste management system in Surau Gadang is planned to use the 3R method with a design period of 8 years. Where the number of underserved populations at the end of the design period (2018) is 20,704 people.
b. Management starts from the source. In disaggregated garbage housing areas are collected door to door by using motorized trash tricycles to TPST. The processing carried out at the source is composting with the Takakura method (high income housing), while the processing carried out at TPST is composting with the method of Bokhasi and sorting of other wastes for sale, recycling and making skills. Residue from the remaining processing that will be disposed of to TPST is 18.43%. Management is carried out by the community. Whereas in the non-residential area, disaggregated waste will be collected towards the TPS by Badan Usaha Ikhlas.
c. Financing consists of investment costs and maintenance operations. Investment comes from the government while financing for operations and maintenance comes from waste retribution from the community and the benefits of processing waste at TPST. The cost of waste management in this area includes an investment fee of Rp. 1,322,000,000 operational and maintenance costs of Rp. 26,678,066 / month and the estimated profit from this waste management is Rp. 102,117,284 / month.
d. Community participation can be seen from the involvement of citizens, namely participating in the construction or supervision of the construction of the TPST, participating in TPST institutions and personnel, separating and sorting waste at the source and processing, paying waste retribution, engaging in processing at TPST, maintaining cleaning of roads and channels, and participating in socialization and training.

References

[1] N. Artiningsih, "Community Participation in Household Waste Management," Semarang, 2008.
[2] Environmental Service Program, "Community Based Waste Management Training Module. Development Alternatives, Inc. for USAID, "Medan, Bandung, Subang and Surabaya, 2006.
[3] Department of Public Works, "Procedures for Waste Management in Settlements, SK-SNI-3242-2008," Jakarta, 2006.
[4] E. Damanhuri, "Final Disposal Technique," Bandung: ITB, 2004
[5] S. Alhusna, "The Generation Unit, Composition and Potential for Domestic Waste Recycling in Padang City," Padang: Andalas University, 2009.

Study of mercury concentration in plants in Traditional Buladu Gold Mining

M Mahmud[1,*], F Lihawa[2], Y Saleh[2], F Desei[1] and B Banteng[1]

[1] Faculty of Engineering, Gorontalo State University, Jenderal Sudirman Street No. 6, Gorontalo City, Indonesia
[2] PSLK, Gorontalo State University, Jenderal Sudirman Street No. 6, Gorontalo City, Indonesia

*marikemahmud@yahoo.com

Abstract. The objective of the study was to examine the levels of mercury concentrations in plants due to the traditional gold mining of Buladu, Sumalata District, Gorontalo Regency. Buladu gold mining was located in Sumalata District, North Gorontalo Regency, Gorontalo Province. Plant samples were taken in 3 locations, namely plants living along the Hulawa River, living near to the tailings and those in residential areas. The number of plants that became the samples along the Hulawa River were 13 samples, 5 samples in the tailings, and 16 samples in residential areas. The total number of plants were as many as 34 samples. The samplings were taken from the roots and leaves so that there were a total of 68 samples. Plant analysis was carried out using mercury analyzer at UGM Integrated Research and Testing Laboratory. The quality standard used as a reference for plants employed the Decree of the Director General of National Agency of Drug and Food Control No: 03725 / B / SK / VII / 89, where mercury levels could not exceed 0.5 mg / kg. Data analysis used tables and graphs and was subsequently interpreted. The results showed that the concentration of mercury in the leaves of plants living along the Hulawa River ranged from 0.00142 mg / kg to 0.41617 mg / l and that in the root ranged from <0.00014 - 14.890 mg / kg. Mercury concentrations in plants living around the tailings in leaves ranged between <0.00014 - 1.30822 and at roots ranging from 0.01058 - 12.59366. In residential areas, the concentration of mercury in leaves ranged from <0.00014-1.44368 mg / kg and in the root ranged from <0.00014-1.94505. Based on these results, the concentration of mercury in plants along the river, around the tailings and in the people's resident area tended to be above the quality standards set by the Decree of the Director General of National Agency of Drug and Food Control No: 03725 / B / SK / VII / 89.

1. Introduction

Buladu traditional gold mining is located in Sumalata District, North Gorontalo Regency. Gold exploration activities in Buladu area by the Dutch Government which began in the Dutch Colonialization era (18th century). There is historical evidence in this area including 3 Dutch graves on Buladu Beach died in 1899, mine pits with rails and lorries, gold ore processing equipment in the form of large sized pots, and solid tailings found around the mine site. Around the 1970s, the exploitation activity was continued with the traditional mining model. The location of the mining was reopened by the local community. At that time gold search activities were carried out traditionally by drilling sand and rock deposits along the Buladu River [1].

The impact of the mining activities can be positive for the mining business area. However, mining activities can be negative towards the local ecosystem. The emergence of positive and negative impacts from mining business occurs during the exploration and exploitation phase including the processing and sale of mining products and post-mining. Mining business known as small-scale gold mining is considered to be the cause of environmental damage. Ore processing is carried out with an amalgamation process in which mercury (Hg) is used as a binding medium for the gold [2].

In this area, the livelihoods of many people depend on mining. The results showed that 70% of people living in locations around mining were living as miners [3]. The people who lived as miners show that 75% graduated from elementary school, 5% from junior high school and 20% from high school. The low level of education caused people in the Buladu gold mine did not manage waste residues properly. A low level of education also triggers people to depend on their life in traditional mining. A low level of education causes them not to be aware of the dangers of mercury poisoning for their bodies [3]. This triggers the public to throw away the waste carelessly. Communities in the mining location are not aware of the dangers of wastewater being discharged into the environment which will be absorbed into the soil and eventually be absorbed by plant roots.

The first plant's main structure is the root, known by its scientific name radix. Roots have less important role than leaves and stems. The function of the root is to absorb water and nutrients, which will then be transmitted to the stem and leaves, resulting in a metabolic process. The nature of the roots is generally contrary to the nature of the stem, among others, is to grow away from the center of the earth, which is known as the term positive geostrophic or go to a water source and grow away from the light [4]. The nature of the roots goes to the water source, causing the roots to absorb liquid waste that is absorbed into the soil. This will cause the roots to accumulate heavy metals from mining waste. The process of gold mining at Buladu gold mining used mercury in the processing steps. The remaining mercury waste was wasted into the environment, especially the soil. It entered to the soil and could eventually be absorbed by plants. This would be dangerous because if the plants accumulated mercury, it could enter humans through food chain relationships. The purpose of this study was to identify mercury concentrations in various plants that lived in the traditional mining area of Buladu, Sumalata District, North Gorontalo Regency.

2. Research Methods

The location of the study was in Buladu gold mining in Sumalata District, North Gorontalo District, Gorontalo Province. Plant samples were taken in 3 locations, namely plants that lived along the Hulawa River, lived close to tailings and lived in residential areas. Sampling location was shown in Figure 1.

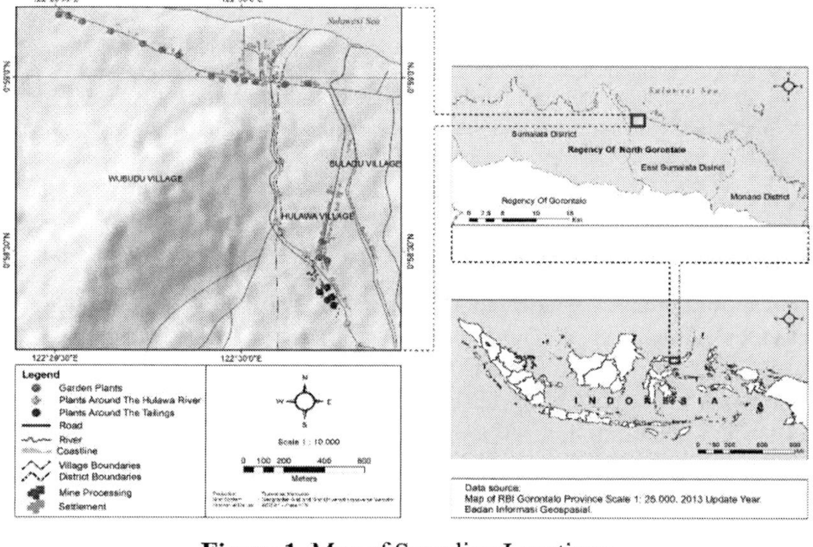

Figure 1. Map of Sampling Locations

The number of sample plants that lived along the Hulawa River were 13 samples, 5 samples in tailings and in 16 samples in residential areas. The total number of plants were as many as 34 samples. Samplings were carried out on the roots and leaves so that it obtained a total of 68 samples. Plant analysis was carried out using mercury analyzer at UGM Integrated Research and Testing Laboratory. The quality standard used as a reference for plants employed the Decree of the Director General of National Agency of Drug and Food Control No: 03725 / B / SK / VII / 89, in which mercury levels cannot exceed 0.5 mg / kg. Data analysis applied tables and graphs and was subsequently analyzed.

3. Result and Discussion

3.1. Mercury concentration in plants along the Hulawa River

3.1.1. Mercury Concentration at Plant Roots. The analysis results of the highest mercury concentration in the roots of plants that lived along the river obtained a value of 14.890 mg / kg, on needle grass plants (*andpagon aciculatus*). The lowest value showed <0.00014 mg / kg in taro plants (*Colocasia esculenta*). The average value of mercury concentration in plant roots along the Hulawa River was 2.387 mg / kg. This result was above the quality standard set by the Decree of the Director General of National Agency of Drug and Food Control No: 03725 / B / SK / VII / 89 that should not be exceeding 0.5 mg / kg. Some types of plants such as *suplir* (*adiantum cuneatum*), grass plants (*andpagon aciculatus*) and taro (*Colocasia esculenta L*) had concentrations above the established quality standards. The analysis results of the mercury concentrations in plant roots around the river were shown in Table 1.

Table 1. Concentration of Mercury in the Roots of Plants Living around the Hulawa River

No	Names	Result mg/kg	Quality Standard mg/kg
1	Suplir *(Adiantum cuneatum)*	0.88752	0.5
2	Taro *(Colocasiaesculenta)*	1.37442	0.5
3	Grass (*Andpagon aciculatus*)	14.890	0.5
4	Needle grass (*Andpagon aciculatus*)	5.11985	0.5
5	*Tapak Dara* (*Catharanthus roseus L.G.Don*)	0.06701	0.5
6	*Suplir (Adiantum cuneatum)*	1.34	0.5
7	Taro *(Colocasia esculenta L)*	3.88	0.5
8	*Suplir (Adiantum cuneatum)*	1.69	0.5
9	*Suplir* (*Pterydophyta)/filiciinae)*	0.65601	0.5
10	Taro *(colocasia esculenta(L) Schott)*	0.82258	0.5
11	*Tapak Dara* (*Catharanthus roseus L.G.Don*).	0.26817	0.5
12	Taro (*Colocasia Esculenta*)	<0.00014	0.5
13	Water hyacinth (*Eichhornia crassipes)*	0.03647	0.5

Source: Primary data (2016)

Kosegeran, [5] conducted research on plants that lived in Talawaan Gold Mining. Samples were taken in 3 locations as many as 9 samples of ferns and 3 samples of the soil. The results showed that ferns did not contain mercury, while the soil from 3 locations contained 0.6 ppm mercury. This result was lower compared to research conducted at Buladu gold mining. Plants in the Buladu gold mining showed that the *suplir* type (*adiantum cuneatum*) had mercury concentrations of 0.88752 - 1.69 mg / kg, grasses (*andpagon aciculatus*) ranging from 5.11985-14.890 mg / kg, and taro (*colocasia esculenta*) ranging from 0.00014-3.88 mg / kg. This result was already very high because it was already above the established quality standard. The increase of mercury concentrations in plants was caused by high concentrations of mercury in sediments. The concentration of mercury in the sediments along the Hulawa River at 3 sampling ranged between 10.8731-55.0680 mg / kg; 0.08995 - 136.70 mg / kg and 0.55255 - 244.16 mg / kg [3]. There was a relationship between elevation of mercury in plants and high concentrations of mercury in sediments. The higher the concentration of mercury in the growing medium, the higher it will be in plants. This result was same as the research conducted by Lona, et al [6]. Research conducted by Lona et al (6) showed that mercury absorbed by roots ranged from 0.112 to

0.2997 mg / g and leaves ranged from 0.0221 to 0.0287 mg / g. The higher the addition of metals to the growing medium, the higher the concentration of mercury accumulated by the organ of the *seruni* plant. Plants can absorb pollutants from a medium so that they accumulate around the roots of plants. Plant roots will absorb pollutants together with the absorption of nutrients and minerals in the media and then be translocated to the leaves [7,8].

Based on these results, mercury pollution in plants at the Buladu gold mining site was already very high and endangering the community because mercury could enter to human body through food chain. Plants get nutrients for the growth needs from the surrounding environment by absorption of the roots. The process of absorbing nutrients and minerals into plants through an active transport mechanism occurs between cells, where nutrients are absorbed in the form of elements or compounds.

The results of research conducted on plants showed that the highest accumulation of pollutants was in locations 0-500 m from pollutant sources [9]. The amount of heavy metal taken by plants from the soil is determined by the availability of pollutant materials. The more pollutant content in the soil, the more easily it is absorbed by the roots. The results of the analysis of mercury concentration in the leaves of plants that live around the river are shown in Table 2.

Table 2. Analysis Results of Mercury Concentration on Leaves around the Hulawa River

No	Plant's Name	Results mg/kg	Quality Standard mg/kg
1	Fern plants *(Pterydophyta) /filiciinae)*	0.03550	0.5
2	Taro *(colocasia esculenta (L) Schott)*	0.09744	0.5
3	Needle grass (*Andropagon aciculatus)*	0.20938	0.5
4	Needle grass (*Andpagon aciculatus*)	0.41617	0.5
5	*Tapak Dara* (*Catharanthus roseus L.G.Don*)	0.05923	0.5
6	Fern plants (*Pterydophyta)/filiciinae)*	0.02168	0.5
7	Taro (*Colocasia Esculenta*)	0.12209	0.5
8	*Suplir* (*Pterydophyta)/filiciinae)*	0.06983	0.5
9	*Suplir* (*Pterydophyta)/filiciinae)*	0.01194	0.5
10	Taro *(colocasia esculenta(L) Schott)*	0.431	0.5
11	*Tapak dara* (*Catharanthus roseus L.G.Don*)	0.02917	0.5
12	Taro (*Colocasia Esculenta*)	0.06377	0.5
13	Water hyacinth (*Eichhornia crassipes)*	0.00142	0.5

Source: Primary data (2016)

The analysis results showed that mercury concentrations were higher in roots than in leaves. The concentration of mercury was very high in plants due to the residual processing of gold being disposed of in the treatment tank. If it is full, it will overflow in the surrounding soil and eventually flow and enter the river. It caused the plants around to be contaminated with mercury. The tendency of mercury accumulation in the roots was higher than the leaves because the root was a part that functions to absorb nutrients from functioning plants and organs that had a direct contact with the planting medium. The absorption results were obtained in the form of elements of mercury to be translocated to other parts which are leaves.

3.1.2. Analysis results of mercury concentration on plant roots around tailings. The analysis results of mercury concentration in plant roots that lived around the tailings showed that the highest concentration was 12.59366 mg / kg in green taro plants (*Colocasia Esculenta*) and the lowest was 0.01058 mg / kg in needle grass plants (*Andpagon aciculatus*). Based on these results, mercury concentration in plants had exceeded the limit set by the Director General of National Agency of Drug and Food Control No. 03725 / B / SK / VII / 89. The analysis results of mercury concentrations in plant roots that lived around tailings can be seen in Table 3.

Table 3. Analysis Results of Mercury Concentration in Roots of Plants that Lived Near Tailings

No	Names	Analysis Result mg/kg	Quality Standard mg/kg
1	*Suplir (Adiantum cuneatum)*	2.37107	0.5
2	Taro (*Colocasia Esculenta*)	12.59366	0.5
3	Needle grass (*Andpagon aciculatus*)	7.35	0.5
4	Soft *suplir* (*Pteridium aquallinum)*	1.73	0.5
5	Needle grass (*Andpagon aciculatus*)	0.01058	0.5

Source: Primary data (2016)

The high concentration of mercury in plants around tailings is related to the high concentration of mercury in sediments. Mercury concentrations in tailings sediments in sampling I, II and III ranged from 10.0643 - 36.4008 mg / kg, 3.31 - 135.55 mg / kg and 104.36 - 236 mg / kg [3]. One of the functions of the root is to absorb water and nutrients from the soil, so the concentration of mercury will be high in the roots.

Direct tailing disposal to the soil without any treatmnet caused the soil to be contaminated with mercury so that there was a possibility of mercury accumulation in the surrounding food plants. In addition, it could also cause the infiltration of Hg to ground water used by residents as a source of clean water. Mercury is a cumulative poison, in the sense that a small amount of mercury is absorbed in the body for a long time will cause danger. The dangers of diseases caused by mercury compounds include damage to hair and teeth, loss of memory and disruption of the system requirements [2].

3.1.3. Analysis results of mercury concentration on plant leaves around tailings. The analysis results of mercury concentration in the leaves of plants that lived around the tailings showed that the highest concentration was 1.30822 mg / kg on needle grass (*Andpagon aciculatus*) and the lowest was 0.00014 mg / kg in taro (*Colocasia Esculenta*). This result was above the quality standard set by the Decree of the Director General of National Agency of Drug and Food Control No: 03725 / B / SK / VII / 89. Based on the decree of the Director General of National Agency of Drug and Food Control No. 03725 / B / SK / VII / 89, the concentration of plants cannot exceed 0.5 mg / l. The analysis results of mercury concentration in leaves living around the tailings can be seen in Table 4.

Table 4. Analysis Results of Mercury Concentration on Leaves of Plants that Lived Near Tailings

No	Names	Analysis Result mg/kg	Quality Standard mg/kg
1	*Suplir ((Adiantum cuneatum)*	0.51563	0.5
2	Taro (*Colocasia Esculenta*)	<0.00014	0.5
3	Needle grass (*Andpagon aciculatus*)	1.30822	0.5
4	*Suplir* (*Pteridium aquallinum)*	0.17754	0.5
5	Needle grass (*Andpagon aciculatus*)	<0.00014	0.5

The analysis results of mercury concentrations in plant roots that lived around the tailings ranged from 0.01058-12.59366 mg / kg higher than the mercury concentration in the leaves of plants ranging from 0.00014-1.30822 mg / kg. Eckenfelder [10] stated that the ability of absorption and accumulation of heavy metals by plants is divided into 3 processes, namely the absorption of heavy metal precipitates by the roots, precipitation of mercury pollutants in the soil mobilized by plant roots by accumulation, and absorption by the root surface and deposited in the root zone. After from the root, mercury is translocated towards certain plant organs such as the stem and leaves and then locate metals in certain tissues to keep them from inhibiting the plant metabolism [11].

3.1.4. Analysis results of mercury concentration on plant roots around settlement area. The analysis results showed that the highest mercury concentration was in the roots of *pandanus* plants with 1.94505 mg / kg (*Pandanus Amarylilfolius*). The lowest concentration was in mayyana medicinal plants (*Coleus Atropurpureus benth*) with <0.00014 mg / kg. Several types of garden plants consumed by the community had been above the quality standard, namely lemongrass (*Cymbopogon citratus*), galangal (*Alpinia galanga*), water spinach (*Ipomea Reptana Poir*), *Pandan* (*Pandanus Amarylilfolius*), and curcuma (*Curcuma Zanthorrhiza*). It will endanger humans who consumed it. The analysis results of mercury concentrations in the roots that lived in the resident yard can be seen in Table 5.

Table 5. Analysis Results of Mercury Concentration on Plant Roots in the resident yard

No	Names	Analysis Result mg/kg	Quality Standard Mg/kg
1	*Pandan* (*Pandanus Amarylilfolius*)	1.94505	0.5
2	Lemongrass ((*Cymbopogon citratus*)	1.21653	0.5
3	Galangal (*Alpinia galanga*)	1.09105	0.5
4	Water spinach (*Ipomea Reptana Poir*)	1.18	0.5
5	*Kumis Kucing* (*Orthosiphon Aristatus*)	0.44447	0.5
6	Maiyana (*Coleus Atropurpureus benth)*	0.04787	0.5
7	*Pandan* (*Pandanus Amarylilfolius*)	0.73919	0.5
8	*Kencur* (*Kaemferia galaga L)*	0.08194	0.5
9	Lemongrass (*Cymbopogon citratus*)	0.03934	0.5
10	Galangal *(Alpinia galangal*)	0.064	0.5
11	Water spinach (*Ipomea Reptana Poir*)	0.26518	0.5
12	Curcuma *(Curcuma Zanthorrhiza)*	1.44368	0.5
13	Rice *(Oryza sativa)*	0.01073	0.5
14	Water spinach (*Ipomea Reptana Poir*)	0.05493	0.5
15	Mayyana (*Coleus Atropurpureus benth)*	<0.00014	0.5
16	Rice *(Oryza sativa)*	0.00950	0.5

Source: 2016 Primary Data

The amount of mercury concentration in plants is directly proportional to the amount of mercury in the growing media; in this case it is the resident area. If the soil is slightly contaminated with mercury, there will be little mercury accumulated in the plant. Research conducted by Khairuddin, [12] showed that the concentration and weight of metal mercury absorbed in the roots of spinach plants is directly proportional to the medium of planting. The results of a study conducted by Santoso et al [13] in Poboya gold mining showed that the concentration of mercury in crops such as peanuts ranged from 0.098-0.45 ppm, corn 0.07-0.43 ppm, rice plants 0.03-0.21 ppm and shallots ranged 0.01 - 0.32 ppm.

This result is lower than the concentration of mercury in *pandanus* plants (*Pandanus Amarylilfolius*) with 0.73919 - 1.94505 mg / kg, lemongrass (*Cymbopogon citratus*) 0.03934-1.21653 mg / kg, galangal (*Alpinia galangal*) 0.064-1.09105, water spinach (*Ipomea Reptana Poir*) 0.05493 - 1.18, rice (*Oryza sativa*) with 0.00950 -0.01073 mg / curcuma (*Curcuma Zanthorrhiza*) 1.44368 mg / kg, mayyana 0.00014 - 0.04787 mg / kg , *kumis kucing* (*Orthosiphon Aristatus*) 0.44447 mg / kg and *kencur* (*Kaemferia galaga* L) 0.08194 mg / kg living in the resident area around Buladu mining location. These plants were planted by residents to be used daily. The high accumulation of mercury concentrations in plant roots consumed by people around Buladu mining location was very dangerous because it will endanger human health.

Soil pollution by mercury (Hg) is often related to the addition of heavy metals from various factors such as fertilizer, lime, mud and manure. The dynamics between the amount of Hg taken from the soil and plants is not linear and depends on several variables such as soil sediment content, carbon exchange capacity, content of oxides and carbonates, redox potential, formulations used and total metal content. In general, mercury uptake in plants is related to pollution levels. If the pollution level is low, the amount of mercury is below the allowable level [14]. Many plants absorb Hg which tends to be accumulated in the roots [15], and some can even accumulate in moderate amounts in the shoots [16]. Research by Suszycynsky and Shann [17] showed that plant exposure of Hg^{o} can be taken and accumulated by the shoots but not transferred to the roots [18].

3.1.5. Analysis Results of Mercury Concentration on Plant Leaves around Settlement Area. The analysis results of mercury concentrations in the leaves of plants that lived in the residential areas showed that the highest concentration was 1.44368 mg / kg and the lowest was 0.00014 mg / kg. The average rate of mercury concentration was 0.27107 mg / kg. Some plants that had been above the quality standard of the Director General of National Agency of Drug and Food Control No. 03725 / B / SK / VII / 89 are *Pandan* (*Pandanus Amarylilfolius*), galangal (*Alpinia galanga*), water spinach (*Ipomea Reptana Poir*) and curcuma (*Cymbopogon citratus*). Research conducted by Ordak, et al [19] showed that herbal plants had higher mercury concentrations in spring than in autumn. Perennial plants have high levels of mercury compared to seasonal plants (monocarpic). Therefore, usually the use of herbal plants had significantly higher levels of mercury compared to others.

Based on these results, cultivated plants consumed by people living around the mining location had been contaminated with mercury. This would endanger humans who consume these plants as food. The toxicity of mercury in plants can affect the antioxidant system, affect photosynthesis, inhibit the growth and production of plants and also affect nutrient uptake [20]. The results of the analysis of mercury concentration in the roots of plants that lived around residential areas are shown in Table 6.

Table 6. Analysis Results of Mercury Concentration on Plant Leaves in Residential Areas

No	Names	Analysis Result mg/kg	Quality Standard Mg/kg
1	*Pandan* (*Pandanus Amarylilfolius*)	0.52674	0.5
2	Lemongrass (*Cymbopogon citratus*)	0.08691	0.5
3	Galangal (*Alpinia galanga*)	0.95055	0.5
4	Water spinach (*Ipomea Reptana Poir*)	0.52858	0.5
5	*Kumis Kucing* (*Orthosiphon Aristatus*)	0.04159	0.5
6	*Maiyana* (*Coleus Atropurpureus benth)*	0.00753	0.5
7	*Pandan* (*Pandanus Amarylilfolius*)	<0.00014	0.5
8	*Kencur* (*Kaemferia galaga L)*	0.08194	0.5
9	Lemongrass ((*Cymbopogon citratus*)	0.03934	0.5
10	Galangal (*Alpinia galanga*)	0.064	0.5
11	Water spinach (*Ipomea Reptana Poir)*	0.26518	0.5
12	Curcuma/ Lemongrass (*Cymbopogon citratus*)	1.44368	0.5
13	Rice *(Oryza sativa)*	0.13247	0.5
14	Water spinach (*Ipomea Reptana Poir*)	0.09708	0.5
15	*Mayyana* (*Coleus Atropurpureus benth)*	0.06938	0.5
16	Rice *(Oryza sativa)*	<0.0014	0.5

Source: 2016 Primary Data

In absorbing heavy metals, plants form reductase enzyme in the root membrane which functions to reduce metal. Heavy metal cycles from roots show that mercury must be transported through transporting tissues, namely xylem and phloem, to all parts of the plant to improve the efficiency of transporting metals in the chelate molecule (binding molecule). In further, mercury is accumulated in all parts of the plant in the roots, stems and leaves [21,22]. Mercury contaminants are very dangerous for health risks for those who consume foods that contain mercury. The entry of mercury in the population, such as vegetables and grains that have exceeded the permitted limit can cause serious risks [23,24].

4. Conclusion

The results of the study showed that the concentration of mercury in the leaves of plants that lived along the Hulawa River ranged from 0.00142 mg / kg to 0.41617 mg / l and at the root ranged from <0.00014 - 14.890 mg / kg. Mercury concentrations in plants that lived around the tailings in leaves ranged from <0.00014 - 1.30822 and at the roots ranged from 0.01058 - 12.59366. In the residential area, the concentration of mercury in the leaves ranged from <0.00014-1.44368 mg / kg and at the root ranged from <0.00014- 1.94505. Based on these results, the concentration of mercury in plants both along the river, around the tailings and in resident areas tended to be above the quality standards set by the Decree of the Director General of National Agency of Drug and Food Control No: 03725 / B / SK / VII / 89.

Acknowledgments
Thank you to the Ministry of Research, Technology and Higher Education for providing financial assistance, and thanks to the Chancellor, the Head of the Research Institute and all parties who had assisted in the completion of the research. The author also would like to thank all the authors of the books and journals as sources for the literature review.

References
[1] BALIHRISTI, 2008. *Laporan Akhir Kegiatan Pengawasan Pelaksanaan PETI.* Provinsi Gorontalo
[2] Mallongi A 2017 *Dampak Limbah Cair dari Aktivitas Institusi dan Industri* (Yogyakarta: Gosyen Publishing)
[3] Mahmud M, Lihawa F, Desei F, Saleh Y dan Banteng B C D. 2016. Model Pengelolaan Penambangan Emas Tradisional Buladu. Laporan Penelitian PUPT. Universitas Negeri Gorontalo
[4] Rosanti D 2013 *Morfologi Tumbuhan* (Jakarta: Erlangga)
[5] Kosegeran A O, Rondonuwu S, Simbala H dan Rumondor M 2015 Kandungan merkuri pada tumbuhan paku (Diplazium Accedens Blume) di daerah tambang emas Tattelu-Talawaan. Kabupaten Minahasa Utara *Jurnal Ilmiah Sains* **15**(1) 59-65
[6] Lona L M, Linda R dan Mukarlina 2015 Pengaruh logam merkuri (Hg) terhadap pertumbuhan seruni (Wedelia trilobata L.Hitchc) *J. Protobiont.* **4**(3) 26-30
[7] ITCR 2001. Phytotechnology Technical and Regulatory Guidance Document. Technical/ Regulatory Guidelines pp 1-78
[8] Wahyuni D. 2016. *Potensi Ipomea aquatic Forsk Sebagai Agen Fitoremediasi Air Asam Tambang Batubara. Skripsi.* Jurusan Biologi. Fakultas Matematika dan Ilmu Pengetahuan Alam Universitas Sriwijaya.
[9] Heriyanto N M 2011 Kandungan logam berat pada tumbuhan, tanah, air, ikan dan udang di hutan mangrove *Jurnal Penelitian Hutan Mangrove* **8**(4) 197-205
[10] Eckenfelder W W 2003 *Industrial Water Pollution Control* (New York: McGraw Hill)
[11] Lembah V A A, Darmawan S dan Isrun 2014 Konsentrasi merkuri (Hg) dalam tanah dan jaringan tanaman kacang tanah (*Arachis hypogaea L*) akibat pemberian Bokashi Titonia (*Titonia diversifolia*) pada tailing tambang emas Poboya Kota Palu *e-J. Agrotekbis* **2**(3) 249-259
[12] Khairuddin, Sikanna R dan Sabaruddin 2017 Kajian kemampuan akar tanaman kangkung darat (Ipomea reptans Poir) dalam menyerap logam merkuri pada tanah tercemar *Jurnal Riset Kimia Kovalen* **3**(3) 303-312
[13] Santoso F J, Wahyudi H I dan Isrun 2014 Evaluasi kandungan logam berat merkuri (Hg) pada beberapa tanaman pangan palawija di sekitar areal pengolahan tambang Emas di Kelurahan Poboyo, Kota Palu *e-J. Agrotekbis* **2**(2) 138-145
[14] Patra M and Sharma A 2000 Mercury toxicity in plants *Botanical Review* **66**(3) 379-422
[15] Lenka M, Panda K K and Panda B B 1992 Monitoring and assessment of mercury pollution in the vicinity of chloralkali plant. IV. Bioconcentration of Mercury in Situ Aquatic and Terrestrial Plant at Ganjam, India *Archives of Environmental contamination and Toxicology* **22**(2) 195-202
[16] Rodriguez L, Rincon J, Asencio I and Rodrıguez-Castellanos L 2007 Capability of selected crop plants for shoot mercury accumulation from polluted soils: Phytoremediation perspectives *International Journal of Phytoremediation* **9** 1–13
[17] Suszcynsky E M and Shann J R 1995 Phytotoxicity and accumulation of mercury in tobacco subjected to different exposure routes *Environmental Toxicology and Chemistry* **14**(1) 61-67

[18] Azevedo R and Rodriguez E 2012 Phytotoxicity of mercury in plants: A review *Journal of Botany* **2012**(848614) http://dx.doi.org/10.1155/2012/848614
[19] Ordak M, Wesolosky M, Radecka I, Muszynska E and Zasdrozny M B 2016 Seasional variations of mercury levels in selected medicinal plants originating from Poland *Biol. Trace. Elem. Res.* **173**(2) 514-524
[20] Kumar B, Smita K and Flores L C 2017 Plant mediated detoxification of mercury and lead *Arabian Journal of Chemistry* **10** S2335-S2342
[21] Ghosh M and Singh S P 2005 A Review on phytoremediation of heavy metals and utilization of its byproducts *Applied Ecology and Environmental Research* **3**(1) 1-18 DOI: 10.15666/aeer/0301_001018
[22] Zulfikah, Basir M dan Isrun 2014 Konsentrasi merkuri (Hg) dalam tanah dan jaringan tanaman kangkung (Ipomea reptans) yang diberi Bokashi Kirinyu (Chromolaena odorata L) pada limbah tailing penambangan emas Poboyo Kota Palu *e-J. Agrotekbis* **2**(6) 587-595
[23] Li R, Wu H, Ding J, Fu W, Gan L and Li Y 2017 Mercury pollution in vegetables, grains and soils from areas surrounding coal-fired power plants *Scientific Reports* **7**(46545) DOI:10.1038/srep46545
[24] Yu H, Li Jing and Luan Y 2018 Meta-analysis of soil mercury accumulation by vegetables *Scientific Reports* **8**(1261) DOI:10.1038/s41598-018-19519-3

Composition of mixing gypsum additives as environmentally friendly material in tropical soil to reduce grounding resistance

Yul Martin[1,2]

[1]Electrical Engineering Department, Faculty of Engineering, University of Lampung, Indonesia
[2]Postgraduate School of Environment Science, University of Lampung, Indonesia

*yul.martin@eng.unila.ac.id

Abstract. Grounding resistance is directly proportional to the amount of soil type resistance. While soil type resistance is influenced by several things, namely soil structure, temperature, the effect of water content (moisture), and the effect of chemical content in the soil. The purpose of this study was to analyze the decrease in the value of earth resistance by adding additives in the form of gypsum without a mixture of soil and gypsum mixed with soil. The results showed that the value of grounding resistance with the addition of gypsum without soil mixtures had a higher resistance value than grounding resistance without the addition of additives. Grounding resistance with the addition of gypsum mixed with soil on average can reduce the value of grounding resistance by 153.56 ohms with 25% gypsum, 157.2 ohms with 75% gypsum and 169.91 ohms with 50% gypsum.

1. Introduction

Indonesia is located in the equatorial lane which has a high level of thunder days ranging from 100-200 days of thunder per year. The distribution of electrical power is transmitted through high voltage tranmission, then distributed through a 20-kV medium voltage to the center of the load. Construction of towers and distribution poles can be struck by lightning due to being open to nature so that lightning currents will reach a certain point of the object being struck which causes disruption of power distribution. The disturbances that occur will have the effect of reducing the voltage and stability of the power system, and will endanger humans and can damage electrical and electronic equipment. This can be maintained with a good grounding system [1].

A good grounding system has soil resistance values ranging from 1-5 ohms, but to get a low resistance value is quite difficult because it is influenced by soil type, temperature and humidity, soil electrolyte conditions and temperature. The grounding system must pay attention to several important things, namely the resistance to the grounding electrode, the resistance between the grounding electrode and the ground, the resistance from the ground around the grounding electrode. Resistance to grounding electrodes is usually ignored because the value of the electrode resistance is usually smaller than that of ground resistance [2]. The value of grounding resistance around earthed ground electrodes needs to be considered because a low soil type resistance value is needed as a support for the grounding system where the fault current will be channeled to the ground.

The type of resistivity value from the soil around the electrode is usually not directly obtained a low value, therefore to design a good grounding system it is necessary to do an assessment first on the grounding place to be used. Areas with high soil type resistance if they want to be used as grounding soil need to be treated on the soil so that the value of the earthen resistance is low. Reducing soil type resistance can be done in several ways, namely by modifying grounding electrodes to be planted in the soil and adding additives to the earthen soil [3].

In our previous research we carried out experiments using bentonite and zeolite additives as additives that can be used to reduce the soil resistivity values of various types, namely sodium chloride (NaCl), magnesium (Mg), copper sulfate ($CuSO_4.H_2O$), and calcium chloride ($CaCl_2$). [4,5] This study uses gypsum (calcium sulfate dihydrate) as an additive. Gypsum is used as an additive to reduce soil type resistance because it can absorb water and improve soil structure.

This study conducted an experiment on adding gypsum as an additive to the soil. Gypsum added to the soil was varied into four variations, namely gypsum without mixed with soil, and gypsum mixed with soil with a composition of 75% gypsum mixed with 25% soil, 50% gypsum mixed with 50% soil and 25% gypsum mixed with 75% soil. This study aims to observe the effect of adding variations in gypsum additives to a decrease in the value of soil type resistance. The expected results from this study are to obtain the best variation of additives which can reduce grounding resistance significantly

2. Literature review

2.1 Tropical Soil
Soil is a material composed of 8 granule-forming mineral elements and the largest percentage is oxygen reaching 46.6%, other elements are Si, Al, Fe, Ca, Na, K, Mg [6]. Mixing between gypsum and soil elements produces a compactness in neutralizing the resistance of stem electrode. So that this research can achieve the actual goals. The nature of the soil itself can be determined by research on the physical properties of the soil in the form of water content, volume weight, specific gravity, behavior of plasticity and elasticity in the soil and the size of its grain. The density of a soil also influences the behavior of the results of the research and compaction of the soil will cause the loss of water and air in the pore [5]. The following is given a table of the physical properties of the soil used before the soil is mixed with gypsum.

Table 1. Physical properties of soil used in research

Description		BOR 1 1,60 – 2,00
Undisturbed Sample		
Water Content	%	37,39
Density	gr/cm^3	1,47
Specific gravity (Gs)		2,57
Percent Lose No. 200	%	33,94
Atterberg Limit		
LL	%	51,33
PL	%	39,59
PI	%	11,75
Direct Shear Test		
Cohesion (c)	kg/cm	0,15
Internal Friction Angle (ϕ)	...^{0}	26,54
Consolidation		
Cv	cm^2/s	0,132
Cc		0,129

Table 1 show the water content at the time of sampling includes a high-water content. Soil samples are taken at a depth between 1.6 m and 2 meters. Taken by undisturbed tube. From the results of this study it was stated that the clay content reached 33.94% and included clay sand.

2.2 Gypsum

Gypsum is a chemical that has the formula $CaSO_4.2H_2O$. Gypsum is claimed to contain 50% to 95% $CaSO_4$ (Calcium Sulfate Dehydrate) and in natural CaSO4 there are 23.5% sulfur and 29.4% calcium. The sulfur content of gypsum does not reduce the alkalinity of the soil (Alkalinity is the property in which substances form chemical salts when combined with acid) and the calcium content in gypsum does not affect the acidity (pH) of the soil [7]. The physical form of gypsum can be seen in Figure 1.

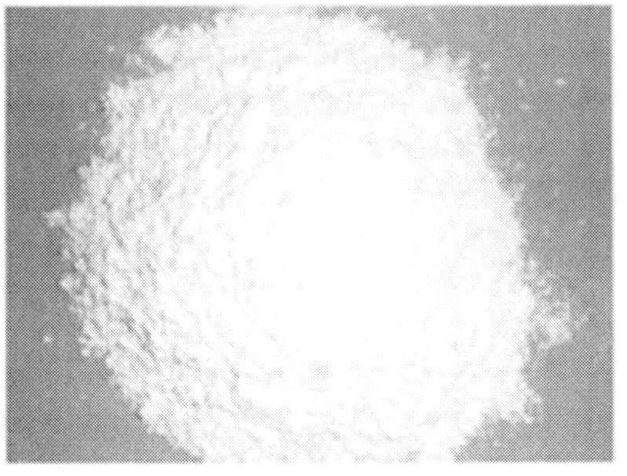

Figure 1. Gypsum powder

Gypsum as an additive used to change the chemical composition of the soil, which has a low solubility value that is not easily lost. Gypsum is a stone formed from the deposition of sea water, a mineral composed of calcium sulfate dehydrate with the chemical formula $CaSO_4.2H_2O$. Composition of Calcium Oxide (CaO) 32.57%, Water (H_2O) 20.93%, and Sulfur (S) 18.62%. This composition is expected to reduce gypsum resistance. Gypsum in the form of rock, formed because of two things, namely evaporation of ground water and deposition of groundwater itself. Climate change and oxidation reactions cause sulfur to turn into sulfuric acid in the soil containing $CaCO_3$ and forming gypsum. Gypsum is dissolved salt, hydrous calcium sulphate $CaSO_4.2H_2O$. The solubility rate is 2.6 grams of dm-3 in water at a temperature of 250 C and a pressure of 1 atmosphere. The arid and semi-arid soil causes gypsum to dissolve during the rainy season and tends to settle when the soil starts to dry out. [8].

The type of resistance (ρ) gypsum has different values depending on the surroundings. The research that has been done shows the resistance value of pure gypsum in the form of rocks in the amount of more than 800 ohm.m whereas for areas with wet soil conditions, it shows that gypsum has a high conductivity with a low resistance (close to 1 ohm.m) This is because the ions dissolved in water from material containing salt. [9]

2.3 Driven Rod Electrode

A good grounding system is very much determined by the type of soil that is good for determining the soil resistivity because not all types of soil have a good resistivity value, which is influenced by the density and contamination of the soil. To see grounding systems can be shown in various soil resistivity values for various types of soil. Table 2 shows the criteria for soil types and their resistivity values.

Table 2. Soil Type and Resistivity [10]

Soil Type	Resistivity (ohm-cm)
Clay, Garden Soil, dl	500-5000
Clay	800-5000
Mixture of clay, sand and gravel	4000-25000
sand and gravel	6000-10000
State, Rocky stone, dll	1000-50000

The results of the study in table 2 classify the criteria for soil types with resistivity values obtained in certain regions. Resistivity values are taken from measurements through samples from certain types of soil but it is possible to get results with conditions similar to samples taken.

The grounding system aims to obtain a low resistance value to channel fault current to the ground through a single rod electrode medium. Single rod electrodes are made of copper, steel profiles, which are influenced by the size, dimensions, and material of the rod electrodes. In general, rod electrodes use cylinders made of pure copper, bare copper rod and coated (copper-clad steel), stainless steel rod (stainless rod), copper wire inserted into galvanized pipe rods, copper plated steel. This study aims to determine the value of grounding resistance on clay to mixing gypsum with various compositions.

3. Experimental setup

The method used to determine the value in ground resistance by punching the ground with a depth of 1 m, where the stem electrodes are planted at a predetermined depth and diameter, then add the gypsum additive into the hole and mix it with the soil with the composition specified. Gypsum is varied before it is used as an additive in soil soils. There were five soil sources in this study, 100% gypsum, 75% gypsum, 50% gypsum, 25% gypsum and no gypsum as a comparison. Each grounding hole made in this study is illustrated as in Figure 2. The grounding electrode used is inserted into the grounding hole with a diameter of 10 centimeters and a depth of 1 meter. Next, each hole earth additive. The distance between the electrodes and the other electrodes is 50 centimeters.

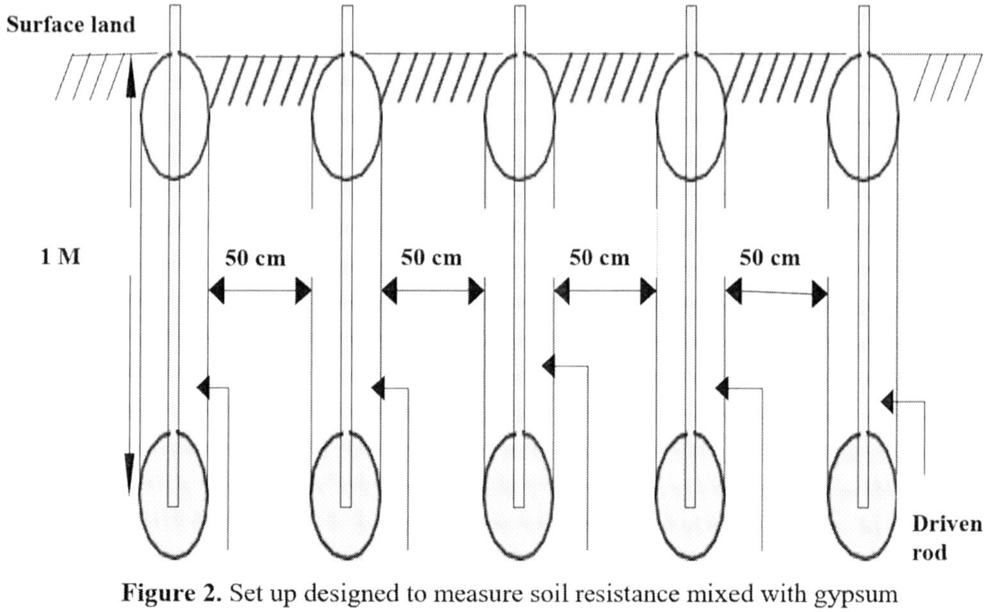

Figure 2. Set up designed to measure soil resistance mixed with gypsum

The composition of the mixture between soil and gypsum and the distance between 50 cm holes can be shown below :

1. Ground: 100%
2. Gypsum: 100%
3. Gypsum: 75%
4. Gypsum: 50%
5. Gypsum: 25%

Electrodes used in the type of electrode rod made from copper with a diameter of 0.015 m with a length of 1 m. The tip of the electrode will be planted into the ground as a point to touch the center of the electrode to the ground. To plant electrodes in a predetermined hole and be marked on holes that have been dug. There are 5 holes dug with a depth of 1 m, then hole 1 is filled with 100% pure soil without mixture and holes 2 to 5 are filled with mixing gypsum with a presentation value of 25% -75% as additives that have been smooth and not clot.

Electrodes used in the type of electrode rod made from copper with a diameter of 0.015m with a length of 1 m. The tip of the electrode will be planted into the ground as a point to touch the center of the electrode to the ground. To plant electrodes in a predetermined hole and be marked on holes that have been dug. There are 5 holes dug with a depth of 1 m, then hole 1 is filled with 100% pure soil without mixture and holes 2 to 5 are filled with mixing gypsum with a presentation value of 25% -75% as additives that have been smooth and not clot.

4. Result and discussion

This research was started by making grounding holes and continued by mixing variations of gypsum with soil with variations of 100% gypsum, 75% gypsum mixed with 25% soil, 50% gypsum mixed with 50% soil and 25% gypsum mixed with 75% soil. The mixed mixture of gypsum is put into the grounding hole with a depth of each hole of 1 m and a diameter of 10 cm.

4.1 Results of Measurement of Grounding Resistance

4.1.1 Variation of Gypsum 100%. The graph in Figure 3 shows the resistance value with the length of the soil measurement as a sample without mixture with gypsum and the measurement of soil resistance values mixed with gypsum 100%.

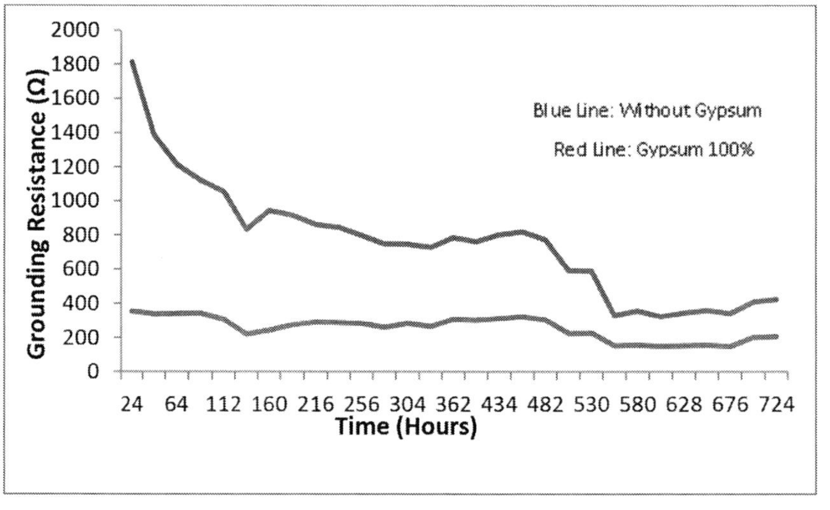

Figure 3. Grounding resistance of bentonite-mixed gypsum (100%)

Figure 3, it is seen that the value of grounding resistance with the addition of gypsum is very high at the beginning of the measurement. Gypsum density in the soil affects (decreases) the ability of gypsum to absorb water. This causes the value of grounding resistance to be high. After the location around the object of research experienced rain, grounding resistance with the addition of 100% gypsum began to decline. This decrease in the value of grounding resistance is because the gypsum in the grounding hole has absorbed the water and minerals found in the surrounding soil. The measurement data in the research from 538 hours to 732 hours showed the value of grounding resistance by adding 100% gypsum not much different from the grounding resistance without the addition of additives. This shows that grounding resistance with the addition of 100% gypsum does not differ much from the value of grounding resistance without the addition of additives

4.1.2 Variation of Gypsum 75%. In Figure 4, it can be seen that the value of grounding resistance with the addition of 75% gypsum is better than the value of grounding resistance without the addition of additives where the grounding resistance with the addition of gypsum is 75% lower than that of grounding without additives. The value of grounding resistance with the addition of 75% gypsum has decreased dramatically at the beginning of the grounding measurement to decrease with a research time of 130 hours. The decrease in the value of grounding resistance is due to the mixture of materials used as additives absorbing soil water and minerals found in the soil around the grounding hole.

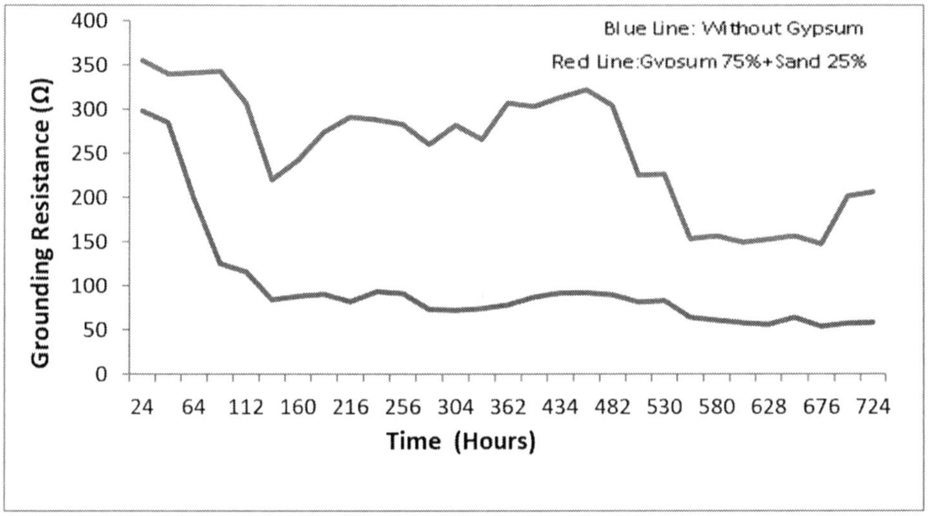

Figure 4. Grounding resistance of bentonite-mixed gypsum (75%)

4.1.3 Variation of Gypsum 50%. Figure 5, it can be seen that the value of grounding resistance with the addition of 50% gypsum is better than the value of grounding resistance without the addition of additives. This is because the mixture of 50% gypsum material and 50% soil is able to absorb water better than the soil without the addition of additives. The measured grounding resistance values from 556 hours to 732 hours of planted electrodes did not undergo significant changes. This shows that the mixture of 50% gypsum and 50% soil has reached a saturation point in reducing ground resistance where the lowest value of grounding resistance is measured at 51.8 ohms.

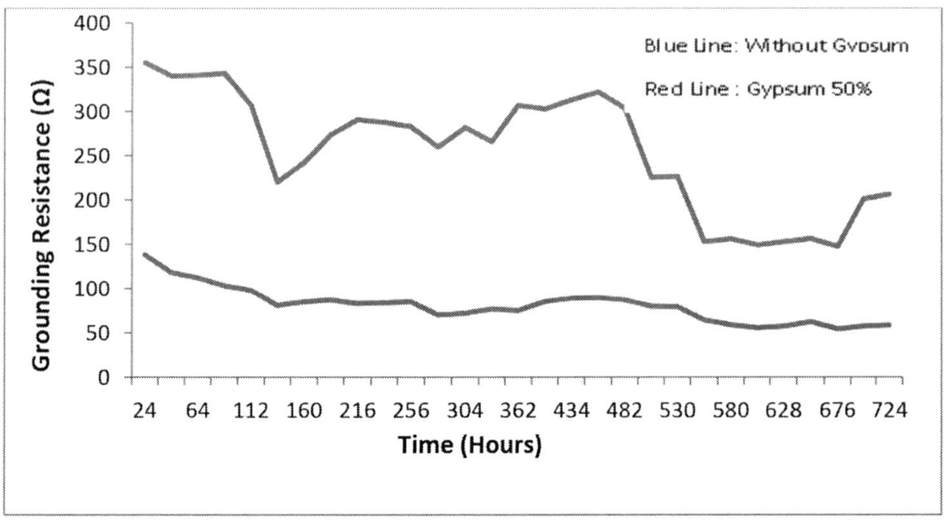

Figure 5. Grounding resistance of bentonite-mixed gypsum (50%)

4.1.4 Variation of Gypsum 25% The graph in Figure 6 below shows the value of soil resistance with a variation of 75% addition of gypsum on the ground from the resistance value on the blue line of measurement of resistance to the soil. There was a significant decrease after measurement of 24 hours as shown in the graph in figure 6. From the various measurements made with the composition of soil and gypsum showing variations in soil and gypsum 25% far more significant the results obtained based on measurements for 724 hours.

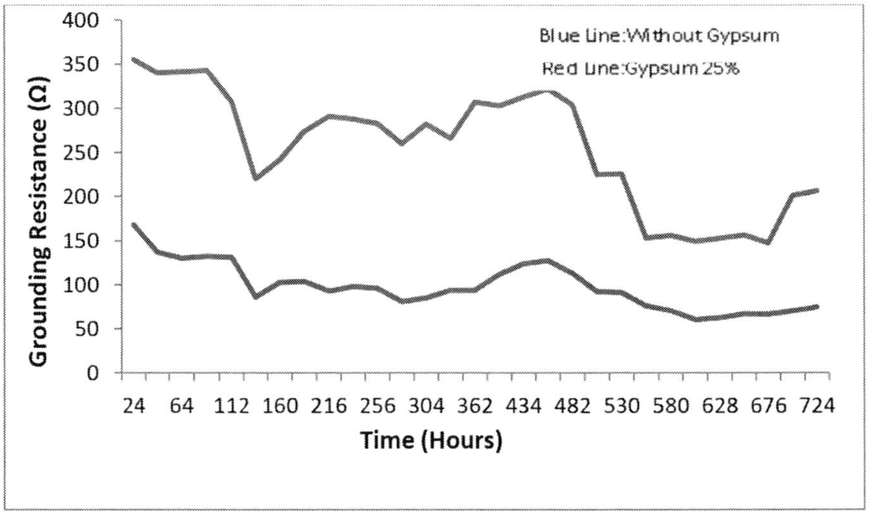

Figure 6. Grounding resistance of bentonite-mixed gypsum (25%)

The value of grounding resistance with the addition of 25% gypsum is better than the value of grounding resistance without the addition of additives where the value of grounding resistance with the addition of gypsum is 25% lower. This is due to the ability of the mixture of gypsum material 25% and soil 75% in absorbing water and minerals in the soil better, compared to soil without additives. The measurement data during the 732-hour study showed the lowest grounding resistance value obtained with the addition of 25% gypsum which was equal to 60.3 ohms. The data shows that 25% gypsum has reached its saturation point in reducing the value of grounding resistance. This is because the ability of 25% gypsum to absorb water and minerals around it has reached the maximum limit.

5. Conclusions

Based on the results of measurements of the value of earth resistance with the addition of gypsum, some conclusions can be obtained as follows:

1. Additional of gypsum as an additive to the soil can reduce the value of soil type resistance. In this study, Gypsum absorbs water and minerals contained in the soil so that the value of the resistance of its species decreases.
2. The value of grounding resistance with the addition of 50% gypsum in this study is the best among the other groundings where the addition of 50% gypsum on average can reduce the value of grounding resistance by 68.24%.
3. The value of grounding resistance with the addition of 100% gypsum has a value that is not much different from the value of grounding resistance without the addition of additives.

Acknowledgements

This work was support by grant of faculty of engineering University of Lampung.

References

[1] Zoro Reynaldo 2013 The phenomenon of lightning, Pertamina Refinary Unit (RU) IV, Cilacap, Central Java

[2] Hutauruk T S 1991 *Neutral Grounding of Equipment and Grounding Equipment* (Jakarta, Indonesia: Erlangga)
[3] G Vijayaraghavan, M Brown and M Barnes 2004 *Practical Grounding, Bonding, Shielding and Surge Protection* (London: IDC Technologies)
[4] IEEE Guide for Safety in AC Grounding Substation, IEEE std 2000
[5] Yul Martin, Samsul Bakri, Neneng K and Ade Damaria M 2018 Response substances of Bentonite Additives Activated Physics and chemistry as environmentally friendly materials to reduce the value of Grounding prisoners *National Conference on PPL UNS and BPPTPDAS.*
[6] Foth H 1990 *Fundamental of Soil Science* (New York: John Wiley & Sons Inc.)
[7] Afriani L 2014 *Soil Shear Strength* (Yogyakarta, Indonesia: Graha Ilmu Publisher)
[8] U Jones 1982 *Fertilizer and Soil Fertility second edition* (Virginia: Reston Publishing Co.)
[9] Food and Agriculture Organization of the United Nations (FAO) 1990 Management of Grypsiferous Soils, Rome
[10] A G Maysounave 2011 *Geoelectrical Characterization of Sulfate Rocks* (Barcelona: Universitat de Barcelona)

IOP Conf. Series: Earth and Environmental Science **314** (2019) 012020 doi:10.1088/1755-1315/314/1/012020

Microbial degradation of batik waste water treatment in Indonesia

Bunyamin Muchtasjar[1,*], H Hadiyanto[2] and Munifatul Izzati[3]

[1] Departement of Health Science, University of Muhammadiyah Purwokerto 53181, Indonesia
[2] Departement of Chemical Enginering, Diponegoro University, SH-Tembalang, Semarang 50239, Indonesia
[3] Departement of Biology, Faculty of Mathematics and Natural Sciences, Diponegoro University, SH-Tembalang, Semarang 50239, Indonesia

*beben_ump@yahoo.com

Abstract The existence of the batik industry is economically enough to provide a great livelihood for the community. The increasing demand of batik by the community has an impact on the growth of the batik industries in various regions in Indonesia. However, this industrial activity has a negative impact in production of liquid waste disposal. In batik production process, the liquid waste was produced by more than 95% which is emitted from process of soaking, heating and rinsing. Liquid waste has the potential to cause environmental pollution and affect the condition of living organism, including humans. One of efforts to reduce pollutants in batik waste water is by using biodegradation. Biodegradation is usually carried out by a consortium of a number of microbes. The objective of this study was to evaluate the potential of microbes for batik waste water degradation. Samples were taken from water and sludge contaminated batik waste water in the sewerage and river at Banyumas. Isolation and purification using the dilution planting method were conducted, to obtain pure culture using the pour plate technique, and maintenance of bacteria using streak culture. About 8 genus of microbes found in water bodies that have been contaminated by batik waste water are *Mesophilobacter, Methylococcus, Agrobaacterium, Neisseria, Xantobacter, Deinococcus, Sporosarcina,* and *Bacillus*. The result showed that BOD degradation in batik waste water by microbial consortium was about 85.71%. Regarding on the result, it can be concluded that bacteria found in the sewerage and river were able to degrade pollutant from batik wastewater.

1. Introduction

The existence of batik industry is economically enough to provide a great livelihood for the community. Increasing demand of batik by consumers has an impact on the growth of the batik industries in various regions in Indonesia [1]. However, this industrial activity has a negative impact due to its waste water disposal [2]. In the production process, the liquid waste produced more than 95% [3].

The waste water produced comes from the process of soaking, heating and rinsing. Waste water has potentially the environment and affect the condition of living organism, including humans. Indicators of polluted waters can be seen from changes that are easily observed, such as temperature, pH, color, odor, sediment, and the presence of certain microorganism [2,4].

Indicators used to determine polluted waste water are BOD (Biological Oxygen Demand), COD (Chemical Oxygen Demand), DO (Dissolved Oxygen), pH, nitrogen, phosphorus, heavy metals, CO_2, suspended solid material organic and inorganic, and total solids both organic and inorganic [5]. These indicators are the standard for polluted or uncontaminated waters [6].

In addition to pollution of aquatic environment, batik industry waste water can also cause odor, especially waste water in the form of liquid. The odor is due to the suspension of solids from coloring material used in the batik process. High temperatures will result in decreasing Dissolved Oxygen (DO) content in water which kill many organism so that it disturbs the balance of the water ecosystem [7,8].

The problem of managing batik waste water in several batik industry centers in Banyumas are to dispose the waste directly into the river without any treatment. Batik waste water discharged into the river without treatment has potentially polluted the environment and disrupt the life of aquatic biota.

The ability of microorganisms in organic matter in batik liquid waste has indeed been recognized by many researchers. Reliability of microbes including bacteria, fungi, and protozoa in waste water treatment and its role in maintaining the ecological balance of the waters has been widely elaborated. One of the commonly used bioremediatory agents is bacteria. Bacteria have an important role in the decomposition of organic matter. In general there are two types of bacteria based on the need for oxygen, it's aerobic bacteria and anaerobic bacteria. The objective of this research was to evaluate the BOD degradation by using microbial isolated from water body contaminated by batik waste water.

2. Material and Methods

Bacteria was isolated from river and sludge contaminated with batik waste. Isolation and purification using the dilution planting method or sample dilution were carried for the identification. In order to obtain pure culture bacteria the pour plate technique for maintain bacteria using streak culture (culture of scratching) was used. Dilution were carried for the identification. In order to obtain of samples was carried out up to level 10^{-3}. Samples from dilutions of 10^{-1}.10^{-2}, 10^{-3} as many as 1 ml to be grown on nutrient agar by pour plate. Then the culture was incubated 24 hours at room temperature.

The colonies are separated and show different features, each of which is grown into nutrient agar medium in a streak quadrant. Culture was incubated for 24 hours at room temperature. Separate colonies that show different characteristics are transferred back to the nutrient agar medium in a streak quadrant, and so on until the colony is obtained with uniform characteristics (pure culture).

Pure isolates obtained from the isolation were grown on nutrient agar medium in a spread plate, and incubated at room temperature for 24 hours. The culture observed characteristics included colonies, colony shape, elevation, colony surface texture, colony color, and colony size. Observation of bacterial microscopic characteristics include: gram staining, endospore staining and bacterial cell measurements. Bacterial identification was carried out by matching its characteristics with the bacterial identification book from Bergey's manual of determinative bacteriology 9th. This is to find out the name of indigenous bacteria produced from the isolation of batik waste water. After identifying the bacterial genus, the next step is to multiply the bacterial culture as needed. The multiplication of bacteria is added to batik waste water to evaluate the growth and BOD degradation.

3. Results and Discussion

Based on the identification activities that have been carried out, bacteria from the genera *Mesophilobacter, Methylococcus, Agrobaacterium, Neisseria, Xantobacter, Deinococcus, Sporosarcina,* and *Bacillus* were obtained. These are types of aerobic or facultative anaerobes which are effective as biological agents in processing organic waste.

The results of identification of bacterial isolates by observing the macroscopic characteristics of bacterial colonies were based on colony morphology which included colony shape, colony color, colony edge, and elevation of bacterial colonies that had been cultured on nutrient agar plates.

Observation of the microscopic characteristics of bacterial cells includes the shape and size of cells, observation of the results of gram staining, and observations of endospores staining results. Based on observations of microscopic characteristics. Biochemical test of bacterial cell activity including carbohydrate fermentation test, TSIA test, methyl red test, indole test, proskauer voges test and citrate use test.

Table 1. Identification of bacteria from river and sludge contaminated with batik waste

No	Code of isolates	Colony shape	Colony edge	Elevation of bacterial	Surfave properties	Colony color	Size of colony	Cell shape	Size of cell	Gram straining	Biochemical test					Name of genus
											TSIA test	MR test	VP test	Sitrat test	Indol test	
1	A1(1)	Irregular	Undulate	Corvex	Wringkled	Opaque	Large	Basil	78,43135 µm	-	-	-	-	-	+	Mesophilobacter
2	A1(2)	Irregular	Entire	Raised	Glistening	Transluent	Large	Cocci	65,18269 µm	-	-	+	-	+	-	Methylococcus
3	A2	Irregular	Undulate	Pulvinate	Rough	Transluent	Large	Basil	8,603492 µm	-	-	+	-	+	+	Agrobacterium
4	A3	Rhizoid	Undulate	Pulvinate	Wringkled	Opaque	Moderate	Cocci	46,94431 µm	-	-	+	-	-	+	Neisseria
5	B2	Rhizoid	Undulate	Pulvinate	Rough	Transluent	Moderate	Cocci	112,1552 µm	-	-	-	+	+	+	Methylococcus
6	B3	Irregular	Entire	Pulvinate	Rough	Transluent	Large	Basil	6,769941 µm	+	+	+	-	+	+	Xanthobacter
7	B4	Irregular	Undulate	Crateriform	Wringkel	Transluent	Large	Cocci	88,16432 µm	+	-	+	-	-	+	Deinococcus
8	C1	Irregular	Undulate	Flat	Wringkled	Transluent	Large	Cocci	113,2664 µm	+	-	+	-	-	+	Sporosarcina
9	C2	Irregular	Undulate	Crateriform	Wringkled	Opaque	Large	Cocci	90,3688 µm	+	-	+	+	+	+	Methylococcus
10	C3	Irregular	Undulate	Flat	Rough	Opaque	Moderate	Basil	16,77075 µm	+	-	+	+	-	+	Bacillus
11	D	Irregular	Undulate	Umbonate	Wringkled	Transluent	Large	Cocci	7,039983 µm	-	-	+	-	-	-	Methylococcus

Determination of bacterial isolates was carried out based on the macroscopic characteristics of colonies, microscopic cells and biochemical tests of bacterial cell activity. Testing the characteristics of bacterial isolates was carried out based on the book Bergey's manual of determinative bacteriology 9th [9]. Bacterial culture is an attempt to increase the amount of bacteria isolated.

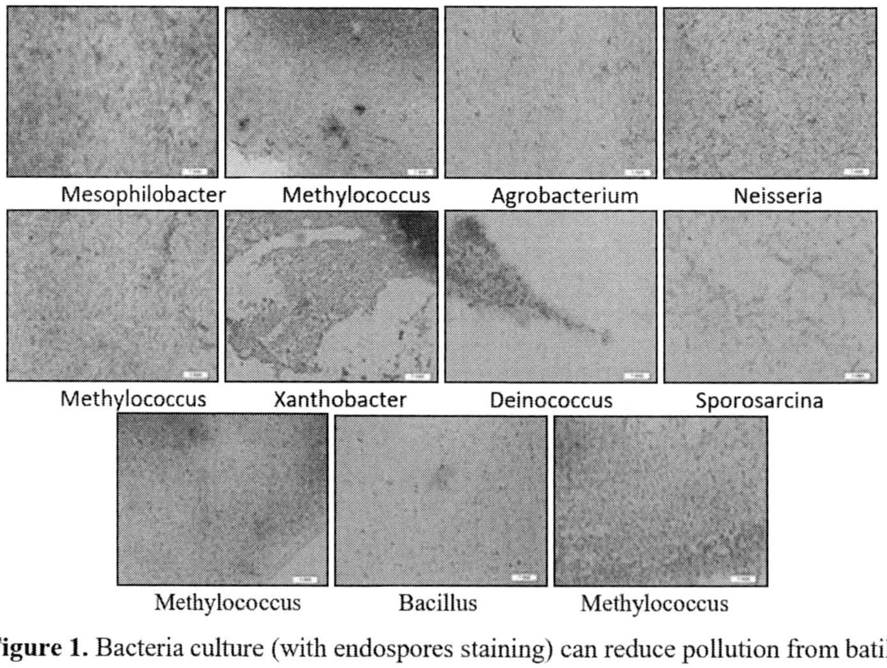

Figure 1. Bacteria culture (with endospores staining) can reduce pollution from batik

Bacterial culture is carried out on nutrient broth medium which has been added with batik waste water as much as 2%. The culture is then in sheaker for approximately 2x24 hours at room temperature.

Observations were made to determine the resistance and ability of living bacteria by looking at the color of the medium, that the medium experienced turbidity which showed bacterial growth in the medium.

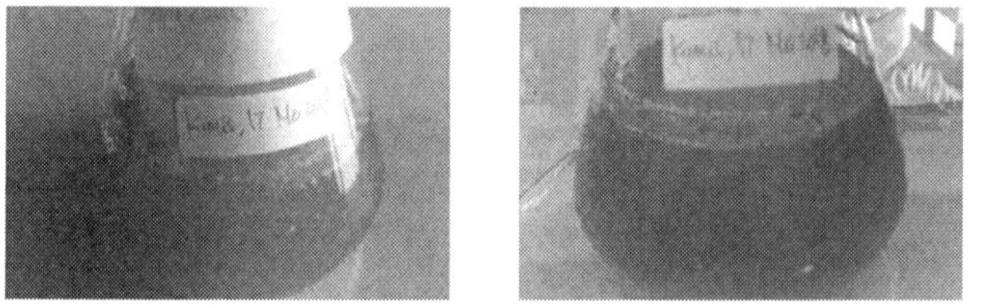

(a) (b)

Figure 2. Observation bacterial growth on medium; (a) before adding bacteria, (b) after adding bacteria (6 days)

Typical oxidation or decomposition of organic matter, especially in some types of industrial waste water containing for example phenols, detergents, oils, etc., bacteria must be given a time of adjustment (adaptation) [10] several days through contact with the waste water (acclimatization) [5], before it can be used as seeds in the water BOD analysis. Conversely some organic and inorganic substances can be toxic to bacteria (such as cyanide, copper, etc.) and must be reduced to the desired extent [5].

The parameters used in the research activities are based on the parameters contained in the Minister of Environment Regulation Number 5 of 2014 Republic of Indonesia concerning waste water quality standards, one of which is BOD. These parameters are used to determine the content of organic matter in batik wastewater [11].

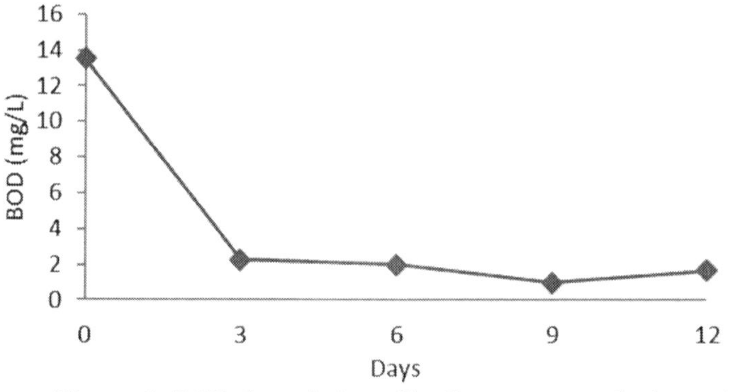

Figure 3. BOD degradation of batik waste water by bacteria

The results of the observation showed that the old treatment of contact had a significant influence on the BOD value, resulting in a decrease of up to 85.71%. The ability of bacteria to decompose organic matter must be supported by other biological agents that can reduce or eliminate residual decomposition of organic matter by bacteria [12].

In decomposing organic matter in batik liquid waste, bacteria release enzymes to decompose organic compounds and produce by-products in the form of carbon dioxide (CO_2), methane (CH_4), hydrogen (H_2) and water (H_2O), and energy to support metabolic activity [13]. Decomposition of azo compounds which are organic pollutants in batik liquid waste by bacteria produce CO_2, H_2O and NH_3 compounds and energy in the form of biomass [13]. Decomposition of azo compounds by bacteria was carried out using the enzyme azo reductase [2]. Enzymatic azo reduction is catalyzed by an enzyme called azo reductase [14]. Inorganic materials resulting from decomposition of organic matter by bacteria will then be absorbed by plants with the content of the enzyme phytochelatin in their roots and leaves [5],12].

4. Conclusion

The study evaluated the use of microbial consortium to degrade batik waste water. From the observation result, we obtained about 8 bacterial in a colony symbiosis of mutualism that take an important role of cleaning pollutants from batik waste water. The degradation of BOD in batik waste water was 85.71% and it also proved that the batik waste water can be degraded by using microbial bioremediation method

Acknowledgment
Sponsored by Indonesia Scholarship Endowment Fund (LPDP), Ministry of Finance – Republic of Indonesia

References

[1] Li Y, Hu C J, and Yao X 2009 Innovative batik design with an interactive evolutionary art system *Journal of Computer Science and Technology* **24**(6): 1035–1047 https://doi.org/10.1007/s11390-009-9293-5
[2] Forgacs E, Cserháti T and Oros G 2004 Removal of synthetic dyes from wastewaters: A review. *Environment International*, **30**(7) 953–971 https://doi.org/10.1016/j.envint.2004.02.001
[3] Sari M M, Hartini S, and Sudarno 2015 Pemilihan desain instalasi pengelolaan air limbah batik yang efektif dan efisien dengan menggunakan metode life cycle cost *Jati Undip* **X**(1), 27–32.
[4] Suprihatin H 2014 Kandungan organik limbah cair industri batik Jetis Sidoarjo dan alternatif pengolahannya *Pusat Penelitian Lingkungan Hidup Universitas Riau* 130–138
[5] Gaur N, Narasimhulu K, and PydiSetty Y 2018 Recent advances in the bio-remediation of persistent organic pollutants and its effect on environment *Journal of Cleaner Production*, **198**(1) 1602–1631 https://doi.org/10.1016/j.jclepro.2018.07.076
[6] Wang Z, Xue M, Huang K and Liu Z 2011 *Textile dyeing wastewater treatment. Huazhong University of Science and Technology China*. https://doi.org/10.5772/22670
[7] Mratihatani A S and Susilowati I. (2013). Menuju pengelolaan sungai bersih di kawasan industri batik yang padat limbah cair. *Diponegoro Journal of Economics* **2**(2) 1–12 استرجع في من http://ejournal-s1.undip.ac.id/index.php/jme
[8] Pande S and Kost C 2017 Bacterial Unculturability and the Formation of Intercellular Metabolic Networks *Trends in Microbiology* **25**(5) 349-361 https://doi.org/10.1016/j.tim.2017.02.015
[9] Bergey D and Holt J G 1994 *Bergey's manual of determinative bacteriology*
[10] Zhong F, Xu M, Metz P, Ghosh Dastidar P and Zhu J 2018 A quantitative metabolomics study of bacterial metabolites in different domains. *Analytica Chimica Acta* **XXX**(1) 1–8 https://doi.org/10.1016/j.aca.2018.02.046
[11] Peraturan Menteri Lingkungan Hidup RI No. 5 Tahun 2014 tentang Baku mutu air limbah (Jakarta)
[12] Jacob J M, Karthik C, Saratale R G, Kumar S S, Prabakar D, Kadirvelu K and Pugazhendhi A 2018 Biological approaches to tackle heavy metal pollution: A survey of literature *Journal of Environmental Management* **217**(1): 56–70. https://doi.org/10.1016/j.jenvman.2018.03.077
[13] Paz A, Carballo J, Pérez M J and Domínguez J M 2017 Biological treatment of model dyes and textile wastewaters *Chemosphere* **181**(1): 168-177 https://doi.org/10.1016/j.chemosphere.2017.04.046
[14] Yang Y, Xu M, Guo J, and Sun G 2012 Bacterial Extracellular Electron Transfer in Bioelectrochemical Systems *Process Biochemistry* **47**(12): 1707-1714 https://doi.org/10.1016/j.procbio.2012.07.032

Sangon and Acacia plant suitability for land reclamation of coal mines in PT. Karbindo Abesyapradhi Sijunjung District

Admizal Nazki[1,*], Nasfryzal Carlo[2], Indang Dewata[2], Fitratul Rahmi[2] and Rihan Efendi[2]

[1] B.Eng, Mining Engineering Faculty, Universitas Negeri padang, Indonesia
[2] Researcher, Mining Engineering Faculty, Universitas Negeri Padang, Indonesia

*adnaz@rocketmail.com

Abstract. The loss of the vegetation indirectly had removed the functions of the forests in regulating water system, controlling erosion and flood, absorbing carbon dioxide, supplying oxygen and controlling temperature. One of the efforts done to restore the quality of the environment around mining areas was by doing reclamation in accordance with the Regulation of Indonesian Government No. 78 Year 2010. This research applied cross-sectional design. The sample was taken in the areas vegetated. The variables studied were the chemical content of the soil (N, P, K, Ca and Mg), pH and Cation Exchange Capacity. The sample spread out in five areas reclaimed in Sungai Tambang PT Karbindo Abesyapradhi. The result of the research showed that the reclamation was done by planting sengon and acacia trees. The fertility of the soil (clay texture, acidity, Ca, Magnesium, Cation Exchange Capacity of the soil, C-Organic and N-total of the soil) in the reclaimed areas was similar. However, partially affecting both plant growth and sengon acacia tree is a cation exchange capacity and Soil Texture. It was recommended to PT Karbindo Abesyapradhi to maintain and care for reclamation has been done, because if based on the evaluation of the report on the implementation of reclamation showed reclamation does not meet the criteria of success, the Minister, the Governor, or the Regent / Mayor in accordance authority may assign a third party to conduct reclamation activities or entirely by using collateral reclamation.

1. Introduction

PT. Karbindo Abesyapradhi, the leading mining business permit owner of coal production operations in West Sumatra has the obligation to carry out environmental management, especially when and after mining in accordance with applicable regulations. The company is obliged to conduct an analysis of the important impacts of its activities, develop a plan for management and monitoring of the environment in its working area, and comply with every provision in EIA, RKL / RPL, and its employment contract. Because in the Amdal agreement PT. Karbindo Abesyapradhi which is located at positions 101020'30'-101022'40' 'East Longitude and 0049'30"-0052'13,8' 'South Latitude administratively this area is in the Kanagarian River Tambang - Kiliran Jao, District Kamang Baru, Sijunjung Regency, West Sumatra Province is done by open pit mining with a total coal reserve of ± 4,405,631 tons of production averaging 200,000 tons per year, so it is estimated that the mine life of ± 22 years is agreed upon the reclamation must be cultivated, including sengon and acacia trees.

Furthermore, the company is obliged to reclaim and revegetate ex-mining land so that the land has carrying capacity and environmental functions in accordance with the provisions. Thus, reclamation activities should not only be seen as a part of mining operations but are stages in the mine closure process.

Based on the description above, the researcher raised this research with the title "The Conformity of Sengon and Acacia Plants for Reclamation of Former Coal Mine in PT. Karbindo Abesyapradhi-Sijunjung Regency". Reclamation activities are emphasized on improving soil quality, in terms of soil organic matter content, macro nutrients N, P, K, Ca, Mg soil and value of Cation Exchange Capacity (CEC) as well as weight volume (BV) of land, from reclamation activities- reforestation which has been carried out since 1993 to research in stages. Land that has been laid out and planted with the main plant species sengon (*Albizia falcataria*) and acacia (*Acasia mangium*). Thus it is expected to describe the role of plants in the status and dynamics of soil fertility. The soil fertility status is determined by the physical, chemical and biological properties of the soil. But this study focuses on observing canopy or tree canopy in terms of soil fertility and soil chemical properties.

2. Coal and Reclamation

Coal is a sediment composed of organic matter and inorganic materials. Organic materials are made up of remnants of plants that experience decomposition and changes in their physical and chemical properties, either before or after being covered by other deposits. Inorganic materials consist of several minerals, for example clay minerals, carbonates, sulfides, silicates, and so on

Coal mines, especially open-pit mines, require large areas to be temporarily disturbed. This raises environmental problems, including soil erosion, dust, noise and water pollution, and impacts on local biodiversity. Actions are carried out in modern mining operations to reduce these impacts. Good environmental planning and management will reduce the impact of mining on the environment and help preserve biodiversity.

In the environmental aspect, mining companies from the outset should pay attention to the Environmental Impact Analysis (AMDAL) that they have made, in accordance with the Decree of the State Minister of Environment No: 17 of 2001 concerning Types of Business Plans and / or Activities Required with EIA. General mining business activities with wide permits for Mining Business Permits (IUP) above 200 hectares or wide open areas for mining above 50 hectares cumulative per year must be completed with an AMDAL. This is very necessary to avoid open land that is too large.

The potential for significant impacts on the environment from general mining businesses includes changing landscapes, ecology and hydrology. Then, the duration of the business activity will also have an important impact on air quality noise, vibration when using explosives, and the impact of liquid waste produced. For exploitation of coal/peat production of more than 250,000 tons/year, primary ore more than 250,000 tons/year and secondary ore/alluvial deposits of more than 150,000 tons/year are all required to use AMDAL.

Reclamation of ex-mining land is an activity that aims to improve or organize the use of disturbed land as a result of mining business activities so that it can function and be efficient in accordance with its designation. In accordance with the prevailing laws and regulations, every Mining Company that engages in mining business activities, except for general investigation and exploration activities, is obliged to reclaim the former mining activities it undertakes. The cost of conducting reclamation is part of the production expense, which is one of the deducting factors of business sales (income derived from the company's mining products) to obtain a gross profit (loss). In order to guarantee the compliance of mining companies to carry out reclamation in accordance with the Reclamation Plan, mining companies are required to provide a Reclamation Guarantee, the amount of which is in accordance with the Reclamation Cost Plan that has been approved by the Minister, Governor and Regent / Mayor according to their authority. Reclamation guarantees can take the form of time deposits, bank guarantees, insurance, and accounting reserves (accounting reserve). The guarantee must be placed by the Mining Company before the Company starts a mining production or exploitation business.

IOP Conf. Series: Earth and Environmental Science **314** (2019) 012021 doi:10.1088/1755-1315/314/1/012021

3. Geological Process of Coal Formation

Coal comes from plant material through the geological process and the influence of pressure and heat so that the physical and chemical properties change. Because of its chemical composition, geologically coal is classified more closely as mineral rock [1].

The main components of coal are Carbon, Hydrogen, Nitrogen, Oxygen, Sulfur. These minerals include Silica (SiO_2), alumina (Al_2O_3), Iron oxide (Fe_2O_3), Magnesia (MgO), phospor pentoxida (P_2O_5) and Potassium oxida (Na_2O dan K_2O) [1].

Indonesian coal has a low rank coal, which is from peat (peat) to subbituminous, the quality is very good, the ash content and its sulfur with adequate heating value for heat and physical generation. Coal in Indonesia is generally young, but the coal is of good quality, because it has experienced very high pressure caused by symptoms of mountain formation (orogenesis).

4. Considerations in Reclaiming Former Mine Land

There are 3 (three) considerations that must be considered in carrying out mining land reclamation, namely consideration of soil chemistry, soil physical and land topography [2]. First, chemically, there are two problems related to spoil material produced by open coal mining, namely the overburden is too alkaline or too acidic. In arid and semi-arid areas with low rainfall, this will be a problem because ground water containing high salt will cause water to be difficult to penetrate plant root membranes so that it greatly affects and disrupts plant growth. The second is related to the physical nature of the soil. The rock blasting process will reduce the amount of rock because it turns into sand and dust and reduces the number of soil particles because it will turn into clay. The stepping technique produces mixing between sand, dust and clay. Third is the topography of the land. The slope of the land will cause problems in revegetating the land. The type and number of plants need to be considered in relation to their ability to control erosion and nutrient deterioration.

5. Research Methodology

The research method used in this writing uses quantitative methods, with the type of crossectional research. Crossectional research is descriptive research, but in certain cases it is used for analytical research or exploratory research combined with descriptive research, or explorative with case control or crossectional [3].

5.1. Description of Sampling Locations

The research location is the Western Dump Area (revegetation from 1993 to 2012). Administratively these locations are in the Mining Business Permit (IUP) area, PT. Karbindo Abesyapradhi in the Jorong area of the Tambang River, Parik Rantang Minute Nagari, Kamang Baru District, Sijunjung Regency, West Sumatra Province.

Sampling locations for each revegetation year from 1993 to the present (see appendices III and IV), are also distinguished by locations with monoculture sengon planting and monoculture acacia planting locations that have been reforested according to the age of growing greening plant communities and conditions control as a comparison.

5.2. Research time

The research was carried out on ex-coal mining land planted with Sengon and Acacia trees in the Mining Business License Area (WIUP) of PT. Karbindo Abesyapradhi located in Sijunjung Regency, West Sumatra Province in July to November 2012

6. Research Parameters

6.1. Physics Analysis - Soil Chemistry

Physics quality parameters - Soil chemistry analyzed in the laboratory includes texture, volume weight (BV), pH, C-organic, N - amount, available phosphorus, available potassium, available magnesium and CEC.

6.2. Profile of Cover of Greening Plants
The intensity of coverage of all stands against the surface of the land below. By looking at the number of trees, the density, the average tree diameter and the type of tillers and growing below which is below the stand, it is expected to further illustrate the adaptability of acacia and sengon plants, as well as see the level of soil fertility.

6.3. Physics Quality - Soil Chemistry
Tools used such as hand drills or hoes and sample bags (plastic bags), while the materials needed are stationery.

6.4. Profile of Cover of Greening Plants
The tools used include meters, while the materials needed such as stationery and paper millimeters.

7. Research Location
Geographically the mining business permit area of the production operation of PT. Karbindo Abesyapradhi is in a position 101020'30''-101022'40'' East longitude and 0049'30''-0052'13,8'' South latitude. Administratively it is in the Sungai Tambang Village area, Rantang Trench, the concern of Sijunjung West Sumatra Regency with the area of mining business permit for production operations ± 2000 Ha (See Figure 1. Map of OP IUP).

The location can be reached from the City of Padang past Solok - Sijunjung Kiliranjao, with a distance of 165 Km. This location can be reached within ± 3 hours by road (Trans Sumatra Road). From Kiliranjao to the location of PT. Karbindo Abesyapradhi is ± 11 Km away, condition of class III paved road (Street of Teluk Kuantan).

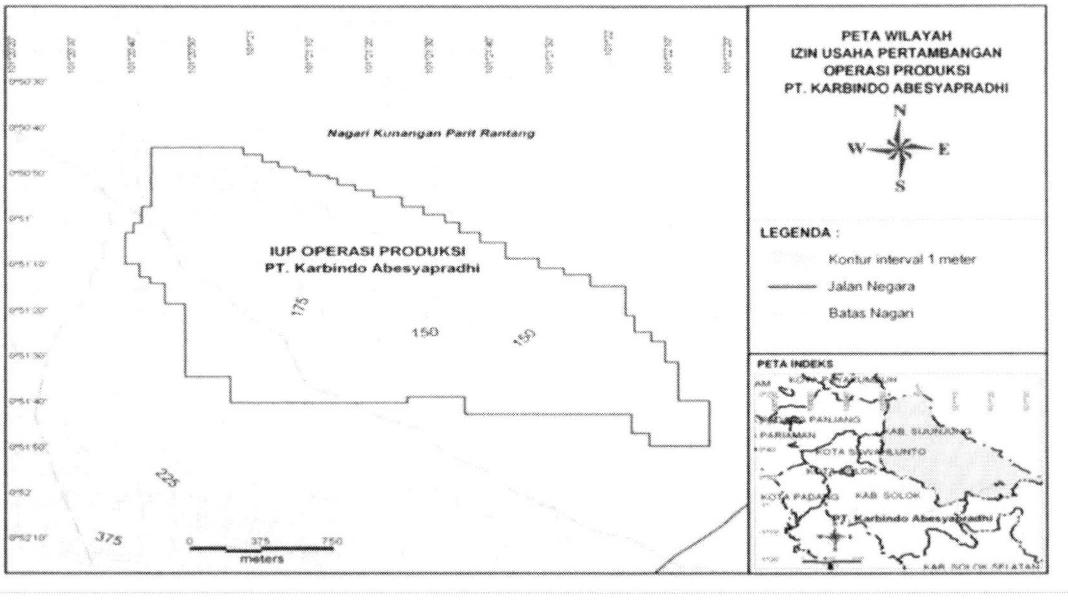

Figure 1. Map of Production Operation IUP of PT. Karbindo Abesyapradhi.
(Source: [4])

8. Coal Mining Activities PT. Karbindo Abesyapradhi
The mining system carried out by PT. Karbindo Abesyapradhi is using a surface mining system with a pit pit method. The application of open mining methods is adjusted to the calculation of layers of coal reserves with slopes, while the back-filling method serves as an effort to reduce the area of land that is open due to mining activities, so that stockpiling activities are in line with the movement of active mining operations.

8.1. Overburden Transport and Stockpiling Disassembly

Land excavation and soil transfer work is the beginning of mining activities. Amdal document PT. Karbindo Abesyapradhi explained the volume of the soil layer that was excavated and moved requires a separate place so that it does not cause problems. Overburden which is dug up will be deposited in a separate place. The location of a waste dump area (outside dump) is between 600 - 1,200 m from the mine area (pit area). Overburden dismantling is done using balsting, because the material is so hard that it cannot be disassembled directly (free). From detailed geological exploration, overburden that requires 100% blasting. Demolition using free digging or blasting still takes into account geotechnical considerations, so that it follows the predetermined pit area configuration.

To maintain slope stability at the present location, there are corner benches (either single bench or multi bench), the northern pit area ranges 55^0-65^0, high bench 10,0- 50,0 m, work width 20-40.0 m, and for slope angles, both single bench and multi bench pit areas in the southern part range from 300-450 (overall 300), bench width 15.0 m, bench height 10,0-20,0 m, working width 20,0-40,0 m. In some parts of the angle the slope is not stable, because it is the floor of the coal seam.

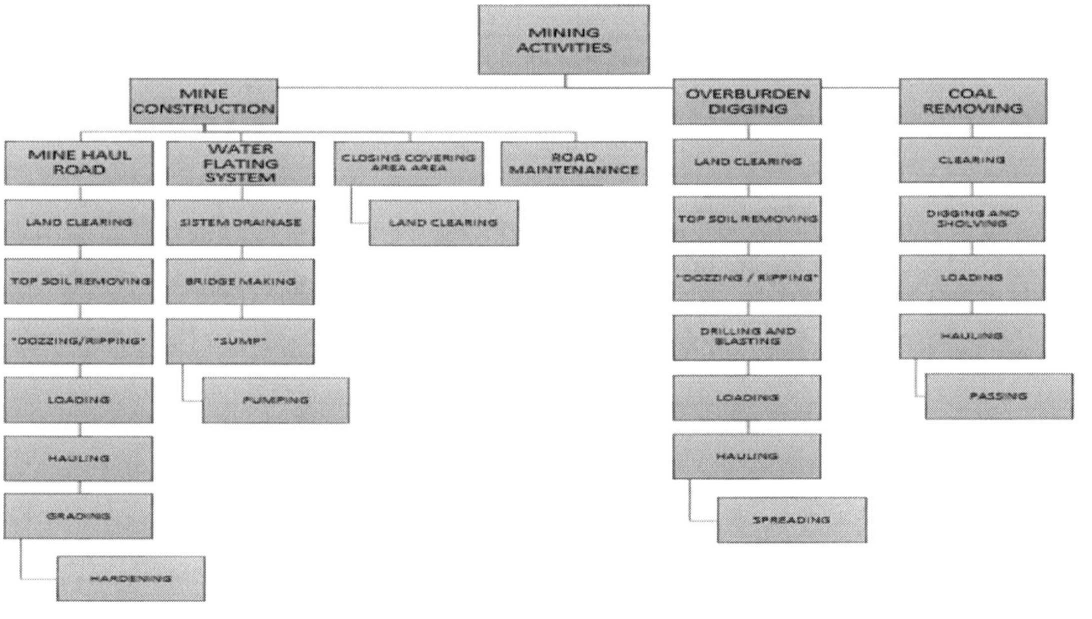

Figure 2. Mining Activities of PT. Karbindo Abesyapradhi.
(Source: PT. Karbindo Abesyapradhi)

8.2. Coal Excavation, Cleaning and Loading

A steep slope coal layer and consists of several partings excavated mechanically using a heavy equipment (excavator). To do this, surface cleaning and coal breaking are done by cutting the top of each parting the act of destroying coal from side rocks or parting which is thick enough can have a negative impact on the quality of coal. This coal cleaning work uses a digging tool with a smaller capacity by a qualified operator. Then the coal is arranged into two piles, clean coal and dirty coal, then ready to be loaded into the conveyance. Coal is mined in a series of horizontal terraces; the overburden is dug first to facilitate the extraction of coal seams. Then, loaded and transported to the run of mine (ROM), then processed dryly in a coal processing plan (CPP) with a capacity of 100 MT per hour and then coal products are stacked in the stockpile area. Coal products are transported to the stock pile of gulf of Bayur Padang port. The utilization of the said land space is rented with PT. PELINDO II gulf Bayur Padang Branch (See Figure 4.5. Flow Chart Mining Process).

9. Type of soil

The results of laboratory analysis of the Land Research Station and Agro-climate of the Balitbang Department of Agriculture, Bukit Tinggi, found two types of soil found in the project area, namely Latosol and Podzolic soils. Soil Type Latosol is found in areas with choppy physiography to bump, even to the top of the hill. From the observation of the physical properties of the soil, it is known that this area has a texture of sandy clay to sandy clay loam (See Table 4.1. Results of Soil Texture Analysis). Soil fertility is very instrumental in supporting the growth of vegetation, especially when land reclamation and natural succession of vegetation that grows on the soil, Assessment of soil fertility is done by taking into account determinants such as pH (H2O), C, N, C / N, NTK and saturation bases, which are seen in locations that are considered to represent the project area.

Table 1. Results of Soil Texture Analysis

NO.	Type of Analysis	LOCATION						
		P1	P2	P3	P4	P5	P6	P7
1.	Sand	24,67	5,5	38	24,5	32	40,25	7
2.	Ash	47,67	42	35,25	32,8	37,25	33	38,6
3.	Clay	31	52,5	23,3	42,7	30,8	26,8	49,6
4.	Organic ingredients	1,7	1,57	0,79	0,92	1,68	0,65	1,07

Information:
- Location P1 = at Pit B
- Location P2 = at Pit A
- Location P3 = Rear Base Camp
- Location P4 = Reserve Forest
- Location P5 = near washing
- Location P6 = Downstream of the River Mine
- Location P7 = Drill well

Figure 3. Flow Chart Mining Process.
Source: In the Mineral and Coal Engineering and Environmental Rectorate, Directorate General of Mineral and Coal, Department of Energy and Mineral Resources

10. Physical and Chemical Properties

This data shows the occurrence of chemical dynamics that are not uniform between parameters one with other parameters. The description of each parameter can be explained below:

10.1. Textured clay content

Some types of land are land planted with acacia trees, sengon trees and natural land which is not planted with vegetation, so the soil pH results are obtained as follows:

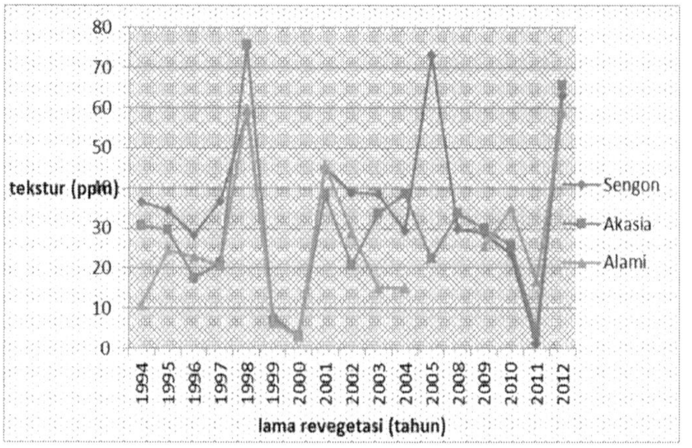

Figure 4. The results of examination of clay texture content at the Coal Mining location of PT. Karbindo Abesyapradhi in 1994-2012.
Source : Author documentation from software Microsoft Excel

The results of the examination of the content of upright clay texture of sengon trees in the reclamation area decreased in 3 years, because the process of draining by rain water, then increased sharply in the 5th year because the tree canopies had begun to tighten so that the sloping process was reduced and decreased in the following years due to decomposition of sarasah occurs. Whereas in the acacia tree stand there was an increase in the 5th year and then sharply declined the following year. Based on the graph above, it can be seen that clay content in natural areas or not planted with vegetation tends to be under the land planted with vegetation.

10.2. Soil pH

Some types of land are land planted with acacia trees, sengon trees and natural land which is not planted with vegetation, so the soil pH results are obtained as follows:

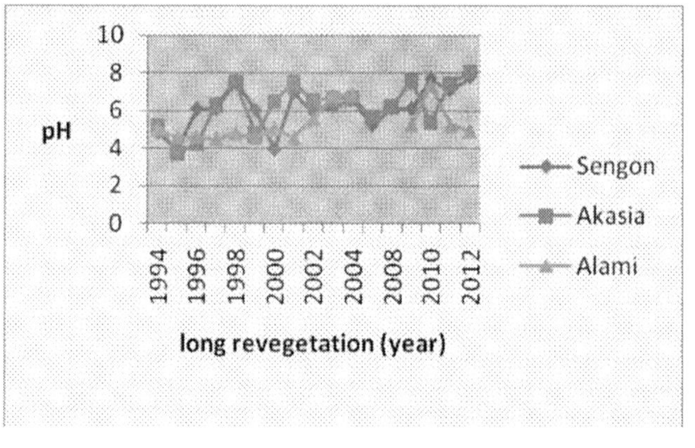

Figure 5. The results of examination of soil pH content at the Coal Mining location of PT. Karbindo Abesyapradhi in 1994-2012.
Source : Author documentation from software Microsoft Excel

The results of soil pH examination in vegetation and natural areas showed that the soil pH content in the reclamation area tended to be alkaline since the 2nd year even though in certain revegetation years there was a decrease in pH both in the stands of sengon and acacia vegetation. Whereas the pH of the soil in natural areas or not planted with vegetation tends to be acidic, this is because the natural soil base is Red Yellow Podsolic or Red Podsolic low pH.

10.3. Ca soil content

In some types of land, namely land planted with acacia trees, sengon trees and natural land which is not planted with vegetation, the resulting Ca soil content is obtained

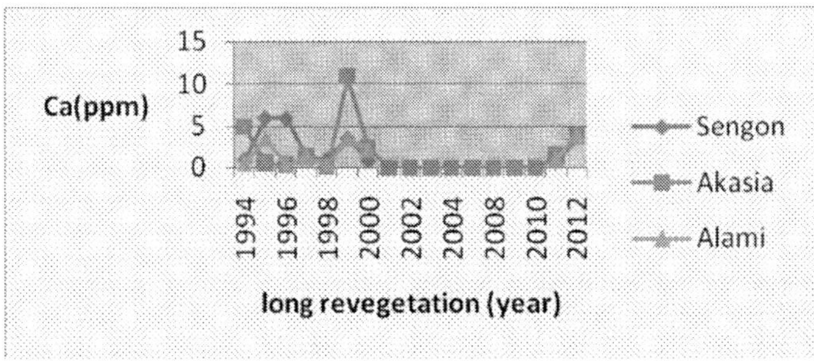

Figure 6. The results of the examination of the Ca content of the land at the Coal Mining location of PT. Karbindo Abesyapradhi in 1994-2012.
Source : Author documentation from software Microsoft Excel

Based on the picture above, it can be seen that the Ca content of the land planted with acacia vegetation has increased in the 6th year, because laboratory analysis results tend to continue to increase, from 2001 to 2011 cation exchange rates such as Ca, Mg, Na, K and CEC monitoring parameters for analysis is not taken anymore. While the Ca content in the natural area is lower than the area planted with acacia and sengon vegetation.

10.4. Mg Soil Content

In some types of land, namely land planted with acacia trees, sengon trees and natural land which is not planted with vegetation, the soil Mg content is obtained as follows:

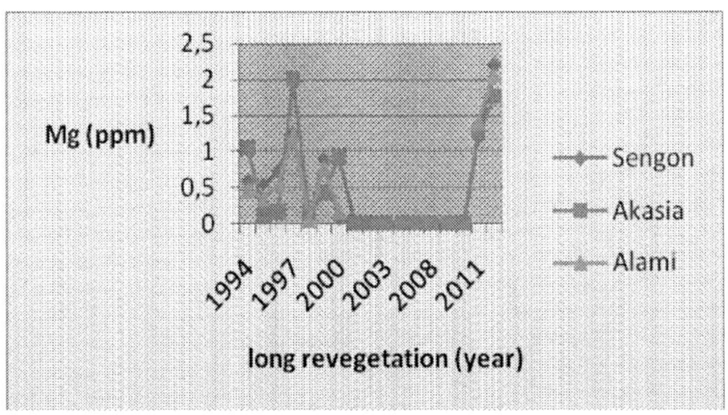

Figure 7. The results of the examination of the soil Mg content at the Coal Mining site at PT. Karbindo Abesyapradhi in 1994-2012.
Source : Author documentation from software Microsoft Excel

The results showed that the soil Mg content in the reclamation area decreased year 2 to year 8 but there was an increase in the 15th year for both sengon and acacia vegetation stands. While the content of soil Mg in natural areas tends to be lower than in areas planted with sengon and acacia vegetation

11. Discussion

11.1. Clay Texture

The results also found that the longer the life of the mine the texture of clay tends to decrease. In this case, the relationship between texture and soil structure to plant growth is very close, there is a reciprocal relationship between components one with the other components. Plant growth can be influenced by texture and soil structure. In soil conditions that have the dominant texture of sand, the binding capacity of the soil to water and other organic materials is small. Soils with the dominant texture of sand tend to easily release the nutrients needed by plants. In a soil condition like this, the growth of plant roots will develop well. Easy root to penetrate into the soil. Drainage and aeration in the texture of the dominant sandy soil is quite good, but the texture of this soil tends to easily release the nutrients needed by plants.

For good plant growth, soil with aeration, drainage, and the ability to store water and good nutrients must have a balanced component of sand, dust and clay. So that plants can grow in optimal conditions. In addition to soil texture, another factor that has a close relationship with plant growth is soil structure.

As a result of plants experiencing growth, it turns out that plants can cause the formation of soil structures. With the presence of plants, aggregation of the soil will be formed into a more stable structure. Plants can reduce soil damage due to rain, so nutrients can be maintained and available to plants and microorganisms that live in the soil. From this description, the relationship between texture and soil structure to plant growth is related to one another. Without the texture and soil structure that is good for plants, plant growth is less optimal. Because, there are factors that limit plant growth due to unfavorable texture and soil structure. If the state of the texture and structure of the soil is in a steady state, then these factors can be overcome. In addition, the presence of plants on the ground seems to be able to help the formation of soil structures. This is caused by the existence of a root system found in the soil that is able to form natural divisions. Thus, the binding capacity of the soil increases with each other.

11.2. Mg content

The results showed that the soil Mg content in the reclamation area decreased in the second year to the eighth year but there was an increase in the 15th year for both sengon and acacia vegetation stands. This is caused by plants that are around acacia and sengon trees that have grown, the plants are rather large and weathering occurs. It means that we can see the availability of magnesium due to the weathering process of minerals containing magnesium. Furthermore, as a result of this process, magnesium will be free in the soil solution. This condition can cause magnesium to disappear with percolation water, magnesium is absorbed by plants or various other living organisms and absorbed by clay particles and deposited into secondary minerals. Availability of magnesium in soils that have high acidity and will decrease due to loss from the soil. Loss of magnesium is the same as calcium, such as erosion, due to washing and transporting plants or other living organisms.

12. Physical countermeasures

The application of plant rehabilitation methods creates a balance between human intervention and ecosystem efforts to design their own environment (self-design). This self-design provides benefits in terms of providing survival in the initial conditions of the growth of rehabilitation, stabilizing the condition of the forest after the initial phase of rehabilitation, and all of this requires costs. The types found in the location of the study are the location of PT. Karbindo Abesyapradhi as there are associated plant species in each greening area, for example, we take the example in the South E area because these plants can indirectly symbiosis with nitrogen fixing bacteria, have strong roots, and produce a lot of litter, such as: Eupatorium conjugatum (kerinyuah), Lee indica, Sida rhombifolia, Piper aduncum (betel-siriah), Canarium sp, Zingiber sp, Merremia sp, Laportea sp (Jilatang), Syzigium jambos (Jambak), Clausena excavata (sicerek), Alocasia esculenta (taleh), Grewia sp (madang), Litsea sp, Mucuna sp (aka lilik), Ficus lepicarpa (aro), Abelmoschus sp, Cyathea sp (nail bareh) and other suitable types (See attachment 2 Data on Observation of Flora).

Besides that, the presence of organic C is used by the function of the soil in meeting energy needs. The function of the soil is the microbes that can best adapt to the changing environment conditions, such as in mining lands that are degraded compared to soil microbes from other groups.

13. Economic countermeasures

Most of the inhabitants of the Nagari region around the location of PT. Karbindo Abesyapradhi works for the average company earning income as shown in table E. 2.1 below:

Table 2. Compilation of Number of Manpower PT. KA (End of December Condition for 2013 Working Year)

NO	LABOR QUALIFICATION	TOTAL (PEOPLE)	INCOME (Rp. Jt)
1.	Operations	147	Rp. 3.500.000,00
2.	Supervisor	34	Rp. 8.000.000,00
3.	Administration - Office Workers	27	Rp.4.500.000,00
	TOTAL	208	

In addition, there are also various service businesses currently developing. The compilation of data collection results can be seen in Table E.2.2 below.

Table 3. Compilation of results of community livelihood data collection

NO	NAGARI REGION	LIVELIHOODS AND AMOUNT OF POPULATION (SOUL)				
		Farmer	trade	PNS/ABRI	Private	Services
1.	Kungan Parik Rantang	3.113	534	71	58	66
2.	Kamang	4.733	381	158	121	72
3.	Muaro Takung	330	216	82	32	45
	TOTAL	8.176	1.131	311	211	183
	PERCENTAGE	81,66	11,30	3,11	2,11	1,83

In addition to the Jorong Sungai Tambang I region (the location of PT Karbindo Abesyapradhi's coal mining business activities), currently various activities or economic activities of the community are developing rapidly. This indication can be seen from the existence of new stalls, new shops and the growth of shop houses (shophouses) that are used for trade services, both in the Parik Rantang Nagari Kunangan Market (the Sungai Tambang market had previously been moved) as well as in the Kiliran Jao road section - Teluk Kuantan (Jorong Parik Rantang and also Jorong Kamang). The development of said economic activities is not fully related to the existence of PT. Karbindo Abesyapradhi, but also the impact of the existence of other businesses such as plantations and palm oil mills of PT. Bina Pratama Sakato Jaya.

14. Social Response Efforts

As the government program returns to Nagari, all the social components of a Nagari community (including Ninik Mamak) play a role in regional development. In accordance with the prevailing government order, the control of all-natural resources in a Nagari including land and permits is the full authority of Ninik Mamak. In addition to settlements and agricultural businesses, land use that has developed around the location of PT. Karbindo Abesyapradhi includes fields, gardens and most of them are still in the form of forests. Field commodities such as corn, soybeans, long beans, peanuts, chillies and so on, while the garden products are dominant types of rubber and oil palm. Compilation of land use can be seen in Table E.3.1 below:

Table 4 Compilation of Land Use According to Nagari Region Around the Location of PT. Karbindo Abesyapradhi (End of Year 2013)

NO.	LAND USE	NAGARI REGION					
		Kunangan Prk Rantang		Kamang		Muaro Takuang	
		Area (Ha)	%	Area (Ha)	%	Area (Ha)	%

1.	Settlement	28,0	0,36	111,0	0,59	43,5	0,77
2.	Rice fields	34,0	0,43	136,0	0,72	168,0	2,98
3.	Field	734,0	9,32	2.043,0	10,85	1.429,0	25,37
4.	Garden	5.733,0	72,83	2.594,0	13,77	1.800,0	31,96
5.	Swamp	1,0	0,01	1,0	0,01	4,0	0,07
6.	Bush	342,0	4,34	655,0	3,48	1.051,0	18,66
7.	Mining	119,0	1,51	305,0	1,62	3,5	0,06
3.	Forest / Thicket	881,0	11,19	12.987,0	68,96	1.133,0	20,12
	TOTAL	7.872,0	100,00	18.832,0	100,00	5.632,0	100,00

In accordance with the ethnicity that grew and developed in the residential area around the coal mining business location of PT. Karbindo Abesyapradhi, the prevailing customs are Minangkabau (more dominant), Javanese and Batak (especially in the social environment of the Tambang River and Tanang River communities). Although the ethnicity that developed in the Sungai Tambang area is heterogeneous, assimilation between ethnic groups - including cultural or traditional assimilation - has taken place in harmony. So, until now there has never been an inter-ethnic social conflict. Furthermore, from the results of monitoring, it is also known that through the local community development program a harmonious relationship between one another has been created between PT. Karbindo Abesyapradhi with members of the local community.

PT. Karbindo Abesyapradhi is committed to making sustainable development very important, this is reflected in what the company does in the form of community empowerment. Moreover, of course it will risk the name of the company to get and maintain permits especially socially in conducting its operations in the community.At this time the mining industry has realized that to gain access and maintain its reputation must certainly be able to demonstrate its ability to conduct mine closure and effectively gain support from stakeholders, especially the communities where the mine operates. The expectations of existing regulations and stakeholders are getting higher, so that to get maximum results, a good and right method is needed.

15. Conclusion

The conclusions obtained from the study of the suitability of Sengon and Acacia Plants for Reclamation of Former Coal Mine in PT. Karbindo Abesyapradhi - Sijunjung Regency is as follows; There is no difference in soil fertility status or chemical content (clay texture, acidity, Ca content, Mg, soil potassium, soil CEC, C-organic soil, N-total soil) on both sengon and acacia tree vegetation growth.Partially affecting the growth of plants, both Acacia and Sengon trees, are Cation Exchange Capacity (CEC) and soil texture.

References

[1] Bambang P dan Lina M J 2005 *Metode penelitian Kuantitatif,Teori dan Aplikasi* (PT. Raja Grafindo Persada)
[2] Afandie R dan Nasih W Y 2002 *Ilmu Kesuburan Tanah* (Penerbit Kanisius)
[3] Bejo S 2008 *Kebocoran Hutan dan Anomali Illegal Logging* (Wana Aksara)
[4] Di Rektorat Teknik Dan Lingkungan Mineral Dan Batubara Direktorat Jenderal Mineral Dan Batubara, Departemen Energi Dan Sumber Daya Mineral
[5] Hanafiah K A 2005 Dasar-Dasar Ilmu Tanah (Jakarta : Penerbit PT. Raja Grafindo Persada)
[6] Iskandar 2008 Staf pengajar Dept. Ilmu Tanah & Sumberdaya Lahan, Fakultas Pertanian IPB, dan Peneliti pada Pusat Studi Reklamasi Tambang, (LPPM IPB) e-mail : issi-iskandar@yahoo.com; issi_iskandar@ipb.ac.id
[7] Kustiawan W 2001 Perkembangan vegetasi dan kondisi tanah serta revegetasi pada lahan bekas galian tambanga batubara di Kalimantan Timur *J. Ilmiah Kehutanan Rimba Kalimantan* **6** (2) 20-31
[8] Mohamad S 2006 *Lingkungan Hidup (The Living Environment)* (Jakarta : CV. Restu Agung Jakarta Pusat)

IOP Conf. Series: Earth and Environmental Science **314** (2019) 012022 doi:10.1088/1755-1315/314/1/012022

Studies breeding places of *Aedes aegypti* and *Aedes albopictus* in dengue endemic Padang area

Aidil Onasis[1,2,*], Eri Barlian[1] and Abdul Razak[1]

[1] Department of Environmental Science, Post Graduate Program, Universitas Negeri Padang, Prof. Dr. Hamka Street, West Air Tawar, Padang-West Sumatera 25131 Indonesia

[2] Padang Health of Polytechnic, Environmental Health Departement, Siteba Street, Padang-West Sumatera 25146 Indonesia

*onassisaidil@yahoo.co.id

Abstract. Suspect vector of Dengue Haemorrhagic Fever (DHF) and Chickungunya transmitted by Aedes aegypti and Aedes albopictus in container as potential breeding places. This study used a descriptive design, data collection in Kuranji, Nanggalo and Padang Timur and Lubuk Begalung district for 5 months in Padang with populations in house as many as 400 larvae index for density analysis to interpretation. The results container at risk of becoming breeding places of Aedes aegypti in Padang Timur districts 14.25%, and Aedes albopictus in Kuranji 20.75%. 3 classification area analysis are Aedes aegypti and Aedes albopictus in the container in the form of supervision that more routine in Nanggalo compared Kuranji and Lubuk Begalung well as in Padang Timur as well as control by getting rid of that is rare in Kuranji and Padang Timur while in Nanggalo and Lubuk Begalung category often get rid of the landfill. Updating data analysis results in the form of survey results, mosquito density monitoring strategy regularly and periodically by the community with assistance by health workers as well as the risk of transmission as well as the anticipation of Extraordinary Events and prevention of dengue outbreaks.

1. Background

Studies potential breeding in surveillance and epidemiological analyzes provide visualization as well as an ecologically, revealing trends, dependencies and inter-relationships, including the risk of diseases transmitted by Aedes aegypti and Aedes albopictus [1,2,3].

The mosquito's infectious disease by the Aedes aegypti and Aedes albopictus is still a public health problem that has both social and economic impact. [2,4] Social losses that occur, among others, caused panic in the family to the death of family members and reduced life expectancy. The immediate perceived economic impact on the patient is the cost of treatment, while the indirect is the loss of working time, school time and other costs incurred in addition to treatment such as transportation and accommodation during patient care. [2,4,5,6]

The prevalence of incidence and risk of DF and Chikungunya disease is transmitted by Aedes aegypti and Aedes albopictus mosquitoes in Padang city based on Padang City Health Office (DKK) report that Padang is an endemic area of dengue fever and sporadic that is almost encountered in 11 districts. This condition provides the spread of dengue cases that are still greater than the national Case Fatality Rate (CFR) that must be below 1%.

Human factors contributed to the growth of mosquito populations, mostly large cities in Indonesia, has grown rapidly with all its implications, such as the growth of slums due to urbanization, limited water supply, and unprofessional environmental management, transmission caused by many potential places as nests used Ae mosquitoes. aegypti and Aedes albopictus to lay eggs such as tanks and water Tubs in WC, buckets, Drum, Bird food place, Overflow Refrigerator and Dispenser and other containers. [2,6,7]

Control programs existing are known as 3 M movement, albitization and Fogging but this condition not perceived yet as a need to protect themselves and the family against the risk of transmission, it is necessary to share information and knowledge about breeding places, to find larvae, able to recognize larva and estimate the density of larvae at each community water reservoir [1,3,6].

Ecological analysis is expected to be able to study the potential of the environment to be breeding places mosquitoes and design form of control that is easily applied by the community. The whole process of surveillance and epidemiological analysis. provides a means for visualizing and analyzing epidemiological data, showing trends, dependencies and risk-related diseases including those transmitted by Aedes aegypti and Aedes albopictus mosquitoes and other mosquitoes as Suspect vectors. [1,4,7,8,9]

2. Materials and Methods

Observations and surveys breeding larvae in August to December 2017. This study population as potential house mosquito breeding Aedes aegypti and Aedes albopictus. Followed by the field of data processing into the data with computer software, through the determination of the risk of transmission and dignity proportionally qualitative namely: CI <11% Free Larvae and CI> 11% area is classified density by Index Larvae.

3. Result

Density of Aedes aegypti larvae and Aedes albopictus. The research results in Table 1 and Table 2 described the density of Aedes aegypti and Aedes albopictus based Container Index shows the number density of more than 10% so the value density Low was given a scoring of 1, the value of the given weight 2 and the highest value.

Table 1. Density of Distribution Aedes aegypti Larvae Breeding by Container Index in Padang 2017

No.	District	Number	Container Index (%)
1.	Kuranji	265 (+35)	13,20
2.	Nanggalo	328 (+40)	12,19
3.	Padang Timur	456 (+65)	14,25
4.	Lubuk Begalung	368 (+45)	12,22

From table 1, explained that the highest density of Aedes aegypti larvae is in the Padang Timur sub-district of 14.25%

Table 2. Density of Distribution Aedes Albopictus Larvae Breeding by Container Index in Padang 2017

No.	District	Number	Container Index (%)
1.	Kuranji	265 (+55)	20,75
2.	Nanggalo	328 (+50)	15,24
3.	Padang Timur	456 (+45)	15,04
4.	Lubuk Begalung	368 (+45)	12,22

From table 2, explained that the highest density of Aedes albopictus larvae is in the Kuranji sub-district of 20.75%

Characteristics of breeding places with mosquito larvae are from 1.417 containers / domestic water reservoirs positive larvae as many as 195 containers with average Density 13.76%. The highest density on the Aedes aegypti larvae in Padang Timur District (14.25%) and Aedes albopictus mosquito larvae in Kuranji district (20.75%).

Trend analysis of density of Aedes aegypti larvae and Aedes albopictus larvae research results can be explain the distribution of the density of area a high density is in Padang Timur district, whereas in Nanggalo, Kuranji and Lubuk Begalung district with the density being in the community landfill. Compared with Kuranji district. The results are consistent with research that says that the area affected by dengue fever in general is a district is high populated. The houses that are close together facilitate the transmission of disease-based breeding places domestic water reservoirs. [2,6,7,8]

4. Conclusion

The density of Aedes aegypti larvae in Padang Timur District 14.25% and Aedes albopictus in Kuranji 20.75%. The distribution of the density of Aedes aegypti larvae in breeding places in Padang Timur district and Aedes albopictus in Kuranji with the highest indices and weights classification.

Larvae density monitoring is conducted periodically, regularly and periodically with the recommendation of at least every month. Update Survey with Periodic monitoring which give container index by Health and Research Program Vectors to be integrated

References

[1] Alto B W et al 2012 Demographic consequences of predators on prey: trait and density mediated effects on mosquito larvae in containers *PLoS ONE* **7**(11) e45785

[2] Baewono D T et al 2012 Distribusi spasial kasus demam berdarah dengue (DBD), analisis indeks jarak dan alternatif pengendalian vektor di Kota Samarinda Provinsi Kalimantan Timur *Media Litbang Kesehatan* **22**(5) 131–137 available at : http://ejournal.litbang.depkes.go.id/index.php/MPK/article/view/2907

[3] Couret J, Dotson E and Benedict M Q 2014 Temperature, larval diet, and density effects on development rate and survival of Aedes aegypti (Diptera: Culicidae) *PLoS ONE* **9**(2) e87468

[4] Higa Y 2011 Dengue Vectors and their Spatial Distribution *Tropical Medicine and Health* **39**(4 Suppl) 17–27 doi: 10.2149/tmh.2011-S04

[5] Hu W et al 2011 Spatial analysis of notified dengue fever infections *Epidemiology and infection* **139**(3) 391–399 https://doi.org/10.1017/S0950268810000713

[6] Luz P M et al 2011 Dengue vector control strategies in an urban setting: An economic modelling assessment *The Lancet* **377**(9778) 1673–1680 Available at: http://dx.doi.org/10.1016/S0140-6736 (11) 60246-8

[7] Onasis A 2012 Hubungan Kondisi TPA Rumah Tangga Dengan Keberadaan jentik Aedes sp Pada Kejadian DBD Kota Padang Tahun 2010

[8] Vanwambeke S O et al 2006 Multi-level analyses of spatial and temporal determinants for dengue infection *International journal of health geographics* **5**(5); doi: 10.1186/1476-072X-5-5

[9] Wen T H et al 2010 Spatial-temporal patterns of dengue in areas at risk of dengue hemorrhagic fever in Kaohsiung, Taiwan 2002 *International Journal of Infectious Diseases* **14**(4) e334-e343; https://doi.org/10.1016/j.ijid.2009.06.006

Decreasing level of heavy metals Fe and Mn use the wetland method at coal open mining PT Bukit Asam South Sumatra Province

Heri Prabowo*, Ali Amran and Ardinis Arbain

Department of Environmental Science, Post Graduate Program, Universitas Negeri Padang, Prof. Dr. Hamka Street, West Air Tawar, Padang-West Sumatra 25131 Indonesia

*heri.19782000@gmail.com

Abstract. PT. Bukit Asam, Tbk is a coal mining company that directly produces waste water. For this reason, it is necessary to treat wastewater to neutralize substances that pollute the wastewater, one of the methods used is constructed wetland. The method used is descriptive quantitative method. Where this study will produce the influence of the constructed wetland method in neutralizing waste water from the coal mining process at PT. Bukit Asam, Tbk. Based on the results of the study it can be seen that the waste treatment caused by the Constructed Wetland coal mining process can increase the Ph value from 3 samples averaging 2.6 mg / l, reducing the Fe metal content from 3 samples averaging 9.98 mg / l , and lowering the metal content of Mn from 3 samples averaging 0.75 mg / l and decreasing the level of Total Suspended Solid from 3 samples averaging 486 mg / l. The constructed wetland method that has been implemented at PT. BA has succeeded in neutralizing Fe and Mn metals and TSS according to the standard quality values that meet the standards set by the government.

1. Introduction

Mining is an activity utilizing minerals and coal so that it can be used to meet human needs, one of the industries that supply domestic and foreign energy needs is coal mining. Coal is widely used in various industries including electricity generation, steel industry, space heating and others, in its production activities the industry requires a lot of coal as raw material, in order to fulfill this, Indonesia is one of the largest coal producing countries in the world, to meet the needs Indonesian and foreign coal must implement good mining practices, in fact the coal mining process in Indonesia is generally still not optimal in the implementation of good mining practices, one of the things that are still largely ignored by coal mining companies, namely the problem of handling wastewater into the environment outside the mine [1].

For waste water management, the government has implemented a standard for wastewater treatment, which is based on the environment minister's decree number 113 of 2003, that every mining business is obliged to process wastewater, so that the quality of wastewater discharged into the receiving environment does not exceed the standard quality of wastewater determined.

The wastewater produced from mining activities is called mine acid water, the waste contains iron metal, manganese and high turbidity levels which can pollute river water, and can be resistant to humans and living things that use the river's functions [2]. "High levels of metal Fe can affect water quality including causing reddish-brown water to smell and cause crusting in the piping system".

Therefore PT. Bukit Asam Tbk. Tanjung Enim Mining Unit needs to conduct acid mine drainage treatment before being channeled to the receiving river. In processing PT. Bukit Asam Tbk. perform processing of acid mine drainage in two ways, namely by active treatment and passively treatment, the treatment system is in active treatment using chemicals in the form of cursive, quicklime and pH adjuster (NaOH solution), and for processing in a passive treatment using constructed wetland methods (artificial wetlands), this method is applied by using water plants that can increase pH, reduce TSS, and absorb heavy metals, especially Fe and Mn in waste water produced from coal mining, to meet environmental quality standards contained in the Minister of Environment Decree Number 113 of 2003, and the Governor of South Sumatra Number 8 of 2012.

2. Materials and Methods

This research is descriptive quantitative research. The purpose of this study was to provide an overview of the effects of the constructed wetland method in neutralizing coal mining wastewater. In this study, to analyze the increase in pH value, decrease in Fe metal content, Mn metal content and suspended residues in acid mine drainage, laboratory tests were carried out and the data were analyzed and compared with wastewater quality standards according to Minister of Environment Decree No.113 2003 and the Governor of South Sumatra Regulation No. 8 of 2012. In this study, samples were taken of each pool in the Stockpile 1 mud settling pond, sampling of each pool was carried out three times with a span of 1 week.

Table 1. Metals and Quality Standards of Fe and Mn Metals in Acid Mine Water

No	Parameter	Unit	Quality Standart
1	Iron (Fe) Total	Mg/L	7
2	Manganes (Mn) Total	Mg/L	4

Reference: Governor of South Sumatra No. 8 of 2012

3. Results and Discussion

3.1. Change in PH value

Constructed wetlands can reduce pH values from 5.89 to 6.21 in the first week, 5.11 to 6.33 in the second week and 4.86 to 6.44 in the third week, the following can be seen in the graph.

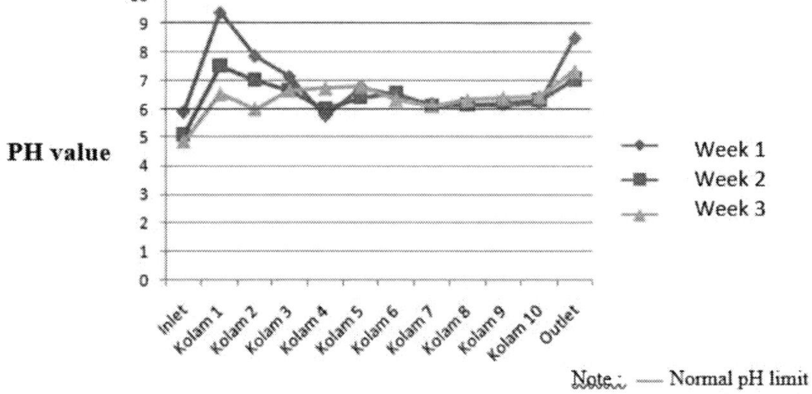

Figure 1. Graph PH value

For the research of metal content is the same as the research for pH value where sampling is done three times with a distance of one week.

3.2. Change in value of Fe

The effect of Constructed Wetland can reduce dissolved metal Fe in mine acid water waste from 10.215 mg / l to 0.5025 mg / l in the first week, and 9.3272 mg / l to 0.6572 mg / l in the second week, and 12.3122 mg / l to 0.621 mg / l in the third week, the following can be seen in the figure image.

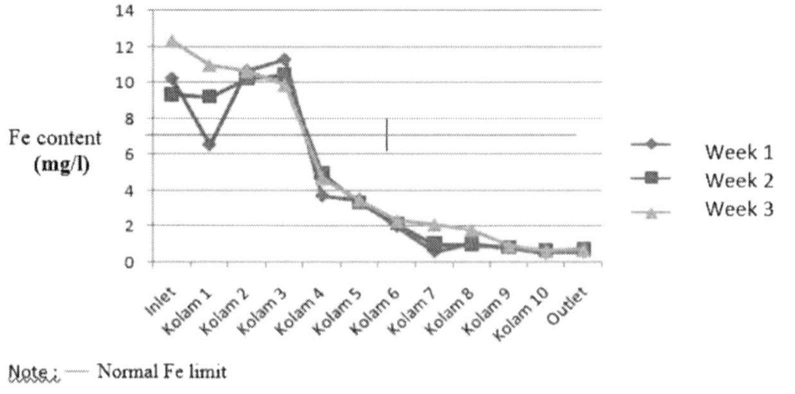

Figure 2. Graph Fe value

3.3. Change in the value of Mn

For the study of metal content the same as the research for pH and Fe values where sampling was done three times with a distance of one week, from the results of the study found the effect of Constructed Wetland can reduce dissolved Mn metal content in mine acid water waste from 0.8586 mg / l to 0.1512 mg / l in the first week, and 0.7735 to 0.1165 mg / l in the second week, and 1.0023 to 0.1516 mg / l at the third week, the following can be seen in the figure.

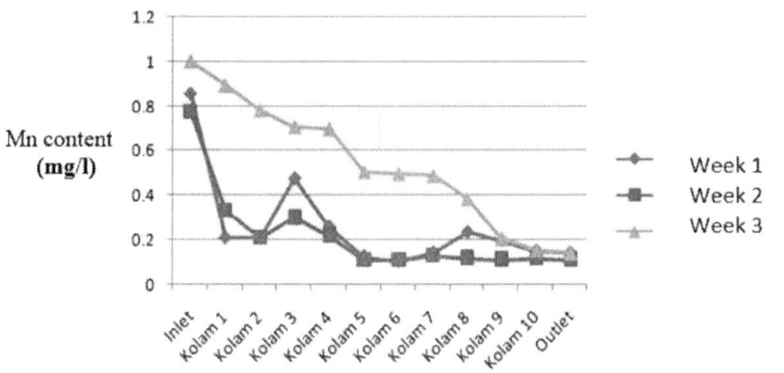

Figure 3. Graph Mn value

3.4. Change in TSS value

For the study of metal content the same as the research for the value of pH, Fe and Mn where sampling was done three times with a distance of one week, from the results of the study found the effect of Constructed Wetland can reduce TSS levels in mine acid water waste from 405 mg / l to 3 mg / l in the first week, and 443 mg / l to 6 mg / l in the second week, and 621 mg / l to 4 mg / l at the third week, the following can be seen in the figure.

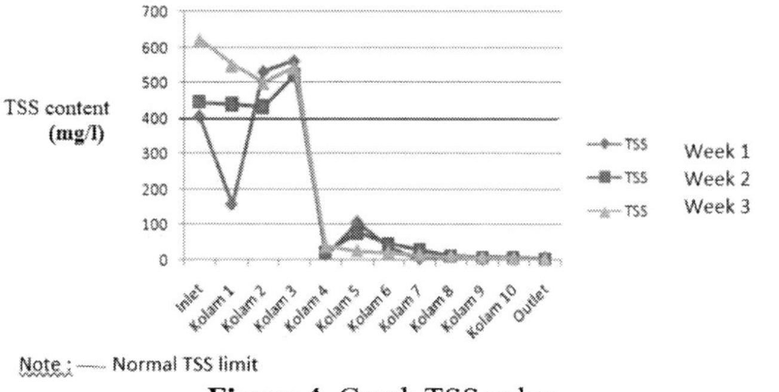

Figure 4. Graph TSS value

3.5. Effect of water plants on pH values

According to Arroyo [3] study that aquatic plants can increase pH values from low to high, this is caused by:

1. The content of alkalinity of compost used as a wetland matrix, because the compost can increase the growth of sulfate reducing bacteria which can increase alkalinity, and then can increase the pH value.
2. Interaction of aquatic plants with the surrounding environment, the presence of aquatic plants floating in the pond causes anaerobic environment under the pond, then the water plants carry out photosynthesis by taking CO2, due to the taking of the CO2 by plants so that the pH value of the water increases.

From the results of testing by mardaniyah the average pH increase caused by aquatic plants is 35%

3.6. Effect of aquatic plants on Fe and Mn levels

Based on Prihatini [4] study that water plants can reduce levels of Fe and Mn, this is caused by:

1. Interaction between sulfides (S^{2-}) produced in the process of reducing sulfates with valentic metals 2 (such as Fe^{2+} and Mn^{2+}) forming precipitated metal sulphides.
2. The process of metal absorption by plant tissue
3. The process of adsorbing metals by organic materials
4. The process of metal biosorption by microorganisms found in wetland environments

From the results of laboratory tests, it has also been proven that these aquatic plants can absorb certain amounts of metal in the following table from the results of tests carried out

Table 2. Accumulation of absorption of Fe and Mn elements by plants

Plant type	heavy metal	Accumulation (pp/gram dry weight)	
		Root	Stem
Kiambang	Mn	7682	516
(Salvania natans)	Fe	13231	2842
Vetiver	Mn	486	119
	Fe	3964	448
Enceng Gondok	Mn	1120	1366
	Fe	5350	394
Tipha	Mn	936	1484
	Fe	15116	1432
Pakis	Mn	1700	1432
	Fe	13923	1432
Mandoan	Mn	526	260
	Fe	62686	7642

From the table it can be seen that the absorption of Fe and Mn content is mostly absorbed in the roots of the aquatic plants. According to Mardalena [1] the spread and composition of Fe metal accumulates a lot at the bottom of the pond or in other words Fe besides absorbed by plants is also deposited by plants, then for Mn metal the spread and composition in the pond shows the same condition so that Mn metal is not easily deposited

3.7. Effect of water plants on levels of Total Suspended Solid (TSS)
According to Risnawati [5] TSS is suspended materials that are not soluble in water. Suspended solids consist of particles whose size and weight are smaller than sediments. According to Hadiwidodo [6] the decrease in TSS levels in ponds because there is a connection with plant roots, the longer the plant roots, the TSS levels will decrease, the length of plant roots is directly proportional to the duration of the experiment and decreases in TSS levels in the pond. According to Arroyo [3], the decline in TSS is because plants have positively charged root hair that attracts opposite-charged colloidal particles such as suspended solids, so that the particles attach to the roots and are slowly absorbed and assimilated by plants and microorganisms.

3.8. Effect of water discharge on decreasing TSS levels
Water flow can affect the contact process between waste and decomposing microorganisms, in the constructed wetland method used here surface flow with the incoming water discharge is not too large, because the water discharge can affect the process of absorption of plants in absorbing metals, From the results of the author's observation, the relationship between water discharge and decreasing TSS levels can be seen in the following table:

Table 3. Relation of water discharge to decreasing TSS levels

Comparison	Research result		
	Week 1	Week 2	Week 3
TSS *Inlet (mg/l)*	405	443	621
TSS *Outlet (mg/l)*	6	3	2
Debit (m^3/s)	0.21	0.14	0.08

Based on table 3 it can be concluded that the slower the water discharge in constructed wetland ponds, the higher the decrease in TSS levels that occur. Conversely, the faster the water discharge in constructed wetland ponds, the smaller the decrease in TSS levels.

4. Conclusion
From research conducted at PT. Bukit Asam, Tbk, the authors draw conclusions:

1. Acid mine water treatment applied by PT. Bukit Asam, Tbk with passive treatment technology, namely constructed wetland method.
2. Plants used with the constructed wetland method are using kiambang, vetiver, water hyacinth and kiapu.
3. Passive treatment of acid mine drainage applied by PT. Bukit Asam, Tbk, using constructed wetland methods is effective in decreasing levels of Fe and Mn.
4. The saturation time of plants in the constructed wetland method is 6 months, so that plants need to be renewed once every 6 months, so that the treatment of acid water with the constructed wetland method is maximal.
5. From the results of testing the pH value, the level of Fe metal, the metal content of Mn and the total residual thesis, it meets the applicable liquid waste quality standards in Minister of Environment Decree Number 113 of 2003, and the Governor of South Sumatra Number 8 of 2012.

References

[1] Mardalena M, Faizal M and Napoleon A 2018 Metal of iron (Fe) and mangan (Mn) from waste water coal mining with fitoremediation techniques with using floating fern (Salvinia Natans), water lettuce (Pistia Stratiotes) and water-hyacinth (Eichornia Crassipes) *BIOVALENTIA: Biological Research Journal* **4**(1)

[2] Puspita L, E Ratnawati, I N N Suryadiputra and A A Meutia 2005 *Lahan Basah Buatan di Indonesia* (Bogor, Indonesia: Wetlands International-Indonesia Programme) available at https://indonesia.wetlands.org/id/publikasi/lahan-basah-buatan-di-indonesia/

[3] Arroyo P, Ansola G and de Luis E 2010 Effectiveness of a full-scale constructed wetland for the removal of metals from domestic wastewater *Water, Air, & Soil Pollution* **210**(1-4) 473-481

[4] Prihatini N S, Nirtha I and Iman M S 2016 Role of Purun Tikus in vertical subsurface flow constructed wetland in treating manganese (Mn) from coal mine drainage *TROPICAL WETLAND JOURNAL* **2**(1) 1-7

[5] Risnawati I and Damanhuri T P 2010 Lindi metal allowance using constructed wetland metal removal from leachate in constructed wetland *Journal of the Civil and Environmental* Faculty Environmental Engineering Study Program, Institute Technology of Bandung

[6] Mochtar H, Oktiawan W, Primadani A R, Parasmita B N and Gunawan I 2012 Pengolahan air lindi dengan proses kombinasi biofilter anaerob-aerob dan wetland *Jurnal PRESIPITASI* **9**(2) 84-95

Density, coverage and biomass of seagrass ecosystem in the Lobam Island, Bintan Regency - Indonesia

Widya Prarikeslan*, **Dedi Hermon, Yurni Suasti and Aprizon Putra**

Department of Geography, Universitas Negeri Padang, Prof. Dr. Hamka Street, West Air Tawar, Padang - West Sumatera 25131 Indonesia

*whe2geounp@gmail.com

Abstract. Seagrass ecosystem is a sea plant that can store carbon in the form of biomass and in sediments known as blue carbon. This research conducted in around coastal waters of Lobam island, Seri Kuala Lobam sub-district, Bintan regency that has seagrass community type *Thalassia hemprichii* and *Enhalus acoroides* which has an important role as a carbon sink. This research aims to determine differences of density, coverage and differences in biomass based on a seagrass tissue at the *Above ground* (leaf *Abg.*) and *Below ground* (rhizoma and root *Blg.*) of each station and the relationship between the density towards seagrass biomass. The research results show that the seagrass density in station 2 was higher than in station 1. While the difference in the seagrass biomass there is in the section of *Abg* (leaf) and *Blg* (rhizoma and root) in station 1 and conversely there is no significant difference in station 2.

1. Introduction

The issue of global warming that has implications on the rate of climate change that occurs today is very worrying, where human activity is the largest contributor of carbon dioxide (CO_2) gases into the air. Human activities that can release CO_2 emissions are land burning, motor vehicle emissions, factory wastes and so on that lead to an increase in greenhouse gas concentrations (GHG). [1] explain this increase causes the radiation balance to change so that the earth's temperature increases. This thermal nature of radiation causes global warming of the atmosphere (global warming) [1,2]. Scientific evidence until now has also proven that there are marine ecosystems that act as carbon sinks.

One mitigation effort undertaken to reduce atmospheric CO_2 concentration is to nurture and develop the ability of forests and oceans to absorb and store carbon. Forest utilization in these mitigation efforts has been widely implemented, while the application of the role of the oceans has not been seen significantly [3,4]. Thus, oceans have a high enough ability to bind and store CO_2. Therefore, it is indispensable for marine ecosystem services in carbon sequestration.

The seagrass ecosystem is an important ecosystem to support the life of aquatic biota and is used as a nurturing, spawning, feeding and enlargement of larvae of aquatic organisms [5,6]. Seagrass ecosystems are important to protect because of their very important functions for the sustainability of fishery resources. The assessment of the seagrass condition becomes something that is needed as a control to see the condition of seagrass beds from time to time. [7] add the function of the seagrass is as a provider of shelter for marine biota living in it, as well as a nursery ground for several species of marine biota.

IOP Conf. Series: Earth and Environmental Science **314** (2019) 012024 doi:10.1088/1755-1315/314/1/012024

There are 4 seagrass families known throughout the world's waters, 2 of which are in Indonesia, i.e *Hydrocharitaceae* and *Potamogetonaceae*. In Indonesia, there are 12 types of seagrasses, from 12 species of seagrass growing in Indonesian waters [6,8] where 10 species can be found in the waters of Bintan Island, Riau Islands [9]. Seagrass ecosystem in Indonesia is often encountered in the inner intertidal and upper subtidal regions. Seen from the pattern of horizontal seagrass zoning, seagrass ecosystem is located between two important ecosystems i.e mangrove and coral reef ecosystems. Seagrass ecosystem is closely related and interacting, as well as the link and as a buffer of the sedimentation rate from mangrove ecosystem on the coast to coral reef ecosystem towards the sea [10,11].

The status of the seagrass condition determines the occurrence of the indication of seagrass damage due to the activity and influence around the coast. Changes in the conditions and status of seagrass beds can be analyzed using a community approach in the form of levels of coverage and density. From this approach, can be seen how far the damage to seagrasses occurred.

Lobam Island is one of the natural islands in Sri Kuala Lobam sub-district, Bintan Islands which has seagrass ecosystems in the category of healthy to moderate [12]. The existence of seagrass which is broad enough in the waters of the island will certainly contribute to the climate balance through the seagrass function as a carbon sink (blue carbon). Based on these evolving issues and given the importance of the role of marine ecosystems in reducing GHG emissions, there is a need for a lot of research that can encourage the continued development of carbon in biomass calculations. This study aims to determine differences of density, coverage and differences in biomass in seagrass tissue in the *Above ground* (leaf) and *Below ground* (rhizoma and root) sections between stations and the relationship between density with seagrass biomass.

2. Research Methods

2.1. Research Location

This research was conducted in the waters of Lobam island, Sri Kuala Lobam sub-district, Bintan regency, Riau Islands province. Primary data collection in the field and analysis in the laboratory was carried out in November 2017. Geographically the location of the research is at coordinates 0°58'10.50" - 1°0'6.36" N and 104°13'9.11" - 104°15'52.42" E (Figure 1) is one of the conservation area plans based on the Decree of the Regent No. 261/VIII/2007 dated August 23, 2007 [13] and also the pre-eminent natural tourism area of Bintan regency and a pilot program on seagrass management in Indonesia, especially in the eastern part of the island of Sumatra.

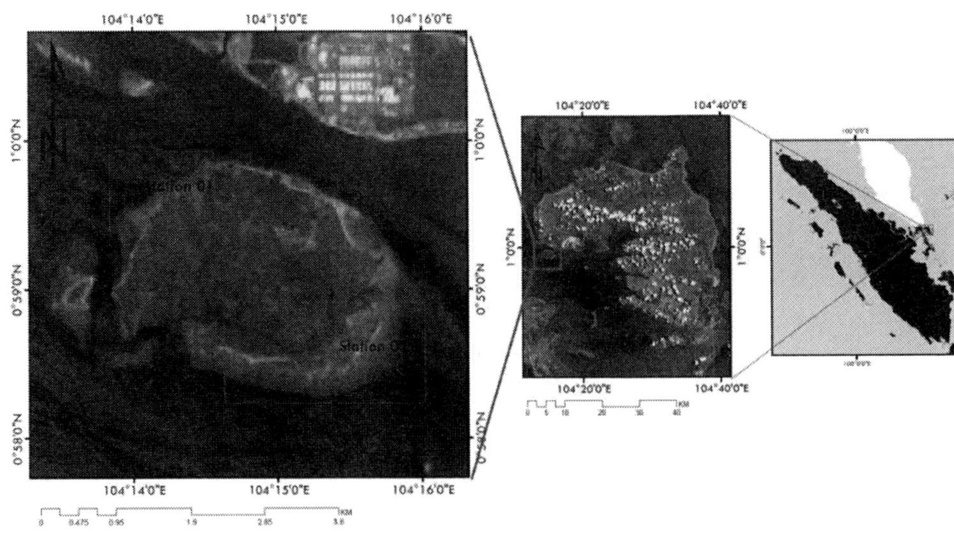

Figure 1. Map of Research Location.

The method used in this research is survey method, i.e observation and sampling directly in the field. Determination of sampling points taken at 2 observation stations where station 1 is located in an area close to anthropogenic activities (catching fish) in the Lobam strait, while station 2 is located in the mangrove area which has little influence on anthropogenic activities.

2.2. Data and Samples

Sampling and observation of the condition of seagrass were carried out using the line transect method using a sized-square (50 × 50) cm^2 [6,14]. The number stand of seagrass beds type of *Thalassia hemprichii* and *Enhalus acoroides* in the square plot was calculated to determine its density (stand/m^2) and the percentage value of seagrass coverage (%) was determined by observing the cover of seagrass leaf of *Thalassia hemprichii* and *Enhalus acoroides* (the other types were ignored) in each small box. The number of sampling points was 8 points scattered in 2 the observation stations.

Biomass sampling is done using a transect measuring (50 × 50) cm^2. Seagrass found on the transect is taken one rod (revoked) on the depth of root penetration. The sample is put into a sample bag after being cleaned of the substrate then put into ice box and taken to the laboratory, then cleaned (from sand, biota and epiphytes) and washed with fresh water to knock out epiphytes or biota attached to the seagrass. Each seagrass is separated between the *Above ground* (leaf *Abg.*) and *Below ground* (rhizoma and root *Blg.*). Then all the seagrass samples were dried by putting them in an oven at a fixed temperature of 60 °C for 72 hours, then weighed to find out the dry weight using analytical scales. After that, calculation of seagrass biomass (g BK/m^2) was carried out using the formula according to [15,16,17].

2.3. Seagrass Biomass Analysis

The calculate of seagrass coverage as follows [18,19]:

$$Seagrass\ Coverage\ (\%) = \frac{Amount\ of\ seagrass\ coverage\ values\ (4\ box)}{4} \tag{1}$$

The calculate seagrass density as follows [18]:

$$Density\ Ea\ \left(stand/m^2\right),\ and\ density\ Th\ \left(stand/m^2\right) = Amount\ (Ea^* \times 4)\ and\ (Th^* \times 4) \tag{2}$$

Information:
Ea = *Enhalus acoroides*
Th = *Thalassia hemprichii*
* = Amount of individuals in squares
4 = Conversion constants (50 × 50) to 1 m^2.

Biomass is an organic material produced through photosynthesis, good in a shape of a product or waste. Seagrass tissue biomass (g) can be calculated by a formula [15,17]:

$$Seagrass\ tissue\ biomass\ (g\ BK/m^2) = \frac{Dry\ weight\ (g)}{Plot\ Size\ (m^2)} \tag{3}$$

The data that obtained from research results were analyzed statistically and discussed descriptively with reference to the literature. Data processing is made with a software of Microsoft Excel and *Statistical Package for Social Science* (SPSS) version 16.0, as well as geographic information data assisted with a software of *Environment for Visualizing Images* (ENVI) version 5.1 and ArcGIS 10.4.

3. Results and Discussion

3.1. Density and Percentage of Seagrass Coverage

Based on the research results than obtained the seagrass density values on the 2 stations (Figure 2[a]), It can be seen that the value of seagrass density as a whole at station 2 (natural of condition waters) with an average total value of 128.00 stands/m^2 is higher than the seagrass density value in station 1 (location of anthropogenic activities) with an average value of 115.00 stands/m^2. While the percentage of seagrass coverage in the research location is presented in Figure 2[b], it is seen that the percentage of seagrass coverage in Station 2 is higher with an average value of 30.03% from in Station 1 with an average value of 27.21%.

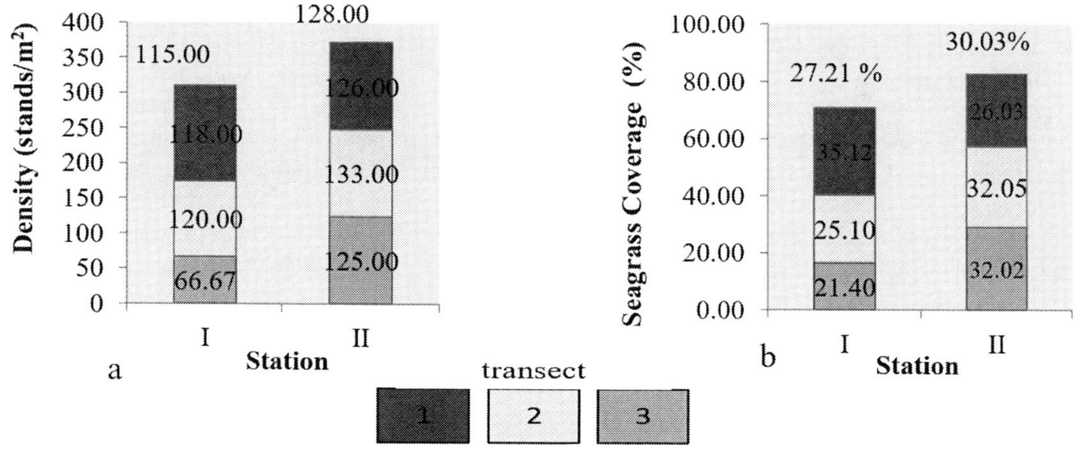

Figure 2. a) Density; and b) Percentage of Seagrass Coverage.

Seagrass type of *Thalassia hemprichii* and *Enhalus acoroides* found in the waters of Lobam island are a type of seagrass that usually lives in shallow waters which are always exposed during low tide [20]. The relatively shallow waters of Lobam island with a high level of brightness strongly support high seagrass density and coverage. On each squared transect of each research station, there are different seagrass density of different *Thalassia hemprichii* and *Enhalus acoroides*. Differences of density and seagrass coverage in each station can illustrate that the spread is varied, this can be caused by differences in environmental conditions and treatment obtained in 2 station representing dense areas from anthropogenic activity (catching fish and coastal tourism) and lack of anthropogenic activity (natural environmental conditions).

The seagrass density of *Thalassia hemprichii* and *Enhalus acoroides* in station 2 was higher than seagrass density in station 1. The average seagrass density value at a distance of 0 m, 10 m, and 20 m at each station explained that the seagrass distribution varies according to the average value of density. In station 1, the average of seagrass density in each plot is different and fluctuating. On the contrary in station 2, increasingly getting towards the sea the value of seagrass density is getting higher.

The difficulty of seagrass developing in Station 1 allegedly due to the relatively shallow waters at low tides and the waters that are relatively much affected by anthropogenic activities that come from household and industrial waste, fishing vessels, ship disposal, and beach tourism [21]. On the contrary seagrass in station 2 is better than in station 1, this is because it grows in waters that are relatively deep despite being shallow and slightly affected by anthropogenic activities (Located in mangrove areas that are still natural). Based on research [21] the growth of seagrass is strongly influenced by tidal patterns, salinity, pH and water temperature, and anthropogenic activities in coastal areas such as fishing activities, berths of small vessels, harbour development and recreation or beach tourism, both directly and indirectly can also affect seagrass growth.

3.2. Seagrass Biomass

Biomass values of seagrass in the waters around Lobam island can be seen in Figure 3. Based on the graph in Figure 3[a] and the calculation of the total value of seagrass biomass in station 2 is higher with a value of 252.64 g BK/m² from in station 1 with a value of 150.17 g BK/m². The average value of seagrass biomass at the research location from 2 research Stations was 201.40 g BK/m².

The value of largest seagrass biomass is on the *Blg* (rhizoma and root) from 2 stations compared to the *Abg* (leaf) section. Seagrass biomass values of *Enhalus acoroides* is higher than *Thalassia hemprichii* in part of rhizoma and leaf, where is the length of the leaf blade of *Enhalus acoroides* can reach 75 cm and width 1.0-1.5 cm with rhizoma width is 1 cm [22].

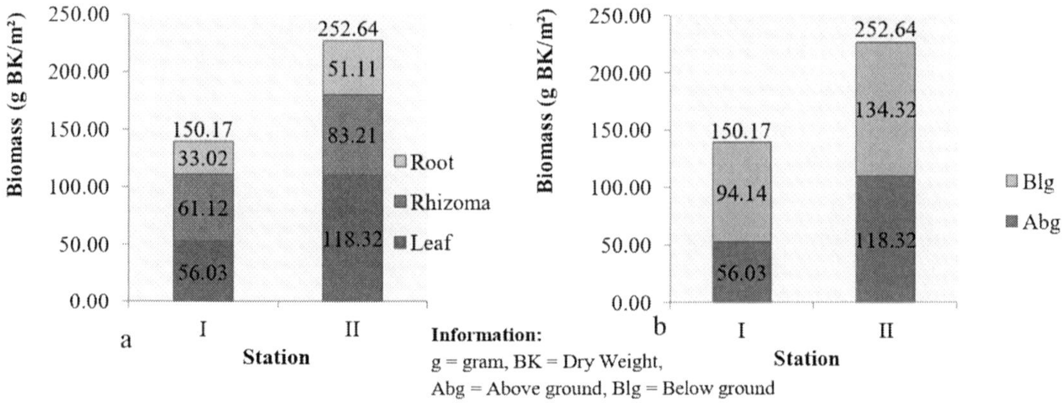

Figure 3. a) Comparison of the average biomass on the tissue; and b) Comparison of the average biomass in *Abg* and *Blg*.

Seagrass biomass in station 2 was higher from station 1. This was because in station 2 seagrass type of *Enhalus acoroides* and *Thalassia hemprichii* was found to have a density value and the percentage of seagrass coverage was greater than station 1. This affects the value of seagrass biomass in every station, [23,24] explains that seagrass meadows that have a type of seagrass that has a larger size leaf and rhizoma and high density will cause a higher biomass. The high biomass of seagrass species of *Enhalus acoroides* and *Thalassia hemprichii* in station 2, reflects that the seagrass type this likes of the environmental conditions in station 2 where it does not get the influence of anthropogenic activity intensively from station 1.

Suspected leaf size that long makes the leaf more easily carried away by the flow so that it gives a burden in the rhizoma to defend itself. Where the seagrass in station 2 is located far from the shoreline and directly affected by the high seas, suspected of making the rhizoma and root at this station strong and large in order to survive from the currents and waves. While seagrass condition in station 2 was found with better conditions than seagrasses at station 1. Longer leaf and larger rhizoma make station 2 have relatively high biomass values compared to seagrasses in station 1 which have shorter (not intact) leaf or are more severed due to human activities and environmental factors, thus making the biomass content at station 1 relatively low both in leaf, root and in the rhizoma. If pay attention overall in each Stations, then it will be seen that the average density and biomass values will be directly proportional, i.e station 2 is the Station that has the highest density and biomass value. For more details, the appearance of the bases waters for the condition of the seagrass ecosystem in the observation station from the analysis of density, coverage and biomass can be seen based on lyzenga analysis using the *Geography Information System* (GIS) approach [25,26,27,28,29].

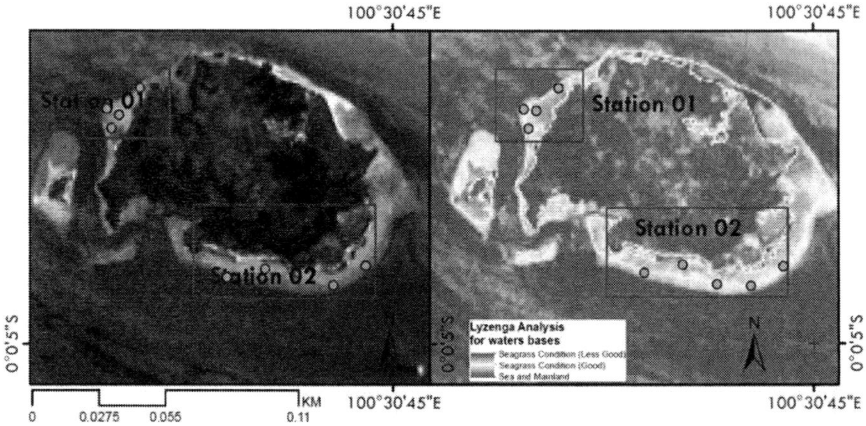

Figure 4. Differences in the colour of coastal waters in the research location for seagrass conditions.

3.3. Relationship of Density with Biomass

In the results of the significance test (Tukey and Benferroni), showed that the biomass of each seagrass tissue had a significant that is a tissue of leaf and root tissue. On the test results of the group similarity, it was found that the biomass between root and leaf had a significant difference and there were no significant differences between root biomass and rhizoma and rhizoma biomass and leaf. Based on Figure 3 and T-test, it is known that in station 1, the amount of biomass of Blg is greater than that of *Abg* ($p<0.05$). In contrast to station 2, the average value of seagrass biomass in part of *Abg* and *Blg* did not differ significantly ($p>0.05$).

Based on the results of simple linear regression test in Figure 4 it is known that there is a seagrass density relationship with the seagrass biomass, indicated by the correlation coefficient (r) that is 0.413, which means that the relationship between density and the seagrass biomass is medium with the regression equation $y = 96.95 + 0.761x$. The coefficient value of determination (R^2) = 0.171, which means 17.1% of the seagrass biomass variation can be explained by the seagrass density variables, while the remaining 82.9% is influenced by other factors outside the independent variable

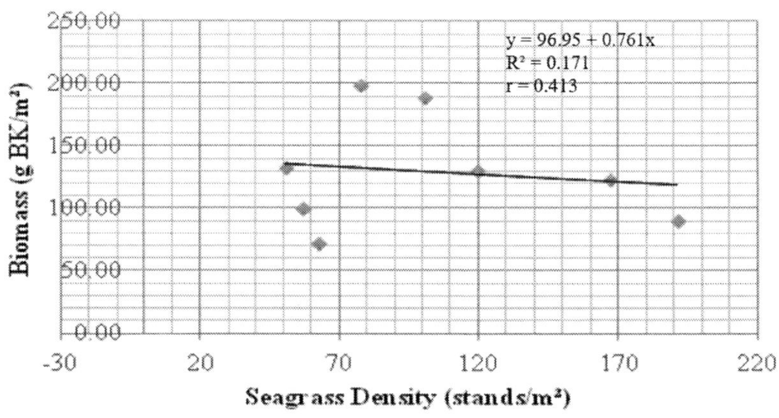

Figure 5. Relationship of density with seagrass biomass.

4. Conclusion

Seagrass density of *Thalassia hemprichii* and *Enhalus acoroides* in Lobam island beach was higher in station 2 than in station 1, while seagrass coverage was classified as poor and included in the medium category with the percentage of seagrass coverage also higher in station 2 than in station 1. The largest biomass value is found in the section *Blg* (root and rhizoma) seagrass type of *Thalassia hemprichii* and *Enhalus acoroides* at both Stations. There is a significant difference in biomass the section of *Abg* and *Blg* in station 1, and conversely there is no significant difference in station 2. The relationship between the seagrass density and the seagrass biomass is medium, because the growth conditions of seagrass stand are not the same or the size of each seagrass stand is different. The high potential of the seagrass biomass illustrates the magnitude of the role of these ecosystems in tackling the adverse effects of global warming so that efforts need to be made to protect and undertake the conservation of the ecosystem.

Acknowledgments
The authors wish to thank Rector of Padang State University - Indonesia for support and we also express our gratitude to all those who have helped with the administration process as well as those who helped us in the field.

References

[1] Hermon D 2016 The changes of carbon stocks and CO_2 emission as the result of land cover change for tin mining and settlement in Belitung Island Indonesia *J. Geography and Earth Sciences* **4** 17-30
[2] Perez-Lopez R, Montes-Hernandez G, Nieto J M, Renard F and Charlet L 2008 Carbonation of alkaline paper mill waste to reduce CO_2 greenhouse gas emissions into the atmosphere *Applied Geochemistry* **23**(8) 2292-2300
[3] Nellemann C 2009 The environmental food crisis: the environment's role in averting future food crises (a UNEP rapid response assessment)
[4] Hermon D 2017 *Climate Change Mitigation* (Jakarta: Rajawali Pers - Radja Grafindo)
[5] Lamb J B, van de Water J A, Bourne D G, Altier C, Hein M Y, Fiorenza E A and Harvell C D 2017 Seagrass ecosystems reduce exposure to bacterial pathogens of humans, fishes, and invertebrates *Science* **355**(6326) 731-733
[6] Tanto T A, Putra A, Hermon D and Damanhuri H 2018 Suitability of seagrass ecosystem for marine ecotourism in Padang City, West Sumatera Province *Forum Geografi* **32**(1) 88-95
[7] Cullen-Unsworth L and Unsworth R 2013 Seagrass meadows, ecosystem services, and sustainability *Environment: Science and Policy for Sustainable Development* **55**(3) 14-28
[8] Erftemeijer P L 1994 Differences in nutrient concentrations and resources between seagrass communities on carbonate and terrigenous sediments in South Sulawesi, Indonesia *Bulletin of Marine Science* **54**(2) 403-419
[9] Heriyanto T and Amin B 2017 Analysis of biomass and carbon reserves in seagrass ecosystem of Malang Rapat Village Bintan District, Kepulauan Riau Province *Applied Science and Technology* **1**(1) 22-28
[10] Putra A, Tanto TA and Ilham 2015 Granules size analysis and sedimentation rate to coral reefs and seagrass in the Bungus Bay waters Padang City *Prosiding Internasional Conference on Green Development in Tropical Region* 171 – 178
[11] Wisesa M M, Bakti D and Fadhilah A 2018 Abundance of sea cucumbers on the ecosystem of seagrasses Inunggeh island, Tapanuli Tengah Regency North Sumatera Province *IOP Conference Series: Earth and Environmental Science* **122**(1) 012107
[12] Wouthuyzen S 2015 *Seagrass Resources and their Management in the East Coast of Bintan Island, Indonesia* (Jakarta: Page LIPI-JSPS Asian Core Program Joint International Seminar on Coastal Ecosystem in Southeast Asia)
[13] Suherman S and Prayitno H B 2012 Examination of the chemical properties of seawater of the Tambelan Islands *Indonesian Journal of Chemistry* **12**(3) 235-241
[14] Quiros T A L, Croll D, Tershy B, Fortes M D and Raimondi P 2017 Land use is a better predictor of tropical seagrass condition than marine protection *Biological conservation* **209** 454-463

[15] Howard J, Hoyt S, Isensee K, Pidgeon E and Telszewski M 2014 *Coastal blue carbon: methods for assessing carbon stocks and emissions factors in mangroves, tidal salt marshes, and seagrass meadows* (Virginia USA: Conservation International, Intergovernmental Oceanographic Commission of UNESCO, International Union for Conservation of Nature. Arlington)

[16] Graha Y, Arthana I W and Astawa I K 2016 Simpanan Karbon Padang Lamun di Kawasan Pantai Sanur, Kota Denpasar *J. Environmental Science* **10**(1) 46–53

[17] Indriani A, Wahyudi J dan Yona D 2017 Cadangan Karbon di Area Padang Lamun Pesisir Pulau Bintan, Kepulauan Riau *Oseanologi dan Limnologi* **2**(3) 1-11

[18] Rahmawati S, Irawan A, Supriyadi I.H and Azkab M H 2014 *Panduan Moniroting Padang Lamun* (COREMAP CTI LIPI: Jakarta. PT. Sarana Komunikasi Utama) p 37

[19] Duffy J P, Pratt L, Anderson K, Land P E and Shutler J D 2018 Spatial assessment of intertidal seagrass meadows using optical imaging systems and a lightweight drone *Estuarine, Coastal and Shelf Science* **200** 169-180

[20] Wu K, Chen C N N and Soong K 2016 Long distance dispersal potential of two seagrasses Thalassia hemprichii and Halophila ovalis *PloS one.* **11**(6) e0156585

[21] Putra A 2017 Evaluasi Kesesuaian Pemanfaatan Ruang Pada Kawasan Pesisir Teluk Bungus Kota Padang *Tesis* (Padang: Pascasarjana Universitas Andalas)

[22] Green E P and Short F T 2003 *World Atlas of Seagrasses* Prepared by the UNEP World Conservation Monitoring Centre (USA: Press Berkeley)

[23] Wahab I, Madduppa H and Kawaroe M 2017 Seagrass species distribution, density and coverage at Panggang Island, Jakarta *IOP Conference Series: Earth and Environmental Science* **54**(1) 012084

[24] Deyanova D, Gullström M, Lyimo L D, Dahl M, Hamisi M I, Mtolera M S and Björk M 2017 Contribution of seagrass plants to CO_2 capture in a tropical seagrass meadow under experimental disturbance *PloS one* **12**(7) e0181386

[25] Lyzenga D R 1981 Remote sensing of bottom reflectance and water attenuation parameters in shallow water using aircraft and landsat data *International Journal of Remote Sensing* **2**(1) 71–82

[26] Putra A, Tanto T A, Farhan AR, Husrin S and Pranowo S P 2017 Pendekatan metode Normalized Difference Vegetation Index (NDVI) dan Lyzenga untuk pemetaan sebaran ekosistem perairan di Kawasan Pesisir Teluk Benoa-Bali *J. Ilmiah Geomatika* **23**(2) 87-94

[27] Putra A and Mutmainah H 2016 The mapping of temporary evacuation site (TES) and tsunami evacuation route in North Pagai Island, Mentawai Islands Regency-Indonesia *IOP Conference Series: Earth and Environmental Science* **47** 1 012020

[28] Hermon D, Ganefri, Putra A and Oktorie O. 2018 The model of mangrove land cover change for the estimation of blue carbon stock change in Belitung Island-Indonesia *International Journal of Applied Environmental Sciences* **13**(2) 191-202

[29] Hermon D, Putra A and Oktorie O 2018 Suitability evaluation of space utilization based on environmental sustainability at the coastal area of Bungus Bay in Padang City *International Journal of Geomate* **14** (41) 193-202

Adaptation model after Tsunami hazard 1994 at South Beach East Java

Ketut Prasetyo[1,2,*]

[1] Department Geography-State University of Surabaya, Indonesia
[2] Secretary of Study Center of Disaster and Environmental-State University of Surabaya, Indonesia

*ketutprasetyo@unesa.ac.id

Abstract. Beached South of East Java passed by tectonic plate have tendency become area which trend of to tsunami disaster. Phenomenon the happening of year tsunami disaster 1994 which have swallowed good victim of soul and also estate do not cause to discourage in as resident to linger on the in place. Tsunami as source of pressure adapt environment live beached resident Beached South of East Java likely not yield is same response sort of environment. On that account, passing approach qualitative and is quantitative of researcher wish to know adaptation strategy model used by resident in Beached South of East Java Province. Research method by use facto expose method, while to collect data by porposive sample technique. The result of the researched hence known that environmental adaptation strategy model at Beached South of East Java after tsunami disaster, that is: First of Pancer model, model Pancer distinguish that its resident stand at bay and back to environment of is ex-tsunami disaster, them believe the happening of eye tsunami disaster because God destiny. Second of Lampon model, characteristic model Lampon differ from characteristic model Pancer. At Lampon model, Tsunami pasca resident leave environment which have swallowed victim, their have of migration to in more peaceful environment. Third model Watu-Ulo, at this model of resident remain to live location but they try to make barrier dam in the form of fence and croopes. Gift of is name of model like Pancer model, model Lampon, and Watu-Ulo model is the name of model which pursuant to coastal environment or location the happening of tsunami knowledge of Resident about very low tsunami later then new tsunami disaster is once experienced of by resident and also mount trust that tsunami disaster is Destiny Which Single The most, hence third this matter equipment Coastal resident characteristic of South East Java in adapting with environment Conclusion result of research is resident response in Coast South East Java differ in accepting and avoiding tsunami disaster, while environmental adaptation model at Lampon society assessed by most peaceful. So that given suggestion that for the resident of as in Pancer and Watu Ulo is immediately given by counselling to increase awareness of resident about level of danger which possibly will befalling it, and also give training of is way of facing tsunami disaster if in a moment happened.

1. Introduction

The tsunami originated from the word Tsu means the place where the ship docked and nami was a wave. Thus, the Tsunami is defined as a natural disaster in the form of waves of sea water that befall the environment where the ship is anchored. In English literature Tsunami is sometimes referred to as tidal wave, and is often translated literally as "tidal wave". This term is actually less precise because it has absolutely no connection with the tides of the sea which is generally determined by the attraction of astronomical objects. In contrast to the waves generated by the wind which only moves the upper sea water. Tsunami waves cause motion in the entire water column, from the surface to the sea floor.

According to RP Corner (1981) in Saroso BS (1996) that based on the source Tsunami can be divided into three, namely: (a) Tsunamis generated by the tectonic earthquake of the ocean floor plate; (b) Tsunamis generated by volcanic activity under the sea volcano; and (c) Tsunamis generated by avalanches on the seabed. Furthermore, it was emphasized that almost 95% of the Tsunami events were due to the tectonic earthquake of the oceanic base plate and until now it has not been predicted when or when the Tsunami will hit an area.

Relating to the position on the active volcanic pathway, historical records of Tsunami disasters often hit the coasts of our country. Since the eruption of Krakatoa in 1883 more than 30 disasters have been recorded. Based on the records of the Meteorology and Geophysics Agency (BMG) that an average of 460 earthquakes a year were recorded in Indonesia, approximately 70% were tectonic earthquakes originating from the seabed which could potentially cause a Tsunami.

The South Coast of East Java is a part of the coastal environment adjacent to the Indian Ocean tectonic plate lane and one of the coastal environments in Indonesia that was hit by the Tsunami disaster, especially in 1994 which has brought many victims of property and lives. Furthermore, according to the records of the Research and Development Agency of the Ministry of Public Works that the victims of property and lives as a result of the 1994 tsunami were as follows: 238 deaths, 15 missing, 789 injured, 992 damaged houses, 340 missing boats. Wave height and range of Tsunami to land in 1994 in the South Coast area of East Java are in Grajagan (Banyuwangi) 6.9 meters high and a range of 300 meters, in Lampon (Banyuwangi) wave height of 11 meters and a range of 300 meters, in Pancer (Banyuwangi) wave height of 11 meters and a range of 300 meters, in Rajegwesi (Banyuwangi) wave height of 14 meters and a distance of 150 meters to the coast, in Watu Ulo (Jember) wave height of 5 meters and a range of 500 meters, in Tambakrejo (Malang) wave height of 4 meters range of 50 meters while in Sine (Tulungagung) wave height of 6 meters with a range of 50 meters. But it seems that the Tsunami disaster has claimed many victims: property, objects and lives did not dampen the guts for residents on the South coast of East Java to remain in the neighborhood.

Against the background that Tsunami disasters are difficult to predict when they occur, and if they occur, they can result in victims of considerable wealth and lives, so with the knowledge that there are still many people who remain in an environment that are classified as vulnerable to Tsunami, it is interesting to reveal how environmental adaptation strategies of a community south coast of East Java. In relation to the discussion of the environment, it is necessary to understand the existence of the Environmental Paradigm Scale paradigm [1]. Broadly speaking, it is understood that New Environmental Paradigm (NEP) is a view that basically places humans' part of the ecosystem, and humans are closely related to the life of the entire cosmos. So, the NEP hopes humans think dialectically. Natural wealth and sustainability of it carrying capacity are always associated with human ecological responsibility and awareness. Thus, it can be understood that the NEP concept basically states that humans are part of their ecosystem and are not separate parts of it.

Regarding the study of cultural and environmental relations there are several views. First-view of determinitic antrophogeography. This view states that the environment can determine culture. The culture that develops is determined by the environment. The environment in question is a natural environment such as climate, soil and topography of Meggering in Sukadana [2]. As part of the ecosystem, human existence is influenced by the environment, such as geography and climate. Different geography and climate cause different patterns and human behavior. People who live in mountainous areas have a different lifestyle than people who are high al in the coastal area. Second possibility. According to this view the environment does not absolutely determine the characteristics of culture, but only provides possibilities that are characteristic of culture. Differences in geographical conditions are only a source of cultural variation. This view is not completely wrong. Many regions have the same physical condition, but the culture of the community is different, as the culture of the Gunung Kidul community in Yogyakarta is different from the Majalengka community in West Java. This can be assessed from the results of research on environmental compatibility by MKLH as cited by Warsono [3]. Third, the view of cultural ecology was initiated by Steward [4]. According to Steward, there is a relationship between culture and environment [4]. This relationship is seen in economic activities, especially in the production sector. Humans must initially adjust to the environment, by developing a culture in harmony with nature. According to Steward culture is not only determined by the environment, but also a creative process in which there are levels in adaptation [4]. The fourth view of cultural ecosystemicism proposed by Cliffort Geertz [5]. This view actually has similarities with Steward's view. The difference is that Geertz emphasizes balance. Geertz stressed that the relationship between humans and the environment is a relationship of interdependence [5]. Therefore, humans must always create and maintain new balances on the environment [6]. According to the results of Geertz's research, it is known that in Indonesia history and politics are factors that determine human adaptation to the five-view adaptive dynamic environment developed by John Bennet [5]. This view still recognizes the existence of human and environmental relations, as well as changes that occur as a result of these mutual relationships. Therefore, humans must adapt. In the process of adapting to these changes, humans have choices, even though the choice contains a problem. Related to environmental adaptation there is a theory put forward by Bell [7]. A model of interaction between individuals and the environment that involves perception, coping, and cumulative effects. A person's perception of the environment is derived from objective physical conditions and individual differences. The perception of the environment produces an optimal stimulus which ultimately forms a hemeostatic, and environmental perceptions appear as optimal limits. Furthermore, stimulation of perception of the optimal range causes stress, or overload and or reactants. Then someone will be coping, if successful will do adaptation or reinforcement, whereas if not successful, pressure (arousal) or stress will continue, and finally will produce cumulative effects. First and second views are more linear by placing the environment as the main cause of a culture. Both of these views do not see any influence between the environment and culture. In the third and fourth views, the recognition of the interplay between environment and culture is acknowledged. Stewards have also shown how human adaptation to the physical environment has influenced cultural characteristics, but Steward's shortcomings do not talk about how the process of mutual influence between culture and the environment and how humans develop new cultures and are related to environmental change [4]. The view of the relationship between culture and environment turns out that Bennet's view is more complete when compared to other views. The reason used to assess the completeness of the relationship model is adaptive dynamics, in addition to a systemic concept, which also includes the process of behavioral adaptation, decision making and choices that are indirectly under regular process control. Adaptive dynamics emphasizes the element of possibilism which is ignored by ecosistemics and also the element of choice of human actions which is ignored by Steward [4].

2. Objectives of research

Research Objectives: 1. Knowing about Tsunami disaster, 2. Knowing the model of enviromental adaptation after the Tsunami disaster, and 3. Evaluating the model of environmental adaptation carried out by the people of the South Coast of East Java after the Tsunami disaster

3. Research methods

The method the reseached is exposure facto, and using quantitative and qualitative approaches. Quantitative approaches are used to obtain data: the distance of the population's residence to the coastline, the form of the settlement layout. Whereas a qualitative approach to uncover community knowledge and perceptions of Tsunami disasters.

4. Result of the reseached

4.1. Knowing about Tsunami disaster

Referring to the results of the study that the level of education of the population is relatively low, and their knowledge of tsunami disaster is low, the condition of the population like this causes obstacles to realizing or changing attitudes and behaviors in dealing with the disaster.

Based on the results of interviews with some people who had been hit by the Tsunami all explained that their knowledge of the Tsunami did not yet exist. Even the term Tsunami was only known after the disaster in 1994 ago. They do not know the origin of the cause of the Tsunami.

Despite the Tsunami disaster in 1994, it turned out that the attitudes of the community towards the location of their settlements that were vulnerable to Tsunami disasters were differences between the people in Lampon Beach and Pancer Beach.

Broadly speaking, the characteristics of the people of the South Coast of East Java who had been affected by the Tsunami were mostly embraced Islam, the level of education that had been taken by most had finished elementary school, and most of the livelihoods were fishermen.

In relation to the culture that is owned, the South Coast community has traditional "*sedekah laut*" ceremonies and they have confidence in the existence of "*Nyai Roro Kidul*" as the South Sea India Ocean . As traditional fishermen, they have knowledge of the existence of the "Southwest Season" danger that occurs from December to March. According to them, during the months they were forced to take a break, they did not go to sea because in that month there was a big wave and fish famine occurred. Most of the settlements that have been hit by the Tsunami disaster are located in tidal areas.

Resignation, relying on natural conditions, the phenomenon of population like this is what we need to increase their knowledge. They need to be developed to try / out to get out or overcome problems that are expected to occur.

4.2. Model of Enviromental Adaptation After the Tsunami Disaster

Model adaptation of the environment of the South Coast of East Java, especially in the coastal communities that have been hit by the Tsunami disaster can be divided into three model : 1).the Pancer model is remain in a location that was hit by the Tsunami with submission, 2). the Lampon model is moving safer residential locations and 3). the Watu-Ulo Model is making protection.

a. The Pancer Model

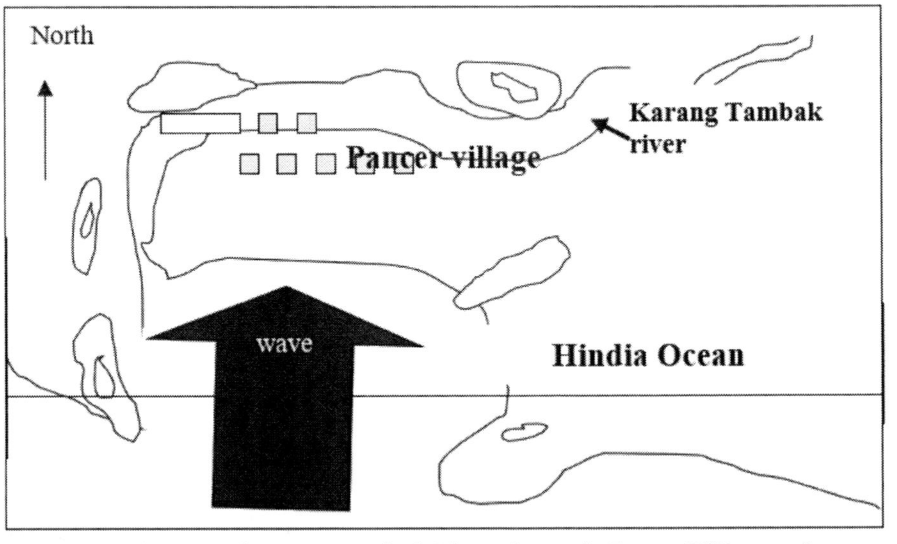

Figure 1. The map of Pancer Beached. The setlemen in Pancer Village at the open beached.

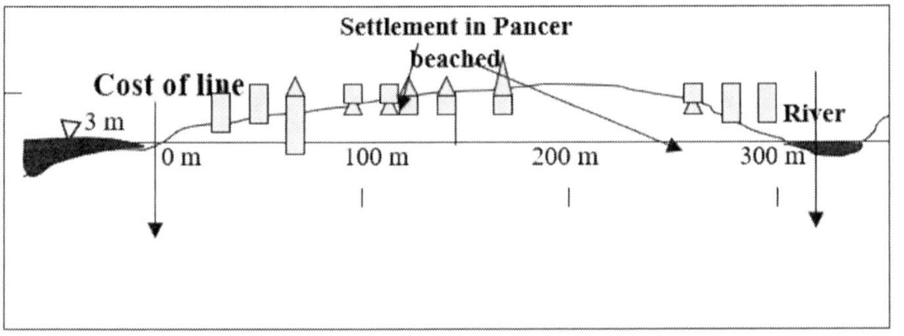

Figure 2. Profille of settlement at Pancer village, Banyuwangi distric

Based on Figures 1 and 2, it is known that the model of spatial settlement in Pancer Village on the vulnerability of Tsunami disasters is the highest. Because the resident houses in Pancer Village directly face the open beach so that in the event of a Tsunami there is no barrier and immediately hit the settlement. In addition, with settlements located on the open bay-shaped beach, in the event of a tidal wave or Tsunami, there will be an accumulation of tide in the bay.

The 1994 tsunami disaster that occurred in Pancer Beach with the death of 120 dead, 526 injured, 704 houses damaged and 200 damaged and missing boats did not seem to be a meaningful lesson for the people in Pancer village. They were back again in the location that had claimed the victim in 1994. The perception of the population in Pancer that the tsunami disaster was God's destiny, so that if Tsunami happened again they resigned, did not have the knowledge to deal with it

b. The Lampon Model

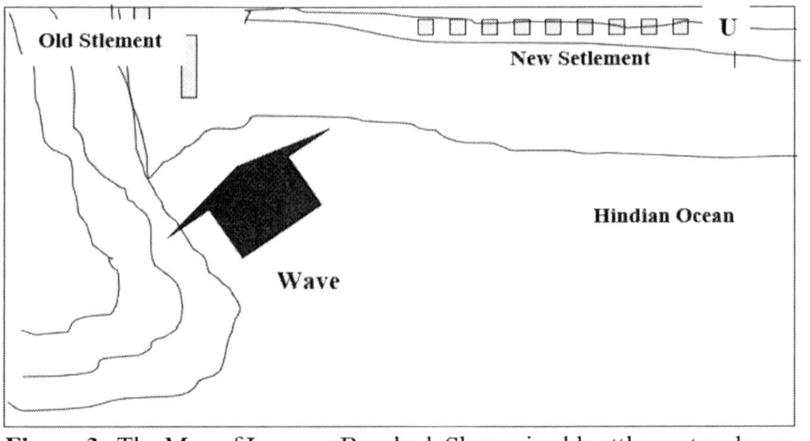

Figure 3. The Map of Lampon Beached. Shown is old settlement and new settlement after tsunami hazar

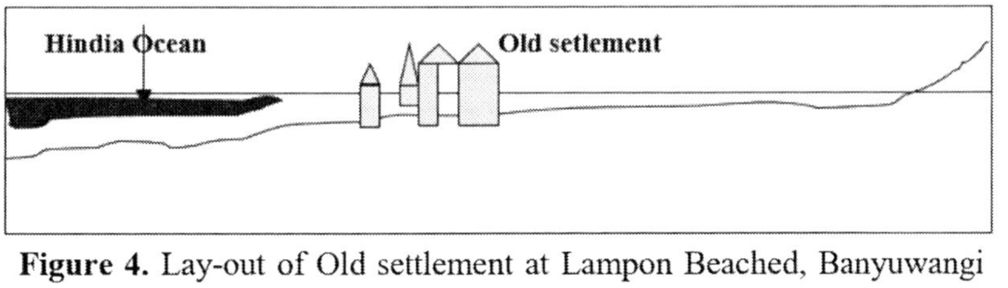

Figure 4. Lay-out of Old settlement at Lampon Beached, Banyuwangi Distri

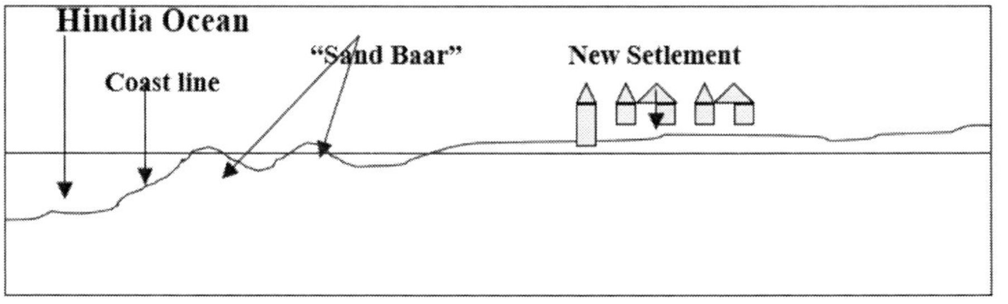

Figure 5. Lay-out of New Setlement at Lampon Village, Banyuwangi Distric

The form of the beach in the form of an open bay as can be seen in figure 4 causes the tidal wave to directly hit the beach. This condition is similar to Pancer Bay. But in adapting to the environment after the Tsunami there are differences. If in Pancer the population chooses to stay / return to live while the coastal lampon community chooses to move in a location that is considered safe from the Tsunami.

c. The Watu Ulo Model

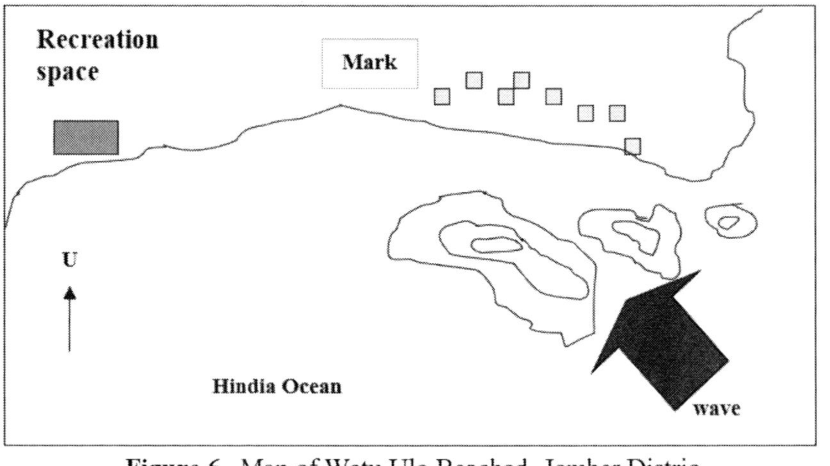

Figure 6. Map of Watu Ulo Beached- Jember Distric

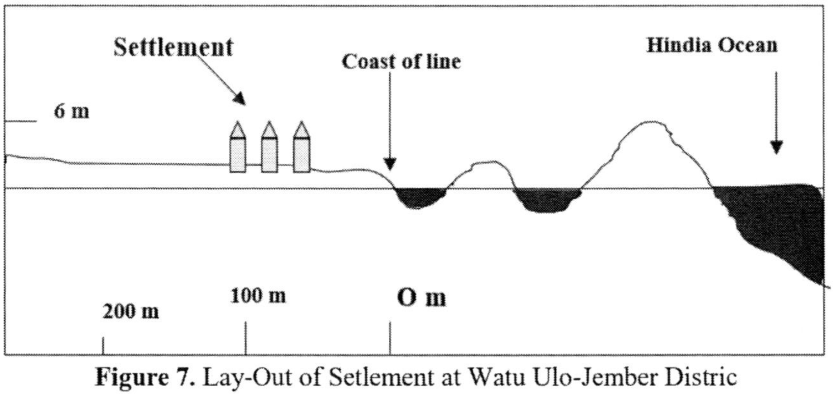

Figure 7. Lay-Out of Setlement at Watu Ulo-Jember Distric

5. Evaluating the model of Environmental Adaptation

The Tsunami safest adaptation model referring to the environmental adaptation model in the communities of the South Coast of East Java after the Tsunami, it is known that the safest adaptation model is the adaptation model as practiced by Watu Ulo Beach community. Then the next allater is the adaptation model in the Pesanggarahan beach community.

As in the case in Pancer and Lampon. In both environments even though both victims and property have fallen, but the Pancer population showed little phenomenon that is different from the population of Pancer who remained settled and carried out economic activities in the location, while in Lampon the location of the former disaster was only used by economic activities namely the port fish catcher.

Phenomenon in Lampon and Pancer proves that the basic needs of employment will override the dangers that always threaten the body and soul. In New Environment Paradigm, the joint economy is a decisive component in quality environmental management. So that at the same time this implies that the economic phenomenon of the population is low, their lives depend on environmental conditions.

Population trust in the research area that the tsunami disaster is the destiny of God Almighty, then this phenomenon seems to identify that the character of the South Coast of East Java population in adapting to the environment was revealed by Chiras who said: "environmental disaster let it happen, then naturally natural will improve it self ." [8]

Population patterns cluster in the coastal plain. Despite the Tsunami disaster in 1994, it was evident that the attitudes of the community towards their settlement locations were vulnerable to Tsunami disasters, there were differences between the communities in Lampon Beach and Pancer Beach.

The people in Pantai Pancer returned to occupy settle in locations that had been hit by the Tsunami, while the Lampon Beach community after the Tsunami disaster largely left their settlements that had been the Tsunami disaster. The reasons put forward by the Pancer beach community to stay in the location because of the Tsunami are disasters and God's destiny, so that if a disaster happens later it is God's destiny too. Whereas the reason stated by the Lampon beach community to move or leave the former site was hit by the Tsunami because the people felt traumatized by the disaster. Conceptually, if we use the theory of environmental psychology, as revealed by Bell, the people on the Pancer coast are not stressed by the Tsunami disaster stimulants, they can copy the Tsunami [7].

The situation is different between the Pancer beach community and the Lampon beach community in receiving Tsunami disaster stimulants. At the Lampon beach community there was a post-Tsunami stress. Therefore they did not want to resettle in locations that had been affected by the Tsunami.

We can observe different environmental adaptation models in the community at Watu Ulo Beach. In both coastal locations the community environmental adaptation model adapts natural conditions to protect from Tsunami disasters. Whereas in Watu Ulo Beach in an environment of adaptation so as not to be affected by the Tsunami in addition to adjusting the natural conditions, they also made retaining waves in the form of artificial likes planted with trees

6. Conclusions

Based on the results of the research can be concluded as follows:

1. The level of education of the population is relatively low, and their knowledge of tsunami disaster is low, the condition of the population like this causes obstacles to realizing or changing attitudes and behaviors in dealing with the disaster.
2. Model adaptation of the environment of the South Coast of East Java, especially in the coastal communities that have been hit by the Tsunami disaster can be divided into three, namely: the Watu-Ulo Model (making protection), the Lampon model (moving safer residential locations) and the Pancer model (remain in a location that was hit by the Tsunami with submission) Variations in the model of environmental adaptation are made possible by the Tsunami events they have only experience
3. Based on the evaluation results of the environmental adaptation model, the Lampon Model is the safest, while the Pancer model which is considered the most risk if there is a tsunami disaster again

Reference

[1] Pelstring L 1997 *Measuring Environmental Attitudes: The New Environmental Paradigm* http/ww.ibs.msu.edu/sca/lbs335/eassy1-Picotte.html.

[2] Sukadana A 1983 *Antropologi-Ekologi* (Surabaya: Airlangga University Press)

[3] Warsono 1992 *Strategi Adaptif Migran Madura di Surabaya Khususnya Bagi Golongan Kenek*

(Jakarta: Tesis Magister Lingkungan, Program Pasca Sarjana UI)
[4] Steward J 1955 *Theory of Culture Change: The Methodology of Multilinear Evolution* (Urbana: University of Illinois Press)
[5] Geertz Cliford 1983 *Involosi Pertanian: Proses Perubahan Ekologi di Indonesia* translated by S Suparmo (Jakarta: Barata Karya Aksara)
[6] Bennet W John 1976 *The Ecological Transition: Cultural Anthropology and Human Adaptation* (New York: Pergamon Press Inc.)
[7] Bell P A et al 1978 *Environmental Psycology* (Philadelphia: W.B. Saunders Co.)
[8] Chiras Daniel D 1991 *Environmental Science. A Framework for Decision Making* (Denver: Benyamin Co.)

Hydrograph separation method and baseflow separation using Chapman Method – A case study in Peusangan Watershed

Ichwana Ramli[1,*], **Sayed Murthada**[2], **Zulkifli Nasution**[3] **and Ashfa Achmad**[4]

[1] Agricultural Engineering Department, Faculty of Agriculture, University of Syiah Kuala, Indonesia

[2] Department of Mining and Energy of East Aceh, Idi Rayeuk, Indonesia

[3] Agroecotechnology Department, Faculty of Agriculture, University of Sumatera Utara, Indonesia

[4] Architecture and Planning Department, Engineering Faculty, Syiah Kuala University, Indonesia

*ichwana.ramli@yahoo.co.id

Abstract. Base flow is the portion of stream that is not run off and results from seepage of water from the ground into a channel slowly over time. Base flow is considered to be the groundwater contribution to stream flow. Estimates of the amount of base flow can be derived from stream records. This study is about groundwater recharge condition at two sub watersheds in Krueng Peusangan: sub watershed of Lut Tawar and sub watershed of Teupin Mane (Krueng Teumbo). The Krueng Peusangan Watershed is one of priority for watershed management because this watershed is in a critical condition. The Krueng Peusangan Watershed is an important water catchment area in Aceh where there are one million people depend on life from the water source watershed, such as dringking water, agriculture, and fisheries. The Krueng Peusangan includes four regencies, an upstream Aceh Tengah, a midstream area in Bener Meriah and downstream area in Bireuen and Aceh Utara. This research aims to get groundwater recharge using Seasonal Recession Method and Base Flow Separation Method. Both of methods were used to estimate groundwater recharge for each peak in stream flow during the period of record (2008-2012). The results showed that the trend pattern of the stream hydrograph could be explained using the exponential function where the dots lowest discharge that is the end of the recession (y) than any period of time on stream hydrograph semi logarithmic (x). The regression formula for the watershed of Krueng Peusangan is: $y=1.3x^{-0.12}$ (Lut Tawar -Wih Nareh) and $y=1.94x^{-0.42}$ (Teupin Mane-Krueng Teumbo). The filtering method (Base Flow Separation Method) by Chapman (1999) gave a lower percentage of the low flow component (base flow) for Sub Watershed of Teupin Mane (Krueng Teumbo): 2008/2009 (18%), 2009/2010 (3%), 2010/2011 (1%), 2011/2012 (1%). For Sub Watershed of Lut Tawar (Wih Nareh): 2008/2009 (1%), 2009/2010 (1%), 2010/2011 (4%), 2011/2012 (1%). Estimates using the two methods define a reasonable range of the base flow contribution to stream flow in the watershed.

1. Introduction

Groundwater is one of the most important natural resources, which supports human health, economic, development and ecological diversity. Challenges in the future are increasingly limited sources of surface water, and the ever increasing needs of water for purposes in supporting development.

Assessment of water resources available for both surface and groundwater flow is very important for optimal use and protection considerations, and also for the prediction of floods and low flows [1].

Based on the depth and density of surface drainage, groundwater flow systems can be classified into three types: (1) shallow, (2) intermediate, and (3) regional flow systems [2]. For water resources management must understand the contribution of groundwater in the river so that good groundwater resource management techniques will reduce problems such as the disruption of the hydrological cycle balance and impact on groundwater contamination. One of the components of groundwater is base flow. It is a component of streams flow derived from natural aquifer storage. Availability of river water is influenced by surface runoff or base flow component directly. Stream flow can also be affected by direct flow disturbances, such as diversion of runoff and water harvesting mechanism.

A popular method by Meyboom (1961) [3] was used to determine the volume of recharge with seasonal recession and the recession curve displacement method by Rorabaugh. The method is a basic method used in the concept of river flow contributions from direct runoff and groundwater base flow [4]. The base flow recession curve is the lower part of the falling limb of a hydrograph and expresses the relation between base flow and time. Recession constants describing the slope of the decline in stream following the dynamic movement of water in the basin fill [5]. Accurate estimation of the groundwater recharge is very important as an indicator of the proper management of the groundwater system [6].

A multitude of methods have been developed, which can be conveniently categorized into four basic approaches: graphical base flow separation, filtering algorithms, frequency analysis and recession analysis. In this research, we will use two methods for base flow analysis in Krueng Peusangan watershed, recession analysis and hydrograph base flow separation. The aim of this study is to determine the volume of the watershed recharge in Krueng Peusangan based on the available data using the both methods. The base flow is assumed to fluctuate (ups and downs) and the river discharge represents the integration of the physical condition of watershed and precipitation.

Hydrograph base flow separation is one of analytical methods developing for separate base flow from total stream flow. Although most procedure are based on physical reasoning, element of all separation technique are subjective. Manual separation of hydrograph stream into surface flow and groundwater flow is difficult and inexact. In this research, analysis of base flow using filtering method (base separation method) was only for two sub watersheds of Krueng Peusangan: downstream (Krueng Teupin Mane-Teumbo) and upstream (Lut Tawar-Wih Nareh).

2. Materials and Methods

The study was conducted in the watershed of Krueng (Figure 1). Precipitation data for the upstream and downstream areas were derived from Stations of Meteorology and Climatology of Lhokseumawe and Bebesan Takengon. The data as used in this study were stream flow data issued by the Office of Water Resources and Headquarter of Krueng Aceh Watershed, Aceh Province (NAD).

The hydrograph stream during the periods when all discharges are derived from groundwater sources is known as base flow recession curve. For analysis groundwater recharge using recession analysis.

Where Q_t is the discharge of the river at time t and e^{-a} is a constant definiting the rate of recession. It is common practice to replace baseflow recession constant e^{-a} with k. Thus, equation (1) becomes (2)

$$Q_t = Q_o e^{-at} \quad (1)$$

$$Q_t = Q_o k^t \quad (2)$$

The pattern of groundwater recharge condition changes was obtained by connecting the dots of lowest discharge (y) in which it was the end of the recession from every period to the time (x) on flow semi-logarithmic hydrograph . The lines that formed as a particular curve which followed the end of the recession flow fluctuation. Trend was selected that had coefficient determination R^2 (8).

For base flow separation method, this study used the one that was proposed by Chapman (1999) [9]. It is given by equation (3).

$$Qd(i) = \frac{3\alpha-1}{3-\alpha}(Qd(i-1) + \frac{2}{3-\alpha}(Qt(i) - Qt(i-1)) \qquad (3)$$

Where:
Qd(i) = base flow (m^3/s),
Qt(i) =total flow and α is coefficient with value 0.925.

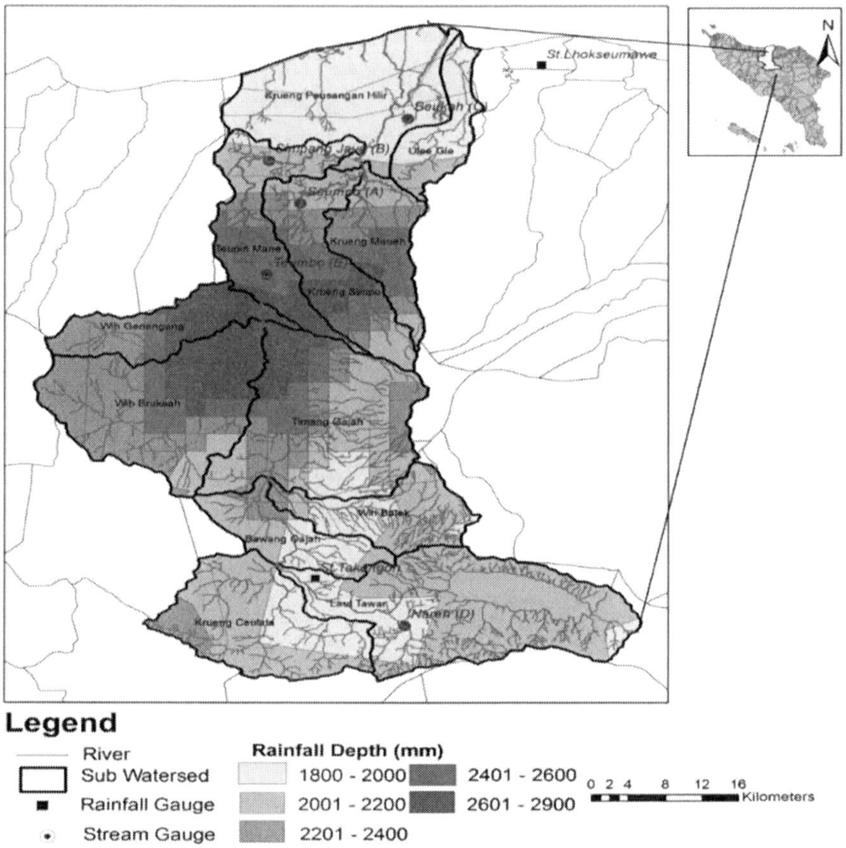

Figure 1. Krueng Peusangan watershed and the observation points of discharge [7]

In this study, the discharge data at point D and point E will be used as observation points for hydrograph Method and Baseflow Separation Using Chapman Method. Point at (D) Wih Nareh representing sub watershed of Lut Tawar, (E) Krueng Teumbo representing sub watershed of Krueng Celala, Lut Tawar, Wih Balek, Bawang Gajah, Timang Gajah, Wih Bruksah, Wih Genengen.

3. Results and Discussion

Infiltration that occurs in an area is influenced by various factors, such as infiltration capacity, rainfall, climate, topography and geological structure. This can be seen from the conservation index of an area that is used to show the ability of an area to absorb rainwater that falls to the surface of the land. The infiltration process can occur if there is an unsaturated zone between the surface of the ground and the surface of the ground, so that water can infiltrate into the soil.

Infiltration can occur if there is no watertight space between the surface of the groundwater, so that the water on the surface can infiltrate into the soil. This infiltration process is influenced by soil structure, land use and climate. The Krueng Peusangan watershed is dominated by brown podsolic soil and yellow red podsolic and medium and rather fine soil texture which is a low absorption of water. In good vegetation conditions, the stability of water absorption can be helped by prisoners from the roots and canopy of vegetation that covers the watershed area. But in this area research, there are about 33% land conversion functions as plantations.

Generally, the Krueng Peusangan watershed has a slope of 0.002 % and the lowest about 43-750 mpl in the middle and lower reaches of the watershed. While the highest point around 800-3000 mpl is located at the head or around the mountain Peutsagoe, Geureundong and mountains Gayo. The impact caused the physiographic Krueng Peusangan have a level slope were categorized grade flat slope-being. This means that the areas with the category of flat to being dispersed in the basin, especially the middle and lower reaches. While the steep watershed slopes to very steep generally found on the edge of the foothills of the upstream Krueng Peusangan. Either directly or indirectly, and topographic effect is most pronounced in setting rock are relatively high [2].

Changes in the distribution of river flow and groundwater recharge over space and time is also determined by changes in temperature, evaporation, and a significant rainfall [10]. One system is highly vulnerable to climate change are water resources. Climate change affect water availability, quality, and potential water damage. Many sectors and systems depend on the availability of water resources, so that changes in the hydrological regime and water quality due to climate change will have an impact on the social and economic fields. To overcome the negative impact of climate change on water availability, land use needs to be done. Rainfall analysis for 20 years (1992-2011) based climate classification of Smith Ferguson in Krueng Peusangan (Station Bebesan) was classified in type B with wet category 28.86%. In the same location had been a shift in climate over the past two decades due to the 1992-2001 was classified in type B category (23.68%), while in 2002-2011 was classified in type C category (34.25%).

The process of absorbing rainwater into the soil is strongly influenced by rainfall and watershed characteristics in the form of watershed morphology and morphometry. Watershed morphology is information on the physical shape of the watershed that is able to characterize watershed characteristics including watershed shape, watershed topography and drainage pattern, while the watershed morphometry states quantitatively the condition of river channel network which includes area, width, slope, river order and density. The extent of a watershed or sub-watershed functions as a reservoir of water flow from the surface water storage site. The wider the sub-watershed, the more water can be accommodated and the longer it takes for rain to reach the outlet.

Characteristics of drainage patterns in sub-watershed sub-locations are dendritic. The results of the analysis show that the Krueng Peusangan watershed has a tree-shaped flow pattern so that it can be classified into the dendritic flow pattern. Dendritic patterns are generally found in areas that have uniform rock and wide distribution

The Krueng Peusangan watershed can be visually categorized into longitudinal watersheds with a northward orientation. The watershed roundness level using the circularity ratio method approach has a value of 0.31. The elongated shape is characterized by the presence of a main river that receives streams and creeks on the left and right sides which then flow into the sea. The Lut Tawar sub-watershed has a non-round shape with a value of circularity ratio 0.42 while the Teupin mane sub-watershed has circularity ratio 0.2 with a non-round category.

The shape of the sub-watershed in the Krueng Peusangan watershed is generally not rounded, indicating that the shape has a debit and centralized velocity is relatively small. This condition causes the water travel time of the tributaries to vary to the center of the flow because it experiences infiltration first. But the chance of water to erode the soil and disperse the soil material is getting bigger. If the effort to rehabilitate the land is not done, the chances of erosion and sedimentation are quite large.

The Krueng Peusangan watershed has a drainage density that can be classified into the medium category with a density index of 0.84. While the density index of the sub-watershed ranges from 0.37 to 1.39. The index of drainage density for Lut Tawar is 1.071 with a moderate category and for the location of krueng teumbo is 0.802 with a medium category. The density of drainage obtained for each watershed is smaller than 1.74, which means that the Krueng Peusangan watershed has features of poor surface drainage, many occurrences of inundation and often flooding is also a zone of sedimentation. Sub-watersheds that have a high drainage density will have better hydrological conditions when compared with watersheds with low drainage density. This is caused by the increasing drainage density, the rainwater will be spread evenly into the creeks and before entering the main river the water has a longer time and more permeated into the soil which in turn will increase the availability of underground water.

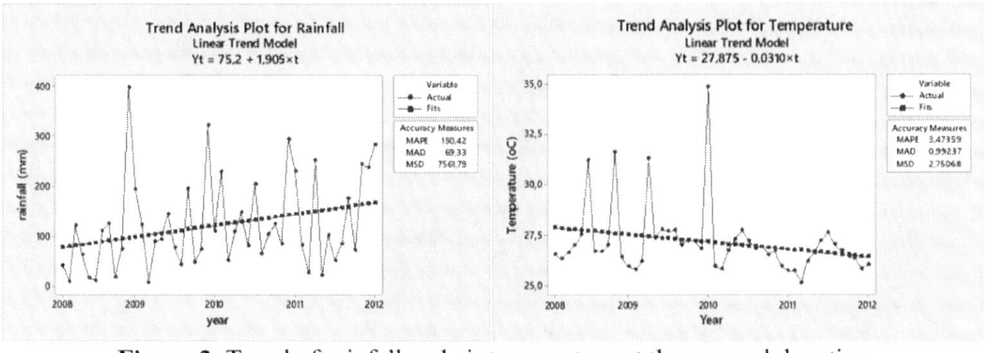

Figure 2. Trend of rainfall and air temperature at the research location

In figure 2 shows the rainfall (2008-2012) continues to increase and the air temperature decreases at the research location. This will certainly cause increased air humidity. In addition to watershed physical factors, rainfall, climatic elements influence the pattern of water availability and temperature, air humidity, wind speed and duration of irradiation. These conditions will greatly influence to reduce river discharge and deposits of ground water.

This study in the Peusangan watershed is very difficult to determine the exact desires of the baseline which can then be analyzed for volume loss because there are no long dry months. Correcting the changing fluctuations from the recession line can change the unstable conditions of the refill area of the Teupin mane sub section at the measurement point in Teumbo. Therefore further analysis is needed for a longer period of time. For sub-watershed teupin Mane, this area is returned by Latosol soil types (33.89%) yellow red podsolic (48.68%), and Andosol (0.19%) (BP Krueng Aceh watershed, 2010) which is a type of soil absorbency. In conditions with agricultural land cover (68.72%) and secondary dryland forest (22.81%) covering forest areas that have been allocated to agricultural land through deforestation to be used as plantation gardens and fields in accordance with what is spent. Increased direct runoff in the rainy season, this causes the catchment volume of watersheds in this area to increase extreme fluctuations (unstable). Changes in the distribution of river currents and recharge soils above and time are determined by changes in temperature, evaporation, and importantly, precipitation [11]. Some of the effects of change on the hydrological process have been revised [12].

The line produces a regression equation (D) sub watershed of Lut Tawar, $y = 1.3x^{-0.12}$; $R^2 = 0.941$, and (E) sub watershed Teupin Mane, $y = 1.94x^{-0.42}$. with $R^2 = 0.917$. (figure 3 and 4). That was showed the pattern of groundwater recharge condition that occurred at some point in Krueng Peusangan watershed. It is obtained by connecting the dots lowest discharge that is the end of the recession than any period of time on stream hydrograph semilogarithmic. The lines that form a particular form of the exponential curve following the end of the recession flow fluctuation of each period of observations at the study site.Here, we can see the good relationship between the lowest discharge recession variable points with the time observation period. the figure also shows that at location E (Teupin Mane watershed) has a higher slope. This means reducing the water absorption pattern compared to the Lut Tawar watershed, while location (D) tends to have a horizontal slope. Instability of recharge that occurred will affect the amount of ground water reserves in the watershed [13]. The base flow recession was a function of topography, drainage patterns, soil type, geology of the watershed.

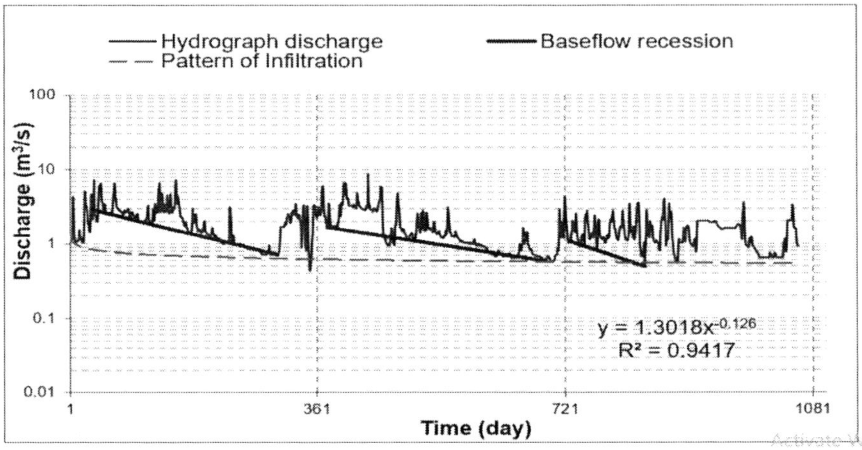

Figure 3. Hydrograph of Discharge, Baseflow recession and Pattern of Infiltration at Location D (Wih Nareh) Period November 2008 - October 2011

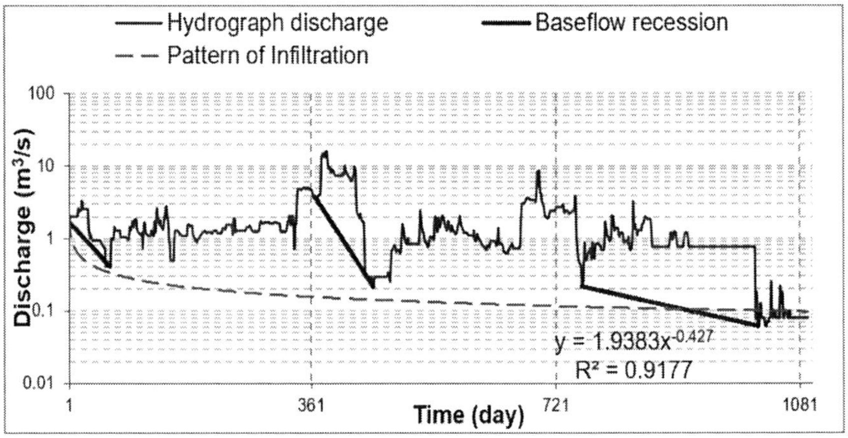

Figure 4. Hydrograph of Discharge, Baseflow recession and Pattern of Infiltration at Location E Period November2008 - October 2011

The condition of absorption volume was also influenced by the characteristics of the rainfall in the region of the area. Conditions that occurred in both sub watersheds indicated that there had been an imbalance fluctuations of water seeped into the aquifer (groundwater catchment layer) on recharge area due to rainfall and high evapotranspiration in mid and downstream sections of Krueng Peusangan. Therefore, a clear direction needed for land use and functions of forests, especially in the recharge area to maintain the balance of the hydrological cycle and the quantity of groundwater properly. In the Upstream Krueng Aceh Watershed there had been some problem with decreasing direct flow because deforestation and sedimentation. Other side, in the downstream (Krueng Tepin Mane-Teumbo) there had also been some problem such as flooding.

3.1 Base Flow Separation Method by Chapman

The purpose of basic flow separation is to distinguish two streams base (use of groundwater inflow) and fast flow (surface runoff and interflow) [14]. Baseflow index (BFI) states the comparison of the base flow to the total discharge on the River. The baseflow index shows the ratio between the base flow to the total flow discharge in the River. According to [15], the baseflow index value that is around 0.9 indicates a permeable base. the high baseflow index value illustrates the bases that have a stable flow regime and the das can provide water supply to the river, even though the dry season is quite long. a low baseflow index value (between 0.15 - 0.2) shows an impermeable watershed with a flashy flow regime.

Figure 5 is present one of sample the results of the base flow and direct flow component of the daily stream flow in Sub Watershed Teupin Mane (Teumbo) during 2008/2009 as calculated by Chapman (1999). According to the result, base flow is not the dominant value in the Sub Watershed Teupin Mane (Teumbo). The filtering method (Base Flow Separation Method) by Chapman (1999) gave Base Flow Index (BFI) of the low flow component (base flow) for Sub Watershed of Teupin Mane (Krueng Teumbo): 2008/2009 (18%), 2009/2010 (3%), 2010/2011 (1%) and 2011/2012 (1%). The Chapman method in this study used the empirical coefficient taken 0.95.

Many emperical methods and formulas for separating the base flow from the stream. Digital filter methods and digital graphic methods are two methods that are sufficient, it's easy to analyze [16]. Digital filter method (recursive digital filter works by processing discharges data for separating the base flow from the hydrograph. Chapman's Method tend to estimate lower base flow

In general, the base flow contribution was very minimum than the direct flow. Based on [17], the Sub Watershed Teupin Mane (Teumbo) showed deteriorating watershed conditions (critical). In February 2016, Teupin Mane River flooded and some houses damaged while dozens of other inundated (Serambi Indonesia, 2016). It showed the sub watershed Teupin Mane (Teumbo) one of critical sub watershed of Kreung Peusangan.

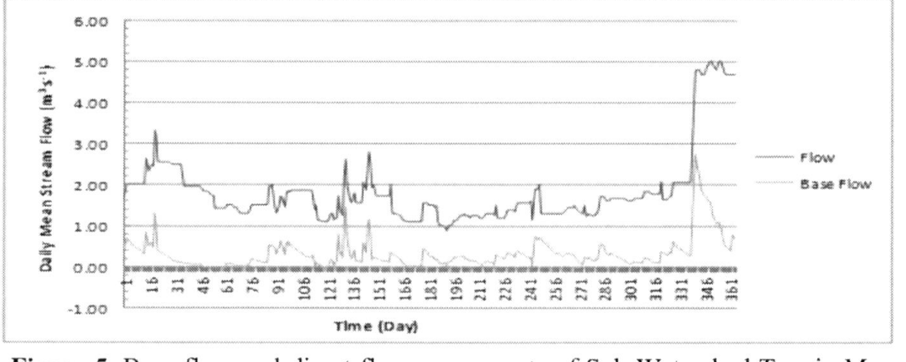

Figure 5. Base flow and direct flow components of Sub Watershed Teupin Mane (Teumbo) in 2008/2009.

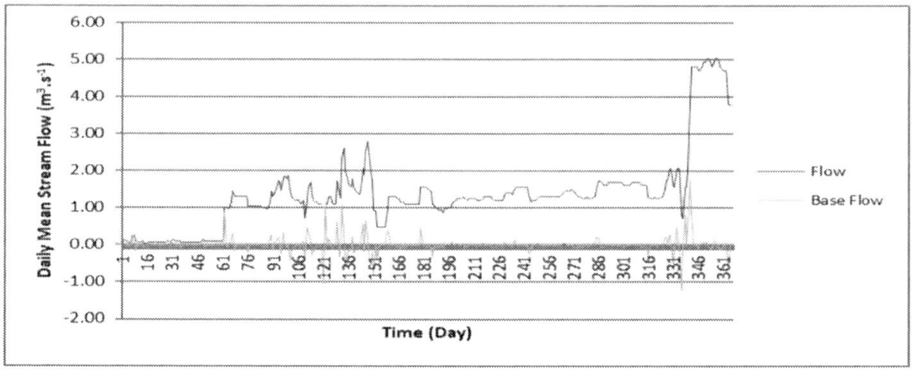

Figure 6. Base flow and direct flow components of Watershed of Lut Tawar (Wih Nareh) in 2008/2009

For upstream Sub Watershed Lut Tawar/ Wihnareh Figure 6 (one of sample), BFI was 2008/2009 (1%), 2009/2010 (1%), 2010/2011 (4%), 2011/2012 (1%) , respectively. In Sub Watershed Lut Tawar, it found that decreasing river flow and sedimentation. where is important hydrological issue to be solved in The Krueng Peusangan Watershed. According to the result of BFI, the upstream was very minimum of base flow, because of deforestation in that area. Based on[10], the largest proportion of deforestation in Krueng Peusangan Watershed occurred in the upper catchment area

The instability in the chapman method is also seen in the slope recession baseflow which tends to be flat (Lut Tawar) and steep for the Teumbo location. High average BFI values indicate watersheds with water availability both in the dry season and the rainfall is quite good, while the low BFI shows the watershed with the base flow contribution is quite low. Many methods for calculating the base flow index that is six methods (two parameters, ihacres, EWMA, Eckhardt, local minimums, and fixed intervals) resulted in an average BFI calculation of more than 0.7 and consistent across all watersheds. While the three methods (one parameter, Chapman, and lyne Hollick) tend to produce low BFI values and tend to be unstable [18]. Each method has a tendency to estimate the base flow higher and others tend to estimate lower. The instability in the chapman method is also seen in the slope recession baseflow which tends to be flat (Lut Tawar) and steep for the Teumbo location.

4. Conclusions

The pattern of absorption that occured in the watershed area of Krueng Peusangan (D) sub watershed Lut Tawar for $y = 1.3x^{-0.12}$,and (E) sub watershed of Teupin Mane for $y = 1.94x^{-0.42}$. The location of E had a higher slope which decreased absorption pattern compared to watershed of Lut Tawar, while location of (D) tended to flat.The location measurement at points E showed infiltration volume fluctuated widely and unstable. Point D showed that the catchment volume increased, it was indicated Positive (good). The volume of absorption in the watershed of Krueng Peusangan midsection fluctuated sharply. Base Flow Separation Method by Chapman gave Base Flow Index (BFI) of the low flow component (base flow) for Sub Watershed of Teupin Mane (Krueng Teumbo): 2008/2009 (18%), 2009/2010 (3%), 2010/2011 (1%), 2011/2012 (1%). BFI was 2008/2009 (1%), 2009/2010 (1%), 2010/2011 (4%), 2011/2012 (1%) In this study, the result gave base flow value smaller than direct flow. In general, Krueng Peusangan watershed needs to do conservation and water resource management to solve the problem in this watershed.

References

[1] Gonzales A L, J Nonner, J Heijkers and S Uhlenbrook 2009 Comparison of different base flow separation methods in a lowland catchment *Hydrol. Earth Syst. Sci.* **13** 2055–2068

[2] Tetzlaff D, Seibert J, McGuire K J, Laudon H, Burn D A and Dunn S M 2009 How does landscape structure influence catchment transit time across different geomorphic provinces? *Hydrological Processes* **23** 945–953

[3] Meyboom P 1961 Estimating groundwater recharge from stream hydrographs *Journal of Geophysical Research* **66** 1203-1214

[4] Hongbing S 2004 Estimating the fate of precipitation from stream discharge: A case study in New Jersey *New Jersey Academy of Science* **49**(2) 9-15

[5] Hector D Rivera-Ramirez, Warner Glenn S and Frederick N S 2007 Prediction of master recession curves and base flow recessions in The Liquillo Mountains of Puerto Rico *Journal of the American Water Resources Association* **38**(3) 693-704

[6] Kyoochul Ha, Dong-Chan Koh, Byoung-Woo Yum and Kang-Kun Lee 2008 Estimation of river stage effect on groundwater level, discharge, and bank storage and its application *Geosciences Journal* **12**(2) 191-204 DOI: 10.1007/s12303-008-002-y

[7] Ichwana, Zulkifli Nasution and Sumono 2013 Determining groundwater recharge from stream flow with seasonal recession method *Aceh International Journal of Science and Technology* **2**(1) 8-16 ISSN:2088-9860

[8] Moriasi D N, J G Arnold, M W Van Liew, R L Bingner, R D Harmel and T L Veith 2007 Model evaluation guidelines for systematic quantification of accuracy in watershed simulations *Jurnal American Society of Agricultural and Biological Engineers* **50**(3) 885-900 ISSN 0001-2351

[9] Chapman T 1999 A comparison of algorithm for streamflow recession and base flow separation *Hydrological Processes* **13**(5) 701-714
[10] Khasanah K, Mulyoutami E, Ekadinata A, Asmawan T, Tanika L, Said Z, van Noordwijk M and Leimona B 2010 *A Study of Rapid Hydrological Appraisal in the Krueng Peusangan Watershed, NAD, Sumatra* Working paper nr.123. Bogor, Indonesia. World Agroforestry Centre. 53p. DOI: 10.5716 WP10339
[11] Chiew F H S 2007 Estimation of rainfall elasticity of streamflow in Australia *Hydrol. Sci. J.* **51**(4) 613–625
[12] Rosenzweig C, Casassa G, Imeson A, Karoly D J, Liu Chunzhen, Menzel A, Rawlins S, Root T L, Seguin B and Tryjanowski P 2007 *Assessment of observed changes and responses in natural and managed systems* In: Climate Change 2007: Impacts, Adaptation and Vulnerability. *Contribution of Working Group II to the Fourth Assessment Report of the Intergovernmental Panel on Climate Change* (ed. by M L Parry, O F Canziani, J P Palutikof, P J van der Linden and C E Hanson) pp 79–131 (UK: Cambridge University Press) http://www.ipcc.ch/pdf/assessment-report/ar4/wg2/ar4-wg2-chapter1.pdf
[13] Fetter C W 2001 *Applied hydrogeology 4th ed* (Prentice Hall Publishing Company)
[14] Eckhardt K 2012 Technical Note: Analytical sensitivity analysis of a two parameter recursive digital baseflow separation filter *Hydrol. Earth Syst. Sci.* **16** 451-455 DOI:10.5194/hess-16-451-2012
[15] Tallaksen L M and H A J van Lanen (Eds.) 2004 Hydrological drought–processes and estimation methods for streamflow and groundwater *Developments in Water Science* **48** (Amsterdam: Elsevier Science B.V.) pp 579 ISBN 0-444-51688-3
[16] Murphy R, Z Graszkiewicz, P Hill, B Neal, R Nathan, To Ladson 2009 Australian rainfall and runoff revison Project 7: baseflow for catchment simulation. Stage 1 report – volume 1 – selection of approach. AR&R Report Number, P7/S1/004, ISBN: 978-085825-9218
[17] Ichwana, Ashfa Achmad and Susi Chairani 2016 Determination of micro watershed model based on ecohydrology for the management at Krueng Peusangan watershed *Proceedings of International Conference on Engineering and Science for Research and Development (ICESred)* Hermes Hotel Banda Aceh, 25-26 Oktober 2016, Banda Aceh
[18] Indarto, Novita E, Wahyuningsih S, Ahmad H and Pudjojono M 2016 Baseflow separation study at watersheds of UPT PSDA Pasuruan, East Java *Jurnal Keteknikan Pertanian* (JTEP) **4**(2) 227-236 available online at http://journal.ipb.ac.id/index.php/jtep/article/view/15911 DOI: 10.19028/jtep.04.2.227-236

Tenurial conflict between parties in the technical implementation unit of the Banjar Regency Forest Management Unit in the Province of South Kalimantan

Joni Saputra* and Indang Dewata

Enviromental Science Study Program, Postgraduate Program, Universitas Negeri Padang, Indonesia

*joni_malay57@yahoo.co.id

Abstract. The main issues in the management of forest resources at present, include: legal uncertainty over forests, high tenure conflicts over forest areas and uncertainty over community management areas over forest areas. One of the root causes of legal uncertainty over forests is overlapping policies governing forest management. The purpose of this study was to find a solution in handling tenurial conflicts at UPT KPH Banjar Regency. In the implementation of tenure conflict research the methodology used was interviews. The results of this study found different perceptions about forest, where the community considers the forest when there is still physical forest, on the other hand forestry based on law 41 of 1999 is not so and there are different perceptions between BPN and Forestry relating to the status of the region. BPN together with the Regional Government are the parties that strengthen the position of the community with the certificate issued and the definitive village decision issued. The existence of FMUs is expected to be streamlined to build synergies between institutions in forest utilization and management as outlined in a work program, therefore integrated coordination of the Forestry Planning Agency, Director General of Forestry Production Development, Directorate General of Forest and Land Rehabilitation, BPHP, BPKH and BPDAS.

1. Introduction

In order to realize sustainable forest management, according to [1] has mandated the establishment of forest management areas that must be implemented at the Provincial, District / City level and management units. The formation of the forest management area is in the form of the smallest forest management unit (KPH) in accordance with its main function and designation so that it can be managed efficiently and sustainably. The implementation of forest management in Forest Management Units (KPH) is the implementation of forest management which includes forest governance activities, preparation of plans for forest management, forest use, forest rehabilitation, forest protection and nature conservation.

According to [2] concerning Forest Arrangement and Preparation of Forest Management Plans, as well as Forest Utilization stated that what is meant by Forest Management Units (KPH), includes: a. Conservation KPH (KPHK), b. Protected KPH (KPHL) and c. Production KPH (KPHP). The FMU area is that all forest areas (conservation, protection and production) are divided into KPHs which are part of strengthening the national forest management system, provincial and district / city governments. The FMU area is defined in one or more main forest functions. If in a KPH area there are more than the main functions of the forest, then the determination of the name of the FMU is based on the function that is predominantly dominant. For example, if a FMU is dominated by production forests, it is called KPHP. The FMU area is stipulated in one administrative area or crossing the administrative administration area depending on the existing legal system and the facts of the field. The Minister of Forestry determines the area of KPH by taking into account the efficiency and effectiveness of forest management.

Development of FMUs includes the formation of regions and the formation of institutions. Establishment of forest management areas in the form of FMU units through the procedure stages, as follows: First, design by the Governor with consideration of the Regent. Second, directives by the Minister. Third, the formation of the Governor and Fourth, stipulation by the Minister. As for the establishment of management institutions in each KPH unit, it is intended to be responsible for doing the following: a. Management planning; b. Organizing; c. Management implementation; and D. Supervision control. The establishment of these institutions must pay attention to the competency requirements of KPH personnel and continue to respect the existence of forest utilization permits (eg IUPHHK, etc.) according to existing legislation.

In the implementation of the establishment of forest management areas to realize sustainable forest management, all forest areas are divided into KPHs. The development of FMU is a priority and the strategic target of forestry development must be achieved. In the 2009-2014 Strategic Plan of the Ministry of Forestry, the strategic goal that must be achieved is the operation of KPH institutions in 15 provinces. For this reason, all the main and supporting elements to achieve these targets must be optimally pursued, including the availability of information data on the development of FMU development.

The strategic role of FMUs in the development of forest resources at the site level includes optimizing community access to forests; conflict resolution, real decentralization of the forestry sector; has strategic value in playing a role in supporting climate change commitments; carry out the implementation and management of the right location, on target, the right activities and the right funding; bridge the funding optimization of climate change handling; supporting facilities in investment in the forestry sector; and increased success in rehabilitation and reclamation.

The main issues in the management of forest resources at present, include: legal uncertainty over forests, high tenure conflicts over forest areas and uncertainty over community management areas over forest areas. One of the root causes of legal uncertainty over forests is overlapping policies governing forest management. At the same time the expansion of the mining and plantation sector was increasingly massive into the forest area. This fact also caused the high incidence of forestry tenure conflicts in various regions.

The uncertainty of the community management area over the forest area is related to the problem of land claims and the return of land rights. The exclusion of indigenous peoples and local communities from their ownership of land and territory has brought a series of problems to the community. They practically do not own land anymore, or have but are very limited and inadequate, to maintain their survival and their culture. Such facts do not reflect the overall problems of indigenous peoples and local communities, but depend on the extent to which they can obtain the land and resources needed to ensure the continuity of economic development and sufficiency in fulfilling living needs.

In general, the causes of conflict can be seen as follows: In connection with conflicts that occur in forest areas, based on [2] concerning Forest Arrangement and Preparation of Forest Management Plans, and Forest Utilization, one of the main tasks of the FMU is to ensure that there is room for resolution of tenure conflicts by providing community-based forest management (HKm), Community Plantation Forest (HTR).) and Village Forest (HD). In addition to access to these forest management, KPH must open a communication room with the parties involved in managing forest resources through a multi-stakeholder forum as KPH partners.

2. Methodology (techniques and tools)

In the implementation of tenure conflict research the methodology used was interviews. Some of the tools used to simplify extracting information are as follows:

Table 1. Technical and analytical tools

Number	Tools	Key Information	Techniques / Information Sources
1.	Village History Flow	• The origin of the formation of the village and • important events / events related • land tenure that has an impact on people's lives	• Focused discussion group (FGD) with the community • Document review (secondary data, such as village profiles or monographs, results of preliminary information)
2.	Social mapping and resources	Village boundaries and settlement • Available natural resources • Problems / challenges faced related to social and resource aspects • Economic potential that can be developed	
3.	Tendency and change	• Changes that occur are related to people's lives in a certain period of time • Efforts that have been made to overcome problems / obstacles • Changes that occur are related to the economy of the community • Security conditions	
4.	Rapid Land Tenure Assessment and Analysis of Dispute Style	Regional history • Area area • Regional appointment letter in the form of SK, Perda, or other documents • Map of area designation Stakeholders in KPH area • Which parties have interests in the PHH area? • What programs / projects exist within the FMU area? • How is the management of each party? • Are there conflicts between parties? • In what form is the conflict? • Are there victims of the conflict? • To what extent is the resolution of the conflict between parties? • Are there maps of function for	Interview and document review

Number	Tools	Key Information	Techniques / Information Sources
		each party? Status of power, governance and governance permission • Is there a BAP of authority, governance and permit arrangements? • Is there a review of BKSDA • Are there supporting documents • Are there any new developments related to the status of the power of attorney, governance and permit arrangements? The style of the parties to the dispute and resolve conflicts: • Who are the actors involved in the conflict? • How far is the role and involvement of women in conflict? • How did the conflict occur? • What strategies do the parties make to resolve conflicts that occur? (must be distinguished between women's groups, male groups and mixed groups) • Are there solutions that have been realized?	
5.	HuMa-Win	HuMa-Win Documentation of data from land tenur assessment	Study of FGD Processes, interviews and supporting documents

3. Results

3.1 KPHP BANJAR

3.1.1 Status of KPH. The Banjar KPH in Banjar Regency, South Kalimantan Province was designated as a Model KPH through the Decree of [3]. The area of the Banjar Model KPHP according to the ministerial decree is 139,958 Ha consisting of 42,090 Ha Protection Forest (HL), 25,354 Ha of Limited Production Forest (HPT), and 72,513 Ha of Permanent Production Forest (HP).

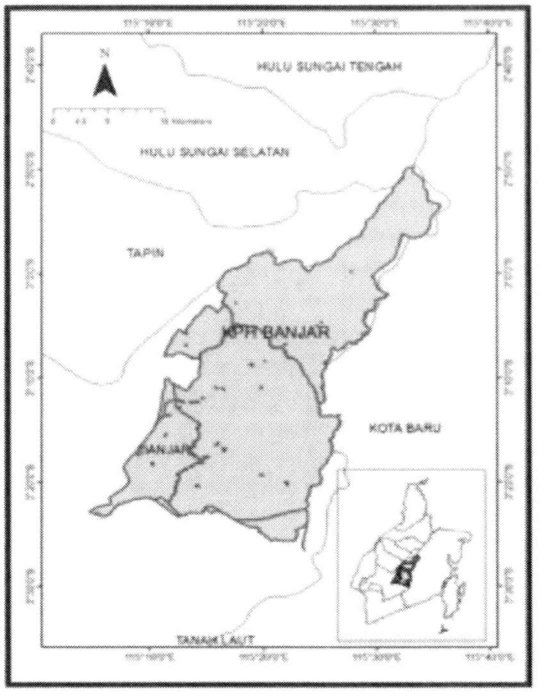

Figure 1. Map of the Model Banjar KPHP area, South Kalimantan

The organization of field operations in the Banjar KPH was determined based on the Banjar Regent Regulation Number 13 of 2009 concerning the Establishment, Organization and Work Procedure of the Banjar Regency Forest Management Unit Technical Implementation Unit. The main tasks of the KPH UPTD that must be responsible for the Head of the Banjar Regency Forest Service are implementing forest governance and preparing forest management plans, implementing forest utilization and forest area use, implementing forest rehabilitation and reclamation, and implementing forest protection and conservation in the Banjar Model KPHP area South KalimantanMore references.

3.1.2 Status of Authority and Governance. 1) Land Authorization. According to the Forestry Service, forest areas in Kupangrejo Village and Pakutik Village are the Banjar Model KPHP working areas of the UPTD of the Banjar Regency Forest Service. Evidence of Forestry Service claims to forest areas in Kupangrejo and Pakutik locations are a) [3] concerning Establishment of Unitary Areas of Production Forest Management (KPHP) Banjar Regency Banjar Regency, South Kalimantan Province, and b) boundary pal as evidence in the field, even though the installation (boundary) has not been fully completed.

While the people of Kupangrejo Village and Pakutik Village claim to have taken control of this land by cultivating it since 1962, even according to residents' accounts, before that year there were already indigenous people. For decades the community has never received information that the village land and its cultivation are forest areas. It was only in 1997 that they learned that the land was a forest area, namely when there was a PT. MHB entered and worked on their land. In addition, in that year there was information that the village government was no longer allowed to issue seals as a statement of land ownership known to the sub-district head.

The community's understanding of the forest is if the area in question is still forested, ie many large trees, many animals, can be used for hiding (taking refuge, the language of the protected local community is invisible). So protected forests are forests that can still be used to hide because they are still thick. With this perception, the community did not really understand when the area which had no forest turned out to be declared as a forest area. In addition, so far there has been no management / management, according to the community, it is the land of their ancestors.

The community feels that the claims to their land are quite strong because of the support from the National Land Agency (BPN) by issuing land certificates through the PRONA program. In 2012, 100 villagers from Kupangrejo received land certificates. The claims of the village community were strengthened by the village being determined as the definitive village in 1978 for Kupangrejo Village and 1982 for Pakutik Village.

Land ownership by the community between 1984 and 1998 was also proven by seals issued by Pembakal (Village Chief). Seal is a certificate of ownership of land that is known by the Camat. But the assessor did not get the seal document. In 1998, land tenure was reduced in intensity, which was only limited to working on it. This is based on the testimony of residents that since 2012 a Declaration of Land Cultivation has begun to be made which is often known as SPORADIK, namely the statement of cultivation of land by cultivators, justified by the witness of the boundary, known by the Chairperson of the RT.

2) Regional Governance. The Forestry Service Office through the FMU feels that at present it is certainly not able to manage an area of nearly 140 thousand Hayang to be managed by the government. Therefore community involvement in jointly managing forest areas is needed. The Forestry Service through KPH is working to implement a government development policy oriented to community based forest management (CBFM), namely the Community Forestry (Hkm) scheme, Village Forest (HD), and Community Plantation Forest (HTR), including the conflict villages. . Regarding the offer of these programs, both residents of Kupangrejo and Pakutik Villages stated that they rejected the HKm, HD and HTR programs. Residents only want the land they have occupied and managed to be owned in full.

The Watershed Management Center (BPDAS) through the Head of Institutional Section said that the BPDAS principally served HKm and HD licenses to the entire community. Of course the priority given to the community feels that the claims to their land are strong enough because of the support from the National Land Agency (BPN) by issuing land certificates through the PRONA program. In 2012, 100 villagers from Kupangrejo received land certificates. The claims of the village community were strengthened by the village being determined as the definitive village in 1978 for Kupangrejo Village and 1982 for Pakutik Village. Land ownership by the community between 1984-1998 was also proven by seals issued by Pembakal (Village Chief). A seal is a certificate of ownership of land that is known by the Camat. But the assessor did not get the seal document. In 1998, land tenure was reduced in intensity, which was only limited to working on it. This is based on the testimony of residents that since 2012 a Declaration of Land Cultivation has begun to be made which is often known as SPORADIK, namely the statement of cultivation of land by cultivators, justified by the witness of the boundary, known by the Chairman of the RT.

3) Regional Permit Arrangement. As a rule that applies to forest areas (countries), licensing in the Banjar KPH area is issued by the Minister of Forestry. There are six Business Permits for Utilization of Timber Forest Products (IUPHHK) issued by the Minister of Forestry at KPHP Banjar, namely: (1) PT. Inhutani III, (2) PT. HR Buana covering an area of 26,745 based on Decree Number 262 / Kpts-V / 1992, (3) PT. Kirana Rimba covering an area of 4,000 Ha, (4) PT. Prima Multi Buana covering an area of 12,690 Ha, (5) PT. Prima Buana Bestari covers 12,195 Ha, and (6) Litbanghut (KHDTK) BPK covering 623.5 Ha.

The use of the Banjar KPH forest area which has not gone through the Ministry of Forestry's licensing process is recorded as many as 34 cases with an area reaching 243.7 ha for residential residents, 770.2 hectares for transmigration, and 170 ha for mining. The three uses are proposed for ratification through the Regional Spatial Review (RTRW) in 2009.

3.1.3. Tenure conflict 1) Community vs Forestry Conflict. Based on information from the conflicting parties, in Kupangrejo Village there was a conflict between the community and forestry as the main actor. Forestry in this case is the FMU as the representative of the Forestry Service and BPKH as the executor of the Ministry of Forestry's boundary. The prominent causes of the conflicts that occurred in the two villages were the factors of unclear land status and differences in values about the forest. There is no control of the status of land for decades, so that when the community controls the land during this period no other party claims before forestry finally states that the land is a forest area. Differences in perceptions about forests are also a factor that causes conflicts to occur between communities and forestry, where the community thinks the forest when there is still physical forest, on the other hand forestry based on law 41 of 1999 is not the case.

2) Forestry VS BPN Conflict. There are differences in perceptions between BPN and Forestry regarding the status of the region. BPN together with the Regional Government are the parties that strengthen the position of the community with the certificate issued and the definitive village decision issued. In Pakutik Village the main party in conflict with Kupangrejo Village, namely the community and forestry. The difference is that the supporters of Pakutik Village were only the Regional Government with definitive village designation in 1982. The community worked on the land and settled during that period until the definitive village was determined by the local government and a certificate of land ownership from the BPN was issued. After all this time, PT. MHB, the Forest Service, and the BPKH that carry out activities as a form of claim that the land is a forest area. This confuses citizens who are in the right status.

3) Conflict Resolution Efforts. In 2009 the local government attempted to accommodate and propose areas that had been in the status of using forest areas that were not yet licensed by the Minister of Forestry in Banjar district through the CBFM scheme. The forestry party tries to accommodate the community to participate in managing forest areas through HKm, HD and HTR, but this effort has not yet been received by Kupangrejo Village and Pakutik Village.

The Kupangrejo village community insists on asking the village of 586 hectares to be enclosed or released from the forest area and become the property of the community. Residents argue logically that the 586 hectares of land with 362 households (head of the family) are very small compared to other generally wider villages. While the people of Pakutik Village who felt they did not have certificates and land were more than 900's with fewer inhabitants, namely 227 households wanted the government to release 454 forest areas for the people so that each household would receive a 2 ha portion.

The government of the two villages acknowledged that the disputed area was a forest area within the forest boundary made by the Planology Agency and boundary arrangements were made. The boundary pal which was taken away by the residents of Kupangrejo Village is probably the function limit between HL and HP. The Pakutik Village Government instructed its citizens not to remove the boundary pales installed by the forestry. This is intended to be able to know the extent that must be proposed to be released from the forest area.

4. Conclusion

a. The existence of FMUs is expected to be streamlined to build synergy between relations between institutions in forest utilization and management as outlined in a work program, therefore integrated coordination of the Forestry Planning Agency, Director General of Forestry Production Development, Directorate General of Forest and Land Rehabilitation, BPHP, BPKH and BPDAS.

b. In preparing the KPH management plan in the future the existence of communities around the FMU area needs special attention. Community governance, especially in the KPH Model area needs to be accommodated in the management of FMU by prioritizing the guarantee of forest functions. Various government schemes such as Community Forestry (HKm), Community Plantation Forests (HTR), and Village Forests are expected to be able to provide tenure certainty for people who access the area. However, caution is needed whether this scheme is in accordance with the form of community tenure at the location.

c. The forms of empowerment schemes which are options for the community vary greatly, such as: Community Plantation Forest (HTR), Community Forestry (HKm), Village Forest (HD),

Community Forest / Rights Forest (HH / HR). Apart from any form of empowerment scheme that will be chosen, it is better to get treatment of assistance, especially in the process of understanding the selected scheme policy and the decision-making process to choose the scheme.

d. Re-socialize the construction of policies that distinguish State Forest and Forest Areas, so that the options are more open and reflect the real conditions of Indonesia's forest area.

e. The weakness of forest planning, forest governance and forest management planning that has been running so far can be seen from the low capacity to carry out these activities. This is due to the orientation of the government and regional government in forest product licensing and commodities and not in the management of forest areas / areas, so that control over the implementation of forest and forest products utilization activities becomes weak.

f. In determining the vision and mission of the FMU it is necessary to place balanced sustainability (social, economic and environmental) as a concentration of forest resource management and wherever possible accommodate the views of the parties.

5. Recommendation

a. General:

1. FMU development policies must be able to internalize KPH development activities to various parties that have different tasks, functions and authorities. The complexity of regulating the role of parties needs to be accommodated in a policy, so that the establishment of FMU development policies requires openness to invite active participation by the parties.
2. The urgency of the FMU development implementation program needs to be understood by all relevant parties in order to reduce barriers that are in the nature of disagreements and coordination barriers that are needed.
3. The parties related to FMU development as an institution must be able to manage workloads and develop procedures and schedules so that there is a common direction in FMU development.
4. Monitoring and evaluation is carried out to formulate the problems faced and the form of intervention needed so that FMU development can run in accordance with the stated objectives.

b. National:

It is necessary to increase the capacity of national KPH development to realize national resource allocation for FMU development continuously, until all FMUs are built for all forest areas.

c. Area:

It is necessary to identify provinces / districts that are relatively ready, at least political support to implement them, especially in the short term, FMU development can actually be a basis for lifting regional economic activity where this can be a trigger factor for national FMU development.

References

[1] Undang-Undang Republik Indonesia 1999 Undang-Undang Republik Indonesia Nomor 41 Tahun 1999 Tentang Kehutanan (Jakarta)

[2] Peraturan Pemerintah Republik Indonesia 2008 Peraturan Pemerintah Republik Indonesia Nomor 3 Tahun 2008 Tentang Perubahan Atas Peraturan Pemerintah Nomor 6 Tahun 2007 Tentang Tata Hutan dan Penyusunan Rencana Pengelolaan Hutar serta Pemanfaatan Hutan (Jakarta)

[3] Keputusan Menteri Kehutanan 2009 Keputusan Menteri Kehutanan Nomor 793/Menhut-II/2009 Tentang Penentuan Area Kesatuan Pengelolaan Hutan Produksi (KPHP) Model Provinsi Banjar Kalimantan Selatan

[4] Kementrian Kehutanan 2011 *Pengembangan KPH: Konsep, Legislasi,Implementasi* (Jakarta)

[5] Kementrian Kehutanan 2012 *Peraturan Terkait Kesatuan Pengelolaan Hutan* (Jakarta)

[6] Putusan Mahkamah Konstitusi Nomor 45/PUUIX/2011 Tentang Judical Review Undang-Undang Nomor 41 Tahun 1999 Tentang Kehutanan dengan Konstitusi Republik Indonesia Tahun 1945, 21 Februari 2012

[7] Peraturan Menteri Kehutanan 2012 Peraturan Menteri Kehutanan Nomor P.33/2012 Tentang Perubahan Atas Peraturan Menteri Kehutanan Nomor P.40/2010 Tentang Organisasi dan Tata Kerja Kementrian Kehutanan (Jakarta)
[8] Suciana D, Gamin G, Widianto, Syaifullah dan Muh A 2013 *KPH, Konflik dan REDD* (Bogor: Working Group on Forest Land Tenure)

The use of sugar-cane waste (Bagasse) energy as substitution of fossil-fuel energy in Krebet Sugar Factory, East Java, Indonesia

Kartika Eka Sari* and Christia Meidiana

Urban and Regional Planning Department, Enginering Faculty, Universitas Brawijaya Jl. MT. Haryono No. 167 Malang, East Java, Indonesia

*kartika_plano@yahoo.co.id

Abstract. Industry is one form of economic activities that process raw material and/or utilize industry resource to produce product with a higher value or benefits. Rapid growth of industrial activities an increase community's welfare. While on other side, industrial development can increase the value of CO2 emission that can influence global warming. Krebet Sugar Factory in one of the largest sugar industries in Indonesia and so far, still using fossil fuels (gasoline, diesel, residue oil) as energy resource in production process. Based on 2011 National Greenhouse Gas Emission Reduction Action Plan, CO2 emission reduction's target in Indonesia, especially in industrial sector for year 2020 is 0,001 Gton CO2 (Scenario 26%). One of mitigation action that can be done to support CO2 emission reduction is using alternative energy resource/ fuel bagasse. The aim of this research is to evaluate utilization of factory waste for alternative energy resource/fuel (bagasse). Krebet Sugar Factory have production capacity of 840 ton/day. Average of Electricity resource provided by PT. PLN is 26,40 KVA/month, while the electricity average usage provided by Krebet Sugar Factory's power plant is 4.320 KVA/month and Krebet Sugar Factory Fuel's consumption for one year is 19.500 ton. Based on production data, average bagasse production of Krebet Sugar Factory is 420.533,34 ton so can produce energy value in the amount of 4.878,19 Terajoule that can be used as an alternative energy resource.

1. Introduction

Krebet Sugar Factory in Malang, which having been operating since 1906 is one of the largest sugar companies in Indonesia, it is located in Bululawang District, Malang Regency. In 2004, Krebet SF produced 6000 tons of sugar per day and in 2010, it increased to 11,500 tons of sugar per day and this factory will continue to increase its production capacity. Greenhouse gases in sugar factory come from process of production, combustion, utilization of electrical energy, and waste of production [1].

The use of large-scale energy at sugar factory can increase greenhouse gases (CO2) production that affect global warming. According to [2], the main gases which are categorized as greenhouse gases and which have the potential to cause global warming are carbon gases (CO2), CH4, and NO2. [3] states that carbon gases, especially CO2, have percentage of 50% in total greenhouse gases.

In accordance to National Greenhouse Gas Emission Reduction Action Plan 2011, Indonesia's GHG emission reduction target by 2020 is 26% by using mitigation actions which one of them is by the development of carbon sequestration. According to [4], one effort to reduce greenhouse gases is by replacing fossil fuels and utilizing industrial waste as a source of alternative energy. Data of previous research show the use of fossil energy sources in Krebet SF is 19,550 tons/year (gasoline, diesel and residual oil). From the use of fossil fuels in this case produce CO2 emissions of 61,568.04 tons.

2. Literature review

2.1. Sugar Production Process

The sugar factory industrial activity basically has 6 stages: milling station, refinery station, evaporation station, crystallization station, separation station, and settlement station [5].

- Milling Station: The process at milling Station can be divided into two stages; they are preliminary process and sugar cane extraction.
- Purification Station: The non-sugar substances which are also contained in sugar nira are separated by controlling the temperature, pH, and remaining time in each equipment, so sucrose which is contained in sugar nira is not inverted. Most of non-sugar substances will be separated, one is called as blotong and the result of this nira is called pure/clear nira.
- Evaporation Station: Pure/clear nira still has high content of water. To make the use of steam in process of crystallization efficient, water clear nira is evaporated until it reaches 30 - 32 degrees Celsius. The evaporation process is carried out vacuum.
- Crystal Station: Thick nira is further evaporated until sugar crystals forms. This crystallization process is also carried out under vacuum. To achieve the desired crystal size, process of cooking is done in various stages. Final result of the Crystal Station is massecuite; sugar crust which surrounding still contains strup layers.
- Separation Station: Sugar crust in massecuite is separated from its strup by utilizing centrifugal force. This centrifugation process is also performed in several stages, depending on the type of massecuite that is being rotated.
- Completion Station: Sugar which is produced in Rotation Station still contains high content moisture, therefore sugar is dried and cooled using Sugar Drier and Cooler (SDC) to obtain sugar with desired water content and temperature.

2.2. Sugar Factory's CO2 Emission Source

Sugar factory industries that using fossil and biomass fuels potentially produce GHG (Green House Gas) emission and contributing global warming. Based on [1], The source of CO2 Emission in Sugar factory comes from burning fuel and solid waste processing. CO2 emission sources based on Research and Development center of Enviroment and Forestry Ministry of Indonesia (2017) are Industrial process and Product Usage, Energy Consumption and Industrial Waste.

Based on [6], CO2's measurement result in the process in milling station is higher that CO2 measurement result in other stations. According to President's Regulation Number 61/2011 in National Action Planning for GHG Emission Reduction, Sugar Factory is clasified in Food and Beverages Industry that has a GHG Emission Reduction's target amount 0,069 Mton CO2 in year 2020.

3. Methods

The variables used in this research are Bagasse Production from Sugar Production Process, Energy Consumption of Bagasse, Energy's Conversion Factor of Bagasse and Co2 Emission Factor of Bagasse. Research method using alternative scenario technique based on existing data. Alternative scenario performed through CO2 emission production from bagasse usage. Calculation of CO2 using CO2 emission conversion which has been recognized by [2].

$$GRKemission = fuelconsumtion \times femissionfactor \quad (1)$$

To identify energy consumption of bagasse is to multiply the amount od bagasse production with the calorie value of bagasse. Based on ICSEFA (2014), the calorie value of bagasse is 1,852 kcal/k or 11,6bTJ/Gg.

Energy consumption of bagasse = bagasse production x 11,6 TJ/Gg (2)

Afterwards to identify CO2 emission production from bagasse usage is to multiply energy value of bagasse with CO2 emission factor. Bagasse is classified as biomass energy, so the CO2 emission factor used in this research is 100,000-ton *CO2/TJ.*

Bagasse CO2 Emission = Energy consumption of bagasse x 100,000 CO2/TJ (3)

Emission factors are obtained from emission factors that have been determined by IPCC *(the default)*. Fuel consumption used by the Krebet sugar factory is Gasoline, Solar and Residue in Ton/Year units. The third factor in emissions from these fuels can be seen in Table 1.

Table 1 Factor of CO2 Emission in Fuel

No	Product	CO2 Emission Factor (Ton/TJ)
1	Gasoline	68,6
2	Solar	73,4
3	Residue oil	76,6

Source : IPCC, 2006

4. Results

Krebet Sugar Factory is a sugar factory owned by PT. The Rajawali I Sugar Factory which is located in Krebet Village, Bululawang District, Malang Regency. This location is same as the location of Krebet Sugar Factory II. Area used by Krebet SF is an area of 24.08 hectares. Production activities in Krebet SF are a neutral sulfitration process which is done physically and chemically, with production capacity of 70,000 tons of cane per day. The process of activities that happen in sugar production are as follows.

1. Production activities
   Stages of sugar production ranging from sugar cane at Krebet sugar factory are as follows.

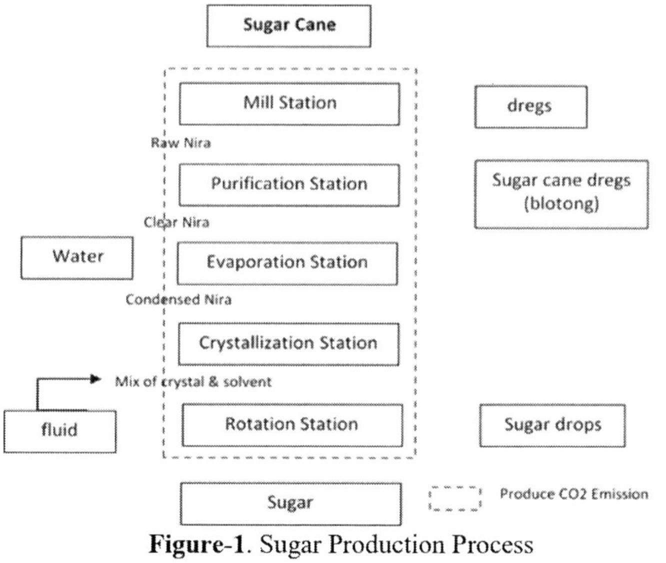

Figure-1. Sugar Production Process

2. Post-production activities
   Post-production activities are period of sugar cane planting, so that the process of sugar production is stopped until harvest time of sugar cane. Activity that occurs during the post-production season is only the mobilization of labor and office activities, so that the only energy source used is electricity.

4.1. Sugar Factory Production Capacity

Sugar production capacity in Krebet Sugar Factory on planting season year 2010-2014 which average is 1,501,904.78 tons of sugar cane with average sugar result of 105,071,742 tons/ year. The final product consists of two types, they are sugar and drops.

Table 2. Sugar Factory Capacity

Plant season	Rendemen (%)	Production (ton)	
		Sugar Cane	Sugar
2011	7,2	1.051.286,9	75.377,27
2012	6,5	1.583.997,8	103.118,26
2013	6,8	1.394.934	94.855,51
2014	6,7	1.715.724,8	114.095,77
2015	7,8	1.763.580,4	137.911,9
	Average	1.501.904,78	105.071,742

4.2. Energy Consumption

Electrical energy used in the Krebet SF operation and activities of lighting is supplied by PT. PLN and electricity coming from power plants owned by Krebet SF. The use of electricity that is supplied by PT. PLN is averagely 26.40 KVA per month, while electricity usage comes from Krebet SF power plants is averagely 4,320 KVA per month.

Operational activities and sugar production in Krebet sugar factory use fossil fuel energy which includes gasoline, diesel fuel, residual fuel and lubricants. The following table shows consumption of fossil fuel usage in Krebet sugar factory per year.

Table 3. Energy Resources Consumption

Fuel Type	Use (Ton/tahun)
Gasoline	1.500
Diesel	5.000
Residue	10.000
Oil	3.000
Total	19.500

Based on Table 3, known that fuel consumptions by Krebet sugar factory are gasoline, diesel fuel, residual fuel, and lubricating oil. Total use of Krebet sugar factory fuel in one year is 19,500 tons of fuel.

4.3. Krebet Sugar Factory's Solid Waste

Solid waste which is generated from the sugar production process consists of bagasse, blotong, and kettle ash. Bagasse comes from residue of milled sugar cane which sap milked at the milling station. Blotong is result of separation of sap and its manure by filtration at purification station. Kettle ash comes from the residue of combustion on the boiler.

4.4. CO2 Emission of Fossil Energy Resources

The total fuel used in Krebet sugar factory is 19.5 Gg/Year,the largest use is residual oil which is 10 Gg/year.

Table 4. Energy Resources Consumption

Fuels	Consumption (Ton/yearn)	Consumption(Gg/year)	Conversion Factor (TJ/Gg)	Energi (TJ/year)
a	b	c = b / 1000	D	e = c x d
Gasoline	1.500	1,5	46,4	69,6
Diesel	5.000	5	43	215

Residu	10.000	10	40,4	404
Oil	3.000	3	43,4	130,2
Total	19.500	19,5		818,8

Energy consumption of Krebet sugar factory in one year is 818.8 Terajoule/year. Then it is calculated with CO2 emissions from fuel energy which amount is shown in Table 5.

Table 5. CO2 Emmision of Energy Resources Consumption

Fuel	Energy (TJ)	Emission Factor (Ton CO2/TJ)	Emission Factor (Gg CO2)	CO2 Emmision (Ton CO2)
a	b	c	d = b x (c/106)	e = d x 1000
Gasoline	69,6	69.300	4,82	4.823,28
Diesel	215	74.100	15,93	15.931,5
Residu	404	77.400	31,27	31.269,6
Oil	130,2	73.300	9,54	9.543,66
		Total	61,568	61.568,04

Based on Table 5, and based on the amount of fuel energy, it is known that CO2 emissions produced by the Krebet sugar factory are 61.568 Gg CO2 per year or 61,568.04 Ton CO2 per year (1 Giga gram = 1000 Ton)

4.5. Use of Sugar Cane Waste (Bagasse)

One way to reduce CO2 emissions is by using bagasse as an alternative fuel to supply energy of Krebet sugar factory. Krebet sugar factory itself produces waste of bagasse. Therefore, the purpose of bagasse which is used as alternative fuel is to apply zero waste, thus there is still leftover bagasse which can be used as raw material for particle board production.

Sugar cane waste (Bagasse) is a side-product in sugar-making process, especially from grinding process (Ministry of Environment, 2005). Bagasse contains mostly lingo-cellulose, 48-52% water, 47.7% fibre. Other elements contained in bagasse are 47% carbon, 6.5% hydrogen, 44% oxygen, and 2.5% ash. It is known that bagasse production is averagely of 280 kg production or 28% per 1 ton of sugarcane (Rabelo, 2011). According to review ministry of environment, calorific value of 1 ton dry bagasse is equivalent to 598 liters of diesel. The following is bagasse production in Krebet Sugar Factory.

Table 6. Baggase Production

Year	1 Year Production (ton)		
	Sugar Cane	Sugar	Bagasse
a	b	c	d = b x 28%
2011	1.051.286,9	75.377,27	294.360,33
2012	1.583.997,8	103.118,26	443.519,38
2013	1.394.934	94.855,51	390.581,52
2014	1.715.724,8	114.095,77	480.402,94
2015	1.763.580,4	137.911,9	493.802,51
∑	1.501.904,78	105.071,742	420.533,34

Assuming that 280 kg or 28% of 1 ton of sugar cane becomes bagasse, the average bagasse production from 2011 to 2015 is 420,533,34 ton. Average value of bagasse production will be used to determine average energy value that can be removed from the bagasse. Conversion calculation of bagasse production into energy uses the IPCC formula, 2006. The following is a calculation of bagasse energy conversion.

Table 7. Total Energy of Baggase

Average Bagasse Production (Ton)	Average Bagasse Production (Gg)	Calorie value (TJ/Gg)	Energy (TJ)
A	b = a / 1000	c	d = b x c
420.533,34	420,53	11,6	4.878,19

If it is compared to other energy sources in Krebet SF, Bagasse has greater energy value compared to any fossil energy sources.

Table 8. Energy Resouces Comparation in a year

Energy Resources	Energy (TJ)
Gasoline	69,6
Solar	**215**
Residue oil	404
Lubricant	**130,2**
Bagasse	4.878,19

Large energy value will result in large CO2 emissions and so does bagasse energy, it has large emission factor which is equal to other biomass fuel emission factor (100,000 tons CO2 / TJ), so this recommendation is done by simulating bagasse energy usage in order to find out that energy is adequate to replace fossil fuels and that it has lower CO2 emission than energy from fossil fuel. Based on the simulation that has been done, there are amount of CO2 emission output. The following is comparison of CO2 emissions results in existing conditions at the time of bagasse usage.

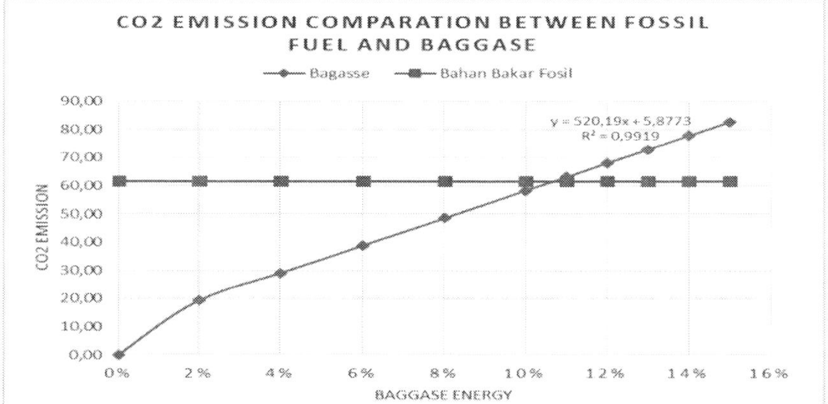

Figure-2. CO2 Emission Comparation Between Existing and Baggase Usage

In graphic above, it is known that point tangent of x and y (10.7; 61,57) with curve equation is y = 5,8773 + 520,19x. From curve equation y, it is known that intersection point between line x and y lies at 5.8773 if x = 0. If bagasse energy values of 0, then value of carbon emission equal to 5.8773. Intercept (5,8773) can also be interpreted as the influence of variables or other factors outside this study that is not researched. Energy value that can be allowed to be used in as substitution of fossil fuel energy based on curve equation y = 5,8773 + 520,19x is maximum of 10.7% from total bagasse energy which produced. Thus, simulation that is used for the recommendation is use of bagasse energy by 10%, since energy value is close to the value of fossil fuel energy which used in Krebet sugar factory and it has lower CO2 emissions value than the existing value of CO2 emissions. The following is energy calculation which uses 10% of bagasse production.

Table 9. Energy Resouces Comparation in a year

Average Baggase Production (Gg/year)	10% of a (Gg/year)	Calorie Value (TJ/Gg)	Energy (TJ)
a	b = a / 10%	C	d = b x c
420,53	42,05	11,6	487,82

From the calculation of energy which is generated by bagasse, it is known total CO2 Emission can be generated through the following calculations.

Table 10. CO2 Emission of Baggase

10% Average Baggase Production (Gg/year)	Energy (TJ)	CO2 Emission Factor (Ton/TJ)	CO_2 Emission (Gg)
a	B	C	d = b x (c/10^6)
42,05	487,82	100.000	48,78

Based on Table 9 and Table 10, it can be seen that value of the energy produced from 487.82 Gg bagasse is amounted to 42.05 Terra Joule. Then it will be calculated along with CO2 emissions from the energy released by bagasse, which generating CO2 emissions of 48.78 Gg CO2. The following is comparison of each existing CO2 emission values with the value of CO2 emissions during its use.

Table 11. CO2 Emission Comparation

Comparation	Existing	Simulation
Gasoline	4,82	-
Diesel	15,93	-
Residu	31,27	-
Oil	9,54	9,54
Baggase (10%)	-	**48,78**
Total (Gg/Year)	61,57	**58,32**
Deviation	3,25	

Based table 11, it can be seen that there are differences in results of existing CO2 emissions during and after the simulation. On simulation, it can reduce CO2 emissions about 3.25 Gg / year or about 5.3% of total CO2 emissions from the use of existing fossil fuels.

5. Conclusion

Bagasse is one of Sugar Factory's solid waste comes from residue of milled sugar cane which sap milked at the milling station and has average production amount 420.533,34 ton/year of bagasse. For sugar production, Krebet Sugar Factory consumed 19,500 tons of fuel per year that content 818,8 TJ/year of energy and produce 61.568 Gg of CO2 per year or 61,568.04 Ton of CO2 per year (1 Giga gram = 1000 Ton).

One way to reduce CO2 emissions is by using bagasse as an alternative fuel to supply energy of Krebet sugar factory. Every year Krebet Sugar Factory produce 420.533,34 ton/year of bagasse that content 4.878,17 Terra Joule of energy. Based on the comparison of bagasse energy and conventional energy that has been used, it can be concluded that Bagasse is very potential for alternative energy in Sugar Factory. Bagasse also produce CO2 and by calculating CO2 emissions produce by bagasse, it can be concluded that the use of 10% bagasse can reduce the amount of CO2 emissions from Krebet Sugar Factory.

Aknowledgement
I would like to thank you to all the academic staf of Urban and Regional Planning Department, Universitas Brawijaya. And also thank you to Auza Ghifary that already assist during data collection period.

References
[1] Sihombing S R A BR 2012 Potensi Penurunan Emisi Gas Rumah Kaca Pada Industri Gula (Studi Kasus PT PG RAJAWALI II Unit PG SUBANG) *Skripsi* (Bogor: Teknologi Industri Pertanian IPB)
[2] IPCC 2006 IPCC Guidelines for National Greenhouse Gas Inventories Energy; Chapter 2: Stasionary Combustion **2** (Washington D.C: USA)
[3] Rukaesih A 2004 *Kimia Lingkungan* (Yogyakarta: Andi)
[4] Fiantisca A 2002 Kajian Inventarisasi Gas Rumah Kaca Pada Sektor Pertanian dan Kehutanan dengan Menggunakan Metode IPCC *Skripsi* (Bogor: Jurusan Geofisika dan Meteorologi-FMIPA, IPB)
[5] Santoso B 2005 Proses Pembuatan Gula dari Tebu Pada PG X (Jakarta: Universitas Gunadarma)
[6] Purwaningsih E 2011 Polusi Udara dan Pengaruhnya terhadap kesehatan Masyarakat (Studi Kasus Pabrik Gula Mojo Sragen) *Tesis* (Yogyakarta: Universitas Gajah Mada)
[7] Peraturan Mentri Perindustrian 2011 Peraturan Mentri Perindutrian Nomor 64/M-IND/PER/7/2011 tentang Jenis-Jenis Industri dalam Pembinaan Direktorat Jendral dan Badan di Lingkungan Kementrian Perindustrian

Effect of amino acids intake of rubber seeds processing as alternative comestible length for infants born against

Ratna Dewi Puspita Sari

Environmental Science, Post Graduate Program, Lampung University, Bandar Lampung, Indonesia

*ratnadps@gmail.com

Abstract. Sumatra and Kalimantan are the region with the area and the highest rubber production in Indonesia, including Lampung province that is also abundant production of rubber seed every year. Rubber seed contains a protein that is high enough. With the high crude protein, it can be used to make rubber seed chips and snacks potential to be used as a source of protein for pregnant ibi amino acids to help meet the mother during pregnancy. Sufficient amino acid intake during pregnancy associated with a body length of birth. This research was observational analytic with cross sectional design. Sampling method used in this study is consecutive sampling. The study was conducted in East Lampung in 2016. The subjects were all pregnant women including rubber farmers in East Lampung region and met the inclusion criteria of the study. The samples obtained were 124 pregnant women and their babies. Data were analysed descriptively. The results obtained indicate that rubber seed chips can help meet amino acid intake during pregnancy and affects the baby's body length at birth. From the sample of the research note that 89.5% of the infants had a body length of> 48 cm (normal) and 10.5% of infants had a body length of <48 cm (short). The nutritional intake of pregnant women is very important to prevent children born with a body length is not normal. The results obtained indicate that rubber seed chips can help meet amino acid intake during pregnancy and affects the baby's body length at birth. From the sample of the research note that 89.5% of the infants had a body length of> 48 cm (normal) and 10.5% of infants had a body length of <48 cm (short). The nutritional intake of pregnant women is very important to prevent children born with a body length is not normal. The results obtained indicate that rubber seed chips can help meet amino acid intake during pregnancy and affects the baby's body length at birth. From the sample of the research note that 89.5% of the infants had a body length of> 48 cm (normal) and 10.5% of infants had a body length of <48 cm (short). The nutritional intake of pregnant women is very important to prevent children born with a body length is not normal.

1. Introduction

Indonesia is the country's second largest natural rubber producer after Thailand, with an area of rubber plantation in the world [1]. Most of the rubber produced from the rubber of the people which covers a total area of 81% of existing dar. Rubber plantation nearly spread all over Indonesia, including Lampung Province. Based on data from the Plantation Office of Lampung Province [2], the land area of rubber in the province of Lampung is 158.999 ha which includes 15 East Lampung district that currently has the potential rubber cultivation increasingly passionate. Rubber is growing every year, both in terms of acreage and production (tonnes). The average production from 2007 to 2014 as many as 46 798 tonnes. The rubber plant is one kind of plant MPTS (Multi Purpose Tree Species) which produces non timber forest products such as gum [3]. Rubber seeds contain nutrients especially proteins that could potentially be used as a raw material of food. Rubber seed as a food ingredient has not been fully utilized by the community, but each plant capable of producing seeds rubber around 0.8 to 1.2 tonnes / ha / year (for plants with over 4 years of age). The abundance of rubber seed in the East Lampung Regency is one of the capital to improve the creative food industry in the District. Rubber seed as a food ingredient has not been fully utilized by the community, but each plant capable of producing seeds rubber around 0.8 to 1.2 tonnes / ha / year (for plants with over 4 years of age). The abundance of rubber seed in the East Lampung Regency is one of the capital to improve the creative food industry in the District. Rubber seed as a food ingredient has not been fully utilized by the community, but each plant capable of producing seeds rubber around 0.8 to 1.2 tonnes / ha / year (for plants with over 4 years of age). The abundance of rubber seed in the East Lampung Regency is one of the capital to improve the creative food industry in the District.

Constraints are less optimal utilization of rubber seeds as an alternative comestible due to the content of HCN. However, HCN reduction techniques related research has been done before and proven to reduce the HCN content until the limit is safe to consume all circles from children, adolescents, adults and pregnant women. Boiling rubber seeds for 15 minutes and soak them in water for 24 hours (the soaking water replacement every 6 hours) has been shown to reduce levels of cyanide (HCN) in the rubber seeds. Rubber seeds that have been safely consumed can be used as raw material for rubber seed chips. Rubber seed has a proportion of parts that can be consumed around 57%. In addition, rubber seed contains nutrients, especially protein. Rubber seed nutrient composition consisting of 27% protein, 32.3% fat, carbohydrates 15.9%, water 9.1% and 3.96% [4]. With the crude protein content is high, then the rubber seed chips containing amino acids and also high potential to be used as a source of protein snacks [4]. Processed rubber seeds can be used to meat the daily protein requirement for pregnant women in addition to meat, fish, eggs, milk and various derivatives in terms of support programs Protein Adequacy Score (AKP). Quality of food proteins is determined by the composition and quantity of essential amino acids [5]. This amino acid plays a role in meeting the nutritional needs of pregnant women. The adequacy of nutrition of pregnant women affect the nutritional status of children in the womb which will further define the development of children, especially during growth (golden age) [6].

Formulation of the problem is how does the influence of amino acid intake from rubber seed treatment as an alternative to body length comestible baby is born?. The purpose of this research is to determines the effect of amino acid intake from rubber seed treatment as an alternative to long comestible birth weight. This research can be used for:

1. As the information to the public and the government about the effect of amino acid intake from rubber seed treatment as an alternative to long comestible birth weight.
2. Can be input to the government or the parties concerned in order to increase the creative food industries in East Lampung regency.

2. Method

2.1. The design, location and research subjects

This research was observational analytic with cross sectional design. Sampling method used in the study is consequtive sampling. The study was conducted in East Lampung in 2016. The subjects were all pregnant women including rubber farmers in East Lampung region and met the inclusion criteria of the study. The samples obtained were 124 pregnant women and their babies. The subject has the following inclusion criteria.

1. Pregnant women who are physically healthy and clinical
2. Not having a birth defect
3. Residing in East Lampung
4. Eating chips during pregnancy rubber seeds
5. Parents babies willing to participate in the study by signing the informed consent

Subjects who met the inclusion criteria measurements of the baby's length at birth by looking at the data in the book MCH (Maternal and Child Health Handbook).

2.2. *Collecting, processing and data analysis*
The data collected were body mass index (BMI), protein level and baby's body lenght. Body lenght measurements carried out by trained personnel using berketelitian-board lenght of 0.1 cm. Analysis of the data used is descriptive analysis and presented in tabular form. While the rubber seed chips consumption data collected by food recall 1 x 24 hours per month during the study and were taken on average consumption. How rubber seed processing into chips [7]:

1. Literature review. Information on reduction techniques cyanide (HCN) contained in the rubber seed and the details of the content of the nutritional value obtained from the literature. The information is used as a reference in the processing of rubber seeds to make it safe for consumption.
2. Selection and extraction of rubber seeds. Sorting is done to get a rubber seed deserves to be further processed as raw material comestible. High-quality rubber seed memantulnya marked with rubber seeds when dropped. Rubber seed extraction is done with the aim of separating the hard seed coat with meat seeds. The extraction process using a hammer or stone tools.
3. HCN reduction process. HCN reduction process is based on past studies as well as information obtained from our library. Boiling rubber seeds for 15 minutes followed by immersion in water for 24 hours and the replacement of the soaking water every 6 hours is used as a method of HCN reduction in this study.
4. Processing of rubber seeds as an alternative comestible. Rubber seed that has passed the HCN reduction process, to be further processed as an alternative comestible grains rubber chips.

3. Results

Table 1. Body Mass Index (BMI) pre pregnancy and third trimester

Classification of	Pre		Third Trimester	
BMI	Amount	(%)	Amount	(%)
Underwight	20	16,1	17	13,7
Normal range	68	54,9	76	61,3
Overweight at risk	17	13,7	12	9,7
Obese	19	15,3	19	15,3
Total	124	100	124	100

BMI of pregnant women in pre-pregnancy and third trimester are generally included in normal BMI.

Table 2. Protein levels at the beginning of second trimester and third trimester

Protein Levels	Beginning of Second Trimester		Third Trimester	
	Amount	%	Amount	%
$\geq$ 6,4 mg/dl	109	87,9	115	92,7
< 6,4 mg/dl	15	12,1	9	7,3
	124	100	124	100

The protein levels at the beginning of the second trimester of pregnancy was 12.1% which lacked protein and decreased to 7.3% in the third trimester.

Table 3. Body length measurement results baby is born

Long Firm	Amount	Percentage (%)
> 48 cm (normal)	111	89.5
<48 cm (short)	13	10.5
Total	124	100

At birth body length measurements of babies born to mothers who regularly eat chips of rubber seed with an average of 300 grams / per month. The results showed 89.5% of infants born with normal body length> 48 cm and 10.5% of infants born with a body length ≤48 cm.

4. Discussion

In the rubber seed processing chips, do boiling water for 15 minutes, and replacement of water every 6 hours for 24 hours. This time is shorter when compared with Widayati [8] and this ensures that the content of amino acids in the rubber seed still available. In order to get the maximum processing, rubber seed must go through perfect processing to eliminate the anti-nutritional substances in it (HCN / Acid Cyanide). Widayati [8] do boiling for 45 minutes and test of hemagglutinin activity against rubber seed analyzed indicate negative values. The processing of rubber seeds will improve the digestibility of the protein, from 85.95% in the seeds of 102.46% of raw rubber into the rubber seeds that have been soaked and boiled. The percentage of amino acids decreased, while the percentage of the proteins varied according to a decrease in protein content. Amino acids in the seeds of the rubber is methionine, cysteine and lysine. Score each seed raw rubber is 32.0 (methionine + cysteine) and 45.6 (lysine). While on the rubber seeds that have been soaked and boiled amounted to 36.9 (methionine + lysine) and 40.4 (lysine).

BMI was calculated in pregnant women to determine the nutritional status of pregnant women related to nutritional intake during pregnancy. From the data obtained shows that pregnant women generally are on a normal BMI. The number of normal BMI of pregnant women increased from 54.9% to 61.3% in the third trimester. The number of pregnant women with protein levels ≥ 6.4 mg / dl also increased in the third trimester from 87.9% to 92.7%. The next step is to measure the length of the baby's birth where during pregnancy the mother has additional protein intake through rubber seed chips (Table 3).

In this study the results obtained indicate that rubber seed chips consumption habits affect the baby's body length. Babies are born with a body length of normal (> 48 cm) more than babies with a body length is not normal / short (≤48 cm). Emawati, et al [6] states that babies are born with a short body length of 6 times risky short stays at 12 months compared to normal birth, while pregnant women with protein intake is less than 58% RDA is 1.6 times greater risk of having a child stunting at the age of 12 months than women with ≥58% RDA of protein intake. Babies born to normal (body length ≥ 48 cm) have a higher survival rate compared with children who were born with a body length is not normal (<48 cm). The length of the body is born with the risk of occurrence of stunting.

Based on research Septiani [9], the rate of maternal nutrient consumption is still low. Only 39% of pregnant women who meet their protein requirements. Amino acid intake of protein obtained from rubber seed chips to help meet the needs of an amino acid in pregnant women. Pregnant women need protein in sufficient quantity and complete amino acid content because during pregnancy increased protein requirements for growth and development of the fetus [10]. The impact of protein intake deficiency is a growth disorder in the fetus, such as intrauterine retardation, birth defects, low birth weight and miscarriage [11]. Nutrition during pregnancy is one of the important factors that influence the development of the embryo and fetus and health status of pregnant women. Pregnancy is a continuous phase, so that deficiency in a period would give a different impact on the outcome of pregnancy. Quality of babies born very dependent on the state of maternal nutrition before and during pregnancy.

Short birth length of the body can lead to stunting. Short stature (stunting) became one of the nutritional problems in developing countries, including Indonesia. Riskesdas data in 2010 showed that 35.6% children in Indonesia are having problems stunting. That is, more than a third of our toddler has a height lower than the standard height toddler age. Stunting be an indicator of chronic malnutrition which describes the history of child malnutrition in the long term and does not rule out the possibility in the womb or have had recurrent infectious diseases after birth. Children who experience severe stunted not only on shorter physical stature, but also on cognitive function. Children who suffer from malnutrition, including short,

Grantham-McGregor [12] states that a failure of linear growth stunting due to malnutrition and infections both before and after birth. Semba [13] also stated that stunting in children early age is associated with cognitive, motor, and a low socio-emotional development and increase mortality. Short children cannot reach full growth potential and will become stunted adolescents and adults. The consequences of stunting will continue in adulthood, with reduced work productivity. In women, there is increased risk of death in childbirth and adverse birth outcomes around the world where about one-third of preschool children is stunted. Therefore, to prevent stunting in early childhood, especially at birth,

5. Conclusion

Rubber seed chips help meet amino acid intake during pregnancy and affect the baby's body length at birth.

Reference

[1] Joshi L, Authority G, Akiefnawati R, Mulyoutami E, Wulandari D and Penot 2006 Diversified Farmer A Smallholder Rubber Agroforestry for Better Alternative to monoculture *Paper presented in Rubber Development in Lao PDR: exploring Improved System for Smallholder Rubber production* (Vientiane: Lao PDR) pp 9-11

[2] The Plantation Office of Lampung Province Area 2014 Rubber Plantation in Lampung (CPM)

[3] Sukmawati W, Arkeman Y and Maarif S 2014 *J. Industrial Engineering* 58-64

[4] Ly J, Chhay Ty and Chiev Phiny 2001 *Livestock Research for Rural Development* pp 13-2

[5] Hardinsyah Hadi Riyadi and Victor Napitupulu 2013 *J. Nutrition*

[6] Emawati Fitrah, Yuniar Rosmalina and Yurista Permanasari 2013 *Effect of maternal protein intake and weigh the baby was born on the incidence of stunting in children aged 12 months in Bogor* (Bogor: Research and Development, Ministry of Health RI)

[7] Rivai R R, F Damayanti and M Handayani 2015 Potential development of rubber seed (Hevea brasiliensis) as an alternative food in North Bengkulu *Pros. Sem. Nas. Masy. Biodiv. Indon.* **1**(2) 343-346 DOI: 10.13057/psnmbi/ m010229

[8] Widayati and Restu Yuni 1988 *Thesis* (Bogor: Institut Pertanian Bogor)

[9] Septiani E 2008 *Planning study of food consumption for pregnant women with the goal programming method* (Bogor: Institut Pertanian Bogor)

[10] Almatsier S 2006 *Basic Principles of Nutrition Science* (Jakarta: Gramedia)

[11] Kristiyanasari 2010 *Pregnant women Nutrition* (Yogyakarta: Nuha Medika)

[12] Sally Grantham McGregor, Y B Cheung, Santiago Cueto, Paul Glewwe, Linda Richter, Barbara Strupp and the International Child Development Steering Group 2007 *Developmental potential in the first 5 years for children in developing countries* (US National Library of Medicine National Institutes of Health) pp 60-70

[13] Semba R D, de Pee S, Sun K, Sari M, Akhter N and Bloem M W 2008 *Effect of parental formal education on risk of child stunting in Indonesia and Bangladesh: a cross-sectional study* (US National Library of Medicine National Institutes of Health) pp 322-8

"One"

Martinus Bambang Susetyarto

Architecture Department, Faculty of Civil Engineering and Planning, Universitas Trisakti, Jakarta, Indonesia

*bambang.s@trisakti.ac.id

Abstract. The robustness of the *One* building as the main building of *Sa'o* traditional house in *Ngada* Regency, Flores, has been proven to be reliable. The experience and testimony of indigenous people proves that when there was a shock in the area due to the earthquake and tsunami, the human victims who lived in traditional houses were the least number. The most severe damage occurred when *Sa'o* was shaken violently, until the *One* building fell to the ground. The *One* building is located on a stage floor made of wooden beams with split bamboo floor coverings. The stage floor is supported by wooden pillars or megalithic stone pillars. Floor and ceiling construction are tightly integrated with wooden columns and plank walls. The construction of the *One* building use wooden pegs, interlocking board walls, without nails or screw bolts, and tied thatched roof coverings. Therefore, inspired by the *One* building is the *Ngadha* architectural identity I designed the "*One*" model as a space box rescue from the earthquake and tsunami disaster. The goal is that when the disaster occurs, each family could save themselves to the "*One*" box rescue, close tightly for several minutes to avoid accidents and security threats from more severe natural disasters. The "*One*" prototype model must be tested for its strength, patented, calculated the technical feasibility, collaborated with industry, the government, and the local government.

1. Introduction

Actually, the *One* building which is the core space inside the *Sa'o* traditional house in *Ngadha* culture, in the Ngada Regency, the Province of Nusa Tenggara Timur, Flores Island, proved to be reliable. Although the earth of Flores is located on the ring of fire, and tectonic earthquake shocks that often hit the region are at risk of coming tsunami waves, but the record of human victims in the *Ngadha* traditional settlement due to these natural disasters are not much in number. Many indigenous people believed that living in the *One* building will be surviving from any disaster and threat. Their ancestors are in side of the *One* building, and would always protect them. Therefore, the *One* is a cube-shaped building, measuring ± 3.60 x 3.60 square meters, height from floor to ceiling is about 3.00 meters, made of wood and bamboo structures, which are constructed strongly, with the connection methods by wooden pegs without using nail or screw bolt.

Referring to the experience and testimony of indigenous people in the event of a shock caused by an earthquake, the damage to the *One* building was the least in number. Likewise, with the safety of humans who lived in the *One* building. The worst damage if the *One* building is shaken violently so that the building is separated from its position on the construction of the stage floor, where the construction is supported by the foundation of wooden pillars or megalith pillars which can also collapse, and finally the *One* building fell on to the ground [1]. The floor and ceiling part which were

IOP Conf. Series: Earth and Environmental Science **314** (2019) 012030 doi:10.1088/1755-1315/314/1/012030

firmly integrated with the construction of the *One*, wooden columns and plank walls, even though there was a strong earthquake shock, the floor and ceiling construction elements would not be released, or broken due to being hit by roof construction. The wooden post connection system and interlocking board walls without screws or screw bolts are made so solid, both in terms of workmanship and in terms of material usage.

Therefore, inspiring to the traditional building of *One* in the traditional house of *Sa'o* which became a peculiarity of *Ngadha* architecture, then I was designing an idea of the "*One*" as a savior space, a savior box from the earthquake (and tsunami). The goal is that in the event of an earthquake, every family must save themselves by entering to the "*One*" for a few minutes to avoid accidents and /or safety threats due to more severe natural disasters. The "*One*" prototype model was patented, and then tested for its strength, its solidity, its technical feasibility study and its market feasibility. Further, by collaboration with the industry, the government and the local government I create it to be a real savior property in the context of earthquake (and tsunami) disaster.

2. Understanding the *Sa'o* traditional house

Traditional houses in *Ngadha* culture are referred to as the *Sa'o*. The *Sa'o* structure of the *Ngadha* traditional house vertically is made up of heads - bodies – feet, or in the local language is called *zeta ulu - zale wekki / zeta wekki - zale wai*, as well as calling the human body parts. *Ulu* is the head, *wekki* is the body, and *wa'i* is the foot. When referring to the roof space, walls, and stage floor poles, the roof space as *zeta ulu*, the wall as *zeta wekki*, the stage floor as *zale wekki*. Space under the floor stage as a foundation pole room with a stage construction called *zale wa'i*. So, vertically, traditional houses are grouped into four vertical spaces, namely: the roof (*ubu sao*), the wall (*kebi ube kedu*), the stage floor (*befa*), and the stage pile (*tere leke / leke*) [2]. The structure of the *Sa'o* traditional house is understood vertically and horizontally as shown in Figure 1

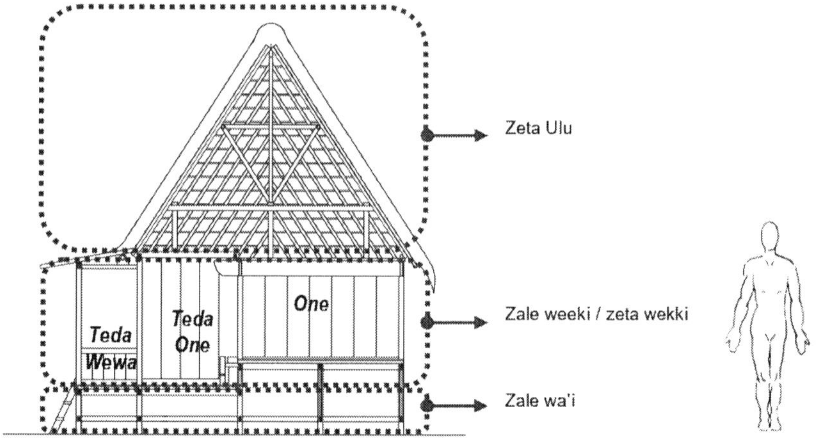

Figure 1. the structure of the *Sa'o* traditional house [2]

Horizontally, the components of the traditional house spaces are the *One*, the *Teda one*, and the *Teda wewa*. Particularly in the *One* room, there are two recitations, namely: the women's room, it is called *papa bhoko*; and the men's room, it is called *papa lewa*. This enlightenment is not interpreted as segregation of room functions, but as a state of mutual complementarities. Whereas, the *Teda one* and the *Teda wewa* are not conceptualized as different zones. However, these two spaces are enlargement of the *One*, as the additional rooms, or *bheli* in local language, at the right or at left of the *One*, where the floor surface is made not as high as the surface of the *One* floor. The functions, benefits, and meaning of these rooms in traditional houses are as follows [3]

2.1. The One

The *One* is as a core space inside the *Sa'o*, the main room of the *Sa'o*, or it is called the real residence. The word of *One* means a space to live in, a place to get togetherness. Multi-use functions of the *One*, namely: accommodating the independent activities of family members/clans, they are called *ana woe* in the local language; resting, cooking, traditional rituals, traditional meetings, giving birth to babies, laying down bodies, and others. The benefits of the *One* room are to get protection, security and safety from the ancestors. By living in the *One* room together in a family/clan entity, then protecting each other, keeping each other safe, and surviving can be realized. In the *One* room, there are ancestral spirits that guard the spiritual life of the inhabitants of the traditional house, they are called *ana sa'o* in the local language; guarding against interference from evil spirits around the area. The ancestor which is called *ebu nusi* in the local language, is symbolized in the form of *sau uwi*, a piece of bamboo *aur* and *bhoka,* the *tuak* traditional drink bottles, which are stored in the *mata raga* on the middle wall of the *One*, is called as *ube kedu*. The meaning of the *One* is a space to bring together the spiritual needs between *ana sao* and *ebu nusi*. In addition, the *One* is the base of their life, which is called *ulu* in the local language and the end room of their life, which is called *eko* in the local language. Thus, the *One* is as the smallest world, or micro cosmos, of the life of *woe* and *ana sa'o* in the indigenous people of the *Ngadha* culture. Therefore, this meaning is articulated more clearly by the *Lima pade*, the name of 'traditional architect' of the *Ngadha* culture, by making the floor surface of the *One* has the highest floor surface among the other rooms in the *Sa'o*. The room of *One* is a single building without a permanent connection, without the use of nails and iron hinges, and has a separate structure and construction from the *Teda one* or the *Teda wewa* building. Besides being used for traditional ritual activities, the *One* room is also used for cooking chicken as a sacrifice for ancestors. Therefore, on the right side near the entrance are three stone stoves as a traditional kitchen, is called *lapu lika* in local language. The right side of the *One* area, there is a traditional kitchen is called a female area, or *papa bhoko ana fai* in the local language, and vice versa it is called a male area, or *papa lewa ana saki.*

Figure 2. the zoning of the *One* [2]

2.2. The Teda one and the Teda wewa

The *Teda one* and The *Teda wewa* have functions as a porch, or additional space for storing devices for daily work activities, for resting, accepting family/clan guests, tourists/researchers, and others. In addition, the function room is used to accommodate general activities, additional activities, or new activities that were previously unthinkable by indigenous people who are generally engaged in agricultural livelihoods, teachers, civil servants and tourism services in recent times. Activities carried out by indigenous villagers, such as: farming activities, weaving, reading, watching television, listening to the radio, and others. So, the function of worldly activities, semi-private and semi-public activities are accommodated in the *Teda one* room and the *Teda wewa* room. The benefits of *Teda one* is to accommodate the abundant activities in the *One* room, especially related to the traditional ritual activities of the *Mosalaki* at the *One* room. The benefit of the *Teda wewa* is to accommodate the abundant activities in space of the *One* room and the *Teda one* room, especially related to the semi-private and semi-public activities.

Indeed, the *Teda one* and the *Teda wewa* actually are as foyer space which is a transitional space from 'the smallest world,' i.e. the *One* to 'small world,' i.e. the *Sa'o* up to the 'outside world,' i.e. the *Loka* or the *Kisa loka*, and vice versa. The *Teda one* and the *Teda wewa* are meant as space to create brotherhood with residents /tourist guests /researchers from outside the traditional village. Another meaning is as a tourist service building, which constructively is not attaching permanently to the *One* construction. The height of the both floors are lower than the *One*. Thirty years ago, the *Teda wewa* and the *Teda one* has the roofs of split bamboo, are called *lenga* in the local language. But at the present, the *One* and the *Teda one* has only thatch roof construction, which is as the modification of the structure and construction of the thatch roof, is called keri in local language. The *lenga* roof changed to be the thatch roof due to the limitations of the *betho* bamboo material in *Ngadha* traditional villages. Meanwhile, the difference in the floor height between the *Teda wewa*, the *Teda one*, and the *One* are as expression of the hierarchy of space values in term of worldly transitional activities in the *Teda one* and *Teda wewa*, and transcendental activities in the *One*.

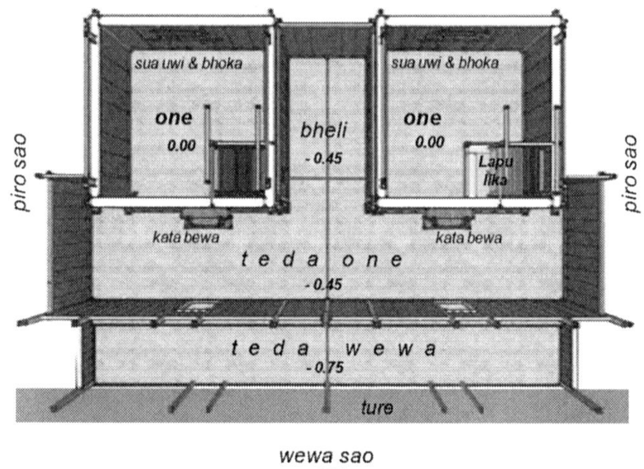

Figure 3. Isometry of the room of *One-Teda One-Bheli-Teda Wewa* [2]

2.3. The Bheli

The *Bheli* is a room that was built as additional room. It could be on the right side or the left side of the *One* room, or in between two *One* rooms if two *Sa'o* building units are continuous into one large house unit with two *One* rooms. The function of the *Bheli* room is a place for resting sleep, to do independent activities, or additional space to store valuables, and others. In addition, the *Bheli* serves to accommodate special activities, such as accepting family/clan guests, or live-in tourists/researchers. The benefit of *Bheli* is to increase the capacity of room around the *One* so that a traditional ritual ceremony can be held which involve many *ana woe* or *ana sao*, then all participants can be accommodated in the *Sa'o*. In addition, the existence of the *Bheli* further develops brotherhood with residents/tourist guests/ researchers from outside the traditional village. Another benefit is as a tourist service room. Therefore, this meaning is articulated by the *Lima pade* by making separate buildings from the *One* building, not attaching permanently to the construction of the *One*, not nailing or installing iron hinges, and making the floor lower than the floor surface of the *One*. Figure 2 Isometry of the room of *One-Teda One-Bheli-Teda Wewa*, where the *Bheli* looks are made in between two *One* of two *Sa'o* units, so that the *Bheli* room is quite large. The surface of the floor is also not as high as the *One* floor, but it is the same height as the floor surface of the *Teda one*. So, the *Bheli* additional room has the same hierarchy of values as the *Teda one*. The production of the *Bheli* is also based on the utilization of the *piro sao,* the space on the right side and on the left side of the *One*. Moreover, after the *One* and the *Teda one* room has one thatched roof, so that the span of the *Soku dolu,* the beam which extends to the front is used while at the same time compensating for the *Soku dolu* span towards the rear, and the space under the roof on the right and the left of the *One* is used for making the *Bheli*. Even though the construction of the building is close to the *One* wall, the construction of this room does not stick to the construction of walls and floors to the *One* permanently.

3. Methodology

The research method which I used to understand the traditional houses in *Ngadha* culture was an inductive method with a scope of problems in the *Sa'o* design in several traditional villages in the Ngada Regency. Further, lesson learnt about the character of the *One* building, how its strengths and weaknesses in accommodating indigenous people's activities, the impact of security and safety of indigenous people, and specifically understanding its rigidity so that it has advantages as a building that can be used as a lifeline for earthquakes and tsunami disaster.

4. The "*One*" a savior of the earthquake and tsunami

Understanding the features of the *One* building from *Ngadha* traditional house in Flores, namely as a main residential building, protecting and rescuing indigenous people from various threatening dangers, and knowing the *One* building is traditionally believed by indigenous people as ancestral dwellings, places of *ulu* and *eko* of their life in the smallest world, then inspired by the power of the *One* I was designing a new *One* building, which has more or less the same meaning, namely a space box rescue from the dangers of earthquakes and tsunami. The new *One* building is not intended to replace the existence of the *One* in *Ngadha's* traditional houses today. However, the new *One* building is intended for villagers in the area of earthquake and tsunami hazards throughout Indonesia. Actually, in fact, this design is the development of ideas from the model of sustainability of vernacular kampongs, the Ngada district, Flores as I have discussed in my previous paper [4]. Further, I wrote the new *One* building as the "*One*" with an apostrophe sign on it.

The building material is selected from high quality and transparent fiberglass material. The structure and construction system used an interlocked joint model, very strong in accepting earthquake and tsunami loads, stable and solid construction, tightly closed and waterproof and shocks. In fact, to maintain the tidiness I add a rubber belt that will automatically lock when the last door is closed. In the building, of course, there is enough space to breathe, for the adequacy of food and drink supplies, so that those who live in it will survive in an emergency. This rescue cube box is also equipped with a number of balloons that automatically expand when stagnant in the water, so that this will lift the "*One*" always above the water level if there is an earthquake followed by a tsunami hazard. This kind of idea should be introduced and socialized to the *Ngadha* traditional people, especially the *Utu Bhou Dia Sa'o*, the forum in construction of Ngadha traditional houses. Perhaps, there will be some wise advice from the forum to make the "*One*" design better for the future. Because, the *Ngadha* traditional people really understand the characters of the earthquake or/and tsunami in the Flores island. More important thing to convey at the forum is that the *Ngadha* traditional people must know the idea of "*One*" is derived from their traditional architecture and is designed for those who really need safety devices from earthquake and tsunami natural disaster. By doing so the sense of pride and confidence would grow in the hearts of the *Ngadha* traditional people to continue to be the people who advance where they are. I know that in the *Ngadha* area, the lagging factor in the human resources and the threat of natural disasters are parts of serious problems that must be addressed immediately in addition to the problem of poverty. Therefore, academics and the government have the time to be responsible for solving these problems.

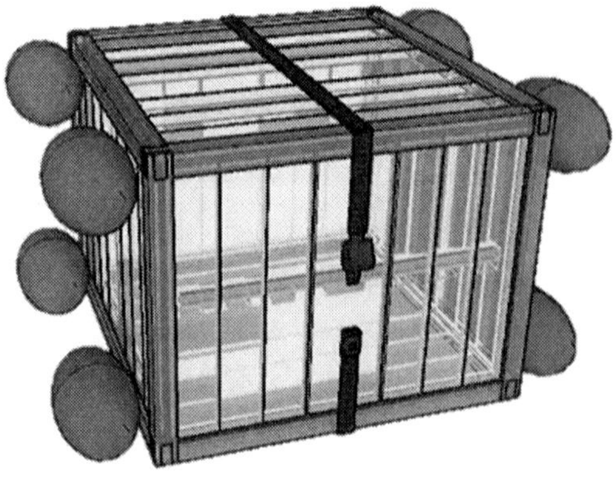

Figure 4. The "*One*" a savior of the earthquake and tsunami [4]

The "*One*" structure and construction system will continue to be tested for strength in civil and water engineering laboratories, tested for its durability, technical feasibility, market feasibility, so that the construction industry is willing to hold it. The concept and schematic of the "*One*" design of the rescue cube box from the dangers of the earthquake and tsunami is currently being handled by my patent, so that it can later be developed by the industry, and then marketed to the government and local governments who need it. This "One" building can be placed in the inner room, on the terrace of the building, on the porch of the house, on the yard, in the green open space, on the sports field, on the beach space, and especially on disaster evacuation routes made by architects so that it can be easily seen, can be reached by families, residents of residential areas, or the general public who are facing the dangers of earthquakes and tsunamis. I will improvise its architectural design towards better product design, more effective and efficient, funny and interesting to have it. Of course, as attractive as any of the "*One*" design must be guaranteed to be robust, their reliability highly tested in accepting earthquake and tsunami loads. As for globally, the intended "*One*" design is as shown in Figure 4 the "*One*" a savior of the earthquake and tsunami.

References

[1] Susetyarto B M 2012 Architecture and Environmental Sustainability: Critical Issues in Vernacular Kampong of Bena, Flores *Applied Mechanics and Materials* **253-255** 22-26; DOI: 10.4028/www.scientific.net/AMM.253-255.22

[2] Susetyarto B M 2013 *Arsitektur Vernakular, Keberlanjutan Budaya di Kampung Bena Flores* (Padepokan Seni Djayabinangun) pp 50-51

[3] Djarot P, Benediktus B, Herman F H, Ignatius H, Kristiana B and Reginaldo C L 2016 *Menelusuri Arsitektur Vernakular NTT* (Padang, Indonesia: Rumah Kayu) pp 89-93

[4] Susetyarto B M 2018 Model of sustainability of vernacular kampongs within Ngadha culture, Flores *IOP Conference Series Earth and Environmental Science* **106** 012076; DOI: 10.1088/1755-1315/106/1/012076; Licence CCBY 4.0; Project: Cultural Sustainability

Hazard analysis and social vulnerability to predict the loss population is affected by natural disaster of landslides in the Pesisir Selatan District of West Sumatra

Triyatno[1,*], Isril Berd[2], Idris[3] and Viola Putra[3]

[1]Graduate Program of Environental Science, Universitas Negeri Padang, Indonesia
[2]Lecturer of Fakultas Teknologi Pertanian, Universitas Andalas, Indonesia
[3]Lecturer Faculty of Economic, Universitas Negeri Padang, Indonesia

*triyatno@fis.unp.ac.id

Abstract. Pesisir Selatan District is one of the areas in West Sumatra that often experiences natural disasters of landslides. Natural disasters of landslides that occur in this area often cause losses in the form of property or human lives and disruption of road access to other areas. The aims of this research were to analyze the hazard of landslides based on land characteristics, analyze the carrying capacity of the region based on landslide hazards, analyze social vulnerability, and analyze prediction of Population loss. The method used in this study is a survey method with a geomorphological approach. Data used in the form of primary data and secondary data. The technical analysis used is in the form of scoring to analyze landslide hazards and GIS used to analyze the carrying capacity of the region, vulnerability, and exposed populations. The results showed that the hazard of landslides in the study area was divided into three, namely low, medium, and high hazards. The hazard of low landslides is generally found in plain areas near the coast. This area is formed due to the results of sedimentation processes from both rivers and the sea. The hazard of moderate landslides is generally found in hilly areas, and the hazard of high landslides is generally found in hilly or mountainous areas. The hazard of landslides is caused by physical characteristics in the form of high rainfall, steep slopes to very steep slope, high rock weathering. Social vulnerability in the research area is divided into two parts, namely low and medium. Low social vulnerability is caused by the ratio of male population more than female population, ratio of vulnerable groups, ratio of poor Population, and a ratio of disabled Population. Social vulnerability is being caused by a greater number of female populations, ratios of vulnerable groups, and disabled Population. The prediction of population loss found in the Lakitan Tengah area is in Lengayang Subdistrict, which is 11,869 Population.

1. Introduction

The West Sumatra region is an area that has the potential to be hit by natural disasters in the form of earthquakes, tsunamis, volcanic eruptions, floods, flash floods, landslides and coastal abrasions. The potential for disasters in the West Sumatra region is due to the fact that this region is located on the west coast of Sumatra Island and crosses the equator, causing this area to have high rainfall, weathered rock conditions, regional morphology from the plains to the mountains. The occurrence of natural disasters from year to year tends to increase both in number and intensity so that losses incurred by an event of natural disasters tend to increase also from the following year, both property losses and human lives.

The West Sumatra region that has the potential for natural disasters is the Pesisir Selatan District. The Pesisir Selatan District area often experiences natural disasters such as coastal abrasion, floods, flash floods, and landslides. The potential for natural disasters in the region of the Pesisir Selatan District is largely determined by physical conditions such as geology, geomorphology, hydrology, soil conditions, land use and land cover. Natural disaster occurrences of landslides in the Pesisir Selatan Regency often occur during the rainy season, high rainfall and supported by easily weathered rocks and thick soil solum cause this area has considerable potential for landslides to occur.

The natural disaster occurrence of landslides that occurred in the Pesisir Selatan District has caused many losses in the form of disruption of transportation routes, loss of property and loss of human life. The high impact of losses caused by natural disasters of landslides in this region is mostly caused by cutting slopes for road construction or settlements. Settlement patterns in this area generally fall on the plain or in mountainous areas such as those found in the Painan and Bayang Utara regions. The aims of this article are to analyze the landslide hazard, carrying capacity of regions based on landslide hazard levels to predict social vulnerability and population loss which can result in high loss of property and human lives.

2. Methodology

This type of research is in the form of quantitative research using a geomorphological approach [1, 2, 3, 4]. The method used in this study is a survey method to analyze the level of hazard of landslides. Landslide hazard levels were analyzed based on geomorphological units in the form of land units and social vulnerability of landslides obtained by analyzing satellite images in the form of Ikonos images to see the distribution of settlements found in areas that have potential or hazard of landslides [1, 5, 6, 7],

The type of data used in this study is primary data and secondary data. Primary data in the form of data obtained directly in the field are in the form of data on land characteristics, namely geological, geomorphological, soil, hydrology and land use conditions [8, 9, 10], Data The secondary data used in this study is population data obtained from BPS data of the Pesisir Selatan District in Figures, basic maps of a 1: 250,000 scale geological map, 1: 250,000 scale geomorphological maps, 1: 150,000 scale maps, scale land use maps 1: 250,000.

The stages of research carried out in the form of pre-field stages, field stages, and post-field stages. The details of the activities carried out in this study are as follows;

2.1. Pre-field stage

Activities carried out in the pre-field stage in the form of preparation of data needed in the field in the form of literature studies on the hazard and social vulnerability of landslides of natural disasters, preparing basic maps for guidance in the field, and determining sample points for field data collection. Analyzing satellite imagery in the form of an Ikonos image to determine the distribution of residential settlements in areas that have the potential for landslides.

2.2.Field stage

The activities carried out at the field stage were in the form of matching sample point maps with actual conditions in the field, and continued with field data collection in the form of land characteristics data that included geological, geomorphological, soil, hydrological and land use data. Taking documentation of research in areas that have the potential for natural disaster landslides.

2.3. Post-field stage

The activities carried out in the post-field stage are to classify field data, tabulate field data, analyze field data, and produce map results and research reports.

3. Data analysis

To determine the hazard of landslides that occurred in the study area, a formula was formulated by [1, 3, 11] as follows;

$$I = \frac{c-b}{k} \quad \text{(1)}$$

Information;
I: large range of class intervals
c: highest number
b: lowest number
k: number of classes desired

Based on the formula above, the interval of the landslide hazard class in the study area is as follows;

$$I = \frac{56-14}{3} = \frac{42}{3} = 14 \quad \text{(2)}$$

information:
the highest number of physical characteristics determining landslide hazards 56 (c)
number of lowest physical characteristics determinant of landslide hazard 14 (b)
desired number of classes 3 (k)
To determine the social vulnerability of natural disasters, landslides in the Pesisir Selatan District area use a formula that refers to Perka No. 2 BNPB 2012, in DDRI 2013 as follows;

$$Ks = 0{,}6 * \left(\frac{\log\left(\frac{KP}{0{,}01}\right)}{\log\left(\frac{100}{0{,}01}\right)}\right)0{,}1 * RJK + 0{,}1 * RK + 0{,}1 * RPC + 0{,}1 * KR \quad \text{(3)}$$

information:
Ks: social vulnerability
KP: population density
RJK: sex ratio
RPC: ratio of disabled population
KR: vulnerable group

To determine the number of loss population, based on analysis of landslide hazard data, social vulnerability of the community, and population.

4. Result and Discussion

Hazard of landslides is an event that often occurs during the rainy season and usually occurs in areas that have a slope of more than 15% with weathered material on it [12, 5, 13, 14], The hazard of landslides that occur in the Pesisir Selatan District is a natural process of slope evolution towards common levels [15, 12, 14]. The hazard of landslides that occur in Pesisir Selatan district can be seen from the characteristics of the land in the form of geological, geomorphological, soil, hydrological and land use conditions [15, 16]. For more details on the characteristics of the land in the research area can be seen in the following Table 1, Table 2, Table 3.

Table 1. Land Characteristics for Landslides Natural Disasters Pesisir Selatan District

		Relief/Topography				Soil				Geology		Groundwater	
No	Location	Slope (%)	Slope Shape	Slope Length (m)	Altitude Relief (m)	Soil Texture	Permeability cm / hour	Depth soil (cm)	Structure Rock	Level Weathering Rock	Depth weathering Rock (cm)	Spring	Depth ground water (cm)
1	Muaro Siri Tapan	95	Ir	203	78	Ld	15.3	2.10	Mr	S	2.35	A	501
2	Padang Kejai Tapan	2	Dt	63	5	Lp	18.9	39	Dt	R	60	Ta	160
3	Ampalu Tapan	98	Ir	98	2	Lb	31.4	73	Mr	R	142	A	53
4	Bukit Subarang Batu Bt. Kapas	30	Ir	144	96	Lb	23.9	40	Mr	R	81	A	87
5	Bukit Tuiak Bt.kapas	63	Dt	134	83	Lp	35.9	22	Mr	R	33	A	76
6	Kampung Jambak Bt.Kapas	2	Dt	57	3	Ld	29.8	25	Dta	R	93	Ta	145
7	Bukit Pulai	2	Dt	53	2	Ld	21.8	79	Dta	R	95	Ta	148
8	Bukit Pincuran Boga painan	98	Ir	141	88	Lp	38.5	31	Mr	R	46	Ta	63
9	Bukit Timbulun Painan	54	Ir	132	73	Ld	15.1	27	Msf	K	59	A	68
10	Bukit Vila Painan	99	Ir	133	86	Lp	25.7	81	Msf	K	141	A	165
11	Bukit Gaung, Pasar Baru	35	Ir	93	91	Ld	13.0	13	Mr	R	130	Ta	143
12	Sibingkeh	48	Ck	15	95	Ld	13.8	117	Mr	S	126	Ta	166
13	Siguntur	99	Ir	80	98	Lp	29.3	30	Mr	S	84	A	30
14	Koto Pulai, tarusan	85	Ir	153	76	Lp	31.4	48	Mr	S	78	A	142

Source: Results of Field Measurement and Data Analysis in 20018
Information:
D = data
H = scor

Slope shape	soil texture	rock stucture	Weathering of rocks	Spring
Ir = irriguler	Ld = dusty clay	Dt = plain	Sgr = fres K = higt	A = There is spring
Dt = plain	Lp = sandy clay	Mr = sloping	S = Medium R = low	Ta = no spring
Ck = concave	Pl = clay sandy	msf = massif		

Land use
Sbl = scrub bush Ld = farm Kc = mixed farm Pm = settlement Ht = forest Sw = paddi field

Table 2. Land Characteristics Scor for Landslides Natural Disasters Pesisir Selatan District

		Relief/Topography				Soil				Geology		Groundwater	
No	Location	Slope (%)	Slope Shape	Slope Length (m)	Altitude Relief (m)	Soil Texture	Permeability cm / hour	Depth soil (cm)	Structure Layer Rock	Level Weathering Rock	Depth weathering Rock (cm)	Spring	Depth ground water (cm)
1	Muaro Siri Tapan	4	4	3	3	4	4	4	4	3	4	4	1
2	Padang Kajai Tapan	1	1	2	1	4	4	2	1	2	2	1	2
3	Ampalu Tapan	4	4	3	1	4	4	3	4	2	3	4	4
4	Bukit Subarang Batu Bt. Kapas	3	4	3	3	4	4	2	4	2	2	4	4
5	Bukit Tuiak Bt.kapas	4	1	3	3	4	4	1	4	2	1	4	4
6	Kampung Jambak Bt.Kapas	1	1	2	1	4	4	1	1	2	2	1	1
7	Bukit Pulai	1	1	2	1	4	4	3	1	2	2	1	1
8	Bukit Pincuran Boga painan	4	4	3	3	4	4	2	4	2	1	1	4
9	Bukit Timbulun Painan	4	4	3	3	4	4	2	2	4	2	4	4
10	Bukit Vila Painan	4	4	3	3	4	4	3	2	4	3	4	2
11	Bukit Gaung, Pasar Baru	3	4	3	3	4	4	1	4	2	3	1	1
12	Sibingkeh	4	3	1	3	4	4	4	4	3	3	1	2
13	Siguntur	4	4	3	3	4	4	2	4	3	2	4	4
14	Koto Pulai, tarusan	4	4	3	3	4	4	2	4	3	2	4	4

Source: Results of Field Measurement and Data Analysis in 2018

Table 3. Land Characteristics of Pesisir Selatan District

No	Location	Land		Rain		Landslide Hazard	
		Land use		Rain fall mm/mounth			
		D	H	D	H	D	information
1	Muaro Siri Tapan	Sbl	2	48	4	48	High
2	Padang Kejai Tapan	Ld	2	38	3	28	Low
3	Ampalu Tapan	Kc	2	40	3	45	High
4	Bukit Subarang Batu Bt. Kapas	Sbl	2	45	4	45	High
5	Bukit Tuiak Bt.kapas	Ht	1	40	3	39	Medium
6	Kampung Jambak Bt.Kapas	Sw	3	35	3	27	Low
7	Bukit Pulai	Kc	2	40	3	27	Low
8	Bukit Pincuran Boga painan	Kc	2	44	3	43	High
9	Bukit Timbulun Painan	Pm	4	48	4	48	High
10	Bukit Vila Painan	Kc	2	46	4	46	High
11	Bukit Gaung, Pasar Baru	Kc	2	39	3	38	Medium
12	Sibingkeh	Ht	1	41	3	40	Medium
13	Siguntur	Ht	1	46	4	46	High
14	Koto Pulai, tarusan	Pm	4	49	4	49	High

Source: Results of Field Measurement and Data Analysis in 2018

Based on the table above the hazard of landslides that occur in the study area can be divided into several parts, namely; high, medium and low. The hazard of landslides in this study area is determined by the characteristics of the land in the form of weathered geological conditions and rock structures that follow the slope direction and the weathering depth of rocks more than 100 cm. The morphometry of the slopes of the study area starts from 0% -> 45% with slope lengths varying from 53 m to 203 m, this causes the shape of slopes that vary from flat to irrigulaler or irregular which can cause the amount of water that can be accommodated in the slope segment which has a concave shape [13, 17, 18]. Land used in the study area is in the form of shrubs, mix farm, rice fields, forests and settlements, this variation in land use causes the hazard of landslides in the study area to also vary such as in residential areas that cut slopes to establish buildings. Slope cutting activity causes the amount of water that enters the soil, causing increased mass of soil mass [14, 17, 19].

The hazard of landslides in the study area can be divided by three classes, namely low, medium and high. The hazard of low landslides is generally found in plain areas located near the coast or alluvial plains. This area is generally formed due to the sedimentation process resulting from the erosion process in the upstream section. The hazard of low landslides is generally found in hilly areas that have slopes of 25% - 45% and spread almost everywhere throughout the study area while the hazard level of high landslides is generally found in areas that have slopes of more than 45% and in areas that have experienced rock weathering levels go on. In the slope segment there are generally springs which trigger landslides. For more details, see the following table 4.

Table 4. Percentage of Area of Hazard of Landslides

Subdistrict	Village	Percentage of Low Hazard Area	Percentage of medium Hazard Area	Percentage of high Hazard Area
Basa Ampek Balai Tapan	Ampang Tulak Tapan	49.38	31.76	18.86
Basa Ampek Balai Tapan	Batang Arah Tapan	77.73	21.28	0.99
Basa Ampek Balai Tapan	Binjai Tapan	61.91	36.97	1.11
Basa Ampek Balai Tapan	Kubu Tapan	99.62	0.36	0.02
Batang Kapas	IV Koto Hilie	35.11	56.72	8.17
Batang Kapas	IV Koto Mudiek	17.27	56.02	26.71
Batang Kapas	Koto Nan Duo IV Koto Hilie	59.93	36.62	3.44
Batang Kapas	Koto Nan Tigo IV Koto Hilie	64.12	32.13	3.75
Bayang	Gurun Panjang	67.16	18.14	14.70
Bayang	Koto Barapak	47.34	46.04	6.62
IV Nagari Bayang Utara	Koto Ranah	7.02	61.04	31.93
IV Jurai	Bungo Pasang Salido	37.46	37.22	25.31
IV jurai	Lumpo	22.39	53.23	24.38
Koto XI Tarusan	Ampang Pulai	68.64	24.99	6.36
Koto XI Tarusan	Barung Barung Balantai	16.02	70.28	13.69
Koto XI Tarusan	Barung Barung Balantai Selatan	24.4	64.75	10.86
Koto XI Tarusan	Batu Hampar	67.18	26.41	6.41
Koto XI Tarusan	Duku	20.52	64.50	14.98
Koto XI Tarusan	Kapuh	73.53	15.28	11.19
Koto XI Tarusan	Kapuh Utara	64.18	30.02	5.80
Lengayang	Kambang	21.98	54.27	23.75
Lengayang	Kambang Timur	15.06	60.77	24.18
Lengayang	Kambang Utara	22.65	65.53	11.82
Lengayang	Lakitan Tengah	24.22	61.52	14.26
Linggo Sari Baganti	Aie Haji	41.48	52.92	5.60
Linggo Sari Baganti	Lagan Hilir Punggasan	80.09	19.67	0.24
Linggo Sari Baganti	Lagan Mudik Punggasan	43.01	53.11	3.88
Pancung Soal	Inderapura Selatan	38.6	37.07	24.34
Pancung Soal	Inderapura timur	49	33.66	17.35
Pancung Soal	Inderapura utara	40.4	47.50	12.10
Pancung Soal	Kudo-kudo inderapura	99.21	0.79	0.00
Sutera	Amping parak	91.49	7.04	1.47
Sutera	Amping parak timur	46.23	47.81	5.96

Source: Data Analysis in 2018

Based on the table above the percentage of the area that has a low, medium and high landslide hazard varies, this is caused by several factors including slope, soil characteristics, geology, morphology and land use. The highest percentage of landslide hazards is highest in the Bayang Utara area, which is 92.98% of the total area. This shows that the Bayang Utara area has a topography in the form of mountains and hills that have a slope of more than 45%, and have areas that have the potential for landslides only around 7.02% of the total area. Areas that have low landslide hazards are found in the Tapan Kubu area, which are 99.62% of the area and Kudo-Kudo Inderapura is 99.21% of the total area. This area is a plain area which is mostly in the form of sediment, for more details can be seen in the following picture.

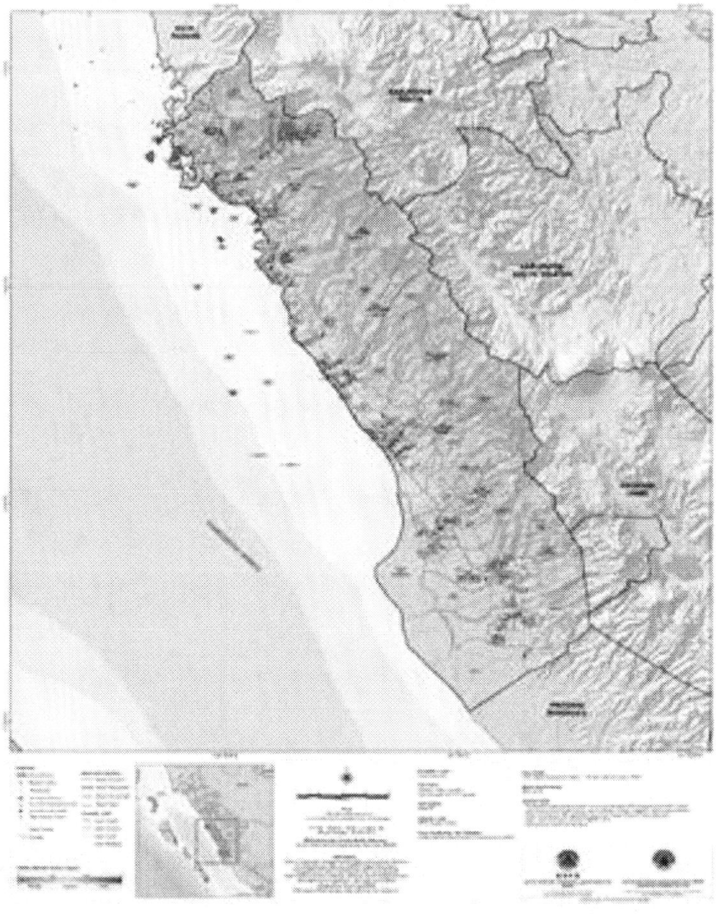

Figure 1. Landslide Hazard Map Pesisir Selatan District West Sumatera

Social vulnerability shows the level of social fragility of a community in the event of a natural disaster in the form of a landslide. The social vulnerability of the study area was seen from the percentage of population density, sex ratio, ratio of vulnerable groups, poor population, and ratio of disabled population. For more details, see the following table 5.

Table 5. Social Vulnerability and Prediction Population loss of Landslide Hazards in Pesisir Selatan District

Subdistrict	**Village**	Sex Ratio Percentage	Percentage of Vulnerable Groups	Percentage of Poor Population	Percentage of Disabled Population	Social Vulnerability	Social class vulnerability	Amount of moderate Population loos Prediction	Amount of high Population loos Prediction	Total of Population loos Prediction
Basa Ampek Balai Tapan	Ampang Tulak Tapan	1.02	0.23	0.1	0.01	15.57	Low	8	3	11
Basa Ampek Balai Tapan	Batang Arah Tapan	0.99	0.14	0.07	0.01	42.57	Medium	255	421	676
Basa Ampek Balai Tapan	Binjai Tapan	0.96	0.2	0.09	0.01	31.94	Low	132	-	132
Basa Ampek Balai Tapan	Kubu Tapan	0.99	0.25	0.11	0.01	0.14	Low	-	-	-
Batang Kapas	IV Koto Hilie	0.97	0.12	0.07	0.00	46.25	Medium	876	314	1,190
Batang Kapas	IV Koto Mudiek	0.95	0.11	0.06	0.00	40.80	Medium	246	270	516
Batang Kapas	Koto Nan Duo IV Koto Hilie	0.97	0.25	0.13	0.01	43.18	Medium	423	318	740
Batang Kapas	Koto Nan Tigo IV Koto Hilie	0.99	0.46	0.19	0.01	36.48	Medium	73	191	264
Bayang	Gurun Panjang	0.91	0.27	0.11	0.02	41.81	Medium	320	280	600
Bayang	Koto Barapak	0.91	0.2	0.09	0.01	20.81	Low	24	-	24
IV Nagari Bayang Utara	Koto Ranah	0.89	0.54	0.33	0.05	38.24	Medium	90	254	344
IV Jurai	Bungo Pasang Salido	0.94	0.53	0.14	0.03	37.13	Medium	199	92	292
IV jurai	Lumpo	0.96	0.16	0.06	0.01	54.36	Medium	1,827	2,307	4,134
Koto XI Tarusan	Ampang Pulai	1.03	0.07	0.03	0.00	52.88	Medium	1,403	1,891	3,294
Koto XI Tarusan	Barung Barung Balantai	0.96	0.2	0.08	0.01	51.48	Medium	2,516	136	2,652
Koto XI Tarusan	Barung Barung Balantai Selatan	0.9	0.23	0.09	0.01	31.79	Low	129	-	129
Koto XI Tarusan	Batu Hampar	0.92	0.26	0.1	0.01	0.13	Low	-	-	-
Koto XI Tarusan	Duku	0.93	0.1	0.05	0.01	45.77	Medium	651	456	1,107
Koto XI Tarusan	Kapuh	0.93	0.16	0.07	0.01	38.21	Medium	81	265	346
Koto XI Tarusan	Kapuh Utara	1	0.5	0.17	0.03	28.07	Low	-	72	72
Lengayang	Kambang	0.93	0.23	0.11	0.01	54.30	Medium	889	3,200	4,089
Lengayang	Kambang Timur	0.94	0.15	0.08	0.01	57.90	Medium	1,572	5,540	7,112
Lengayang	Kambang Utara	0.94	0.16	0.08	0.01	55.68	Medium	1,543	3,517	5,060
Lengayang	Lakitan Tengah	1.02	0.37	0.17	0.01	61.27	Medium	2,394	9,475	11,869
Linggo Sari Baganti	Aie Haji	0.96	0.64	0.24	0.01	36.55	Medium	197	69	266
Linggo Sari Baganti	Lagan Hilir Punggasan	0.99	0.63	0.24	0.01	0.19	Low	-	-	-
Linggo Sari Baganti	Lagan Mudik Punggasan	0.98	0.93	0.29	0.01	35.48	Medium	224	-	224
Pancung Soal	Inderapura Selatan	1.01	0.02	0.08	0.01	46.58	Medium	420	833	1,252
Pancung Soal	Inderapura timur	1	0.01	0.05	0.00	33.33	Low	115	49	164
Pancung Soal	Inderapura utara	0.99	0.02	0.03	0.01	42.52	Medium	485	187	672
Pancung Soal	Kudo-kudo inderapura	0.98	0.03	0.12	0.01	29.21	Low	87	-	87
Sutera	Amping parak	0.98	0.27	0.13	0.01	32.89	Low	14	138	153
Sutera	Amping parak timur	1.04	0.6	0.22	0.01	44.21	Medium	393	467	861

Source: Data Analysis in 20018

IOP Conf. Series: Earth and Environmental Science **314** (2019) 012031 doi:10.1088/1755-1315/314/1/012031

Based on the table above the highest percentage of sex ratios is found in the Ampiang Parak Timur which is 1.04 area and Inderapura Timur which is 1.02, indicating that in this area the male population is more than the female population. The ratio of the percentage of male population and the smallest percentage is found in the Barung-Barung Balantai Selatan area, which is 0.9, which means that in this area there are more female residents, this is if a landslide occurs, this area must be priority in action mitigation.

The percentage ratio of the highest vulnerable groups in the Punggasan Langan Mudik area in Linggo Sari Baganti District is 0.93, indicating that this area has a population that is less than 12 years old and a population of more than 60 years old, which causes this area to be more vulnerable in the event of a natural disaster landslides. The highest percentage ratio of poor Population is found in the Koto Ranah area of Bayang Utara District which is 0.33, this indicates that this area has more poor Population than other regions. The high rate of poverty in this area is due to the fact that some Population depend on agricultural and forest products. The percentage ratio of the disabled population in the study area is very small, ranging from 0.00 to 0.05, this indicates that in terms of the disabled population the study area does not have a high value. This means that if a natural disaster occurs, landslides in the research area do not cause a burden to the community, especially when they carry out the evacuation process

The level of social vulnerability in the research area can be divided into two parts, namely low and medium social vulnerability. The level of social vulnerability is generally found in regions that have a low percentage ratio of male and female population, namely the male population is less than the female population and has a high percentage ratio of vulnerable groups of more than 0.1, this is shows that in this area the population is less than 12 years and more than 60 years is higher and the percentage ratio of the poor is higher. Low social vulnerability in the study area shows a greater male and female population ratio, namely the male population is greater than the female population and has a lower ratio of vulnerable groups and has a smaller ratio of poor Population.

The Population loss is a prediction of the number of Population living in the area who have the hazard of moderate and high landslides. The highest prediction of Population loss is in the Lakitan Tengah of Lengayang District, which is 11, 869 Population, because most residents choose a place to live in a hilly area that has a hazard of moderate and high landslides. In general, Population in this area have livelihoods as farmers, so they are closer to going to work in the landfarm or pady fields. Population who live in this area generally choose a place to stay near the hilly foot area by cutting slopes. The Population loss prediction to landslide hazards in the lowest research area is 0 Population, this means that in areas that have a moderate and high landslide hazard there are no residents. In this area generally in the form of forests or mixed farm, the community generally lives in the plains area.

5. Conclusion

Based on the description on landslide hazards in the study area, some conclusions can be drawn as follows;

1. The landslide hazard in the study area can be divided into three parts, namely the hazard of low, medium and high landslides. This difference in the level of landslide hazard is caused by the physical characteristics of an area in the form of; geological conditions, morphology, soil, hydrology, and land use,
2. The level of social vulnerability in the research area can be divided into two parts, namely low and medium. Low social vulnerability is generally found in regions that have a higher percentage ratio of males than females, the composition of vulnerable groups that are low or have few vulnerable groups, and the ratio of the percentage of disabled Population is small and has a small ratio of poor Population,
3. The highest number of settlements in the Lakitan Tengah area is 11,869 Population and the lowest is 0 in the Lagungg Hilir Punggasan, Batu Hampar and Tapan Kubu areas.

6. Suggestion

Based on the description above about the landslide hazards in the area of Pesisir Selatan District to reduce the impact of losses that will be caused by a landslide natural disaster event, the suggestions that can be taken are as follows;

1. to reduce the hazard of landslides in this area, it should be noted that the causes of landslides are in the form of physical characteristics of the area.
2. to reduce the hazards of landslides in this area it is necessary to increase the social vulnerability of the community, especially in areas of high landslide hazards that have residents,
3. to reduce the number of Population loss to the research area, there should be a regulation on land use for settlements, especially in areas that have the hazard of moderate and high landslide.

References

[1] Ainon N O, Wan M, Naim W M and Noraini S 2012 GIS Based Multi-Criteria Decision Making for Landslide Hazard Zonation. Asia Pacific International Conference on Environment-Behaviour Studies, Salamis Bay Conti Resort Hotel, Famagusta, North Cyprus, 7-9 December 2011 *Procedia - Social and Behavioral Sciences* **35**: 595 – 602

[2] Mardiyatno 2001 Risiko Longsoran Daerah Samigaluh *Tesis* (Yogyakarta: Fakultas Geografi, Universitas Gadjah Mada)

[3] Federica C, Francesca S, Piernicola L, Claudia V, Francesco C, Osvaldo B 2016 A Geomechanical Approach to Landslide Hazard Assessment: the Multiscalar Method for Landslide Mitigation. VI Italian Conference of Researchers in Geotechnical Engineering – Geotechnical Engineering in Multidisciplinary Research: from Microscale to Regional Scale, CNRIG 2016. *Procedia Engineering* **158**: 452 – 457

[4] Yueping Y, Bin L, Wenpei W, Liangtong Z, Qiang X, Yang G, Nan Z, Hongqi C, Tiankui L and Aiguo L 2016 Mechanism of the December 2015 Catastrophic Landslide at the Shenzhen Landfill and Controlling Geotechnical Risks of Urbanization *Journal Engineering* **2**: 230–249

[5] Henky N, Dhandhun W, Gilang A D, Ahmad C, Bachtiar W M and Ajeng L 2015 Geomorphometric Characteristics of Landslides in the Tinalah Watershed, Menoreh Mountains, Yogyakarta, Indonesia The 5th Sustainable Future for Human Security *Procedia Environmental Sciences* **28**: 578 – 586

[6] Carlo T S, Samuele S, Nicola C, Filippo C 2016 Geomorphic Indexing of Landslide Dams Evolution *Journal Engineering Geology* **208**: 1–10

[7] Filippo C, Veronica T and Daniela L 2016 Spatial Patterns of Landslide Dimension: A Tool for Magnitude Mapping *Journal Geomorphology* **273**: 361–373

[8] Marten G Andrée B S, Eva K, Brian M, Jeremy G. Venditti , Alain G , and Kelsey W 2018 Sensitive Clay Landslide Detection and Characterization in and Around Lakelse Lake, British Columbia, Canada *Journal Sedimentary Geology* **364**: 217–227

[9] Raymond W.M. Cheung, K.Y. Choi 2013 Landslide Disaster Prevention and Mitigation Through Works in Hong Kong *Journal of Rock Mechanics and Geotechnical Engineering* **5**: 354–365

[10] Ryley A B, and W Andy T 2016 Loss of Slope Support due to Base Liquefaction:Comparison of 1g and Centrifuge Landslide Flume Experiments. Soils and Foundations **56**(2): 251–264

[11] Triyatno dan Erna J 2012 Pemetaan Zonasi Bahaya dan Risiko Longsoran di Daerah Ngarai Sianok *Jurnal Pelangi* **4** (2) ISSN: 2252-7168

[12] Ari T, Caterina M and Kerstin J 2015 A Fast and Efficient Algorithm to Map Prerequisites of Landslides in Sensitive Clays Based on Detailed Soil and Topographical in Formation *Journal Computers & Geosciences* **75**: 88–95

[13] Maryam I 2011 A Comparative Study of Fuzzy Logic Approach for Landslide Susceptibility Mapping Using GIS: An Experience of Karaj DAM Basin in Iran. The 2nd International Geography Symposium GEOMED 2010 *Procedia Social and Behavioral Sciences* **19**: 668–

676

[14] Hasali H, Rangali R S Sb, N L Deshapriyac and Lal S 2018 Landslide Susceptibility Mapping Using Logistic Regression model (a Case Study in Badulla District, Sri Lanka)7th International Conference on Building Resilience; Using scientific knowledge to Inform Policy and Practice in Disaster Risk Reduction, ICBR2017, 27 – 29 November 2017, Bangkok, Thailand. *Procedia Engineering* **212**:1046–1053

[15] Keren D, Zhenhong L, Roberto T, Guoxiang L, Bing Y, Xiaowen W , Haiqin C ,Jiajun C and Julia S 2016 Monitoring Activity at the Daguangbao Mega-Landslide (China) Using Sentinel-1 TOPS Time Series Interferometry *Journal Remote Sensing of Environment* **186**: 501–513

[16] Suzuki, Tetsuo F, and Takeshi T M 2014 Peak and Residual Strength Characteristics of Cement-Treated Soil Cured Under Different Consolidation Conditions *Journal Soils and Foundations* **54**: 687–698

[17] Shruti N, K.S. Sajinkumar, Thomas O, V.J. Anuja, Rinu A. Samuel , and C. Muraleedharan 2017 Early Warning System for Shallow Landslides Using Rainfall Threshold and slope stability analysis *Journal Geoscience Frontiers* **xxx**: 1- 12

[18] Takashi T, A.M.R.G.Athapaththua, Takashi H, and Masaki K 2015 Investigation of Landslide Calamity due to Torrential Rainfall in Shobara City Japan *Journal Soils and Foundations* **55**(5): 1305–1317

[19] Takashi T, Seiji K, Shota N, Masahiro K, Shinji N and Nao Y K, 2014 Landslide and Mudflow Disaster in Disposalsite of Surplus Soil at Higashi-Hiroshima due to Heavy Rain Fall in 2009 *Journal Soils and Foundations* **54**: 621–638

IOP Conf. Series: Earth and Environmental Science **314** (2019) 012032 doi:10.1088/1755-1315/314/1/012032

The endurance test of *Nannochloropsis* sp. paste isolated from Lampung Mangrove Centre (LMC)

Tugiyono[1,*], Agus Setiawan[2], Emy Rusyani[3] Suharso[4], and Siti Nurjannah[1]

[1]Departmet of Biology, Faculty of Mathematics and Natural Sciences University of Lampung, Indonesia
[2]Department of Forest Management, Faculty of Agriculture University of Lampung
[3]Lampung Marine Culture Development Centre, Indonesia
[4]Departmet of Chemistry, Faculty of Mathematics and Natural Sciences University of Lampung, Indonesia

*tugiyono.1964@fmipa.unila.ac.id

Abstract. *Nannochloropsis* sp. has been used as natural feed because it has high nutrition that is good for fish larva growth and development. However, the *Nannochloropsis* sp. availability continually is in insufficient amount and this often becomes a problem in culturing because this is difficult to culture in mass volume. The objective of this research was to make *Nannochloropsis* sp. paste and to test the paste quality based on the life endurance of *Nannochloropsis* sp. cells from isolate of Lampung Mangrove Centre in intermediate scale culture by using different fertilizer combination and NaOH doses. This research used factorial completely randomized design with two treatments and three repetitions. The first treatment was administration of agricultural fertilizer combination of (P) urea 40 ppm, ZA 20 ppm, and TSP 5 ppm, and Conway (C) technical fertilizer as control. The second treatment is the administration of NaOH doses of: 100 ppm, 125 ppm, 150 ppm, and 175 ppm. Data were analyzed by using One Way Analysis of Variance (ANOVA), and Least Significant Difference test with $\alpha = 0.05$ was performed whenever any significant difference was found. The research result showed that *Nannochloropsis* sp. paste from isolate of Lampung Mangrove Centre had highest life endurance level with Conwy technical fertilizer and dose of NaOH 100 ppm with population density of 18066666.67 x 10^4 cell/mL.

1. Introduction

Based on Decree of East Lampung regent No.660/305/04/SK/2005/1546/J. 26/KL/2005 in 10 May 2005, Lampung Mangrove Centre is one of mangrove ecosystems in Lampung province, and it is located in Margasari village of Labuhan Maringgai sub district in East Lampung district with 700 Ha area width [1] (monography of Margasari village, 2005). Lampung Mangrove Centre as an ecosystem becomes one of feed sources for living creatures such as fish and shrimp larvae that contain of phytoplankton and zooplankton.

Currently, phytoplankton as a natural feed source is very needed in culturing, especially in fish hatchery activities. However, feed necessity often becomes a problem because this feed availability is often less than sufficient. There are two plankton kinds for natural feed; phytoplankton and zooplankton. Phytoplankton is the organism producing micro algae [2,3]. There are three kinds of phytoplankton most often found in Lampung Mangrove Centre water for fish feed; *Nannochloropsis* sp. *Tetraselmis* sp. and *Nitzchia* sp. [4].

Nannochloropsis sp. has been used for natural feed because it has high nutrition that is good for fish and shrimp larva growth and development, and in addition, this micro alga is easy to grow in varying environment conditions [5]. The *Nannochloropsis* sp. availability in continual manner often becomes a problem in culturing, because of environmental change factors and less sunrays in rainy season, so that *Nannochloropsis* sp. is difficult to be mass cultured [6].

The effort to provide phytoplankton population is performed by producing *Nannochloropsis* sp. paste. This paste making is a practical way to provide natural feed stock. The paste making is performed by precipitating *Nannochloropsis* sp. by adding NaOH in culturing media, so that pH value in water can improve [7,8]. Therefore, a research concerning paste making by using fertilizer and NaOH with different doses with an objective to find out the life endurance of *Nannochloropsis* sp. isolate from Lampung Mangrove Centre was performed.

2. Research method

This research was conducted from August to November 2018 in zooplankton laboratory of Living Feed Division, in Ocean Fishery Culture Office in Lampung, located in Jalan Yos Sudarso, in Hanura village of Teluk Pandan sub district, Pesawaran district, Lampung province.

Equipment to use in this research included aquarium, aquarium jar, table, droplet pipette, hemocytometer, satin cloth, microscope, glass cover, Erlenmeyer flask, measuring cup, scale, spatula, plastic spoon, plastic disc, beaker glass, plastic hose, hand counter, plastic funnel, filtering cloth, magnetic stirrer, plastic container, thermometer, DO meter, spectrophotometer, pH meter, and refractometer. Materials to use in this research included *Nannochloropsis* sp. seeds from Lampung Mangrove Centre, agricultural fertilizer (urea, ZA, TSP), Conwy technical fertilizer, Vitamin B12, NaOH, sterile sea water, vidone, citric acid, aquabidest, alcohol 70%, fresh water, disk washing detergent and chlorine solution.

This research used experimental method with factorial completely randomized design with two treatments and each treatment was repeated three times. The first treatment was 12 aquariums administered with combined agricultural fertilizer (P) or Urea 40 ppm, ZA 20 ppm, and TSP 5 ppm; and other 12 aquariums were administered with Conwy technical fertilizer (C) 1 ppm as control. The second treatment was administration of NaOH doses; 100 ppm, 125 ppm, 150 ppm, and 175 ppm.

Data obtained in this research were presented in tables and graphics. Data of cell density and cell daily growth rate of *Nannochloropsis* sp. isolate from Lampung Mangrove Centre were presented in tables and graphics and were analyzed by using descriptive analysis. Data analysis was done by using One Way Analysis of Variance (ANOVA) and Least Significant Difference test with 95% level (α = 0.05) was conducted whenever any significant difference was found.

2.1. Preliminary research

2.1.1. Cluturing Nannochloropsis sp. Isolate from Lampung Mangrove Centre. The culturing of *Nannochloropsis* sp. isolated from Lampung Mangrove Centre (LMC) was done in intermediate scale (semi-mass). Stages in culturing *Nannochloropsis* sp. were as follows: the *Nannochloropsis* sp. seeds from LMC was cultured by using Erlenmeyer flask that would be used in the intermediate scale culturing. The *Nannochloropsis* sp. seeds were removed into intermediate scale by using 1000 liter volume aquarium. The seeds of *Nannochloropsis* sp. from LMC was cultured in the aquarium and re-cultured again up to 24 aquariums. The seeds of *Nannochloropsis* sp. from LMC were removed from aquarium into one container of 1 m^3 and harvested. The initial density of *Nannochloropsis* sp. seeds were counted by using hemocytometer. The initial seed density for spreading and volume of sea water in each aquarium were calculated by using the following formula [9]:

$$\mathbf{V1 \times N1 = V2 \times N2} \qquad (1)$$

Note:
V1 = seed volume for initial spreading (mL)
V2 = desired volume of *Nannochloropsis* sp. culture media (mL)
N1 = seed density / stock of *Nannochloropsis* sp. (cell/mL)
N2 = desired seed density of *Nannochloropsis* sp. from LMC (cell/mL)

24 aquariums were filed with sterile sea water and Conwy technical fertilizer was administered into 12 aquariums and agricultural fertilizer (urea, ZA, and TSP) was administered into other 12 aquariums. Testing for measuring salinity, temperature, DO, and light intensity were performed. Water quality test was performed in the beginning and at the end of treatment. Water quality testing was performed in water quality laboratory in Ocean Fishery Culture Office in Lampung. The measurements were performed to measure: nitrite, nitrate, phosphate, ammoniac, and pH. Cell density of *Nannochloropsis* sp. from LMC in 24 aquariums were counted

2.1.2. Nannochloropsis sp. Paste Making. Stages in *Nannochloropsis* sp. paste making were as follows. NaOH doses were weighed; 100 ppm, 125 ppm, 150 ppm and 175 ppm. NaOH was dissolved in the bottle and then poured into aquariums containing *Nannochloropsis* sp. from LMC. The aquarium content was stirred up slowly until homogenous and it was helped with aeration for precipitating process. The aquariums were covered with tarpaulin to prevent sunrays and then left aside for 24 hours. After 24 hours, the *Nannochloropsis* sp. was harvested by filtering it. Precipitated deposit of *Nannochloropsis* sp. was removed from aquarium into filtering container for 24 hours to obtain *Nannochloropsis* sp. paste that was free from water. The *Nannochloropsis* sp. paste was then entered into plastic bags and weighed for each bag, and then they were labelled according to NaOH doses.

2.2. Main Research
Stages in the main research were as follows. Sterile sea water, 24 aquariums and *Nannochloropsis* sp. paste seeds from LMC in the preliminary research were prepared. 5% citric acid was weighed and dissolved into the sterile sea water. *Nannochloropsis* sp. paste seed was entered into a container and filtered by using filtering cloth and then dissolved by using sterile sea water. Each sample of *Nannochloropsis* sp. seed was administered with 5% citric acid concentration. 24 culture containers were arranged and randomized by using factorial completely randomized design with two treatments and each treatment was repeated three times. The containers were filled with sea water and aerated, given with agricultural fertilizer, Conwy fertilizer, and Vitamin B12 2 mL. measurements for salinity, temperature, DO, and light intensity were performed. Water quality test was performed in the beginning and at the end of treatment. Containers were covered with covered with aluminum foil to prevent evaporation and then cell density was counted every 24 hours.

3. Result and Discussion

3.1. Nannochloropsis sp. Paste weight
Data of *Nannochloropsis* sp. paste weight in each treatment can be seen in Figure 1.

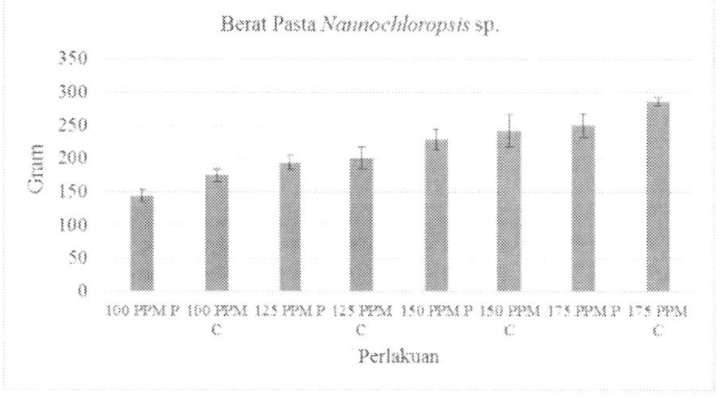

Figure 1. Graphic of *Nannochloropsis* sp. paste in each treatment

Figure 1 above shows that the highest average of *Nannochloropsis* sp. paste weight in this research is produced by treatment of 175 ppm of C (286 grams).

Note:
100 PPM P = treatment of agricultural fertilizer and NaOH 100 ppm dose
100 PPM C = treatment of Conwy technical fertilizer and NaOH 100 ppm dose
125 PPM P = treatment of agricultural fertilizer and NaOH 125 ppm dose
125 PPM C = treatment of Conwy technical fertilizer and NaOH 125 ppm dose
150 PPM P = treatment of agricultural fertilizer and NaOH 150 ppm dose
150 PPM C = treatment of Conwy technical fertilizer and NaOH 150 ppm dose
175 PPM P = treatment of agricultural fertilizer and NaOH 175 ppm dose
175 PPM C = treatment of Conwy technical fertilizer and NaOH 175 ppm dose

3.2. Nannochloropsis sp. population density

In the endurance life test of paste of *Nannochloropsis* sp. isolate from Lampung Mangrove Centre (LMC) in the intermediate scale, culturing was performed for 5 research days and the population density was counted from 4 September to 8 September 2018 in every 24 hours. The research result shows that the highest *Nannochloropsis* sp. population density for each treatment is obtained in third day of research (Figure 2). The highest average population density is in treatment 10 ppm of C with population density of 18066666.67 x 10^4 cell/mL, then followed by treatment 125 ppm of P with population density of 17950000.00 x 10^4 cell/mL, treatment 125 ppm of C with population density of 16883333.33 x 10^4 cell/mL, treatment 150 ppm of P with population density of 14166666.67 x 10^4 cell/mL, treatment 100 ppm of P with population density of 13783333.33 x 10^4 cell/mL, treatment 150 ppm of C with population density of 12783333.33 x 10^4 cell/mL and treatment 175 ppm of P with population density of 9516666.67 x 10^4 cell/mL, and the lowest is in treatment 175 ppm of C with population density of 8233333.33 x 10^4 cell/mL.

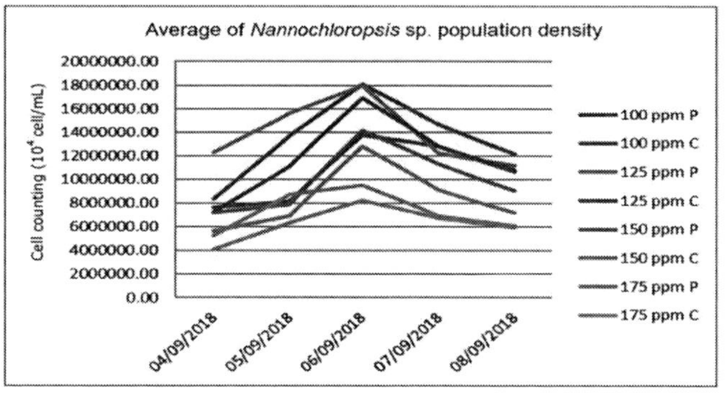

Figure 2. Graphic of average of *Nannochloropsis* sp. density in each treatment

note:
100 ppm P = treatment of agricultural fertilizer and NaOH 100 ppm dose
100 ppm C = treatment of Conwy technical fertilizer and NaOH 100 ppm dose
125 ppm P = treatment of agricultural fertilizer and NaOH 125 ppm dose
125ppm C = treatment of Conwy technical fertilizer and NaOH 125 ppm dose
150 ppm P = treatment of agricultural fertilizer and NaOH 150 ppm dose
150 ppm C = treatment of Conwy technical fertilizer and NaOH 150 ppm dose
175 ppm P = treatment of agricultural fertilizer and NaOH 175 ppm dose
175 ppm C = treatment of Conwy technical fertilizer and dosis NaOH 175 ppm dose

Graphic in Figure 2 shows that the first and second day of treatment show slow adaptation phase toward administration of combination of fertilizer and NaOH. It is suspected that the seeds underwent longer adaptation phase before coming into exponential phase. In addition, the adaptation phase went longer because extreme temperature in the first and second day of treatment that caused death some of cells while some others proceed to make adaptation by stopping cell splitting. [10] explain that too high sunrays intensity can inhibit phytoplankton growth rate. In addition, [11] also explains that *Nannochloropsis* sp. micro alga culture in optimal temperature shall produce good population development.

In the third day, the population density kept on increasing or entered the exponential phase where in this phase the *Nannochloropsis* sp. performed cell splitting and reached peak of population density or exponential phase. This can occur because nutrition and water quality of culture media are still supportable for the growth and cell splitting of *Nannochloropsis* sp. [12].

In the fourth dan fifth day of treatment, *Nannochloropsis* sp. growth underwent decrease, and it may occur because worsening water quality and nutrition runs out so that *Nannochloropsis* sp. cannot do more metabolism to support its growth. Cell population density decreases fastly because the death rate of phytoplankton is higher than its growth rate [13]. This was affirmed with color change in culture media into clear and *Nannochloropsis* sp. precipitation in the bottom of aquarium had sandy texture and was unable to dissolve again when being stirred up [14]. The graphic in Figure 2 does not show stationary phase, and it is suspected that the stationary phase occurred in less than 24 hours while observation of population density was performed in each 24 hours. Therefore, the stationary phase was unobservable and death phase was directly observed after exponential phase.

The graphic in Figure 2 above shows that the average cell life endurance of *Nannochloropsis* sp. isolate from Lampung mangrove Centre (LMC) that is able to produce highest average of 18066666.67 x 10^4 cell/mL population density is treatment 100 ppm of C, while the lowest is treatment 175 ppm of C with 8233333.33 x 10^4 cell/mL population density.

3.3. Growth rate

Data of *Nannochloropsis* sp. specific growth rate calculation result can be seen in Table 1. Table 1 shows that the highest *Nannochloropsis* sp. growth rate occurs in treatment 100 ppm of C with 0.15 cell/mL/day, and then followed by treatment 150 ppm of C with 0.14 cell/mL/day, treatment 175 ppm of C with 0.14 cell/mL/day, treatment 125 ppm of C with 0.13, treatment 175 ppm of P with 0.12 cell/mL/day, treatment 150 ppm of P with 0.11 cell/mL/day, treatment 100 ppm of P with 0.08 cell/mL/day, and the last treatment 125 ppm of P with 0.06 cell/mL/day.

Table 1. Average of *Nannochloropsis* sp. specific growth rate in each treatment

Treatments	Specific growth rate (cell/mL/day) *Nannochloropsis* sp. (Mean ± Standar Eror)
100 ppm P	0.08 ± 0.02[a]
100 ppm C	0.15 ± 0.03[a]
125 ppm P	0.06 ± 0.02[a]
125 ppm C	0.13 ± 0.02[a]
150 ppm P	0.11 ± 0.02[a]
150 ppm C	0.14 ± 0.04[a]
175 ppm P	0.12 ± 0.02[a]
175 ppm C	0.14 ± 0.05[a]

Note:
The same superscript letter in the same column states no significant difference at $\alpha = 0.05$

Note:
100 ppm P = treatment with agricultural fertilizer and NaOH 100 ppm dose
100 ppm C = treatment with Conwy technical fertilizer and NaOH 100 ppm dose
125 ppm P = treatment with agricultural fertilizer and NaOH 125 ppm dose
125ppm C = treatment with Conwy technical fertilizer and NaOH 125 ppm dose
150 ppm P = treatment with agricultural fertilizer and NaOH 150 ppm dose
150 ppm C = treatment with Conwy technical fertilizer and NaOH 150 ppm dose
175 ppm P = treatment with agricultural fertilizer and NaOH 175 ppm dose
175 ppm C = treatment with Conwy technical fertilizer and NaOH 175 ppm dose

3.4. Generation time

Generation time is the time required by micro alga to split cell, from one cell into two cells [15]. The calculation for generation time of each treatment can be seen in Table 2.

Table 2. Average generation time of *Nannochloropsiss* sp. in each treatment

Treatments	Generation time (hour) *Nannochloropsis* sp. (Mean ± Standar Eror)
100 ppm P	11.75 ± 2.64[a]
100 ppm C	16.39 ± 5.45[a]
125 ppm P	7.68 ± 2.41[a]
125 ppm C	6.37 ± 1.08[a]
150 ppm P	10.38 ± 2.75[a]
150 ppm C	5.47 ± 0.83[a]
175 ppm P	16.75 ± 12.99[a]
175 ppm C	17.31 ± 11.71[a]

Note:
The same superscript letter in the same column states no significant difference at $\alpha = 0.05$
100 ppm P = treatment with agricultural fertilizer and NaOH 100 ppm dose
100 ppm C = treatment with Conwy technical fertilizer and NaOH 100 ppm dose
125 ppm P = treatment with agricultural fertilizer and NaOH 125 ppm dose
125ppm C = treatment with Conwy technical fertilizer and NaOH 125 ppm dose
150 ppm P = treatment with agricultural fertilizer and NaOH 150 ppm dose
150 ppm C = treatment with Conwy technical fertilizer and NaOH 150 ppm dose
175 ppm P = treatment with agricultural fertilizer and NaOH 175 ppm dose
175 ppm C = treatment with Conwy technical fertilizer and NaOH 175 ppm dose

The fastest generation time is obtained by treatment 175 ppm of C (treatment with Conwy technical fertilizer and NaOH 175 ppm dose) that reaches 17.32. This indicates that the time required by one cell of *Nannochloropsis* sp. to split into two is 17.32 hours. The second fastest generation time is 16.75 hours by treatment 150 ppm of C, followed by 16.39 in treatment 125 ppm of P, 11.76 hours by treatment 100 ppm of P, 10.38 hours by treatment 100 ppm of C, 7.69 hours by 150 ppm of P, 6.37 hours by treatment 175 ppm of P, and the last 5.48 hours by treatment 150 ppm of P.

3.5. Water quality

Water quality data were measured in the beginning and in the end of *Nannochloropsis* sp. culturing. Measured parameters included temperature, DO, pH, salinity, light intensity, nitrite, nitrate, phosphate and ammoniac. Water quality data during the research are presented in attachment 5.

Water quality data indicated that parameters influencing the *Nannochloropsis* sp. growth were temperature, salinity, light intensity, DO, pH, and ammoniac [16,17].

Temperature is an important factor influencing micro alga growth. In this research, temperature change in media culture was influenced by light intensity. The optimal range of temperature for *Nannochloropsis* sp. micro alga is 25 – 35 °C [18]. The temperature measurement during the research showed 29 ^{0}C, so that this temperature level became one of *Nannochloropsis* sp. growth inhibitors.

According to [18], the optimal salinity range for *Nannochloropsis* sp. growth is 25-35%. The salinity during research was 28%. It means that the salinity during the research is still proper for optimal growth of *Nannochloropsis* sp., and shows that salinity does not become inhibitor factor.

According to [19], light is the main energy source for micro alga to do photosynthesis. In a water ecosystem, micro alga can do assimilation process of organic materials provided that light necessity is fulfilled properly. Optimal light intensity for *Nannochloropsis* sp. micro alga growth is 2000-8000 lux. Too high light intensity can inhibit photosynthesis. Light intensity during this research was 9000 lux, and this was inconsistent with [19].

During the research, the pH value changes in culture media range 8.27-8.35, and these were still proper for the *Nannochloropsis* sp. growth, because it was in accordance with [16] suggesting that *Nannochloropsis* sp. could live in pH range of 8.0-9.5. In the condition of range pH 6-8 *Nannochloropsis* sp. and other microalgae generally can be applied as adsorbent for heavy metal ions [20,21,22,23,24,25,26].

4. Conclusion

Based on the research results, the research conclusions are as follows:

1. The treatment of Conwy technical fertilizer and NaOH 175 ppm dose produces the most paste.
2. The paste of *Nannochloropsis* sp. isolate from Lampung Mangrove Centre shows highest life endurance in treatment of Conwy technical fertilizer and NaOH 100 ppm dose with population density of 18066666.67 x 10^4 cell/mL.

Acknowledgements

We thank to the Directorate of Research and Community Services, Directorate General of Higher Education (DIKTI), Ministry of Research, Technology and Higher Education of the Republic of Indonesia (Kemenristekdikti) for the research grant through the institutional national strategic grant scheme with the contract number:393/UN26.21/PN/2018. We also thank the Head of Lampung Marine Culture Development Centre for permitting this research.

References

[1] Monografi D M 2005 *Potensi Desa Margasari, Kecamatan Labuhan Maringgai* (Kabupaten Lampung Timur, Provinsi Lampung)

[2] Fulk and Main 1991 Rotifer and Microalgae Culture System. *Proc. of a U.S. Asia Workshop* Honolulu Hawai

[3] Tugiyono, Setiawan A, Rusyani E, Gustina E, and Suharso 2018 Combination of agriculture

fertilizer for intermediate cultivation of isolate Nannochloropsis sp. of the waters of Lampung Mangrove Centre as live feed *Asian J Agri & Biol* **6**(4) 587-593

[4] Tugiyono, Master J, and Suharso 2017 Isolation and identification of phytoplankton from aquatic ecosystemsof Lampung Mangroves Centre (LMC) as biological feed. *Asian J. Agri & Biol* **5**(4) 188-194

[5] Martosudarmo dan Wulani 1990 *Petunjuk Pemeliharaan Kultur Murni dan Massal Mikroalga* (Proyek Pengembangan Budidaya Udang Situbondo)

[6] Muliono A 2004 Pengaruh Suhu dan Lama Penyimpanan Terhadap Kondisi Sel *Nannochloropsis* sp *Skripsi* (Bogor: Fakultas Perikanan dan Ilmu Kelautan IPB)

[7] Kokarkin C Dan Kusnendar E 2000 Marine Microalgae Engineering With a Special Emphasis on *Chlorella* sp. and its Potensial Use in the Future *Proc. Int. Symp on Marine Biotechnology* Ancol Jakarta 29-31 Mei 2000 Jakarta.

[8] Sukarni, Sudjito, Hamidi N, Yanuhar U, Wardana I N G, 2014 Potential and properties of marine microalgae Nannochloropsis oculata as biomas fuel feedstock *Int.J Energy Environ Eng* **5** 279-290

[9] Villegas C T 1995 *Production Natural Food Organism* 9 (Philippines : Southeast Asian Fisheries Development Centre)

[10] Hendersen-Seller B and Markland H R 1987 *Decaying lakes The origins and control of cultural eutropication* (Willey and Sons Chichester)

[11] Ismi S 1996 Perkembangan Populasi Nannochloropsis oculata pada Suhu dan Salinitas yang Berbeda *J. Pendidikan Perikanan Indonesia* **2**(2) 71-75

[12] Pujiono A E 2013 Pertumbuhan Tetraselmis chuii pada medium air laut dengan intensitas cahaya, lama penyinaran, dan jumlah inokulan yang berbeda pada skala laboratorium *Skripsi* (Jawa Timur : Universitas Jember)

[13] Chan E C, Pelczar M J Jr and Krieg N R 1986 *Microbiology* (Singapura: McGraw-Hill Book Company)

[14] Ayustama A L S and Sari E A W 2011 Proses Produksi Mikroalga dalam Photobioreaktor Mini Pond Secara Batch Untuk Bahan Bakar Biodesel *Artikel* (Jurusan Teknik KimiaFakultas Teknik Universitas Diponegoro)

[15] Fardiaz S 1992 *Mikrobiologi Pangan* (Direktorat Jenderal Pendidikan Tinggi Pusat Antar Universitas IPB Bogor)

[16] Taw N 1990 *Petunjuk Kultur Murni dan Massal Mikroalga* (Proyek Pengembangan Udang UNDP FAO)

[17] Fogg G E 1987 *Algal Cultures and Phytoplankton Ecology* (London : The University of Wisconsin Press)

[18] Isnansetyo A dan Kurniastuty 1995 *Teknik Kultur Fitoplankton dan Zooplankton* (Kanisius) (Yogyakarta)

[19] Laven P and Sorgeloos P 1996 *Manual on The Production and Use of Live Food for Aquaculture* (Fisheries Technical Paper Rome)

[20] Buhani, Musrifatun, Pratama D S, Suharso, Rinawati 2017 Modification of Chaetoceros sp. Biomass with Silica-Magnetite Coating and Adsorption Studies towards Cu (II) Ions in Single and Binary System *Asian J. Chem* **29**(12) 2734-2738

[21] Buhani, Rinawati, Suharso, Yuliasari DP and Yuwono S D 2017 Removal of Ni (II), Cu(II), and Zn(II) Ions from Aqueous Solution Using Tetraselmis sp. Biomass Modified with Silica-Coated Magnetite Nanoparticles *Desal. Wat. Treat* **80** 203-213

[22] Buhani and Suharso 2009 Immobilization of Nannochloropsis sp. Biomass by Sol-Gel Technique as Adsorbent of Metal Ion Cu (II) from Aqueous Solution *Asian J. Chem* **21**(5) 3799-3808

[23] Buhani, Suharso and Satria H 2011 Hybridization of Nannochloropsis sp. Biomass-Silica through SolGel Process to Adsorb Cd (II) Ion from Aqueous Solutions *Eur. J. Sci. Res.* **51** (4) 467-476

IOP Conf. Series: Earth and Environmental Science **314** (2019) 012032 doi:10.1088/1755-1315/314/1/012032

[24] Buhani, Suharso and Sembiring Z 2006 Biosorption of Metal Ions Pb (II), Cu (II), and Cd (II) on Sargassum Duplicatum Immobilized Silica Gel Matrix. *Indo. J. Chem* **6** (3) 245-250

[25] Buhani, Suharso and Sembiring Z, 2012 Immobilization of Chetoceros sp. Microalgae with Silica Gel through Encapsulation Technique as Adsorbent of Pb Metal from Solution. *Orient. J. Chem* **28**(1) 271-278

[26] Buhani, Suharso and Sumadi, 2012 Production of Ionic Imprinted Polymer from Nannochloropsis sp Biomass and Its Adsorption Characteristics toward Cu (II) Ion in Solutions. *Asian J. Chem* **24**(1) 133-140

IOP Conf. Series: Earth and Environmental Science **314** (2019) 012033 doi:10.1088/1755-1315/314/1/012033

Relationship of determination pesticide doses with horticultural farmers health complaints in Cikajang, Garut

Suyud Warno Utomo[1,*], Rizqy Fauzi[1] and Haryoto Kusnoputranto[2]

[1] Environmental Health Department, Faculty of Public Health, Universitas Indonesia, Depok 16424, Indonesia

[2] School of Environmental Science, Universitas Indonesia, Salemba, 10430, Indonesia

*sw_utomo@yahoo.com

Abstract. The role of pesticides in agriculture is crucial. Pesticides are needed to protect plants from pests and plant diseases. The opinions among farmers about more doses of pesticides used in plants will be more effective in protecting plants. The pesticides use that do not comply with the rules of use indirectly cause health complaints. This study aims to identify the correlation between the behaviour of farmers in determining the dose of pesticides with health complaints caused to farmers in Cikajang District, Garut Regency. This study uses primary data with 105 subjects and also sprayed plants using pesticides. The study design was cross sectional and the data were analysed bivariately with chi-square. This study there was no significant correlation between the behaviour of farmers to determine the dose of pesticides with health complaints that occurred (P = 0.079 with OR = 2.053 95% CI 0.936 - 4.502). Farmers behaviour in determining the dosage of pesticides to be applied in plants is not related to health complaints that occur in farmers. This is probably caused by other factors that affect health complaints.

1. Introduction

Most farmers still consider the purpose of using pesticides merely to control and eradicate animals that are risk to the plants. Whereas the understanding of pesticides is a chemical substance used to control and expel plant-disturbing animals and diseases that attack agricultural crops [1]. The use of pesticides is influenced by two important factors, namely the size of the dose and the concentration of pesticides used on agricultural land [2]. There are six principles that must be applied in the use of pesticides by farmers, including on target, right quality, right type, timely use, right dose of use, and appropriate method of use [3]. If these six principles can be applied by farmers and other pesticide users, the effectiveness of the pesticide's active energy and active ingredients can be felt.

Cikajang District, Garut Regency, is one of the areas with the largest agricultural centers in West Java Province. The area of Cikajang Subdistrict reaches 12.790,78 hectare and 46.48% of the total area is agricultural land. Agricultural crops that are commodities in the region are cultivated vegetables or better known as horticulture. Horticultural plants are divided into four types, namely based on the stature of plants, plants that can shed leaves, plant life cycle, and plant use. While vegetable commodities which are among the top three are in Cikajang District, including potatoes 1.442-hectare, cabbage 1.033 hectare, and chili 548 hectare [4]. All of the plants must require pesticides for protection of pests.

Based on research conducted by [5] the behavior of farmers in using pesticides is still classified as poor (63%) and causes health problems. Percentage of poor behavior of farmers in pesticide use is shown by the large percentage of farmers who did not read the rules of use first when using pesticides, which is 47.26% and 61.81% of respondents used pesticides not according the rules stated on the pesticide packaging label [6]. Farmers used pesticides in plants based on the situation and condition of the garden on the day their plant. If the weather conditions are dry, spraying plants using pesticides is done every three days or even once a week, but if it is found wet, humid, and rainy, spraying pesticides is done every two days. This is done based on the pesticide grain will be carried by the rain and cannot stick to the plant when exposed to rainwater. Farmers in determining the dose of pesticides are also based on the many pests found in the garden. If the number of pests is increasing and the risk of causing financial losses, the farmer will quickly increase the dose of pesticide used so that the pest can die in the opposite direction. If a little pest is seen by the eyes of farmers, then the dose of pesticide is reduced or not sprayed by farmers waiting for pests in amounts that are at risk of causing financial losses.

The farmers behavior who are less concerned with pesticide doses to protect plants will have an impact on various types of agricultural problems. According to Adriyani [7] there are seven negative impacts of pesticide use that are not in accordance with the usage rules, such as causing water pollution, agricultural soil pollution, air pollution, causing resistant pest species, causing secondary pest explosion, resurgence, damage ecosystem balance, and causing health problems for humans. The discovery of pest resistance due to excessive use of pesticides was found in Bogor, West Java, with *Helopeltis antonii* pests in cocoa plants [6]. Based on several studies that have been conducted, the negative impact of the use of pesticides on human health can be form of symptoms of poisoning and chronic health problems. Symptoms of pesticide poisoning, for example epileptic seizures, nausea, vomiting, headache, anxiety, tremor, loss of consciousness, ventricular fibrillation, and respiratory depression. Then, chronic health problems, such as disorders of the human nervous system and inhibition of cholinesterase enzymes that play a role in the body's metabolic system.

Bad behavior of farmers determines that the dosage of pesticides is not in accordance with the rules and health complaints that arise need to be carried out further research. So, based on the description, this research was conducted to find out the relationship between the behavior of farmers in determining the doses of pesticides in plants with the emergence of health complaints.

2. Method

This study uses a cross sectional study design with independent variables behavior of farmers determine the dose of pesticides for plants and the dependent variable health complaints experienced by farmers after using pesticides. The location of this study is in Cikajang District, Garut Regency. Determination of the number of samples in this study using the proportion estimation formula. The value of CI used is 95% and the deviation value is 10%. As well as the estimated proportion value using the results of BPS data from Garut Regency at 38.18%. The formula for estimating the proportion equation used is as follows:

$$\text{n} = \frac{Za^2 p\,(1-p)}{a^2} \tag{1}$$

Information:
n : size of sample
Za : confidence interval (CI 95%)
P : estimated proportion
d : deviation (10%)

The results of this equation obtained a sample size of 91 respondents, to avoid the dropout of respondents the results were added 10% of the total sample so that the total sample was 105 respondents. Respondents are people who live in Cikajang District who work as farmers or farm laborers and usually spray plants using pesticides. The selection of respondents uses purposive sampling method in each village located in Cikajang District.

In this study univariate and bivariate analyzes were conducted. Univariate analysis was carried out by means of frequency analysis to see the number of farmers who determined the dosage of pesticides for plants according to the rules and not according to the rules and to know the number of farmers who experienced health complaints while using pesticides for plants. Bivariate analysis is done by means of kai squared test by considering the p value and the magnitude of Odds Ratio (OR) resulting from cross tabulation calculations between these variables.

3. Result

Based on interviews, the number of farmers who have good and bad knowledge can be seen in Table 1, as follows:

Table 1. Distribution of Farmer's Behavior in Using Pesticide on Plants

No.	Category	Frequency (Respondent)	Percentage (%)
1	Good	39	37.1
2	Less Good	66	62.9
	Total	105	100

Table 1 shows that more than half (62.9%) of respondents had bad behavior in using pesticides for plants while farmers who had good behavior were 37.1%. The results of the analysis of the behavior of farmers in using pesticides in Table 1 are directly proportional to the percentage of the number of farmers who determine the dose of pesticides for plants not in accordance with the rules of use. These results can be seen in Figure 1, as follows:

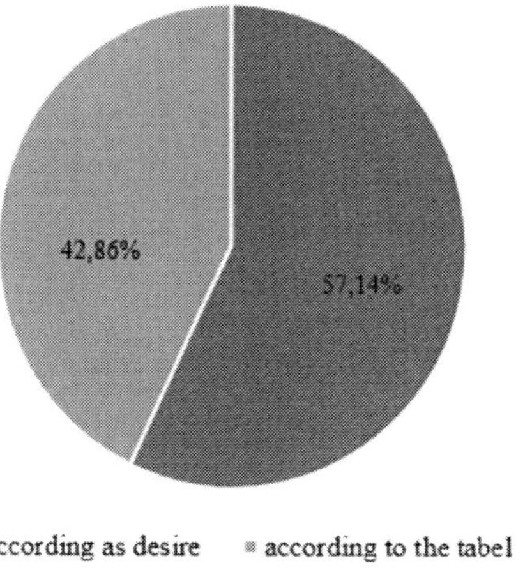

Figure 1. Determine of pesticide doses

Diagram in figure 1 shows that 57.14% of farmers determine the dosage of pesticides is not in accordance with the rules, while 42.86% of farmers determine the dose of pesticides in accordance with the rules contained in the packaging label of pesticide products.

More than 50% of farmers said that they experienced health complaints experienced by farmers while using pesticides for plants. The following are the results of the analysis of the number of farmers who experience health complaints and types of health complaints that are often experienced by farmers.

Table 2. Number of Farmers Who Experience Varied Health Complaints

No.	Health Complaints	Frequency	%
1	Itchy	7	6.67
2	Poignant eyes	5	4.76
3	Limp body	1	0.95
4	Blurred vision	1	0.95
5	Nausea and dizziness	39	37.14
6	Shortness in the chest	2	1.90
7	Do not experience	50	47.62

Table 2 shows that there were 7 (6.67%) respondents of itching on the skin, as many as 5 (4.76%) eye respondents were irritable, 1 (0.95%) respondents felt weak and blurred, as much as 2 (1.9%) respondents experienced tightness in the chest, and health complaints most experienced by respondents were nausea and dizziness as many as 39 (37.14%) respondents. While as many as 50 (47.62%) respondents did not experience health complaints.

The square test of the behavior variables of farmers determining the dose of pesticides with health complaints experienced by farmers while using pesticides can be seen in Table 3, as follows:

Table 3. Results of Chi-Square Analysis Determining Pesticide Doses and Health Complaints

			Health Complaints		Total	P value	OR (95% CI)
			No complaints	There are complaints			
Determination dosage of pesticides	As per Label	n	26	19	45	0,079	2,053 (0,936-4,502)
		%	52.0%	34.5%	42.9%		
	Not Based on Label	n	24	36	60		
		%	48.0%	65.5%	57.1%		
Total		n	50	55	105		
		%	100.0%	100.0%	100.0%		

Based on Table 3 it is known that there are 65.5% of respondents determine the dose of pesticides for plants that do not fit the label and experience health complaints. Whereas the respondent who determined the dose of pesticide for the plant according to the label and experienced health complaints was 34.5%. P value of 0.079 means there is no significant relationship between the behavior of farmers to determine the dose of pesticides with health complaints experienced by farmers while using pesticides on plants. OR value of 2.053 means that farmers who determine the dose of pesticide in plants that do not fit the label 2.053 times the risk of experiencing health complaints compared to farmers who determine the dose of pesticides in plants according to the table.

4. Discussion

Most (62.9%) farmers have poor behavior in the use of pesticides. The results of farmer behavior analysis were obtained from interviews with farmers using a questionnaire with five questions. Among the behaviors of farmers reading the label first before using pesticides, farmers use personal protective equipment when mixing pesticides before spraying on crops, farmers re-examine the packaging of pesticide products before reuse, farmers pay attention to the direction of the wind when spraying plants using pesticides, and farmers save pesticide products in a special room. Of all the questions, the behavior of farmers to re-examine pesticides before being reused obtained results that did not re-examine the most compared to the number of farmers who re-examined pesticides. The things that are examined are the state of pesticide formulations and instructions for use on the label of pesticide products. Based on the results of an interview with a head of the Girijaya Village farmer group in Cikajang District, things that need to be re-examined when they want to use pesticides such as the state of pesticide formulations have changed color and shape or not, and the quantity of pesticide formulations remaining to be used so that farmers do not need to make sense so that you can still use it.

The reason the farmers stated that they did not re-examine the pesticides when they wanted to use it again, they felt that they had understood the method and formulation of the pesticide. In fact, according to the results of observations and targeted interviews, most of the ways farmers determine the dose of pesticides in plants by using the principles of weather forecasts and the number of pests contained in plants in other words farmers determine the dose of pesticide use according to their feelings. And many of the farmers claimed to have obtained a way to determine the dosage and mixing pesticides on the plants from fellow farmers and personal experiences in the field where the information is chain-based to previous peers without any reference material as a source that can be accounted for.

Whereas the definition of the dosage in question is the number of pesticides in liters or kilograms that are used to control pests per certain area or each plant that is carried out in one application or more. Usually a large dose of a pesticide is listed on the label of a pesticide product, not derived from the mere estimates and feelings of the farmer [2]. This condition is illustrated in the analysis results of Diagram 1 which states that there are 57.14% of respondents determine that the dosage of pesticides does not match the label. Determination of pesticide doses that do not match the usage label by farmers sometimes during rainy weather and many pest conditions on plants, they increase the dosage according to the efficacy of pesticides in disabling plant pest pests. These results are in accordance with the research of [8, 9] that to reduce the risk of farmers production routinely and intensively spraying pesticides with very high frequency of spraying, even most farmers spray pesticides without taking into account the presence or absence pest attack. The results of other studies conducted by Basuki [10] state that the majority of farmers in Brebes and Cirebon spray insecticide regularly and intensively, with high doses and short duration of spraying with one another to prevent pests from developing worse.

Negative impacts can occur, one of which is the high level of pesticide residues found in plants. Factors that influence the high levels of pesticide residues in plants, including the types of pesticides used, such as the easy or difficult pesticide active ingredients dissolved in water, the large dose of pesticides when applied to plants, and the timing of the application of pesticides to plants [11].

Based on Table 2 about health complaints experienced by farmers due to the use of pesticides it is known that most respondents experienced complaints of nausea and headache as many as 39 (37.14%) respondents and 7 (6.67%) respondents experienced skin irritation in the form of hives due to exposure to liquid pesticides after spraying. And there were also respondents who experienced symptoms of severe poisoning, namely asphyxiation as much as 2 respondents and blurred vision by 1 respondent.

Health complaints that occur to farmers are the initial symptoms of poisoning that occur within 6-12 hours after exposure. The use of organophosphate, organochlorine and carbamate pesticides is the main cause of the initial symptoms of poisoning [12,13]. The initial symptoms of poisoning include irritation of the skin, nausea, vomiting, weakness, headaches, and health problems. While health complaints in the form of shortness of breath experienced by 2 respondents are a continuation of the same type of pesticide [14].

The initial symptoms of poisoning that occur due to various kinds of factors that influence [3], such as the physical condition of farmers who are less healthy at the time of spraying, not breakfast at the time of spraying, not using PPE recommended for spraying pesticides, especially masks, gloves, and closed clothes. Farmers who do not pay attention to the direction of wind gusts can also cause the mixture of pesticides as well as the behavior of farmers when spraying, such as smoking, scratching the body, rubbing the eyes, and not cleaning the limbs after spraying plants [2].

Pesticide residues that are swallowed into the human body and then deposited with fat tissue in the body can disrupt the process of metabolism in the human body. For example, exposure to organochlorine pesticides against the human body. Organochlorine stimulates the nervous system and causes paresthesia, is sensitive to stimulation, irritability, disruption of balance, tremor and convulsions. Several chemicals including the organochlorine group induce facilitation and hyperexcitation in synapse links and neuromuscular linkages that result in repeated disarmament in central neurons, sensory neurons, and motor neurons. Organophosphates and carbamates inhibit AChE. The functions of acetylcholinesterase (AChE) include hydrolysis of acetylcholine (ACh), inhibited the neurotransmitter can be increased in number and function [15]. The availability of ACH in the brain causes neurons to be easily transported and improve cognitive function of the brain. If AChE is interrupted, it will cause a disturbance in the central nervous system that affects the nervous system failure of the human. Continuous exposure in a long duration will cause health problems that are acute and chronic and lead to death.

The results of the kai squared analysis on the determination of pesticide doses with health complaints stated that there was no significant relationship ($p = 0.079$) between farmers who determined pesticide doses according to the label and did not match the label. While the OR value of 2.053 means that farmers who determine the dosage do not match the label of 2,053 times the risk of experiencing health complaints compared to farmers who determine the dose of pesticide according to the label. The statement that there is no significant relationship between farmers who determine the dosage of pesticides according to the label and does not match the label for health complaints experienced by farmers is predicted to have other factors that affect health problems that occur to farmers, such as the psychological level of farmers affected by various kinds of social, economic, educational and political fields of life. As well as predicted food intake factors received by respondents who were not examined in this study. Although there is no significant relationship, a persuasive and preventive approach is needed so that poor behavior in determining the dose of pesticides in plants does not cause wider environmental and health problems.

5. Conclusion and Suggestion

Based on the results of the study there is no significant relationship between the behavior of farmers who determine the dose of pesticides in plants with health complaints experienced by farmers but still at risk of causing health problems twice as much as farmers who determine the dose of pesticides is not in accordance with the label. Communication, information and education needs to be carried out by the Cikajang District Agriculture Office in collaboration with farmer groups in each village about the use of good and correct pesticides in order to protect farmers from acute and chronic health problems.

References

[1] Fait A, Iversen B, Tiramani M, Visentin S and Maroni M 2001 *Preventing Health Risks From The Use Of Pesticide In Agriculture* (Italy: World Health Organization)
[2] Sudarmo S 1991 *Pestisida* (Jakarta: Kanisius)
[3] Pusat Penelitian dan Pengembangan Hortikultura *Teknik Penyemprotan Pestisida* (Jakarta: Kementerian Pertanian Republik Indonesia)
[4] Pemerintah Kecamatan Cikajang 2017 *Profil Kecamatan Cikajang Kabupaten Garut Jawa Barat Tahun 2017* (Garut: Jawa Barat)
[5] Fauzi D 2018 Analisis Tingkat Keuntungan Petani Kentang Merah di Kabupaten Solok *Menara Ilmu* **12** 32-39
[6] Utami U C 2016 *Hubungan Pengetahuan, Sikap, dan Tindakan Penggunaan Pestisida Dengan Tingkat Keracunan Pestisida pada Petani di Desa Kembang Kuning Kecamatan Cepogo* (Surakarta: Universitas Muhammadiyah Surakarta)
[7] Adriyani R 2006 Control of Environmental Pollution caused by Pesticide in Agricultural Process *Jurnal Kesehatan Lingkungan* **3** 95-106
[8] Saptana, Daryanto A, Daryanto H K dan Kuntjoro 2010 Analisis Efisiensi Teknis Produksi Usahatani Cabai Merah Besar dan Perilaku Petani dalam Menghadapi Risiko *Jurnal Agro Ekonomi* **28** 153-188
[9] Badrudin U dan Jazilah S 2013 Analisis Residu Pestisida pada Tanaman Bawang Merah (*Allium ascalonicum* L.) di Kabupaten Brebes *Pena Jurnal Ilmu Pengetahuan dan Teknologi* **24** 75-86
[10] Basuki R S 2009 Pengetahuan Petani dan Keefektifan Penggunaan Insektisida oleh Petani dalam Pengendalian Ulat Spodoptera Pada tanaman Bawang Merah di Brebes dan Cirebon *J-Hort* **19**(4) 459-747
[11] Asfawi S dan Hartini E 2013 *Kontaminasi Residu Pestisida Dalam Buah Melon Di Kabupaten Grobogan* (Semarang: Universitas Dian Nuswantoro)
[12] Departemen Kesehatan RI 1984 *Pengenalan dan Penatalaksanaan Keracunan Pestisida.* (Jakarta: Direktorat Jendral PPM&PLP Departemen Kesehatan Republik Indonesia)
[13] Hallenbeck W H and Cunningham-Burns K M 1985 *Pesticides and Human Health* (New York: Springer-Verlag)
[14] Utami A Nurmansyah A dan Wayan LI 2017 *Helopeltis antonii (Hemiptera: Miridae)* Resistance Level in Cacao Plantation Against Three Types of Synthetic *Insecticide Journal of Industrial and Beverage Crops* **4** 89-98

[15] Wu C K, Thal L, Pizzo D, Hansen L, Masliah E and Geula C 2005 Apoptotic signals within the basal forebrain cholinergic neurons in Alzheimer's disease *Experimental Neurology* **195** 484-96

Determination of the water quality of Panasen River as a source of irrigation water

Sofia Wantasen*, **Jooudie N Luntungan and Annie E Tarore**

[1]Faculty of Agriculture, Universitas Sam Ratulangi, Indonesia

*swantasen@unsrat.ac.id

Abstract. The Panasen River is located in the upper part of Tondano Watershed. This river has an important role including being used as an irrigation water source. The purpose of this study was to assess the water quality of the river as a source of irrigation water in terms of the value of Sodium Adsorption Ratio (SAR) and the quality standard according to Government Regulation No. 82/2001 class III.The method used was field observation and laboratory analysis for river water samples. The measured parameters were Na, Ca, Mg, as well as other chemical parameters. The measured data were compared to the Quality Standards according to the regulation and using the SAR formula. The results showed that the SAR of river water varied from 0.10me/l to 0.40me/l. The results of the chemical analysis of 30 water quality parameters indicated that they generally met the requirements, except the chlorine (0.03-0.26 mg/l) that has exceeded the quality standard according to Government Regulation No. 82/2001 class III (0.03 mg/l). The high concentrations of chlorine caused by domestic sewage that goes into the body of water. It can be concluded that the river water still meets quality standards for irrigation water except chlorine.

1. Background

The river is the main body of aquatic ecosystem that has an important role in the hydrological cycle and functions as a catchment for the surrounding area. Consequently, the condition of a river is strongly influenced by the characteristics of the surrounding environment [1].

With the disposal of various types of waste and rubbish containing diverse types of pollutants into the river, both of which may decompose or which cannot be broken down, will lead to increasingly heavy burden which was accepted by the river. If the load is received by the river exceeds the thresholds that are set based on the raw quality, then the river is said to be contaminated, either by physical, chemical, or biological [2]. With the disposal of various types of waste containing various types of pollutants into the river, both biodegradable and non-biodegradable, it will cause more weight to be borne by the river. If the load received by the river exceeds the threshold set based on quality standards, then the river is said to be polluted, both physically, chemically, and biologically [2].

Irrigation water is water that is important for the growth and production of rice plant. The function of the irrigation water is as a solvent in which the minerals, and other dissolved elements can enter into plant cells [3].

The high activity around River Panasen has the potential to change the water conditions. Declining water quality is caused by contaminant sources that enter the water body. The estimation unit for water pollution can be classified in physical, chemical, and biological parameters. Physical parameters include increase in temperature, suspended solids, dissolved solids, DHL, and salinity. Chemical parameters include an increase and decrease in water pH, reduced DO value, increase in BOD and the presence of dissolved heavy metals. Biological parameters are the increasing content of pathogenic bacteria in water [4].

The management of River Panasen as a source of irrigation water can be done properly if its water quality as a source of irrigation water is known. In this case, the water of River Panasen river needs to be analyzed against the parameters of the quality of irrigation water required in accordance with the Government Regulation No. 82 of 2001 Class III and the value of Sodium Absorption Ratio (SAR).

The quality of irrigation water for plant needs is relies heavily on the substances contained in it. In principle, irrigation water must not contain substances that harm plants, such as sulfur or chlorine. The water that is good for plants must contain nitrogen, phosphorus, potassium, and calcium, among others [5]. Physically, the irrigation water can also affect soil conditions. If the irrigation water contains Fe or Na which exceeds the quality standard, it can cause the soil to become dense, thereby reducing circulation of air in the soil. Likewise, if the irrigation water is contaminated with chemical elements originating from residential waste, the congestion will reduce fertility rates and plant growth and productivity. Therefore, this research is important to obtain the value of Sodium Adsorption Ratio (SAR) as one of the factors in determining the suitability of water for irrigation and determining the quality status of River Panasen as irrigation water. This research aims to :

1. Determine the water quality of River Panasen as a source of irrigation water in terms of the value of Sodium Absorption Ratio (SAR).
2. Determine the status of River Panasen water quality as a source of irrigation water in terms of physical, chemical and biological parameters according to the Government Regulation of the Republic of Indonesia Number 82 Year 2001 Class III.

This research was conducted on paddy fields that received water from River Panasen in the upstream region of the Tondano Watershed, Minahasa Regency. This research lasted for one year. Data collection in the field is carried out for six months. Water samples are taken twice, each every three months.

2. Materials, Tools and Methods

a. The materials used for this study are water samples of River Panasen, aquadest, tissue, name label to mark containers and writing instruments.
b. Tools for taking water samples (water sampler, ice box, GPS, pH meter, water thermometer), tools for in situ measurements (pH meter, DO meter, thermometer), and equipment in the laboratory, i.e., spectrophotometer.
c. Primary data for water quality were obtained by taking data directly in the field, i.e. water samples from River Panasen. The parameters observed were pH, temperature, salinity, SAR (Na, Ca and Mg), DO, BOD, COD, Boron, NO3, PO4, arsenic, cobalt, selenium, cadmium, chromium (VI), copper, lead, mercury, zinc, cyanide, fluoride, nitrite, free chlorine, sulfur as H2S, oil and fat, surfactant, fecal coliform, and total coliform. Sampling water is done using the composite sampling method. Taking the water sample was prepared in the field and analyzed in the laboratory. In situ measurements were carried out on the parameters of pH, Temperature, DO, DHL, and Salinity.
d. Water quality sampling uses the composite sampling method and refers to the Indonesian National Standard [6, 7].
e. Sampling points and samples of river water are determined based on river water flow [8].
f. Secondary data on River Panasen water quality which was taken periodically (time series data for three years).
g. Compare the measurement data for each water quality parameter with the quality standard, in accordance with the water class [9]
h. Calculate SAR that is more closely related to the percentage of sodium that can be exchanged in the soil using the formula as follows:

$$SAR = \frac{NA^+}{\sqrt{\frac{Ca^{2+} + Mg^{2+}}{2}}} \quad (1)$$

where concentrations of Na_+, Ca_{2+}, and Mg_{2+} are expressed in milliequivalents per litre, as totals.

Water classification is based on SAR value; below 10 is very good, 10 to 18 is good, 18 to 26 is sufficient, above 26 is bad [10]. Low sodium water can be used for irrigation in almost all types of soils with the least risk to develop sodium exchange. Water containing moderate sodium content will pose a considerable risk because sodium in fine-textured soils have a high cation exchange capacity, except when there are gypsum in the soil. However, such water can be used properly on granuler or organic soils with good absorption. Water with high sodium content can cause sodium exchange which is harmful in most soils and requires special soil handling for good water flow, high meltdown, and the addition of some organic substances. Water that is very high in sodium is usually unsatisfactory for irrigation purposes, except when it has low and possibly moderate salinity, where liquefaction of soil calcium or use of gypsum or other repairs makes it easy to use the water.

3. Data Analysis

The data obtained in this study were analyzed using the Sodium Absorption Ratio (SAR) formula, then compared with Quality Standards according to Government Regulation No. 82/2001 Class III.

4. Results and Discussion

4.1. Results of Analysis of Irrigation Water Quality using Sodium Adsorption Ratio (SAR) Formula

Laboratory analysis of river water obtained the following results: the value of parameter Sodium in the locations of River Panasen 1, River Panasen 2, River Panasen 3, River Panasen 4, and River Panasen 5 were 0.46 mg/l, 1.57 mg/l, 1.54 mg/l, 1.51 mg/l and 1.67 mg/l, respectively. The value of parameter Calcium in the locations of River Panasen 1, River Panasen 2, River Panasen 3, River Panasen 4, and River Panasen 5 were 0.31 mg/l, 0.36 mg/l, 0.38 mg/l, 0.34 mg/l and 0.35 mg/l, respectively. The value of parameter Mg in the locations of River Panasen 1, River Panasen 2, River Panasen 3, River Panasen 4, and River Panasen 5 were 0.78 mg/l, 0.58 mg/l, 0.87 mg/l, 0.69 mg/l and 0.58 mg/l, respectively. The value of parameter SAR in the locations of River Panasen 1, River Panasen 2, River Panasen 3, River Panasen 4, and River Panasen 5 were 0.62 mg/l, 2.26 mg/l, 1.93 mg/l, 2.09 mg/l and 2.44 mg/l, respectively. All the paramaters value in milliequivalents per litre are presented in Table 1 and Figure 1.

Table 1. Sodium, Calcium, Magnesium and Sodium Adsorption Ratio (SAR) Concentrations

No	Location	Parameters			
		Sodium (Na) (me/l)	Calcium (Ca) (me/l)	Magnesium (Mg) (me/l)	SAR (me/l)
1	River Panasen 1	0.020	0.015	0.068	0.10
2	River Panasen 2	0.068	0.018	0.048	0.37
3.	River Panasen 3	0.067	0.019	0.072	0.31
4.	River Panasen 4	0.066	0.017	0.057	0.34
5.	River Panasen 5	0.073	0.017	0.048	0.40

Note : SAR (*Sodium Absorption Ratio*) is the ratio of natrium absorption.

According to [10], the value Sodium Adsorption Ratio (SAR) can be interpreted as follows. The value smaller than 3 me/l is good, from 3 me/l to 9 me/l is medium and above 9 me/l is bad. Excess sodium in irrigation water relative to calcium and magnesium or relative to the total soluble salt content can adversely affect soil structure and reduce the rate at which water moves into and through the soil (infiltration, permeability), as well as reduce soil aeration [11].

Sodium Adsorption Ratio (SAR) in the diagram below (Figure 2) shows that it has not crossed the threshold or the specified standard value. According to Effendi [4], the high SAR value shows that sodium in irrigation water replaces calcium and magnesium ions in the soil and changes the structure of the soil which will eventually make the plant leaves burn and the tissue dies. It can be seen in the diagram above; the SAR value is still below the threshold so that the concentration of sodium is still less concentrated than calcium and magnesium ions. As comparative, Yusuf [12] obtained the SAR values of the Bening Upper River (0.61) and Bening Lower River (0.59) as a rice field irrigation source in South Mopuya II Village, North Dumoga Subdistrict, Bolaang Mongondow District, are still in good quality.

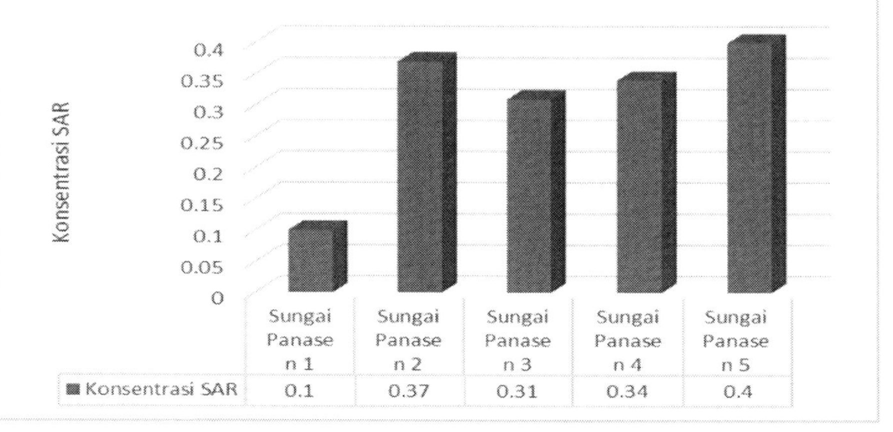

Figure 1. Sodium Adsorption Ratio (SAR) Concentration

Figure 2 shows that the concentration of Sodium Adsorption Ratio (SAR) in Panasen River 1, Panasen River 2, Panasen River 3, Panasen River 4, Panasen River 5 were 0.1 me/l, 0.37 me/l, 0.31 me/l, 0.34 me/l, and 0.4 me/l, respectively. Panasen River 5 has the highest SAR concentration and the lowest concentration was in Panasen River 1. Those five locations generally met the requirements because they were still below the threshold (Quality Standard <3 mg / l).

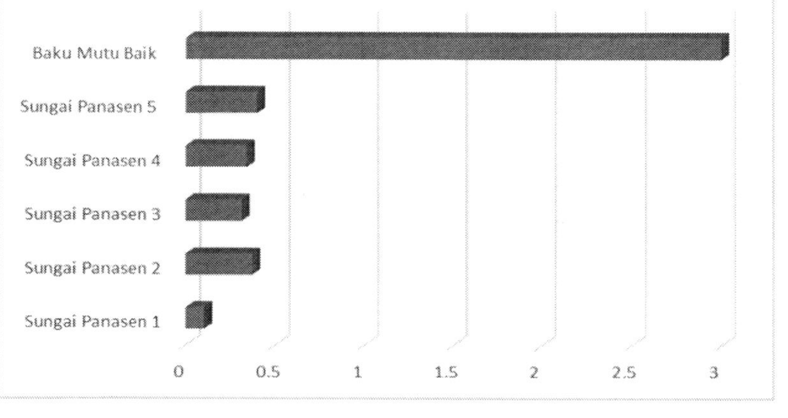

Figure 2. Sodium Adsorption Ratio (SAR) Concentration and Quality Standards

4.2 Results of Analysis of Irrigation Water Quality (Government Regulation No. 82 of 2001)

The results of the analysis of River Panasen water quality on 30 parameters indicate that the water generally meets the requirements according to quality standards, except the chlorine parameter that has exceeded the quality standard as stated in Government Regulation No. 82/2001. As presented in Figure 3, the quality standard for chlorine is 0.03 mg / l, while data from the analysis were in the range of 0.03 to 0.26 mg / l.

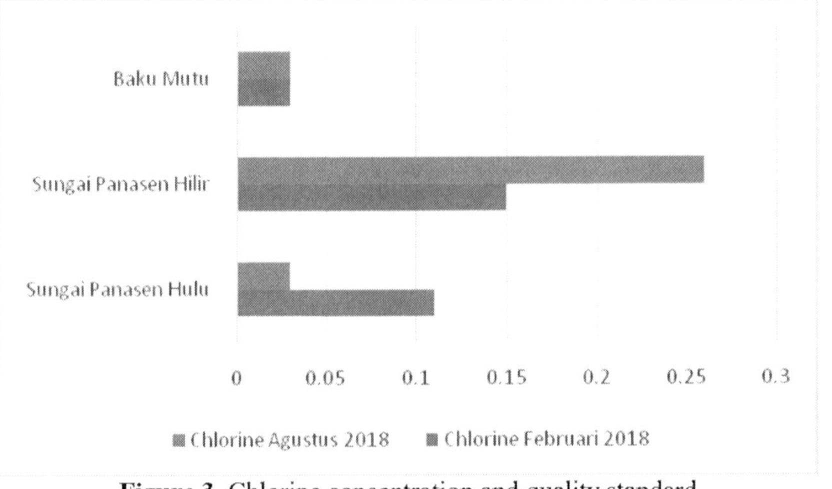

Figure 3. Chlorine concentration and quality standard (Government Regulation No. 82/2001 Class III)

The high concentration of chlorine was caused by domestic waste entering the water body. Besides being beneficial to humans, chlorine can also be toxic to the environment. The nature of chlorin as a strong oxidizer makes it easier for chlorine to bind to other compounds and is carcinogenic [13].

5. Conclusion

1. River Panasen as a source of irrigation water in terms of the value of Sodium Absorption Ratio (SAR) was considered good quality.
2. River Panasen as a source of irrigation water in terms of physical and chemical parameters according to the Government Regulation No. 82/2001 Class III was considered good quality but it has a high chlorine concentration was considered be toxic to the environment.

References

[1] Asdak C 2004 *Hidrologi dan pengelolaan daerah aliran sungai* (Yogyakarta: Gadjah mada univ. Press)
[2] Sinaga J dan Mukhlis J 2013 Kualitas air irigasi di desa air hitam kecamatan limapuluh kabupaten batubara, *Jur. online agroekotek* **2**(1) p 186-91
[3] Mawardi M 2016 *Irigasi asas dan praktek Bursa ilmu* (Yogyakarta)
[4] Effendi H 2003 *Telaah kualitas air bagi pengelolaan sumber daya dan lingkungan perairan* (Yogyakarta: Kanisius)
[5] Bardan M 2014 *Irigasi Graha ilmu* (Yogyakarta)
[6] Standar nasional indonesia (SNI) 6989.57: 2008 *Metoda pengambilan contoh air permukaan* Badan standarisasi nasional.
[7] APHA 2005 *Standard methods for the examination of water and waste water* American public health association (APHA) *21st ed.* Method 10200H and 4500-NO2-B.
[8] Hadi A 2015 *Pengambilan sampel lingkungan* (Jakarta: Erlangga)
[9] Pemerintah republik indonesia 2001 Peraturan pemerintah republik indonesia no 82 thn 2001 tentang pengelolaan kualitas air dan pengendalian pencemaran air (Jakarta)
[10] Ayers R, Westcot D 1985 *Water quality for agriculture. FAO Irrigation and drainage paper No*

29 (Rome)

[11] Canadian council of ministers of the environment 2008 *Canadian water quality guidelines* (Ontario) p 204

[12] Yusuf H, Wantasen S dan Lumingkewas A 2018 Kajian kualitas air sungai bening sebagai sumber air irigasi persawahan di desa mopuya selatan II kecamatan dumoga utara kabupaten bolaang mongondow *Jur. Cocos* **1** (3)

[13] Manahan S 2005 *Environmental Chemistry (8th edition)* (Florida USA CRC Press LLC) p 812

Preliminary study on socio-economic aspect towards Jakarta climate resilient (case study: Cengkareng District, West Jakarta and Penjaringan District, North Jakarta)

Tusy Augustine Adibroto[1,*], Pini Wijayanti[2], Rizky Pratama Adhi[1], and Rudi Nugroho[1]

[1] Center for Environmental Technology, Agency of the Assessment and Application of Technology, Indonesia
[2] Bogor Institute of Agriculture, Bogor, Indonesia

*tusy.adibroto@bppt.go.id

Abstract. This paper is the preliminary of guidance development for mitigation and adaptation efforts to overcome annual flood, which is exacerbated by the land subsidence in the north Jakarta. In addition to the physical studies of hydrodynamic modeling, a survey on socio-economic aspect was also conducted to the affected communities. In order to establish decisions for the option of flood disaster mitigation and adaptation efforts, the results of scientific studies that discuss both technical aspects and socio-economic and cultural characteristics should be gained. The aim of this study was to assess the social, economic, and cultural aspects of the communities in selected flood vulnerable locations. This paper is part of the entire study "Hydrodynamic Modeling for Flood Reduction and Climate Resilient Infrastructure Development Pathways in Jakarta". The results of the socio-demographic analysis indicated that the study area is densely populated and flood vulnerable. Thus, it requires immediate mitigation efforts to reduce the flood risk in the future. The relevant and appropriate conceptual framework in the context of social resilience in 3 (three) phases of disaster; before the occurrence of floods, when facing floods, and during recovery after a flood are analysis of public perceptions of flooding, analysis of exposure, and vulnerability.

1. Introduction

1.1 Background

At present, flooding is still a major environmental problem for Jakarta in the future. This phenomenon is caused by natural factors such as increased intensity of rainfall in upstream areas, local rainfall, and increased sea levels; also, by human factors such as excessive groundwater extraction and urbanization followed by expansion of urban areas and land conversion [1]. Especially North Jakarta, which is the most vulnerable to climate change in Southeast Asia [2], is severely threatened by coastal flooding, land subsidence and sea level rise [2,3].

The Jakarta flood risk value is expected to continue to increase in the future [4]. In general, flood risk values are estimated by multiplying the chance of flood hazard in a number of return periods with potential losses caused by flooding on these opportunities [5]. The research shows that due to a combination of land subsidence, increased rainfall and increased sea levels, the risk of flooding in Jakarta in 2030 is expected to increase 1.8 times higher than in 2015 [4]. The main factor increasing risk this is a land subsidence, which will last for the next decade even though groundwater extraction is stopped [6]. The most significant reduction in land surface in Jakarta in the period 1997-2005 occurred in the north-west [3]. The study showed that from 2007 to 2010, the average land subsidence in 3 (three) stations in North Jakarta varied from 11-19 cm / year, in addition to sea level rise reaching 2-4 mm / year in 1992-2009 [3]. The value of flood losses represents a decrease in people's welfare. Although most people who live in flood-prone areas are accustomed to flooding and suffer losses due to flooding, the decline in welfare is something that should be avoided. This is because the continued decline in welfare will have an impact on the decline in health, a decrease in decent living standards, and cause other social problems.

The DKI Jakarta Provincial Government has tried to make a number of mitigation efforts to reduce the adverse effects of flooding, carried out through dredging, cleaning and widening of rivers (otherwise known as normalization programs), increasing pump capacity, and improving river and sea dikes. Adaptation efforts will be carried out through the creation of giant sea walls and the construction of reservoirs in the upper Ciliwung. However, these mitigation and adaptation efforts will not necessarily reduce the risk of flooding in the future, given that changes in natural factors (such as rainfall and sea water increase) and human factors (population increase, urbanization and land-use change) are not uncertain and very much related to changes that occur in the area around Jakarta (Bodetabek). Therefore, various mitigation and adaptation efforts are still needed, besides being adjusted to the availability of government budgets as well as the socio-cultural characteristics of the community in flood-prone locations and the principles of sustainable development. In-depth scientific studies that discuss the technical aspects of the choice of mitigation and adaptation efforts and the characteristics of socio-cultural and community culture are needed as input for decision making.

1.2. Objective, Target and Scope

1.2.1. Objective. The objective is to conduct a detailed study of the social, economic and cultural aspects of communities in flood-prone locations, which include socio-economic conditions and community demographics, community perceptions of flood risk, potential losses that will be caused by flooding and willingness to be relocated to support mitigation efforts or adaptation from the government.

1.2.2. Target. While the target of this activity is to obtain data for socio-economic conditions, level of perception, vulnerability, resilience, and willingness to be relocated from the community in the study area.

1.2.3. Scope. The scope of Socio-economic and Culture Study Activities for Flood-Susceptible Communities in Jakarta are:

a. Literature study: conducted by examining the contents of scientific journals and previous studies to build a conceptual framework to prepare detailed components of questions in the questionnaire in surveys and focus group discussions. To describe the characteristics of the socio-economic and impact of floods, a study was carried out in a number of agency reports in the study area, and in publications a number of agencies (BPS: West Jakarta, North Jakarta, DKI Jakarta, and Provincial BPBD).
b. Focus Group Discussion (FGD): conducted in January 2017 by involving local government officials, such as the Village Head, Sub-District Head, BPBD, and representatives of the flood task force. And in April 2017 with more comprehensive participants came from various stakeholders (academics, central and provincial governments and local communities).
c. Field survey: a survey with interviews conducted in January 2017 against 342 households after a study from the modeling team (technical aspects) obtained results in the area of flooded area and depth of flooding in various flood scenarios and their mitigation. Sampling frames of research respondents were all households that were routinely affected by flooding in Kapuk Muara Village, Penjaringan District - North Jakarta and Kapuk Village, Cengkareng District - West Jakarta. The workflow of this study is illustrated in Figure 1.

d. Data were analyzed using qualitative and quantitative analysis. Qualitative analysis was applied to describe the characteristics of the demographic description and the impact of the flood in the study area. Quantitative analysis is carried out after survey and FGD data are obtained by statistical and econometric methods.
e. Compilation of Final Report.

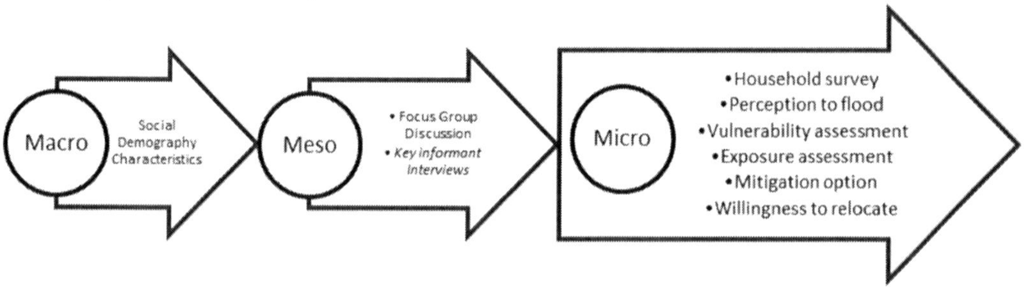

Figure 1. Study Workflow

2. Characteristics of the Study Area

This chapter discusses the results of the literature study to describe the social demographic characteristics and prepare the conceptual framework of socio-economic studies. The selected locations are in Penjaringan District - North Jakarta and Cengkareng District - West Jakarta. The illustration of the study location is shown in Figure 2.

Figure 2. Research Sites in Penjaringan District - North Jakarta and Cengkareng District - West Jakarta

2.1 Socio Demographic characteristics

The study location is a densely populated and densely populated urban village. In 2015, the total population was 309,215 people and the number of households was 101,086 households (KK), where the Kapuk village was the village with the highest population of 148,197 people and the largest number of households was 48,034 households [7]. In 20 years, the population and household in 4 (four) urban villages showed a rapid increase. The population in the Kapuk village, Kedaung Kali Angke, Kapuk Muara, and Pejagalan increased by 2.9; 1.8; 3.5; and 1.7-fold, and the number of households increased by 2.4; 1.7; 4.4; and 2.4-fold.

Cengkareng Sub-District is the most populous sub-district in West Jakarta City. Of the nine sub-districts in West Jakarta, 23% of the population lives in the Cengkareng Subdistrict area [8]. The population in this sub-district has increased 2.4-fold over a period of 20 years, and its population density exceeds the average population density of DKI Jakarta in 2015, which is 15,374 people per km^2 [7]. Demographic details of Cengkareng District in Table 1 below

Table 1. Characteristics of Socio Demography in Cengkareng District, West Jakarta

Year	Sq. (Km^2)	Neighbourhood (RT)	Upper Neighbourhood(RW)	Households	Inhabitants	Density (Inhabitants/km^2)
1995	27,93	768	71	65.402	209.391	7.497
2005	30,09	974	83	96.208	230.623	7.664
2010	26,54	998	84	101.108	304.945	11.490
2015	26,54	1.013	84	159.108	495.850	18.683

Source: [9][10][11[12][13]

Of the 6 (six) sub-districts in North Jakarta, 19% of the population lives in the sub-district of Penjaringan [14]. The population in this sub-district has increased 1.2-fold over the past 15 years. The demographic details of Kecamatan Penjaringan are presented in Table 2 below.

Table 2. Socio Demographic Characteristics of Penjaringan District, North Jakarta

Year	Sq. (Km^2)	Neighbourhood (RT)	Upper Neighbourhood(RW)	Households	Inhabitants	Density (Inhabitants/km^2)
2000	35,49	768	63	47.372	251.633	7.090
2005	35,49	800	65	19.487	178.026	5.016
2010	35,49	841	70	66.531	216.842	6.110
2015	45,41	867	72	102.777	305.121	6.390

Source: [15,16,17,18,19]

When viewed from the age group, the population of the 2013 projection for Kecamatan Penjaringan shows that the population is dominated by productive age, where the age group 15-64 reaches 241,229 people or 74.65%. The population of women is more than men with a ratio of 97.8 [20]. While based on the 2014 projection, Cengkareng Subdistrict was also dominated by productive age which reached 395,827 or 71.18%. Male population is more than women with a ratio of 105 [13].

Figure 3. One of the Conditions of Settlements in the Study Area

This study focused on the Kapuk Sub-District, Cengkareng Sub-District - West Jakarta and Kapuk Muara Sub-District, Penjaringan District - North Jakarta. The realization of the APBD for 2014 in the

Kapuk Kelurahan 9.3 billion [13] while the Kapuk Muara village was 3.06 billion [20]. Picture of the study location is shown in Figure 3.

Table 3. Comparison of Socio-Demographic Characteristics of the Study Area, 2015

Area	Sq. (km^2)	Neigbourhood (RT)	Upper Neighbourhood (RW)	Households	Inhabitants	Density (Inhabitants/km^2)	Sex Ratio
Kapuk	5.63	222	16	48.034	148.197	26.323	104,81
Kapuk Muara	10.06	101	10	11.907	36.578	3.637	104,14

Source: [13][19]

Based on population, Kapuk Village is more than eight times denser than Kapuk Muara Village (Table 3). This is in line with the very high changes in the composition of the population that occurred in Kapuk Village (Table 4).

Table 4. Changes in Population in the Study Area

Area	Birth (Inhabitants)	Mortality (Inhabitants)	Migration In (Inhabitants)	Migration Out (Inhabitants)
Kapuk (2015)	2.178	518	1735	1777
Kapuk Muara (2013)	551	92	1043	854

Source: [13][19]

2.2 Characteristics of Land Use

Land use in both kelurahan is dominated by housing, industry and offices as shown in Figure 4. The condition of housing and environmental facilities in a region can illustrate the level of welfare and health of the region. In Kapuk Village 46.54% of the buildings are permanent [13], while in the Kapuk Muara village 81.58% of the buildings are permanent and there are no apartments [19].

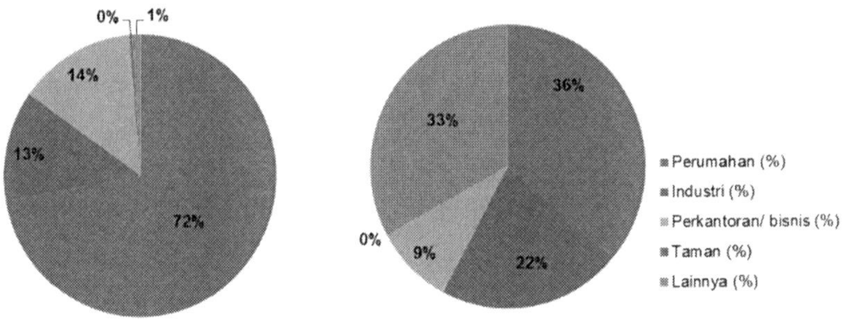

Figure 4. Percentage of Land Use in the Study Area

Note: *Perumahan* (in English means Settlement), *Industri* (in English means Industry), Perkantoran/Bisnis (in English means Office/Business), *Taman* (in english means Park/Green Open Space), and *Lainnya* (in English means Others)
Source: [13,18]

Although dominated by land use for housing, the existence of processing industries is very influential on the economic life of the people in Cengkareng Subdistrict, especially in the Kapuk Village (53%) [21], because this industrial business can absorb labor and is one of the eyes livelihoods of the surrounding community. Industrial activities in the study locations are dominated by large and medium industries which include the textile industry and textile products (convection, apparel, and garments). The comparison of the number of large and small industries in the study area is shown in Table 5.

Table 5. Number of Large and Medium Industries and the Number of Workers in the Study Area

Area	Medium and Large Industries	
	Industries (unit)	Labor (peoples)
Kapuk (2014)	62	3848
Kapuk Muara (2013)	55	6382

Source: [13][19]

2.3 Flood Disaster History

The study area is a flood-prone area. Table 6 and Table 7 show the number of flood events and their impacts during the 2015-2016 period. During this time, Kapuk Village was more frequently affected by flooding and more affected communities.

Table 6. Impact of Floods in Kapuk Village, Cengkareng District - West Jakarta, in 2015-2016

Flood	Number of Affected				Flood Height	Duration
	Neighbourhood	Upper Neighbourhood	Households	Inhabitants	(cm)	(Date and Day(s))
January 2015	4	0	1468	5538	20 - 40	23 – 24, (Two Days)
February 2015	14	114	10637	41924	20 - 100	02, 09 – 14, (Seven Days)
Desember 2015	1	1	0	0	10 - 20	13 (One Day)
February 2016	14	37	0	0	5 - 30	26 - 27, 29, (Three Days)
March 2016	3	10	0	0	5 - 15	1 (One Day)

Source: [22][23]

Table 7. Impact of Flood in Kapuk Muara Village, Penjaringan District - North Jakarta in 2015-2016

Flood	Number of Affected				Flood Height (cm)	Duration (Date and Day(s))
	Neighbourhood	Upper Neighbourhood	Households	Inhabitants		
Januari 2015	5	0	0	0	20 - 30	23 - 24, (two days)
Februari 2015	5	3	8	24	20 - 100	02,09 - 11, (four days)
Februari 2016	1	0	0	0	20	26, (one day)

Source: [22][23]

3. Literature Study

3.1 Resilience Concept

Resilience comes from the Latin word "resilio", meaning "to jump back" [24]. A number of literatures interpret the concept of resilience in a variety of ways. Resilience as adaptive flexibility in responding to disasters [25]. Resilience as the ability of people to recover from adversity [26]. Disaster resilience is the ability of the community to overcome a number of shocks, social losses, and changes in the political environment [27]. In general, resilience is defined as a flexible response to the dangers that occur and shows the ability to "bounce back" to the previous state after the danger occurred and caused damage. Resilience depends on changes or adjustments that require a fast and flexible response [28] in all phases of the disaster and can be conceptualized as a link between 4 (four) interrelated dimensions namely technical, organizational, social, and economic [29], as shown in Figure 5 below.

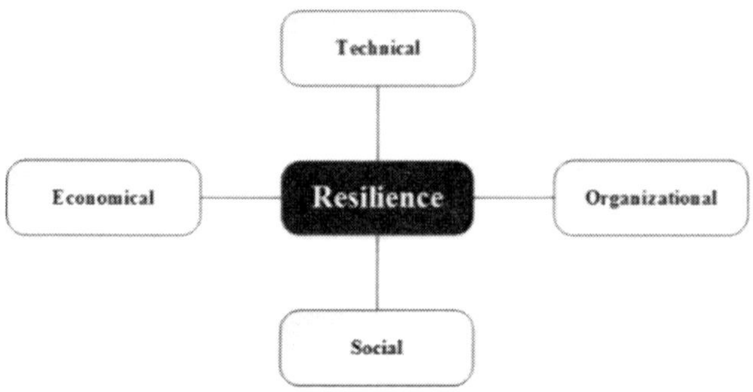

Figure 5. Resilience Framework [29]

3.2 Concept of Perception of Flood Risk

Community perceptions of flood risk refer to the risks felt by the community towards flooding in the context of limited and uncertain information about flooding in their area. Perception will motivate people to mitigate flood risk and also encourage demand for government efforts to mitigate and protect against flooding. Therefore, the perception of flood risk that is "felt" can provide useful information in flood risk management [30], where the perceived risk will affect the response and risk management of a community in the future [31].

Table 8. Indicators in Measuring Social Resilience

	Social Resilience Indicators		
High	Community participation Education Exchange of Information Learning Social support Sense of community Level of Trust	Community participation Education Exchange of Information Social support Sense of community Level of Trust	Community participation Exchange of Information Learning Social support Sense of community Level of Trust
Medium	Demography information	Coordination	*Community efficacy*
Low	Improvisation, inventiveness coping style Ability to cope with emergency situation/disaster Leadership	Ability to cope with emergency situation/disaster Leadership	Improvisation, inventiveness coping style Ability to cope with emergency situation/disaster Leadership
	Pre-Disaster	**Response (During Disaster)**	**After Disaster (*recovery*)**

Source: [32]

The key to success in flood management is related to how communities and local leaders respond to the risk of flooding and how they work together [33]. However, everyone in the community element may have a difference in assessing the risk of flooding, because they do not have the same information about the possibility of flooding [34].

Many factors will influence people to take flood mitigation measures, such as experience of past flooding [33][35], lack of dependence on government protection efforts, emotions, and fear [36]. Losses due to flooding can be minimized by mitigating and adapting both those carried out by the government (public) and personally. Self-motivation to reduce flood losses can be affected by several obstacles such as limited time, money, knowledge, or social support [36].

3.3 The concept of Flood Vulnerability Analysis (Flood Vulnerability Analysis)

Two types of exposure indicators in floods [34]. First, exposure related to the element of risk (usually related to elements of the flood location), such as: elevation, distance to the river, and return period. Second, exposure that focuses on the characteristics of floods in general, namely: duration, depth, and current strength.

The actual damage from flooding depends on the vulnerability of the socio-economic system and the affected ecological system [37]. Indicators of vulnerability measure how sensitive elements are exposed to risk will react or change when threatened with several types of hazards [34]. Flood vulnerability analysis can be demonstrated by the flood-damage function), which connects the value of damage to elements exposed to risk to indicators of exposure and vulnerability. Information about flood damage, exposure indicators and vulnerability indicators are important for building stage damage curves. Figure 6 shows how this analysis is related.

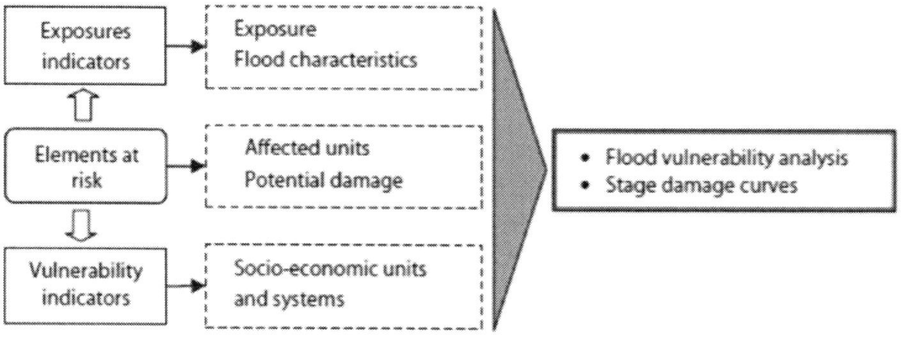

Figure 6. Indicators in Flood Vulnerability Analysis
Source: [34]

4. Analysis

The FGD with representatives of the community and industry produced fairly comprehensive information to describe the socio-economic conditions of the community. The FGD results showed a good understanding from community representatives on the causes of floods in Penjaringan and Cengkareng sub-districts. Some of the causes of flooding described in the FGD include; 1) high intensity of rainfall with frequent frequency, 2) siltation of rivers, 3) limited green areas, 4) rivers full of garbage, 5) land subsidence, and 6) pump usage that is not optimal. Some of the causes of the flooding have now been addressed by both the government and the community. However, the potential for future floods due to natural factors must still be watched out.

Social resilience is measured based on a total score of 8 (eight) attributes of social security. From the results of the questionnaire, the assessment of social security from survey respondents was higher than the assessment of FGD participants. Respondents in Kapuk Muara have the highest social resilience compared to other kelurahan. From the point of view of the community, the attribute of high social resilience is a sense of togetherness, a level of mutual trust and effectiveness. There are 3 (three) indicators that have the highest average value, namely sense of togetherness (4,10), trust (4.01) and learning (3.99). In other words, there are 3 (three) things that make the people in the study locations have social resilience, namely feeling part of the community, knowing each other and trusting the people around, and when facing flooding, they take experience from the events of previous floods.

Community perceptions at the location of the study on flood hazard varied between kelurahan. Although the community in the study location is between two major rivers, namely Cengkareng Drain and Angke Drain, their "perceived probability" is lower than "perceived consequences". It can be seen in Figure 6.2 that they were not too worried about the possibility of a larger flood in the future, but they were more worried about the impact of the flood and were worried about staying in that location. On a scale of 1 to 5, the average perceived probability is 1.6 and "perceived consequences" is 2.5. The highest value of "perceived risk" is given by households in Pejagalan Village. The community in the study location is accustomed to flooding and is aware of the risk of flooding that might occur. The response of the community when they learned that there would generally be a flood was to move items upstairs. After the floods arrived, and the height of the floods increased, a number of respondents said they would decide to evacuate. However, 40% of respondents said they would not evacuate even though the floods were rising. This is done because the community considers that the items in the house need to be maintained, besides that the community also thinks that the refugee sites are relatively limited so they are more comfortable to stay in their homes when there is a flood. One of the refugee locations is SD Kapuk 05 Pagi in Kapuk.

From the results of the FGD in April 2017 some of the inputs obtained from academics, the Central Government (i.e Bappenas) were about the need to highlight the findings of the survey about the refusal of the grass-root community towards the relocation plan by taking a participatory approach. To determine the location of relocation it is necessary to study the pattern of community movements for economic activities so that the maximum distance of the location can be obtained. In addition, the relocation building should be no more than 4 floors so that the community that had lived in the land-site still feels comfortable. Also, new economic activities should be sought based on the potential of the local community which is enhanced through empowerment activities that are integrated with economic policy. There is input that suggests the emphasis on collaboration and synchronization between various relevant stakeholders in locations such as the Ministry of Public Works and the relevant SKPD of the Provincial Government.

5. Conclusion

From the results of the study and FGD conducted, the following conclusions were obtained:

a. Socio-demographic analysis shows that the study area is a flood-prone area, experiencing high and densely populated land subsidence, requiring immediate mitigation efforts to reduce the risk of flooding in the future.
b. The relevant and appropriate conceptual framework for use in the next stage of study is social resilience in 3 (three) phases of disaster, namely before a disaster occurs, when facing disasters and after a disaster; and need deep vulnerability analysis such as community perceptions of flooding, as well as analysis of exposure and vulnerability.
c. Regarding the tendency to refuse relocation, a participatory approach and study of community movements related to economic activities are needed so that the maximum radius for the relocation location can be estimated.
d. It is necessary to think about new types of economic activities that are in accordance with the potential of the community and that appropriate community empowerment is carried out.
e. There needs to be continuous socialization of the community affected by the development plan in order to produce better conditions.
f. In the context of implementing relocation there is a need for various stakeholder collaboration and synchronization approaches related to flooding and management of water resources.

References

[1] Brinkman, J J and M Hartman 2009 *Jakarta flood hazard mapping framework, in International Conference on Urban Flood Management* edited (Paris A reference) pp 1-9
[2] Yusuf, A. A., and H. Francisco 2009 *Climate change vulnerability mapping for Southeast Asia* (Singapore : Economy and Environment Program for Southeast Asia (EEPSEA), pp 10-15
[3] Abidin, H. Z., H. Andreas, I. Gumilar, Y. Fukuda, Y. E. Pohan, and T. Deguchi 2011 Land Subsidence of Jakarta (Indonesia) and its relation with urban development *Natural Hazards* **59**(3): 1753-1717

[4] Budiyono, Y., J. C. J. H. Aerts, D. Tollenaar, and P. J. Ward 2016 River flood risk in Jakarta under scenarios of future change *Nat. Hazards Earth Syst. Sci.* **16**(3): 757-774
[5] Meyer, V., D. Haase, and S. Scheuer 2009 Flood risk assessment in European river basins-concept, methods, and challenges exemplified at the Mulde river *Integrated Environmental Assessment and Management* **5**(1): 17-26
[6] Bakr, M. 2015 Influence of Groundwater Management on Land Subsidence in Deltas: A Case Study of Jakarta (Indonesia) *Water Resources Management* **29**(5): 1541-1555
[7] BPS (Badan Pusat Statistik) DKI Jakarta 2016 *Jakarta dalam Angka 2016*, BPS Provinsi DKI Jakarta (Jakarta)
[8] BPS (Badan Pusat Statistik) Jakarta Barat 2016a *Jakarta Barat dalam Angka 2016*, BPS Jakarta Barat (Jakarta)
[9] BPS (Badan Pusat Statistik) Jakarta Barat 1996 *Kecamatan Cengkareng dalam Angka 1996*, BPS Jakarta Barat (Jakarta)
[10] BPS (Badan Pusat Statistik) Jakarta Barat 2001 *Kecamatan Cengkareng dalam Angka 2001*, BPS Jakarta Barat (Jakarta)
[11] BPS (Badan Pusat Statistik) Jakarta Barat 2006 *Kecamatan Cengkareng dalam Angka 2006*, BPS Jakarta Barat (Jakarta)
[12] BPS (Badan Pusat Statistik) Jakarta Barat 2011 *Kecamatan Cengkareng dalam Angka 2011*, BPS Jakarta Barat (Jakarta)
[13] BPS (Badan Pusat Statistik) Jakarta Barat 2016b *Kecamatan Cengkareng dalam Angka 2016*, BPS Jakarta Barat (Jakarta)
[14] BPS (Badan Pusat Statistik) Jakarta Utara 2016c *Jakarta Utara dalam Angka 2016*, BPS Jakarta Utara (Jakarta)
[15] BPS (Badan Pusat Statistik) Jakarta Utara 1996 *Kecamatan Penjaringan dalam Angka 1996*, BPS Jakarta Utara (Jakarta)
[16] BPS (Badan Pusat Statistik) Jakarta Utara 2001 *Kecamatan Penjaringan dalam Angka 2001*, BPS Jakarta Utara (Jakarta)
[17] BPS (Badan Pusat Statistik) Jakarta Utara 2006 *Kecamatan Penjaringan dalam Angka 2006*, BPS Jakarta Utara (Jakarta)
[18] BPS (Badan Pusat Statistik) Jakarta Utara 2011 *Kecamatan Penjaringan dalam Angka 2011*, BPS Jakarta Utara (Jakarta)
[19] BPS (Badan Pusat Statistik) Jakarta Utara 2016a *Kecamatan Penjaringan dalam Angka 2016*, BPS Jakarta Utara (Jakarta)
[20] BPS (Badan Pusat Statistik) Jakarta Utara 2016b *Statistik Daerah Kecamatan Penjaringan 2016*, BPS Jakarta Utara (Jakarta)
[21] BPS (Badan Pusat Statitik) Jakarta Barat 2016c *Statistik Daerah Kecamatan Cengkareng 2016*, BPS Jakarta Barat (Jakarta)
[22] BPBD (Badan Penanggulangan Bencana Daerah) Jakarta 2015 *Rekap Rinci Banjir Bulanan 2015* BPBD Provinsi DKI Jakarta (Jakarta)
[23] BPBD (Badan Penanggulangan Bencana Daerah)Jakarta 2016 *Rekap Rinci Banjir Bulanan 2016* BPBD Provinsi DKI Jakarta (Jakarta)
[24] Klein, R. J. T., R. J. Nicholls, and F. Thomalla 2003 Resilience to Natural Hazards: How Useful is this Concept? *Environmental Hazards* **5**(1): 35-45
[25] Cutter, S. L., L. Barnes, M. Berry, C. Burton, E. Evans, E. Tate, and J. Webb 2008 A Place-Based Model for Understanding Community Resilience to Natural Disasters *Global Environmental Change* **18**(4): 598-606
[26] Buikstra, E., C. Rogers-Clark, H. Ross, D. Hegney, C. King, P. Baker, and K. McLachlan 2011 *Ego-resilience and psychological wellness in rural communities, in Continuity versus Creative Response to Challenge: The Primacy of Resilience and Resourcefulness in Life and Therapy, edited (*18): 273-290 e-ISSN 9781619427075
[27] Adger, W. N. 2006 Vulnerability *Global Environmental Change* **16**(3) 268-28

[28] Berkes, F., and H. Ross 2013 Community Resilience: Toward an Integrated Approach *Society & Natural Resources* **26**(1): 5-20

[29] Bruneau, M., S. E. Chang, R. T. Eguchi, G. C. Lee, T. D. O'Rourke, A. M. Reinhorn, M. Shinozuka, K. Tierney, W. A. Wallace, and D. v. Winterfeldt 2003 A Framework to Quantitatively Assess and Enhance the Seismic Resilience of Communities *Earthquake Spectra* **19**(4): 733-752

[30] Bubeck, P., W. J. W. Botzen, L. T. T. Suu, and J. C. J. H. Aerts 2012 Do Flood Risk Perceptions provide useful insights for flood risk management? Findings from central Vietnam *Journal of Flood Risk Management* 5(4): 295-302

[31] Adger, W. N., S. Dessai, M. Goulden, M. Hulme, I. Lorenzoni, D. R. Nelson, L. O. Naess, J. Wolf, and A. Wreford 2009, Are there social limits to adaptation to climate change? *Climatic Change* **93**(3-4): 335-354

[32] Khalili, S., M. Harre, and P. Morley 2015 A temporal framework of social resilience indicators of communities to flood, case studies: Wagga wagga and Kempsey, NSW, Australia *International Journal of Disaster Risk Reduction* **13**: 248-254

[33] Fatti, C. E., and Z. Patel (2013), Perceptions and responses to urban flood risk: Implications for climate governance in the South *Applied Geography* **36**: 13-22

[34] Messner, F., and V. Meyer (2007), *Flood damage, vulnerability and risk perception - challenges for flood damage research, in Flood risk management - hazards, vulnerability and mitigation measures, edited by J. Schanze, E. Zeman and J. Marsalek,* (Springer: Dordrecht) *pp 149-168*

[35] Siegrist, M., and H. Gutscher 2008 Natural hazards and Motivation for mitigation Behavior: People Cannot Predict the Affect Evoked by a Severe Flood *Risk Analysis* **28**(3): 771-778

[36] Grothmann, T., and F. Reusswig 2006 People at Risk of Flooding: Why Some Residents Take Precautionary Action While Others do not *Natural Hazards* **38**(1-2): 101-120

[37] Cutter, S. L. 1996 Vulnerability to Environmental Hazards *Progress in Human Geography,* **20**(4): 529-539

Science and technology for the Alert Group Disaster "Tumik Singgalang" East Malalak Nagari, Malalak District, Agam Regency, West Sumatra

Ansosry, Eri Barlian, Nurhasan Syah and Abdul Razak

Department of Environmental Science, Postgraduate Program, Universitas Negeri Padang, Prof. Dr. Hamka Street, Air Tawar Padang-West Sumatera 25131 Indonesia

*osh5161@ft.unp.ac.id

Abstract. In the implementation of devotion in Jorong Saskand obtained a commitment on the part of districts, villages and ellipse as a service area. The commitment is obtained based on some identification problems faced by people who are important to be solved collectively. Society needs an explanation in terms of the academic science related to the potential danger to human settlements that are prone to landslides. The survey was conducted using a test to determine the type of layer geoelectric lithology / rock. Based on the results of the interpretation of electrical resistivity tomography. Wenner configuration showed layers of clay and pumice tuff at this track. Scientific analysis is translated in simple language to convey to the public. The solution to the problems facing society is the improvement of information systems at the community level while considering the potential and local wisdom. Public understanding of scientific terms that are directly involved in the technical processes related to geological conditions in disaster areas and this becomes the basis for an early warning system to collect and effective mitigation mechanisms to do. Improvement of early warning information system for landslides that involve elements that exist in society. To strengthen the information system related to early warning, then the required agreement among participants so that generated Mechanism of Landslide Mitigation Participative Approach. This method is outlined in the form of Standard Operating Procedures (SOP) early warning system for landslides Jorong Saskand, East Malalak, Agam Regency.

1. Introduction

Jorong Saskand, East Malalak Nagari, Malalak District is one part of Agam district. The geographical conditions of Agam Regency consist of flat areas, wavy flat areas, wavy and wavy flat areas and mountainous hills. The flat area owned by Agam Regency is at a slope between $0^0 - 3^0$ with 662 Km^2. While the wavy flat area which has a slope between $3^0 - 8^0$ has a total area of 8o 153 Km^2 [1]. This region is also located between two mountains, namely Mount Singgalang with an altitude of 2,877 meters and Mount Tandikat. The location of the study can be seen from Figure 1.

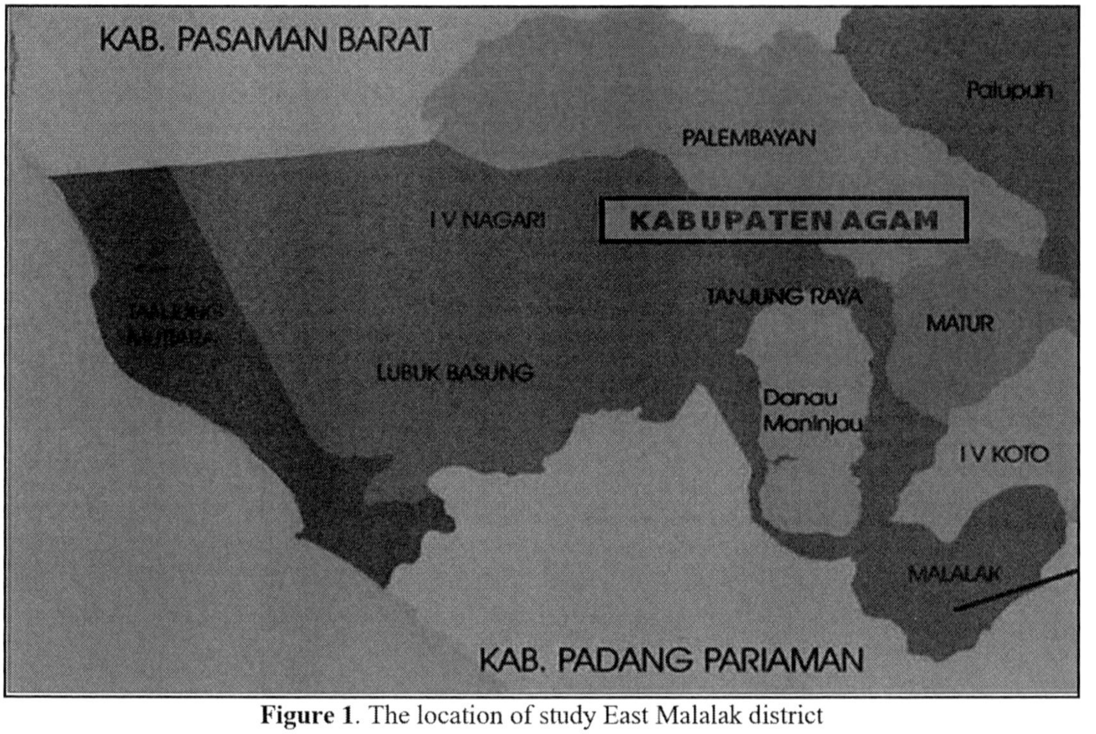

Figure 1. The location of study East Malalak district

To see the condition and record of events that have occurred can be known according to the Indonesian Disaster Data and Information (DIBI). The official information system issued by the National Disaster Management Agency (Badan Nasional Penanggulangan Bencana (2012), recorded 10 times landslides in the last 10 years in Agam District, which also occurred in several Malalak sub-districts [2].

Therefore, to reduce the risk of landslides, an early warning is needed. An early warning system is a series of systems to notify of the occurrence of natural events, can be disasters or other natural signs. Early warning to the community on disasters is an act of providing information in a language that is easily digested by the community. The ultimate goal of early warning is that the community can live and move safely in an area and the order of an area.

Geologically, the Jorong Saskand region, East Malalak Nagari is located under a steep hill, located in geological alignment, with loose lithology which has the potential to occur in landslides [3]. Based on the experience of the community that there was once a landslide accompanied by rocks which in the local language was called galodo. In addition to causing environmental damage, loss of property, the galodo also caused casualties. The galodo incident originates from a hill that landslides in an area about 200 m from the settlement. But now what people are worried about is the hill in their area of residence. According to observations the field has experienced *fractures* due to the earthquake that occurred in previous years.

From the results of a survey in early January 2012 at Jorong Saskand, the research team identified that the community in Jorong Saskand through the Disaster Preparedness Group (KSB) formed by an NGO had tried to make a document about the vulnerability analysis known as the *Hazard Vulnerability Capacity Assessment*. The results of the document show that the level of vulnerability and disaster risk is quite high [4]. This conclusion was obtained from the study and analysis regarding the economic, social and knowledge conditions of the community in the face of a landslide that threatened Jorong Saskand. Another important concern is because the people living in vulnerable areas have nearly 500 people and have farming livelihoods around the hills which are categorized as prone to landslides.

While the evacuation route and the safe route for the local community are insufficient. This is because access to open and safe areas is constrained by river flows that are likely to be the main stream if landslides accompanied by rocks (galodo) occur. So that the isolation of the area becomes the worst condition experienced by the people in Jorong Saskand if a landslide disaster (galodo) occurs.

According to information from the KSB members, as long as community assistance in improving preparedness for landslide hazards carried out by an NGO has not touched the landslide disaster early warning system and mechanism. The information obtained is only limited to knowledge about preparedness and testing (simulation) of events based on the tasks and functions of the group itself.

Therefore, based on the elaboration of the situation and the conditions above, it can be concluded that the community in Jorong Saskand, East Malalak Nagari, Agam Regency experienced quite complex and dilemma problems. But with the research program to be conducted in Jorong Saskand, East Malalak Nagari, Malalak District, Agam Regency is expected to be able to obtain a method of integration of early warning systems and landslide mitigation mechanisms.

2. Research Methods

2.1. Method of approach

Approach and direct assistance to the community in a participatory manner is a method for solving problems during program implementation. The community will be asked to participate directly in the implementation later through a series of agreed activities.

To carry out the initial steps in addition to field survey / mapping, it was carried out as well as the screening of commitments with the government, nagari leaders and other parties who would be involved during the implementation of the program. The research team with a background in geology and civil science will conduct a survey / mapping of the geological conditions of the area to determine the geological structure, rock, landslide fields and sampling to further analyse the development of models / estimates of landslide fields for vulnerable areas and affected assets. The results of the research team will be synchronized with the system and mechanism of the early warning tools that BPPT has made.

To carry out the integration method of the early warning system and the landslide mitigation mechanism that will be developed by the research team demands the role of the community involved in the program. Communities that know the area, try to agree together on disaster mitigation and especially implement and ensure the sustainability of existing systems and mechanisms. While the team is only an intermediary and researcher to ensure the condition scientifically and recommend systems and mechanisms for early warning and landslide mitigation to partners / KSB "Tumik Singgalang".

2.2. Plan of Activities

Plans for activities in solving problems faced by KSB "Tumik Singgalang" are the initial steps to provide counselling to explain as completely as possible the technical problems related to regional geological conditions and scientific disaster. After this group correctly understands the problem, the research team explains the stages of implementing the integration method of the early warning system and the landslide mitigation mechanism.[5]

The stages of mapping the condition of the area are the first step to know and understand the geology of the real conditions of the threat and the modelling / analysis of the affected areas while trying to understand the concept of early warning tools with partners. The next stage is mentoring by the research team as well as collective and participatory community agreement on the method of integration of early warning systems and landslide disaster mitigation mechanisms [6]. After the two stages above are carried out properly, the next stage is the agreed stage of testing / implementation of the system simulation. If the achievement of the test results is not maximized, improvements will be made until the early warning system can be implemented and understood by all levels of society.

The success of this research program needs support from many parties including the active participation of the Disaster Preparedness Group (KSB) "Tumik Singgalang" as the main partner of this research program. The forms of participation of partners that have been supported include providing access to the research team such as providing a separate place / room, access to information and the widest opportunity for the research team to come at any time in the process of identifying problems, dismantling, replacing and reassembling and landslide early warning system testing. With such broad access support, the community research team can freely express their ideas in a broad, complete, optimal and responsible manner. With the beam diagram the solution to solving the problems faced by the "Tumik Singgalang" Disaster Preparedness Group can be seen in the appendix.

3. Result and Discussion

Geographically, Jorong Sasai Kandang borders Jorong Subarang, the length of the North, South borders Jorong Toboh, on the east borders on Mount Singgalang, and on the west borders on Jorong Limo Badak. Jorong in the hilly area, as seen in Figure 2, generally has a potential danger that threatens the people living in the area.

Figure 2. General topography of the Saskand region

Based on Figure 2, it can be seen that the Sasai Kandang area is near the line of structural alignment. Geographically, Jorong Sasai Kandang borders on Jorong Subarang, the length on the North, South borders on Jorong Toboh, on the east borders on Mount Singgalang, and on the west borders on Jorong Limo Badak. Jorong in the hilly area as seen in the picture generally has a potential danger that threatens the people living in the area. The land contour is in the form of hills, this makes the Sasai cage area the most potential area for landslides (in regional terms is galodo). The results of recording these events can be seen in table 1 below;

Table 1. Historical events in Jorong Saskand and impacts

Year	Disaster events	Impact
1926	Earthquake	Houses damaged
1979	Galodo	4 people died, houses damaged, bridges got washed away
2006	Earthquake	Home damaged, damaged rice field
2007	Earthquake	12 houses damaged, 19 homes were damaged, facility public (schools and mosques) damaged
2008	Galodo	5 destroyed houses, 6 people died, bridges broken, farm land damaged 30 ha
2009	Earthquake	13 houses badly damaged, 9 medium damaged houses, mosques and mushalla damaged, hills cracking

From the historical plot of Jorong Saskand we can see a series of events disaster that has hit this area. the characteristics of the disaster that hit Jorong Saskand were the earthquake and galodo, but the biggest threat to the Saskand community was Galodo, because from the field survey and mapping several locations had been found which had cracked on the hillsides surrounding Jorong Saskand which would eventually break and cause landslides / banjir bandang / galodo, moreover rainfall in the area is quite high. The Malalak area is the most vulnerable to earthquakes, because this area is an area traversed by the earthquake fault line, as evidenced by the 2009 West Sumatra earthquake, Malalak was one of the worst affected areas. One of the drivers in the Malalak sub-district is Jorong Saskand who has felt the impact.

Jorong Sasai enclosure is located in a high degree of slope so it also has a very high threat of landslides. A number of landslides occur several times, affecting physical, property and life losses. At present the threat of landslides is still haunting the people of Jorong Sasai Kandang. As a result of the 2009 West Sumatra earthquake there were cracks at several points in the slopes of Mount Singgalang which at times could threaten the people of the Jorong Sasai cage especially in conditions of very high rainfall. From the historical path of Jorong Saskand disaster we can see a series of disasters who had hit this area. characteristics the disaster that hit Jorong Saskand was earthquake and galodo, but it became the heaviest threat to the people of Saskand is Galodo, because of the survey and field mapping obtained several locations who had cracked the slope the hills surrounding Jorong Saskand which one day will break and cause landslide / flash flood / galodo, let alone bulk rain in the area is quite high. Area Malalak is the most vulnerable area against earthquake, because of the area this is the area passed by the lane earthquake fault, proven in the earthquake 2009 West Sumatra earth, malalak is one of the worst affected areas. One of the drivers in this Malalak sub-district is Jorong Saskand who feels it the impact.

Jorong Sasai Kandang area located at a high degree of slope so that also has a very landslide threat high. Several landslides have occurred, impact on physical loss, property and soul. Currently the threat of landslides is still on haunt the people of Jorong Sasai Kandang. As a result of the 2009 West Sumatra earthquake occurred cracks on several points of the mountain slope Singgalang can at any time threatening the people of Jorong Sasai cage especially in very heavy rainfall conditions high.

Figure 3. Illustration of lithology and geological problems in Jorong Saskand

Signs of landslides in this area have been studied and known simply by the Saskand community. It is characterized by shrinking river water when it rains with high rainfall and a very loud thud. At this time the community was asked to stand by. However, early warning according to local wisdom needs to be refined by developing a systematic early warning system.

The following are the results of identification of community problems in the Sasai Kandang in a participatory manner in assessing vulnerabilities and capacities in their area. Assessment of these vulnerabilities is done using the PRA (Participatory Rural Assessment) method. The PRA tool used in assessing the vulnerability and capacity of the Sasai Kandang drive includes, transects, disaster maps, season calendars, varnish diagrams, historical flows, and the daily schedule of the community.

1) **Physical Vulnerability.** Among the physical vulnerabilities possessed by Jorong Saskand are:
   a. Residential settlements are on hillsides.
   b. Residential settlements near the river flow
   c. Many forest conditions are deforested
   d. Hills many cracks (fractures)
   e. Evacuation paths are inadequate Evacuation
   f. locations do not yet exist
2) **Social Vulnerability.** Among the social vulnerabilities possessed by jorong sankand are:
   a. Lack of population knowledge about how to deal with and disaster management
   b. Lack of youth living in the village
   c. Lack of Ulama
3) **Economic Vulnerability**. Among the economic vulnerabilities possessed by jorong sankand are:
   a. Sources of livelihoods (field rice fields) on the hillside
   b. There are no alternative jobs
   c. Many families that are less able
   d. The planting season is not simultaneous
   e. Prices of agricultural products often fall on the market
   f. Far from urban
   g. Transportation inadequate
   h. Erratic weather, which affects agricultural yields and planting patterns
   i. Plant pests (rats, pigs)
   j. Scarcity of Fertilizers

In addition to having many vulnerabilities, the Jorong Saskand community also has the capacity to reduce disaster risk. With the existing capacity if realized and maximized, of course the risk that will be caused by the disaster will decrease. Among these capacities are physical, social and economic capacity.

1) **Physical Capacity**
   a. The existence of a mosque as a temporary evacuation site for
   b. smooth irrigation channels
2) **Social capacity**
   a. Disaster Preparedness Groups (KSB)
   b. Strong relationships between institutions
   c. There is a concern of migrants to their home villages
   d. People know the signs of a disaster High
   e. solidarity and public awareness
3) **Economic Capacity**
   a. existence of farmer groups and Gapoktan
   b. Presence of Water User Farmers Association (P3A)
   c. The existence of nagari-level BMTs
   d. There isSPP of
   e. Sufficient Agricultural Products

Identification of vulnerability issues and capacities that have been carried out needs to be supported by hazard assessment so that the magnitude of the threat risks is identified. Through the PRA method, a general overview of maps based on HVC (Hazard, Vulnerability, Capacity) is obtained. 80% of the area jorong Sasai Kandang is a disaster-prone area, the disaster in question is landslides and flash floods. The red shaded area above is the area around the hillsides and river flow. Whereas the pattern of residential settlement of 60% is on hillsides and follows the river flow.

There are 3 points of vulnerability which are hilly cracks which can threaten people's lives and livelihoods at any time with landslides. From the map, it was also illustrated that only one road used as an evacuation route was in a disaster-prone area because it followed the river flow.

The evacuation area of the Sasai Kandang jorong is centered in the open area of Jorong Lomo rhinos, and temporary evacuation at the Sasai Kandang mosque. To take the two evacuation sites, you can only pass one road, the jorong road, Sasai Kandang. The capacity of the main evacuation site can accommodate approximately 300 people, while the temporary evacuation site in Saskand Jorong mosque can accommodate 100 people.

3.1 Coordination for screening commitments

Initial coordination of the activities is carried out after obtaining a letter issued by LP2M UNP. Research permit is given to the sub-district, nagari and jorong. At the time of coordination, a persuasion approach was carried out with stakeholders and policies in the Sub-Districts, Nagari and Jorong. When the initial coordination was obtained agreement and support in the implementation of the activities later. This is because the activities carried out are needed by the community. The condition of vulnerable areas and the condition of the community that requires attention related to disaster risk reduction efforts.

3.2 Geological survey of the area

The research area and its surroundings are plain and hilly areas with steep slope. Morphology in the eastern part (upstream) of the research area is hills which are composed of deposited material in the form of lumps of tuff, which is easily released due to weathering. Morphology in the western part (downstream) of the study area is a plain area.

The Ampera hill area in Jorong Saskand which has the potential for landslides has a variety of slope angles, namely slopes with an angle of ± 30° found in the middle to the foothills and slopes with angles of ± 35°-40 ° found in the middle to the top of the hill. The middle area of the hill that has the potential for landslides with soil thickness reaches ± 20-30 meters. Analysis of the data was obtained from more detailed mapping based on the geological map of the Malalak area as seen in Figure 4.

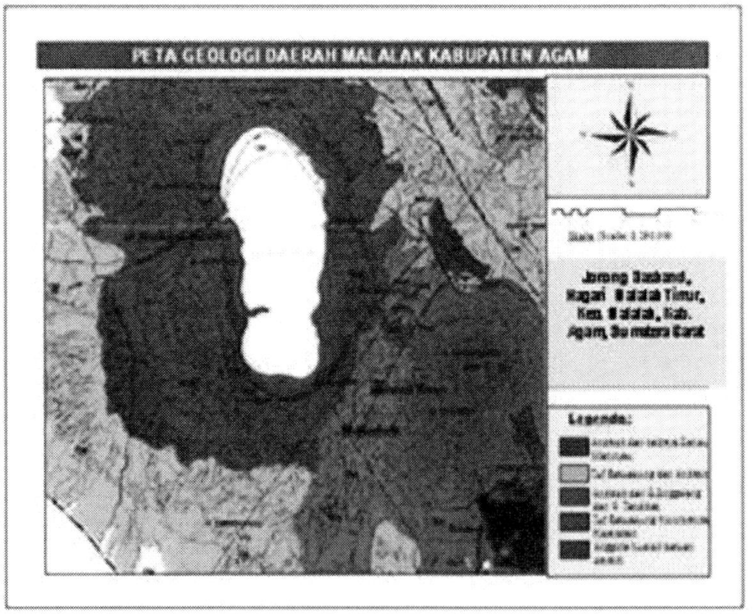

Figure 4. Malalak geological map of land Use

Land use in the middle area of potentially landslide hills in the form of shrubs and at the foot of hill settlements and rice fields. The condition of paddy fields is always planted with seasonal plants such as rice, corn and crops. Water that continues to be held in the middle of the hill due to soil cracks results in increasing the intensity of the water entering the soil. This condition will reduce soil shear strength significantly and increase slope load.

The sloping morphology of the study area leads to a vertical force that pulls rocks downward higher, this supports the occurrence of landslides in the study area. The river in the study area occupies the southern part of the research area, at the foot of the hill which has the potential for landslides there is a small channel used to drain water to the rice fields. This causes the water to seep into the soil continuously. This event accelerates avalanches in the research area as shown in Figure 5.

Figure 5. Survey location and field data collection points field

Based on observations this area consists of volcanic material from the Maninjau caldera eruption in the form of hiperstin hornstone tufa (Qhpt) (purple) which is entirely in the form of 2-10 cm lapilli stone containing horblenda and hiperstin or biotite, rather compact white or grayish yellow for fresh and brown ingredients from decaying and pumice and andesite (Qpt) (pink) in the form of fibers glass 5-80% white pumice fragments measuring 1-20 cm, rather compact. The description of the lithology that appears in the field can be seen in Figure 6.

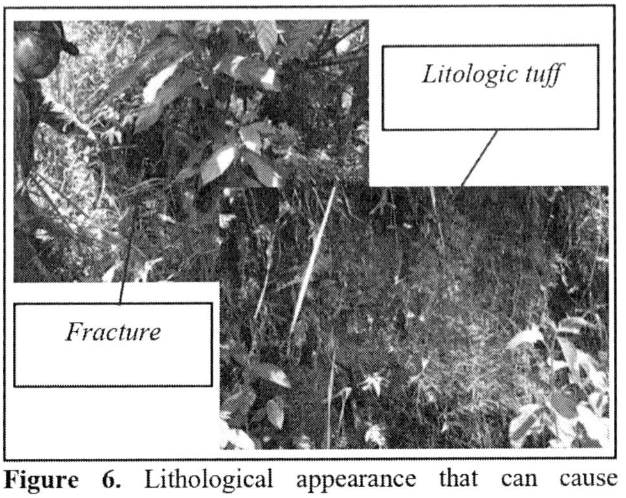

Figure 6. Lithological appearance that can cause landslides

Based on the results of the survey and lithological sightings which have several cracks, it can be concluded that the hill above the settlement has a high potential for landslides.

3.3 Analysis of the results of the field survey

Processing the results of the analysis of the geoelectric survey carried out can be seen from the 3 generated trajectories namely:

Track 1.
This track is located 26° from the north of the measurement location, at the coordinates S 00° 23' 17.2" and E 100° 17' 01.3". Based on the results of electrical resistivity tomography, the Wesner configuration shows the presence of colloidal deposits and pumice tuffs on this track. The clay (colivium deposits) occupy the top of the track and the pumice tufa is located at the bottom of this track with varying thicknesses.

Track 2.
This track is parallel to line 1 on the coordinates S 00° 23' 17.3" and E 100° 17' 01.0". Based on the results of electrical resistivity tomography, the Wesner configuration shows the presence of colloidal deposits and pumice tuffs on this track. The clay (colluvium deposits) occupy the top of the track and the pumice tufa is located at the bottom of this track with varying thicknesses.

Track 3.
This track is located 48° from the north of the measurement location in the coordinates 00° 23' 17.3" and E 100° 17' 00.9". Based on the results of electrical resistivity tomography, the Wesner configuration shows the presence of colloidal deposits and pumice tuffs on this track. The clay (colluvium deposits) occupy the top of the track and the pumice tufa is located at the bottom of this track with varying thicknesses. The track is a test sample to find out more in detail the lithology of the area. To see more clearly through trajectory mapping can be seen in Figure 7.

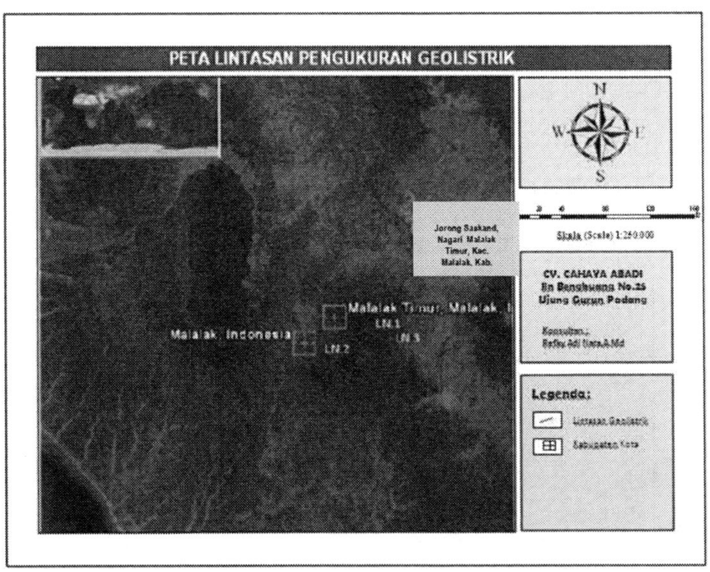

Figure 7. Geoelectric measurement path map in Jorong Saskand

Based on the picture above, testing has been done by producing several things related to the lithology of the area. The results of mapping and testing will be the basic material in conducting socialization and assistance to the community.

3.4 Dissemination of the results of regional mapping

After the mapping results were obtained, a meeting was held with community groups incorporated in the KSB. This is done in order to provide basic knowledge in terms of science related to regional conditions. Submission of information from the mapping survey needs to be done so that people can understand what the research team has done.

In general, the delivery of information to the public is more fundamental in understanding the potential of landslides. To make it easier for the public to understand, the research team made a simple description of the condition of the hill as a landslide point as shown in the following figure;

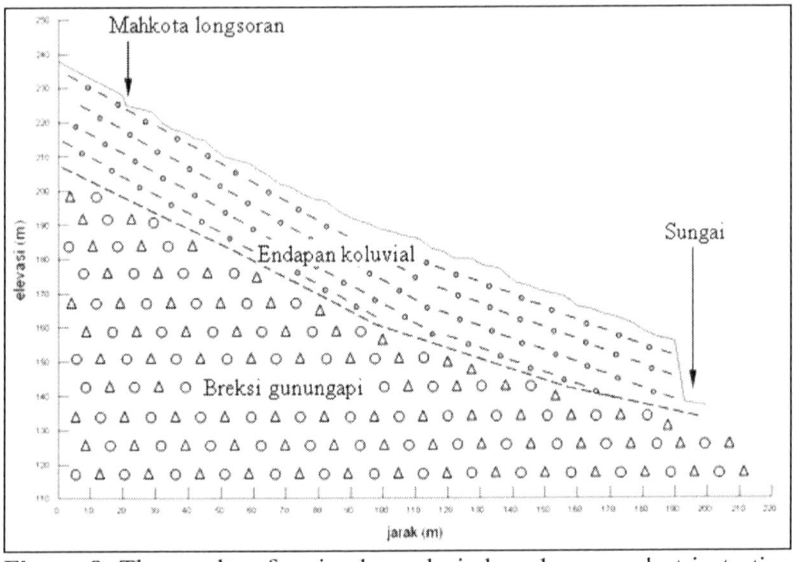

Figure 8. The results of a simple analysis based on geoelectric testing based on geoelectric testing

Socialization to the community in providing an explanation of the picture above has been done. Communities belonging to the KSB tumik Singgalang will pass on the knowledge provided by the research team to another general public. The role of KSB is more empowered as a group that has been formed in an effort to reduce disaster risk, especially landslides.

3.5 Focus Group Discussion (FGD) the role of the community in mitigation and prevention

Before entering into the FGD activities material was provided as an introduction to the FGD activities. First, the Implementation Team explained that Risk is the most important thing in a disaster, namely the risk of losing lives and assets. Therefore, the most important action that must be taken is to minimize the risk of small risks if we can face vulnerability and sponsored hazards. Provision of material prone to landslides in Jorong Saskand, East Malalak Nagari, Malalak District when rainfall is more than standard especially in the watershed. In particular this activity focuses on landslides. This is inseparable from the frequent frequency of landslides that occur in East Malalak District.

One that is needed by disaster-prone communities is the early warning system (EWS). This is absolutely necessary in order to create a society that is ready, alert and fast in facing disasters. In landslide-prone areas, such as in Jorong Saskand, East Malalak District is the most important part of disaster management. With the application of good and true will be able to protect and save the public from the threat of disaster.

The community can make various efforts to save lives and property. Early warning (EWS) is the key to effective risk reduction. It will be effective if it involves the participation of all components of society, besides that it can be understood and reaches all levels of society and must be followed by a systematic Standard Operating Procedure (SOP). Therefore. The IbM Program Implementation Team

"Singgalang Tumik Disaster" Group Faced Landslide Disaster, Jorong Saskand, East Malalak Nagari, Agam District Organized a Focus Group Discussion (FGD) on an appropriate method to integrate early warning systems with optimal landslide mitigation mechanisms, efficiency and agreed upon by the community in a participatory manner. The active involvement of the community is highly demanded to produce a method that is simple and easy to implement by the community.

4. Testing and monitoring of community systems and mechanisms in landslide handling

Anticipating the high risk of landslides in East Malalak Nagari, Malalak District, Agam Regency, which is between two mountains, namely Singgalang Mountain with an altitude of 2,877 meters and Mount Tandikat, Disaster Preparedness Group "Singgalang Tumik", In Jorong Saskand, East Malalak Nagari, Malalak Subdistrict, trials and monitoring of systems and mechanisms of the community in handling landslides. One of them is a landslide handling simulation that is centered in the local village field. In addition to these activities, to anticipate and handle the disaster, the role of the "Tumik Singgalang" Disaster Group was also increased.

"Tumik Singgalang" Disaster Preparedness Group has a community-based disaster management function. Starting from pre or disaster mitigation, emergency response, and recovery after a disaster. The "Tumik Singgalang" Disaster Preparedness Group not only moves to one village area but can also help surrounding villages within a sub-district. The formation of the "Tumik Singgalang" Disaster Alert Group is equipped with various disaster management facilities such as social barns for logistics, shelter, tents and other equipment. After empowerment, development is then carried out with the preparation of a roadmap (vulnerability map), disaster management SOP, and need assessment.

"This simulation is for SOP trials. With the "Tumik Singgalang" Disaster Preparedness Group equipped with logistics, they can move on their own, not depending on the government especially if there is a disaster when the office holidays are closed, they can independently move on their own. The Chairman of the "Tumik Singgalang" Disaster Preparedness Group, Jorong Saskand, Nurwan St Mangkudun said that before the simulation of the Lonsor land handling, his party had made sponsored anticipation efforts by draining the cracks of the land and fixing the trenches. The "Tumik Singgalang" Disaster Preparedness Group has 60 members consisting of the rapid reaction team (TRC) division, shelter, evacuation, public kitchen, communication and transportation. "With this simulation it is very positive, so that in the future we can anticipate and handle all possible disasters. The Jorong Saskand area, East Malalak, Agam Regency because of the form of cliffs so that it is very likely to occur landslides.

5. Conclusion

Based on the activities carried out in the Science and Technology program for the Community in Jorong Saskand, several conclusions and suggestions were obtained. These results are conclusions from several processes that have been running so the results are in the form of;

1. Mapping
   Presents visual information about the level of vulnerability of geological natural disasters in an area, as input to the community and / or district / city and provincial governments as basic data to carry out regional development in order to avoid disasters.
2. Examinations
   Conduct investigations during and after a disaster, so can know the causes and ways of controlling it.
3. Monitoring
   Monitoring is carried out in disaster-prone areas, in strategic areas economically and services, so that the danger level can be identified early, by users and the people who live in the area.
4. Socialization
   Provides understanding to the general public, about natural disasters of landslides and their consequences. Dissemination is carried out in various ways including sending posters, booklets, and leaflets or can be directly to the public and government officials

5. Landslide disaster inspection
   Aims to learn the causes, processes, disaster conditions and disaster management procedures in an area affected by a landslide
6. Community role Focus Group Discussion (FGD)
   Provide understanding to participants (KSB) about the importance of an early warning system, know what equipment is important in a landslide early warning system, and know what steps should be taken in the early warning system, even though not well conceptualized.
7. follow-up plan will be made public
   presence of repair / strengthen early warning information system landslide disasters, the required agreements between the public / participants KSB thus produced Procedures (SOP) early warning system for landslides Jorong Saskand
8. Tests and monitoring system
   In order to anticipate the and handle all possible disasters

References

[1] Badan Pusat Statistik., 2016, Kabupaten Agam Dalam Angka, BPS Kabupaten Agam.
[2] Data dan Informasi Bencana Indonesia (DIBI)., 2012, Sistem informasi resmi yang dikeluarkan oleh Badan Nasional Penanggulangan Bencana (BNPB), Sumatera Barat.
[3] Kastowo dan Leo. 1973., Peta Geologi Lembar Padang, Sumatera, Skala 1:250.000. Direktorat Geologi, Ditjen Pertambangan Umum, Departemen Pertambangan
[4] Kurniawan, Budi 2011, Laporan Akhir Program AIR SUMBAR, MERCYCORPS – Padang
[5] Mukhlis T, Fathani T F dan Sudarno I 2008 Perencanaan sistem peringatan dini bencana tanah longsor di Dusun Lucu Palongan Desa Campoan Kecamatan Mlandingan, Kabupaten Situbondo, Jawa Timur *Forum Teknik Sipil* **Vol. 18** No. 3 (2008) 888-898 available at http://puslit2.petra.ac.id/ejournal/index.php/cef/article/view/17509
[6] Karnawati, D., dan Fathani, T. F., 2007, Extensometer Otomatis dengan Anglemeter, Kompas dan Alarm, Extensometer Manual dengan Anglemeter, Kompas dan Alarm, Alat Pengukur Curah Hujan Otomatis, Petunjuk Manual, Fakultas Teknik Universitas Gadjah Mada bekerjasama dengan Kementrian Percepatan Pembangunan Daerah Tertinggal Republik Indonesia, Yogyakarta.

Waste Management in Kurai Taji Market Sub-District South Pariaman, Pariaman City

Hendra Arifin[1,*], Nurhasan Syah[2] and Eri Barlian[2]

[1] Graduate Program of Environmental Science, Universitas Negeri Padang, Indonesia
[2] Lecturer of Environmental Science, Universitas Negeri Padang, Indonesia

*hendraarifin38@yahoo.com

Abstract: Waste problems will not be separated from behavioral problem and lifestyle. Increased activity in the market greatly influences the quantity of waste piles in the market environment. This research began from the findings of poor waste management at Kurai Taji Market. Here, Waste management is not in accordance with the standard as according to Cunningham (2004) which is the stage of modern waste management consists of 3R (Reduce, Reuse, Recycle) before being destroyed or destroyed. The purpose of this study is to see how the appropriate form of management and strategiess are used in waste management at the Kurai Taji Market. The methodology used is descriptive qualitative. Waste Management in Kurai Taji Market, Pariaman City is as follows; Waste management in the Kurai Taji Market with the concepts of Reduce, Resue, and Recycle (3R) has not been implemented by market managers because the traders as a whole and the communities around the market have not reduced waste generation and reused market waste and they also do not have a habbit how to care the environment at Kurai Taji market.

1. Introduction

The market based on the Housing Estate Planning Guidelines is a regional shopping center whose main function as an environmental shopping center that sells daily necessities including vegetables, meat, fish, fruits, rice, flour, clothing materials, grocery items, school supplies, other household appliances [1]. Pariaman city runs the economy in the trade sector, followed by the service sector, where there are five traditional markets in this city. Traditional markets are a place where someone conducts buying and selling transactions in the form of goods and services to fulfill their daily needs, especially food and other primary household needs. Traditional markets in Pariaman City are Balai Nareh Market, Padusunan Market, Jati Market, Pariaman Market and Balai Kurai Taji Market.

Management of traditional markets such as Kurai Taji Market must be considered. Not only the management of its development will have an impact on increasing income but also on environmental hygiene management so that the distribution of goods in the market continues to run smoothly.

Kurai Taji Market is classified as a traditional market. Kurai Taji Market has an area of 1 hectare with 200 traders. Kurai Taji market has always seemed to be shabby, dirty, and chaotic. The stigma attached to traditional markets is generally motivated by the behavior of market traders, visitors or buyers and market managers. Kurai Taji market was built by the government and self-help community. Currently Kurai Taji Market is a significant contributor of waste in Pariaman Cty in daily life-day.

According to [2] "Waste is material that has no value or is not valuable for reprocessing purposes, waste is material that is wasted or disposed of from the results of human activities and natural processes that do not have economic value." Market waste has characteristics that slightly different from residential waste. Market waste composition is predominantly organic waste. Plastic waste on the market is fewer in number than household waste.

Waste problems will not be separated from behavioral problems and lifestyle. Increased activity in the market greatly influences the quantity of waste piles in the market environment. Waste management can be carried out efficiently and directed if functional relationships between waste elements can be clearly identified and understood. In order for the waste management system to be efficient, each element, both individuals and collectively must optimally consider various limitations such as costs, technology, education and community behavior.

At present the community in Pariaman City, especially the people who live around the Kurai Taji Market, still sees waste as a useless waste, not giving value as a resource that needs to be utilized. The community in managing waste still relies on the end-of-pipe approach, garbage is transported and disposed of to the final waste processing site.

Waste management at Kurai Taji Market is not so good. Such as the availability of one garbage truck. Ideally the waste before stacking in the trash can is separated according to the type. In order to facilitate preparation before processing. But the fact is that all market traders dispose of all types of waste in a garbage truck that is available in only one unit. Moreover, residents around the Kurai Taji Market also dispose of household garbage to a shelter truck at Kurai Taji Market along with restaurant merchants as well as dispose of their kitchen waste to the Kurai Taji Market garbage collection. The smell caused by the piles of garbage is very polluting the air around the market area.

This problem really requires serious handling of extras by the manager of the market environment. Market waste alone cannot be fully handled by the market manager, coupled with the addition of waste from the surrounding community. In order for waste management to undergo a better transformation, ideally market managers must educate and guide traders and the public about the obligation to clean up their respective trading areas in order to minimize the slum environment. But in fact this has never been done by market managers. Naturally, if the traders are not so concerned about the garbage around it. Traders only hand over the garbage to five cleaners.

Based on the description of the problems that exist in the Kurai Taji Market, the researchers felt that they needed to carry out research with the title "Waste Management at Kurai Taji Market, Pariaman City."

1.1. Problem Formulation
As for the formulation of the problem in the study of Waste Management in Kurai Taji Market, Pariaman City are as follows;

1. How is the waste management at Kurai Taji Market, Pariaman City?
2. What is the right strategy to deal with the garbage problem in the Kurai Taji Market in Pariaman City?

1.2. Research Purpose
The objectives of this study are:

1. To find out the right waste management at Kurai Taji Market, Pariaman City.
2. Looking for the right strategy to deal with the waste problem in the Kurai Taji Market, Pariaman City.

2. Literature Review

2.1. Waste
According to [3]"waste is a material that has no value or is not valuable for ordinary or primary purposes in fabricating or using damaged or defective goods in manufacturing or over-material or

rejected or discarded material". In Law Number 18 concerning Waste Management stated the definition of waste as human activities and/or natural processes in the form of solids. According to [4] re things that are no longer desired by those who have and are solid. While in the Academic Script Garbage Draft is mentioned that waste is left over a business or activity in the form of solid or semi-solid in the form of organic or an organic matter which can be decomposed or unraveled which is considered to be useless and discarded into the environment.

"Solid waste or waste include objects in the form of plastic, aluminum, iron, cans, glass, etc., while liquid waste is the seepage of liquid from decaying waste, commonly called *lindir* and can contaminate groundwater and rivers" [5] hen according to [2] waste is material that has no value or is not valuable for reprocessing purposes, waste is material that is wasted or disposed of from the results of human activities and natural processes that do not have economic value".

2.2. *Market*

Market is people who have the desire to be satisfied, for shopping, and packaging to spend it (6) While the same opinion says that the market is people who have the desire to be satisfied, money to shop, and willingness to tax [7].

Traditional markets are places where sellers and buyers meet to conduct transactions, in the case where existing market organizations are still very simple, low levels of efficiency and specialization, dirty physical environment and narrow building patterns [8]. Traditional markets are built and managed by the government, regional government, private sector, State-Owned Enterprises and Regionally-Owned Business Entities.

2.3. *Waste management*

[9] argues that waste management is an effort to create beauty by processing waste that is carried out harmoniously between the people and managers or the government together. Meanwhile, according to [2] waste management is an activity that includes the collection, transportation, processing, recycling or disposal of waste material. In [10] oncerning Waste Management, waste management is a systematic, comprehensive and sustainable activity that includes the reduction and handling of waste that aims to improve public health and environmental quality and make waste as a resource. Management of household waste and household-like waste consists of: (a) waste reduction; and (b) handling waste. Where the reduction of waste in question includes activities: (a) limitation of waste generation; (b) recycling of waste; and / or (c) reuse of waste. According to [11] the stage of modern waste management consists of 3R (Reduce, Reuse, Recycle) before finally being destroyed or destroyed. The following are described in the waste management cycle:

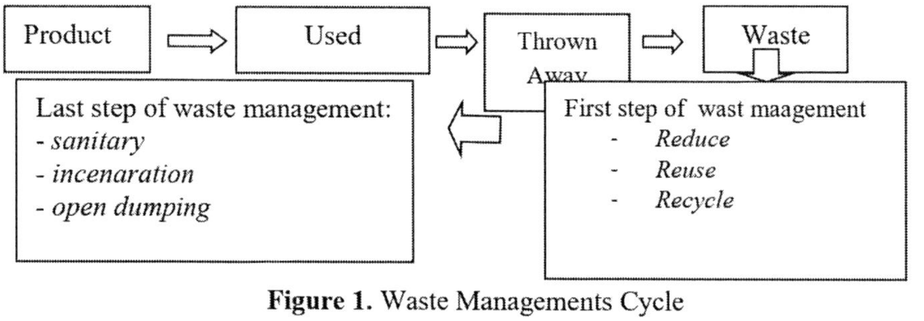

Figure 1. Waste Managements Cycle
Source: [11]

2.4. *Conceptual Framework*

Concept Problems related to the above phenomenon occur at Kurai Taji Market, Pariaman City. Researchers felt it was important to evaluate the waste management process through research on Waste Management at the Kurai Taji Market. The following can be described in the conceptual framework in Figure 2:

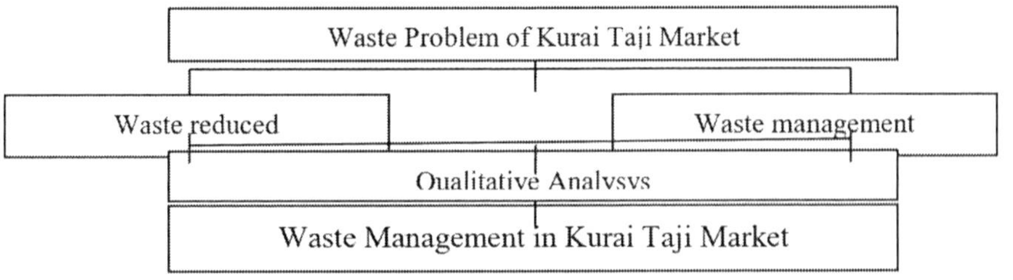

Figure 2. Conceptual Framework for Waste Management at Kurai Taji Market

3. Methodolgy Research

This study uses a qualitative description research method. Informants in this study were market traders, communities, market managers, and local governments. The researcher used three data collection techniques, namely through observation, interviews and documentation studies. Data collected with various techniques above, then carried out data analysis. [12] explains that in qualitative research there are three stages of analysis, namely data reduction, data presentation and conclusion drawing.

4. Result and Discussion

4.1. Result

4.1.1. Reduce Analysis. Waste is solid waste consisting of organic and inorganic substances which are considered useless anymore and must be managed so as not to endanger the environment and protect development investment. Waste needs to be managed maximally because it has been regulated in the Waste Management Act. waste management efforts can be carried out by reducing efforts first. The point here is that everything is trying to reduce the cause of garbage.

Based on input from traders, the public and market managers described above, it can be concluded that efforts to reduce the amount of waste in the Kurai Taji Market have not been carried out in every layer that has accumulated garbage in the Market. Like traders and the surrounding community, as well as market managers, there has been no program implemented so that the amount of garbage at Kurai Taji Market decreases.

4.1.2. Reuse Analysis. The principle of reuse is done as much as possible to sort out items that can be reused, and also avoid using items that are only disposable. This can extend the time spent on the item before it becomes garbage. the result of the interview with WR regarding the reuse of waste is "wet waste is processed into organic fertilizer and dry waste is made into crafts".

Based on the results of interviews, it can be concluded that a small portion of the community has used wet waste for compost, and dry waste is burned and only a small proportion of traders do this. Market managers only stack wet and dry garbage in one container and transport it to a landfill. The garbage is left buried just like that without being reprocessed.

4.1.3. Recycle Analysis. Waste now continues to be a problem because of the low awareness of traders, communities and market managers in Balai Kurai Taji to process and recycle waste (recycle). Recycling is a process to make used materials into new materials with the aim of preventing the actual waste become something useful, reduce the use of new raw materials, reduce energy use, reduce pollution, land damage and greenhouse gas emissions when compared to making new goods.

Explanations from all community representatives provide information that is not much different from the information given by traders and market managers. It can be concluded that the collected waste in the container is transported to disposal and just piled up without doing something like reprocessed into useful items such as compost and handicrafts.

4.2. Discussion

Waste management along with the development of the city has become the main agenda agenda faced by almost all cities, including in Pariaman City. The problem of waste management will never be resolved if it only relies on the government without any involvement from the community as a source of producing waste itself.

In [10] it was said that waste problems cover many aspects, therefore management must be carried out comprehensively and integrated with new innovations that are more adequate in terms of all aspects, both social, economic and technical aspects from upstream to downstream in order to provide economic benefits, be healthy for the environment, and can change people's behavior, meaning that waste handling needs to be done from the source. Community-based waste management with the 3R concept aims to reduce waste from the source, reduce environmental pollution, provide benefits to the community, and can change people's behavior towards waste. The 3R concept is actually very simple and easy to implement, but difficult to implement. Because the success of the 3R concept is largely determined by community participation by changing its behavior which is generally influenced by the socio-cultural character and socio-economic character that characterizes people's lives.

4.2.1. Reduce. Limitation of waste generation or waste reduction (reduce) is interpreted as an activity to reduce usage or behavior patterns that can reduce waste production and not make excessive consumption patterns. Based on the data analysis conducted, it was found that efforts to reduce the amount of waste in Kurai Taji Market had not been carried out by every layer of individuals who disposed of garbage in the market such as traders and surrounding communities. Also, with the market manager, no program has been implemented to reduce the amount of waste at Kurai Taji Market.

The sources of waste are based on observations of not only traders, surrounding communities, and visitors. But also, seasonal street vendors who sell when Monday at the peak of the market crowd or called "hall". As a result, the garbage will never decrease, on the contrary there will be more garbage accumulation in the container. If the market is in a hall on Monday, a pile of garbage exceeds one container collected.

The problem above can be found a solution starting from each officer, both from the market manager and from the Pariaman City Environmental Agency, must coordinate together in terms of waste management, so that the garbage problems that arise will be resolved immediately. The absence of waste reduction efforts can be overcome by involving every element of society, traders, and managers. This is in line with the opinion [13] "The involvement of all elements in waste management is one of the key factors in overcoming waste."collected in a forum. Then given direction so that public awareness, especially the traders to always maintain the cleanliness of the market, including disposing of garbage in containers that have been provided. For market managers, it is best to provide a dump/garbage container that meets established health criteria as well as make slogans in every corner of the market that contains an appeal to maintain the cleanliness of the market environment.

4.2.2. Reuse. Reuse of waste according to the Explanation of Article 11 Paragraph (1) letter c Government Regulation [10] concerning Household Waste Management and a Kind of Household Waste is an attempt to reuse waste in accordance with the same function or different function and / or reuse parts of garbage that are still useful without going through a processing process first. Reuse is the activity of reusing materials or materials that are still suitable for use, for example by using hand over paper towels, using used mineral bottles for flower pots or using used bottles for cooking oil containers. Another example that can be done is to wear back clothes such as used clothes to orphans. But the closest thing is to giving clothes that are too small to younger siblings or relatives in need.

Based on the results of the study, some people use wet waste for compost, and dry waste is burned. Only a small number of traders do this. Market managers only stack wet and dry garbage in one container and transport it to a landfill. The garbage is left buried just like that without being reprocessed.

From the results of the study, the resulting market waste has economic potential that is feasible to be processed, especially in terms of composting. So, in this case, given a solution so that the waste is reduced and more beneficial for the market owner or the manager can open opportunities for investors who want to do processing in that market.

4.2.3. Recycle. Garbage is an item not useful for ordinary or primary purposes in the use of damaged goods in manufacturing or overloaded or rejected materials. Waste is divided into two organic and organic waste. Organic waste is made up of plants and animals that are taken from nature or produced from agricultural, fishery or other activities. This garbage is easily described in natural processes. The rest of the market results in the form of flour, vegetables, fruit skin etc. are classified into organic waste. Furthermore, inorganic waste, which comes from natural resources cannot be renewed such as minerals and petroleum, or from industrial processes.

The two types of waste above can be recycled or often called recycle. Recycle according to the Explanation of Article 11 Paragraph (1) letter b Government Regulation [13] 81 of 2012 concerning Waste Management is an effort to use waste to become useful items after going through a management process first. The results of the research on recycle garbage in the Kurai Taji Market are garbage collected in containers transported to dumps and piled up without doing anything like reprocessed into useful items such as compost and handicrafts. The point here is that there is no recycling process of waste carried out at the Kurai Taji Market.

Related to the results of the above research, it can be indicated that things that become obstacles or obstacles in the waste management process in the Market are the importance of maintaining the cleanliness and health of the environment around the market by minimizing the remaining waste which the reduce method uses. Then there is still an assumption in the community that garbage is material / goods that have no economic value anymore. Furthermore, there is still no workforce who is willing to assist in carrying out the activities of implementing the 3R (Reduce, Resue and Recycle).

5. Conclussion

The research findings obtained through interviews, observation and documentation have obtained the, then conclusions can be drawn regarding Waste Management in Kurai Taji Market, Pariaman City as follows; Market management in the Kurai Taji Market with the 3R concept has not been implemented by the market manager. The traders as a whole and the people around the market have not yet made a reduction in waste generation, and the use of market waste again.

Reference

[1] Departemen PU 1987 *Petunjuk Perencanaan Kawasan Perumahan*
[2] Alex S 2012 *Sukses Mengolah Sampah Organik Menjadi Pupuk Organik* (Yogyakarta : Pustaka Baru)
[3] Gunawan S G 2007 *Farmakologi dan Terapi* (Jakarta : FKUI)
[4] Artiningsih N 2008 Peran Serta Masyarakat Dalam Pengelolaan Sampah Rumah Tangga *Tesis* (Semarang : UNDIP)
[5] Slamet J S 2002 *Kesehatan Lingkungan* (Yogyakarta: Gadjah Mada Universty Press)
[6] Basu S 2002 *Manajemen Pemasaran Edisi 3* (Yogyakarta: BPEE Yogyakarta)
[7] Stanton W J 2000 *Manajemen Pemasaran Edisi Revisi* (Jakarta: Erlangga)
[8] Agustiar 2015 *Peran Pasar Tradisional yang Mulai Ditinggalkan* (Solo : Harian Informasi)
[9] Neolaka A 2008 *Kesadaran Lingkungan* (Jakarta : Rineka Cipta)
[10] Undang-Undang Republik Indonesia 2008 Undang-Undang Republik Indonesia Nomor 18 Tahun 2008 Tentang Pengelolaan Sampah (Jakarta)
[11] Cunningham D J 2004 Mind, Culture, and Activity **11** (2)
[12] Moleong J Lexy 2005 *Metodologi Penelitian Kualitatif EdisiRevisi* (Bandung: Remaja Rosda Karya)
[13] Yarianto 2005 *Perlu Paradigma Baru Pengelolaan Sampah* (Jakarta)

[14] Peraturan Pemerintah Nomor 81 Tahun 2012 tentang Pengelolaan Sampah Rumah Tangga dan Sampah Sejenis Sampah Rumah Tangga (Jakarta)

Effect of school head integrity and communication climate on productivity through teacher work discipline in state vocational Padang City

Ariswan*, Rusdinal , Muri Yusuf and Gusril

Universitas Negeri Padang, Padang Indonesia

*arispengawas@gmail.com

Abstract. The phenomenon in the study illustrates the lack of productivity of teachers in implementation of the task. This study aims to show the influences; (1) the integrity and the communication climate of principal affect the work discipline of teachers in vocational schools, (2) the principal integrity towards the work discipline of teachers in Vocational High School of Padang, (3) work discipline affects the work productivity of teacher in Vocational School. Type of research is Path analysis. The population of this study was all the teachers in vocational high school of Padang with a status of civil servants (PNS), at least one year and still active teaching. The total of populations are 795 people. The sample of the study was 266 people who chosen by using the stratified proportional random sampling technique. The research data was collected using a Likert scale questionnaire model. Data were analyzed by Pathway analysis of moderating variables. The results of the study showed that principle integrity and communication climate directly and indirectly affect productivity through teacher work discipline.

1. Introduction

Educators and education personnel have the right to obtain adequate and adequate income and welfare guarantees, awards that are appropriate to their work and work performance, and career coaching in accordance with the demands for the development of quality professionalism. Educators in this case teachers, occupy a strategic position to realize all these expectations and goals, because teachers as educators are directly involved in the learning process in the classroom and the entire process of education in schools [1]. Teachers are required to have academic qualifications, competencies, educator certificates, physically and mentally healthy, and have the ability to realize national education goals. If a number of these requirements are met, then the role of these will be able to determine the progress and decline of organizational productivity.

Accroding to [2] said that the fulfilment of the right work by concentrating human resources and effort on work can affect organizational productivity Productivity in organizations depends largely on the motivation of its members, where the actions of members are directed towards achieving organizational goals. Correspondingly, productivity is a function of motivation. Motivation can be positive or negative, this shows that if motivation is positive it can encourage increased work productivity. Conversely if decreased motivation can reduce work productivity.

School organizational productivity is largely influenced by teacher work productivity. Dangkua's research in [3] reports that teacher work productivity can increase between 35% through the influence or encouragement of the principal, and around 60% -65% is determined by the teacher's personal abilities which include teacher work motivation, teacher work discipline, teacher interpersonal relations and teacher work environment. Therefore, the product of the teacher's work activities must be the concern of the principal as the leader of the organization because the high and low productivity of the teacher's work can affect the achievement of the overall school organization goals.

Based on the data obtained from the education office of Padang city; 1) For the year 2013/2014 there were no Padang City Vocational High School teachers who fulfilled the requirements to become outstanding teachers at the level of West Sumatra Province, 2) Teachers who had obtained IV / b rank above and only 24 people, out of a total of 795 teachers, 3) Academic qualification of S.2 level of education is only 120 people from 795 teachers, 4) The average score of the results of teacher performance assessment carried out by assessor is 45 of the maximum value of 56, 5) teachers who have conducted classroom action research and wrote a new Journal of 46 teachers from 795 teachers, 6) The average results of the acquisition of the National Vocational High School level Examination in 2013 only reached 6.54. From the ideal value of 7.00, 7) graduates of vocational schools who work independently and absorbed by the business and industry world are only 30% of the total graduates, 8) graduates of vocational schools are accepted at state universities in Padang city only 10%.

The above conditions are thought to reflect the work productivity of teachers in the Padang City State Vocational School as a whole. The low teacher work productivity is thought to be strongly related to the principal's leadership integrity variables, teacher's work discipline, and the communication climate at school [4]. If this is allowed then it will have a negative impact on the teacher's work productivity in Padang City State Vocational School. Based on the symptoms that occur, it is necessary to examine what factors influence the work productivity of the teacher. This study is expected to produce smart solutions as recommendations for the implementation of preventive and corrective policies on the quality of schools in the future, especially at vocational high school in the Padang.

2. Research Methods

The type of research is quantitative research using Path analysis. This research was conducted in all state vocational high schools in the City of Padang as many as ten school. The subjects of this study were the teachers of vocational school in Padang City. The population in this study were 795 people. The sample amounted to 266 people. Data collection was carried out by giving questionnaires about integrity of the principal, communication climate and work discipline on productivity of teacher work in vocational high school of Padang. The implementation have been done in 2 months after the trial is carried out by taking questionnaire data. Research data were analyzed using path analysis techniques, description analysis, hypothesis testing [5]. The table above show the result of the questionnaire analysis.

Table 1. Level of Achievement of the Response of each Teacher's Work Productivity Indicator

Indicator	Score ideal	Average	% Level of achievement	Category
Planning, implementation, evaluation and follow-up	160	138.9	86.82	High
Academic achievement	20	10.8	54.00	Less
Professional Development Work	35	21.6	61.80	Less
Participation in the Scientific Forum	10	7.71	77.70	Medium
Teacher's Overall Work Productivity Score	225	179.0	79.57	Medium

Source: Primary data processed

Based on the table above, each indicator of teacher productivity can be interpreted, the first indicator is planning, implementing, evaluating and following up with an achievement level of 86.82% including in the high category, the second indicator of achievement is 54.0% including low category, indicator of professional development level achievement 61.80% is included in the sufficient category, the next indicator is to participate in scientific forums, the level of achievement is 77.70%, including the adequacy category, ideal score of teacher work productivity79.57% is included in the category enough.

2.1. Principal Integrity

Variable Principal integrity (X1) was measured through 45 items in the questionnaire with 266 respondents' affordability. Empirically found the lowest score range of 45 and the highest score is 255. Based on the results of data analysis can be stated that the average score is 185.31 with a standard deviation of 24.22, median 186.0 mode 206, with the number of class 9 and class 12. The level of achievement of each indicator can be seen in Table 2 below:

Table 2. Achievement Rate of Response of each Principal Integrity Indicator

Indicator	Score ideal	Average	% Level of achievement	Category
Commitment to Service	50	43.8	87.62	High
Values	50	42.5	85.00	High
Consistent in action	50	40.3	80.76	High
Firm in establishment	75	58.6	78.16	Enough
Overall Principal Integrity Score	225	185,3	82.36	High

Source: Primary data processed

Based on the table above can be interpreted each indicator of principal integrity, the first indicator of service commitment with the achievement level of 87.62% including high category, the second indicator of achievement levels of 85.00% is included in the high category, indicators are consistent in action with achievement level of 80.76% with a high category, the indicator acts on a value system with a level of achievement of 78.7%, improving ability continuously with a level of achievement of 77.5%, firm indicators in the achievement of achievement 78.16% including sufficient categories and overall indicators of integrity of principals with achievement levels of 82.36% including high category.

2.2. Communication Climate

The communication climate variable (X2), empirically found the lowest score range 44 and the highest score 220. Based on the results of the data analysis it was found that the average score was 175.06, with standard deviation 21.33, median 176.50, mode 135 and number of class 9. and length class 12, based on the results of these calculations it was found that the climate level criteria for communication with distributional trends. The level of achievement of each indicator can be described in the following Table 3.

Table 3. Achievement Rate of Response of each Principal Integrity Indicator

Indicator	Score ideal	Average	% Level of achievement	Category
Planning, implementation, evaluation and follow-up	160	138.9	86.82	High
Academic achievement	20	10.8	54.00	Less
Professional Development Work	35	21.6	61.80	Less
Participation in the Scientific Forum	10	7.71	77.70	Medium

Teacher's Overall Work Productivity Score	225	179.0	79.57	Medium

Source: Primary data processed

Based on Table 5, the above can be interpreted as each indicator of the communication climate, the first indicator of trust with a level of achievement of 85.15% including good category, the second indicator of joint decision making level of achievement 77.65% is included in the category enough, honesty indicator achievement level 77.64% with enough categories, Downward communication indicator of achievement level 78.00% with adequate category, communication openness indicator above the level of achievement 76.08% with sufficient categories and indicators of attention to high-performance goals achievement level of 80.86% with good category. Overall communication climate score79.57% with sufficient categories.

2.3. Description of Variable Work Discipline

The variable Teacher discipline variable (Y1), as many as 43 questionnaire statement items with 266 respondents' affordability empirically the range of cooperative discipline scores with the lowest score of 45 and the highest score 215. The level of achievement of teacher work discipline can be seen in Table 4 below:

Table 4. Level of Achievement of the Response of each Indicator of Teacher's Work Discipline

No	Indicator	Score ideal	Average	% Level of achievement	Category
1	Obedience of teachers to carry out tasks	70	56.59	80.85	High
2	Awareness of teachers advancing teaching tasks	40	33.90	84.7	High
3	Teacher responsibility in the field of student teaching and coaching	50	43,06	86.11	High
4	Exemplary teacher in teaching	55	46,66	84,84	High
5	Teacher's Ideal Work Discipline Score	215	180.1	83.811	High

Source: Primary data processed

Based on Table 4 above, each indicator of work discipline can be interpreted, the first indicator of obedience of teachers to carry out tasks with the level of achievement of 80.85% with a good category of second indicators teacher awareness of advancing teaching tasks achievement level 84.7% with high category, third indicator of teacher responsibility in the field of student teaching and coaching the achievement level of 86.11% is in the high category, and the teacher's exemplary indicators in the teaching level of achievement are 46.66% in the high category, overall the score of teacher work discipline with 83.81% achievement in the high category.

3. Testing Requirements

3.1. Normality

Normality testing aims to test the assumption that the distribution of samples from instrument the damage samples comes from a population that is normally distributed, the normality testing technique used in this study is Kalmogorov Smirnov, with the criteria used is if Asymsig obtained is greater than 0.05 then the null hypothesis is accepted means the population is distributed normal.

Table 5. Summary Analysis of Normality Tests (N = 266)

No.	Variables	KS	Sign.	Conclusion
1.	Productivity	1.273	0.078	Normal
2.	Principal Integrity	1.281	0.075	Normal
3.	Communication Climate	1.304	0.067	Normal
4.	Work discipline	0.796	0.551	Normal

Notes: $\alpha = 0{,}05$

Data homogeneity testing was conducted to find out whether the research sample came from homogeneous population. To test homogeneity was done by Levene Statistic analysis. Analysis of variant homogeneity is done to determine whether the assumption that all three variables have the same (homogeneous) variant. The results of homogeneity can be seen in Table 6.

Table 6. Test of Homogeneity of Variances

	Levene Statistic	df1	df2	Sig.
Principal Integrity	1,623	4	261	,169
Communication Climate	,812	4	261	,518
Work Discipline	,863	4	261	,487
Productivity	1,017	4	261	,399

Source: Primary data processed

Table 6 can be seen that the magnitude of the Levene Statistic integrity of the principal (X1) is 1.623 while the probability or significance is 0.169, the communication climate (X2) is 0.812 while the probability is 0.518 which means greater than 0.05, the work discipline is 0.863 with a probability of 0.487, is accepted which means the assumption that the population variance is identical (homogeneous) is acceptable.

3.2. Hypothesis

The findings illustrate that the integrity of the principal influences work productivity through work discipline and teacher achievement motivation in Padang City State Vocational School. Communication climate influences productivity through teacher work discipline in Padang City State Vocational School. Principal integrity, communication climate, work discipline and achievement motivation together have a direct effect on teacher work productivity in Padang City State Vocational School.

3.3. Path Analysis Model

The direct influence of each variable, namely the variable principal integrity (X1) of teachers in Padang City State Vocational High School, Communication climate (X2) influences teacher achievement motivation (Y2) at Padang City State Vocational School. The principal's integrity (X1) affects the productivity (Y) of teacher work in the State Vocational School of Padang City. Communication climate (X2) affects the productivity (Z) of teacher work in City Vocational Schools Padang, Work Discipline (X3) influences the work productivity of teachers in Padang City State Vocational School.

Indirect influence principal integrity (X1) of schools affects work productivity (Z) through Padang work discipline (Y1). What is the integrity of the principal (X1), Principal Integrity (X1) communication climate (X2) work discipline (Y1) (Y2) jointly influence the work productivity (Z) of teachers in Padang City State Vocational School.

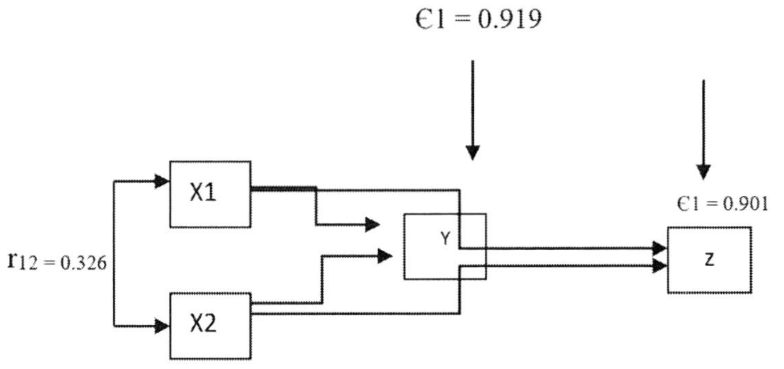

Figure 1. Conceptual framework

The findings show that the principal's integrity affects work productivity through work discipline and teacher achievement motivation in Padang City State Vocational Schools because the F count > F table means that integrity through work discipline affects the productivity of the principal. Integrity is the principal capital of the principal, but at the same time the capital that is rarely owned by the leader. This is the biggest tragedy in leadership. Lead researcher [6], in their book entitled Credibility: How Leaders Gain and Lose It, Why People Demand It, reports the results of their research for almost twenty years from surveys of thousands of professionals from four continents that there are two tens of characteristics of admired leaders.

Brophy features a list of motivational strategies used by the teacher to give students stimuli to be productive in learning 1) linkages to environmental conditions, which contain sportsmanship environmental conditions, conditions of level of difficulty, meaningful learning conditions, and meaningful strategy makers, 2) expectations to succeed, to contain the success of the program, the purpose of teaching, remedial socialization from outside awards that can contain gifts, positive competencies, value of learning outcomes [7,8].

4. Conclusion

Based on the results of the study, some conclusions can be drawn, namely the integrity of the principal influences on the work discipline of teachers in the City Vocational School the value of t count > t table. The integrity of the principal influences the work productivity of teachers in the State Vocational School of Padang City the value of t count > t table. Communication climate affects the work discipline of teachers in the State Vocational School of Padang City, the value of t counts > t table. Work discipline affects the work productivity of teachers in the State Vocational School of Padang City the value of t count > t table.

References

[1] Erianjoni 2018 Pengembangan Materi Ajar Sosiologi Tentang Mitigasi Bencana Berbasis Kearifan Lokal di Kota Padang *SOCIUS* **4**(2) 97-108
[2] Handoko T H 2012 *Manajemen Personalia dan Sumber Daya Manusia* (Yogyakarta : BPFE)
[3] Tolla I 1992 Kajian tentang Peranan Supervisi Kepala Sekolah dalam Meningkatkan Produktivitas Kerja Guru SMA di Sulawesi Selatan *Laporan Penelitian* (Ujung Pandang : IKIP Ujung Pandang)
[4] Erianjoni 2016 *Kesulitan dalam Belajar Sosiologi* (Jakarta: UNJ)
[5] Riduan 2010 *Cara Menggunakan dan Memaknai Path Analysis* (Bandung : Alfabeta)
[6] Kouzes J M and Posner B Z 2011 *Credibility, How Leaders Gain and Lose It. Why People Demand It* 2th ed USA : Josey Bass Terjemahan Adiwiyoto A (Jakarta : Professional Books)
[7] Hamzah B U 2011 *Teori Motivasi dan Pengukurannya* (Jakarta : PT Bumi Aksara)
[8] Erianjoni and Elviati 2015 *Implementasi Kurikulum 2013 pada Mata Pelajaran IPS di SMP SBSNP Kabupaten Agam* (LPMP Sumbar) pp 49-59.

IOP Conf. Series: Earth and Environmental Science **314** (2019) 012039 doi:10.1088/1755-1315/314/1/012039

Strategy of ecotourism development in Pariaman City

Lucyanel Arlym[1,*] and Dedi Hermon[2]

[1] Postgraduate Program of Environmental Science, Padang State University, Indonesia
[2] Lecturer of Environmental Science, Padang State University, Indonesia

*lucyanel95@yahoo.com

Abstract. This study is to select strategies for the development of ecotourism in Pariaman City. The process of this research starts with identifying and analyzing strategic values, internal and external environments and supporting and inhibiting factors. The next process is to determine strategic issues using SWOT Analysis techniques, SPACE Matrix, QSPM Matrix, Alternative Strategy Formulation, Strategy Selection, Strategy Selection Implications. This research is descriptive qualitative. The main goal of this research is solving the problems faced/at the study location. The approach for the study is a qualitative descriptive approach. Data collection techniques used were observation, interviews and literature study. Interview activities carried out on a number of speakers who were considered competent in one matter. Library Study is a method used to trace historical data or review literature and report reports relating to the research title. The strategy in accordance with the conditions for the development of ecotourism in the Pariaman City are: 1) Intensive Strategy, a strategy with intensive efforts if the competitive position of Pariaman City with existing tourist destinations and tourism services wants to improve; 2) Diversification Strategy: Diversification strategy is Pariaman City government strategy with the adds tourist destinations and new tourism services to help increase regional income.

1. Introduction

Tourism and sustainable development are two things that complement each other. Sustainable development is a development concept that emphasizes the balance of economic, ecological/environmental and social approaches. The environment that is maintained will be able to encourage an increase in tourist visits. Therefore good environmental/ecological safeguards can be a good tourist destination as well so that it further develops into an eco-tourism concept [1]. [2] adds tourism is one sector that is expected to provide a major contribution to the effort to maintain economic stability and improve people's welfare. Therefore the development of tourism needs to be continued and improved by using the resources and potential of tourism to become economic and non-economic forces that can be relied upon to support the implementation of regional autonomy. Along with the development of the times accompanied by increasing awareness and concern for the preservation of the environment that became the object of natural attractions, then began to emerge the thoughts and ideas for holding tourism activities that are environmentally sound.

One of the most popular tourism activities today is ecotourism activities. According to The International Ecotourism Society (TIES), [3],[4] is a journey made to unspoiled areas to preserve the environment and improve welfare. Ecotourism has, in essence, three core tenants. [5] holds that, done well, it can 1) protect and enhance the environment; 2) respect local cultures and provide tangible benefits to host communities; and 3) be educational as well as enjoyable for the traveller.

Conventional tourism impacts the environment both physically and culturally. Tourism has increased by more than 100% between 1990 and 2000 in the world's biodiversity hotspots, regions richest in species and facing extreme threats [6]. [7] explains the ecotourism is one of the promising tourism segments in the world today. Many countries in the world build their economic progress by using ecotourism. Ecotourism begins when there is a negative impact on conventional tourism activities. This negative impact is not only expressed and proven by environmental experts but also cultural experts, community leaders and tourism business actors themselves. The impact of environmental damage, uncontrolled influence of local culture, reduced the role of local communities and business competition that began to threaten the environment, culture and economy of the local community, as well as many other negative effects.

The vision of Pariaman City as a tourist destination city is a form of the desire to make tourism service activities as one of the activities that play an important role in driving the economy of the city of Pariaman. One effort to utilize local resources is to develop Ecotourism [8]. In this case, the tourism carried out has an integral part with conservation efforts, empowering the local economy and encouraging higher respect for differences in habit or culture. The educational aspect is a major part of ecotourism management because it carries a social mission to make people aware of their existence, the environment and the consequences that will arise if there is an error in the management of global environmental empowerment [9].

Another thing that must be considered is the development of culture in indigenous communities around the ecotourism area which is different from the culture of tourists [10]. Whether we realize it or not, gradually there will be a cultural shift that might eliminate indigenous culture. Ideally in an ecotourism area arises an attachment and mutual respect between indigenous communities and tourists. To minimize the bad impact that arises in the future, it requires integrity, quality, loyalty and managerial capabilities in carrying out supervision. The study was conducted to see how the strategy for managing ecotourism development in Pariaman City in the future.

2. Research Method

The approach used in the study is a qualitative descriptive approach. Descriptive method is a method that is carried out to find out and be able to explain the characteristics of the variables under study in a situation [11]. The qualitative approach places more emphasis on describing variables in terms of either definitions, conceptual explanations, notes or other forms that describe field conditions.

The research location is in Pariaman City. The study location was purposively selected with several considerations, first. The study location is a developing ecotourism activity path. Second, the potential location that tends to be natural with small natural damage is an attraction to be developed, third, the potential diversity of tourist attractions starting from turtle breeding, diving, along with the beach, island tours. This research was conducted from February 2015 to April 2015.

To support data collection activities, activities were previously sought to identify the types and sources of data in the location identification process. First, qualitative data is data in the form of words or sentences, including data on the physical condition of the study location, social conditions, a number of policy descriptions, data on perceptions and aspirations of both stakeholders and the community. This data is generally presented in the form of descriptions, and photographs. Both quantitative data are data in the form of quantified numbers or codes. Quantitative data used include data on the physical condition of study locations such as area.

There are 2 types of data sources obtained, primary data and secondary data. The first is the primary data source. The primary data source is direct data obtained from the object. The resource persons used in this research are:

1) Mayor and Vice Mayor of Pariaman,
2) Head of planning and development agency of Pariaman City,
3) Head of tourism Pariaman City,

4) Head of a marine of Pariaman City,
5) Head of the subsection of Turtle Conservation Area
6) Citizen from the tourist area. The sampling technique used was purposive sampling technique. The number of resource persons around the area that became the resource persons in this study was determined by a number of 5 resource persons for each area so that there were 10 resource persons representing the community.
7) Community leaders. Information from community leaders is determined based on people in the desa/"kelurahan" who are expected to know a lot about the history, events and religious uniqueness of the study location.

The secondary data source comes from agencies or parties that have published data or information that is useful in the research process. These data sources are the head of the village office, and the planning and development agency of Pariaman City. Data collection in this study was carried out by: Observation; Interview; Library Study. Methods and Data Analysis Techniques carried out by qualitative descriptive analysis, namely by giving a review or interpretation of the data obtained so that it becomes more clear and meaningful compared to just the numbers.

Data Analysis Framework. There are two kinds of descriptive analysis used in this study, namely descriptive explorative and descriptive comparative analysis. Descriptive explorative emphasizes exploration/excavation of information in more depth and focuses on the objectives of the analysis that you want to achieve. Comparative descriptive aims to compare a description/description with a particular variable. Several types of strategy terms are defined by several experts, among others: [12,13,14,15] Tourism according to some experts [16,17,18] concerns travel activities with the aim of not seeking income. Ecotourism formulation was first introduced by proposed by Hector Ceballos-Lascurain (1987) [19] as follows: *"Nature or ecotourism can be defined as tourism that consist in travelling to relatively undisturbed or uncontaminated natural areas with the specific objectives of studying, admiring, and enjoying the scenery and its wild plantas and animals, as well as any existing cultural manifestations (both past and present) found in the areas"* .

The International Ecotourism Society (1990), "Ecotourism is a way of life that conserves the environment and improves the welfare of local people" [3,4]. [20] defines ecotourism as a sustainable form of natural resource-based tourism that focuses primarily on experiencing and learning about nature, and which manages ethically low impact, non-consumptive and locally oriented (control, benefits and benefits and scale) [15]. [9] summarize the basic concepts of ecotourism into five core principles, including those that are nature-based, ecologically sustainable, educational environment, and useful local tourists and produce satisfaction [20,21].

3. Result

Important strategy formulation techniques can be integrated into a three-stage decision making framework using the model developed by [12,22]. As shown in Table 1. The tools displayed in this framework can be applied to all sizes and types of organizations and can help strategy makers identify, evaluate and choose strategies.

Table 1. Analytical Framework for Strategy Formulation

1. Input Stage				
(EFE)		Competitive Matrix of Profile		EFI
2. Matching Stage				
SWOT	BCG	IE Strategy	SPACE	Grand Strategy
3. Decision Making Stage				
		Quantitative Strategic Planning Matrix (QSPM)		

Source: [12]

Stage 1 of the formulation framework consists of the External Factor Evaluation Matrix (External Evaluation Factor-EFE), Internal Factor Evaluation Matrix (Internal Evaluation-IFE Factor) and

Competitive Profile Matrix. Phase 1, the input stage, contains basic input information needed to formulate a strategy. **Stage 2**, the matching stage, focuses on creating alternative strategies that make sense by taking into account the main external and internal factors. Stage two techniques include: Strength, Weaknesses, Opportunities and Treat-SWOT Matrix, Strategic Position and Action E-Matrix. **Stage 3**, the decision stage, involves only one technique, the Quantitative Strategic Planning Matrix (QSPM). QSPM uses input information from stage 1 to objectively evaluate alternative strategies identified in stage 2. The QSPM shows the relative attractiveness of various alternative strategies and thus provides an objective foundation for the selection of alternative strategies. Based on the results of the analysis using the EFE and IFE Matrix, SWOT Matrix, SPACE Matrix and using QSPM, the strategy that fits the current conditions for ecotourism development in Pariaman are:

3.1. Intensive Strategy
Intensive strategy is a type of strategy that requires intensive efforts if the competitive position of Pariaman City with existing tourist destinations and tourism services wants to improve.

a. Market development
   Market development is a type of strategy that introduces current tourist destinations and tourism services to new geographical areas.
b. Development of tourist destinations and tourism services
c. Development of tourist destinations and tourism services is a type of strategy that seeks to increase revenues through the improvement of current tourist destinations and tourism services or the development of tourist destinations and tourism services.

3.2. Diversification Strategy
The diversification strategy is a type of strategy where the Pariaman City Government adds tourism destinations and new tourism services to help increase regional income. Related diversification is a type of strategy where the Pariaman City Government adds new tourist destinations and tourism services but still relates to old tourist destinations and tourism services.

4. Conclusion
Pariaman City is a tourist area that can be developed into an ecotourism area that is attractive to domestic and international tourists who want to enjoy the concept of ecotourism. Based on the results of the analysis using the EFE and IFE Matrix, SWOT Matrix, SPACE Matrix and using QSPM (Quantitive Strategic Planning Matrix), then the strategy that fits the current conditions for ecotourism development in Pariaman City are: 1) Intensive strategy is a type of strategy that requires intensive efforts if the competitive position of Pariaman City with existing tourist destinations and tourism services wants to improve; and 2) the diversification strategy is a type of strategy where the Pariaman City Government adds tourism destinations and new tourism services to help increase regional income. Related diversification is a type of strategy where the Pariaman City Government adds new tourist destinations and tourism services but still relates to old tourist destinations and tourism services.

Acknowledgments
The author to express our gratitude to Pariaman city government and all those who have helped with the administration process as well as those who helped us in the field.

References
[1] Hermon D 2016 The Strategic Model of Tsunami Based in Coastal Ecotourism Development at Mandeh Regions, West Sumatera, Indonesia *J. Environment and Earth Science* **6**(4) 40-45
[2] Putra A, Tanto TA, Pranowo W S, Ilham, Damanhuri H, Suasti Y and Triyatno 2018 Suitability of Coastal Ecotourism in Padang City-West Sumatera: Case Study of Beach Recreation and Mangrove *J. Segara* **14**(2) 87-94

[3] Blangy S and Mehta H 2006 Ecotourism and ecological restoration *J. Nature Conservation* **14**(3-4) 233-236
[4] Tanto T A, Putra A, Hermon D and Damanhuri H 2018 Suitability of Seagrass Ecosystem for Marine Ecotourism in Padang City, West Sumatera Province *Forum Geografi* **32**(1) 88-95
[5] Honey M 2006 *Foreword in ''Le guide des destinations indige`nes'' Indige`ne editions* (France: Montpellier)
[6] Christ C, Hillel O, Matus S and Sweeting J 2003 Tourism and Biodiversity *Mapping Tourism's Global Footprint, Conservation International Washington, DC US*
[7] Bhuiyan M A H, Siwar C, Ismail S M and Islam R 2011 The role of home stay for ecotourism development in east coast economic region *American Journal of Applied Sciences* **8**(6) 540
[8] Hermon D 2019 Evaluation of Physical Development of The Coastal Tourism Regions of Tsunami Potentially Zones in Pariaman City Indonesia *International Journal of Geomate* **17** (59) 189-196
[9] Page S J and Dowling R K 2001 *Ecotourism* (Pearson Education Limited)
[10] Scheyvens R 1999 Ecotourism and the empowerment of local communities *Tourism management* **20**(2) 245-249
[11] Sekaran U and Bougie R 2016 *Research methods for business: A skill building approach* (John Wiley & Sons)
[12] David Fred R 2011 *Strategic Management: Concept and Cases* (New Jersey: Prentice Hall, Upper Saddle River)
[13] Hunger and Wheelen 2000 *Strategic Management & Business Policy, 8th edition* (New Jersey: Prentice Hall, Upper Saddle River)
[14] Hamel G, Prahalad C K 1996 *Competing for the Future* (Cambridge : Harvard Business Press)
[15] Porter M E 1996 *What is strategy* (Published November)
[16] Kodhyat H 1996 *Sejarah Pariwisata dan Perkembangannya di Indonesia* (Gramedia Widiasarana Indonesia Untuk Lembaga Studi Pariwisata Indonesia)
[17] Burkart A J, Medlik S 1981 *Tourism: past, present and future. Tourism: past, present and future. 2th edition.*
[18] Inskeep E 1991 *Tourism planning: an integrated and sustainable development approach* (Van Nostrand Reinhold)
[19] Ceballos Lascurain H 1996 *Tourism, ecotourism, and protected areas: The state of nature-based tourism around the world and guidelines for its development* Iucn
[20] Fennell D A and Dowling R K 2003 *Ecotourism Policy and Planning* (Oxon and Cambridge : CABI Publishing)
[21] Hermon D, Putra A and Oktorie O 2018 Suitability Evaluation of Space Utilization Based on Environmental Sustainability at The Coastal Area of Bungus Bay in Padang City *International Journal of Geomate* **14**(41) 193-202
[22] Putra A and Mutmainah H 2016 The Mapping of Temporary Evacuation Site (TES) and Tsunami Evacuation Route in North Pagai Island, Mentawai Islands Regency-Indonesia *IOP Conference Series: Earth and Environmental Science* **47** 1 01202

Environmental study and development planning center of tomato production

Zul Azhar[1,*], Hasdi Aimon[1], Idris[1] and Elida[2]

[1] Faculty of Economics, Universitas Negeri Padang, Prof. Dr. Hamka Street, West Air Tawar, Padang-West Sumatera 25131 Indonesia

[2] Hospitality and Tourism Faculty Universitas Negeri Padang, Prof. Dr. Hamka Street, West Air Tawar, Padang-West Sumatera 25131 Indonesia

*zulazhar@fe.unp.ac.id

Abstract. This article from the research of [1] states that the average total production of tomatoes in one planting period is 9.74 tons and is equivalent to a monetary value of Rp. 10.17 million. It means that tomato farmers in one planting period for 3-4 months get income between Rp. 9.77 million - Rp. 10.60 million or one month between Rp. 2.44 million - Rp. 2.65 million. Socio-culture of tomato farmers must abandon traditional farming culture that is less supportive, needs to increase production. In addition, [1] stated tomatoes during the harvest season, their production exceeded market needs, so that very much production was not utilized, because the average tomato production at harvest time reached 16.74 tons per farmer in one planting period. In connection with that case, it requires a downstream planning of more production (over production). This, if not considered, will make tomato farmers suffer losses and make demotivation to grow tomatoes. Tomatoes are wasted to rot causing flies, creating an unhealthy environment. Quantitative research method, data collection by distributing questionnaires to tomato farmers at the research site. Measurement The research instrument used a Likert scale with a weight of 1-5 values to assess each statement contained in the questionnaire. The variables of this study include economic, socio-cultural and environmental factors. Data analysis was performed using the Partial Least Square. The results of this study have a significant effect between the social fields on the economic field and economic factors on a significant environment while there is no significant influence between the social and environmental fields. So that environmental studies must be analyzed before project activities (business) will be carried out. If the project (business) activities are carried out then it can improve the welfare of tomato farmers who are environmentally friendly.

1. Background

Economic development in several countries has succeeded in raising the standard of living of the people in general. Economics with a series of theories that have been born has played an important role in shaping the flow of thought and intuition of economic planners [2]. But along with the progress of economic development, several problems and environmental impacts arise both in externalities and internalities. Depleted resources, increasingly advanced technology and the emergence of various types of pollution are beginning to be considered to hinder economic progress, because they damage the environment [3].

Agriculture is one sector that plays an important role in the economy [4] in Solok District in particular and West Sumatra in general. This is due to the contribution to Solok Regency's Gross Regional Domestic Product (GRDP), which is relatively large at 26.9 percent (BPS, West Sumatra 2017). Based on the research findings of [5], it shows that the average total production of tomatoes in one planting period in Tanang Bukik Sileh Nagari Salayo is 9.74 tons and is equivalent to a monetary value of Rp. 10.17 million with a standard deviation of 0.43. This means that farmers in one planting period for 3-4 months will get income between Rp. 9.77 million - Rp. 10.60 million. So, income in one month is between Rp. 2.44 million - Rp. 2.65 million. Therefore, tomato farmers from Nagari Selayo Tanang Bukik Sileh still need to increase their production capacity, economically profitable and will improve the welfare of farmers.

In addition, [3] stated that tomato production during the harvest season exceeds market needs, so that large quantities of production are not utilized, because the average tomato production at harvest time reaches 16.74 tons per farmer in one planting period. In connection with that, it requires a downstream planning of more production (over production). It is necessary to plan the down streaming of tomato products, if not thought out, it will make tomato farmers suffer losses and make demotivation to grow tomatoes.

Therefore, an Environmental Assessment and Development Planning Center is needed for tomato production in Nagari Bukit Sileh which is analyzed by Partial Least Square by looking at the socio-cultural local linkages with the economic field and its impact on the environment.

2. Theoretical Review

Socio-cultural; Everyone learns patterns of action in interacting with others, from individuals who occupy various social roles. Socialization means the learning process of community members to recognize and appreciate the culture of the community in their environment. Culture is very closely related to society. Melville J. Herskovits and Bronislaw Malinowski suggest that everything contained in society is determined by the culture that belongs to society itself. According to Andreas Eppink, culture contains the whole notion of social values, social norms, science and overall social, religious, and other structures, in addition, all intellectual and artistic statements that are characteristic of a society [6].

In everyday life a person will not be separated from the environment [7]. A person's personality is formed from his environment, because that personality leads to positive attitudes and behavior which of course must be supported by a norm (social culture) about its truth and adhered to as a guideline in acting. Glaser et al. (1987), Organizational culture is often described in the sense of being shared. [8], Kotter and Heskett (1992); Culture has power that influences the environment. The Effect of Culture on the Environment, culture is created or realized as a result of the interaction between humans and all the contents that exist in this universe. human beings were created by God Almighty with the mind that they are able to work on this earth and in essence human beings are caliphs on this earth. In addition, humans are also given advantages in terms of reason, intelligence, will, and imagination. Culture that has existed in a community environment is not possible to avoid the culture that comes from other groups, because of the contact and also the interaction between the two. A social group will adopt a certain culture when the culture is useful to overcome the problems and demands that exist at that time.

Economy; The environment is closely related to economic activities in the fields of production, consumption and distribution. Along with this, there is concern that the quality and quantity of resources is decreasing due to economic activities which are feared to threaten the continuity of economic progress in the next stage. For this reason, a benchmarking tool is needed to determine profit losses from various activities that have an impact on the environment.

Taking into account the possible impact of a problem when a business activity is running or at the end of a project, then the protection and control measures are also an important part of environmental studies for the development of a business or project [3]. So that it can minimize the impact of negative externalities due to economic activities.

IOP Conf. Series: Earth and Environmental Science **314** (2019) 012040 doi:10.1088/1755-1315/314/1/012040

Environment; Environmental problems and the use of natural resources both for agriculture, industry, mining and so on should be deemed necessary as an issue of environmental economics. Economic activity will have an impact on the environment and even cultural problems have links with the environment. More and more technological innovations will develop new products. Innovation is a process of renewal from the use of natural resources, energy and capital, new arrangements of labor and the use of new technologies which will all lead to the existence of production systems, and the creation of new products [9]. The process of innovation is very closely related to technology and economics. In a new invention usually requires a long social process and through two special stages namely discovery and invention. Discovery is an invention of a new element of culture, both in the form of a new tool, a new idea, created by an individual or a series of individuals in the community concerned [10]. Discovery becomes invention only if the community has recognized, accepted, and applied the new invention.

The case that happens a lot in the economic activity sector is very dilemma, on the one hand it needs economic growth, to increase growth, sacrifice resources, while resources are not closely watched to damage the environment, because social culture is closely related to the economy and environment. To see the interrelationship between social, economic and environment using the SEM [11] (triple bottom line) model in the construction of the Nagari Bukit Sileh tomato plantation, a thought map was created as shown in Figure 1. below;

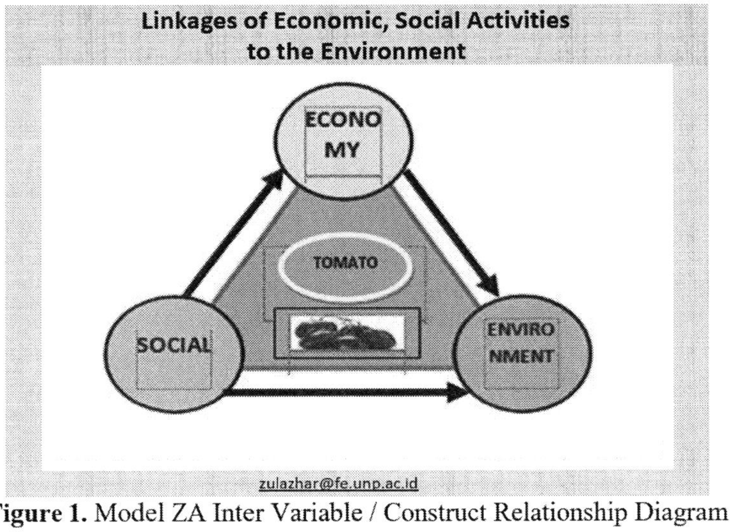

Figure 1. Model ZA Inter Variable / Construct Relationship Diagram

3. Research Methods

The research method is quantitative, the location of data collection in Nagari Bukit Sileh in 2018. The primary data obtained by questionnaires and interviews with tomato farmers or traders in the study location were 98 people. Data retrieval technique is done by sample selection method by accidental sampling.

There are three constructs as Latent variables and 20 manifest variables which consist of socio-cultural 4 indicators, Economy 6 indicators and Environment 10 indicators (see attachment 1). Each construct is measured by manifest variables or influencing indicators. The relationship between constructs (SOCIAL, EKO and LING) and indicators is formative because indicators affect constructs [12].

Then it is weighted with a Likert scale value that is value 1 (strongly disagree), value 2 (disagree), value 3 (agree), value 4 (strongly agree), and value 5 (strongly agree). The direction of the diagram is the main model of the relationship between the constructs / research variables can be seen in Figure 2 below.

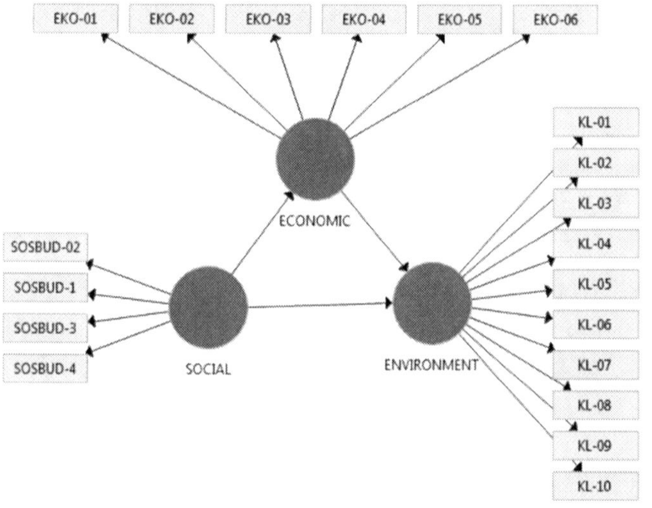

Figure 2. Interconstruct Relationship Diagram

4. Results and Discussion

After the questionnaire data was tabulated, then carried out the analysis resulting in path coefficients as shown in Figure 3 below.

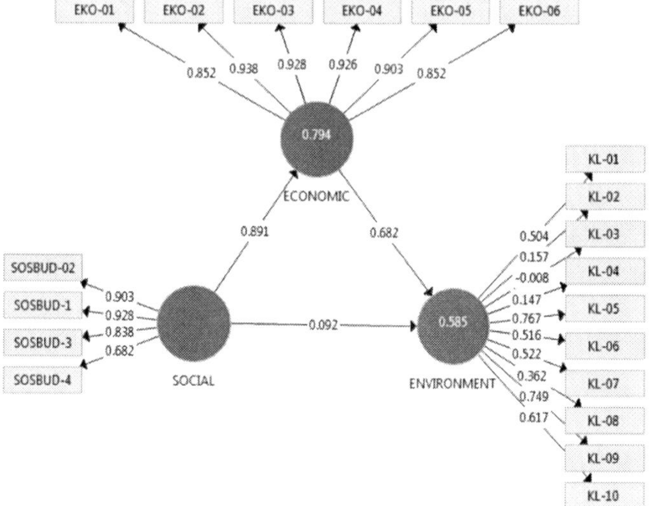

Figure 3. The output model of the path coefficient output

Based on the output of the path coefficient above the indicator which does not have a loading factor coefficient below 0.7, it is dropped from the research diagram (Haryono, 2017), namely Social Culture 04, KL1, KL2, KL3, KL4, KL6, KL7, KL8, KL10. So that the Fit Line output is obtained.

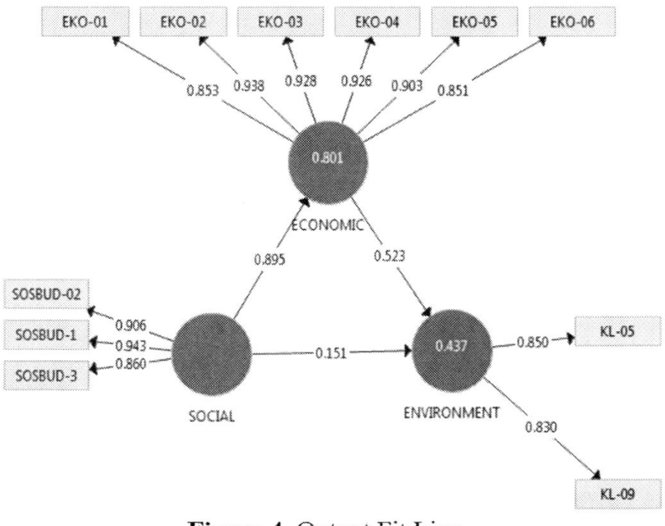

Figure 4. Output Fit Line

Next, the Calculate-Bootstrapping command produces a T Statistic coefficient as a test of the research hypothesis as shown in the following picture.

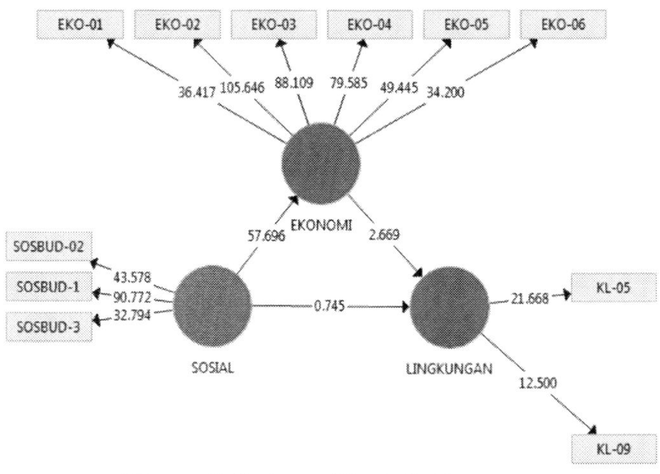

Figure 5. Output of Statistic T.

To test construct validity, the indicator validity is seen from the Loading Factor value (LF) must be equal to 0.7, which is said to be valid [13]. Based on the results of the LF in table 2 (see attachment 2) it turns out that all ≥ 0.7, then all indicators on the model are fit. Next the results of outer weights in table 3 (see attachment 3) from the calculation of calculate Boost rapping has a T Statistic value ≥ 1.96 said to be valid besides that the indicator is also valid because it has P Value ≤ 0.05.

So that the results of calculation of variable models that are already fit can be used to test hypotheses at the stage of structural measurement. According to [12] for testing construct reliability is measured by the value of Cronbach's Alpha and Composite Reliability. The Cronbach's Alpha value of all constructs must be equal to 0.7. Based on the results of the construct Reability based on Composite Reability, it can be done by looking at the AVE to show the magnitude of the variant indicators contained by the construct [14,12]. The big limit value is 0.5. It turns out that the AVE results in table 4 are all above 0.5. (see attachment 4). Furthermore, comparing the root AVE with the correlation between constructs where the latent variable correlation can be explained that the AVE Root for Economy is 0.657 while the correlation is maximally economic with the environment

The next step is to evaluate the structural model by looking at the significance of the construct relationship shown by the t value by looking at the output option calculate Bootstrapping in table 5 (see attachment 5). Based on the T Statistic value, the magnitude of 1.96 is said to be valid because it has a small P value of 0.05. It turns out that there is a social field to the environment that is invalid or insignificant because of the large P value of 0.05. Finally, based on the Path Coefficients above, hypothesis testing can be carried out as follows;

First hypothesis; the Social field has a positive and significant effect on the economic field. The result of t value is 52.403 96 1.96 or P Value 0,000 ≤ 0.05. So it can be concluded that there is a significant influence on the social field of the economic field.

Second Hypothesis; Social field has a positive and significant effect on the Environment. The result of t value is 0.764 96 1.96 or P Value 0.447 5 0.05. So that it can be concluded that there is no significant influence on the social field on the Environment.

Third Hypothesis; Economics has a positive and significant effect on the Environment. The result of t value is 2.405 96 1.96 or P Value 0.000 ≤ 0.05. So that it can be concluded that there are significant economic effects on the environment.

Based on the results of testing the hypothesis above the relationship between constructs according to the map thought the study because the indicators are formative, the indicators affect the construct. The Social Sector has a positive and significant influence on the economic field. The result of t value is 52.403 ≥ 1.96 or P Value 0,000 ≤ 0.05. So it can be concluded that there is a significant influence on the social field of the economy. The better the socio-cultural interaction to do tomato plantations will support the production of tomatoes which ultimately improve people's welfare [5]. Effect of Organizational Culture on Motivation, Job Satisfaction and Performance of employees, especially employees in the production section. In a positive way a person's behavior will influence their performance [7].

Furthermore, the Social Sector has a positive and significant effect on the Environment. The result of t value turns out to be 0.764 ≤ 1.96 or P Value 0.447 ≥ 0.05. So that it can be concluded that there is no significant influence on the social field on the Environment. This is predicted because this area is at the foot of Mount Talang which is suitable for crops for secondary crops. Nagari Bukit Sileh is an area that is almost evenly distributed as a tomato farmer and the quality of tomato production is very good farmers also use liquid organic fertilizer in accordance with the results of research [15].

Economics has a positive and significant effect on the Environment. The result of t value is 2.405 ≥ 1.96 or P Value 0.000 ≤ 0.05. So that it can be concluded that there are significant economic effects on the environment. To improve the quality of tomato production [16], it will have an impact on the production of high yields of low tomatoes and even farmers not taking fruit from the stem, because picking costs are more expensive than the sale of tomatoes. The results of sorting from the sale of tomatoes are wasted in the field and in the streets which eventually rot bring flies that damage the environment.

5. Conclusion

The findings from the results of research on environmental studies and the planning of the development of the tomato production center in Solok Regency West Sumatera can be summarized as follows; There is a significant influence between the social sector on the economic field in planning the development of tomato production centers. In this case there is a socio-cultural interaction between tomato farmers and the community in Nagari Bukit Sileh and other Nagari as forward linkage and linkage backwash. There is no significant influence between the social field and the environment in planning the development of tomato production centers. This is due to the quality of land due to the use of fertilizers, soil erosion and landslides. However, floods rarely occur in Nagari Bukit Sileh. There is a significant influence between the economic sector on the environmental field in planning the development of tomato production centers. This together with the progress of economic development arises several problems and environmental impacts. Depleted resources, increasingly advanced technology and the emergence of various types of pollution are beginning to be considered to hamper economic progress, because they damage the environment. This is due to the excessive use of fertilizers so that the reduced species of other species including eels are almost difficult to find.

Suggestions for policies in the Solok district government to build infrastructure, increase the related industries so that there is a guaranteed market for tomatoes produced, so that economic activities are smooth. The local government should accommodate to create industrial businesses related to tomato plantations as a leading industrial sector. So that an integrated tomato production center is formed which can finally absorb labor in rural areas [14]. Besides maintaining environmental health so farmers reduce the use of fertilizer pesticides.

Appendix A:

Table 1. Variable and Indicator Names and Codes based on the Liker Scale

Variable	Indicator	Code	Likert Scale
STUDY OF THE ECONOMIC FIELD	a. Business development	EKO-1	Likert Scale 1 s/d 5
	b. Farmer's income	EKO-2	
	c. Farmer's lifestyle	EKO-3	
	d. The pattern of consumption of farmers	EKO-4	
	e. Infrastructure	EKO-5	
	f. The results of development are enjoyed by the community	EKO-6	
SOCIAL AFFAIRS STUDY	a. Social interaction between farmers	SOS-1	Likert Scale 1 s/d 5
	b. Social interaction between farmers and other Nagari communities	SOS-2	
	c. High-income product farmers	SOS-3	
	d. Courtesy and hospitality of extension staff	SOS-4	
ENVIRONMENTAL STUDY	a. Land quality due to fertilizer use.	KL-1	Likert Scale 1 s/d 5
	b. Soil erosion or landslide	KL-2	
	c. Flood	KL-3	
	d. Environmental Conservation Plants.	LK-4	
	e. Decreased species	KL-5	
	f. Health Quality	KL-6	
	g. Affected by lung disease, cancer and poisoning	KL-7 KL-8	
	h. Decreasing Water Quality	KL-9	
	i. Excessive Use of Fertilizers	KL-10	
	j. Products rot and damage the environment		

Appendix B:

Table 2. The outer loadings

Outer Loadings

Matrix

	EKONOMI	LINGKUNGAN	SOSIAL
EKO-01	0.853		
EKO-02	0.938		
EKO-03	0.928		
EKO-04	0.926		
EKO-05	0.903		
EKO-06	0.851		
KL-05		0.850	
KL-09		0.830	
SOSBUD-02			0.906
SOSBUD-1			0.943
SOSBUD-3			0.860

Appendix C:

Table 3. Outer Weights

Outer Weights

Mean, STDEV, T-Values, P-Va... | Confidence Intervals | Confidence Intervals Bias C... | Samples | Copy to

	Original Sampl...	Sample Mean (...	Standard Devia...	T Statistics (|O...	P Values
EKO-01 <- EKO...	0.178	0.177	0.007	25.160	0.000
EKO-02 <- EKO...	0.196	0.196	0.006	35.634	0.000
EKO-03 <- EKO...	0.196	0.196	0.005	36.745	0.000
EKO-04 <- EKO...	0.190	0.190	0.006	29.605	0.000

Appendix D:

Table 4. AVE, AVE Root, Composite Reability, Cronbach's Aplha dan R^2

Constructs	AVE	AVE Root	Composite Reability	Cronbach's Aplha	R^2
Economy	0,811	0,657	0,963	0,953	0,801
Environment	0,706	0,498	0,823	0,584	0,437
Social	0,817	0,667	0,931	0,887	-

Source: Data Processed 2018

Appendix E:

Table 5. Path Coefficients

Path Coefficients

Mean, STDEV, T-Values, P-Va... | Confidence Intervals | Confidence Intervals Bias C... | Samples | Copy to

	Original Sampl...	Sample Mean (...	Standard Devia...	T Statistics (|O...	P Values
EKONOMI -> L...	0.523	0.545	0.217	2.405	0.018
SOSIAL -> EKO...	0.895	0.896	0.017	52.403	0.000
SOSIAL -> LIN...	0.151	0.130	0.197	0.764	0.447

Source: Data Processed 2018

References

[1] Zul Azhar and Hasdi Aimon 2017 Model Perencanaan Pembangunan Berbasis Lingkungan Sentra Produksi Tomat Nagari Salayo Tanang Bukik Sileh Kabupaten Solok *Ikraith-Humaniora* **2**(121) 31–40

[2] Z Azhar 2012 *Teknik Perencanaan Pembangunan Berbasis Komputer, Edisi Pertama* (Padang, Indonesia: Fakultas Ekonomi UNP)

[3] Z Azhar 2017 *Kajian Lingkungan dan Perencanaan Pembangunan, Buku Ajar* (Padang, Indonesia: Fakultas Ekonomi Universitas Negeri Padang)

[4] Azhar Zul 2014 Analisis Perencanaan Anggaran Pembangunan Kabupaten / Kota Sumatera Barat Untuk Pembangunan Sumberdaya Manusia *Nasional, 2014*

[5] Z Azhar, H Aimon and Z Ardi 2018 Achieving optimal tomato production levels using the downstream of production *PROCEEDING | ICESST* 2018

[6] Wijayanto 2012 *http://newijayanto.blogspot.co.id/2012/04/pengaruh-budaya-terhadap-lingkungan.html.*

[7] Koesmono H T 2006 Pengaruh budaya organisasi terhadap motivasi dan kepuasan kerja serta kinerja karyawan pada Sub Sektor Industri Pengolahan Kayu Skala Menengah di Jawa Timur *Jurnal Manajemen dan Kewirausahaan* **7**(2) 171–188

[8] Molenaar K, Brown H, Caile S and Smith R 2002 Corporate culture: a study of firms with outstanding consideration safety *Professional Safety* **47**(7) 18-27

[9] Adams R, Jeanrenaud S, Bessant J, Denyer D and Overy P 2016 Sustainability-oriented innovation: a systematic review *International Journal of Management Reviews* **18**(2) 180–205

[10] Suddaby R, Bruton G D and Si S X 2015 Entrepreneurship through a qualitative lens: Insights on the construction and/or discovery of entrepreneurial opportunity *Journal of Business Venturing* **30**(1) 1–10

[11] Ghozali I 2014 *Structural Equation Modelling Metode Alternatif dengan Partial Least Square, edisi 4* (Semarang: Badan Penerbit Universitas Diponegoro)

[12] Haryono S 2017 *Metode SEM untuk Penelitian Manajemen AMOS LISREL PLS* (Jakarta: Luxima)

[13] Chin W W 1998 *The Partial Least Squares Approach to Structural Equation Modelling* In: G A Marcoulides (ed.) *Modern Methods for Business Research* (London: Lawrence Erlbaum Associates, p.) pp 295-236

[14] Chin W W and Newsted P R 1999 *Structural equation modeling analysis with small samples using partial least square* in Hoyle R H (ed.) 1999 *Statistical Strategies for Small Sample Research* (Sage Publication) pp 307–341

[15] Wijaya A S et al 2015 Produksi dan kualitas produksi buah tomat yang diberi berbagai konsentrasi pupuk organik cair *e-J. Agrotekbis* **3**(6) 689–696

[16] Sita B R and Hadi S 2018 Produktivitas dan faktor-faktor yang berpengaruh terhadap produksi usahatani tomat (Solanum Lycopersicum Mill) di Kabupaten Jember *JSEP (Journal Soc. Agric. Econ.* **9**(3) 67–78

Pattern of palm-based agroforestry the Bugis etnic community in the Regency of Kolaka Indonesia

W K Baka[1], I S Rianse[2*], U Rianse[2], W G Abdullah[2] and Zulfikar[2]

[1]Faculty of Cultural Science, Halu Oleo University, Kendari, Indonesia
[2]Faculty of Agriculture, Halu Oleo University, Kendari, Indonesia

*ilma.rianse@gmail.com

Abstract. The study aimed to identify patern of palm-based agroforestry Bugis etnic community in Kolaka Regency. This research was conducted in Kolaka Regency. Respondents were determined by purposive sampling. The method data analysis used is qualitative analysis. The results showed that there were fourteen palm-based agroforestry patterns that were cultivated in a subsistence manner in Kolaka Regency. The agroforestry pattern of coconut-palm-cacao-clove is the most agroforestry pattern cultivated by community farmers (28.5%). The variation plant of agroforestry patterns is can be describe the welfare level of farmer.

1. Introduction

Native Indonesian agroforestry gardens show characteristics that deserve attention in the framework of agricultural and forestry development, especially for infertile areas. In these areas only annual crops can produce sustainably, while for food crops and other seasonal crops it is only possible through massive fertilization [1]. Meanwhile, Non-Timber Forest Products of palm-based are one of the forest products that have advantages and are most in touch with the surrounding forest communities [2].

One of the most effective models of sustainable agriculture is agroforestry, according to the conditions of farmers [1]. Agroforestry is generally considered only as a "kitchen garden" which is nothing more than a complement to other farming systems, where production is only devoted to own consumption by producing by-products. Economically the application of agroforestry contains four important things, namely: (1) competition, (2) diversified diversification, (3) economic rentability, and (4) sustainability. Associated with aspects of ecosystem sustainability, prosperity and economic prosperity, food security and resilience of the community in the face of economic crisis, it is suspected that the choice of farming by applying agroforestry patterns is the right solution [3].

Many benefits from the agroforestry cropping systems, both in terms of economic, social, and environmental. Palm and coconut plants have very good conservation benefits [4]. Therefore the pattern of agroforestry with palm-based and coconut-based is one of the main supporters of the sustainability of an agroforestry pattern. The pattern of agroforestry has become one of the sources of community livelihoods in Southeast Sulawesi.

Agroforestry patterns are developed with simple and complex agroforestry patterns. There are various patterns of agroforestry based on coconut and sugar palm which are cultivated by the Bugis Etnic community in Kolaka Regency. This is because besides the conservation benefits of coconut and palm sugar, it is also due to the economic value of palm and coconut sap. This agroforestry pattern has become part of Bugis people's lives. Thus, this research is important to identify patterns of palm-based agroforestry carried out by the community.

2. Theoritical Background

2.1. Social Culture of Agroforestry

Decision making of farmers in agroforestry exploitation is not always based on financial considerations, or in other hand financial considerations are not always the number one aspect of decision making, but there are more dominant socio-cultural aspects. Although farmers have the highest income from one commodity, farmers do not cultivate the commodity monoculture which is actually more profitable. One of socio-cultural factor that encourage people to develop the agroforestry, if someone can bequeath on agroforestry farm to their children and grandchildren, than the agroforestry are the sense of pride for them, so the agroforestry has bequeath value [5]

Feeding long-term needs is one of the reasons farmer plant trees. Tree production that can be taken continuously is very suitable as a 'old age guarantee plant'. The existence of these plants makes them more confident, because they will not depend on others in their old age. Considering the decreasing power and physical strength, they tend to choose annual plants that do not require intensive and heavy maintenance, but provide continuous income [5]

Systems of land utilization that are applied individually must be in harmony with local culture and vision of people's on their location and relationship with nature. The land utilization landscape and its development form are part of the identity of the people who live in there. Farmers usually have a strong need to side with local culture. History and tradition play an important role in the life, ways and systems of their land utilization. Improvement which doesn't sinergy with their social, cultural, and spiritual values can create stress and power be at cross purpose [6].

3. Method of the study

The study was conducted in the regency of kolaka. Kolaka is well know by its agroforestry based on coconut and Palm sugar. Respondents were determined by purposive sampling. Number of respondents are as much 30 people. In the study, qualitative analysis was conducted to describe pattern of palm-based agroforestry the bugis etnic community in Kolaka Regency.

4. Results and Discussion

4.1. The Characteristics of Bugis Ethnic Farmers

Age of etnic bugis farmers who applied pattern of agroforestry in this was dominated by farmers in productive (83,3 persen). In their age, we assumsed farmers could make the agroforestry production on maximum level. Farmers whose were in productive age criteria will be willing to keep running the palm agroforestry business, because only requires their energy while the raw materials and auxiliary materials were still available in nature around their residence, abundantly. Good education is needed for existence and sustainability a business [7]. Farmers who non productive level keep their agroforestry farming, because its give continous income and has bequeath value for their children.

Table 1. Distribution of Characteristics Agroforestry Bugis Etnic Farmers

No.	Age (years)	Number (People)	Percent (%)
1	15-54 (Productive)	25	83,3
2	> 54 (Non Productive)	5	16,7
	Total	11	100,0
	Average	45,6	
No	Education Level		

1	Elementary School	3	10
2	Junior High School	3	10
3	Senior High School	19	70
4	Bachelor Degree	3	10
	Total	30	100,0
	Average	SLTA	
No.	House Conditicion		
1	Emergency	3	10,0
2	Semi Permanent	19	63,3
3	Permanent	8	26,7
	Total	30	100,0
	Average	Semi permanent	
No.	Land Area (ha)		
1	< 1	11	33,7
2	1 -1,50	10	33,3
3	> 1,51	9	30
	Total	30	100,0
	Average	1,08	
No	Land of Right Status		
1	Owner	30	100,0
2	Rent	0	0,0
	Total	30	100,0
	Average`	Owner	

Education level is one of supporting factor for the succses of farmers in doing thier bussines, cause it influance farmers on make decisions to adopted the inovation. We founded that education level of farmers was dominated by farmer at Senior High School Level (70 percent). The low education level of respondents caused a lack of knowledge and the application of management concepts in the practice of agroforestry business. Next, their business scale was hardly ever increase since the beginning do [7].

4.2. Patterns of Palm-Based Agroforestry The Bugis Etnic

Agroforestry has several forms in Indonesia, such as home garden (*pekarangan*) and traditional garden (*kebun*). Home gardens to some degree resembles natural forest in structure and species diversity [8]. It is encouraging that many villagers maintain their traditional home gardens and gardens. Those traditional agricultural lands as well as forest still provide high diversity of plants to rural community for many purposes [9].

The agroforestry system has long been carried out by farming Bugis Etnic communities in Kolaka Regency. In practice they did not realize that the farming practices they had carried out were an agroforestry system. The practice of agroforestry is subsistence, so that there are no significant efforts to improve the quality and quantity of products. The agroforestry system in Kolaka Regency is only limited to the diversification of long-term crops in one land area. Farmers only do it as a routine in the techniques of cultivation, harvesting, and marketing that are often without the processing of crops. Farmers also only use the main plant parts, namely the fruit from coconut, cacao, clove, *durian*, and lansium, as well as sap from the palm tree. However, farmers have not made much use of other plant parts, such as palm sugar has many benefits (leaves, stems, roots, waste from the main product) [6], both for the needs of their families especially for commercialization. Thus, the social, economic and environmental benefits of an agroforestry system have not yet been fully realized.

The patterns of agroforestry found in the communities in Kolaka Regency are very diverse. Based on the biophysical conditions of the land, coconut and or palm sugar grow in almost every plantation area of the farming community in Kolaka Regency. Some of the identified agroforestry patterns can be seen in table 2.

Table 2 shows that there are fourteen agroforestry patterns based on the combination of types of crops cultivated in each respondent's land area. Coconut and palm sugar plants are quite dominant together in a land area. It is shown that there are eight (8) agroforestry patterns, two of which are coconut and palm sugar.

Table 2. Pattern of coconut and palm sugar based in the Regency of Kolaka, 2016

No	Pattern of agro foresrtry	Total Respodents (%)	Averages of plants composition (%)	Averages of land hectare (Ha)	Averages of number of trees per Ha
1	Coconut-Palm sugar	7.14	60 : 40	0.75	133
2	Coconut-Palm sugar-Cacao	17.86	20 : 15 : 65	1.10	799
3	Coconut-Palm sugar-Cacao-Clove	28.57	4 : 2 : 74 : 20	1.33	1.015
4	Coconut-Palm sugar-Clove-*Durian-Lansium*	3.57	2 : 11 : 22 : 55 : 3 : 7	1.50	303
5	Coconut-Clove-*Durian*-Lansium	7.14	6 : 45 : 30 : 13 : 6	2.00	104
6	Coconut-Cacao-Clove	7.14	1 : 86 : 13	2.50	729
7	Coconut-Palm sugar-Cacao-Lansium	3.57	1 : 1 : 49 : 49	0.50	2.030
8	Coconut-Palm sugar-Cacao-Clove-Lansium	3.57	1 : 1 : 94 : 2 : 2	0.50	936
9	Coconut-Palm sugar-*Durian-Lansium*	3.57	20 : 12 : 34 : 34	1.00	148
10	Palm sugar-Cacao-*Durian*-Lansium	3.57	10 : 75 : 5 : 10	0.24	1.667
11	Coconut-Palm-Cacao-*Durian*-Lansium	3.57	1 : 1 : 94 : 2 : 2	1.00	1.012
12	Palm sugar-Lansium	3.57	88 : 12	3.00	57
13	Coconut-Kakao-Cengkeh-*Durian*	3.57	4 : 70 : 23 : 3	1.00	356
14	Palm sugar-Kakao	3.57	2 : 98	0.50	820

Based on table 2, it is further known that the combination of palm sugar without coconut is more common in agroforestry patterns, which are as many as five (5) patterns, while the combination of coconut without sugar palm is only one (1) pattern. This means that sugar palm plants are very potential to develop in Kolaka Regency, although they have not been cultivated. The diversity of plant species serves many purposes for community, especially those living in rural areas. Each traditional community may have specific knowledge and use of plants found in their environment. Having many plant species and tribes, Indonesia is rich in biological and cultural diversity [9].

The coconut-based and palm-based agroforestry patterns which aplied Bugis etnic farmers that are dominantly cultivated by respondents are the third pattern, namely Coconut-Palm sugar-Cacao-Clove (28.57%). The four types of plants are planted with irregular distances, are spots. The different thing about sugar palm plants is because these plants are not planted intentionally but grow on their own and then are allowed to grow large and then are used by sap and or palm fiber, leaves, stems. In this third pattern, the main crops are cacao, which is 74% of the total plant. Even so, the results of the clove plant provide the greatest benefit compared to coconut, palm sugar, or cocoa. Cloves have very high economic value, reaching IDR 90,000 per kg. Even, clove give farmer the highest economic value than palm trees but they still keep the palm trees because it give them continous income each day.

The function of traditional garden (forests of community) which was previously conservation, now functions as production forests in the form of wood and non-timber. The average traditional garden is in the form of mixed forests or seasonal forests [10], such as palm sugar that functions as source of palm sugar raw materials (sap) and than provide environmental benefits in favor of soil and water conservation, carbon absorption, flood protection, and water transportation [11].

The average number of trees per ha is still very varied among respondents, with very wide variations. The average number of trees per hectare in pattern 12 is only 57 trees while in pattern 7 it reaches 2,030 trees. This is due to the insignificance of the distance of the plants, especially because of the sugar palm trees that grow wild so without the spacing arrangement. Planting trees is determined by the level of wealth (according to local size) and land status. The number of poor households (controlling narrow land) that plant fewer trees than rich households, as well as the number of trees planted by poor households is less than the number of rich household trees (controlling large areas). Poor households that control narrow land are more likely to use their land for food crops or trade crops than for tree crops [12].

5. Conclusions and Suggestions

5.1. Conclusions

Based on the results and discussion, it can be concluded that:

- We founded there are 14 (fourteen) pattern of palm-based agroforestry Bugis etnic in Kolaka Regency. The 3 pattern of agroforestry (coconut - palm sugar – cacao - clove) were cultivated by farmers more than 25 percent
- The variation plant of agroforestry patterns is can be describe the welfare level of farmer.

5.2. Suggestions

Things that can be suggested from the results of this study include: so that farmers always maintain sugar palm and or coconut plants and even cultivate them better in their agroforestry systems because of the large environmental, economic and social benefits of these crops. Optimizing the economic benefits of agroforestry systems based on coconut and palm sugar crops can be done by planting more cloves, as well as adding more land. Furthermore, it is recommended that the government, private sector, and tertiary institutions collaborate to assist and facilitate farmers in promoting high economic and financially sound agroforestry systems

Refferences

[1] Rianse U and Abdi 2010 Agroforestri: *Solusi Sosial dan Ekonomi Pengelolaan Sumber Daya Hutan* (Bandung : Alfabeta)

[2] Hutapea TOA, Rahmawaty and Irawati A 2016 Potensi Hasil Hutan Non Kayu Kelompok Palmae Berbasis Sistem Informasi Geografis (Sig) Di Kecamatan Lumban Julu KPHL Model Unit XIV Tobasa *Peronema Forestry Science Journal* **5**(1)

[3] Sanchez PA 1995 Science in Agroforestri (Nairobi: P.O. Box 3067)

[4] Abdullah WG, Rianse U, Iswandi M, Taridala SAA, Rianse IS, Zulfikar LZ, Baka LR, La Abdi A, Cahyono E, Widayati W and Baka WK 2015 Potency of Natural Sweetener: Brown sugar *WSEAS Transaction on on Business and Economics* **12** 99-110

[5] Suharjito D, Sundawati L and Utami SR 2003 *Aspek Sosial Ekonomi dan Budaya Agroforestri* (Bogor: World Agroforestry Centre (Icraf))

[6] Reijntjes C, Haverkort B and Waters-Bayer A 1992 *Pertanian Masa Depan: Pengantar untuk pertanian berkelanjutan dengan input luar rendah* (Yogyakarta: Terjemahan Penerbit Kanisius)

[7] Abdullah WG, Rianse U, Muhidin, Widayati W, Mihrad ES, Taridala SAA, Rianse IS and Baka WK 2018 Farmer's Motivation in Aren Sugar Processing Business. *International*

Conference on Agriculture, Environment, and Food Security **122**(1) 012007 doi :10.1088/1755-1315/122/1/012007

[8] Soemarwoto O. 1983. *Ekologi, Lingkungan hidup dan Pembangunan* (Jakarta : Djambatan)

[9] Wiryono and Lipranto 2013 The Diversity of Locally Useful Plants in Batu Ampar Village Near Bukit Raja Mandara Protected Forest Area in South Bengkulu District. *Jurnal Manusia dan Lingkungan* **20** (2) 119-128

[10] Setiaji, Sadono R, Hartono dan Machfoedz MM 2016 Analisis Perkembangan Wilayah, Pemusatan Perubahan Penggunaan Lahan Hutan dan Kaitannya dengan Kehidupan Sosial Budaya Masyarakat Merapi *Jurnal Manusia dan Lingkungan* **23** (1) 92-101

[11] Rianse IS, Abdullah WG, Midi LO, Hartono S, Suryantini A, Jamhari and Widayati W 2016 Financial, Economic and Environmental Feasibility Analysis of Palm Sugar Domestic Industry in Kolaka Indonesia *International Journal of Economics and Management Systems* **1** : 104-109.

[12] Brokensha D and Riley BW 1987 *Privatization of Land and Tree Planting in Mbeere, Kenya. In Raintree JB* (ed). *Land, Trees and Tenure* 187-192 (Nairobi and Madison: ICRAF and The Land Tenure Center).

IOP Conf. Series: Earth and Environmental Science **314** (2019) 012042 doi:10.1088/1755-1315/314/1/012042

Pukung pahewan: The effort of natural resources conservation in Dayak Ngaju community

Bulkani[1,*], Ilham[2] and Saiffullah Darlan[3]

[1] Universitas Muhammadiyah Palangkaraya, Palangka Raya, Indonesia
[2] Universitas Muhammadiyah Palangkaraya, Palangka Raya, Indonesia
[3] Universitas Palangka Raya, Palangka Raya, Indonesia

*bulkaniardiansyah@gmail.com

Abstract. In this article, I will discuss about *Pukung Pahewan*. It is a name of the prohibition area with certain terms and extents. *Pukung Pahewan* comes from the Ngaju Dayak language. *Pukung* means island or region, while *pahewan* means prohibition. The determination of *Pukung Pahewan* refers to certain objects such as trees, stones, and other objects that are sacred to a particular region, which should not be disturbed or destroyed, including the surrounding area. Moreover, *Pukung Pahewan* is determined by customary agreement with certain characteristics, i.e. the *Diwung/Handiwung* tree, *Lunuk*/Banyan tree or other large trees, the presence of sacred, and sometimes required by the Bird's Nest/*Antang*. The concept of *Pukung Pahewan* by the Dayak Ngaju community in many areas of Central Kalimantan proves to eliminate the notion that a shifting cultivation system can damage natural resources.

1. Introduction

Pukung Pahewan is a name of prohibited area in Central Kalimantan. The name of *Pukung Pahewan* comes from Dayak Ngaju language, *Pukung* means Island (*pulau*) and *Pahewan* means prohibition (*larangan*). *Pukung Pahewan* is determined based on customary agreements with certain characteristics, including the presence of *Diwung/Handiwung* trees, *Lunuk/Beringin* trees or other large trees, the existence of *Keramat* and sometimes the existence of eagle / *Antang* bird nests. In the custom of Dayak Ngaju community who inhabit the Central Kalimantan, the establishment of the *Pukung Pahewan* area is usually in conjunction with the determination of the area for settlement and farming. At the time of searching for a new region more fertile for farming, the Dayak community always divide their territory into three zones, namely residential areas, cultivated areas, and *Pukung Pahewan*. The determination of the *Pukung Pahewan* area is intended as a place to relocate spirits that inhabit the forest, who's their residence has been made into settlements and fields. The spirits are moved to the area *Pukung Pahewan* with certain ceremonies and the area should not be disturbed by the community as well as sacred.

Pukung Pahewan is a form of local wisdom of Dayak community, especially on Dayak Ngaju to conserve nature and the environment. By the establishing of the *Pukung Pahewan* area, which is basically the implementation of the concept of zoning and relocation, the flora and fauna in the area are not disturbed by their habitat because they have been moved. The *Pukung Pahewan* concept applied by the Dayak Ngaju community in many regions of Central Kalimantan has proven to eliminate the assumption that a shifting cultivation system can damage natural resources. When the forest is cleared for fields and settlements, there are still certain areas that are left as a prohibited area and sacred as a place of relocation. Furthermore, after the community moves to other more fertile areas, the fields and settlements that are left will become forest again, so that the flora and fauna that have been relocated in the *Pukung Pahewan* area and will get a wider area to develop again.

2. *Pukung Pahewan* and the Distribution of Resources

2.1. The pattern of division of space between humans and other organisms in the tradition of the Dayak community

The culture and traditions of the Dayak community cannot be separated from the context of the beliefs that they held i.e. *Kaharingan* belief or religion, or rather the *Hindu Kaharingan* religion. Most of the traditions of the Dayak community, derived from traditions in the *Hindu Kaharingan* e.g. variety of traditional rituals, clothes, and habits and abstinence or *pali*. Moreover, its weapon, has a religious meaning that is closely related to *Kaharingan.*

Dayak culture was originally based on *Kaharingan* culture that has known and encompasses the concepts of human relations with nature, and human relations with God called *Ranying Hatalla.* Furthermore, as a culture, *Kaharingan* has encompassed cultural system boundaries as stated by anthropologists Clyde and Florence Kluckhohn [1], i.e:

- Conception of the nature of human life
- Conception of the nature of human work
- Conception of human position in time and space
- Conception about human relations with the surrounding environment
- Conception of human relations with each other

According to Lewis KDR, a *Hindu Kaharingan* religious figure, the *Kaharingan* people had known the harmony and balance system since the beginning which was implanted through the three pillars supporting the life or existence and belief system called *Garing Hatungku Tungket Langit*. The three supporting pillars consist of *Ranying Hatalla* which represents the concept of divinity, *Kayu Erang Tingang* represents the concept of life with custom (*belom bahadat*), and *Kayu Kumpang Saribu* represents the concept of human intelligence. In the concept of *Hindu Kaharingan* religion, to achieve peace and prosperity of life in the world and the hereafter *Saran Danum Sangiang*, humans must be devoted to God, respect and uphold customs and live pursuing comprehensive intelligence (*harati*). Thus, the three concepts above must be carried out in a balanced and harmonious manner.

In the belief of the Dayak community, the universe consists of the upper world, and the underworld. The upper world, inhabited by *Ranying Hatalla*, who is the highest ruler. The rulers of nature, who argues by Hermogenes Ugang in Nila Riwut [2] called as God. This concept of God is similar to the divine concept in other religious systems and beliefs in the world. *Ranying Hatalla*, living in the sky, so it is often referred to as *Ranying Hatalla Langit*. In addition, the word of *langit* describes its position in the upper realm.

Ranying Hatalla lives in the upper realm called *Sangum Saran Danum*. This place is described as a place that is full of balance, peace, calm and prosperity, and has a harmonization of relationships between various elements and elements of nature, where the inhabitants share space and time in an interactive unity. This concept is similar to the concept of heaven or nirvana in other religions and beliefs in the world. The harmony and peace like heaven which *Ranying Hatalla* duplicates to the surface of the earth. Consequently, *Ranying Hatalla* send the first man, named *Raja Bunu* (King Bunu), to come down to earth.

Raja Bunu (King Bunu) and his family descended to the surface of the earth, which represented the underworld and called the *Saran Danum Kalunen*, bringing the main mission is to invite people to obey *Ranying Hatalla* to form a good and balanced community so that the harmonization of relations between humanity and nature is well preserved as happened to the world above. The ideal situation in the upper world, *Saran Danum Sangiang*, should be able to be replicated and recreated on the surface of the earth. Furthermore, it is this philosophy that underlies the purpose of sending *Raja Bunu* (King Bunu) to earth, as the concept of sending Adam in the belief of the major religions in the world.

According to Nila Riwut [2], *Ranying Hatalla* has provided *Raja Bunu* (King Bunu) with a variety of rules and procedures for life, including procedures for humans to return to the upper realms or *Saran Danum Sangiang*. This concept also has many similarities with the concepts of the major religions in the world, namely in time when humans will return to the place where humans originally came from. To be able to return to *Saran Danum Sangiang* which everlasting. Therefore, humans must obey all the rules that have been set. Humans can also feel the pleasure of living in the underworld, *Saran Danum Kalunen*, if they are able to maintain harmony and balance of nature as happened in the *Saran Danum Sangiang*.

Thus, in the belief of the Dayak community, the earth and all its living systems must reflect the ideal system in the world above, which mean, if the system in the world is believed to be a well ordered and harmonious system, then the system on earth must be like that.

Furthermore, humans and the environment is something that cannot be separated and each others, e.g Humans who destroy the environment will effect to what they have done. Conversely, if humans treat the environment very well then it will get some good benefits for themselves. According to Law [3] which states that the environment is a unity of space with all objects, the power of circumstances and living things. In short, environmental science is called the biotic component and the abiotic component. Rachmad [4] also stated that humans are unique and different from other creatures because they get cultural heritage. Thus, Dayak community is having a culture to take care of the environment, namely its manifestation by protecting the environment.

This replication of philosophy, then underlies and influences the entire life system of the Dayak community. The emergence of a ban, *pali*, or *pamali* is the implementation of that philosophy, because the ban function is to maintain the balance and harmony of nature, so that the earth is maintained in harmony as harmony in the *Saran Danum Sangiang*.

According to Nila Riwut [2], the main mission of *Kaharingan* is to invite people to the right path by worshiping and glorying *Ranying Hatalla* in every attitude and deed, including the task of maintaining harmony in the universe.

The concept of replicating the situation in the *Saran Danum Sangiang*, which is the upper world also influences the pattern of the division of space between humans and other organisms in the Dayak community, because that is the situation in the *Saran Danum Kalunen*. The pattern of space distribution must refer to the harmony of relations, which by Tjilik Riwut [5], it can be divided into 3 types of relations, namely the relationship of the Dayaks with *Ranying Hatalla*, human relations with humans, and human relations with the surrounding environment.

Maintaining the balance of the relationship means maintaining the cosmic balance. It is also reflected in the balanced relationship between nature above the underworld, *Ranying Hatalla* who lives in the upper realms, and *Jata* who live in the lower realms, and the balance of human relations with nature.

Thus, for Dayak community, the obligation to share space with fellow human beings and other organisms in the natural environment is not just a cultural message, but rather a religious message, in order to maintain the quality of the relationship with *Ranying Hatalla*.

The way the Dayak community shares space with other organisms in the surrounding environment can be seen in various forms e.g. the existence of *pahewan* which are prohibited areas with certain conditions and areas. More specifically, the stipulation of *pukung pahewan* which referred to certain objects such as trees, stones, and other sacred objects, which must not be disturbed or damaged, including the surrounding area. It is illustrated the ability of the Dayak community to share space with other organisms in the natural surroundings.

In several places in Central Kalimantan, we can see the abilities of the Dayak community to maintain harmony with nature which symbolizes the concept of willingness to share space with other organisms that live in nature e.g. the placement of yellow flag and *ancak* or places of offerings on trees that grow on the outskirts of a lake or river. Therefore, the surrounding area will look haunted.

The haunted impression around the area will cause the community is avoiding the area. By the avoidance of the area from interaction of humans, it will give more space to the flora and fauna around the area to develop better.

3. *Pukung Pahewan* among human habitation needs

Central Kalimantan is the second largest province after Papua. With land that is still very large and a relatively smaller population compared to other regions in Indonesia, the relative need for development of land for housing does not experience problems in the past. However, with the increasing economic needs of the community, the need for additional land is not only aimed at expanding the land for housing, but more towards meeting economic needs, for example for farming, gardening, farming and business land. This causes the growth of opportunities for conflict between people and nature. The chances of this conflict are getting bigger in line with the rapid growth of oil palm plantations and the mining sector.

Consequently, as a result of the growing need for land for this economic sector, the area that was slowly allocated for *pukung pahewan* also experienced a reduction. In fact, the existing of *pukung pahewan* many of them were lost and displaced by land expansion e.g. the presence of *pukung pahewan* in the *Terusan* river area which is a tributary of *Kahayan* River, Banama Tingang sub-district, in Pulang Pisau district. Additionally, according to information from traditional leaders around the area, in the past there was a number of *pukung pahewan*, both stone and *Handiwung* trees, but since the year of 2000, the *pukung pahewan* has been lost because it was encroached on by the expansion of community-owned gardens and the presence of traditional mining.

Actually, the existence of *pukung pahewan* in the Dayak community was originally a direct or indirect local wisdom, aimed at preserving nature. With the existence of certain areas that are used as *pukung pahewan*, the area and its surroundings will be preserved. Thus, the placement of *pukung pahewan* can also be referred to as the Dayak community's efforts to share places with other organisms in an area. *pukung pahewan* positions are generally outside residential areas, not only because the *pukung pahewan* area is considered a haunted area and there is a waiting area, but also proves that the Dayak community in the past has known the concept of zoning.

The zoning system can be seen from the separation between the areas designated as safe zones for residence and farming, and other zones that cannot be penetrated for settlement and farming. Certain areas are haunted, generally because there are *Handiwung* trees and large trees such as *Beringin* trees which are considered inhabited by spirits, then set as *pukung pahewan*. The term "set" must not be formally interpreted, because in reality, the set of an area as a *pukung pahewan* is more of an informal consensus. While, other areas that are not considered haunted, can be used by the community for shelter and farming.

The *pukung pahewan* area is generally also established almost simultaneously when people open a new field or village. The opening of new fields and villages certainly requires a place, thus disrupting the balance of the ecosystem that already exists in the region. For this reason, the *pukung pahewan* area was established, which was intended as a place to conserve or protect new flora and fauna around the newly opened area. This sorting concept is a concept of zoning, namely where zones are for farming, and where zones are for nature conservation.

The choice of large trees as the *pukung pahewan* which gives meaning that the survival of the flora will be guaranteed, because large trees also produce seeds that are ready to grow again. Large trees in the *pukung pahewan* area also become fauna sanctuaries because their habitat has been used for farming or shelter.

The existence of prohibitions or *pali*, which is then continued by speaking and acting to posterity. The stories about the awesomeness of *pukung pahewan* are passed on from generation to generation, so the existence of *pukung pahewan* in a period of time is maintained. It helps to preserve the surrounding area of *pukung pahewan*. Sustainable conservation concepts are inadvertently inherited through the existence of *pukung pahewan*. Moreover, until the year 2000, we could see the existence of *pukung pahewan* in many places, which was marked by the large number of large trees still standing around the area. After the year 2000, in line with the increasing need for land for economic activity, and the changing concept and perception of the public about the meaning of awesomeness in a place, then the existence of *pukung pahewan* slowly disappeared.

4. Respect for the rights of other organisms in *Pukung Pahewan*

As explained above, *pukung pahewan* is a form of local wisdom of the Dayak community to conserve natural resources, one of which is through a zoning system. The existence of *pukung pahewan* is an effort to relocate evicted organisms due to increased land requirements for settlements and fields. Farming with a rotating system, requires a temporary relocation site for flora and fauna. Thus, that its sustainability is maintained. For this reason, certain zones are needed to relocate. When the field is abandoned to move to another location, the flora and fauna that were originally relocated in the *pukung pahewan* area, they can occupy the former fields that have become forest again.

The existence of restrictions for the community to disrupt and damage the flora and fauna in the *pukung pahewan* area is also evidence of respect for the rights of organisms. This is closely related to the Dayak community's belief system, that every organism has a spirit, has a soul, and has something to do with the supernatural. There are still many Dayak people who believe that the prey with certain types i.e. deer or *Karahau*, are often the incarnation of spirits so people must be careful before hunting them, which mean, don't kill animals that are actually spirits, especially animals that are in the vicinity of the sacred area like *pukung pahewan*.

The abundance of natural resources in the past with a small number of human populations, causes people to use natural resources wisely and sustainably. Based on parents' philosophy, first is not to use natural resources excessively, and take them as needed. This habit is evidence of respect for other organisms in the natural surroundings. Sometimes, due to considerations of access, the use of natural resources around the village and fields has increased, which has resulted in excessive utilization, the *pukung pahewan* concept is an alternative solution that is smart enough to preserve nature.

5. *Pukung Pahewan* as part of the Zoning system

5.1. Determination of the area of Pukung Pahewan

As explained in the previous section, *pukung pahewan* can be interpreted as a sacred area, or a prohibited area because the place is considered sacred, haunted, and inhabited by spirits. *Pukung*, which in Dayak Ngaju language means island or archipelago, and *Pahewan* which means the area of prohibition.

The term of *pukung* in *pukung pahewan* is actually an assertiveness word, because it also means *Purak* which means island, thick, lots of dense forests. Word of *Pukung* or island in the context of language in Central and South Kalimantan, more towards the plural meaning, for example, if it is called a rubber island, it means the number of rubber trees that grow in the region.

Thus, *pukung pahewan* can also be interpreted as a place where there are many trees, objects, or similar organisms in certain areas, which are forbidden to be disturbed. Therefore, one of the characteristics of the *pukung pahewan* area is the large number of trees that live around it. Furthermore, if when we pass in an area, we see that there are many large trees that collect in a place, while in other places it is not like that, it can be assumed that the area is *pukung pahewan*.

Determination of an area as *pukung pahewan* does not follow a definite indicator, either from the terms, characteristics, or the determination process. In general, the main characteristics of the area that can be determined as *pukung pahewan* include:

a. There are banyan trees or other large trees that are considered haunted, such as Durian trees, Kasturi trees, and so on. It is based on the belief of the Kaharingan community that large trees have a watchman or are inhabited by spirits. According to Nila Riwut [2] in the Kaharingan belief, there are gods (*ilah-ilah*) who are the subordinates of *Ranying Hatalla* and *Jata*, which are located between the heavens and the earth, and there are also gods that roam the earth and under the earth. These gods included inhabiting large trees and other places that were sacred.

Thus, the existence of a banyan tree or other large tree as a condition of *pukung pahewan's* location, is more of a physical requirement, behind that condition there is a belief in the existence of a guardian god of the trees. In ancient times, large trees like this were also used as a place to hang of *Raung*, it is for *Kaharingan* coffin. Sometimes, there is a lot of *Raung* hanging on the branches of large trees. This of course will add to the magical atmosphere of the surrounding area.

b. There are *Handiwung* wood. This tree is a kind of palm or palm tree with the Latin name *Oscosperma tigillarium*. The tree usually grows wild, grows clumped like bamboo, with 5-30 tillers. Tree trunk height can reach 30 meters, straight and thorny, with stems up to 20 cm in diameter. This tree also has spines on the leaves and stems. This tree trunk is hard, and can last hundreds of years. This tree grows on marshy land, especially in peat swamp areas.
   In some areas, for example in Riau, this *Nibung* or *Handiwung* tree is considered a symbol of unity, because of its strength.
   For Dayak community, especially inland and coastal communities, *Handiwung* is a sacred tree because it is considered capable of driving away various black magic, because it has a hard texture. The wood from *Handiwung* can also be used as arrows, spears, or small knives. There is trust in the Dayak community, that a person's immunity to a sharp weapon can be destroyed if he uses weapons from the *Handiwung* wood. The shamans and warlords of the past always kept weapons made of *Handiwun* trees.
   Thus, *Handiwung's* position is quite important in the culture of the Dayak community, consequently, its existence is sacred, and is used as an indicator of the need for the establishment of *pukung pahewan* in its growing areas. On the other hand, the entry of the *Handiwung* tree into the *pukung pahewan* area is an effort to preserve its sustainability.

c. There is a *Tajahan Antang*. In the Indonesian Dayak Dictionary, the word of *manajah* means to call, while *tajah* means call. While *antang* means a bird in the air, which is always connoted as a powerful eagle and believed to come from an invisible human. This eagle usually gives a special sign when *manajah antang* ceremony is held. The Dayak community generally perform certain ceremonies to ask for instructions about something, and use the arrival of the *antang* as a messenger from *Ranying Hatalla*. The ceremony is called *manajah antang*, and the results, place, or mark obtained are called *Tajahan Antang*.
   In general, when going to determine the location of a field or village, *manajah antang* ceremony is performed with certain rituals. At the time of the ceremony, there will be eagles or *antang* circling in a location, therefore, that is the location chosen as a farm (*malan*) or village.
   Thus, the word *Tajahan Antang* in the context of the *pukung pahewan* location, it can be interpreted as a specific place designated or given by a magical eagle, which is worthy of being used as a *pukung pahewan*. Sometimes, in the *pukung pahewan* area, or the place that will be used as *pukung pahewan*, there is an eagle nest. The eagle's nest can also be referred to as *Tajahan Antang* and a sign that the area is worthy of being used as *pukung pahewan*.
   The presence of *Tajahan Antang* in the physical sense, namely the presence of eagle nests around the *pukung pahewan* area is also marked by the existence of various living creatures around it. The Dayak community believe, if at any place there is *Tajahan Antang*, then the area is fertile, can be overgrown by large trees, and inhabited by various types of animals from worms, snakes, to bears.

d. There is a *Karamat* (sacred*)* or *Karamat* place (sacred place). *Karamat* is a sign that the area is sacred, generally marked as a small house where offerings are offered.
   Karamat locations are generally located in dense but haunted forests that are close to the settlement area, making it easier for residents to provide offerings. In order to preserve the *Karamat* location, a ban on cutting or destroying trees, hunting, and even utilizing and collecting forest products in the vicinity was made.
   The existence of *Karamat* is not something absolute for the terms or characteristics of *Pukung Pahewan*, because sometimes the *Karamat* is not yet available when a place is designated as

Pukung Pahewan, it is mean, *Karamat* is sometimes built after a place is designated as *Pukung Pahewan*.

In other circumstances, it is often found that *Karamat* is already in place when people will open fields. The sacred place is the place where the previous people put up offerings, then left in line with the move of the village or the move of the people to another place.

This is as an evidence that the Dayak community has known how to farm with a moving rotation pattern. Old and abandoned fields will grow new vegetation, including fruit trees. This old field will return to forest, which the next generation finds and cultivates as a field. In such locations, old *Karamat* is often found.

Based on the characteristics above, the *Pukung Pahewan* area is set. Determination of *Pukung Pahewan* is based on the agreement of traditional leaders and the surrounding community because of the mystical experienced or the characteristics and conditions for determining the location as *Pukung Pahewan* have been fulfilled. Thus, the area of *Pukung Pahewan* also does not have a definite standard. *Pukung Pahewan* area can cover a large area if the area has the characteristics above e.g. in a large area there are many large and old trees that are considered sacred or inhabited by spirits. On the other hand, the area of *Pukung Pahewan* can also only cover a few large trees which are sacred.

However, the conditions above are not binding and absolute conditions for the establishment of *Pukung Pahewan* locations. In some cases, the determination of the *Pukung Pahewan* area is only based on phenomena, experiences, and events experienced by the community. Therefore, it's more nuanced of trust or mutual trust are experienced by each other e.g. if a person or several residents have experienced mystical events in a location, then they agree to establish the location as *Pukung Pahewan*.

Determination of an area as *Pukung Pahewan* spread by word of mouth and some mystical stories related to the region by people. In the end, the *Pukung Pahewan* area became a kind of prohibited area. Sometimes, in the *Pukung Pahewan* area it is also marked by the presence of ceremonial attributes such as *ancak* or offerings that are hung on tree branches which can add to the magical and haunted atmosphere around the area. Consequently, the *Pukung Pahewan* area is always preserved by its flora and fauna and it is characteristic of the area. The existence of large and tall trees, the growth of *Handiwung* trees, and the presence of several ceremonial attributes, the presence of *tajahan antang* and *karamat* are characteristic of the *Pukung Pahewan* area.

5.2. The position of Pukung Pahewan in customary forest areas

In 2012 Constitutional Court ruling, changing the definition of customary forests, as stated in article 1 paragraph 6 changes in Law No. 41/1999 concerning Forestry, which states that customary forests are forests within the territory of customary law communities. Customary forests refer to an area where a group of indigenous people live, in which the forest functions primarily to sustain the lives of indigenous peoples around the forest.

Customary forests are part of social forestry. According to Ministry of Environment and Forestry Regulation Number [6], it is affirmed that social forestry is a sustainable forest management system implemented in state forest areas or customary forest / forest rights carried out by local communities or customary law communities.

Furthermore, judging from the conditions an area can be referred to as a customary forest if there is an indigenous area which is mostly forest. In addition, there are customary law communities or customary rights in the area, which according to Ministry of Environment and Forestry Regulation [7], the *ulayat* rights have been recognized by the local government through regional legal products.

In Central Kalimantan Province, there are many areas that are recognized as customary forests by local indigenous people, the majority of whom are Dayaks. Examples are the *Tabalien* forest or *Ulin* wood in the Mungku Baru area, around the border of Palangka Raya city with the area of Gunung Mas Regency. It is covering an area of around 500 hectares. Another example, the Tumbang Bahanei customary forest in the West Rungan District, Gunung Mas Regency, which covers 8,900 hectares.

Conversely, community rights to customary forests are still formally recognized by the government. The Central Kalimantan Provincial Government through the Forestry Service targets to the establishment of around 1.7 million hectares of land as social forestry in 2019, although until 2018 not one hectare of customary forest areas have been established. The main obstacle is the lack of local government regulations on this matter. In addition, the lack of clarity in determining the area and customary forest area of indigenous peoples. By there is no formal recognition of customary forests in Central Kalimantan, the chances of customary forest encroachment by individuals and corporations will be even higher. It will also ultimately increase the potential for damage to the entire contents of the customary forest including the forests that are sacred and *pukung pahewan* in it.

An example is the customary forest area in Mungku Baru Village, Rakumpit Subdistrict, in Palangka Raya, which is overgrown by *Ulin* or *Tabalien* wood in Dayak Ngaju language. Untill now, the *Ulin* forest is quite awake. It can be seen from the large number of *Ulin* wood trees. The *Ulin* forest area is recognized by the community as a customary forest area. Conversely, it has not received recognition from the government as social forestry. While in the vicinity of the forest area, there are logging companies operating, thus increasing the potential for conflicts of interest between companies and indigenous peoples.

Ulin forest area in Mungku Baru can be well maintained because of the traditional prohibition to cut down *Ulin* trees in that location. According to the information from the local community, there has always been a ban on cutting *Ulin* trees in the area. There is advice or *peteh* from ancestor that who enters the customary forest without permission or destroys the forest the guardian spirit of the forest will be devoured. The people who are known to violate the prohibition also imposed customary fines in the form of *kati*. One *kati* worth Rp. 100,000. The amount of the *kati* will be adjusted to the size of the prohibition ban.

The prohibition to destroy customary forests, as exemplified above, is another form of local wisdom of the Dayak community, even though not all regions are designated or referred to as *Pukung Pahawan*, or at least set as *Pahewan* area. There are several customary forest areas that do not have *Pukung Pahewan* in the explicit meaning, but it has same prohibition meaning. Generally, customary forests contain certain locations which are prohibited areas for certain activities, although not all regions are then called *Pukung Pahewan*.

It is not all of *Pukung Pahewan* are in the customary forest area owned by the Dayak community. We can see from the general conditions for the determination of the *Pukung Pahewan* area, there is no requirement that *Pukung Pahewan* be in a customary forest. There are *Pukung Pahewan* which located in customary forests, but some are outside the customary forest area. *Pukung Pahewan* area can cover a very large area, but can also cover only a few square meters, it is depending on the type of tree or object which is sacred.

In some locations, *Pukung Pahewan* is located in a forest area on the outskirts of a lake or river. It is related to Dayak pattern and life system that cannot be kept away from rivers or lakes.

The existence of *Pukung Pahewan* on the outskirts of rivers or lakes not only can help preserve the vegetation of rivers / lakes, but can help preserve aquatic organisms around the area. It has become a habit for Dayak fishermen to avoid catching fish on the banks of rivers or lakes that are found in *Pukung Pahewan*.

5.3. The type and stand composition of vegetation for Pukung Pahewan

In terms of vegetation, around of *Pukung Pahewan* area, there is a lot of vegetation. It is judging from the conditions for determining the *Pukung Pahewan* area as stated in the previous section, the types of vegetation in the *Pukung Pahewan* area are *Handiwung* trees and large trees such as Banyan (*beringin*) trees. The *Handiwung* tree symbolizes the power to resist reinforcements and antidotes to black magic, while large trees symbolize strength as well as a place for spirits to dwell. In *Kaharingan* religion, large trees are the place of gods, both good and evil.

6. The commitment to keep *Pukung Pahewan*

6.1. The obligation of the Dayak community to Pukung Pahewan

As a place that is sacred, then the *Pukung Pahewan* must be preserved. For this reason, the Dayak people who live around the *Pukung Pahewan* area have several obligations, including:

a. Give offerings (*sajen*). The offerings were intended to the spirits in the *Pukung Pahewan* area did not disturb the people in the villages or fields around the area. Furthermore, offerings are sometimes also placed by someone as a thanksgiving for the achievement of their desires or vows.
b. Maintain the surrounding area from interference from other parties. *Pukung Pahewan* is an area that is sacred, thus, that its sanctity must be maintained from human disturbances.
c. Avoiding the *Pukung Pahewan* area when hunting, looking for wood, or looking for fish. Although *Pukung Pahewan* location is still not too far from fields and villages, the *Pukung Pahewan* area is considered a haunted area where spirits live. Because of its awesomeness, the area tends to be shunned from daily activities, unless there are special needs such as giving offerings.

6.2. The prohibition imposed on Pukung Pahewan

The prohibition that applies to a community is generally intended to maintain the preservation of tradition and custom in that community. Likewise, with the Dayak community.

As explained in the previous section, the Dayak community believes in the balance of relations in a harmonious unity of the system. The balance between the upper and lower realms, the balance of relations between fellow humans, and the balance of human and natural relations. Dayaks believe that if the balance of the relationship is damaged, a disaster will arise.

According to Tjilik Riwut [8], the Dayak people believe that natural disasters occur because of human error themselves, partly because they violate a prohibition on the place of fear, e.g. example, pahewan and other places that are considered to be residents.

The *Pali* concept was born from the belief that every organism has its watchmen. Thus, if we are breaking the *Pali* means disturbing the guardian spirit of the organism. According to Tjilik Riwut [8], Dayak people believe that everywhere there are *ilah-ilah* called by various names such as *gana* or watchmen who have their respective strengths and levels. Disturbing this *ilah* is breaking *Pali*.

Pali in *Dayak Ngaju*, or *Pamali*, which means prohibition, something that must not be violated or not allowed, if carried out would be disastrous for violators. If *Pali* is broken, it will break the balance of the various relationships in the universe. The Calamity received by *Pali* offenders, more as a rage and punishment from the rulers of the universe because humans have damaged the balance of the relationship. The implications of the *Pali* include, among others, the existence of community compliance in preserving the natural environment around it. Likewise, the *Pali* or ban that must be obeyed by the Dayak community against *Pukung Pahewan*.

There are several restrictions on *Pukung Pahewan*, namely:

a. Communities are prohibited from damaging, cutting, or utilizing wood, plants or trees in the *Pukung Pahewan* area. In some cases, limited use is still permitted under the permission of the area's traditional guard elders. In urgent circumstances, permission can also be made by asking or saying permission to spirits who are believed to be waiting for the area. An example is, when we urinate near a large tree around the *Pukung Pahewan* area, we can convey the excuse word first before.
   The Dayak community is strongly believe that every living organism and inanimate matter, such as places, villages, houses, timber trees (especially large ones), mountains, rocks, rivers and lakes, have their own waiting. All have spirits. That is, all living and inanimate objects have their respective values, have a certain role in the system of relationships in the universe, so that all must be respected and guarded.
b. Communities are also prohibited from disturbing organisms around the *Pukung Pahewan* area, for example catching or killing animals that live in the vicinity. Some Dayaks believe that

IOP Conf. Series: Earth and Environmental Science **314** (2019) 012042 doi:10.1088/1755-1315/314/1/012042

animals that live in the vicinity of the sacred area, including the *Pukung Pahewan* area are the incarnation of the region's spirits, so that capturing let alone killing them is the same as disturbing the existence of these spirits. The Dayak community believe that disturbing the existence of spirits will result in anger and retaliation from them.

c. Give offerings in the *Pukung Pahewan* area. The offerings can be hung using *ancak* in large trees or placed in *Karamat* around the *Pukung Pahewan* area. These offerings are intended as offerings to the spirits of the region's guardian spirits. Offerings are intended as a thank you, there is a vow that comes true, a request for something, or that spirits do not disturb humans. The contents of the offerings are also varied, depending on who the offer is intended. For spirits that are considered not too important, offerings can only contain eggs, *Cucur* cakes, sticky rice cakes, betel leaves, and a cup of coffee. But, for spirits that are considered important and of high status, offerings can even be in the form of a head, heart and a little pork or buffalo.

References

[1] Koentjaraningrat 2004 *Kebudayaan, mentalitas, dan Pembangunan* (Jakarta: PT Gramedia Pustaka Utama)
[2] Riwut dan Nila 2011 *Senjata tradisional & pakaian adat dayak kalteng* (Palangka Raya: Badan Perpustakaan dan Arsip Daerah Provinsi Kalimantan Tengah)
[3] Undang-Undang Nomor 32 2009 Perlindungan dan Pengelolaan Lingkungan Hidup (Jakarta)
[4] Rachmad K 2012 *Sosiologi lingkungan* (Jakarta: PT. Rajagrafindo Persada)
[5] Riwut T 2003 *Maneser panatau tatu hiang menyelam kekayaan leluhur* ed Riwut N (Palangka Raya: Pusaka Lima)
[6] Indonesia MLHdKR 2016 Perhutanan Sosial di Peraturan Menteri Lingkungan Hidup dan Kehutanan (Jakarta: Sekretariat Kementerian KLH) p. 83
[7] Menteri Lingkungan Hidup dan Kehutanan Republik Indonesia 2015 Hutan hak di Indonesia MLHdKR Peraturan Menteri Lingkungan Hidup dan Kehutanan (Jakarta: Sekretariat Kementerian KLH) p. 32
[8] Riwut T 2007 *Kalimantan membangun alam dan kebudayaan* (Solo: NR Publishing)

Effectiveness of the Indonesian literacy school program in improving the quality of basic education for marginal communities in the Indonesian border area

Sarjon Defit[1], Zefriyenni[2], Yosa Novia Dewi[3,*] and Yarman[2]

[1] Departement of Computer, Universitas Putra Indonesia (UPI) YPTK, Indonesia
[2] Departement of Economics, Universitas Putra Indonesia (UPI) YPTK, Indonesia
[3] Departement of Education, Universitas Putra Indonesia (UPI) YPTK, Indonesia

*yosa_novia@upiyptk.ac.id

Abstract. This study aims to compile basic data, develop and implement the Indonesian School of Literacy program in Improving the Quality of Basic Education for Marginal Communities. The program aims to establish and improve school performance through a mentoring process for school principals and teachers. In addition, this program also involves parents in the learning process of students while at home. Parents are involved through parenting training held at school. The Indonesian Literacy School is a program that seeks to realize a community-based school model that concentrates on improving the quality of instructional systems and the development of school culture, with a distinctive literacy culture. This program focuses on improving school governance by referring to indicators of improving school performance based on the Uswah Method Tools. The stages in the development of this program are: program planning, program implementation, and the end of the program. This research was conducted in 2018 to make program design and development of program tools. After that, an initial assessment of the area and school was carried out for program intervention. Program interventions are carried out in the Indonesian verandah, namely: Kepulaun Meranti (Riau Province) and Nunukan (North Kalimantan). The results showed that there was an improvement in school governance through improving the quality of instructional systems and the development of school culture, with a distinctive literacy culture.

1. Introduction

Global competition in the era of free trade requires the availability of quality human resources. These human resources are required to have high professional, technical and managerial skills and sensitive creativity for various changes. Based on data from the *United Nations Development Program* (UNDP) (2016) the Indonesian Human Quality Index *(Human Development Index*) is ranked 113th out of 188 countries. Meanwhile, data from the 2014-2015 World Economic Forum, Indonesia's global competitiveness has indeed increased to 34th out of 144 world countries, but in Basic Requirement Indonesia is ranked 46th (including meeting basic education and health only at 74th) [1].

Based on data from the UNDP (2016) the Indonesian Human Quality Index (*Human Development Index*) is ranked 113th out of 188 countries. The *Indonesian Human Development Index* (HDI) in 2015 amounted to 0.698 and was included in the middle human development category. Nevertheless, Indonesia's HDI value dropped sharply to 0.563 (18.2 percent) if the gap was taken into account. The education gap and life expectancy at birth in Indonesia are higher than the average of countries in East Asia and the Pacific, even though this is the field that receives the most funds from the National Budget. In the 2015 APBN, there were 10.5% of the total state expenditure allocated to the education budget [1].

The principal leadership factor is one of the strategic factors that can improve teacher performance. This influence is the creation of a conducive school climate and culture that encourages teachers to work better. Other efforts that can be done to improve teacher performance can be done in the teacher's view by complementing learning facilities and infrastructure that are still lacking [2,3].

Meanwhile, the quality of education is also determined by the literacy skills of students. According to the 2015 OECD, the average assessment of Indonesia's performance in literacy skills needs attention. Early education for schools outside Jakarta is still a serious problem, as many as a quarter of second grade students still cannot read well. To find out the efficiency of the government can be seen in terms of teacher working hours and student-teacher ratios, especially in remote schools. The teacher's time spent working on assignments, teaching in the classroom, and teaching quality are very diverse in each school. Only 44 percent of teachers do teach according to the minimum working hours needed. Some teachers such as in rural areas (53%) and in remote areas (59%) worked less than 18 hours in a week, still lower than teachers in urban areas (37%) [4].

Based on the conditions outlined in the background, it seems that there is a great need for the participation of all levels of society to help improve the quality of education, especially the quality of education for marginal communities. [5] Quality education is expected to be able to reach all students, whether in urban, rural, urban areas, or verandas. Therefore, this research needs to be done to be able to formulate the Indonesian School of Literacy model in Improving the Quality of Basic Education for Marginal Communities with a comprehensive approach from the point of view of educators, typologies and communities, as well as the role of education providers so that families and schools are able to play an optimal role in creating young people a nation that is intelligent in literacy, character and intellectually competent.

2. Literature Review

According to the national development objectives as stated in the SDGs that education is one of the main targets in development through education for all who are qualified. For this reason, this literature review will reveal scientific evidence and opinions of experts that are relevant to the quality of Indonesia's basic education which includes the condition of Indonesia's basic education, student literacy skills, and school management.

2.1. Marginal Society

In KBBI online, marginal words mean relating to the boundary (edge), or being on the edge. Whereas the word community, still based on the KBBI, is defined as a number of people in the broadest sense and bound by a culture that they consider the same. So, it can be concluded that marginal communities are a number of people who are bound by a culture that they consider the same, who are in the suburbs, remote areas, or remote [6].

Marginals include the poor who are characterized by poverty in terms of food, economy, education and low levels of health. The marginal community referred to in this study are people who come from low economic groups, as well as groups who have not received education properly, both in terms of access, quality and cost.

2.2. Education Access for Marginal Society

The emergence of the stigma of marginal communities that poor people are prohibited from school because of the absence of school alignments for them. The education system applied by policy makers that does not favor marginal groups and poverty makes schools a luxury item. It is not only a poor education policy, other government policies also make it harder for people to get education, the government policy indirectly is prohibiting poor people from being banned from school [7,8].

There are two characteristics of the educational dimension of marginal people. First, alienation from self and its environment. Second, self-depreciation, feeling stupid, knowing nothing. Marginal groups of people, generally trapped in poverty because of lack of capital and access to funding sources and business opportunities that do not favor him. The results obtained from the business are minimal on average, where the income earned is only enough for a simple life. In this case the informal sector in general is an alternative economic activity of millions of small people in cities. Problems faced by children who attend school or not go to school can be solved if the local potential for the development of education in marginal communities can be identified [9].

2.3. *School Quality and Management*
In 2003, the Indonesian government at that time was one of the countries that became the center of attention in the world by decentralizing the primary and secondary education system as part of a wider change designed to increase the responsibility of education in both districts and cities. This form of school management decentralization is often referred to as School-Based Management (SBM) which is needed to make major changes to the capacity building of principals, teachers, and the community in managing leadership, developing alternatives to meet local education needs involving parents and communities in school management [10,11].

2.4. *Development of Student Literacy Skills in Indonesia*
A study conducted in 2012 found that a 10% increase in the number of students with basic reading skills increased the country's annual economic growth by 0.3%. The study also found that a 10% increase in the number of students with further reading ability was associated with a higher annual economic growth of 1.3%. Improving basic reading skills combined with an increase in the number of students with advanced reading skills proved to have a very strong impact, because in order to implement the imitation and innovation strategies, a country needs to have a workforce that is least able to read basic levels [10].

Reading skills especially for elementary school children are obtained gradually and through the same stages. In the first stage, the basic skills of children in developing their reading skills by developing sound-knowledge, word recognition and giving simple letter codes become sounds. In the second stage, children begin to find content in written texts that can strengthen and help children in internalizing the complexity of language including in terms of rules and patterns of language that they know verbally [12].

2.5. *Primary School Literacy*
Literacy is the core of all competencies that must be taught in elementary school. Basically, literacy is focused on developing reading, writing, listening and communicating skills. These four skills are the core skills of elementary school literacy and also as a differentiator between elementary schools and other levels of formal education. According to Government Regulation No. 19 of 2005 concerning National Education Standards, the primary focus of an elementary school is the development of the ability to read, write, count, and communicate. Literacy skills at the elementary school level must be taught and developed in all subjects.

Institutionally, the application of literacy will form an effective and enjoyable learning environment. Literacy-based learning habits will be more effective when applied to all subjects considering literacy is an interdisciplinary skill. The ability and learning outcomes of students will also be more optimal if this culture of literacy is developed continuously. Therefore, an engineered learning environment is needed so that Indonesian children have the habit of reading and loving books. The habit of reading books from an early age makes it easier for them to learn anything, including lessons in school, which has an effect on increasing academic achievement [13].

3. Research Method

3.1. *Framework*

The research framework is presented in Figure1 which is a flow chart of the variables to be studied, which are arranged in two stages, namely: the identification stage (collection of baseline data; the analysis phase of the model and intervention in the form of applying the model.

1. Identify the condition of education in rural, urban, veranda and urban areas of Indonesia in selected locations.
2. Identifying the condition and quality of schools in rural, urban, verandah, and urban areas of Indonesia in selected locations prior to program intervention
3. Analyzing student literacy levels in rural, urban, verandah and urban Indonesia in selected locations.
4. Analyze the Indonesian Literacy School model
5. Provide interventions in the form of the implementation of the Indonesian Literacy School model in rural, urban, veranda and urban areas in selected locations.
6. Companion Placement (School Consultant) at selected locations
7. Monitor and evaluate the implementation of the Indonesian Literacy School model in rural, urban, veranda and urban areas in selected locations.

↓

- Base line data as material for modeling, model implementation intervention, and preliminary data before the model is applied in the selected school. This data can also be used as comparative data between before and after the intervention
- Base line as initial data to determine: (1) School Assistance Program and (2) School Initiation Program
- Indonesian Literacy School Program

Figure 1. Research Flow

3.2. Population, Sample Research, and How-to Withdrawal Sample

The population of this study is marginal schools with low quality in the Indonesian verandas. Region selection is carried out purposively. Then, 2 (two) provinces were selected in each region with random sampling, with the provision that the selection of regions was based on the quality index of primary education (APS level / percentage of literacy / teacher quality / literacy figures). Region selection is based on Regional Education Balance Sheet data [14]. The selected provinces are: Nunukan (North Kalimantan Province) and Kepulauan Meranti (Riau Province).

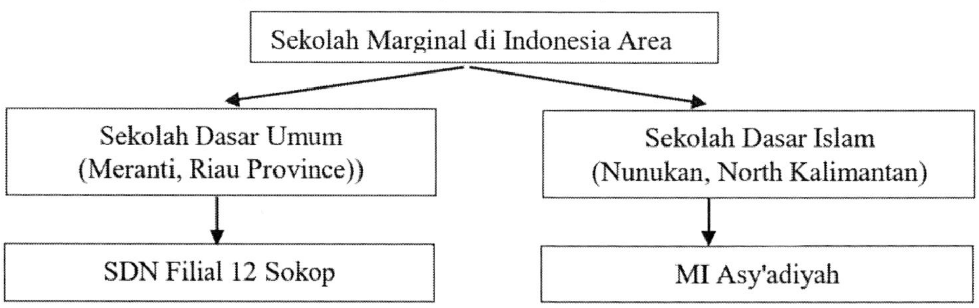

Figure 2. Research Sites

3.3. Types, Techniques, and Data Collection Methods

The data collected in this study consisted of primary data and secondary data. Primary data includes: 1) Quality (Performance) of schools pre and post programs, including: (a) School Leadership, (b)

School culture, (c) Learning Systems. In addition to primary data, this study also uses secondary data from relevant agencies to support the primary data analysis produced in this study.
Data collection was carried out through direct interviews, Focus Group Discussions and In Depth Interviews. The instrument of this research consists of: 1) questionnaire which is a structured question to collect profile data and satisfaction of the Principal, Teacher, and students towards the program; 2) interview guide for indepth interview; 3) FGD guidelines to guide the course of the FGD.

3.4. Data Analysis
The data obtained will be processed through the process of editing, coding, scoring, data entry to a computer, data cleaning, and data analysis. Data will be analyzed with descriptive methods and verification or inferencing methods, after all data is entered into the computer, then processed using SPSS for Windows version 20.0. Data will be analyzed using three methods, namely: 1) descriptive analysis; 2) qualitative analysis; 3) quantitative analysis. Descriptive analysis was used to describe the variables in this study, which consisted of regional profiles, school characteristics, descriptions of the Indonesian Literacy School Program.

4. Research Result

4.1. Regional Profiles and Border Schools

4.1.1. Meranti, Riau Province. Remote Class SDN 12 Sokop is located in Dusun Bandaraya, Sokop Village, Rangsang Pesisir District, Kepulauan Meranti Regency, Riau Province. This Far Class School is located between Sokop village and Tebun Village.

Geographical conditions in Bandaraya Village are lowland areas [15]. Whereas access to these locations can be traversed in two directions, namely from the direction of Batam and from Pekanbaru. The nearest access can now be traversed from Batam direction with an estimated travel time of about 6 hours sea travel. The condition of the road to the location is quite difficult to access because there are a number of heavily damaged road accesses.
Road conditions around schools and residents' homes are still land. If it rains, the road will be muddy and make it difficult for children to get to school. When it rains, children's shoes are always wet. Roads and school grounds are flooded.

This school experiences various difficulties in accessing. Because of its remote location between Sokop Village and Tebun Village. Education in the village is ignored by the local government. There are approximately 6 Km of dirt road that has not been touched by asphalt / concrete. If you want to go to Sokop Village you have to go through a forest that is still on the ground and very difficult to go through.

If we want to go to the next village again, Tebun Village is also very difficult. We also pass the road that has not been touched by asphalt / concrete. Obviously, it will be very difficult to pass if the rainy season arrives. To go to the center of the Meranti Islands Regency we can only cross with Speed Boat for more than 2 hours.

The social conditions of the people in the Meranti Islands Regency, especially in the Hamlet of Bandaraya are nomadic communities. Most of them are people who are from the Akit tribe, Malay and some others are from ethnic Chinese. Most of the income earners are farm laborers or sago laborers with an average income of 1 million / month. With the diverse conditions of the community, this has a direct impact on school life. Thus, every school has a socio-cultural system that is local wisdom.
The condition of education in Bandaraya Hamlet is relatively low with school enrolment rates up to 6 elementary schools. Cadres for educators are very difficult to obtain because most of the population is still illiterate.

The majority of people still adhere to animism. The community also does not have understanding and awareness of a clean lifestyle. Likewise, with public awareness of education, they are not yet aware of the importance of education for children. This is a major factor for related parties, especially the government, in providing access and improving the quality of education for marginal communities in Meranti Islands.

4.1.2. Nunukan, North Kalimantan.
The social conditions of the people in Sebatik are mostly transmigrants from Sulawesi and a small part of Java. Sebatik District consists of 4 villages, Balansiku Village, Tanjung Karang, Sei Manurung, and Padaidi. Distribution of each Kk is Balansiku Village 345 HH, Tanjung Karang Village 612, Sei Manurung Village 229, and Village In 198 HHs, so the total number of HHs in Sebatik District is 1384 HH [16] .

The majority of the Sebatik District community, especially the Tanjung Karang Village, work as farmers, fishermen and traders. This is because the geographical location is close to the beach and full of oil palm proof. The level of education of the community ranged from not graduating from elementary school to master's degree with details: graduating from 3 students, S1 18 people, Academy 4 people, high school 39 people, junior high school 37 people, elementary school 147 people, not graduating 112 people and still 42 blind script. The high rate of not graduating from school and illiteracy in the village of Tanjung Karang is one of the conditions of the community that is very necessary to be followed up [17].

'Emergency' schools located in Sebatik, especially in Sugai Batang, Tanjung Karang Village, are MI As 'Adiyah, with the aim of being able to accommodate children who are difficult to access the nearest school. Meanwhile, the geographical condition of Tanjung Karang Village is the high land behind the hill. Sebatik itself is the outermost island of Indonesia which is directly adjacent to Malaysia. The area is surrounded by beaches, hills, banana gardens and palm oil. To reach Sebatik we start from Tarakan Airport, after that crossing to Sebatik Island with a speed of approximately 4 hours drive, from the port to Tanjung Karang Village takes approximately 30 minutes by motorcycle. Access to the main road is quite good, but access to the entrance to the houses is still very poor.

4.2. Design and Development of the Indonesian Literacy School Program
The Indonesian Literacy School Program is a program that seeks to realize a community-based school model that concentrates on improving the quality of instructional systems and the development of school culture, with a distinctive literacy culture. This program focuses on improving school governance by referring to indicators of improving school performance based on the *Uswah Method Tools*.

4.2.1. Program Goals
This program is present to help the formation and improvement of school performance through the process of mentoring for principals and teachers. This program develops school independence in 6 types of excellence, namely: Literacy Skills, Learning Effectiveness, Instructional Leadership, Conducive Learning Environment, Character Building / Morals, and Educational Leadership.

This program is expected to enable schools to become models for other schools in improving students 'academic and non-academic achievements, as well as the formation of their students' character. One way to achieve this is by providing integrated training that is programmed and periodically held and has continuity in terms of practical implementation in the field. In addition, the program also fixes the school library as a center for learning resources, so that it is expected to be able to facilitate students' access to reading, which in turn will improve reading skills and culture among students.

4.2.2. Program Target
The target of the Indonesian Literacy School Program is that schools experience a minimum increase of 100 scores (in 1 year) from the previous score based on the Uswah Method Tools. The targets of this program are: Principals, all teachers at the assisted schools, school librarians, school committees / community leaders, and student parents.

4.2.3. Program Forms.

The forms of the Indonesian Literacy School program are: Early and Final School Assessment based on the *Uswah Method Tools*, *School Strategic Discussion* (SSD), Student Competency Tests, Training and workshops for teachers, parents and the community, Mentoring (Coaching and Consulting) and program monitoring.

4.2.4. Program Stages.

In achieving school quality improvement, an integrated, comprehensive and leverage program phase is designed to motivate teachers and principals so that the objectives of the program can be realized. One way to achieve this is to provide integrated training that is programmed and periodically held and has sustainability in terms of practical implementation in the field.

4.3. Program Implementation Results

The Indonesian Literacy School Program is divided into 2 types, namely: School Assistance Program and School Initiation Program. The School Assistance Program is focused on schools that have been established by providing mentoring through principals and teachers with the aim of improving school quality / performance. Meanwhile, the School Initiation Program is focused on the establishment of new schools in selected locations, provided there are no educational institutions that can be accessed by the surrounding community.

The program implemented in the Meranti and Nunukan regions is a School Initiation Program by establishing educational institutions that can be accessed by the surrounding community. Several stages were carried out for the School Initiation Program until finally the school stood.

The results of the School Initiation Program are measured using tools to assess school performance. Measurements using this tool are done twice, namely: before the program is implemented and after the program is implemented.

Measurement results show that there is an increase in school performance / quality points in pre and post implementation programs, both in the Meranti region (100 points increase) and in the Nunukan area (374 points increase). This means that the implementation of the Indonesian Literacy School Program has succeeded in improving the quality of schools in the Indonesian Home region (Figure 3).

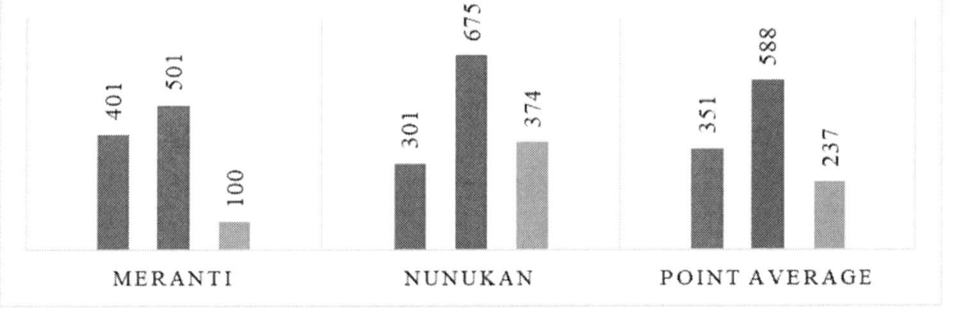

Figure 3. Points of Performance Measurement for Meranti and Nunukan Regional Schools

Figure 4 shows the results of the measurement of school performance in pre and post intervention programs based on 6 criteria. The results showed that the highest score increase was in the Learning Leadership Criteria (22 points increase). Learning Leadership Criteria consist of 3 indicators, namely: School Strategic Planning, Resource Management, Synergy and Networking. Meanwhile, the lowest score increase was in the Literacy and Mathematical Criteria (8 points increase).

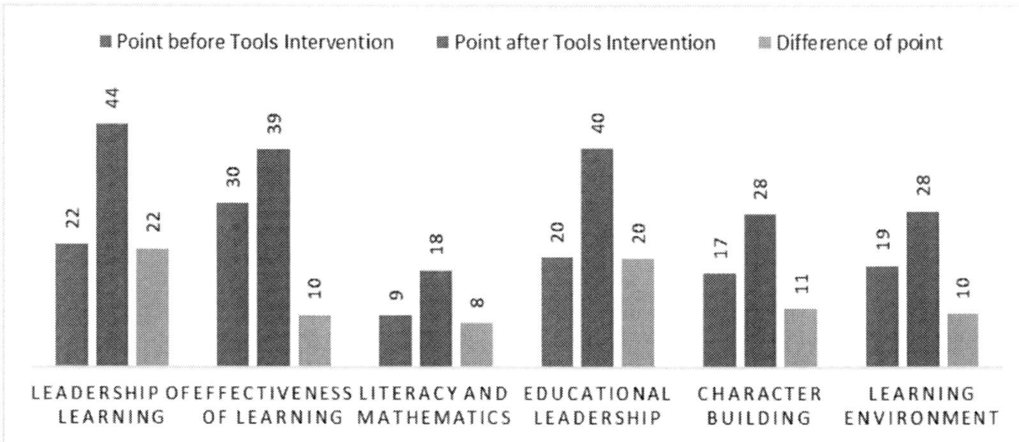

Figure 4. Points of Performance Measurement for Meranti and Nunukan Regional Schools Per Criteria

5. Conclusion

The conclusions of the results of this study, namely: 1) Based on the model that has been prepared, it can be concluded that the stages of the Indonesian Literacy School program are divided into 3, namely: Program Preparation Stage, Program Implementation Phase, and Program Final Stage. The details of each stage are as follows: a) The Program Preparation Phase consists of: school selection and regional potential assessment, signing of an MoU with schools, program launching / socialization, and initial measurement of school performance using the Uswah Method, b) Stage of Program Implementation. This phase is an intervention phase of the Indonesian Literacy School model in eight (8) selected schools. The activities at the implementation stage of this program are: School Strategic Discussion (SSD), training, mentoring (coaching and consulting), monitoring and evaluation, and consultation and mentoring, c) The final stage of the program consists of: the final measurement of school performance using the Uswah Method, the Customer Statistics Index (CSI) measurement / program beneficiary satisfaction, and the final program report.2; 2) The results showed that there was an increase in school performance / quality points in pre and post implementation programs, both in the Meranti region (100 points increase) and in the Nunukan area (an increase of 374 points). This means that the implementation of the Indonesian Literacy School Program has succeeded in improving the quality of schools in the Indonesian Border region.

Acknowledgments

Thanks to KEMENRISTEK DIKTI who has funded this research process. And Special thanks to Yulya Srinovita from Makmal Pendidikan as representative of this research partner.

References

[1] United Nation Development Program 2016 Human Development Report 2016 Human Development for Everyone *UNDP Publishing* http://www.id.undp.org/content/dam/indonesia/2017/doc/INS Indonesia_Country%20Explanatory%20Note_HDR2016.pdf

[2] Purwanto 2012 Kinerja Kepala Sekolah dalam Peningkatan Mutu Proses Pembelajaran dengan Menggunakan Mpmbs *Thesis* (Surakarta : UMS)

[3] Erianjoni E 2018 Pengembangan Materi Ajar Sosiologi tentang Mitigasi Bencana Berbasis Kearifan Lokal di Kota Padang *SOCIUS* **4**(2) 97-108

[4] Organization for Economic Cooperation and development 2015 Education in Indonesia Rising to the Challenge *OECD Publishing* https://www.adb.org/sites/default/files/publication/156821/education-indonesia-rising challenge.pdf

[5] Yanti I, Erianjoni and Firdaus 2014 Deviant Behavior of SMA PGRI 4 Student in the Learning

Process *Jurnal Pendidikan Sosiologi* **1**(1)
[6] KBBI 2017 Pengertian Kaum Marginal https://kbbi.web.id/pengertian-kaum-marginal
[7] Prasetyo E 2009 *Orang Miskin Dilarang Sekolah* (Yogyakarta : Resist Book)
[8] Hermon D and Dalim Y 2005 Penggunaan Media Audio Visual untuk Meningkatkan Kreatifitas Beajar *Jurnal Pembelajaran* **28**(3) 266-276
[9] Supeno 2007 *Model Pendidikan Bagi Masyarakat Marginal* (Sumbawa : DBEP-ADB Kabupaten Sumbawa)
[10] Vemes G, Karam R and Marshall J H 2012 Transforming Indonesia's Centralized Education System to School Based Management *RAND Publishing*
[11] Putra A 2010 Penerapan Pembelajaran Kooperatif Model Group Investigation pada Mata Pelajaran IPS Geografi (Laporan Penelitian Tindakan Kelas)
[12] Hanushek E Hanushek E A and Woessmann L 2012 Do Better School Lead to More Growth? Cognitive Skills, Economic Outcome and Causation *J Econ Growth* **17** 267-321 D10.1007/s10887-012-9081-x
[13] Gove A and Waterberg A 2011 The Early Grade Reading Assessment Aplications and Interventions to Improve Basic Literacy *RTI Publishing* D 10.3768/rtipress.2011.bk.0007.1109
[14] Tim Peneliti Makmal Pendidikan 2013 *Besar Janji Daripada Bukti Kebikajan dan Praktik Pendidikan Indonesia di Era Transisi Demokrasi* (Bogor : DD Makmal Pendidikan)
[15] Kementrian Pendidikan dan Kebudayaan 2016 Neraca Pendidikan Daerah Tahun 2015 http://npd.data.kemdikbud.go.id
[16] Pemerintahan Kepulauan Meranti 2018 Profil dan Letak Geografis Wilayah http://news.merantikab.go.id/web/profil/Letak%20Geografis
[17] Kementrian Kelautan dan Perikanan 2018 Profil dan Letak Geografis Sebatik http://www.ppk-kp3k.kkp.go.id/direktori-pulau/index.php/public_c/pulau_info/297

IOP Conf. Series: Earth and Environmental Science **314** (2019) 012044 doi:10.1088/1755-1315/314/1/012044

Components engaged in the development of school culture in Padang Panjang State High School

Wisma Endrimon*, Rusdinal, Sufyarma Marsidin and M Zaim

Universitas Negeri Padang, Prof. Dr. Hamka Street, Air Tawar Padang, West Sumatera 25131 Indonesia

*wismaendrimon@yahoo.com

Abstract. Improving the quality of education is not enough by just improving the learning process, but it must also be followed by the process of implanting cultural values to students. One school that has successfully developed a school culture is Padang Panjang State High School 1 (SMAN 1 Padang Panjang). The school has successfully developed cultural main values of religion, mutual respect, discipline, and willingness to excel. This type of research is qualitative research. This research was conducted at SMAN 1 Padang Panjang. The informants of this study were the Principal, the teacher assembly, employees, students, and other informants outside the school. Data collection was done through interviews, observations, and documents analysis. Data analysis techniques are done by data reduction, data presentation, conclusion drawing/verification. The study found that the components involved in the development of the school culture in SMAN 1 Padang Panjang covered internal components, including the Principal, teacher assemblies, employees, imtaq (faith and piety) mentors/coaches, and students, as well as external components, including Padang Panjang Mayor, DPRD (Regional House of Assembly), Education Office and School Committee.

1. Introduction

Various efforts continue to be made in order to improve the quality of education both concerning curriculum improvement, development of learning processes, quality improvement, and development of a learning culture.

Habibie stated, "the importance of the synergy between education and the cultivation of cultural values to be implanted in students in schools" ... "The process of education and the civilization process must run in synergy and together" [1]. Suherman stated that "school as a system has three main aspects, which are closely related to the quality of schools, namely the teaching and learning process, school management leadership, and school culture"[2]. Deal and Peterson in Wijayanti (2012) define school culture as, "A set of values that underlies behaviors, traditions, daily habits, and symbols that are practiced by principals, teachers, administrative officers, students, and communities around the school. School culture is a characteristic, character or personality, and the image of the school in the wider community "[3].

Schools as places and means for achieving educational goals, have values and norms that are held firmly by every school member, and are always adjusted to various changes that might affect school performance and school culture. The goal of changing school organizations is basically part of an effort to maintain the existence of the organization to stay alive and meet the needs of the community and the school community. Without these changes and adjustments, the school will slowly "die", because of the inability to deal with the pressures and demands of environmental changes.

There are two main strategies that can be done in improving and developing school quality according to Prihantono, namely strategies that focus on (1) structural dimensions and (2) cultural dimensions with pressure on real behavior changes in the form of actions [4].

One school that is considered to have been able to develop a school culture is State High School 1 of Padang Panjang (SMAN 1 Padang Panjang). This is evidenced by various achievements obtained by students such as schools with the best graduate grades in West Sumatra province (2000 to 2006), Khatam memorization of the Koran to 10 juz, silver medalist (biology) in international Olympiad (in 2004), student exchange participants to Japan, Germany, the United States, winning various academic achievements at the provincial, regional and national levels, and being able to place graduates in various favorite universities in Indonesia (ITB, UI, ITS, UNPAD, UGM, IPB) in addition to various state universities such as UNAND, USU, UNSRI, UNP and others. Almost every year (until 2016) SMAN 1 Padang Panjang is able to place more than 90% of its graduates into state universities in Indonesia. This resulted in a high public interest to send their children to the school. This demand made the Regional Government continue to open two special classes (25% of the acceptance quota) to accommodate students from outside Padang Panjang, and this is different from the two other senior high schools in the same town, namely SMAN 2 and SMAN 3 which only receive 5% of their students for the Padang Panjang outskirts.

The high public interest in sending their children to SMAN 1 Padang Panjang is inseparable from the success of the school in building a school culture that excels based on the main values of education, including honesty, order, discipline, cooperation and togetherness, achievement and based on faith and piety. This research explores further and describes the components involved in developing SMAN 1 Padang Panjang.

2. Research Methodology

This type of research is qualitative research. This research was conducted at SMAN 1 Padang Panjang. The informants of this study were all elements of the school leadership (Principal and Deputy Principal), teacher assemblies, administrative staff, and students and other relevant informants. Data collection is done by (1) interview, (2) observation, and (3) document analysis. Data analysis techniques refer to Bogdan and Biklen (1982) [5] and Miles and Huberman (1992) [6] including (1) analysis of field data, (2) data reduction, (3) data presentation, (4) drawing conclusions/verification.

3. Results

As a system, schools in the whole process involve various components. There are several main components that are directly related to the development of school culture that can be grouped into (1) internal components, and (2) external components.

3.1. Internal Components

When viewed from the internal components of the school, the elements involved cover all elements of the school community in various roles starting from the (1) Principal/Deputy Principal as the school manager, (2) teachers, (3) employees, (4) faith and piety mentors/coaches to (5) students as "raw input" which is due to the education process, students do not only as a recipient but at the same time an analyst and executor of the values developed in the education process, thus making it an object and at the same time the subject of education. Viewing from the "customer" side, because the initial position is "raw input", students after completing their studies can be categorized as "external primary customers", that is individuals who receive and enjoy the results of the educational process they directly experienced.

The internal component elements involved in the development of the school culture in SMAN 1 Padang Panjang turned out to be wider if it refers later to Minister of education and culture regulations No. 28 of 2016 concerning the Education Quality Assurance System, article 11 paragraph (5), where the elements of the School Education Quality Assurance Team consist of at least (a) representatives of education unit leaders, (b) teacher representatives, (c) education staff representatives, and (d) representative of the school committee. In terms of components, the school committee in SMAN 1 Padang Panjang is an external element of the school, but in carrying out the function of education quality assurance, the school committee according to Minister of education and culture regulations No. 28 of 2016 is part of the internal education quality assurance team (school) [7].

3.2. External Components

Regarding the role of external components, the research findings reveal that there are several external components involved in determining the development of school culture in SMAN 1 Padang Panjang, namely (1) Mayor, (2) DPRD/Local parliaments, (3) Education Office, and (4) School Committee, each of which plays a role in accordance with their respective duties and functions.

When viewed at a glance, this finding may not be something different from other schools, but if examined further through the interviewing carried out by the writer with the involved external components such as the Mayor, Chairperson of the DPRD, Head of the Education Office, and School Committees, the very active roles from all external components starting from giving directions, planning, to supervision. Even at the beginning of the appointment of SMAN 1 Padang Panjang as a model/superior school, the Padang Panjang City Government formed a school development team under the control of the Regional Secretary of Padang Panjang City who was assigned to assist the Principal in designing and formulating various steps for school development. Although in the next stage (after regional autonomy and along with school independence) the school development team no longer plays a special role, but in certain steps such as budget planning, budget implementation and oversight of its implementation, the Regional Government always gives special attention to SMAN 1 Padang Panjang. The writer experienced, when being the Head of Planning Section at the Padang Panjang City Education Office (2006-2009), that when discussing the education budget at the City level (Local Government Budget Team/TAPD) chaired by the Regional Secretary, or in work meetings on budget discussions with DPRD (Budget Body Team) when discussing the budget for the scope of education, there were always specific questions about the budgetary requirements for special programs at SMAN 1 Padang Panjang.

4. Discussion

Related to the internal components that influence the development of this school culture, Mortimore (1995) in the concept effective school put forward the roles of: (1) Professional leadership, (2) Shared vision and goals, (3) A learning environment, (4) Concentration on teaching and learning, (5) Purposeful teaching, (6) High expectations, (7) Positive reinforcement, (8) Progress monitoring, (9) Pupil rights and responsibilities, (10) Home-school partnerships, (11) A learning organization [8]. This illustrates that all elements in the school's internal environment, starting from the Principal/Deputy Principal, teacher assembly, supervisors, administrative staff, school security officers, janitors, including students will determine the effectiveness of the school, including in the development of school culture. If there are among these internal components that do not support the school culture development program, then this program will certainly not work effectively. In this case Saleh and Myint see the role of the Principal is indeed more dominant in terms of developing the school culture by involving and working in full with the teacher and all employees, "School leaders are expected to act as agents of change and facilitators who improve the school culture and its effectiveness by transforming the professional learning community. "They have to build human capital by working collaboratively with every teacher and employee in the school" [9].

The findings about the external components involved in the development of school culture in school are basically in line with Gaffar's view in Hadiyanto (2004) which argue that the role of district/city governments in decentralizing education management is no less important than the role of provincial government, namely developing the format of management of education in a comprehensive and integrated manner, developing network with school principals and teachers, developing communication and cooperation with various elements in the community and those in the ranks of district/city governments, including education managers (the term used by Hadiyanto for the education office) as policy makers and accommodating regional needs in accordance with the characteristics of regions, translating central policies according to regional potentials by considering the cultural, economic, sociological, psychological and geographical aspects of the regions concerned [10].

In connection with the role of the school committee, (which before the Minister of Education and Culture No. 044/U/2002 concerning the Education Council and School Committee was BP3), the development of school culture in SMAN 1 Padang Panjang was certainly in line with the objectives of the establishment of School Committees in schools. The Appendix of Decree of the National Education Minister No. 044/U/2002 states that the purpose of establishing a School Committee is (1) To facilitate and channel the aspirations and initiatives of the community in giving birth to operational policies and educational programs in educational units; (2) To increase the responsibility and participation of the community in the implementation of education in educational units; (3) To create transparent, accountable and democratic atmosphere and conditions in the implementation and provision of quality education services in education units [11]. Given the importance of the role of the School Committee in school development, Minister of education and culture regulations No. 28 of 2016 concerning the Education Quality Assurance System, article 11 paragraph (5) places the School Committee as part of the Internal Quality Assurance Team in addition to representatives of the leadership of the education unit, teacher representatives, representatives of education staff, who (a) coordinate the implementation of quality assurance in education unit level; (b) provide guidance, assistance and supervision of education actors in education units in the development and quality assurance of education; (c) carry out the mapping of education quality based on education quality data in the education unit, (d) conduct monitoring and evaluation of the implementation process of quality fulfillment that has been carried out, and (e) provide recommendations for quality improvement strategies based on the results of monitoring and evaluation to the head of the education unit [7].

The research findings and discussion illustrate that the cultural development in schools cannot be separated from 2 (two) main components, namely (1) the internal components of the school including all existing elements which are directly involved in the overall activities of the school starting from the school leader (Principal/Deputy Principal), teachers including counselors (both dormitory advisers for schools that have dormitories, faith and piety mentors for those who have special programs related to faith, librarians, laboratory assistants, and other program supervisors), administrative staff and all staff and school support staff (including janitors, security guards if available), as well as students as the subject and object of education itself, (2) external components including all elements outside the school institution that are directly or indirectly involved in giving direction, supervising or giving support so as to give color to development of school culture, starting from the Regional Head, DPRD, Institution of Trustees (Office of Education with its devices including supervisors of educational units), as well as the community, especially parents of students through the organization of the School Committee. These two components cannot be separated from the whole series of school culture development programs and fully determine the direction and success of the development of school culture.

5. Conclusion

Components involved in the development of school culture in SMAN 1 Padang Panjang, based on the results of the research can be concluded as consisting of two main components, namely (1) internal components including the principal/vice principal, assembly of teachers, employees, mentors/faith and piety mentors, and students, (2) external components including the Mayor of Padang Panjang, DPRD, Education Office and School Committee.

Reference

[1] KOMPAS Februari 19th 2014 *Konvensi PGRI, Pendidikan Nasional Tak Tentu Arah* p 1

[2] Suherman Ade *Pengaruh Budaya Sekolah dan Motivasi Guru terhadap Mutu Pendidikan* available at http://adesuherman.blogspot.com. Retrieved on February 16th 2014.
[3] Wijayanti Rizki *Makalah-Budaya-Sekolah* available at http://kikyuno blogspot.com/2012/05/html, Retrieved on September 26th 2013.
[4] Prihantono Rudi C 2010 Pengembangan Kultur Sekolah Sebagai Upaya Meningkatkan Mutu Sekolah *Jurnal Guru, Pembelajaran di Sekolah Dasar dan Menengah* **2**(7) 149-157
[5] Bogdan Robert C and Biklen Sari Knopp 1982 *Qualitative Research for Education: An Introduction to Theory and Methods* (Boston: Allyn and Bacon, Inc.)
[6] Miles Mathew B and Huberman A Michael 1992 *Analisis Data Kualitatif, Buku Sumber tentang Metode-Metode Baru* (Translation) (Jakarta: UI-Press)
[7] Peraturan Menteri Pendidikan dan Kebudayaan Republik Indonesia Nomor 28 Tahun 2016 Tentang Sistem Penjaminan Mutu Pendidikan Dasar dan Menengah
[8] Mortimore Peter 1995 *Effective Schools: Current Impact and Future Potential* (London: Institute of Education University of London)
[9] Saleh Isa M and Myint Swe Khine 2014 *Reframing Transformational Leadership. New School Culture and Effectiveness* (Rotterdam: Sense Publisher)
[10] Hadiyanto 2004 *Mencari Sosok Desentralisasi Manajemen Pendidikan di Indonesia* (Jakarta: Rineka Cipta)
[11] Keputusan Menteri Pendidikan Nasional Republik Indonesia Nomor 044/U/2002 Tentang Dewan Pendidikan dan Komite Sekolah

Wali Nagari women's leadership style in the development of nagari government in *Nagari Tigo Balai*, Matur sub-district, Agam regency

A Fitri[1,*], A Frinaldi[2] and Erianjoni[2]

[1] Student at Magister of Public Administration Department University of Padang, Indonesia
[2] Lecturer at Faculty of Social Science, University of Padang, Indonesia

*annisafitri196@yahoo.com

Abstract. The research focused on how Walinagari Women's Leadership in Tigo Balai Village, Matur District, Agam Regency. The results of the study showed that women leaders at Tigo Balai Nagari were considered capable of managing the Nagari government and had progressed from before, as evidenced by the improvement in service to the community and development in Tigo Balai Nagari. Walinagari Tigo Leadership Tigo Hall is a leader who is close to subordinates in making decisions involving subordinates or called democratic and not authoritarian leaders, responsible, communicates well, has a high attitude of participation. Continue to seek a balance between personal and social life. The female head of Nagari Tigo Balai can find a solution in the division of tasks that her staff will do. So that the task mandated to them can run optimally to serve social affairs.

1. Introduction

The existence of Village in West Sumatra has its uniqueness since the beginning of the Reformation of West Sumatra Government has declared that the Village Government System will return to the Nagari Government System. This policy is more popular with the term "*Babaliak Kanagari*". Nagari (local government) in West Sumatra is regulated in Regional Regulation No. 2 of 2007 Article 1 paragraph 7 concerning the points of the Nagari Government [1,2]. This policy is followed by all regions in West Sumatra, except for Mentawai who have different customs and cultures from other regions in West Sumatra. Nagari is the most important political unit in the Minangkabau realm, which considers itself to be an autonomous unit in the administration, finance, and other household arrangements.

The Nagari government explained that the implementation of the Nagari government was to regulate and take care of the interests of the community and to organize governmental affairs in Nagari. In its form, the Nagari Government is based on local wisdom which provides support for the re-emergence of the nagari government with its essence, considering the increasingly tenuous kinship, tolerance and concern in the lives of the community [3,4].

Nagari has a set of customary mechanisms to regulate all forms of social relations, such as the system of government, the economic system, the relationship between humans and the relationship between humans and nature. Nagari is governed by the principle of the tigo sapilin rope: namely the link between customary law, Islamic shari'ah and state law (or the Law). Local Kearian values that also deserve to be understood are the relationship between adat and Islam, the Minang people adhere to the principle of *Basandi Syara' Customary, Syara, Basandi Kitabullah,* to regulate and create a balance between human relations, nature, and God. Because indeed the influence of Islamic religion has brought changes in the life of ethnic Minangkabau. [5,6]

Nagari is led by a Walinagari, in running his government. Walinagari was chosen by the nagari children (nagari residents) democratically with direct elections for a 6-year term and then can be re-elected for the next term of office. Usually the chosen one is the person who is considered to be the most masterful about all aspects of life in Minangkabau culture, so that the nagari guardian is able to answer all the problems faced by the nagari children.

After the *"babaliak ka nagari",* the reality that occurred in the selection of the Nagari guardian in West Sumatra was the Nagari trustee selection process dominated by men, but after several Walinagari (Chief of local government at Nagari) elections there were several female candidates who participated, although the results obtained were still inferior to men. man, considering the number of nagari in West Sumatra is 802 nagari in 11 regencies, women's participation in local politics (Nagari government) women's participation in local politics (Nagari government) is still low (data from the regional secretariat of the province of West Sumatra) [7]

An interesting phenomenon in this writing is the victory of female candidates in the nagari tigo balai (Agam district), of course this is the nagari's first victory during the implementation of the walinagari election process in the district, amid an unrelenting focus on low participation and advancement in women's political leadership . the nagari community turned out to be able to display female guardian nagari as a result of people's choice.

Jurniwati's victory as Wali Nagari in the Nagari Tigo Balai 2014 - 2020 period initially led to pros and cons. There are several parties such as traditional leaders who consider that Jurniwati is not worthy of being a Wali Nagari because she is a woman. This is because there are still thoughts that it is inappropriate for men to be led by a woman because women do not understand or do not know what to do in the park, in making decisions often experience obstacles, the difficulty of reaching an agreement in running a good system of government government programs that have not been implemented as a whole, the nature of leaders who are seen as too aggressive, higher stress levels

Eagly and Johnson in Jumiati who have determined that women's leadership style tends to take an approach that invites subordinates to come forward to develop in thinking and leaders take part in carrying out tasks to achieve goals, while different from men who have style leadership that tends to be only superior and subordinate relationships where subordinates do what they are told by their superiors without an emotional approach between subordinates and superiors. Women's leaders tend to involve more emotional ties with the people they lead, so that all affairs between leaders and people led can be carried out in ways that show more feelings of concern and more democratic where female leaders will invite the community they lead to participate involved in various activities and not authoritarian where the leader will tend to give orders to subordinates [8]. For this reason, the writer is interested in discussing about the Women Leadership Style in the administration of Nagari government in Nagari Tigo Balai, Matur Sub-district, Agam Regency.

2. Theory framework

2.1. Leadership style

According to Thoha that leadership style is a behavior norm used by someone when the person tries to influence the behavior of others as he sees it [9]. While Rivai stated that Leadership Style is a set of characteristics used by leaders to influence subordinates so that organizational goals can be achieved or it can also be said that leadership style is a pattern of behavior and strategies that are preferred and often applied by a leader [10]. A leadership style that shows, directly or indirectly, about a leader's confidence in the ability of his subordinates. This means that leadership style is behavior and strategy, as a result of a combination of philosophy, skills, traits, attitudes, which are often applied by a leader when he tries to influence the performance of his subordinates.

Leadership style in understanding the behavior of a leader who needs his ability in leadershipAccording to Kartono [11] states as follows: "The style of leadership is the nature, habits, temperament, character and personality that distinguishes a leader in interacting with others".
Based on the notions of leadership style above, it can be concluded that leadership style is a leader who performs his leadership functions with all his skills and norms of behavior.

According to Sutikno leadership style or leadership behavior or often called Leadership Type [12]. The types of leadership that are widely known and recognized for their existence are as follows:

2.1.1. Autocratic type. This type of leadership considers that leadership is his personal right (leader), so he does not need to consult with other people and no other person should interfere. An autocratic leader has a set of characteristics that are usually seen as negative characteristics. An autocratic leader is a selfish person. An autocratic leader will show an attitude that stands out for his being, and always ignores the role of subordinates in the decision-making process, refusing to accept the suggestions and views of his subordinates.

2.1.2. Type of Free Control or Bodo Period (Laisez Faire). This type of leadership is the opposite of an autocratic type of leadership. In this type of leadership, the leader usually shows passive behavior and often avoids responsibility. A leader with free control tends to choose a passive role and let the organization run according to its own time. Here a leader has free faith by giving the broadest possible freedom to subordinates so that all his efforts will be successful quickly.

2.1.3. Paternalistic type. A paternalistic leader's perception of his role in organizational life can be said to be colored by subordinate expectations of him. The expectation of subordinates is in the form of desire so that the leader is able to act as a father who is protective and worthy of being used as a place to ask questions and to obtain guidance, pay attention to the interests and welfare of his subordinates. The paternalistic leader hopes that the legitimacy of his leadership is an acceptance of his dominant role in organizational life.

2.1.4. Charismatic type. A Charismatic leader has special characteristics, namely his attractiveness which is very attractive, so that he is able to obtain a very large following and his followers cannot always explain concretely why the person is admired. Until now, experts have not managed to find the reasons why a leader has charisma. What is known is that such leaders have enormous attractiveness.

2.1.5. Militaristic Type. A militaristic type leader is different from a military organization leader. A militaristic type leader is a leader in moving his subordinates to use the command system more often, happy to depend on his rank and position, and happy with excessive formality. Demanding high and rigid discipline from his subordinates, and difficult to accept criticism from his subordinates.

2.1.6. Pseudo-democratic type. This type is also called manipulative or semi-democratic leadership. This type of leadership is characterized by the attitude of a leader who tries to express his desires and after that makes a committee, by pretending to negotiate but the truth is nothing but to validate his suggestions. Leaders like this make democracy a veil to get a certain victory. Pseudo-democratic leaders only seem to be democratic when in fact they are autocratic. This leader adheres to pseudo-democracy and leads more to the activities of an authoritarian leader in a subtle, coarse form.

2.1.7. Democratic type. The democratic type is a type of democratic leader, and not because they are elected democratically. The type of leadership where the leader is always willing to accept and appreciate suggestions, opinions, and advice from staff and subordinates, through a discussion forum to reach an agreement. Democratic leadership is active, dynamic and directed leadership. Control activities are carried out in an orderly and responsible manner. The division of tasks is accompanied by clear delegation of authority and responsibility, allowing each member to actively participate.

2.2. Women's leadership

Women are part of civil society which is a key element in any changes both social and political. by providing opportunities for women to act as leaders, governments and organizations this is the power of emancipation itself. there are five characteristics of female leaders [13]

- the ability to persuade, generally female leaders are more persuasive than men, empathy will not disappear
- prove wrong criticism. women have lower levels of ego power than men, meaning they can still feel difficult due to rejection and criticism. but the high level of temperance, friendliness and flexibility they recover quickly and learn from mistakes and move forward with a positive attitude
- female leaders tend to apply leadership style comprehensively when solving problems and making decisions
- have strong charisma, persuasive, confident and strong-willed to complete tasks energetically
- dare to take risks and dare to break the rules

3. Research methods

The research method used in this study is a qualitative descriptive research method that is used to solve or answer the problems faced in the current situation. According to [14] Judging from the type of data the research approach used in this research is a qualitative approach. As for what is meant by qualitative research is research that intends to understand phenomena about what the research subject experienced holistically, and in a way description in the form of words and language, in a specific context natural and by utilizing various scientific methods.

Ali the said this research was carried out by taking steps to collect, classify, and analyze or process data, make conclusions and reports with the main objective of making an objective description of a situation in a description of the situation [15].

4. Results of research and discussion

Structurally the *nagari* government institution is led by a *nagari* guardian as the chief executive. The reality that occurs in the selection process of *walinagari*, is always dominated by men. But lately women have also begun to show their existence in the nagari government, even though they are a *bundo kanduang*, as a wife for their husbands and a mother of their children.

The phenomenon that occurred in Agam district, the mature sub-district precisely in *Nagari Tigo Balai*, namely the emergence of a female leader, after the implementation of the nagari general election. This woman was the first woman to become the nagari guardian in Agam district.

Being a *walinagari* in the sub-village of Nagari must have a great responsibility to the community. Nagari which is a form of social constitution from the Minangkabau community in implementing political regulation, highly upholds "*adat basandi syarak, syarak basandi Kitabullah*". and is governed by the customary foundation called the customary mutual cooperation ".

The appointment of *walinagari* experienced pros and cons, some people could accept Jurniwati as a leader, but some did not, because they were considered not in accordance with the prevailing customs. What the Chairman of *Kerapatan Adat Nagari* (abbreviated as KAN) is a traditional Minangkabau institution at the nagari level which is the guardian and preserver of the Minangkabau customs and culture. KAN is under the supervision of the Minangkabau Natural Density Institute (LKAAM) starting from the sub-district to the provincial level.

Revealed when interviewed about the leadership of the female nagari guardian who said: *".......enough for this time a woman becomes a nagari leader, she hopes that in the future mamak will choose a male leader because after all the top leaders in Minangkabau are the head and the head there is no woman."*

Which means that *ninik mamak* in *Nagari Tigo Balai* does not want to be missed in the next election. For *walinagari*, it is then returned to men. Now let the mother of the democratic system be fully lived for 6 years

The election of Jurniwati as *walinagari* cannot be separated from her experience of being an important person in society. It is this experience that has become more value for the community in making choices in choosing the Tigo Balai walinagari. Jurniwati had served as The Village Community Resilience Institute (LKMD) is one of the community institutions in the village Andaleh Village in 1994-1995, to andaleh village in 1995 to 2000, PJS nagari tigo balai in 2000-2001 and chaired the Bundo Kanduang institution in 2002-2006. This was reinforced by the statement of the head of Jorong Surau Lubuak, Mr. Rio Dasril, who said:*"Mrs. Jurniwati for me is not just a leader but she is like a mother who is valued who is respected and respected, in the leadership system she runs is a system of security where we as guardians of jorong do not have to enter the office but are given the responsibility of fully administering administration in the nation and we also as supervisors ongoing nagari implementation program. Throughout the journey, a nagari guardian had many good things received by the community, ranging from assistance, for example, the assistance of houses that were not suitable to be dissected, the provision of achievement scholarships, but less able and many others. The success of Walinagari today is inseparable from its experience of being the leader of the Andaleh village head and only making the Andaleh suspension bridge as a connecting path by the community when he served as the village head. "*

In the concept of leadership, walinagari is a formal category, because it is appointed and there is legality from the district government. While the leadership style of women in this study is the democratic leadership style applied by the Tigo Balai walinagari, which is seen from the leadership and assessment of subordinates and the community in the organization of Nagari government.

Through interviews with a number of communities and their tools, it was concluded that female guardians cannot be underestimated. Although many do not like his leadership, he really proves that a female leader must be able to carry out his duties as a leader, firm, brave, professional, responsible for having high integrity and discipline.

During his tenure there were many programs implemented starting from facilities and infrastructures such as roads, bridges, improvement of apparatus and community resources through education and training, program socialization in the implementation of nagari governance, improvement of services, village fund transparency, improvement of welfare and others -other.

In line with the results of the interview with Ms. Jurniwati, the writer concluded that there had been a lot of changes because indeed she was truly committed to keeping the mandate of the top leadership, namely the regent and the Tigo Balai Nagari community itself. he did try his hardest and his mind to advance the tigo balai nagari and prove that women could become leaders. He told how his long journey before and after his election as a walinagari were all inseparable from the pros and cons that occurred so that the inner conflict he felt did not deter his enthusiasm to run the nagari government.

All that can be seen from his performance as a leader where all his subordinates and nagari devices including guardians are involved in collaborating to build nagari in accordance with their respective duties and functions, the work culture that he instills is a family work culture without having to impose a will but slowly embrace like his children obey his parents without being forced. The interesting thing from the female leadership is trying to instill the nagari customs which is to foster a love of the community for customs where he has begun to apply it in the nagari guardian's office that every Thursday the nagari wali office employee is dressed as official, then Friday is required to wear kuruang clothes for women and koko clothes for men.

Not only employees but also the nagari people who deal with the nagari guardian's office must kuruang dress (long dress) clothes for women and clothes that are polite for men even though they are not sounded in an order but verbally have been able to influence the community to participate in preserving nagari customs. Likewise, in wedding parties, women also have to wear kuruang clothes, not to be robe, or to use basimpuah for women and baselo for men. Kuruang's dress plan will be used also when it wants to market because indeed, he wants to implement the 12th discord in the Tigo Balai village.

Every person who comes to the walinagari office must greet each other and say goodbye to each other. In the management of the organization according to the author, he is a professional, good management because all related matters must be completed at the office, he does not want to deal at home because he has his own activities such as taking care of his family, gardening and other household activities. In addition, he also maintained the view of people that he was a woman who had left her husband for death.

So great is Jurniwati's dedication in leading from the village head to her election as the nagari guardian, with a supportive and charismatic background that she is able to organize the nagari government, find solutions in the division of tasks to be carried out by her employees and the tools. So that the tasks mandated to employees are able to run optimally to serve community affairs. The effectiveness of the system of division of tasks towards service to the community is felt to have a positive impact from the community. Where people feel the service performance of employees can run quickly and does not result in waiting too long to be served.

In making a decision the leader doing it in a democratic way often involves subordinates in decision making and not only siding with his own desires, so as to provide an opportunity to express opinions that may also be different and challenging with others. But all of that in the end with a joint decision is not just one party. In this case open communication is established, encouraging the participation of all organizational members and leaders close to subordinates. The success of leadership depends on its ability to engage good subordinates when making decisions. Decisions taken by involving subordinates will be carried out by subordinates with good and happy and the similarity of expectations between superiors and subordinates so as to cause morale, intimacy, harmony.

So from the observation of the author, it can be seen that quickly and responsibly employees do work in serving the needs of the results of the analysis conducted on female subjects as leaders found that the subject has a democratic leadership style, which describes high moral individuals, can cooperate, prioritize work quality and can direct yourself. Individuals who have a democratic leadership style prioritize the maintenance of interpersonal relationships, tend to be selfless and care for others.

This trait is classified on the nature of leaders who have a democratic leadership style, plus the nature of motherhood that is very attached to Walinagari so that it tends to have a democratic leadership style because Walinagari feels called to give something better to the environment where he leads. Individuals with a democratic leadership style, generally do not appear on the outside of the individual's nature, but if we dig deeper then the basic characteristics of the individual we can recognize. The hard character outside is only used as a protector in order to survive in an environment full of men making women who as leaders rather tend to have an autocratic leadership style, but after being explored more deeply, the truth of the nature of feminine gender roles can be seen clearly and this is the nature of democratic leadership style.

5. Conclusion

Walinagari woman in Tigo Balai Nagari is considered capable in running the Nagari government, a lot of progress from before seen from the implementation of their functions and functions in the nagari government structure, improving service to the community in decision making always involves the nagari device so that governance runs optimally.

References

[1] Perda (Peraturan Daerah Sumatera Barat) 2007 Peraturan Daerah Sumatera Barat Nomor 2 Tahun 2007 Tentang Pokok-Pokok Pemerintah Nagari (Padang)

[2] Erianjoni 2016 Adoption and Adaptation of Minangkabau Cultural Values for Teaching Sociology Materials in Secondary Schools in Padang City *National Sociology Conference* **V** 1599-1507

[3] Arifin B 2012 *Tribal Management* (Jakarta: Solok Saiyo Sakato) p. 52

[4] Eriyanti F, Desmaila F dan Erianjoni 2018 Membantai Kabau Nan Gadang: the localwisdom of Nagari Alam Pauh Duo *Conference: International Conferences on Educational, Social Sciences and Technology*

IOP Conf. Series: Earth and Environmental Science **314** (2019) 012045 doi:10.1088/1755-1315/314/1/012045

[5] Frinaldi A dan Muhammad A E 2011 Pengaruh Budaya Kerja Etnik Terhadap Budaya Kerja Keadilan dan Keterbukaan PNS dalam Membangun Masyarakat Madani dan Demokrasi *Humanus* **X**(1): 52-61
[6] Fatimah S, Eriyanti F dan Erianjoni 2018 Mapping factors that support social integration of minangkabau community in Luhak Agam *Conference: International Conferences on Educational, Social Sciences and Technology*
[7] The regional secretariat of the province of West Sumatra, Indonesia
[8] Jumiati S. Male and female leadership Surabaya: *proceedings of the 6th anniversary and doctoral degree of colloquium towards the new Indonesian business architecture "strategic leadership: a necessity for prosperity and sustainability"* (Faculty of Business and Postgraduate Unika Widya Mandala) p. 234
[9] Thoha M 2004 *Organizational Behavior, Basic Concepts and Its Applications* (Jakarta: Raja Grafindo Persada) p. 56
[10] Rivai V 2004 *Leadership and Organizational Behavior second edition*. (Jakarta: RajaGrafindo Persada) p. 64
[11] Kartono K 2016 *Leaders and Leadership what is abnomal leadership*? (Jakarta: RajaGrafindo Persada) p. 51
[12] Sutikno, sobry M 2014 *leader and leadership style* First edition, (Lombok: Holistica) p. 35
[13] Fitriani A 2015 Women's Leadership Style *Journal TAPIS* **11**(2):1-24
[14] Moleong dan Lexy J 2013 *Qualitative Research Methods. Revised Edition* (Bandung: PT. Teenager Rosdakarya) p. 3
[15] Ali D 2013 Bringing Local Wisdom-Based Local Government into Practice: A Quadrant Strategy Analysis *Journal Public Policy and Administration Research* **3**(5) ISSN 2224 -5731 ISSN 2225-0972 (Online)

Build management strategies & maintenance *Silokek* Village potential based on local wisdom

Haris Satria, Heldi, Endrizal and Budi Prayitno

Department of Environmental Science, Postgraduate Program, Universitas Negeri Padang, Prof. Dr. Hamka Street, West Air Tawar, Padang-West Sumatera 25131 Indonesia

*satriaharis@student.unp.ac.id

Abstract. *Silokek* village is rich in potential who are at Sijunjung Regency, West Sumatra Province. This area, known as the *MUSIDUGA* (Muaro Silokek & Durian Gadang), has the natural resources in the form of the river which flows between steep limestone hills and beautiful, so that the area proposed to become one of geopark. Problems are found namely the limitations of the community in the management of potential of Natural Resources Company Silokek Village. Yet its optimal strength/ ability society (hope/ the local wisdom, yet high in the management of community spirit). In this research, the author discusses about how to build a strategy management and maintenance potential of silokek village – based on local wisdom. As for the method does is SWOT, internally and externally. On strategic Management, divided into: Actuating (A), Planning (P), Organization (O), Controling (C). At the end of the study, obtained by the method of development potential Silokek Village based on local wisdom with the concept of sustainable development (Ecologic, Social Culture & Economic) with goals "sustainability" through improved skills human resources and Maintenance Management Strategic (MMS).

1. Introduction

Silokek is a beauty with the charm and *nagari* has a strong cultural value, the area of Musiduga located in *Kanagarian Muaro, Silokek, Durian Gadang, Sijunjung,* West Sumatra [1]. The region has an area of ±□41.158 with the State of the region is mostly mountainous, rolling hills, and the River flowed by the Kuantan River and some nifty little. Geopark *Silokek Sijunjung* is one realm of Emergency Government of the Republic struggle base of Indonesia (PDRI) inseparable keberdirian chain as a unitary State of the Republic of Indonesia [2]. Aspirations do not stop on the plan so that the Geopark *Silokek* registered in the UNESCO Geopark due *Silokek* badly needed by society so rich natural potential *Sijunjung ranah minang* culture can prosper and society [3].

Musiduga usually crowded tourist areas visited by tourists on weekends i.e. Saturdays and Sundays as well as public holidays. The management of the tourist Musiduga is under the Tourism Office of Arts and culture and sports (Parsenibudpora) and working with Guardian Nagari (village chief), *Silokek*, and *Muaro Durian Gadang* [4]. According to data from the guardian *Nagari Muaro, Silokek, Durian* and *Gadang* (year 2010), there are 8 *jorong* (RT) i.e. one *jorong* in *Nagari Muaro*, two *jorong* in *Nagari Silokek*, and five at *jorong Durian Nagari Gadang* with a population of 4,113 people [5].

2. Research Method

The research was conducted at *Sijunjung* Regency, West Sumatra Province of Indonesia. The location of the study is presented in Figure 1.

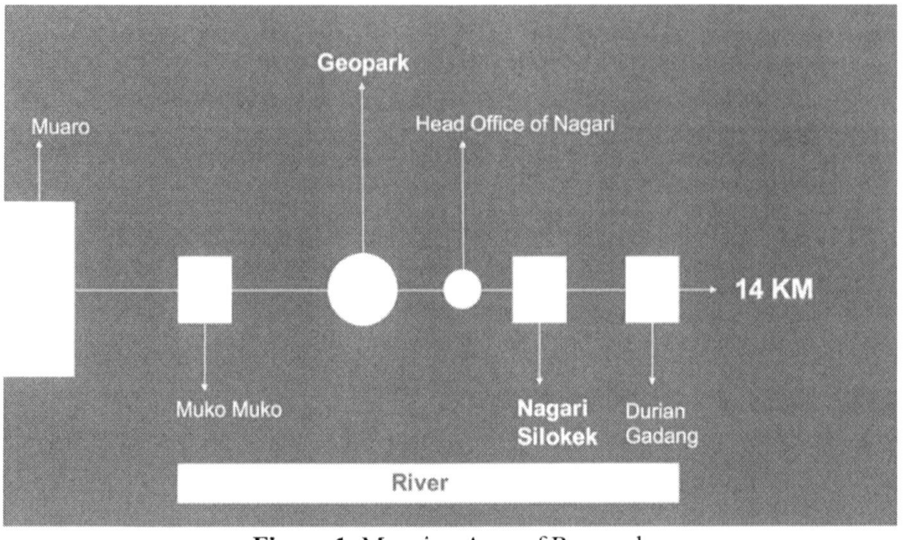

Figure 1. Mapping Area of Research
Source: Satria, Haris 2018

MUSIDUGA (Muaro Silokek & Durian Gadang), has the natural resources in the form of the river which flows between steep limestone hills and beautiful, so that the area proposed to become one of geopark.

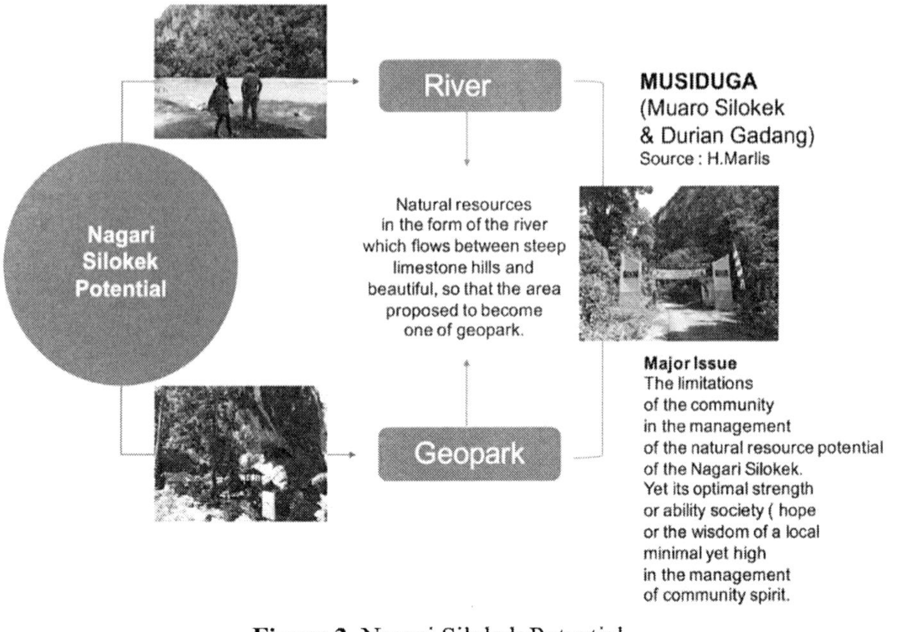

Figure 2. Nagari Silokek Potential
Source: Satria, Haris 2018

Figure 3. SWOT Analysis
Source: Satria, Haris 2018

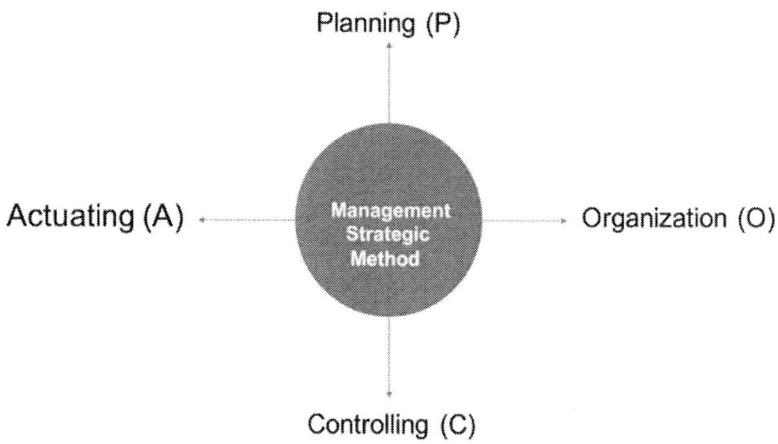

Figure 4. Management Strategic Method
Source: Satria, Haris 2018

POAC is a management process. Which consist of planning, organizing, actuating and controlling. POAC still connected with POSDCORBE (management function). Both are different, but in every element of POAC there always be management function.

The first element in POAC is planning. Planning is a process where a manager makes a decision, decided some strategy to reach the goals, allocate the responsibility to someone else, to measure between success and the goal.

To make a good plan, there are several principles, such as the contribution principle and premising consistence. In Planning process, a manager also needs to make a decision. There are methods is planning, such as elementary methods. When planning is made, manager also need to decided the priority. They can use "Deciding methods".

The second element is organizing. Organizing is a process about how a strategy and tactics which have been created in planning process, we well designed onto an organizational structure. Also, to ensure that all people in organization can work effectively and efficiency which aims to reach the organizational goals. To do a good organizing, manager should not miss the principle of organizing.

IOP Conf. Series: Earth and Environmental Science **314** (2019) 012046 doi:10.1088/1755-1315/314/1/012046

The third is actuating. Actuating is some actions to seek everyone in organization to work for reach the goals. So, actuating aimed is to seek everyone so they want to work with awareness, some kind like realizing a plan. Actuating also need a good leadership too, because the manager is managing a lot of people here.

The fourth is controlling. Controlling is a decide process like, what an organization must reached, what should organization do, to assess the activity and compare it with the standard. Controlling is done in every time, before a process, while the process is working and also after the process.

3. Result

Management means many things to many people. To a layman management means an impressive person occupying an air-condition chamber with an over stacked table and cushioned chair. Some people suggest management as commanding other. To many others, management is nothing more than clerical work and putting fancy signatures. But truly management is the process of planning, actuating, organization and controlling. It is also described as the economic, ecologic and social culture.

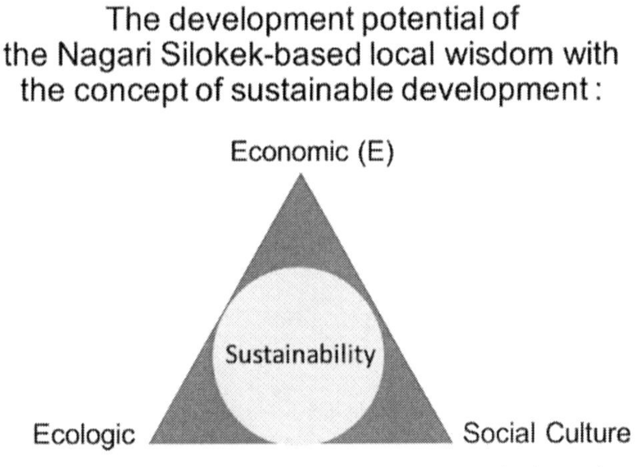

Figure 5. Maintenance Management Strategic (MMS).
Source: Satria, Haris 2018

4. Conclusion

Based on the findings, observations, interviews, information from stakeholders (Regional Governments, Practitioners, Academics and Local Communities: The potential that can be constructed from Nagari Silokek based on local wisdom, from current conditions (existing) is not yet optimal. Therefore, other alternatives need to be built through improved HUMAN RESOURCE capabilities and Maintenance Management Strategic (MMS).

References

[1] Indang Dewata and U Iswandi 2018 *IOP Conf. Ser.: Earth Environ. Sci.* **149** (2018) 012010
[2] Holt Douglas and Douglas Cameron 2010 *Cultural Strategy* (Jakarta: Kompas Gramedia)
[3] Parsons Wayne 2008 *Public Policy: pengantar teori dan praktik analisis kebijakan* (Jakarta: Kencana Prenada Media Group)
[4] Kholil, Intan E K P dan Listyarini S 2014 *Pendekatan Sistem* (Jakarta: Universitas Terbuka Press)
[5] Todaro Michael P 2017 *Pembangunan Ekonomi* (Jakarta: Penerbit Erlangga)

Local wisdom of *Aek Latong* society for mitigation and adaptation of soil movement disaster in North Sumatra, Indonesia

Zainal A Haris[1,2], Agus Irianto[1], Heldi[1], Dedi Hermon[3] and Yulnafatmawita[4]

[1] Department of Environmental Science, Post Graduate Program, Universitas Negeri Padang, Prof. Dr. Hamka Street, West Air Tawar, Padang-West Sumatera 25131 Indonesia
[2] Universitas Putra Indonesia, Lubuk Begalung, Padang Indonesia
[3] Geographical Department, Universitas Negeri Padang, Prof. Dr. Hamka Street, West Air Tawar, Padang-West Sumatera 25131 Indonesia
[4] Soil Science Department, Andalas University, Limau Manis, Padang Indonesia

*zainalabdulharis@yahoo.com

Abstract. Lately, due to advance technology, some of natural disasters could be forecasted, therefore the people can decide where to go to avoid the disasters. However, people in *Aek Latong* North Sumatra did different thing. They just stayed in their villages or regions without moving to anticipate the impact that might happen. This research was aimed to identify the local wisdom of *Aek Latong* society in facing, mitigating, and adapting the disaster, especially soil movement disaster in their region. This research was conducted through 3 steps (preparation, main survey, and data analyses). At the first step (preparation) was conducted to get secondary data about *Aek Latong*, such as geographical position, natural condition, type of natural disaster, society and the tradition. The second step (main survey) was conducted by directly interviewing the people in *Aek Latong* how they face, mitigate, and adapt the disaster happening. The third step was data analyses. Based on data resulted, *Aek Latong* society has a strong belief inherited from their ancients that everything come from the almighty God, therefore they have to face it. They believe that God has specific purposes to do so for them, they do not have to easily give up. Then, they also mitigate and adapt the disaster by earlier checking the signs of disaster and following the advice given, besides they did not construct building and do farming activities in sloping area and in the river bank. It could be concluded that *Aek Latong* society was able to face, mitigate, and adapt to the natural disaster in their region by their belief, knowledge, as well as by following the advice given.

1. Introduction

Indonesia is known as a multiethnic and multicultural country. There are hundreds of original ethnics living in Indonesia, one of which is Batak ethnics living in North Sumatra Province. This ethnics is considered as very strong and very brave toward living challenges. There are many Batak people going away from their hometown to look for some jobs without any knowledge, but they just put their hope into their strength.

On the other hand, Indonesia is considered as a country getting many disasters along the year, either natural or man-made disasters. Sumatra is an island having many disasters, this is due to the fact that it has an a geological fault across the island from the north (Aceh Province) to the South end (Lampung Province). Therefore, potential natural disasters that can happen among which are soil movement under soil surface (tectonic process). This soil movement will cause earthquack, soil surface dislocation, soil surface reduction or subsidence, and some others. One of areas in Sumatra island having high soil movement is *Aek Latong*, North Sumatra Province, Indonesia.

The area being affected by soil movement in *Aek Latong* was about 60% of the total area of 5.42 km^2 [1]. This location is considered as an area (± 60%) having highly susceptible soil movement. This is due to the geologic factors, such as topography. This condition actually had been mapped, however, it was not yet used as an basic information in building either some infrastructure such as streets, bridges, green space, or for other public infrastructures.

Indonesian Government Law No. 24, 2007 [2] on Natural Disaster Control defined disaster as a series of events affecting society life either due to human itself or nature around them causing death, environmental damage, loss of materials, and psychology impact. Disaster is simply defined as a result of hazard due to earth quake, taifun, landslide, meet the vulnerable condition in a community. Main component of disaster is due to interaction between vulnerability and hazard.

Land sliding happens if a movement force is higher than the resisting force. The last force is determined by the the strength of the lithosphere as well as the soil solidification. Slope, water, burden, soil or rock densities affect soil movement force. Landsliding is a kind of soil movement.

Based on planning guidance for landslide susceptible area in 2007 [2], process accelerating landslide was due to water absorption during heavy rainfall and steep slope. The water absorbed was accumulated above the sliding area causing the soil above the area easy to slide down. Furthermore, land slide is also caused by slope instability.

Generally, land slide is caused by 2 factors, controlling and enforcing factors. Factors controlling soil movement is divided into 2 parts, natural (geology, slope, and vegetation condition) and non-natural (especially drainage canal, buffer building and hill, as well as terracing). Factors enforcing land slide are rainfall, steep slope, unstable and thin soil profile, less lithosphere strength, land use planning, tremor, water surface reduction of lake or water body, overburden, erosion, accumulated materials on sloping area, ex landslide, forest clearance, or geology discontinuity, or waste accumulation place [3].

Mitigation conducted on natural disaster could be done through some steps. Somantri (2008) [4] explained some steps of mitigation could be mapping the susceptibility level of the disaster, study the cause and effect of the disaster, identification the process happening during and after disaster, monitoring on susceptible regions especially on strategic areas, and socialization of the disaster and the impact to government or society in the specific regions through poster, booklet, leaflet, or directly through road show.

Lately, natural disasters going to happen can be predicted by using sophisticated information technology. However, Aek Latong society still keeps the local wisdom inherited by the ancestors. The local wisdom given and taught by the parents is in form of knowledge, the way of comprehending and acting on each disasters. What types of local wisdom being kept by the Aek Latong society are interested to study.

2. Research Methodology

This research was conducted by survey method through 3 steps from January to July 2018. The first step was preparation by collecting secondary data (map, rainfall, lithosphere, soil types, land use types, soil movement zones) from literatures and news. The second step was survey by visiting the location to collect primary data (the way of *Aek Latong* society to face, mitigate, and adopt the soil movement disaster) through qualitative-descriptive approach in May 2018. Some of the society and the heads of the tribes were interviewed about the natural phenomenon happened in their area. They were asked on

IOP Conf. Series: Earth and Environmental Science **314** (2019) 012047 doi:10.1088/1755-1315/314/1/012047

what kind of natural disasters they have, how they act during the disasters, how they anticipate, and how they adapt the condition after the disaster. The third step was data analyses and writing report.

Figure 1. Impact of soil movement in *Aek Latong*, *Tapanuli Selatan*, North Sumatra Province

Figure 2. *Aek Latong* Village Area, *Tapanuli Selatan*, North Sumatra Province

3. Resuts and Discussion

3.1. Research Site

Location of the research was in Aek Latong village, South Tapanuli Selatan Region, North Sumatra Province, Indonesia. This village is an area being passed by cross country road in Sumatra island, especially from Medan toward West Sumatra, Riau, Jambi etc. However, the soil is very susceptible to degradation especially during rainy season. This is due to the topography of the area having steep slope. In 2000, a natural disaster happened on which the village of Aek Latong was collapsed and the people houses were felt down (Figure 1). The phenomenon is called "Suhul" (gempa) by the society. The frequency of the "Suhul" phenomenon increased by time.

3.1.1. Lithosphere and Soil Order. Lithosphere in North Sumatra Province ws dominated by Bahorok and Toba tuff, alluvium, and Klue and Kuantan formation. Bahorok formation was dominated by brecci conglomerate rocks which formed Lithosol at the first step, and then Podzolic soil as they were extensively weathered. Conglomerate materials formed Regosol having medium to coarse textured soil. Toba tuff formation was dominated by volcanic ash which developed into Andisols. This soil has fine to coarse texture. Alluvial landscape found along the east coast formed Alluvium, Regosol, Organosols. Klue and Kelantan formation dominated by sasak, tubidite, sand, and gamping stones formed Lithosol, Podzolic and Lithosol having generally coarse soil texture. Other soil types found in this area were Ultisol (16.35%), hidromorphic, glei humic, and Regosol (11.54%) [5].

3.1.2. Slope. It was about 50% of the research site having mountainous topography which lied down between 800-1,915 m above sea level with > 40° slope [1]. The research area as a whole was a a-steep high land. If there was a flood disaster, then it used to be followed by landsliding phenomenon. The classification of slope in North Sumatra Province is presented in Figure 3.

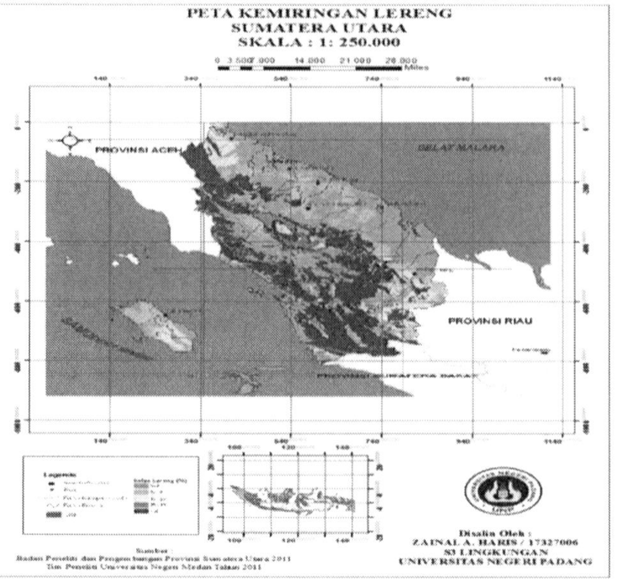

Figure 3. Slope degree of North Sumatera (*Source: Research and Development Institute of West Sumatra Province)*

Based on Figure 3, approximately 50% of North Sumatra Province has steep slope (red color). An area having steep slope was very susceptible to degradation, especially under wet region. As found by Yulnafatmawita *et al* (2013) [6] that the steeper the slope was the higher soil loss due to water movement (erosion) under super wet tropical region in West Sumatra.

3.1.3. Land Use. Land use in Aek Latong region is divided into residential, agricultural (seasonal crop farming, plantation), and public facility (road/street, drainage) types of land use [1].

3.1.4. Geological Structure. Geological map of Aek Latong region in Sipirok Subdistrict is presented in Figure 4. Due to some cracks formed after earthquake, water is easily getting into the rock system causing the rocks are intensively weathered. This causes the geological formation becomes soft and easily moved as the small force is performed. Therefore, the soil movement is readily happened in this region. Based on Table 1, the total rainfall received by Aek Latong region was approximately 2,625 mm/y, with the difference between the driest and the wettest month was 171 mm. The variation of temperature was 0.9°C along the year. According to Smith and Ferguson (1959) [7], the climate of Aek Latong is classified into Af (=very wet). Sloping area under very wet climate is highly susceptible to degradation, especially if the land is not managed based on conservation rules.

3.1.5. Demoghraphy of Aek Latong. Based on survey conducted, Aek Latong is a sub-sub district in South Tapanuli Region, North Sumatra Province. In 2014, Aek Latong had 237 people, with low education level. Almost half (46%) of the society was just elementary school, 25% was uneducated. There was only 29% having junior and senior high school degree, and there were only 2 people having bachelor degree. According to their occupation, there was 80% of family working as farmers, the rest were as pensioners and the civil servants, non-government job, as well as non-job people. The majority of the society is moslem [8].

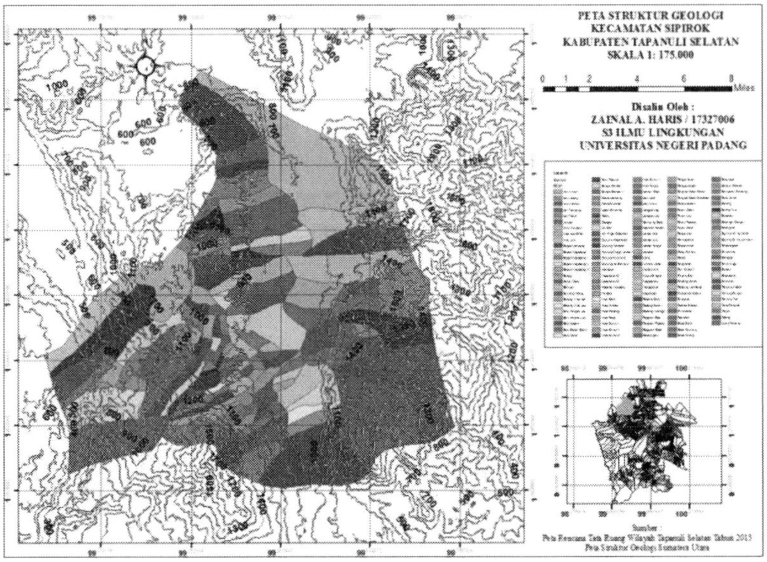

Figure 4. Geological structure of Sipirok, Tapanuli Selatan Region (*Source : Elevation Analyses of North Sumatera Province, 2010)*

Table 1. The climate condition of Aek Latong*

Parameter	Jan.	Feb.	March	Apr.	May	June	July	Aug.	Sept.	Oct.	Nov.	Dec.
Avg. Temp (°C)	22.5	22.5	22.8	23	23	22.4	22.2	22.1	22.1	22.1	22.3	22.3
Min. Temp (°C)	17.1	17	17.5	17.9	17.8	17.1	16.9	16.9	17.2	17.5	17.6	17.4
Max. Temp. (°C)	28	28.1	28.2	28.1	28.2	27.8	27.6	27.3	27.1	26.8	27	27.3
Avg. Temp. (°F)	72.5	72.5	73.0	73.4	73.4	72.3	72.0	71.8	71.8	71.8	72.1	72.1
Min. Temp. (°F)	62.8	62.6	63.5	64.2	64.0	62.8	62.4	62.4	63.0	63.5	63.7	63.3
Max. Temp (°F)	82.4	82.6	82.8	82.6	82.8	82.0	81.7	81.1	80.8	80.2	80.6	81.1
Precip. (mm)	247	217	263	280	179	125	115	152	208	276	277	286

Source: Iklim Aek Latong

3.1.6. Zone of Soil Movement in Aek Latong. Based on evaluation conducted toward slope, lithology, geological structure, and land use planning, Aek Latong in Sipirok subdistrict was classified as an area having high soil movement, with the scoring data were 22. Zone map of soil movement in Aek Latong is presented at Figure 5. Based on Figure 5, all part of Aek Latong area is considered susceptible on soil movement disaster.

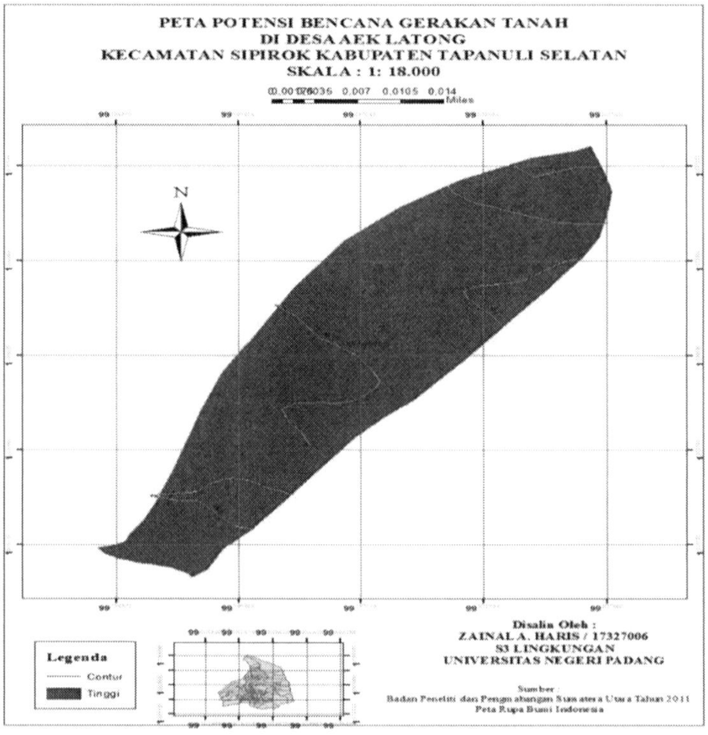

Figure 5. Zone map of soil movement in Aek Latong (*Source: Secondary data analyses, 2017)*

3.2. Mitigation and Adaptation

Mitigation and adaptation are conducted in Aek Latong in order to reduce natural disaster due to soil movement below ground surface. Mitigation is any efforts conducted to anticipate or to lessen the impact of disaster going to happen. However, effective effort that could be done before disaster was not well done yet. The action, actually, is very important to minimize more victims in susceptible area of disaster. Mitigation paradigm in controlling disaster was meant as efforts by introducing disaster susceptible areas, and preparing self-sufficient and self-reliance society. In addition, mitigation was also defined as efforts to increase capacity of society inhabiting in disaster susceptible area to anticipate or to lessen the impacts level.

Among types of mitigation which can be conducted are by changing the labile society behavior through settlement planning, building requirements, building structure having earthquake resistance, and spatial planning having perspective for mitigating disaster impacts. Furthermore, awareness for the disaster impact must be developed and activated among the society. Therefore, it is hoped that the society will not only be the target, but they also become the helper for themselves as well as for the environment during disaster. People awareness being developed among society is a kind of social awareness including earlier notice system, disaster anticipation, response at disaster happening, and ability to handle post disaster. These types of awareness are a kind of social capital to develop mitigation habit in its life of each individual of the society [9].

Land sliding area in Aek Latong is matched with the map of soil movement susceptibility in Sipirok, on which the soil movement is highly active and never stops. Soil movement types in Aek Latong are considered as soil slow movement and fast movement (land sliding) which were happened in two points in Purba Tua sub-subdistrict [10].

There were some factors affecting and accelerating soil movement in Aek Latong. The first factor is steep slope. Since it is on high land having mountainous topography, the slope become steep to very steep.

The second was geo-hydrology factor. Based on data derived from geo-hydrology regional map, it was known that Sipirok region has highly productive water resource. The lithology arrangement was easy to transmit water. The water was easily to come out due to uncompressed aquifer during heavy rainfall. Water functions as lubricant to faster soil degradation and as energy source to push big materials that can open hill wall producing several types of soil movement.

The third factor was earthquake tremor. This factor is really important to consider in Aek Latong for building some public facilities such as streets, housings, offices, etc. Based on pedology map, the soil materials which arranged the road foundation of new Aek Latong road was a kind of residue of vulcanic rock weathering from Nabirong, Sibual-Buali, and Toba Purba which was proved by none hard rock found.

The fourth factor was rock stratigraphy and elastic rebound landslides. This factor was due to continuous rock degradation zone, therefore the stable zone becomes under stressed. The old soil movement needed to be covered, so it causes soil movement in Renun Toru segmen, especially in Aek Latong fault. Then, if earthquake happens, it will shake soil into mud. As a consequence, there will be always areas being increased, disrupted, weathered, or cracked in Aek Latong and the surrounding.

The fifth ecological factor of soil movement is due to no vegetation along the street in Aek Latong. This condition causes that water easily comes out of the soil. Therefore, the wall to resist landslide is susceptible to be cracked. Soil and water movement cannot be separated. Therefore, to conserve soil and water in Aek Latong was conducted by planting some vegetation. For this condition, it is suggested to plant vertifer grass having long roots along the wall sides of the streets or hills.

3.3. Local Wisdom

Local wisdom belonged to Aex Latong society was derived from disasters coming all the time to the society. Since the ancestors did not move or leave their village they had been living for hundreds of years, they had to find out the best way to mitigate and to adapt to the condition. Therefore, they always seek the appropriate methods to solve the problem. The methods are always improved by the time to fix the specific condition.

3.3.1. Types of Local Wisdom for Mitigation and Adaptation to Disaster. Aek Latong society had tried some efforts in controlling soil movement to reduce the impact on their life and their environment. Among the efforts are:

1. Planting vegetation. Planting vegetation having long roots in bare area for buffering the susceptible soil movement, or cutting down the sliding area of the usual soil movement. This is aimed to strengthen the soil by using the vegetation roots. Then, the roots are also able to transmit more water to deeper soil profile. This is due to the fact that the dead roots provide soil macropores continuing from the soil surface into deeper profile. This will increase soil water recharge and reduce water movement on soil surface (runoff) causing soil movement. Furthermore, more vegetation on soil surface causes more water can be transpired from the soil, as a consequence, the soil surface becomes drier. As the soil is drier, it becomes more solid and strong. Therefore, if there is tremor the soil is not easily moved.
2. Constructing Dam. Constructing dam is aimed to cut the sliding area or to reduce area impacted by the soil movement. The type of dams can be derived from stones arranged in a wire or sand put in container and then arranged together to create a buffer to the soil movement.
3. Avoiding very Susceptible Area. The Aek Latong society did not want to leave the village but they just moved to the least risky area, and used land for agriculture which is away from the very susceptible one.
4. Building Strong House against Earthquake. The Aek Latong society used to build their houses from wood in order to get less impact during earthquake.
5. Praying (Istighosah). The Aek Latong society were dominated by moslems. They believed that they could not go any where to avoid the disaster except they have to be closer to the Almighty

God. Therefore, besides conducting physical efforts, they also prayed to the Creator to be out of the disaster.

6. Socializing and Educating the New Generation. Since they did not want to move from their own village, the elder socialized the types of disasters they were used to face along the year. Then, young generation will be educated how to recognize indicators of disasters and what types of impact will come out, and what the action or how to solve the problem. Those types of knowledge are always delivered to the new generation all the time.

3.3.2. The types of Local Wisdom in Facing the Disaster. Aek Latong society had some knowledge from nature indicating natural disaster phenomenon. There are some notices generally being recognized by people in Aek Latong before the disaster happens, such as:

1. Strange sound like thunders. If the Aek Latong inhabitants hear some strange sounds, they have to take some action to save their life directly.
2. Unusual animals comes out of forest or their habitat. Animals are much more sensitive than human being in recognizing the natural phenomenon. Therefore, if people see some unusual animals around them, it means that something will happen.

Some Actions conducted by the Aek Latong society in facing the disaster are

1. The leader of the tribe announces to the society that something will happen
2. They all together runs out of the villages to remote and safe areas for about 6-10 km away
3. They will come back to their villages or houses if they do not see the unusual animals anymore around them, in other words, the animals have come back to their habitats such forest.

4. Conclusion

Aex Latong society conduct the disaster mitigation based on the local wisdom. This wisdom is created due to the local condition they are facing along the time. They always look for the best way to anticipate as well as to adapt with the condition. As they do by time, from one generation to the next, the people habitual turns to be the local wisdom. The wisdom is different for the different location having different natural problems. Socialization about disaster as well as the impact being faced after the disaster, has made the society to become the self-reliance people all the time.

Types of local wisdom of Aex Latong society for mitigation and adaptation to soil movement disaster are planting some vegetation, constructing dam, avoiding highly susceptible area, building wood houses, praying, socializing and educating young generation.

The types of local wisdom of Aex Latong society in facing the soil movement disaster is divided into 2 parts, their knowledge in recognizing (predicting) the disaster will happen (such as strange sound and unusual animal in the surrounding) and their way to act against the disaster (such as announcing to all the society, being away from the susceptible area, and staying in the remote and safe area otherwise the unusual animals have come back to their habitats).

Beyond all of the local wisdom types they have, the spirit left by their ancestor has strengthened their principal that they will never leave their village.

References

[1] Badan Pusat Statistik 2017 *Kecamatan Sipirok Dalam Angka* available at https://tapanuliselatankab.bps.go.id/.../09/.../kecamatan-sipirok-dalam-angka-2017

[2] PM-PU 2007 Peraturan Menteri Pekerjaan Umum No.22/PRT/M/2007 Tentang Pedoman Penataan Ruang Kawasan Rawan Bencana Longsor Departemen Pekerjaan Umum. Direktorat Jenderal Penataan Ruang, Jakarta

[3] Djamal H 2008 Kajian longsoran tebing Ngarai Sianok dan pengelolaan bencana pasca gempa bumi Padang maret *J. Kebencanaan Indonesia* **1**(4) Departemen Pekerjaan Umum

[4] Somantri L 2008 Kajian mitigasi bencana longsor lahan dengan menggunakan teknologi penginderaan jauh *Makalah Seminar Ikatan Geografi Indonesia di Padang* available at https://www.academia.edu/4493377/KAJIAN_MITIGASI_BENCANA_LONGSOR_LAHA

N_DENGAN_MENGGUNAKAN_TEKNOLOGI_PENGINDERAAN_JAUH

[5] BPP PROVSU 2011 Badan Penelitian dan Pengembangan Provinsi Sumatera Utara *J. Inovasi Juni* 2011

[6] Yulnafatmawita Nasution S F and Adrinal 2013 Short term dynamics of soil erosion and nutrient loss during corn growth in ultisols Limau Manis Padang *11th Int' Conf. on The East and Southeast Asia Federation of Soil Sci. Soc. Bogor, Oct 2013*

[7] Schmidt F H and Ferguson H A 1951 *Verhandelingen No. 42 Rainfall types based on wet and dry period ratios for Indonesia with Western New Guinea* (Jakarta: Kementerian Perhubungan Djawatan Meteorologi dan Geofisik)

[8] Badan Pusat Statistik 2017 Kabupaten Tapanuli Selatan Dalam Angka available at https://tapanuliselatankab.bps.go.id/.../kabupaten-tapanuli-selatan-dalam-angka-2017

[9] Imanda A 2013 Penanganan permukiman di kawasan rawan bencana gerakan tanah studi kasus: Permukiman sekitar Ngarai Sianok di Kelurahan Belakang Balok, Kota Bukittinggi *Jurnal Perencanaan Wilayah dan Kota* **24**(2) 141-156 available at http://journals.itb.ac.id/index.php/jpwk/issue/view/373

[10] Heston Y P and Rianto N 2010 *Kebijakan penataan ruang desa Aek Latong akibat longsoran di jalan Tarutung – Sipirok* available at https://pracastino.blogspot.co.id/2010/11/kebijakan-penataan-ruang-desa-aek.html

Socio-economic characteristics of farmers on the existence of floating-rice cultivation demonstration plots in flood prone area in Bojonegoro, East Java

H Irianto[1,3,*], Mujiyo[2], A Qonita[1] and E W Riptanti [1,3]

[1] Department of Agribusiness, Faculty of Agriculture, Universitas Sebelas Maret, Jl. Ir. Sutami 36 A Surakarta, Indonesia
[2] Department of Soil Science, Faculty of Agriculture, Universitas Sebelas Maret, Jl. Ir. Sutami 36 A Surakarta, Indonesia
[3] Study Center and Mentoring of Cooperation and Micro, Small and Medium Enterprises (CMSMEs), Universitas Sebelas Maret, Jl. Ir. Sutami 36 A Surakarta, Indonesia

*radenrara@staff.uns.ac.id

Abstract. Some parts of Bojonegoro Regency along the Bengawan Solo River experience flooding every year due to its downstream area. This leads to a drowning in the rice field areas for several months in the rainy season. As the consequence, the rice fields cannot function well in the agricultural cultivation process. One of the ways to adapt to this condition is cultivating floating rice. However, the further research on the social economic characteristics of the farmers in their accordance to the technological advancement needs to be conducted. This article aims to study the social economic characteristics of the farmers on the existence of the demonstration plots of floating rice cultivation. The method used in this research includes the demonstration plots and farmers survey. The surveyed respondents were 200 farmers living 3 km away from the demonstration plots. The findings show that the characteristics in which respondent farmers are dominated by unproductive ages, low education, low income, limited land tenure, and the lack of information shall be the obstacles in developing floating rice cultivation. This is due to the relatively high operational cost for making floating medium and caring plants different from others. The farmers' mindset tendency to give up to the natural condition is difficult to change when adapting to flood conditions.

1. Introduction

The world's climate is changing and will continue to change in the coming century at rates projected to be unprecedented in recent human history. The risks associated with these changes are real but highly uncertain. Adequate attention must be given to respond to the impacts of climate change that are already occurring, while at the same time preparing for future impacts. It is important to understand the nature of those risks, where natural and human systems are likely to be most vulnerable and what may be achieved by adaptive responses. It is most urgent to ensure adequate and rapid support to the most vulnerable countries and communities. Adaptation is one significant measure that reduces vulnerability to actual or expected climate change effects. There is an urgent need for adaptation assessment in the short-term, as witnessed by the increasingly high costs of extreme weather events, compounded by rising population densities, eroding natural protection systems and aging infrastructure [1].

Agriculture plays a main role in ensuring food security and contributing to economic growth. Currently, this sector is strongly impacted by natural disasters (drought, flood and increasing insect pests and diseases) caused by climate change influencing farmer livelihoods [2]. Rainfed production systems are hypothesized to be most vulnerable to direct effects of climate change (e.g. changes in rainfall patterns) [3]. Bojonegoro regency occupies the largest flood-prone rice fields of about 14,198 hectares, in East Java province. Floods commonly occur due to Bengawan Solo River over burst, particularly in rainy season. The fields are potential for cultivating rice, but floods lasting for months cause these areas to be unproductive for about four months. For farmer households living in flood prone areas, flood will cause a loss of agricultural production [4]. According Ramakrishna and Harwood [5,6], floods undermine farm yields and the national harvest, reducing household and national food availability, and agricultural income derived from crop sales.

Managing the flood prone ecosystem for rice production needs to evaluate the reasons and a comprehensive appropriate technology through research efforts for better rice production under such harsh ecology [7]. Empowering farmers with all kinds of activities can reach sustainable food security in flood prone areas for farmer households. There are activities such as introducing deep-water rice technology, optimizing food consumption, increasing the number of villages, which can produce their own food, building food barns on farmer households, living harmony with flood and providing early warning system [8]. In addition, technology of floating rice cultivation is developed. This is a rice cultivation technique that uses rafts as cultivating media as an effort of adaptation to flood [4].

Floating rice cultivation is an innovation introduced in the flood prone area in Bojonegoro. Potentialities of this locally innovated 'floating rice agriculture' practice need to be assessed for adaptation as a profitable farming with changing climate [9,10,11]. The practice of floating agriculture also helps supplement the income of local communities and contributes to the alleviation of poverty [12]. It, moreover, provides greater food security by increasing the land output and supporting capacity for poor and landless people [13].

The introduction of floating-rice technology requires training, demonstration plots and socialization to farmers. Demonstration plot (demplot) is a method of agricultural extension services to farmers, by making a pilot area, so farmers can see and prove to the object demonstrated. Pilot group should carry out demonstration plots. If pilot project is done successfully, it will drive force to public who are interested in doing similar thing [8]. Before becoming government policy, the floating-rice cultivation needs to be studied regarding the farmers social economic characteristics to prevent failure of cultivation process. This is crucial, given that the cost of floating-rice is relatively high and there is a different treatment procedure in the cultivation. This article aims to study the social and economic characteristics of the farmers on the presence of floating-rice cultivation demonstration plot.

2. Methods

The research location, which was selected through purposive technique, includes Bojonegoro Regency which is a flood prone area experiencing flood every year [4,8]. The locations of floating-rice demonstration plot (demplot) are rice fields near by the residential roads, for instance, in Balungdowo Hamlet, Karangdayu Village, Baureno Subdistrict (Figure 1 and 2). The floating-rice demplot uses floating medium made from bamboo and planting media from the rice field soil and husk charcoal in a ratio of 1:1. The range of the plantation is 20 x 20 cm with Ciherang varieties as the selected planting seeds. This floating-rice cultivation demplot is the development of the previous research finding in determining the highest and the lowest production costs. The rice production resulted in the research scale reaches 13.78 ton/ha. This area was selected as demplot location due to its proneness to flooding for about 4 months during the rainy season. Meanwhile, the survey was carried out to the farmers living 3 km away from the demplot. Based on the mapping using purposive sampling technique, the farmers chosen were from Karangdayu and Pomahan Villages in Baureno Sub-district.

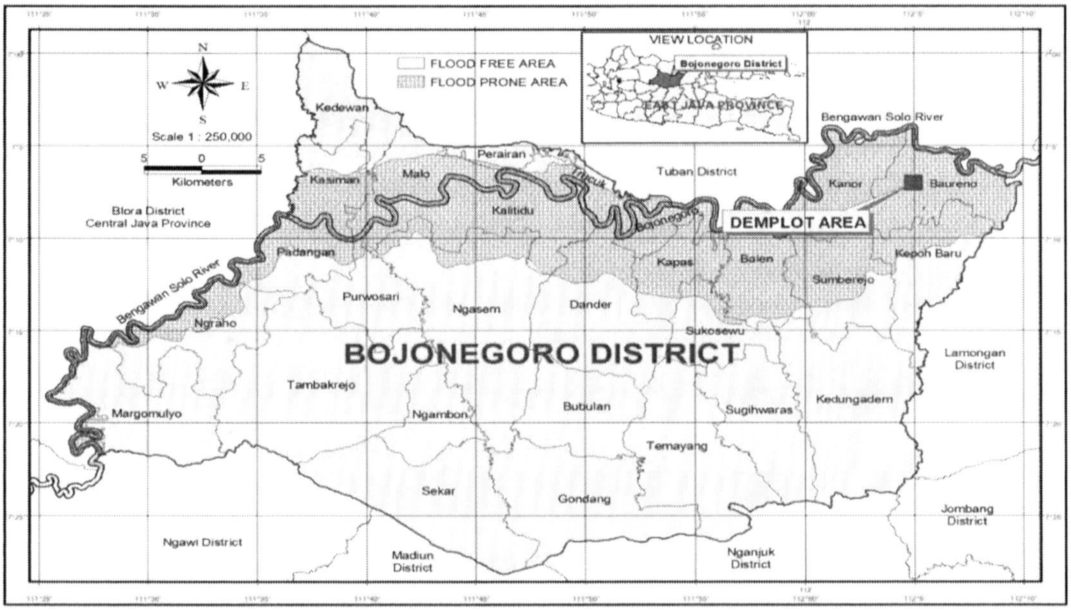

Figure 1. The Location of Floating-Rice Demonstration Plots

The population in this research include the farmers living or owning rice fields near by the floating-rice cultivation demplot, with the maximum radius of 3 km far from the location. The research samples taken were 200 farmers selected in purposive way.

Figure 2. Floating Rice Demplot

The design of this study is exploratory research, in which a research aims to obtain description or study on the condition of local food barns in Central Java. Explanatory research seeks explanations to the observed phenomena, problems, or behaviours [14]. The data were collected through in-depth interview, observation, and recording. The data were analysed using analytical descriptive method.

3. Result and Discussion

The respondent farmers were members of farmer groups in both villages, Pomahan and Karangdayu. The farmer group usually consists of 20 – 25 members from one stretch group. This institution has a fundamental role in deciding the plantation time, irrigation schedule, pest control, agricultural insurance, and the regulation of input subsidies from the government.

Demplot needs to be disseminated to public so the results provide more benefits to them. Demonstration plot can be applied/replicated to other areas that have the identical condition.

Socialization of demonstration plot aims to improve public knowledge about the results from farmers group. Socialization can be done through demonstration plots and promotion through various media both print and electronic media. If public response about the demonstration plot is negative, they will not adopt the results. They do not support the results of the introduction. Therefore, we need to change their response through other agricultural extension activities [8].

Understanding the social economic characteristics of the farmers is significant in order to gain the depiction of the farmers' capacity as the main subject in the agricultural activity. The success of a technology adoption in agriculture cannot be separated from the farmers' social economic condition. Thus, the social economic condition needs to be explored as the input material for the policy makers in applying the technology introduced. Farmers in the surrounding area of floating-rice cultivation demplot are from Pomahan and Karangdayu Villages. The characteristics of social economic condition of the farmers lie in the age, education, number of family members, income, land tenure, type of work, and mastery of information sources.

a. Farmers' age represents their ability in adopting new technology and information in their agricultural business. According Davis [15] proposes that 35 year is the maximum age of young farmer category. In the research location, the age of farmer is described as follows:

Table 1. Farmer Age

Age groups	Number (person)	Percentage (%)
35	30	15
36 – 59	40	20
>59	130	65
Total	200	100

Data source: Primary data analysis (2018)

Table 1 shows that the farmers are mainly in unproductive ages. This phenomenon happens not only in the research location but also in other regions in Indonesia [16]. Young farmers with the maximum age of 35 are very small in number percentage in the research location. This is due to the youth's tendency to work outside the agricultural sector. The dominant number of old farmers and the decreasing number of youths interested in working in the agricultural sector are also experienced by other countries outside Asia, such as European countries and Canada [17,18,19]. This shows that the characteristics of farmer's household are considered less supportive to agricultural cultivation management. In the older age category, farmers tend to be closed to accept innovation such as the application of new technology. The dominating number of the old farmers will lead to the scarcity of farmers in the following years, thus later contributes to the scarcity of labors.

Saragih [20] proposes that age affects the working productivity of any activity requiring physical energy. On the other side, young farmers are considered to be more open minded and bold in trying to apply new technology for the sake of increasing the agricultural productivity [21].

b. In the development of any business, human power is the fundamental factor affecting the success. Although the other factors determining the success have been met, without the support of the adequate quality and quantity of human power, the business will experience various obstacles, even failure. The formal education required for the farmers is the education they have taken since the elementary school until the higher education, as depicted in the following table:

Table 2. Farmer Education

Education Groups	Number (person)	Percentage (%)
Not graduated from elementary school	28	14.0
Graduated from elementary school	83	41.5
Graduated from junior high school	52	26.0

Graduated from senior high school / vocational high school	33	16.5
Graduated from College (D3, S1, S2)	4	2.0
Total	200	100

Data source: Primary data analysis (2018)

The level of formal education of the farmers is important because it determines their capacity in counting, assessing, and analysing a business. The better level of education will result in the better ability to analyse a business [22]. The table above shows that the education level of the farmers is relatively low. This proves that their ability in mastering and applying technology is also limited. Thus, it will later lead to the quality and the quantity of the production result. The old age and the low level of education will affect the farmers' manner and mindset about their interest on the floating-rice cultivation technology.

The high level of education supports the nationalism and insight of the farmers towards the advancement and the adoption of new technology in managing their business. For this reason, farmers with the relatively higher education will be more rational in running their agribusiness by comparing the results gained from the use of new technology to the flooded rice fields. The education has a positive impact to farmers' skills in managing their agribusiness [23,24,25].

Learning from the experience and the knowledge can improve the farmers' skill in making decision wisely [26]. To support the skills and capability of the farmers in mastering and applying the floating-rice cultivation technology, the trainings on cultivation is required.

c. Working as a farmer or any other profession as the source of livelihood can be categorized as the basic and side jobs. As the basic job, the source of family earnings is mostly from farming, as depicted by the table below:

Table 3. Type of Jobs

Type of Jobs	Number (person)	Percentage (%)
Basic jobs	127	63.5
Side jobs	73	36.5
Total	200	100

Data source: Primary data analysis (2018)

Some people with farming as the basic job do not own agricultural land, but only as farm laborers (24. 5% of the total samples) and as cultivators. For some people in the research location, farmer is also as the side job of civil servants, village apparatus, traders, and entrepreneurs.

In the flood prone area, adapting to the work outside the agricultural sector is very possible, given that the rice fields cannot be used for 4 months. Farmers' adaptation activity is mainly by fishing in the inundation areas to get fish for food and for sale. Besides, they can work as daily laborers in the construction work. The adaptation in the agricultural sector has not been common because they regard the risk of cultivating the rice while flooding is relatively high.

d. The number of family members acts as two sides of a coin. In one side, it acts as one of the sources of income when the family members are in the productive age, so that they can help the family finance. Thus, it will give a positive impact in the food security of the farmer household. On the other side, the number of the family members can be the burden of the family itself when they are not in productive age [22]. The number of family members who still become the family burden is as follows:

Table 4. The Total Family Members

Number of dependents	Number (person)	Percentage (%)
1 – 4	120	65

5 – 6	70	35
>6	10	5
Total	200	100

Data source: Primary data analysis (2018)

The more complex the family burden indicates the more increasing the family expenses. To gain great family income, various efforts are done, such as getting income from outside the agricultural sector. This family burden is indeed source of potential workforce, which can be empowered in adopting the floating-rice cultivation. However, if the family burden is under or over the productive age, it is no longer potential workforce to reduce the cost of adopting the cultivation.

e. The family income of the respondent comes from both inside and outside the agricultural sectors. The income from farming is considered to have a high risk of loss when the rice planted and ready to be harvested fails due to the flood. In the conversion to month, the average income of farmer families can be elaborated as follows:

Table 5. Amount of Family Income

Amount of income (IDR/ month)	Number (person)	Percentage (%)
< 1.000.000	118	59
1.000.000 – 3.000.000	74	37
>3.000.000	8	4
Total	200	100

Data Source: Primary data analysis (2018)

The highest household income group is in the income less than IDR 1.000.000/month. This comes from the job as farm laborers who do not work every day. At certain times they are unemployed, for instance when the rice plants are under preservation. Meanwhile, the farmers whose main source of livelihood is not from farming, such as civil servants, village apparatus, drivers, and entrepreneurs, gain the income above IDR 3.000.000/month. Compared to the regional minimum wage, which is between IDR 1.350.000 – IDR 1.400.000, most of the farmers are under the minimum wage. The farmers' income can increase through land expansion, farm productivity increases through optimal utilization of land potential, and the application of agribusiness concept in farming [27].

Seen from the low income, the farmers are likely to find difficulty in funding the floating-rice cultivation due to the relatively high operational cost in making the floating medium. Generally, farmers with low income have no savings to invest because their income can only meet their daily needs, even less than expected.

f. Land tenure. One of the approaches in promoting the farmers' welfare and free them from poverty is by increasing the access of land tenure by the farmers [28]. The main factor of the poverty of villagers who are mainly working as farmers is that most of the farmers belong to small-scale farmers with the average number of land tenure less than 0.5 hectare. In Indonesia, the definition of small farmer commonly refers to the land tenure. According Sayogyo [29] classifies the farmers in East Java into three categories, namely: small-scale farmers with land tenure <0.5 ha, middle-scale farmers with land tenure 0.5-1.0 ha, and big-scale farmers with land tenure >1.0 ha. The land tenure, which comes from the self-owned and rented land as much as 10 percent from the total number of farmers renting lands from other farmers, is depicted as follows:

Table 6. Land Tenure

Land area (m^2)	Number (person)	Percentage (%)
0	49	24.5
1 – 2.500	130	65.0

2.501 – 5.000	18	9.0
>5.000	3	1.5
Total	200	100

Data Source: Primary data analysis (2018)

The biggest number of farmers' land tenure is between 1 – 2.500 m2, while the smallest number is more than 0.5 ha. The system of inheritance division results in narrower land owned by farmers so that the individual management is considered inefficient. Rich farmers who have not yet inherit the land to the heirs own the land tenure above 0.5 ha. With the assumption that the total income of the farmers is only from rice field farming in two seasons in a year, in order to promote the welfare or to free them from the poverty, or in other words, to gain income as much or more than the poverty line by BPS, the land tenure needed by each rice farmer household has to be at least 0.65 ha/capita/year [30].

From the of rice field area cultivated by the farmers, it can be said the farmers living near by the demplot are not categorized as prosperous farmers because their land tenures are under 0.65 ha/capita/years. Moreover, almost 25 percent of the respondents are only farm labours. This comes out as an obstacle in the adoption of the floating rice. They tend not to have the initiative to adopt a technology, except on the orders of the land owner. Because of it, land owners are most potential target to adopt the floating rice technology.

g. The mastery over the information source is reached using several sources to gain the information about the floating rice cultivation technology. The accurate and actual information will pique the farmers' interest in the process of adopting the technology. Furthermore, when the information is delivered by the reliable figures, the channel and the stream of the information will spread out faster to the farmers as the consideration for taking action. The information mastery by the respondent farmers can be elaborated as follows:

Table 7. The Mastery over Information Source of the Floating Rice Cultivation

Information source mastery	Number (person)	Percentage (%)
Self-observation	128	64.00
Information from peer friends	68	34.00
Mass media	4	2.00
Total	200	100

Data source: Primary data analysis (2018)

Most of the farmers get the appropriate information about the floating-rice technology by directly observing the floating rice demplot in the research location. The farmers passing by the demplot location are enthusiastic about the floating rice cultivation. Meanwhile, only some of the respondents know the floating rice cultivation technology from the mass media (online media) because most of the farmers are not conscious about the multimedia technology. The role of extension officers and community leaders also can increase farmers' awareness by inviting them to utilize multimedia technology and thus the floating-rice technology can be implemented successfully.

The more complete the information gained from the floating rice cultivation is, the more it will give positive impact on the increase of motivation and enthusiasm of the farmers in adopting the technology. This is so for by cultivating the floating-rice, farmers can still gain production results when the rice fields are drowned by flood. Meanwhile, inadequate information will give a negative impact in the motivation of the farmers due to the relatively high operational cost. Moreover, old-aged farmers are reluctant to make an innovation in running their agricultural business because they tend to surrender to the natural condition.

4. Conclusion

One of the ways to adapt to the floating-rice cultivation in the flooded area is by developing the floating rice cultivation technology. Before applying this technology in the wider scale, it is very crucial to know the social economic condition of the farmers who directly experience the flood impacts. The characteristics such as the dominating unproductive age, the low level of education, the low income, and inadequate information, become the obstacles in developing the floating-rice cultivation. The floating-rice cultivation technology is something newly introduced in Bojonegoro Regency so that the strong effort is required to attain the success in the application.

Acknowledgements
We would like to express our sincere gratitude to Universitas Sebelas Maret for funding this research. Our sincere appreciation is also extended to all parties who have provided information to complete this article.

Conflicts of Interest. The authors declare there is no conflict of interest

References

[1] Khan A S, Ramachandran A, Malini P and Usha N 2012 Adopt adaptation to changing climate: realizing the need of the hour *Ecology Environment & Conservation* **18**(1) 83-90
[2] Keo S 2015 *Impact of Climate Change on Agricultural Production in North West Cambodia* Conference on International Research on Food Security, Natural Resource Management and Rural Development September 16-18 Berlin Germany
[3] Rodenburg J and Meinke H 2010 *Adapting weed management in rice to changing climates* Second Africa Rice Congress Innovation and Partnerships to Realize Africa's Rice Potential 22–26 March Bamako Mali
[4] Irianto H, Mujiyo, Riptanti E W and Qonita A 2018 The land use potential of flood-prone rice fields using floating rice system in Bojonegoro regency in East Java ICSAE *IOP Conf. Series: Earth and Environmental Science* **142** 012072
[5] Ramakrishna G, Gaddam S R and Daisy I 2014 Impact of Floods on Food Security and Livelihoods of IDP Tribal Households: The Case of Khammam Region of India *International Journal of Development and Economics Sustainability* **2**(1) 11-24
[6] Harwood M R, Hacker J B and Mott J J 1999 Field evaluation of seven grasses for use in the revegetation of lands disturbed by coal mining in Central Queensland *Aust. J. Exp. Agric.* **39** 307-316
[7] Khan A R and Sing S S 2002 *Managing Flood Prone Ecosystem for Rice Production in Bihar Plains* Internal Report United Nations Educational Scientific and Cultural Organization and International Atomic Energy Agency
[8] Riptanti E W, Suprapti S and Qonita A 2016 Sustainable Food Security on Farmer Households in Flood Prone Areas *International Journal of Applied Environmental Sciences* **11**(6) 1435-1450
[9] Nguyen V, Vo O and Huyuh D 2015 Comparing the cost and benefits of floating rice-based and intensive rice-based farming system in the Mekong Delta *J. Asian Economic and Social Society* **5**(9) 202-207
[10] Kien N V 2016 The values and recovery progress of floating rice-based agro-ecological systems for adaptation to climate change in the Vietnamese Mekong Delta *J. Earth Sci. Clim. Change* **7** 9 (Suppl)
[11] Hoque M Z, Haque M E, Afrad M S I and Islam M N 2016 Effectiveness of Floating Agriculture for Adapting Climate Change in Southern Bangladesh *International Journal of Economic Theory and Application* **3**(1) 14-25
[12] Saha S K 2010 *Soilless Cultivation for Landless People: An Alternative Livelihood Practice through Indigenous Hydroponic Agriculture in Flood-prone Bangladesh* Ritsumeikan Asia Pacific University: 139-152

[13] Irfanullah H M, Adrika A, Ghani A and Khan Z A 2007 Introduction of Floating Gardening in the North - eastern Wetlands of Bangladesh for Nutritional Security and Sustainable Livelihood *Renewable Agriculture and Food Systems* **23**(2) 89-96
[14] Nueuman W L 1994 *Social research methods: Qualitative and quantitative approach* (Boston: Allyn and Bacon)
[15] Davis J, Caskie P and Wallace M 2013 How effective are new entract schemes for farmers? *Euro Choice* **12**(3) 32 – 37
[16] Badan Pusat Statistik 2013 *Laporan Hasil Sensus Pertanian 2013* Badan Pusat Statistik Jakarta, Indonesia
[17] Murphy D 2012 *Young farmer finance* Nuffield Australia Project No. 1203 Australia
[18] Wang J H 2014 Recruiting *Young Farmers to join small scale farming: a structural policy perspective* FFTC-RDA 2014 International Seminar on Enhanced Entry of Young Generation into Farming, Oct 20-24 Jeonju Korea
[19] Uchiyama T 2014 *Recent trends in young people's entry into farming in Japan: an international perspective* FFTC-RDA 2014 International Seminar on Enhanced Entry of Young Generation into Farming, Oct 20-24 Jeonju Korea
[20] Saragih B 2000 Agribisnis *Berbasis Peternakan* USESE Foundation dan Pusat Studi Pembangunan IPB Bogor Indonesia
[21] Dewi H, Sudi N dan Rini W 2008 Kajian Sosial Ekonomi Pelaksanaan Inseminasi Buatan Sapi Potong Di Kabupaten Kebumen *MEDIAGRO* **4**(2) 1 -12
[22] Muhammad H 2012 Karakteristik Sosial Ekonomi Rumah Tangga dan Tingkat Ketahanan Pangan Rumah Tangga Petani di Kabupaten Barito Kuala *Jurnal Agribisnis Perdesaan* **02**(04) 320 – 331
[23] Kalirajan K 1990 On Measuring Economic Efficiency *J Appl Economet* **5** 75 – 85
[24] Battese G and Coelli TJ 1993 *A Stochastic Frontier Production Function Incorporating A Model for Technical Inefficiency Effects* No 69
[25] Parikh A, Farman A and Shah M K 1995 Measurement of Economic Efficiency in Pakistani Agriculture *American Journal of Agricultural Economics* **77** 675-885
[26] Othman N and Kutty F 2010 Entrepeneurship behavior amongst Malaysian University Students *Pertanika Journal of Social Science Humanities* **19**(1) 23 – 32
[27] Ira M dan Mujiburrahmad 2014 Karakteristik Petani dan Hubungannya dengan Kompetensi Petani Lahan Sempit (Kasus: Di Desa Sinar Sari Kecamatan Dramaga Kabupaten Bogor Jawa Barat) *Agrisep* **15**(2) 58 – 74
[28] Sri H S dan Mohammad M 2012 Luas Lahan Usahatani dan Kesejahteraan Petani: Eksistensi Petani Gurem dan Urgensi Kebijakan Reforma Agraria *Analisis Kebijakan Pertanian* **10**(1) 17-30
[29] Sajogyo 1977 Golongan Miskin dan Partisipasi dalam Pembangunan (Poor Household and Their Participation in Development) *Prisma* **VI** (3) 10-17
[30] Susilowati S H, Hutabarat B, Rachmat M, Purwoto A, Sugiarto, Supriyati, Supadi, Zakaria A K, Winarso B, Supriyadi H, Purwantini T B, Elizabeth R Hidayat D, Nurasa T, Muslim C, Maulana M, Iqbal M dan Aldilah R 2010 *Indikator Pembangunan Pertanian dan Pedesaan: Karakteristik Sosial Ekonomi Petani Pad* Pusat Sosial Ekonomi dan Kebijakan Pertanian Badan Litbang Pertanian Bogor

The effect of infrastructure development on the development of tourism area and living environment of *Carocok* Beach, Painan

Erasukma Munaf, Syamsul Amar, Idris, Damel Van Wanda and Zikri Alhadi*

Evirontmental and Development Studies, Universitas Negeri Padang, Indonesia

*zikrialhadi@fis.unp.ac.id

Abstract. Carocok beach tourism destination is one of the integrated tourism development areas in Pesisir Selatan district, West Sumatra province which already has a Master Plan. The purpose of the study was to investigate the effect of infrastructure development on the development of tourism and the environment in the *Carocok* beach tourism area. This study is a causal method survey research involving 180 respondents, stakeholders of the *Carocok* beach tourism area including the Government, private parties, non-governmental organizations, tourists, and local residents. The sample selection is done in a convenient purposive sampling manner. Data analysis used path analysis which was preceded by test data requirements: normality, multicollinearity, autocorrelation, and heteroscedasticity. The research error rate was set at 5%. The results of the study show that infrastructure development has a significant direct effect on the development of tourism and the environment. Tourism development also has a significant direct effect on the environment.

1. Introduction

Tourism is an important component of the global economy. Tourism creates billions of dollars of income and millions of jobs worldwide. Tourism is considered by most people, especially in developing countries as the only means of development, and is the only opportunity to improve the quality of life. But at the same time, tourism is starting to show the downside. The activities of investors and tourists have a negative impact on socio-cultural values and natural environmental wealth of people around the world

Carocok Beach tourist destination is one of the integrated tourism development areas of South Pesisir Regency which already has a Master Plan. *Carocok* beach is located west of Painan city, approximately 2 km from Painan Market. This beach is very famous in West Sumatra and Indonesia, according to some people, *Carocok* Beach is not inferior to the Kra Peninsula Beach in Malaysia or Sanur Beach in Bali.

Basically, Ndulu explained the development infrastructure can be divided into: (1) economic infrastructure namely physical infrastructure that is used both in the production process and used by the community, covering all public infrastructure such as electricity, telecommunications, transportation, irrigation, clean water and sanitation and waste disposal (2) social infrastructure, namely social infrastructure such as health and education. Infrastructure is the driving force of economic growth. Insufficient infrastructure is one of the key obstacles to faster economic growth [1].

Investment in infrastructure investment in tourist areas will not only have a positive influence on economic growth, and employment, but can also have a negative influence from a socio-cultural and environmental point of view. Economic, socio-cultural and environmental impacts of tourism activities are important considerations in the planning and economic development of a country, regional region and for the community. Therefore, the community needs to understand the relative importance of the tourism industry to their region, including its contribution to the development of economic, socio-cultural and physical environment of the tourist area.

The aim of the study was to investigate the effect of independent variables on infrastructure development on the dependent variable of tourism and environmental development in the tourist destinations of *Carocok* Beach West Sumatra.

2. Literature review

2.1. Living environment

Holden [2] states that to discuss more accurately the relationship of tourism with the environment, it is necessary to see actors who play a role in tourism activities such as central and regional government authorities, the private sector (investors, contractors), non-governmental organizations that care about tourism, local residents, as well as tourists. In general, actors who play a role in the development of the tourism area mentioned above are human. So if you want to discuss the relationship between tourism and the environment, you need to discuss human relations with the natural environment itself; how humans respect their natural environment, cultural stereotypes about nature, and human ethical relationships with nature.

In the case of tourist areas in this study, what is meant by environmental impacts is the perception of local residents regarding changes in the physical environment that occur due to tourism activities.

2.2. Tourism development

WCED explained the sustainable development is development that meets the needs of the present without neglecting the ability of the next generation to meet their needs [3]. And Weaver explained, sustainable tourism is defined as the application of sustainable development in the tourism sector, or the development of tourism that uses wisely and conserves natural resources for long-term tourism feasibility [4].

2.3. Infrastructure development

Adebayo explained, tourist infrastructure is defined as a physical element that is designed and built for the needs of tourists. A theoretical strong relationship between tourism development and infrastructure development has been put forward by various researchers [5]. Tourism infrastructure includes the main facilities and equipment, equipment, systems, processes and resources needed for the functioning of tourist destinations. Mainly includes roads, railways, airports, ports and others that can facilitate access to tourist destinations. As an addition, infrastructure also includes health care systems, public services and services. Seetanah et al [6] in his research found that tourists are very sensitive to infrastructure or in other words infrastructure provides a positive contribution to the arrival of tourists, mainly from Europe / America and Asia.

Jovanović [7] in his research concluded that there is a strong relationship between infrastructure development and tourism development. Ghulam Rabbany et al [8] in his research described the impact of tourism on natural resources, environmental pollution on a global scale, and its contribution to environmental conservation. INTOSAI [9] revealed that infrastructure development can significantly impact the local environment and community and more widely and use a lot of natural raw materials in its development.

3. Methodology

The research design used is a survey with a causal design. Study respondents totalling 180 stakeholders *Carocok* beach tourism destinations include local government authorities, investors and tourism entrepreneurs, local residents, tourism observers and non-governmental organizations. Data was collected using a questionnaire instrument designed as a closed question with five choices of Likert scale answers. The infrastructure development questionnaire totalled 22 items, the tourism development questionnaire numbered 13 items and 9 environmental questionnaires. Data analysis used path analysis which was preceded by test data requirements: normality, multicollinearity, autocorrelation, and heteroscedasticity. The research error rate was set at 5%.

Based on the description of relevant research and research concepts, the research framework can be identified as follows:

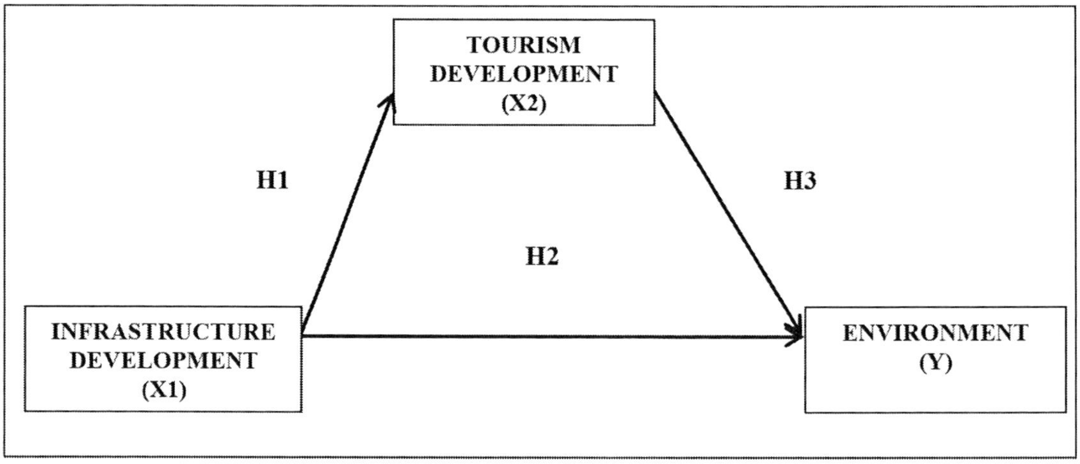

Figure 1. Research Framework

Hypothesis
H1 Infrastructure development (X1) directly affects tourism development (X2)
H2 Infrastructure development (X1) directly affects the environment (Y)
H3 Tourism development (X2) directly affects the environment (Y)

4. Result

4.1. Instrument validity and reliability test

Test the validity and reliability of the instrument carried out on 30 respondents with a summary of the results of the instrument validity and reliability testing as follows:

Table 1. Validity & Realibilty Test

Variable	r-count	r-table	Conclusion	Alpha Cronbach	Conclusion
P1 (XI)	0.878-0.925	0.361	Valid	0.990	Realiable
PP (X2)	0.839-0.928	0.361	Valid	0.980	Realiable
LH (Y)	0.783-0.889	0.361	Valid	0.966	Realiable

The summary of the results of the validity and reliability of the instrument above shows that all research instruments are valid and reliable so that they can be disseminated to collect data.

4.2.Descriptive statistics
The results of research data collection if described statistically are as follows:

Table 2. Descriptive Statstitcs

	N	Range	Minimum	Maximum	Sum	Mean	Std. Deviation	Variance
PI	180	72	38	110	13320	74.00	13.404	179.665
PP	180	45	20	65	7650	42.50	8.465	71.648
LH	180	32	13	45	5220	29.00	6.028	36.335
Valid N (listwise)								

4.3. Test data requirements
Data collected successfully before being used to test the research hypothesis needs to be tested for data requirements including normality, multicollinearity, autocorrelation, and heteroscedasticity. The data requirements test results show:
a. All research variable data are normally distributed
b. The research model has no conclusions about autocorrelation
c. The research model is free from the symptoms of multicollinearity and heteroscedasticity

4.4. Hypothesis testing
The results of hypothesis testing are as follows:

4.4.1. Substructure LH = f (PI, PP)

Table 3. Model Summary

Model	R	R Square	Adjusted R Square	Std. Error of the Estimate
1	.911[a]	.830	.829	2.496

a. Predictors: (Constant), PP, PI

Table 4. Coefficients[a]

Model	Unstandardized Coeficeints		Standardized Coeficients		
	B	Std. Errors	Beta	t	Sig.
1 (Constant)	-.653	1.049		-.622	.543
PI	.180	.029	.399	6.203	.000
PP	.385	.046	.541	8.398	.000

4.4.2. Substructure PP = f (PI)

Table 5. Model Summary

Model	R	R Square	Adjusted R Square	Std. Error of the Estimate
1	.877[a]	.769	.768	4.081

a. Predictors: (Constant), PI

Table 6. Coefficients[a]

	Unstandardized Coefficients		Standardized Coefficients		
Model	B	Std. Error	Beta	t	Sig.
1 (Constant)	1.524	1.711		.891	.374
PI	.554	.023	.877	24.334	.000

a. Dependent Variable: PP

Summary of hypothesis test results are as follows:

Table 7. Result

Relation	Path Coefficient	p-value	p-critical	Significant	Determination
PI → LH	0.399	0.000	0.05	Significant	0.830 (83%)
PP → LH	0.541	0.000	0.05	Significant	
PI → PP	0.877	0.000	0.05	Significant	0.769 (76.9%)
PI → PP → LH	0.877 x 0.541 = 0.475				

a. Infrastructure development (X1 = PI) has a significant direct effect on tourism development (X2 = PP)
b. Infrastructure development (X1 = PI) has a significant direct effect on the environment (Y = LH)
c. Tourism development (X2 = PP) directly affects the environment (Y = LH)

5. Discussion

The results of the above analysis are in line with the research of Seetanah et al. [6] and Jovanović [6] which states that tourists are very sensitive to infrastructure or in other words infrastructure makes a positive contribution to the arrival of tourists; and there is a strong relationship between infrastructure development and tourism development.

The results of the analysis are also in line with the research of Ghulam Rabbany et al [8) which describes the impact of tourism on natural resources, environmental pollution on a global scale, and its contribution to environmental conservation, and in line with INTOSAI [9] which states that infrastructure development can significantly impact the environment and the local community and more broadly and use a lot of natural raw materials in its development.

Jovanovic explained, for the development of successful tourism, the need for more intensive investment in infrastructure modernization is increasingly emerging as a necessary condition. Higher levels of tourism infrastructure development can contribute to increasing production efficiency and distribution of tourism services, and, in some cases, such as remote destinations, increase the supply of tourism services. For the presence in the tourism market, which is becoming more dynamic and demanding, the question of increasing competitiveness is very important. In this case, investment in developing tourism infrastructure is an important component of tourism competitiveness [7].

The construction of new infrastructure can bring significant benefits, including opportunities to build priorities to meet the challenges posed by change and to improve sustainability. However, it can also use significant costs for the environment, both in construction and in its use. Some costs can be overcome and discussed through the planning and design process and opportunities needed to obtain direct and indirect benefits.

Tourism creates disaster situations throughout the world and this condition of destruction can quell regional collaboration and communication. Everyone must be aware of the negative impacts of tourism and take appropriate steps to reduce the problems, especially of each state government and international authority in the tourism industry.

6. Conclussion

Infrastructure development has a significant direct effect on tourism development and the environment. Tourism development also has a significant direct effect on the environment. For this reason, it is recommended that the development of tourist areas need a concept of infrastructure development that is measurable and takes into account environmental sustainability

References

[1] Ndulu B, Kritzinger-van Niekerk L and Reinikka R 2005 Infrastructure, Regional Integration and Growth in Sub-Saharan Africa *Africa in the World Economy*, 101

[2] Holden A 2016 *An introduction to Tourism Environment Relationships Ecotourism and environmental sustainability* (Routledge: Taylor & Francis Group)

[3] WCED (World Commission on Environment and Development) B C 1987 Our Common Future *Report of the world commission on environment and development*

[4] Weaver D B 2006 *Sustainable tourism: Theory and practice* (Routledge: Taylor & Francis Group)

[5] Adobayo K A dan Iweka C O A 2014 Optimizing the Sustainability of Tourism Infrastructure in Nigeria through Design for Deconstruction Framework *American Journal of Tourism Management* **3**(1A) 13-19

[6] Seetanah B, Juwaheer T D, Lamport M J, Rojid S, Sannassee R V and Subadar A U 2011 Does Infrastructure Matter in Tourism Development *University of Mauritius research Journal* **17**(1) 89-108

[7] JOVANOVIĆ S and Ivana I L I Ć 2016 Infrastructure as Important Determinant of Tourism Development in the Countries of Southeast Europe *Ecoforum journal* **5**(1)

[8] Ghulam R M, Afrin S, Rahman A, Islam, F and Hoque F 2013 Environmental Effects of Tourism *American Journal of Environment, Energy and Power Research* **1(7)** 117-130

[9] INTOSAI Working Group on Environmental Auditing WGEA 2013 *Environmental Issues Associated with Infrastructure Development*

The effect of vocational learning strategy and knowledge of sustainable development in increasing traditional cattlemen's skill in making bio-digester

Nadiroh[1,*], Nur Fadli Hazhar Fachrial[2] and Erry Utomo[2]

[1] Environmental Education, Universitas Negeri Jakarta, Indonesia
[2] Educational Research and Evaluation, Universitas Negeri Jakarta, Indonesia

*nadiroh@unj.ac.id

Abstract. The objective of this research is to find out the effect of vocational learning strategy and knowledge about sustainable development to increase traditional cattlemen skills in making Bio-digester. The research design used experiment of 2 x 2 factorial design. The sample were 80 traditional cattlemen at District Sukawangi who were divided in two groups. The result showed as follows 1) There is significant difference on the cattlemen group who was taught with vocational learning strategy between school based and work place based; 2) There were interactions found between two factors there are vocational learning strategy and knowledge about sustainable development; 3) The cattlemen group who have high knowledge about sustainable development and taught with work place-based strategy, the skill in making Bio-digester was higher than the group who taught with school-based strategy; 4) The cattlemen group who had low knowledge about sustainable development and taught with work place-based strategy, the skill in making Bio-digester was higher than the group who taught using school-based strategy. Vocational learning strategy using work place based can be more effective to increase the level skill of cattlemen in making Bio-digester by considering their knowledge about sustainable development.

1. Introduction

The ability of Indonesia produces renewable energy from different types of Biomass of no more than 1 percent of the potential of 49,810 MW. Tropical Indonesia, with fertile soil and the agrarian society should be able to answer the challenges the needs of the energy in the future. The potential of energy from Biomass is very large, Deublein states: "In theory all the amount of space, including the surface of the water can be used to produce biomass" [1].

Human factors have a great influence towards the application of alternative energy production in addition to natural resource and technology factors. The ability of human resources, lack of education and training (training and education), as well as the lack of information on Renewable Energy devoted to the community is the key to answer the challenges of meeting the energy needs for Indonesia in the future [2].

The Skill of making simple Biogas is attempted because it is in line with the interests of society in Indonesia in fulfilment of environmentally friendly energy. The fact that energy is increasingly expensive in the future as well as the efforts of fostering self-reliance and improved the welfare of society, especially society in the sector of household which is far from the energy access.

The population of Indonesia 60 percent resides in rural areas, therefore energy self-reliance in rural areas becomes top priority. The program of Energy Independent Village (EIV) has proven to be able to improve the welfare of poor villagers/isolated/border through the provision of energy access independently developed by utilizing energy for productive activity. Although it has been a priority, yet the utilization of Biogas as an energy source remains still feels a little bit. From 633 EIV which had formed by the end of 2009, spread all over in 27 provinces only 14 EIV using Biogas.

The skills of a person will be different to each other. This is because of the Cognitive Performance. Cognitive performance is the ability/achievements of someone who can clearly visible running cognitive tasks. The cognitive performance influenced by genes, general ability, experience (life experiences) as well as knowledge they have learned.

Richard E. Mayer [3] defines the ability as a potential person in learning gain knowledge. Experience is defined as a person's interactions with the environment and one of the factors that greatly affect the experience is the presence of specialization/specificity of an exercise. Where the existence of special exercises then formed a specialized expertise.

Skills according to John R Schermerhorn, Jr. [4], is the ability of a person in translating the knowledge into a real action as well as indicated by performance. Skills must be poured into a work or act that gives more value or achievement.

Someone who has a particular skill, will look into the personality obviously eager to obtain the knowledge. This opinion was expressed by Philip L. Ackerman [5]. Added that to form a capable private can only be obtained through investments in the form of business throughout a certain period of time.

Bio Digester is a tool that is designed to serve as the digest organic materials by bacteria parser with airtight conditions (anaerobic). The result of the decomposition of organic material it is a gas that is generally called Biogas. Biogas is a gas that is produced from the anaerobic biological processes in the form of the ingredient composition of Methane gas (CH_4), carbon dioxide (CO_2), water (H_2O) in the form of steam, Nitrogen (N_2) a little Hydrogen Sulfid (H_2S). Literacy in Biogas is a gas formed from living beings. The bacteria produce biogas just do in the absence of air (an means no, aerobic means air) [5].

Mazumdar mentions that Biogas is a gas composed of methane, carbon dioxide, hydrogen sulfide and some other gases produced by anaerobic fermentation of organic material such as animal and human feces, foliage, grasses. The presence of methane gas in the biogas gives the ability as a fuel suitable for cooking and lighting [6].

Deublin explains that the formation of methane gas is the biological processes which form when organic material (biomass) decomposes in airtight conditions helped by a kind of microorganisms (methane bacteria) [1]. Naturally this methane gas can form in the stomach or digestive tract of herbivore animals, wet plants experience the compost process and rice paddies inundated. All organic materials (biomass) which can be fermented is called the substrate. This several process of making bio digester also same with previous research held in Kenya and Rwanda [7].

The skills of making Bio Digester is a person's ability to work to produce a Bio Digester tool coupled with knowledge and experience in its making so that one has special skills that is recognized by the community.

A Vocational learning referred as the vocational learning is all aspects required by learners in vocational study including socialization theory, activity theory, collective learning, practices learning, organizations learning and the development of psychology.This combination requires an active approach of pupils to learning. Active, in this sense, means pupils have to integrate their knowledge of facts, concepts and procedures with new facts, concepts or procedures in such a way that they construct their own new meaningful knowledge [8].

Benjamin S. Bloom [9] divides the ultimate goal of learning to be 3 (three) domains namely 1) the realm of cognitive, 2) affective and 3) Psychomotor. Cognitive including the activity associated with listening and considering the knowledge, thinking, creating, and solving the problems.

The meaning of sustainable development is a paradigm in optimizing the environment, economic and social aspects to ensure the integrity of the environment, safety, capacity, quality of life and well-being of the present and future generations by holding on to the values and principles of sustainability [10].

Over the last two decades sustainable development (SD) has become an intensively and widely discussed theoretical concept. The Decade of Education for Sustainable Development (DESD), which was launched by the United Nations in 2005, can therefore be seen as a seminal moment that vividly emphasizes the specific significance that SD bears for the educational sector [11].

Education is a key factor in developing public knowledge and awareness about issues that affect the future of a nation and, subsequently, the world. Environmental problems have become issues of great concern to many parties [12]. A concern for the environment has now become part of the concept of sustainable development, including to make a simple Bio Digester.

Vocational Learning Strategy, one of which is the School Based Course or Training, is a vocational learning strategy-oriented to study reviews of the basic theories, concepts and models related to the world of work which will be learned. The learning strategy of vocational school-based intended in order that theories and concepts related to the world of real work can be transferred to participants well and effectively [13].

Some efforts to improve the skills of making simple Bio Digester using the vocational School-based learning strategy is a strategy of the introduction and transfer of knowledge about the basics of the theory of creation as well as knowledge about Bio Digester theories and concepts of Sustainable development to use the location where the learning tools, supplies, curriculum practice in the classroom [14].

The approach of work based is the vocational learning strategy that creates competence through learning experiences that experienced process of themselves (experience and experimental learning). Pioneer of theories about work-based training expressed by Jacob Mincer and Gary Becker i.e. increased skills can be done either inside or outside of the institution. The process that occurs during the learning done at work is doing the impersonation work (imitation) exemplified by workers who are trained and professional. Learning at work describes the orientation process and lifelong learning which contributes in the development of the individual and the professionalism and support innovation.

A work-based vocational learning strategy in an effort to increase the skill of making Bio Digester is a strategy of approaching participants with the location of the workplace in order to acquire knowledge through the work experience done by themselves, looking at practical examples from professional trainers and learning curriculum oriented work process.

2. Methodology

This research uses Experimental methods by using factorial design 2 x 2. The context of these research studies is traditional cattlemen where the target population is the entire traditional cattlemen in District Sukawangi, County of Bekasi. While the reached population is traditional cattlemen in the village of Setiamekar. It has been chosen the traditional cattlemen as the object of research because traditional cattlemen generally have a number of cows which are not more than 5 cows would utilize directly the Bio Digester being made. Sampling procedure is done with a few steps, namely 1) set District Sukawangi as the area of research, 2) by using cluster random sampling of 8 existing villages in District Sukawangi chosen the village of Setiamekar, 3) selected by random 80 traditional cattlemen in the village of Setiamekar, 4) chosen by random members of the treatment A and B respectively of the Group amount to 40 people, 5) perform the measurement of knowledge about sustainable development using a written test to each member's treatment A and B. 6) written test results later sorted. The value of the order from the results of researcher taking a 27% of every group of 40 cattlemen, then retrieved each group numbered 11 people who have high knowledge about sustainable development and 11 people who have low knowledge about sustainable development.

2.1. Measurement

2.1.1. The Skills of Making Bio Digester. The instruments used to collect the data of skills of making Bio Digester measured using multilevel scale instrument and three assessments: Skilled (S) is given a score of 3, Skilled Enough (SE) is given a score of 2, Less Skilled (LS) is given a score of 1. The highest score for the assessment of skills is 90 and the lowest score is 30.

2.1.2. Knowledge about Sustainable Development. The instrument test of knowledge about sustainable development is used in the form of reserved option is true (T) and false (F) that add up to 30 items. If the correct answer is given a score of 1 and if one is given a score of 0, so that the highest score is 30 and lowest score is 0. The reason the use of test True and False selection because of the goal of the test is the average cattlemen accustomed to thinking of practical, so the revelation of problems is made as simple as possible, the process of score and results more modest correction, and time worked on the test is relatively short.

2.2. Treatment

2.2.1. Vocational School Based Learning Strategy. Coach reveals the basic theoretical studies that describe the functions of Bio typical analysis for digester, recognize the basic concept of the formation of bio gas, individual practice with existing equipment assisted with explanation and demonstration by coaches, and then shows the results of the utilization of gas from the typical analysis for Bio digester.

2.2.2. Vocational Work Place Based Learning Strategy. The delivery of the basic theory that describes the functions of this typical analysis for Bio digester done entirely in the work place. Participants will be mentored and monitored in every activity for workplace-based learning takes place. The delivery of the material, ask and discuss fully carried out directly on-site work, including demonstration and manufacture of Bio digester.

3. Results

Validity Test of using the formula of Correlation of Product Moment. After obtained the validity coefficient (r) or r count then compared with the table, n = 30 i.e. 0312. If r count > r table then the item is said to be valid. On the variable of 30 items of questions submitted to 30 respondents gave the results of the 3 items of questions revealed Drop. The Reliability test to the overall items of instruments expressed to be valid using the formula of α= 0.63.

Independent variables from 30-questions given to 30 respondents gave results 3 items of questions revealed Drop. The Reliability test to the overall items of instruments that expressed to be valid using Kuder-Richardson formula with result = 0.82.

Normality Test of the data in this study using a test of Kolmogorov-Smirnov (K-S) and to test its homogeneity test of population variances using Bartlett both at level of significance α = 0.05.

4. Discussions

The results of ANOVA calculations between groups shows the value of Fcount = 9.46 which turns out to be greater than the value of the Ftable = 4.31 on 99 percent significance level. This means that HO denied and accept H1. This indicates there is a significant difference between the application of school based and work place based vocational learning strategies towards the skills of cattlemen to make Bio Digester.

The results of ANOVA calculations about Knowledge of Sustainable Development showed Fcount = 18.49 which turns out to be greater than the value of Ftable = 7.31 on 99 percent significance level. This means that Ho is denied and accept H1. This indicates there is a difference between the level of high and low knowledge about sustainable development towards the skills to make Bio Digester.

The results of ANOVA calculations showed the presence of Interactions shown on Finteraction = 5.2 which turns out to be greater than the value of Ftable = 4.08 on level of significance of 95 percent. This means the interaction between independent and dependent variables have the Interactions, it can be shown in Figure 1.

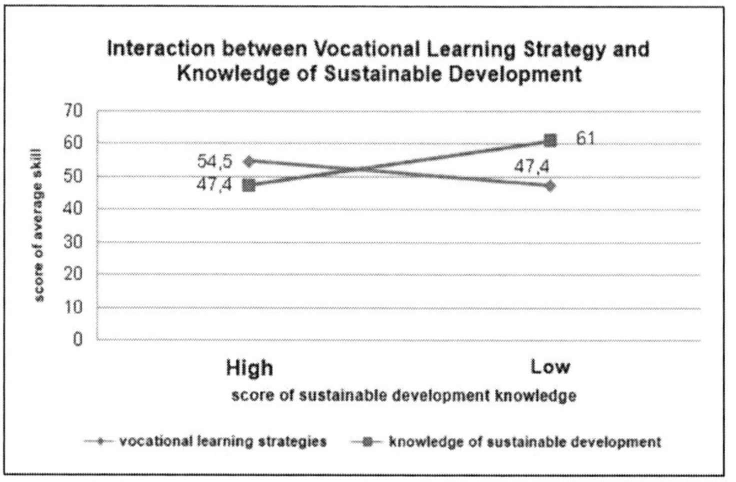

Figure 1. Interaction between Vocational Learning Strategy and Knowledge of Sustainable Development

When the results of the variance analysis showed there were major influences from non-variable against variable and there is an interaction between A and B, then the test is continued with the Tuckey method to know the meaningfulness of such interaction. Tuckey test is used as the number of samples of each cell is the same and the hypothesis testing criteria used in standard significance. Tukey test was used to compare groups of cattlemen who have high knowledge about the Sustainable Development of vocational school based [A1B1] and workplace based [A2B1] learning strategies towards the skills of making Bio Digester results Qcount = 4.64 which turns out to be greater than the value of Qtable = 4.26 on significant level $\alpha = 0.05$.

Based on Tukey Test showed that the skills of the cattlemen make a Bio Digester with high knowledge of Sustainable Development learning strategies that follow either a vocational school-based learning strategy or vocational workplace-based have significant differences. The Calculation analysis of variance with Tukey test stage to compare a group of cattlemen who have knowledge of Sustainable Development with vocational School-based learning strategies [A1B2] and workplace-based [A2B2] towards the skills of making a Bio Digester results Qcount = 4.77 which turns out to be greater than the value of Qtable = 4.26 on significant level $\alpha = 0.05$. Based on Tukey Test showed that the skills of the cattlemen make a Bio Digester with low knowledge of Sustainable Development following the vocational School-based and workplace-based learning strategies have significant differences.

5. Conclusion

This research uses experimental method that aims to find out whether there is a difference or influence learning strategies implementation i.e. vocational School-based and workplace-based learning strategies towards the skills of cattlemen to make the Bio Digester. Based on the results of hypothesis testing can be summed up as follows:

First, the overall skills of cattlemen making a bio digester was given a treatment of vocational workplace-based learning strategy are higher than the cattlemen who have given the vocational school-based learning strategy. Second, there is an interaction between the factors of vocational learning strategies with the cattlemen' knowledge about factors of sustainable development towards the skills of cattlemen in making make the bio digester. Third, on a group of cattlemen who have high knowledge about sustainable development, the skills of the cattlemen make a bio digester by applying workplace-based learning strategies are higher than those applying vocational school-based learning strategies.

Fourth, on a group of cattlemen who have low knowledge of sustainable development, the skills of the cattlemen make a bio digester by applying work place-based learning strategies giving a higher value than the school-based learning strategies. Based on the findings above, in this case can be drawn the conclusion that the vocational learning strategies can effectively improve the skills of the cattlemen make a bio digester considering of the cattlemen' knowledge about sustainable development. For a group of cattlemen who have low or high knowledge about the sustainable development will be given effectively about the vocational work place-based learning strategies.

Acknowledgments
We would like to thank the editorial manager of this publication. We are also immensely grateful to the reviewers for their comments on an earlier version of the manuscript.

References

[1] D Deublein and A Steinhauser 2011 *Biogas from waste and renewable resources: an introduction* (John Wiley & Sons)
[2] A S Goudie 2018 *Human impact on the natural environment* (John Wiley & Sons)
[3] Mayer R E 2003 What causes individual differences in cognitive performance? In R J Sternberg and E L Grigorenko (Eds.) *The psychology of abilities, competencies, and expertise* (New York, NY, US: Cambridge University Press) pp 263-273
[4] Schermerhorn J R et al 2009 *Organizational Behavior, 11th Edition* (New Jersey: John Wiley & Sons)
[5] Ackerman P L 2009 *Personality and Intelligent* on Philip J. Corr and Gerald Matthews *The Cambridge Handbook of Personality (by) Psychology* (Cambridge: Cambridge Press University)
[6] Mazumdar B and Chaudhari P K 2015 Electrochemical treatment of biodigester effluent of maize based starch industry *Indian Journal of Chemical Technology* **22** (September 2015) 201-209
[7] A D Tigabu, F Berkhout and P van Beukering 2015 The diffusion of a renewable energy technology and innovation system functioning: Comparing bio-digestion in Kenya and Rwanda *Technological Forecasting and Social Change* **90**(PA) 331–345
[8] Bijsterboscha E, Scheeb J and Wilmad 2017 Meaningful learning and summative assessment in geography education: an analysis in secondary education in the Netherlands *International Research in Geographical and Environmental Education* **26**(1) 17-35
[9] Bloom, B.C. et al. (1956). Taxonomy of Educational Objectives the Classification of Educational Goals-Handbook 1, Cognitive Domain. Michigan: Logmans.
[10] Nicolescu B 2018 The transdisciplinary evolution of the university condition for sustainable development in *Transdisciplinary Theory, Practice and Education* (Springer) pp 73–81
[11] Agustini C B *et al* 2018 Biogas production for anaerobic co-digestion of tannery solid wastes under presence and absence of the tanning agent *Resources, Conservation and Recycling* **130** 51–59
[12] W A Rosenbaum 2016 *Environmental politics and policy* (Washington DC: CQ press)
[13] J Wellington 2015 *Educational research: Contemporary issues and practical approaches* (Bloomsbury Publishing)
[14] U Balasubramaniyam, L S Zisengwe, N Meriggi and E Buysman 2008 Biogas production in climates with long cold winters *Doc. Prep. by Wageningen Univ. Netherlands Women Eur. a Common Futur. WECF Netherlands* pp 1–68

The geodiversity potential of *Tanah Datar* District developing into a geotourism asset as a geopark in Indonesia

Osronita[1,*], Syafri Anwar[2], Heldi[1] and Eri Barlian[1]

[1] Department of Environmental Science, Postgraduate Program, Universitas Negeri Padang, Prof. Dr. Hamka Street, Air Tawar, Padang-West Sumatera 25131 Indonesia
[2] Department of Geography, Faculty of Social Science, Universitas Negeri Padang, Prof. Dr. Hamka Street, Air Tawar, Padang-West Sumatera 25131 Indonesia

*osronita53@gmail.com

Abstract. This study aims to determine the geodiversity potential of Tanah Datar District, developing into a geotourism asset as a Geopark in Indonesia. This type of research is qualitative descriptive research. The research data acquisition technique was carried out by observation, literature study, interview, FGD (Focus Group Discussion), the method used to determine the delineation of the geoheritage area using the Geographic Information System (GIS) using GIS Arc Software 10.5. The results of geodiversity potential research in Tanah Datar District consist of 3 geosites; 1) Mount Marapi, Mount Singgalang-Tandikek, Straightness of Five Fault Lake (5) Fault Lake and Active Sectional Tectonic Lake 2) Waterfall from hot spring 3) Straightness of Andesite Intrusion. The Geo Area Tabek Patah Salimpaung Normal Fault includes geosite; 1) Straightness of Fault and Fault Talent 2) Lineage of Telaga Line Saga Telaga Air Pakis and Telaga Air Taranang. Geo Lintau Buo Karst Area with Geosite 1) Eksokarst 2) Sinkhole 3) Underground River 4) Goa / Ngalau with Stalactite-Stalagmite 5) Dolina. The potential of this geodiversity is Geotourism Assets and the potential for the development of Indonesian Geopark.

1. Introduction

Geodiversity or geological diversity is a description of the existence and spread of various geological components found in an area, whose conditions can represent the process of diving geological evolution a available time space [1]. The Sumatera fault zone consists of a number of major depression which extends from the northeast of the gorge. This geological diversity will be a geotourism asset.

Geotourism is sustainable tourism with the main focus is the evolution of the earth and geological features that encourage understanding of environment and culture, appreciation and conservation and benefit local communities, geotourism also involves geological and geomorphic features that contribute to the formation of a sense of place for each geosite. specifically these features include various types of landscapes, landscape, rock outcrops, rock types, sediments, soil as well as rock crystals [2]. For geotourism development has several indicator of feasibility and redness (site evaluation) for geotourism that is 1) physical geological of geology site and naturally formed geographic features that are the focus of geotourism planning and development. That Have access that can be taken by vehicles to geotourism (accessibility) 2) Economic sustainability, encouraging economic feasibility of the community and management of the area with sustainable principles, and conservation can develop the capacity of tourism with sustainable principles in the geotourism area without damage the environment, there is not destroying authenticity and the beauty of the area and is responsible for the use of the geotourism area. 3) Educative, informative in the presence of interesting and educative interpretive media, which can increase the awareness of tourists about educational environmental conservation 4) Community participation is the involvement of local communities in the management of the area. 5) Tourist satisfaction is a special impression that tourists get when visiting the geotourism area. If all indicators can be met, then a new area is feasible to be used or developed geotourism activities [5,6].

The types of geotourism activities and attractions include; 1) Sightseeing geo-sites recreation or sightseeing by enjoying the landscape from the unique geography formation 2) Geo-Sport sports related to the earth's topography 3) Geo-Study activities in the outdoors, including geological heritage observation, geo-landscape photography, field trips for geological interests 4) Geo-conservation and Geo-education, conservation programs on earth potential for the sake of education or preservation 5) Geo-festival events created for the continuation of geological resources or promotional platforms for conservation programs 6) Geo-Tours facilities are independent interpretations (Geotourism Map) or tour guide facilities 7) Health and Wellness Geotourism in the form of health or relaxation facilities such as spa therapy, stone therapy, and mud therapy 8) Geo-Trek travels to explore the potential of the earth by observing and understanding the process of forming the earth.

Geopark is built in Buttom-Up which is an implementation of appreciation of natural, non-biological and cultural natural resources which are the essential wealth of the Regional Government that aims to respect the earth with all its contents in full, as a tool to develop regions sustainably based on the pillars of conservation, education and growth local economic value, Geopark is built and developed to glorify the earth's heritage and prosper society [3]. Geopark uses geological heritage that is associated with aspects of natural heritage and cultural heritage for educational purposes, increasing awareness and understanding of problems faced by the community such as how to recognize and make use of earth resources (geo, bio and culture) in a sustainable manner, reduce the impact of geological disasters and how to improve people's welfare by safeguarding the earth's heritage. In carrying out the Geopark concept, a supporting activity for sustainable economic activities is needed in the form of Geotourism activities [7,8]. Geotourism is one of the important activities for take place of a Geopark because geotourism is created and developed based on the principle of sustainable development because the object and attractiveness can be derived or inherited for the future generations.

2. Research Methods

This research method uses qualitative descriptive data analysis method. This research was conducted in three stages; first, the data collection and inventory stage by collecting secondary data then formed a field working group and provided assistance in the inventory of data followed by work inline using social media, interviewing respondents is the community at the geodiversity potential location. Second stage of field observation on the geological site (geosite). The third stage of data analysis and assessment is descriptive by conducting FGD involving agencies related to geological sites to develop the geodivesity potential of Tanah Datar District, the community, stakeholders. Then do regional delineation mapping using ArcGis 10.5 software.

3. Results and Discussion

Geology Map Sheet Padang, Sumatra [10], maps the Great Sumatera Fault (Great Sumatra Fault) in Tanah Datar District in the southwest foot of Marapi mount with coordinates 100025'00 "LS and 0024'30" LS around Nagari or Koto Baru Village, then the fault appears again in Nagari / Paninjau Village with coordinates of 100026'30'BT and 0027'15 "LS continues to merge with the river valley of the Sumpur River, continuously passing the east wall of Lake Singkarak (second picture). Chronologically the fault is categorized as active. Not all of these fault faults can be identified. To the south of the New Koto Nagari to Nagari / Aia Angek Village, some activity from secondary deposits from Mount Marapi is thought to have buried traces of the natural fault phenomenon. The fault that crossed at the foot of Marapi westward passed through the Nagari / Batipuh Village, Mount Rajo, Pitalah and entered Lake Singkarak.

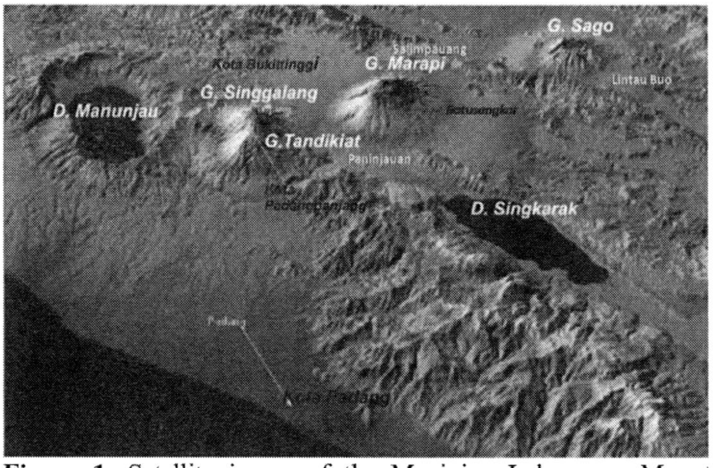

Figure 1. Satellite image of the Maninjau Lake area, Mount Marapi, Mount Tandikek, Mount Singgalang and Singkarak Lake, (Source: SRTM Image)

Based on the results of field observations found the Geodiversity Potential of Tanah Datar Regency including; 1) Fault Lake, there are five (5) Lake Faults which are in a straight line towards Northwest - Southeast in the direction of the Sumatera Fault [2]. The Lake Fault / Sagpond referred to are Talago Batupalano / KotoBaru, Talago Kayu Tanduak, Talago Aia Angek and Talago Paninjau. 2) Sumatrea fault faults, 3) Identified River Flow and hill ridge curvature formed from Quaternary-aged Volcanic sedimentary rock units, the earth's surface age was hit by a thick layer of ice so called the ice age (Diluvium era) [4]. Can be seen in first picture below;

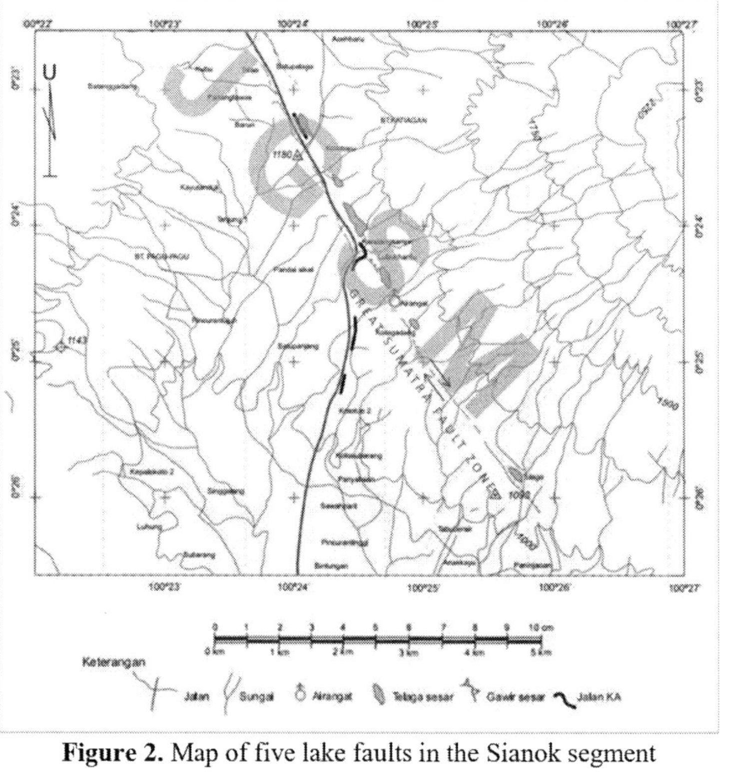

Figure 2. Map of five lake faults in the Sianok segment

Solok Sumatera [11] Geological Map Sumatra [11] shows Salimpaung and Sungayang Subdistricts passed normal / down faults with the azimuth N320E north side lifted (Horst) composed by Filit and Flake Member rocks from the Ku-Perm (Carbon) Formation of Perm-Carbon consisting of shale, filt, slate, quartzite, silt stone, flint and lava flow. The south side has decreased (Graben) compiled by Quaternary Andesite rock products of the Quaternary Mountain consisting of breccias, lumps of lava, lapilli, tuff and sediment lava along a fault line formed a depression (Depression) extending in the direction of a fault line which partially filled with water and became lake or Telaga and called Telaga Sesar. In Tabek Patah Salimpaung area, Fault Lake is formed, namely Lake Air Pakis and Telaga Air Taganang. This fracture process produces landscapes with hills overgrown with evergreen plants to be a panoramic tourist attraction Tabek Patah with steep slopes of Gawir Sesar while the Lowlands part is used as agricultural land and settlements. For the geology of Tanah Datar Regency can be seen there is the third picture following;

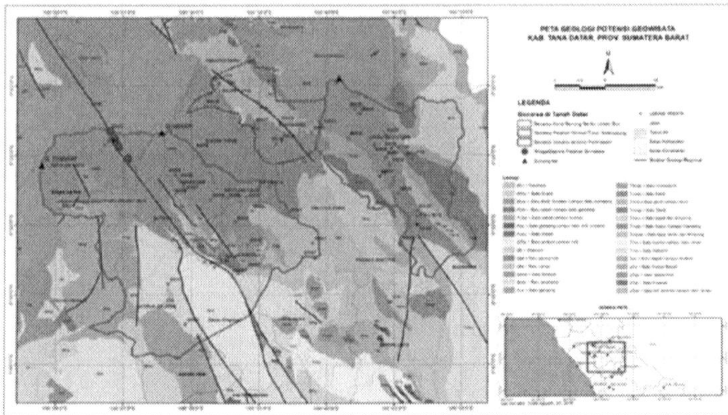

Figure 3. Geological map, Tanah Datar Districts

Morphology of the Karst Region in Lintau Buo Subdistrict was identified by Remote Sensing analysis on satellite imagery, aerial photographs and topographic maps of towering small hills (Menara Karst) with tight and circular contour karst features. This area is composed by limestone rocks of the Kuantan Formation (PCkl) consisting of limestone, filite, shale, and quartzite. This karst area consists of the morphology of Eksokarst, Sinkhole, Underground River, Goa / Ngalau with Stalactites, Dolina. In this karst area there are two Ngalau / Goa which have Archaeological sites. In Ngalau Jambu found fossils of human teeth estimated to be 60-70 thousand BC, Ngalau Kapalo Lakuak or Ngalau Bundo Kanduang contained ancient carvings patterned with Tumpal, Rhombus, Arrows, and Anthrophomorphic / human motifs with an older age than megalithic civilization or the great stone age is thousands of years after the age of human life in caves and niches [12].

Geotourism assets consisting of the Tanah Datar geo-bio-cilutureddiversity district are currently supported by amenities, accommodation, adequate accessibility while the activities and attractions related to Geotourism are still very limited. Cultureheritage flat land is Istano Basa better known as Istana Pagaruyung and Kampung Minang Nagari Sumpur which show the nuances of the old Minangkabau life with the Rumah Gadang building which is also used as a home stay in addition to various historical heritage menhirs which are generally spread in Sungayang District. The next geodiversity aspect is bioheritage namely Andaleh tree and Bilis fish or Bilih fish is an endemic fish found only in Singkarak lake.

The whole potential that has been inventoryed and assessed through FGDs into several geoheritage, bioheritage and cultureheritage potentials in geospatial maps is mapped using ArcGis 10.5 software which produces a geoheritage delineation map of Tanah Datar District. Can be seen in fourth picture the following;

IOP Conf. Series: Earth and Environmental Science **314** (2019) 012051 doi:10.1088/1755-1315/314/1/012051

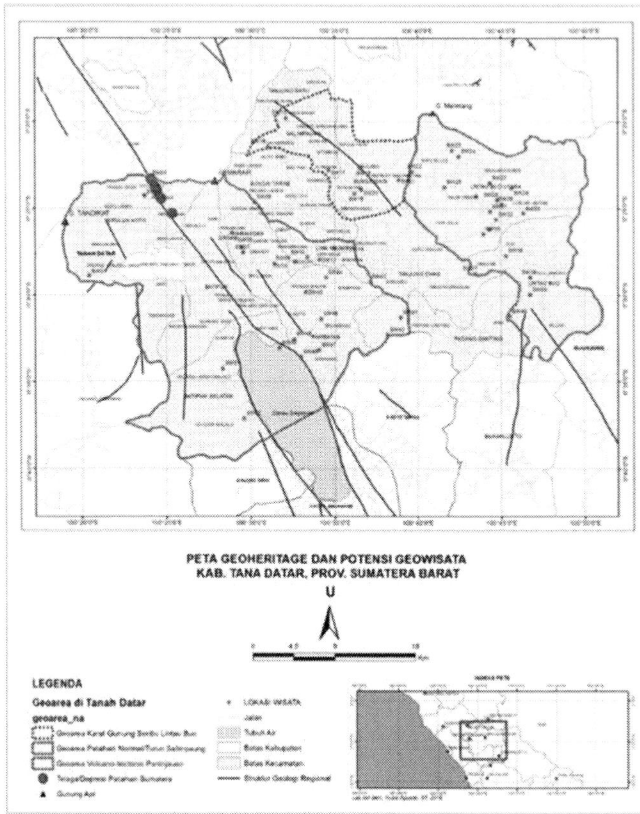

Figure 4. Map of regional delinazation, Tanah Datar districts

Based on the Regional Delineation map in Figure4 Tanah Datar District Geodiversity is divided into three Geoarea, namely Geoarea Paninjau Volcano Tectonic, Normal / Lower Fault Salimpaung Tabek Patah, Lintau Buo Karts area. This marriage delegation will be a Geoarea proposal for the development of the Tanah Datar District Geopark.

4. Conclusion

a. Geodiversity Potential Tanah Datar District as a whole consists of three geo areas; Paninjau Geoarea Vulcano Tecktonic, Geoarea Tabek Patahan Normal Patah Salimpaung, Geoarea Karst Lintau Buo. Geological sites are supported by potential culturediversity potential biodiversity
b. From the results of assessments through the Tanah Datar District Geodiversity Potential FGD supported by biodiversity and specific culturediversity Minangkabau culture is a geotourism asset with facilities, accommodations and accessibility is quite good, while aspects of activities and activities are still very less.
c. Geodiversity potential supported by biodiversity and culturediversity in Tanah Datar Regency, including one of the earth's heritage sites that has the potential to be submitted as a Geopark Candidate in Indonesia.

Acknowledgments

Thank you to the Geologists, Ir. Ade Edward, Progress of Geospatial Data and GIS Yusra Agusti S. T and Data Inventory Team Geopark Tanah Datar, Chair of the Indonesian Geopark Network Dr. Yun Yunus, and all parties involved in this research.

Reference

[1] Samodra H 2016 *Guidelines for Building and Developing Geopark* (Bandung: Geological Agency, Ministry of Energy and Mineral Resources)
[2] Poedjoprajitno 2007 *Morphotectonic and Sumatran Fault Reactivity in Padang Panjang West Sumatra* (Bandung: Center for Geological Survey)
[3] Sriyono 2014 *Geologi dan Geomorfologi Indonesia* (Yogyakarta: Ombak)
[4] Newsome D and Dowling R (eds.) 2010 *Geotourism and Ecotourism* (Good Fellow Publishers)
[5] Newsome D and Dowling R (eds.) 2006 *Geotourism, sustainability impacts and Management* (Butterworth Heinemann, Oxford: Elsevier)
[6] Indrayati A and Wahyu S 2016 *Community Assistance in Preparing Rembang Regency as a Tourism Destination Geographical Education: Community Service Report DIPA* Faculty on Environmental Science, Semarang State University
[7] Rosana M F et al 2015 Dossier Geopark Ciletuh Sukabumi-West Java *Proposal documents to become a National Geopark, Sukabumi: not published*
[8] Rosana M f et al 2016 Geowisata Geopark Ciletuh: geotrek mengelilingi keindahan Mega Amfiteater Ciletuh (The Magical of Ciletuh Ampitheater) *Bulletin of Scientific Contribution* **14**(1) 75-88 p-ISSN: 1693 – 4873
[9] Oktariadi Oki et al 2014 *Geological Heritage of the Minang Islands Guide to Global Geopark* (Bandung: Badan Geologi)
[10] Kastowo and Gerhard W Leo 1973 *Geology Map of Padang Sheet, Sumatra* (Bandung: Directorate of Geology, Ministry of Mining of the Republic of Indonesia)
[11] Silitonga P H and Kastowo 1995 *Geological Map of Solok Sheet, Sumatra* (Bandung: Directorate of Geology, Ministry of Mining of the Republic of Indonesia)
[12] BPCB 2017 *The Batu Sangkar Cultural Preservation Hall* (Batu Sangkar)

Implementation strategy character building of care and environmental culture in school

Agus Teguh Prihartono*, **Aulia Azhar, Yunhendri Danhas, Rusdinal and Nurhasan Syah**

Graduate Program of Universitas Negeri Padang, Prof. Dr. Hamka Street, West Air Tawar, Padang-West Sumatera 25131 Indonesia

*a2006teg@gmail.com

Abstract. Adiwiyata program has been launched in schools. The character building in schools, especially the caring and cultured environment is deemed necessary to be studied, because considering the broad and long-term impact on environmental conditions between generations. To this day, related to the environmental attitudes and culture. The point is to form caring and cultured characters for school citizens through the education system implemented in the education level unit. Therefore, it needs to be measured to what extent the implementation is running and how the improvement strategy is going forward. The number of schools that become the Adiwiyata School, in Padang Town are reaching 243, out of 646 totality of schools. To increase the number of adiwiyata schools in the Padang Town to the future, an improvement strategy is needed based on current conditions. The current condition in question is all the strengths, weaknesses and opportunities and threats that exist. This research is classified as qualitative research. The aim is to increase participation of school into Adiwiyata Program in Padang Town. The method used tin this research is SWOT analysis (strengths weaks threats opportunities). The results of the research, the strategy direction is obtained in the form of growth with the main strategy in the form of SO Strategy (Strenghts - Opportunities), namely "utilizing the power to seize the opportunities that exist", or "increasing the opportunities for existing strengths". The derivatives of this main strategy are 10 (ten) priority strategies.

1. Background

The Adiwiyata program is a national program. Formally juridically, this program is based on (1) the 1945 Constitution Article 28 paragraph 1 and Article 33 Paragraph 4, (2) Law No. 32 of 2009 concerning Environmental Protection and Management, (3) Law No. 18 of 2008 concerning Protection and Management Garbage, (4) Minister of Environment Regulation No. 05 of 2013 concerning Implementation Guidelines for the Adiwiyata Program

In an effort to protect and manage the environment, the Adiwiyata Program is one of the early prevention efforts to anticipate environmental pollution and destruction. That is, this program was not born to answer the problem of pollution and environmental damage that has occurred in an area. But it is precisely to avoid this from happening in the future, by providing environmental education and manners that care and love the environment to students.

Judging to the function of the school as a place of learning for students, the Adiwiyata Program also aims to make the school an ideal place and place to study. Thus, the Adiwiyata Program physically seeks a comfortable, ideal and healthy environment for school citizens in addition to long-term goals as an effort to protect and manage the environment in the concept of sustainable development.

The target of the adiwiyata program is students at schools, ranging from elementary school, junior high school and equivalent to high school and equivalent. In practice, all school residents and committees as well as the community around the school will directly participate. Thus, the principle of involvement of all components in the effort to protect and manage the environment can be realized through the Adiwiyata Program.

There are 4 (four) indicators of a school that can be said or become Adiwiyata School, namely (1) the existence of environmentally sound policies, (2) the curriculum contains efforts to protect and manage the environment, (3) the participation of school members in environmental management and protection efforts (4) facilities and infrastructure are available to answer environmental problems. if the indicator above has been fulfilled by a school, then the school can be declared as a school that cares about the environment and becomes a comfortable and ideal place to study.

However, the number of schools as a Adiwiyata School has only reached 246 schools out of a total of 646 schools. This means that there are 400 schools that have not become Adiwiyata Schools. It can be seen that the percentage of Adiwiyata School has only reached 38 % with the remaining 62 % being schools that do not participate in the Adiwiyata Program. This shows that the school's concern for this program is still low.

As a program ideally the goals and objectives must be achieved, the Adiwiyata Program that has been implemented in the Padang Town for the past 3 (three) years cannot be said to be going well. It is feared that there will be a decline in the number of schools proposed by each year, followed by a decrease in the value of adiwiyata values in schools that have Adiwiyata School predicate at the Adiwiyata School. This condition is very concerning and needs to be addressed immediately.

Therefore, a strategy is needed to increase the number of adiwiyata schools in the West Sumatra Province. The strategy that was born was a strategy formulation based on current conditions with all the existing strategic factors.

The formulation of the problem in this paper is how to increase participation of school to become an adiwiyata school in Padang Town.

2. Aim and Goal

The aim to be achieved is to increase the number of adiwiyata schools in Padang Town. While the targets to be achieved are (1) so that schools issue policies related to environmental protection, (2) integrate environmental protection and management material into the curriculum, (3) participate in environmental protection and management and (4) provide facilities and infrastructure environment.

3. Theoretical Basic

Adiwiyata (Green School) is one of the environmental and forestry state ministry programs that has the aim to encourage the creation of knowledge and also awareness of school citizens in environmental preservation. The Adiwiyata program is basically to make the school a good and ideal place where all knowledge and various norms and ethics that can be the basis of human beings towards the creation of life prosperity in the concept of sustainable development [1]. It can be emphasized that the target of the Adiwiyata Program is the school community and the school's physical environment. The long-term goal is the welfare of life in the concept of sustainable development. According to Soemarwoto [2] sustainable development must contain an increase of 3 (three) aspects in its function, namely ecological, social and economic functions. Thus, the adiwiyata program is an environment-based program for improving social and economic welfare in the future.

The principle of adiwiyata program is participatory and sustainable [1]. Participatory means the school community is involved in school management which covers the entire planning, implementation and evaluation process according to their respective responsibilities and roles.

Sustainability means that all activities must be carried out in a planned and continuous manner comprehensively.

For a comprehensive, sustainable activity that includes planning, implementation and evaluation and participation is clearly not easy. This requires a strategy.

Fathurrohman and Sutikno [3] states that language strategies can be interpreted as tactics, tips, tricks or ways. While in general the strategy is defined as an outline of direction in acting for a predetermined purpose. Referring to the Large Indonesian Language Dictionary [4] the strategy is defined as a careful plan of activities to achieve specific goals. Added by Marrus [5] that the strategy is defined as a process of determining the plans of the top leaders who focus on the long-term goals of the organization, along with the preparation of a way or effort how to achieve these goals. Thus, the goal of giving birth to a strategy is for the long term in achieving a predetermined goal. This is also not simple, because it requires a formulation process.

To formulate a strategy requires a process related to strategic management. According to Griffin [6] strategic management is a comprehensive and sustainable management process aimed at formulating and implementing effective strategies. An effective strategy is a strategy that encourages perfect harmony between the organization and its environment and with the achievement of its strategic objectives. So strategic management or strategic management is the way (strategy) taken in making decisions and actions to respond to opportunities and challenges, so that they still exist and achieve goals. David [7] added that strategic management is the art and knowledge to formulate, implement, and evaluate cross-functional decisions that enable the organization to achieve its objectives. Thus, to create a strategy, it is necessary to have art and knowledge that can contain the method of formulation, how to implement it and at the same time evaluate the decision. A good strategy is not shown by the many strategies that are prepared, but one strategy that contains all possibilities and accommodates the harmony between internal and external resources.

In order for the strategy to be prepared effectively, information is needed about strengths, weaknesses, opportunities and threats related to the conditions and situations that exist. One method for knowing the strengths, weaknesses, opportunities and threats of a company or organization is a SWOT analysis. According to Griffin [6], SWOT analysis is an evaluation of the strengths (strengths) and internal weaknesses (weaknesses) that are carried out carefully, and also an evaluation of opportunities (opportunities) and threats (threats) from the environment. In the SWOT analysis, the best strategy for achieving an organization's mission is to (1) exploit the opportunities and strengths of an organization, and at the same time (2) neutralize its threats, and (3) avoid or correct its weaknesses.

Rangkuti [8] introduced SWOT analysis techniques to produce alternative strategies. Where the resulting strategy is the result of a combination of strengths, weaknesses and opportunities and threats. To combine these strategic factors, they are arranged in the form of a SWOT Matrix. Added by David [7] which has laid the foundation for determining general strategies based on the EFE / IFA Matrix (external factors evaluation / internal factors evaluation). So that it can be compared with the SWOT and Quadrant SWOT Matrix before determining the main strategy chosen.

4. Research methods

The analysis used is SWOT Analysis. Starting with an inventory of each aspect that is related to the implementation of adiwiyata. Furthermore, aspects of these aspects are identified according to external and internal factors. Each factor will be explained in the form of strength (S), weakness (W), opportunity (O) and threat (T) factors. Weakness and strength are internal factors while opportunities and threats are external factors.

Each factor is weighted based on the results of the score of each factor with its rating. Then converted into EFA / IFE Matrix. From the conversion of EFA / IFE Matrix, continued with the conversion to SWOT Matrix to give birth to 4 (four) parent strategy choices which are then chosen to produce a priority strategy.

5. Results and Discussion

5.1 Identification of strategic factors

There are several strategic factors that determine the implementation of the Adiwiyata program. These factors are internal and some are external. Internal factors are factors that originate from within the school, which can be categorized as strengths and or weaknesses. While external factors are factors that come from outside the school, which are categorized as opportunities and threats. These factors influence the target of this study, as well as an indicator of the success of adiwiyata at school. The results of factor identification are presented in Table 1.

Table 1. Strategic Factor Identification Matrix

Factors	Internal factors		External factors	
	S	W	O	T
1. School Motivation		X		
2. Goverment supports such (Bapedalda/BLH/KLH, Diknas, Kemenag, Bappeda dan Dinkes)			X	
3. Program socialization			X	
4. Finance		X		
5. Human resourches at school.	X			
6. School still not yet as Adiwiyata School			X	
7. School as Adiwiyata School			X	
8. The Others program				X
9. Appriciated and reward			X	
10. School environment	X			
11. Head Master commitments and stake holder		X		

11 (eleven) strategic factors above, related to each other actively interact with each other. The results of the accumulated effects of each factor will determine how the Adiwiyata Program will be implemented.

Each factor can be broken down, namely (1) school motivation as a factor within the school personnel that gives effect to whether or not the school led by the Principal follows the program. In fact, schools that have not participated in the Adiwiyata Program are indeed principals and their staff are not motivated. This is a weakness. The second factor (2) is support from the city / district government, namely from Bapedalda / BLH / KLH, the Ministry of Education, the Ministry of Religion, Bappeda and the Health Office in the form of guidance, material support and moral encouragement that are still very weak. The third factor (3) program socialization is still lacking. Effective and efficient socialization will certainly motivate the school. Effective and efficient socialization is also determined by the implementation stages and intervals. The fourth factor (4) financing for program implementation that has not been allocated proportionally so that it becomes a weakness. The fifth factor (5) is human resources in schools, which is a strength factor for program implementation. The sixth factor (6) is the number of schools that have not participated in the program. This factor is an opportunity. Where in the future it is hoped that this school will become an adiwiyata school in the West Sumatra Province. The seventh factor (7) is the school that has become the Adiwiyata School. This factor is an opportunity also to increase the success of the program. Because it can be used as a reference for school schools that have not become adiwiyata schools. The eighth factor (8) is that there is a program that is not synergistic, which has the potential to be a challenge and a threat to the Adiwiyata Program. The 9th factor (9) appreciation and appreciation from the government is a motivator for the school so that this factor is an opportunity . The tenth factor (10) is the school environment which still has the potential to be developed in order to the Adiwiyata School. This is an opportunity. The last factor for eleven (11) is the commitment of principals and stakeholders who are still weak and categorized into internal weaknesses.

11 (eleven) strategic factors, both internal and external in the implementation of the Adiwiyata Program that have been identified, are factors that interact strongly in determining the hue of implementation up to the achievement of the results of the Adiwiyata Program. Schematically, how interactions between factors can be seen in Figure 1.

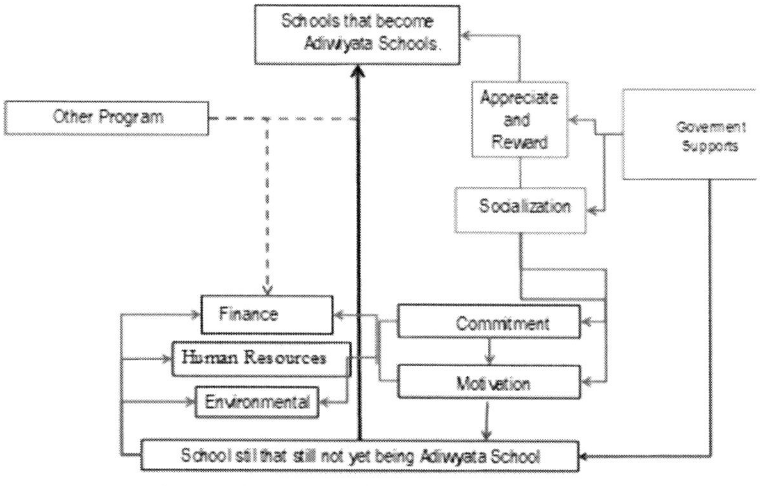

Figure 1. Scheme of Factors Influences to the Adiwiyata Program (Analysis Results, 2018)

In Figure 1, it can be seen that schools that have not become adiwiyata schools will pass through improvements to the physical environment, the ability of human resources in school residents, and funding. Meanwhile, the 3 (three) factors are strongly influenced by motivation. Motivation can come from the support of the city / district government, both in the form of socialization and appreciation to schools that are already adiwiyata.

However, there is also a red broken arrow that blocks schools from going to the Adiwiyata School. This factor is a program program that exists in schools but is not synergistic with the Adiwiyata Program substantially and the mechanism.

From the scheme of factors that make up the interaction with the school process that is not yet adiwiyata towards Adiwiyata School in Figure 1, a more detailed study of these factors is needed. Further study of these factors is related to the nature and quantitative factors that exist. That is, an evaluation of each factor is needed, so that it can determine the current position of the Adiwiyata Program, based on the strategic direction according to the EFE / IFE Matrix

5.2 Evaluation of Internal Strategic Factors and External Strategic Factors

Factor evaluation aims to assess the extent and potential of a factor to the development of adiwiyata program implementation. This evaluation is based on the weight of each factor. Factor weight is obtained from the results of the score with a rating. The ultimate goal is to determine the direction of the strategy which is a general strategy going forward. The direction of this strategy is obtained from the conversion results into the IFE / EFE (Internal Factors Evaluation / External Factors Evaluation) matrix presented in Table 2.

Table 2. Internal and External Factor Evaluation Matrix

Factors (1)	Weight (2)	Rating (3)	Score (4=2x3)
Internal Factors			
1. Motivation	0.20	3	0.60
2. Finance	0.20	3	0.60
3. Human resources.	0.20	3	0.60
4. Environmental	0.20	3	0.60
5. Headmaster commitment and stake holders	0.20	3	0.60
Total	1.00		3.00
External Factors			
1. Goverments support such: kabupaten (Bapedalda, Diknas, Kemenag, Bappeda dan Dinkes)	0.17	2	0.34
2. Socialization	0.17	1	0.17
3. Schools that still not yet as Adiwiyata School	0.17	3	0.51
4. School that being Adiwiyata School	0.17	2	0.34
5. The other program	0.17	3	0.51
6. Appreciate and reward.	0.17	3	0.51
Total	1		2.38

5.3 Strategy Direction

Strategy direction is determined by converting scores obtained from external factors and scores from total internal factors. Conversion was carried out in the EFE-IFE Matrix introduced by David [7]. The results are presented in Table 3.

Table 3. Determination of General Strategies with EFE / IFE Matrix.

EFE / IFE	**Strong (Score:3 – 4)**	**Ordinary (Score:2– 2,99)**	**Weak (Score:1 – 1,99)**
High (Score 3 – 4)	*I. Growth*	*II. Growth*	*III. Turn Around*
Medium (Score:2– 2,99)	*IV. Stability*	*V. Survive*	*VI. Retrenchment*
Low (Score :1 – 1,99)	*VII. Growth diversification*	*VIII. Growth diversification*	*IX. Retrenchment*

Results of Analysis, 2018

From the results of factor evaluation, it is found that the direction of the future strategy is in coordinate II on the IFE / EFE matrix, which is growth. This means demanding the direction of the strategy in the form of growth. Growth means increasing the amount by utilizing the current situation with all the existing strategic factors, both internal factors (strengths and weaknesses) and external (opportunities and threats / challenges).

5.4 Determination of Alternative Strategies

Alternative strategies are derivatives of strategic direction. Determination of alternative strategies is to use the SWOT Matrix. The results are presented in Table 4.

Table 4. Matrix SWOT

	S	W
O	1. Increasing city governments (Bapedalda / BLH / KLH, Diknas, Kemenag, Bappeda and Dinkes) towards human resources in schools. 2. Increase socialization from the provincial government to the city / regency and from the district / city to human resources in schools. 3. Utilizing human resources in schools that have not become adiwiyata schools to become adiwiyata schools 4. Utilizing human resources in schools to refer to schools that have become adiwiyata. 5. Utilizing adiwiyata appreciation and appreciation for human resources at school 6. Increasing the support of city / district governments (Bapedalda / BLH / KLH, Diknas, Kemenag, Bappeda and Dinkes) in utilizing the school environment to go to adiwiyata schools 7. Increase socialization from the provincial government to the city / regency and from the district / city to the school to utilize the school environment in going to adiwiyata schools 8. Utilizing a school environment that has not become an adiwiyata school to improve its environment into an adiwiyata school 9. Utilizing schools that have become adiwiyata as a reference in the use of the school environment 10. Utilizing the potential of the school environment to get adiwiyata appreciation and appreciation	1. Increasing support from city / district governments (Bapedalda / BLH / KLH, Diknas, Kemenag, Bappeda and Dinkes) to motivate the school 2. Socialization from the provincial government to the city / district and from the district / city to the school to motivate the school 3. Schools that have not yet become adiwiyata schools are motivated to become adiwiyata schools 4. The motivation of the school is to make schools that have become adiwiyata schools as a motivation for schools that have not yet become adiwiyata. 5. There is appreciation and appreciation as a tool to motivate school 6. Support from the city / district government in terms of funding to go to adiwiyata school 7. Funding from the school to carry out socialization of the adiwiyata program 8. Funding from schools for schools that have not participated in the adiwiyata program 9. Utilizing schools that have become adiwiyata schools in terms of funding allocation references in schools that have not yet become adiwiyata schools 10. Funding from school to become an adiwiyata school because of appreciation and appreciation 11. Increasing the commitment of school principals with support from the city government 12. Increasing the commitment of school principals to the dissemination of information from the government of city 13. Increasing the commitment of school principals whose schools have not yet become adiwiyata schools, to become adiwiyata schools 14. Increasing the commitment of school principals whose schools have not become adiwiyat schools through schools that

		have become adiwiyata schools. 15. Increasing the principal's commitment through appreciation and appreciation.
T	1. Quality human resources in schools synergize all existing programs with the adiwiyata program 2. Synergizing all existing programs in managing the school environment.	1. Increasing school motivation towards adiwiyata schools and synergizing existing programs with adiwiyata programs. 2. Increase the allocation of costs for adiwiyata programs in proportion to programs that are not synergistic 3. Increasing the commitment of school principals to synergize existing programs with the adiwiyata program

From the SWOT matrix there are 4 (four) types of alternative strategies. Strategy I is an SO strategy (utilizing existing strengths to get opportunities), strategy II is a WO strategy (increasing weaknesses to get opportunities), strategy III is an ST strategy (utilizing the power to avoid threats) and strategy IV which is a WT strategy (minimizing weaknesses and avoid threats).

5.5 Determination of the Main Strategic Direction

Based on the direction of the previous strategy (see Table 3) in the form of a growth strategy, then the choice of strategies presented in the SWOT matrix (Table 4) is then converted into the SWOT Quadrant introduced by Rangkuti [8], as presented in Figure 2 .

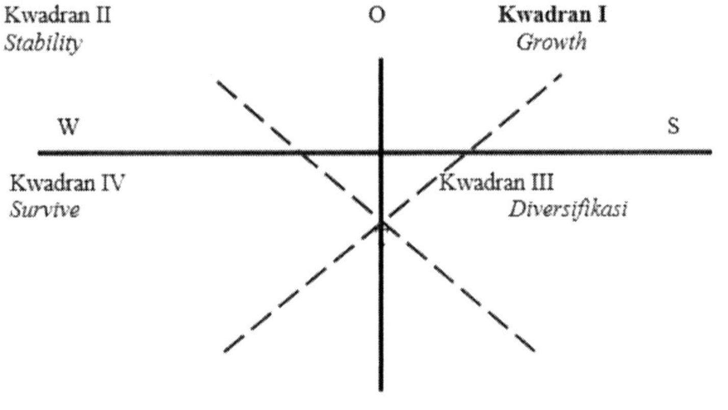

Figure 2. Direction of Staregy Position in Kwadran Strategyc

This provides the basis for establishing the SO strategy as the main strategy chosen, namely:

1. Increase the support of the city / district government (Bapedalda / BLH / KLH, Diknas, Ministry of Religion, Bappeda and DHO) on human resources in schools.
2. Increase socialization from the provincial government to the city / district and from the district / city on human resources in schools.
3. Utilizing the quality of human resources in schools that have not become adiwiyata schools to become adiwiyata schools.
4. Utilizing human resources in schools to refer to schools that have become adiwiyata.
5. Utilizing adiwiyata's appreciation and appreciation of human resources at school.
6. Increase the support of the city / district government (Bapedalda / BLH / KLH, Diknas, Ministry of Religion, Bappeda and DHO) in utilizing the school environment to go to adiwiyata schools.

7. Increase socialization from the provincial government to the city / district and from the district / city to schools to utilize the school environment in going to adiwiyata schools.
8. Utilizing a school environment that has not become an adiwiyata school to improve its environment to become an adiwiyata school.
9. Utilizing schools that have become adiwiyata as a reference in utilizing the school environment.
10. Utilizing the potential of the school environment to get adiwiyata appreciation and appreciation

6. Conclusions, Implications, Recommendations

It was concluded that in order to increase the number of adiwiyata schools in the Padang Town, a strategy for growth was needed. Growth means an increase in quantity based on quality.

For this reason, a strategy is needed in the form of utilizing the power to obtain opportunities (SO strategy), which consists of 10 (ten) priority strategy items, namely (1) increasing city / district government support for human resources in schools, (2) increasing government socialization province to city / district and from district / city to human resources in schools, (3) utilizing the quality of human resources in schools that have not become adiwiyata schools to become adiwiyata schools, (4) utilizing human resources in schools to refer to schools that have become adiwiyata, (5) utilizing adiwiyata's appreciation and appreciation of human resources in schools, (6) increasing the support of city government in utilizing the school environment to go to adiwiyata schools, (7) increasing socialization from the provincial government to the city / district and from the district / city to school to take advantage of the school environment in testing u Adiwiyata school, (8) utilizes a school environment that has not become an adiwiyata school to improve its environment to become an adiwiyata school, (9) utilizing schools that have become adiwiyata as a reference in utilizing the school environment, (10) utilizing the potential of the school environment to get appreciation and awards adiwiyata.

The implications of the results of this study are the ideal model of the implementation of the adiwiyata program. This model is a reference in every year the implementation of the adiwiyata program. Schematically, can be seen in Figure 3.

Strategy Model for Implementing the Adiwiyata Program

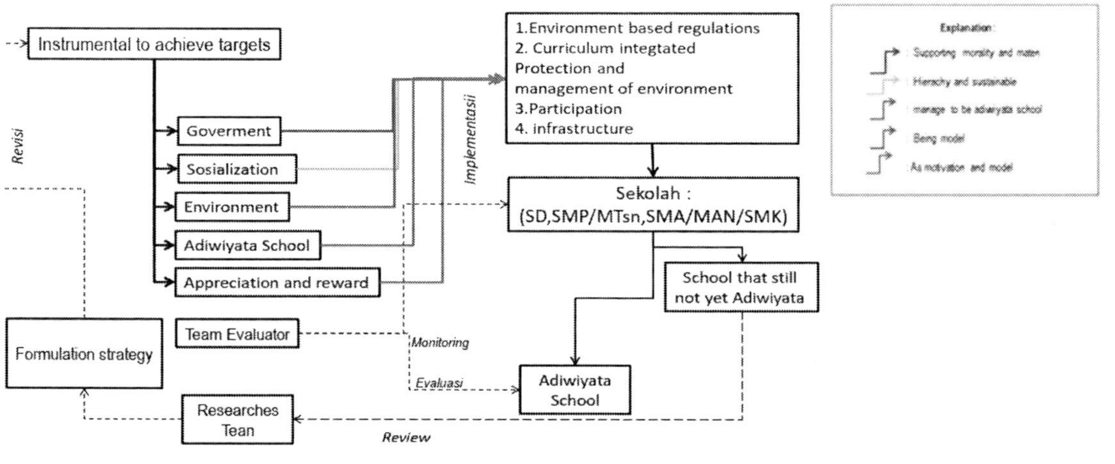

Figure 3. Implementation Strategy Model for Adiwyata Program Increasing

From the results of the study, it is recommended that policies will oversee the process and mechanism for implementing the selected strategies and refer to the adiwiyata program strategy model.

References

[1] Peraturan Menteri Lingkungan Hidup Republik Indonesia No. 5 Tahun 2013 tentang Program Adiwiyata
[2] Somarwoto Otto 2001 *Pembangunan Berkelanjutan* (Jakarta: Rhineka Cipta)
[3] Fathurrohman Pupuh and Sutikno M Sobry 2007 *Strategi Belajar Mengajar* (Jakarta: Refika Aditama)
[4] Kamus Besar Bahasa Indonesia 2005
[5] Marrus Stephanie K 2002 *Building The Strategic Plan: Find Analyze, And Present The Right Information* (USA: Wiley)
[6] Griffin Ricky W 2004 *Management Strategic* (New Jersey: Prentice Hall)
[7] David Fred R 2009 *Management Strategic Concepts* (Jakarta: Salemba Empat)
[8] Rangkuti Freedy 1997 *Teknik Analisis SWOT* (Jakarta: P.T Gramedia Pustaka Jaya)

The influence of green open space and tourism-conscious work culture on the happiness of the people in Solok Regency

Vivi Yulistia Rahayu*, Aldri Frinaldi and Afriva Khaidir

Faculty of Social Science, Universitas Negeri Padang, West Sumatra, Indonesia

*vivirahayu93@gmail.com

Abstract. In an effort to improve the happiness of the community, there is a need for green open space (RTH) and a tourism-conscious work culture. But in reality RTH is not a development priority in Solok Regency. This is seen by not being maintained and not well managed by the green open space. In maintaining green open space, it is necessary to have a tourism-conscious work culture from all relevant parties, but not all parties understand the importance of the tourism-conscious work culture. This study uses quantitative methods with associative research types that aim to determine the effect or relationship between two or more variables. The population in this study is the people of Solok Regency. Determination of the number of samples using the Slovin formula with a total of 400 people, and sampling techniques using Cluster Sampling and Incidental sampling techniques. The instrument used is a questionnaire. Research hypothesis testing uses regression analysis techniques with T Test (Partial) and F Test (Simultaneous). The results showed that: 1) Green Open Space had an effect on community happiness by 21.4%, 2) The influence of Tourism Awareness Work Culture on community happiness was 28.3%, and 3) the influence of green open space and shared tourism conscious culture - same effect on community happiness by 25.0%. 4) Green open space is in a very high category with the level of achievement of respondents at 81.0%. 5) Tourism conscious work culture is in the very high category with the level of achievement of respondents at 81.4%. 6) The happiness of the community is in the high category with the level of achievement of respondents at 77.4%.

1. Introduction

Based on the Solok District Regent Regulation No. 50 of 2017 concerning the Determination of the Location of Green Open Space in Solok Regency, the newly recorded areas as RTH are the areas of Arosuka, Alahan Panjang, Muaro Paneh and Sumani. The focus of green space management's Solok Regency is Alahan Panjang. Alahan Panjang's Green Open Space is a Green Open Space that is the pride of Solok Regency. In addition of being a pride of Solok Regency, Alahan Panjang's Green Open Space is also prepared as a means of tourism for the community to enjoy the natural beauty of Solok Regency. Alahan Panjang is a green open space designated for all types of visitors, both teenagers, families and also groups. The policy to fulfill 30% of green open space in the district / city is a form of green space benefits as the lungs of the city so that the creation of fresh and clean air.

Along with development of the Alahan panjang's Green Open Space, the active role of the community is also needed to maintain the beauty, cleanliness, comfort and foster public awareness in maintaining the green space. Public awareness to maintain the green space environment is very important. In maintaining the green space, the working culture of the people who visit will be reflected.

The work culture that is intended is a tourism-conscious work culture. A conscious tourism work culture is a form of participation and support from the community in encouraging the creation of a conducive climate for tourism growth in an area and aims to improve people's welfare. Tourism awareness in this case is described as a form of public awareness to play an active role in two roles, namely: 1) the community is aware of its role and responsibility as a good host for guests or tourists; 2) the community is aware of their rights and needs to become tourism agents. A tourism-conscious work culture is the behavior / attitude of local people who can support tourism in an area, because it is the tourist community that interacts directly with tourists who come to the tourist attraction.

Open space is a significant place in the lives of many people. The effect of the diversity of open spaces in the city is a pleasant experience. This happens because open space is an area that allows users to participate in different active pleasures such as walking, playing sports or doing passive or informal pleasures such as enjoying the environment, breaking away from all problems, and engaging in social activities such as attending a joint event. Green Open Space has a direct impact on the happiness of the people who treat it. Green Open Sapce based on research results by [1] found that more than 40% of people use local parks every day for various activities.

Based on preliminary observations made on Saturday and Sunday July 7 and 8 2018, Alahan Panjang's Green Open Sapce is still in the stage of development so that the green open space is not well organized. Implementation in the field is also not in accordance with the theory put forward. Alahan Panjang' Green Open Sapce which is the pride of Solok Regency is not well maintained. This is reflected in the amount of garbage scattered on the location. The tourism community visiting Green Open Sapce does not reflect a good attitude in safeguarding tourism objects so that the Green Open Sapce is not well maintained. Traders who also act as tour operators do not participate in maintaining the beauty of green open space, these traders sell in places that are not appropriate such as on the shoulder of the road, near the park and in front of the entrance to the green open space.

2. Study of theory

2.1. Happiness

Happiness is a different concept of satisfaction. The happiness in the Kamus Besar Bahasa Indonesia is a pleasure, tranquility of life (born and inner) while satisfaction is a matter of being satisfied and relieved. Satisfaction is a positive response from customers which is indicated by things such as feelings of pleasure, fulfillment of expectations for a performance and service.

The concept of happiness, according to [2], can be seen from two corners of the field, namely the definition of moral-laden and morally-neutral. The definition of morality requires that the benchmark of happiness is moral values, which in essence is happiness centered on the implementation of virtue. On the other hand, the definition of happiness neutrally emphasizes subjective well-being in the form of full satisfaction with life or achievement of high enjoyment.

Happiness is a goal for every human being. Happiness is something that is fun, joyful, bringing pleasure and achieving a goal. Happiness in each person is different, because happiness is subjective. The happiness of each individual is different from each other despite experiencing the same event. Happiness in each individual depends on meaning and understanding happiness. Happiness according to [3] is the feeling or state of mind of a person which is characterized by a sense of comfort, peace and an urge to be more positive.

One country that is famous for its concept of happiness is the State of Bhutan [4]. The king of Bhutan, Jigme Singye Wangchuck in 1972 introduced the concept of Gross National Happiness (GNH) as the concept of measuring happiness. He introduced this concept as a form of his commitment to be able to develop the Bhutanese economy based on the values of spirituality and culture of the Bhutanese people. At first this concept was just ordinary speech, not a concept that was considered so serious. However, it turns out that the Center of Bhutan Studies, under the leadership of

Kharma Ura, developed the concept and every instrument in GNH so that it eventually became the measuring tool for the happiness level of the Bhutanese population. According to Karma Ura, this measurement concept covers nine aspects / dimensions and consists of 33 indicators taken from 124 variables. These nine aspects / dimensions, among others:

- Psychological wellbeing, consisting of four indicators, namely life satisfaction, positive emotions, negative emotions, and spirituality.
- Health, consists of four indicators, namely health status, number of healthy days, inability to engage in long-term activities, and mental health.
- The use of time, consists of two indicators, namely work time and sleep time.
- Education, includes four indicators, namely literacy, educational qualifications, knowledge and values.
- Cultural diversity, consists of four four indicators, namely language, artistic ability, socio-cultural participation, and Driglam Namzha. Driglam Namzha is a manners and behavior code. In Driglam Namzha also arranged how the Bhutanese people dress, both everyday and formal.Driglam Namzha also regulates a number of cultural assets, such as art and architecture. In english, "driglam" means "order, discipline, custom, rules, regimen". While "namzha" means "system".
- Good governance includes four indicators, namely political participation, political freedom, public services, and government performance.
- Community vitality, including four indicators, including social support, inter-community relations, families and victims of crime.
- Ecological diversity, there are four indicators, namely pollution, responsibility for the environment, nature reserves, and urban issues.
- Living standards, consisting of three indicators, including household income, assets, and housing quality

Happiness according to [3] is the feeling or state of mind of a person which is characterized by a sense of comfort, peace and an urge to be more positive. Based on the results of research conducted by [3] on Building a Model of Tourism Awareness Work Culture among Apparatus and Community of Tourism in an Effort to Increase Community Satisfaction Visiting Tourism Sites in West Sumatra explained that someone who visits tourist sites can relieve stress, fatigue and tired so that it creates a feeling of relaxation, comfort and calm which is expressed by various expressions that reflect happiness [5]. Happiness can be measured using a toga (3) dimension consisting of 8 indicators. These three dimensions,

- Comfortable, is a way / perspective of a person to enjoy a beauty. The indicators are: interesting, fresh, and expressive.
- Peace, is a state of the environment that can produce a relaxed and relaxed atmosphere. The indicators are relax and infrastructure.
- Positive action, is a reflection of positive thinking / positive thinking. The indicators are happy, big heart and enthusiasm.

2.2. Green open space

Green open space is an elongated area / pathway and / or grouping, whose use is more open, a place to grow plants, both those that grow naturally and those that are intentionally planted. Based on Law No. 26 of 2007 concerning Spatial Planning, the spatial planning of the city area must contain a plan for the provision and use of green open space with a minimum area of 30% of the city area. The plan for the provision and utilization of green open space in addition to being included in the City RTRW, City RDTR, or City Strategic Area RTR, is also included in the Urban Area RTR which is a detailed plan of the Regency's spatial layout.

Green open spaces (green open spaces) are planned spaces because of the need for joint meeting and activity places in the open air. Green open spaces (green open spaces), in general the meaning of space is dominated by natural environments outside and inside the city, in the form of parks, courtyards, city recreation areas and green lanes, and facilities that make an important contribution in improving the quality of the settlement environment, as an important element in recreational activities.

Green Open Sapce according to Law No. 26 of 2007 concerning Spatial Planning has the following functions:

- The main function (intrinsic) is the ecological function: to guarantee the procurement of green space to be part of the air circulation system (city lungs); micro climate regulators so that the air circulation system and water can naturally run smoothly; as shade; oxygen producer; rainwater absorbent; animal habitat provider; absorbing air, water and soil pollutants, and; windbreak.
- Extrinsic functions, namely:
  Social and cultural functions: describe the expression of local culture; is a communication media for the city; Recreation areas; container and object of education, research, and training in studying nature.
  Economic function: source of products that can be sold, such as flowers, fruit, leaves, vegetables; can be part of agricultural, plantation, forestry and other businesses.
- Aesthetic functions: increase comfort, beautify the city environment both from the micro scale: home yard, neighborhood, and macro: the city landscape as a whole; stimulates the creativity and productivity of the city; forming architectural beauty factors; creating a harmonious and balanced atmosphere between the built and unconscious areas

Based on government regulations Green Open Space has an understanding: Budihardjo and Sudanti [6] asserted that to develop an environmentally friendly city (ecopolis), green open space is needed that spreads in urban environments, considering that the ecopolis arranges the pattern of population settlements into the environmental carrying capacity. There are 6 indicators that can be used as a basis for assessing the elements of the physical environment of green open spaces, namely:

1. The role of green open space as a city environmental identity;
2. Green Open Space role in traveling destination orientation;
3. The role of green space in creating the beauty of the green system;
4. The role of green open space in enhancing the harmony of the surrounding buildings
5. The role of green space in increasing the comfort of the city;
6. The role of green space in enhancing social interaction.

2.3. *Tourism-conscious work culture*

A tourist-aware work culture can be intended to a concept that can broaden tourism's view not only to consider short-term economic impacts, but with the development of this tourism-aware work culture, tourism perspectives can be developed so that there is a sustainable development and improvement of tourism in line with popular economic growth. significant with the development of natural and cultural sustainability in community life. This can also be an effort to respond to the expectations of the people who visit so that they gain experience and impressions that can improve the quality of satisfaction and happiness. A tourism-conscious work culture can also be a form of mental character that gives meaning to expectations, values, and experiences that are useful for enhancing the capabilities and capacities of human resources and other resources [3].

The development of this tourism-conscious work culture when related to Kotler's opinion [3] is related to a tourist area, it can be considered related to the need for efforts that are able to develop products which are a combination of components, including physical goods, services, experiences, events, people , place, property, organization, information and ideas. Therefore, on the one hand this tourism-aware work culture can mean a development of interpretation of social interactions and human actions needed in development both physically and psychologically as well as expressive actions in an effort to increase the satisfaction of visiting people. Tourism-conscious work culture in the opinion of [3] suggests that what is meant by tourism-conscious work culture is the values that become life views then become a habit, as well as reflected in the form of attitudes, behavior, and concern in social interaction in an effort to realize growth the development of tourism which can encourage the growth and prosperity of the socio-economic life of the community.

Accroding to [3] formulated 10 indicators of tourism-aware work culture that are safe, orderly, clean, cool, beautiful, friendly, memorable, polite, polite and responsive.

- Safe
  A condition in a tourist destination environment that provides a sense of calm, free from fear and anxiety for tourists to travel to these tourist destinations.
- Orderly
  A tourist environmental condition and service in a tourist destination reflects high discipline and physical quality and consistent and articulated and efficient service so as to provide a sense of comfort and certainty for tourists.
- Clean
  An environmental condition and the quality of products and services in tourist destinations reflect a healthy state so as to provide a sense of comfort and pleasure for tourists.
- Cool
  A condition in a tourist destination that reflects cool and shady conditions that will provide a comfortable feeling for tourists.
- 5. Beautiful
  A condition in the tourist destination area that reflects the beauty, attractiveness and provides a deep feeling for tourists so as to realize the potential for repeat visits and encourage promotion to a wider tourism market.
- Friendly
  An environmental condition in a tourist destination that comes from the attitude of the community that reflects a familiar and open atmosphere.
- Memories
  A form of memorable experience in tourist destinations that will provide a sense of pleasure and beautiful memories for tourists.
- Polite
  A condition in a tourist destination that comes from the attitude of the community that reflects good behavior and adheres to the customs that apply in these tourist destinations.
- Courteous
  A condition in a tourist destination that reflects a smooth and good attitude in terms of language and behavior.
- Responsive
  A condition in a tourist destination that reflects a fast attitude in giving a response to tourists who visit.

3. Research methodology

Methodology that used in this study is a quantitative method with associative research, which aims to reveal the influence of both independent / independent variables, namely, green open space, as well as tourism-conscious work culture on the dependent / dependent variable, namely the happiness of the community, both influence and influence simultaneously.

3.1. Population

The population is visitors to the Arosuka green open space. The population of visitors in the Green Open Space of Solok Regency was taken based on the total population of 366,213 people as shown in table 1.

Table 1. Population and district distance to the district capital

No	Sub-district	Distance (Km)	Grouping	Total population
1	Gunung Talang	10,00	Near	50.024
2	Bukit Sundi	21,00	Near	23.421
3	Kubung	22,00	Near	59.414
4	Danau Kembar	26,00	Near	19.768
5	Lembah Gumanti	36,00	Not too far	58.866
6	Payung Sekaki	35,00	Not too far	8.249
7	Lembang Jaya	34,50	Not too far	26.778
8	IX Koto Sungai Lasi	37,00	Not too far	9.696
9	X Koto Singkarak	39,00	Not too far	31.767
10	Hiliran Gumanti	50,00	Far	16.901
11	X Koto Diatas	59,00	Far	18.508
12	Junjung Sirih	45,00	Far	12.137
13	Pantai Cermin	66,00	Far	20.561
14	Tigo Lurah	95,00	Far	10.123
Total population				366.213

Source : data is processed by researchers 2018

3.2. Sample

Sample that used is this study is Cluster Sampling and Incidental Sampling. Accroding to [7] To determine the number of samples from a population in the opinion of Slovin in Prasetyo dalam sugiono with the degree of inaccuracy due to sampling errors that can be tolerated at 5%. From the results of the formula, the number of research samples was 399.56 rounded up to 400 people visiting the green space.

4. Discussion

The results of the discussion in this study are as follows:

4.1. Green Open Space Influence on community happiness

Based on the results of the first hypothesis test, it was found that the partial green open space variable had a positive and significant influence on community happiness variables. This is evidenced by the results of data processing using SPSS (statistical program for social science) software version 17, it found that the significance value of the green space variable on Community Happiness of 0.000 <0.05. The magnitude of the influence of the Green Open Space variable on the happiness of the community in Solok Regency is partially 21.4% where this value is based on the adjusted r square value obtained after the t test of the first hypothesis of 0.214.

The results above in accordance with those expressed [1] found that more than 40% of people use local parks every day for various activities. This concludes that open space is a significant place in the lives of many people. The influence of the diversity of open spaces in the city is a pleasant experience. This is because open space is an arena that allows users to participate in different active pleasures such as walking, playing sports or doing passive or informal pleasures such as enjoying the environment, breaking away from all problems, and engaging in social activities such as attending a joint event.

4.2 Influence of Tourism-Conscious Work Culture on Community Happiness in Solok Regency.

Based on the results of the second hypothesis test, it was found that the variable tourist-aware work culture has a significant positive effect. From the results of the research data conducted using the SPSS (statistical program for social science) software version 17, it was found that tourism-conscious learning activities have a significant effect on the happiness variable of the community with a significance value of 0.000 <0.05 and adjusted r square value of 0.283 which means that the amount of

IOP Conf. Series: Earth and Environmental Science **314** (2019) 012053 doi:10.1088/1755-1315/314/1/012053

influence that is owned by tourism conscious work culture variable is partially to the happiness of the community in Solok Regency is 28.3%.

The findings of this study also prove the truth of the theory put forward by [3] that tourism-conscious work culture influences visitor satisfaction. The effect is 61.1%, which means that a tourism-conscious work culture has a significant effect on increasing the community's capacity. The final achievement of visitor satisfaction is the creation of a feeling of happiness over what has been felt and gained. The Influence of Tourism-Conscious Work Culture on Community Happiness in Solok Regency.

Based on the results of the second hypothesis test, it was found that the variable tourist-aware work culture has a significant positive effect. From the results of the research data conducted using the SPSS (statistical program for social science) software version 17, it was found that tourism-conscious learning activities have a significant effect on the happiness variable of the community with a significance value of 0.000 <0.05 and adjusted r square value of 0.283 which means that the amount of influence that is owned by tourism conscious work culture variable is partially to the happiness of the community in Solok Regency is 28.3%.

The findings of this study also prove the truth of the theory put forward by [3] that tourism-conscious work culture influences visitor satisfaction. The effect is 61.1%, which means that a tourism-conscious work culture has a significant effect on increasing the community's capacity. The ultimate achievement of visitor satisfaction is the creation of a feeling of happiness over what has been felt and gained.

4.3 Effect of Green Open Space and Tourism-Aware Work Culture on Community Happiness in Solok Regency

Based on the results of the third hypothesis test, after the f test was found a significant effect jointly from green open space (X1) and tourism conscious work culture (X2) to community happiness (Y) in Solok Regency with a significance value of 0.000 <0, 05, and also found the adjusted r square value of 0.283 which means that together the two independent variables of this study have an influence of 28.3% on the dependent variable (community happiness) in this research location.

Conclusions from the findings of this study prove that efforts to increase community happiness can be done by building green open spaces and creating a tourism-conscious work culture. In practice, these two factors have a mutually influential relationship to happiness. Green open spaces and a tourism-conscious work culture have reciprocal and mutually influential relationships, in the sense that green open spaces can affect tourism-conscious work cultures and conversely tourism-conscious work cultures can also affect green open spaces. So that these two variables become factors that influence the happiness of the people in Solok Regency.

5. Conclusions

The purpose of this study is to find out whether there is an influence of green open space variables and a tourism-conscious work culture on the happiness of the people in Solok Regency. After doing the research and testing the research hypothesis, the conclusion of the study is as follows: 1) There is an effect of green open space on the happiness of the people in Solok Regency. The significance value of the effect of green open space on community happiness is 0,000 and the magnitude of the effect of green open space on people's happiness at the research location is 21.4%; 2) There is a significant influence on the work culture of tourism to the happiness of the community in Solok Regency. The significance value of the effect of tourist conscious work on the happiness of the community is 0,000 and the magnitude of the influence of tourist conscious work on the happiness of the community at the research location is 28.3%; and 3) There is an influence of green open space and tourism-conscious work culture together on the happiness of the people in Solok Regency. With a significance value of 0,000 and the magnitude of the influence of the two independent variables on the happiness of the community at the study location is equal to 28.3%, the rest is influenced by other variables not included in this study by 71.7%.

References

[1] Liu K 2013 Happiness and Tourism *International Journal of Business and Social Science* **4**(15) 67-70
[2] Awal P 2017 No Title 13 Konsep Kebahagiaan Martin Seligman Sebuah Penelitian Awal pp 359-372
[3] Frinaldi A and Afriva K 2017 Membangun Model Kebijakan Budaya Kerja Sadar Wisata di Kalangan Aparatur dan Masyarakat Pariwisata dalam Upaya Peningkatan Kepuasan Masyarakat Berkunjung ke Lokasi Wisata di Sumatera Barat (Padang : Lembaga Penelitian dan Pengabdian Kepada Masyarakat Universitas Negeri Padang)
[4] Nilson O 2015 Urban Happiness Using Public Space as a Too; for Reating Happiness *Thesis* (Human Ecology)
[5] Frinaldi A 2014 Pengaruh Budaya Kerja PNS Terhadap Pelayanan Publik di Dinas Catatan Sipil dan Kependudukan Kota Payakumbuh Provinsi Sumatera Barat *Humanus* **13** (2) 180-192
[6] Samsudi 2010 Ruang Terbuka Hijau Kebutuhan Tata Ruang Perkotaan Kota Surakarta *Journal of Rural and Development* **1**(1) 11-19
[7] Sugiyono 2018 *Metode Penelitian Kuantitatif dan R&D* (Bandung: Alfabeta)

IOP Conf. Series: Earth and Environmental Science **314** (2019) 012054 doi:10.1088/1755-1315/314/1/012054

Environmental education and disaster mitigation through language learning

S Ramadhan[1,*], E Sukma[2] and V Indriyani[3]

[1] Faculty of Language and Art, Universitas Negeri Padang, Prof. Dr. Hamka Street, West Air Tawar, Padang-West Sumatra 25131 Indonesia
[2] Faculty of Education, Universitas Negeri Padang, Prof. Dr. Hamka Street, West Air Tawar, Padang-West Sumatra 25131 Indonesia
[3] Doctoral Program of Faculty of Language and Art, Universitas Negeri Padang, Prof. Dr. Hamka Street, West Air Tawar, Padang-West Sumatra 25131 Indonesia

*syahrulramadhan.532@gmail.com

Abstract. This study aims to determine the knowledge of the environment, attitudes towards the environment, and skills in designing learning that contains the environment of prospective language teachers. In addition to the environment, this research also links environmental learning with disaster mitigation. This study uses a survey method. The instrument used was a questionnaire in the form of a question sheet consisting of twelve questions relating to the attitudes and knowledge of prospective teacher students regarding environmentally charged learning and disaster mitigation. The results of the study indicate that students still need to improve knowledge about the environment and disaster mitigation to be able to design language learning with environmental education and disaster mitigation. When becoming a teacher, scholar have been able to integrate environmental education and disaster mitigation to provide students with an understanding of it and foster an attitude of environmental awareness and care for the environment.

1. Introduction

Worldwide, everyone faces serious environmental problems, such as global warming, acid rain, ozone layer destruction, environmental pollution, natural damage, and loss of biodiversity that can seriously threaten the lives of current generations, especially future generations [1,2]. Environmental problems are caused by careless and intensive human activities in natural areas to improve their quality of life [3,4]. A more threatening aspect is the unconsciousness of human influences in their environment [5]. Environmental damage results in an ecological crisis that will cause a series of problems with natural resources, climate, pollution, food, and the economy [2,6]. This is because the environment determines the dynamics of life, public health, spiritual and moral development [7].

Humans living today are faced with environmental problems [2]. Humanity has long recognized that its existence depends on the status of its environment [8]. Everyone has the same responsibility to overcome these problems [1,9]. Because environmental damage is not caused from a certain place, but it can come from the surrounding environment. The damage not only affects the future life, but also the present.

Around the world many methods have been applied to solve environmental problems, such as government policies, appeals to society, nature lovers' movements, and strengthening government regulations [1,10]. However, this method is not effective enough to reduce environmental problems. Developing conscious and environmentally responsible individuals is needed to deal with these problems so that a sustainable environment and quality of life are better [5,11,12]. This can be realized through education, more specifically environmental education [5,11,13,14].

Environmental problems are inherently inseparable from environmental awareness, values, and attitudes of people. This encourages countries to start assessing environmental education [2]. Environmental education is a joint effort that is consciously organized to teach or attract human attention about how the natural environment functions and how these humans can manage and protect it [9]. This education takes place in the practice of seeking creativity and discovery, and the exploration of reality without realizing it [15]. Developing knowledge about scientific and social significance of the environment is the first step in the formation of an environmentally friendly society. Through knowledge and awareness, positive values and attitudes emerge that will encourage actions to make adjustments to lifestyle and consumption habits that will reduce the burden we place on the environment [16,17].

The aim of environmental education is to make students participate in protecting the environment [18]. In order for students to participate in protecting the environment, it is important to develop knowledge about the environment, environmental awareness and changes in behaviour toward the environment [19,20,21,22,23]. Individuals should be encouraged not to view the environment as something that should be exploited for their purposes, but as a valuable asset that is worth protecting for human survival [22].

Environmental education can help students re-think the correlation between humans and the environment, begin to understand their environment, be aware of environmental problems and consider environmental issues related to their lives [2,24]. It can also change the overall system of consumer mentality, worldview and education [25]. In addition, it also produces certain behavioural skills and rules in relation to the environment, and develops optimization and harmony between humans and nature [26].

The National Environmental Education Foundation explains the three levels of environmental education described below. First, namely environmental awareness, general familiarity with key environmental issues. Second, knowledge of personal behaviour, which translates awareness into actions to preserve the environment. Third, environmental literacy, which allows students to learn basic principles and acquire skills to carry out their own projects [27].

Educational institutions as key stakeholders are obliged to increase student environmental awareness. In Indonesia, environmental education is not a compulsory lesson in school. Therefore, environmental education can be integrated into other subjects, one of which is language learning. Research on environmental education in language learning has been carried out by other researchers before [9,15,28,29,30]. Using environmental themes in text learning can trigger interest in students learning languages and implicitly can increase students' knowledge of the environment [12,28,31,32]. It can also open a window of opportunity to develop students' thinking skills and communicative skills by using materials that are challenging on global issues, especially about the environment [29]

The themes related to the environment in Indonesia need to be simulated. Indonesia is one country that has many environmental problems. Based on the Sindo newspaper R & D survey in 2018, there were ten major environmental problems in Indonesia. These problems include waste (40%), flooding (20%), polluted rivers (11%), global warming (10%), air pollution (6%), difficulty in clean water (3%), forest destruction (2 %), abrasion (2%), and soil contamination (2%). To foster students' understanding and awareness of the environment, education has a big role. Through language learning, issues regarding this matter need to be integrated [33].

Based on research conducted by previous researchers, it shows that the level of knowledge and environmental awareness of students is low [25,34,35]. This is very worrying, because students are future generations who must be able to maintain the environment. Students need to have experience in taking initiatives relative to urgent requirements in the field of global challenges in society [36]. In environmental education, students need to develop environmental advancement, increase capacity, have experience on environmental issues and develop a vision for a sustainable future [22].

To overcome this, teachers play an important role in teaching environmental education and fostering student awareness about the environment [9,13,28,29,37,38,39]. In addition, the teacher must also develop the ability of students to understand, criticize, and participate rationally in every discourse on environmental issues [8,39]. To make this happen, the teacher must have the necessary knowledge and readiness about the environment in order to have efficient activities in the classroom [40].

As the subject of pedagogical activities, the teacher must produce knowledge, beliefs, and skills in environmental preservation and protection of nature [8]. The teacher views environmental science as an aspect of nature protection that is applied. This attitude does not raise awareness about harmony with nature and knowledge of the relationship between humans and nature [41]. To achieve this, the teacher must have an interest in using themes related to global issues regarding the environment, having a qualified understanding, and a positive attitude towards the environment. However, it was reported that teachers were inefficient in terms of introducing environmental education goals to students, possessing the necessary information and skills, and having self-confidence in relation to environmental education [40,42].

As language teachers, we are the luckiest teachers, all subjects are ours. Whatever our students want to communicate, whatever they want to read, is our subject matter [43]. Language teachers are in a unique position to promote environmental awareness [28]. Many educators today understand the importance of protecting the environment, but language teachers generally do not see themselves as science teachers, nor do they, as the general public understands environmental issues [44]. By integrating environmental education into language classes, educators can increase students' interest in contemporary issues that might directly affect their future; teach students how to contribute to a healthier and more sustainable world; and promoting meaningful language learning and communication [28].

The decrease in environmental carrying capacity is characterized by the number of natural disasters that occur due to human behaviour. Forest fires, floods and landslides that occur in Indonesia are evidence of a crisis of environmental awareness. To prevent further damage, it needs to be anticipated early through environmental education and disaster mitigation [4]. Many natural disasters are caused by environmental damage from humans. Creation of awareness and sensitivity about environmental changes related to mitigation as fraudulent natural hazards needs to be done [45]. An understanding of disaster mitigation can reduce loss and loss of life. Thus, students need to know this through language learning.

To integrate environmental education in language learning is not easy. The teacher must have the necessary knowledge and readiness about the environment in order to have efficient activities in the classroom [40]. Based on these explanations, this study aims to determine the knowledge of the environment, attitudes towards the environment, and skills in designing learning that contains the environment of prospective language teachers. In addition to the environment, this research also links environmental learning with disaster mitigation.

2. Method

This study uses a survey method. The instrument used was a questionnaire in the form of a question sheet consisting of twelve questions relating to the attitudes and knowledge of prospective teacher students regarding environmentally charged learning and disaster mitigation. The selection of question sheets as a research instrument, because the questionnaire cannot accommodate the knowledge and opinions regarding the problems under study and interviews require a considerable amount of time to obtain information. In addition, students also write environmentally charged learning designs and disaster mitigation for language learning. it aims to see the skills of students as prospective language teachers to try to integrate environmental education in language learning. The correspondent is an Indonesian language student who is studying at Universitas Negeri Padang consisting of 30 people.

3. Results

The results of this study are elaborated based on the research objectives which consist of three, including the following. First, knowing the understanding of prospective language teachers about environmental education and disaster mitigation. Second, knowing the attitudes and actions taken by prospective language teachers for environmental education and disaster mitigation. Third, looking at the skills of prospective language teachers in designing learning with environmental education and disaster mitigation. The results of these studies are described as follows.

3.1. Understanding of environmental education and disaster mitigation

The Result of the study on the understanding of prospective language teachers about environmental education and disaster mitigation are described as follows. First, based on students' understanding of environmental education and disaster mitigation, it shows that 19 students (63%) understand and 11 students (37%) do not understand about environmental education and disaster mitigation. This understanding is evidenced by the student's explanation of the nature of environmental education and disaster mitigation and can explain with examples related to it. Students who understand this can explain that environmental education includes cognitive and attitudinal aspects of the environment and knows that disaster mitigation is a human effort to prepare for disasters.

Second, based on information that students have known about environmental education and disaster mitigation, it shows that 21 students (70%) have received information and 9 students (30%) have not received information on environmental education and disaster mitigation. Information obtained by students regarding environmental education is through the government's call for environmental preservation with policies, public service advertisements and posters. In addition, it is also known through counselling about environmental hygiene and practicing how to deal with disasters. Information is also obtained from news about disasters caused by environmental damage. Of the 14 students who knew the information, there were 2 students who had been directly involved in disaster mitigation.

Third, based on students' understanding of environmental impacts, it shows that 30 students (100%) know the environmental impact on humans. The general answer is that the environment has positive and negative effects on humans. Positive impact if the environment is well maintained and has a negative impact if the environment is not maintained. The environment has an impact on human health; besides it also affects the economy of the community. An environment that is not maintained can also cause disasters that will harm humans for their survival.

Fourth, based on student opinions regarding the importance of understanding disaster mitigation, it shows that 29 students (97%) stated that disaster mitigation is important and 1 student (3%) stated that disaster mitigation was not important for humans. Student opinions regarding the importance of understanding mitigation so that people are more aware of disasters that occur and minimize environmental damage resulting in disasters. If a disaster occurs, disaster mitigation can be useful for knowledge about saving oneself in the event of a disaster. This can reduce the losses incurred and especially reduce the loss of life.

Fifth, based on the opinions of students regarding the role of language teachers to provide environmental education and disaster mitigation, it shows that 30 students (100%) stated that language teachers play an important role in delivering environmental education and disaster mitigation. However, the explanation of the role that language teachers can play in environmental education and disaster mitigation is that not all students can provide a detailed explanation of this. In general, students argue that the role of language teachers in environmental education and disaster mitigation uses these themes in language learning, especially in text learning. Language teachers can use environment-themed texts on learning.

Sixth, based on the student's explanation of the approaches / methods that are effectively used in implementing language learning by applying environmental education and disaster mitigation, it shows that 6 students (20%) suggest using environmental, picture, video and audio-visual media, about environmental damage, and disaster mitigation in language learning. 5 students (17%) suggested using the direct method, namely inviting students to directly observe the environment (observation), whether the environment around the house, or school. 6 students (20%) suggested using environment-based (text-based) texts in learning that, through text, students gained knowledge about the environment. 6 students (20%) suggested using the discussion method, learning activities carried out by discussing the environment and the impact of environmental damage, and how to deal with the damage through class discussions. In addition, 2 students (7%) suggested using a contextual approach and 2 students (7%) suggested using a problem-based approach. 3 students (10%) did not know the effective approach / method used in implementing language learning by applying environmental education and disaster mitigation.

Seventh, based on the knowledge that students have known about environmental education and disaster mitigation shows that 29 students (97%) stated that they did not have sufficient understanding to explain environmental education and disaster mitigation in language learning and 1 student (3%) stated that they had sufficient understanding. The insufficient understanding of students is because this knowledge is known through some reading and knowledge that is known to be still in personal interest. Students also stated that environmental education and disaster mitigation were not obtained through formal education and it was suggested that presenters who understand this be brought to prospective language teachers.

3.2. Attitudes towards environmental education and disaster mitigation

The Result of the study regarding the attitude of prospective language teachers regarding environmental education and disaster mitigation are described as follows. First, based on the importance of environmental education and disaster mitigation in language learning shows that 30 students (100%) stated that it was important. The importance of this is because in Indonesia there have been many environmental damages caused by humans. Because the damage to nature causes natural disasters that can disrupt human activities. In addition, Indonesia is also a country prone to natural disasters. So that teaching about environmental education and disaster mitigation to students is very important thing to do to provide understanding and caring attitude towards the environment.

Second, based on the readiness to be a language teacher to shape students' awareness of the environment and disaster mitigation, it shows that 23 students (77%) are ready, while 7 students (23%) are not ready. Of the 23 students who are ready, only 3 students can present actions taken to foster student awareness about the environment and disaster mitigation. What teachers can do is to show that caring for the environment is important, this is done by doing small things such as properly disposing of garbage and carrying out activities related to environmental sustainability.

Third, based on students' thinking about improving environmental knowledge and disaster mitigation, it shows that 30 students (100%) are ready to improve their knowledge because it is important. Environmental problems are a shared responsibility. Students advise that environmental policies need to be socialized and disaster mitigation needs to be simulated so that people know. The most important thing is to grow public awareness, starting with ourselves, to pay more attention to this problem.

Fourth, ways that can be done to increase students' interest in studying environmental education and disaster mitigation show that 27 students (90%) know how to do it and 3 students (10%) do not know how to do it. Students suggest ways that can be done to increase student interest by showing the news or text that relates to this, that the current environment has suffered a lot of damage and can cause disasters. These disasters can affect human life.

Fifth, based on students' interest in studying environmental education and disaster mitigation, 30 students (100%) study it. Because students realize that learning this matters both as human beings and as prospective teachers who must be able to integrate environmental education in learning.

Sixth, based on students' interest in researching environmental education and disaster mitigation, 12 students (40%) were interested, while 18 students (60%) were not interested in doing the research. However, students who are interested in conducting research do not yet know how to start and link their research with the field of education that they undergo. The students' interest regarding this research was motivated by the fact that environmental problems were a topic that could benefit humans.

3.3. Language learning designing skills containing environmental education and disaster mitigation

This research data is a learning design written by students to integrate environmental education and disaster mitigation in language learning. Based on the learning plan, the following points are summarized. First, in the introductory part of learning, the learning design written has not yet seemed to motivate the teacher so that students are interested in studying environmental education and disaster mitigation. Second, in the learning activities section, the themes related to environmental education are still limited, students provide examples of garbage, floods, earthquakes, and tsunamis. Even though the environmental theme is very broad to be given to students. Third, students have not been able to write a design or learning scenario using a particular method, in general students use a lot of learning using media images, audio, or audio-visual. Fourth, the text used in learning to integrate environmental education and disaster mitigation has varied, such as exposition texts, explanations, procedures, poems, and observational reports. Fifth, reflection on the closing part of learning does not appear to foster student attitudes towards the environment.

4. Discussion

Every individual has the same obligation to protect the environment. Lack of knowledge and environmental awareness has caused environmental damage. This incident certainly cannot be left alone. There is still hope to improve it by increasing environmental awareness among the younger generation. School is one of the important roles to make this happen.

Environmental education is important to improve students' attitudes and awareness of the environment. Environmental education must be designed to contribute to the development of cognitive, emotional, linguistic and psychomotor students, and enable them to develop positive attitudes towards the environment [24]. However, environmental education in the school curriculum in Indonesia is not a compulsory lesson that must be learned. To overcome this, environmental education can be integrated in other subjects, namely language learning.

Language teachers play an important role in integrating environmental education in language learning. Teachers can use global issues as a context for developing language skills [46]. This context can be integrated into the text being studied. In accordance with the language learning curriculum, namely text-based language learning. Through the texts learned, students gain knowledge about the environment and are expected to have an impact on environmentally conscious behaviour. Environmental education can be seen through the learning design used by the teacher.

Based on the results of the research that has been done, it shows that the knowledge of students (prospective teacher) to teach environmental education in language learning is not enough. Their understanding is still around general issues about the environment, even though as a 21st century teacher, the world faces serious global issues and must be known by all individuals, especially teachers [13]. The results of the study show that there is a lack of understanding of students (prospective teacher) about the environment, because the students in question rarely get information about the environment. The lack of reading about environmental issues and lack of interest and caring for the environment is one of the factors causing this. Knowledge about the environment is still about understanding concerning matters and personal interests.

Based on the learning design written by students (prospective teacher), it shows that students have been able to integrate environmental education through the texts learned. Students are able to determine diverse texts for environmental education. However, students have not been able to use diverse learning methods. Learning is still about the use of media such as image, or videos that are useful for providing information to students about the environment.

Many methods or approaches that can be used in language learning contain environmental education, such as genre-based learning [30], task-based learning [9,29], and content-based learning [29]. Learning is done by integrating environmental education in language learning. In addition to these methods, teachers can also use other appropriate approaches such as contextual based learning, problem-based learning, project-based learning, and etc.

Based on student (prospective teacher) attitudes towards environmental education shows that students agree to improve their ability to understand the environment, because they realize that environmental education is important to know. The importance of environmental education is useful for oneself and as a teacher whose job is to convey information to students. Teachers need to grow the ability of students to understand, criticize, and participate rationally in every aspect of the environment (human, cultural, physical). It can foster a holistic approach that addresses social, cultural and ethical [39]. If teachers do not have expertise in knowledge, skills, and commitment to the environment, they may not be able to lead effectively to environmental changes [8].

5. Conclusions and Suggestions

Based on the research conducted, that everyone has the same obligation to protect the environment. To be able to maintain the environment requires understanding and knowledge about the environment. Teachers play an important role in conveying this knowledge to students, because students are the next generation who are in charge of protecting the environment.

Environmental education can be integrated into other lessons, one of which is language learning. integration can be done by using themes about the environment in text learning. Through text that is learned, students can learn the language and get environmental knowledge through the contents of the text being read.

Based on this, it is important for teachers to have environmental knowledge and have an environmentally caring attitude. If the teacher does not have the knowledge and attitude, then language learning will be less meaningful. Teachers must be able to design language learning by integrating various topics, one of which is environmental education. You can use text related to the environment or use the appropriate learning method.

This research can be used as an evaluation for all parties, especially prospective language teachers to be able to increase their knowledge about environmental education. By increasing this knowledge, it can foster an attitude of environmental awareness and can improve skills in integrating environmental education in language learning. Teaching environmental education to students to form a generation that understands the environment and has a caring and loving attitude towards the environment.

References

[1] Mitsuki I 2017 Beyond the limitations of environmental education in Japan *Educational Studies in Japan: International Yearbook* Number 11 pp 3-14

[2] Lai C S 2018 A study of fifth graders' environmental learning outcomes in Taipei *International Journal of Research in Education and Science* **4**(1) 252-61

[3] Tung C Y, Huang C C and Kawata C 2002 The effects of different environmental education programs on the environmental behaviors of 8th grade students and related factors *Journal of Environmental Health* **64**(7) 24-9

[4] Wihardjo S D, Hartati S, Nurani Y and Sujarwanta A 2017 The effects of green schooling knowledge level and intensity of parental guidance on the environmental awareness of the early age student *Educational Research and Reviews* **12**(5) 251-57

[5] Erdogan M 2015 The effect of summer environmental education program (SEEP) on elementary school students' environmental literacy *International J. Sci. Env. Ed.* **10**(2) 165-181

[6] Uttara S, Bhuvandas N and Aggarwal V 2012 Impacts of urbanization on environment *International Journal of Research in Engineering & Applied Sciences* **2**(2) 1637-45

[7] Kibbe A, Bogner F X and Kaiser F G 2014 Exploitative vs appreciative use of nature-Two interpretations of utilization and their relevance for environmental education *Studies in Educational Evaluation* **41** 106-12

[8] Nazarenko A V and Kolesnik A I 2018 Raising environmental awareness of future teachers *International Journal of Instruction* **11**(3) 63-76

[9] Nkwetisma C M 2011 EFL/ESL and environmental education: towards an eco-applied linguistic awareness in Cameroon *World Journal of Education* **1**(1) 110-18
[10] Nieblas-Ortiz E C, Acrcos-Vega J L and Sevilla-Garcia J J 2017 The construction of an environmental management model based on sustainability indicators on a higher education institution in Mexico *Higher Education Studies* **7**(1) 15-22
[11] Wang J, Zhu M, Tang X, He M, Xu S, Gao Y and Gu J 2010 Opportunities and challenges for environmental education at yunnan's institutions of higher learning *Chinese Education & Society* **43**(2) 82-93
[12] Uyar Y and Ensar F 2016 Does mother tongue education support development of environmental literacy in Turkey? An analysis of Turkish course book *International Journal of Environmental & Science Education* **11**(1) 1-8
[13] Yumusak A, Sargin S A, Baltaci F and Kelani R R 2016 Science and mathematics teacher candidates' environmental knowlegde, awareness, behavior and attitudes *International Journal of Environmental & Science Education* **11**(6) 1337-48
[14] Akman O and Alagoz B 2017 A study on environmental-oriented knowledge, attitude and behavior level of teacher candidates *Journal of Education and Practice* **8**(1) 229-45
[15] Castillo R and Rojas M del P 2014 Sensitizing young english language learners towards environmental care *Gist Education and Learning Research Journal* Number 9 pp 179-95
[16] Kwan F W B and Stimpson P 2003 Environmental education in Singapure: A curriculum for the environment for adolescents *Journal of Environmental Psychology* **27** 242-51
[17] He X E, Hong T, Liu L and Tiefenbacher J 2011 A comparative study of environmental knowledge, attitutedes and behaviors among university students in China *International Research in Geographical and Environmental Education* **20**(2) 91-104
[18] Jacobs G M and Cates K 2012 Global education in second language and teaching *International Journal of Physical and Social Sciences* **2**(8) 1-22
[19] Kollmuss A and Agyeman J 2002 Mind the gap: Why do people act enviconmentally and what are the barriers to pro-environmental behavior? *Environmental Education Research* **8**(3) 239-60
[20] Schultz P W, Shriver C, Tabanico J J and Khazian A M 2004 Implicit connections with nature *Journal of Environmental Psychology* **24**(1) 31-42
[21] Xuehua Z 2004 An overview of the environmental knowledge system for elementary school students *Chinese Education and Society* **37**(4) 45-7
[22] Çokçaliskan H and ÇelikÖ 2017 Investigation of pre-service classroom teachers' environmental awareness and attitudes *International Electronic Journal of Environmental Education* **7**(2) 73-83
[23] Okur-Berberoglu E 2017 Ecological dynamics model and ecopedagogy-based outdoor experiential education *International Electronic Journal of Environmental Education* **7**(2) 134-151
[24] Buldur A and Ömeroglu E 2018 An examination of the relationship between pre-school children's and their teacher' attitute and awareness towards the environment *Journal of Education and Learning* **7**(2) 221-9
[25] Ponomarenko Y V et al 2016 Students' environmental competence formation as a pedagogical problem *International Journal of Environmental & Science Education* **11**(18) 11735-50
[26] West S E 2015 Understanding participant and practitioner outcomes of environmental education *Environmental Education Research* **21**(1) 45-60
[27] Paterson J 2010 Integrating environmental education *Education Digest* **75**(7) 38-42
[28] Hauchild S, Poltavthenko E and Stoller F L 2012 Going green: Menging environmental education and language instruction *English Teaching Forum* Number 2 2-13
[29] Omidvar R and Sukumar B 2013 The effects of globar education in english language conversation classroom *English Language Teaching* **6**(1) 151-57
[30] Setyowati L and Widiati U 2014 Integrating environmental education into a genre-based EFL

writing class *English Teaching Forum* Number 4 pp 20-7
[31] Napoli M 2011 Going green: Empowering readers to change the environment *Kappa Delta Pi Record* **47**(2) 76-9
[32] Ramos A M and Ramos R 2011 Ecoliteracy through imagery:A close reading of two wordless picture books *Children's Literature in Education* **42**(4) 325-39
[33] Sindo newspaper R&D survey2018 *https://nasional.sindonews.com* accessed on 7 November 2018
[34] Quimby J L, Seyala N D and Wolfson J L 2007 Social-cognitive predictors of interest in environmental science: Recomendations for environmental educators *Journal of Environmental Education* **38** (3) 388-96
[35] Rahman H A 2010 The involvement of the native people in environmental issue: A survey on the Mah Meri Tribe in carey Island, Kuala Langat, Selagor *Human Sciences Journal* **17** 114-34
[36] Obasoro C B, Oyinloye O A and Ilensami A A 2013 The importance of environmental education to secondary education level in Akure South local goverment area, Ondo State Nigeria *African Research Review* **7**(4) 298-306
[37] Kim C and Fortner R W 2006 Issue-specific barriers to addressing environmental issue in the classpoom: An exploratory study *The Journal of Environmental Education* **37**(3) 15-22
[38] Stern M J, Powell R B and Hill D 2014 Environmental education program evaluation in the new millennium: What do we measure and what have we learned? *Environmental Education Research* **20**(5) 581-611
[39] Bergman B G 2016 Assessing impacts of locally designed environmental education projects on students' environmental attitudes, awareness, and intention to act *Environmental Education Reseach* **22**(4) 480-503
[40] Fidan N and Ay T S 2016 Acquisition of operational environmental literacy in social studies course *International Journal of Environmental & Science Education* **11**(13) 5951-68
[41] Liu S Y, Yeh S C, Liang S W, Fang W T and Tsai H M 2015 A National investigation of teachers' environmental literacy as a reference for promoting environmental education in Taiwan *The Journal of Environmental Education* **46**(2) 114-32
[42] Loubser C P, Swanepoel C H and Chacko C P C 2001 Concept formulation for environmental literacy *South Africal Journal of Education* **21**(4) 317-323
[43] Rivers W 1976 *Speaking in many tongues: Essay in foreign language teaching (2nd edition)* (Rowley: MA Newbury House) p 96
[44] Cotton D R E 2006 Teaching controversial environmental issue: Neutrality and balance in the reality of the classroom *Educational Research* **48**(1) 223-41
[45] Manase J 2016 The adequacy of environmental education technique and strategies employed in uluguru mountains hotspot *International Journal of Education & Literacy Studies* **4**(4) 65-70
[46] Gürsoy E and Sağlam G T 2011 ELT teacher trainers' attitute towards environmental education and their tendency to use in the language classroom *Journal of International Education Research* **7**(4) 47-52

IOP Conf. Series: Earth and Environmental Science **314** (2019) 012055 doi:10.1088/1755-1315/314/1/012055

Study on sustainability status of smallholder oil palm plantations Jambi Province, Sumatra Indonesia

Rosyani[1], Edison[2] and Asmadi[3]

[1] Department of Master Environmental Science, University of Jambi, Indonesia
[2] Department of Agribusiness Agricultural Faculty, University of Jambi, Indonesia
[3] Department of Agroecotech Agricultural Faculty, University of Jambi, Indonesia

*rosyani-pertanian@unja.ac.id

Abstract. The growth of oil palm plantations in Indonesia is growing rapidly. Issues of oil palm plantations that are not environmentally friendly are continuously exposed. The purposes of this research are, 1) To determine the role of smallholder in managing oil palm plantations, 2) To analyse the status of sustainability of oil palm plantations, 3) To find sensitive attributes that affect the sustainability of smallholder oil palm plantation management. Analysis method is to answer the first objective which is done descriptively, the second objectives was conducted with MDS approach, with Rappalm, third objectives are to analyse attribute lever was conducted with Leverage analyse. The result of the research is that smallholders can cultivate oil palm plantations. The status of sustainability indicated a very sustainable status (80.41), meaning that economic factors and social factors leverage ecological factors to be sustainable. This means that smallholder should improve the pattern of oil palm crop farming, with lever attributes, namely tree planting among islands of oil palm (plot scale). The integration of sensitive attributes, with the economic dimension and social dimension, produced a research output which have provide solutions sustainability of the smallholder oil palm plantation.

1. Introduction

Palm oil is one of the main commodities of the plantation sub-sector that attracts serious attention from the government, investors and farmers, especially since the 1990s. The actual development of oil palm industry in Indonesia began in 1970s, the period of the 1980s to the present was the era of quick development. Originally, the oil palm plantation actors consisted of the State Plantations (PBN), but at the same year also opened a room for the Private Large Plantations (PBS) and Smallholding Estates (PR) through the form of PIR (Perkebunan Inti Rakyat/ Nucleus Estate and Smallholder Scheme), and later developed into the private plantation forms [1].

The development of agribusiness-based palm oil plantation is a rural-oriented economic development. The target of development of the plantation sector should be to increase the income of rural communities and improve the welfare of farmers. Thus, the expected number of poor people, especially in rural areas can be reduced as even distribution of development; thus, making sustainable palm oil development for rural development that considers aspects of economic, social and environmental balance [2].

Data on the development of oil palm area from 1999 to 2016, the growing land area of oil palm plantations in Indonesia. The area of oil palm plantations from 1999 increased rapidly until 2016. In

1999, the area of oil palm plantation was 3.901.802 hectares. In the sixteen-year period it increased dramatically to 11.118.795 hectares. The total area managed by private companies covered 5.754.718 ha, 4.656.648 ha was managed by smallholders, and the remaining 707.429 ha was managed by state-owned enterprises [3]

Related to commodities, oil palm has a strategic role in supporting the growth of the Indonesian economy since the export of palm oil in recent years has continued to increase. Cumulative value for GDP of plantations continued to increase significantly, from Rp. 81.66 trillion in 2007 grew to Rp. 153.731 trillion in 2011 and continued to soar through the Rp. 159.73 trillion in 2012, or an average annual growth of 14.79% in 2015 [4].

Although the development of the plantation sector, especially oil palm plantations, is encouraging, there are many challenges related to environmental issues. There are many issues within the management and maintenance of oil palm plantations that are considered not environmentally friendly [5] although economically oil palms have provided an increased level of income for welfare. [6]. Oil palm crops have also led to changes in natural forests [7]. Furthermore, the behaviour of farmers who tend not to practice intensive plantation management, results in low productivity and tends to back rumours that oil palm crops are grown in monoculture at the farmers' garden level.

While it is believed that oil palm plantation development contributes greatly to rural development and farmer welfare, the management of oil palm plantations by farmers should pay attention to aspects that ensure environmental sustainability and social responsibility of the community. The hallmark of sustainable development is development that embodies current needs without diminishing the ability of future generations to realize their needs [8]. The concept of sustainable development in a region aims to fulfil the needs of people living independently based on the ability of natural resources, human resources, capital and technology without diminishing the ability of future generations [9]. To achieve that, it is necessary to know the status of sustainability. Sustainability can be categorized into levels of non-sustainable, less sustainable, adequately sustainable, and sustainable. To recognize the status, it is necessary to study the ecological, social and economic factors that are interrelated and can support the development of the region through sustainable development.

From the above, it is interesting to know the sustainability of the development of oil palm plantations by analysing the index and sustainability status. This can be formulated as follows:

1. What is the role of farmers in palm oil cultivation?
2. What is the status of sustainable management of smallholder oil palm plantations?
3. Any sensitive attributes that affect the sustainability of palm oil plantations managed by the smallholders.

The purposes of this research are:

1. To know the role of farmers in the oil palm plantation business?
2. To analyse the status of sustainability of smallholder palm oil plantation management.
3. To analyse sensitive attributes to the sustainability of smallholder palm oil plantation management.

2. Research method

This research was conducted in Mekar Jaya Village, Tabir Selatan sub district of Merangin, Jambi Regency, this village obtained RSPO certification in 2014. The sampling approach was independent self-reliant palm oil plantations with 44 households, respectively. Simple Random Sampling was applied.

2.1 Data Analysis

To answer the first objective of knowing the role of farmers used the analysis of categories: data analysis was done descriptively and conducted to determine the role of oil palm farmers in each dimension; that is ecological, economic and ecological dimensions were analysed by using the following percentage formula:

$$P = \frac{\sum F}{\sum N} \times 100\% \quad (1)$$

Note:
$\sum F$ = Number of frequencies that appear
$\sum N$ = Total number of respondents
P = Percentage

Subsequently, the measurement results of the role of oil palm growers were categorized according to the following:

Table 1. Category of role of farmer group

No	Percentage	Criteria
1.	89% - 100%	very high
2.	61% - 88%	high
3.	41% - 60%	medium
4.	12% - 40%	low
5.	< 12%	very low

Source: [10]

To answer the second objective of analysing sustainability status was done using Multi-Dimensional Scaling (MDS) approach, with the Program Rapfish (Rapid Assessment Technique for Fisheries), technique that was developed by the Fisheries Center, University of British Columbia [11] Based on Rapfish, modification of techniques was tailored to this study and in this study is called Rap-palm.

The data used to determine the status of sustainability is included in the primary data. These data include attributes related to the sustainability of community palm oil management, and include ecological, economic, and social dimensions.

The stages of the implementation of the analysis of oil palm oil sustainability are as follows: conduct review and determine attributes on each dimension of sustainability, i.e. ecological, economic, and social dimensions.

1. The ecological dimension consists of attributes:
   a. Land management without burning
   b. Conservation of undulating land / basin
   c. Use of pesticides
   d. Fertilization
   e. Planting tree crops among palm crops
   f. Selection of seeds
   g. Utilization of river border in the garden
   h. Leaving the tree plant in the basin location
   i. Utilization of midrib, harvest and post-harvest
   j. Use of seeds
2. The economic dimension consists of attributes:
   a. Marketing
   b. Marketing facilities
   c. Potential labour
   d. Area of land exploitation
   e. Market availability
   f. Market structure
   g. Production
   h. Availability of production inputs

i. Acceptance and prosperity of farmers
j. Pricing of palm oil
3. The social dimension consists of attributes:
a. Level of education
b. Experience of oil palm farming
c. The social value of plants
d. Gender
e. Oil palm farming activities in community social activities
f. How to clear land
g. Role in the group
h. Institutional village
i. Tradition in palm oil management
j. Knowledge of riparian area
[12]

The score value which is the index value / sustainability status of each dimension can be seen in Table 2.

Table 2. Category of the status of sustainability of smallholder oil palm plantations

Category	Status of sustainability
0.00 – 25.00	Bad (unsustainable)
25.01 – 50.00	less (less sustainable)
50.01 – 75.00	enough (enough sustainable)
75.01 – 100.00	Good (sustainable)

Source: [13]

MDS is a multivariate analysis method that can be used to analyse metric data (ordinal or nominal scale). This method is also known as one of the ordination methods in an ordination in dimension (ordination in reduced space). Ordination itself is a process of plotting the point of an object along the axes arranged according to an ordered relationship or in a graph consisting of two or more axes.

Through the MDS method, the sustainability point position can be visualized in the form of a triangle diagram, answer the purpose to the three used leverage analysis. The leverage analysis is to determine the sensitive/leverage attributes of each dimension. Leverage analysis is shown by Root Mean Square (RMS). The greater the RMS value changes, the more sensitive the role of the attribute will be to the sustainability status in the relevant dimension [14].

This MDS analysis will also generate stress values, R2, and Monte Carlo analysis. The stress value determines the Goodness of Rapfish. Low stress values indicate good fit, and vice versa. A good model is indicated by a stress value that is smaller than 0.25 ($S < 0.25$) while a good R2 value is approaching a value of 1. Monte Carlo analysis is determining the error in order to assess the uncertainty in MDS. If the MDS value difference with Monte Carlo value is small, then MDS analysis to determine sustainability is good enough with 95% confidence level. Uncertainty can be caused by errors in scoring that is a repetitive process. All these analyses form a package with the MDS program in Software/Rap-palm.

3. Result and Discussion

3.1 The Role of Oil Palm Farmers

The result of the role analysis shows that the farmers in the research location with the value of the role of 73.33 with the high category, the role that is done by the farmers is in the management of oil palm cultivation, and the role in the institution of the palm oil group and role in the effort to stimulate the increase family income and farmer welfare.

3.2 Status of Sustainability of Palm Oil Plant Management and Attribute

In each ecological dimension, the economic dimension and social dimension as well as the calculation of sensitive attributes are described in the subject below;

3.2.1 Ecological Dimension. The results of the MDS analysis on the Ecological dimension are shown in Figure 2 and the leverage analysis in Figure 3. The value of the sustainability status is 60.41 (quite sustainable). The sensitive attribute is attribute number 5, which is that farmers plant other trees among palm oil plants. Thus, other attributes will be supported by sensitive attributes.

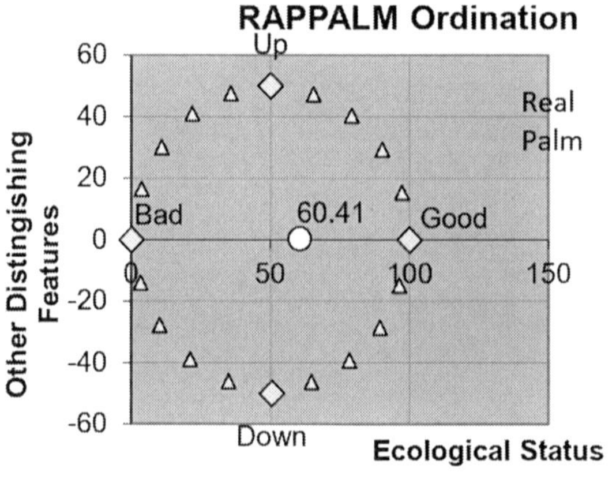

Figure 1. Ecological Dimension Analysis Results

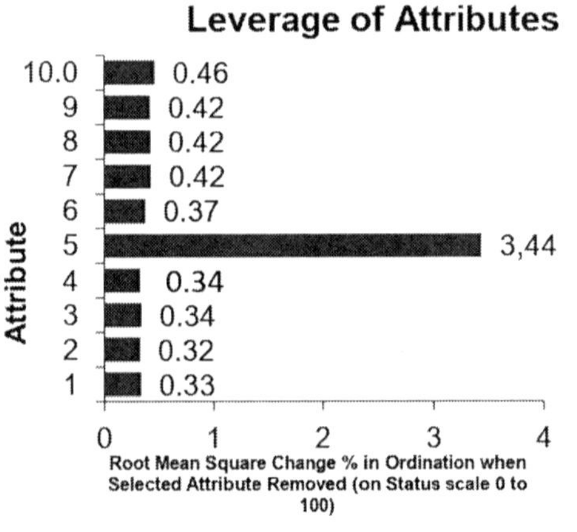

Figure 2. Result of Sensitive Attribute Leverage Analysis

Other attributes that need to be supported are, in the order of the highest value are:

1. The use of seeds according to the seedlings recommended by the BPDPKS, with the use of superior seeds such as Tenerra, Dura and Pisifera, the community at the research location has mostly used the type as recommended, because people realize that the use of good seed types will increase the level of productivity.
2. At harvest and post-harvest time, midribs and litter are collected, then covered around the garden to be used as organic fertilizer. Some farmers do this.
3. Plant protection from pest attacks, not yet using biological protection, not yet using owls (for example *Tyto alba*) as predators. Farmers still use mousetraps manually.
4. Hydrological protection, such as making water collection pits and monitoring wells in the garden has not been done by many farmers. The wells are made only for the benefit of the household. But the irrigation in the garden remains guarded by farmers.
5. Allow pioneer plants and other forest plants to grow in the riparian area. Most of the farmers do it for the protection of the water system in the plantation and for the protection of the moisture of the land.
6. Some farmers use pesticides, use it to control pests, for example borer insects (*Oryctes rhinoceros*).
7. Land management and maintenance done without burning. At the time of the research, farmers no longer burn the land, because of the strict action against the burners by the police.
8. Conservation activities at the location of land with high concentration and not planted with oil palm crop, but with trees so that the water system is maintained. A small portion of the farmers do it.
9. Utilization of midrib, harvest and post-harvest.
10. Use of seeds, smallholders use seedlings as recommended.

3.2.2 Economical Dimension. The results of the MDS analysis of the economic dimensions shown in Figure 3 and the leverage analysis in Figure 4 show the value of the sustainability status of 95.51 (continuous). Sensitive attributes are number 9, namely income from oil palm farming that can support welfare of the farmer families. This means that this attribute becomes a key attribute to increase the position of other attributes. This attribute will be a multiplayer effect for other attributes. Other attributes in the analysis results are still below the average value of MDS.

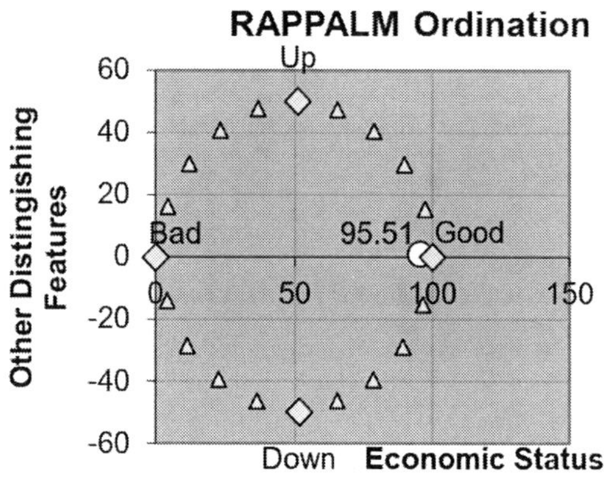

Figure 3. Rappalm Result of Economic Dimension Ordination

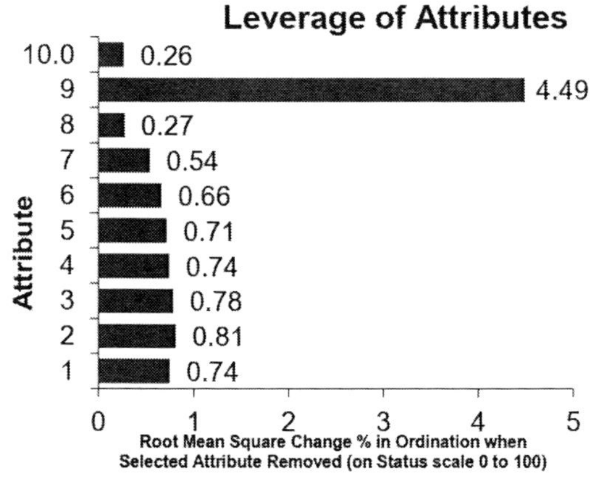

Figure 4. Leverage Analysis Result on Economic Dimension Attribute

Starting from the highest results of the average results are:

1. Oil palm cultivation equipment are available at the site of the garden or at the village level. Farmers usually buy them at the sub-district level. They are found in the village but in limited numbers and types.
2. Potential of family labour in cultivating oil palm crops; available family labour, available enough to process oil palm. However, harvest time still needs additional labour, such as neighbours and families.
3. Convenience in the marketing of palm products. Marketing TBS (oil palm fruit fresh bunches) oil palm plants is always done in the village, some are sold individually and not in groups at lower selling prices.
4. Extent of oil palm plantations at the level of the research location varies. There is a maximum of 4 hectares, there are 1- and 2-hectare plantations. Because the location of the study is the location of self-help farmers, so the area varied.
5. Farmers' strength in competing with price, some farmers can determine where to market their crops, without any connection to middlemen.
6. A small part of the market is concentrated in a centrally located village and is organized by an oil palm fruit collector/buyer at village level (toke)
7. Production results can overcome the economic problems of farmers, although the level of productivity is still low compared to the level of ideal productivity. However, if prices improve farmers can send their family members to school.
8. Market information is always available at the production plant level and in the village. Usually the information is given by groups with interests at the provincial level. Market price comes from the information of Jambi Provincial Plantation Office.
9. Income from oil palm farming that can support welfare of the farmer family.
10. Determination of the price of oil palm at the village level is still controlled by the middlemen, but not much. Some farmers have been able to master this linkage to middlemen.

3.2.3 Social dimension. The results of the MDS analysis of the social dimension shown in Figure 5 and Figure 6 are analyses of sensory attributes, namely the social dimension of the sustainability status value of 91.54 (continuous). Sensitive attributes/levers are attributes of numbers 1 and 9, education level farmers, because with high education, farmers can overcome other attributes that still produce a small value from the average value. Furthermore, the second social dimension whose attributes are above average are the attributes of tradition in the work done by farmers in the research village.

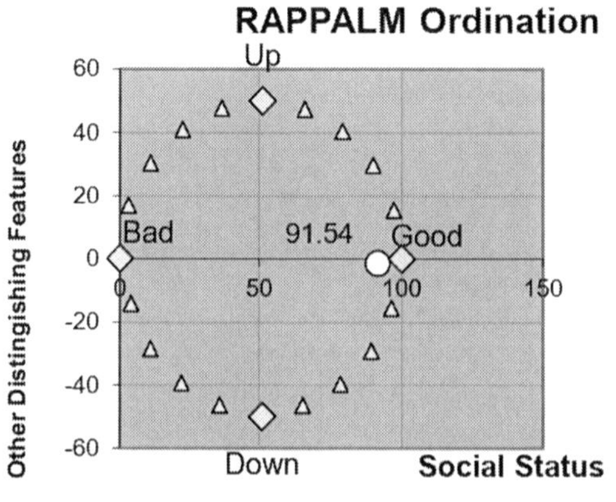

Figure 5. Rappalm Result of Social Dimension Ordination

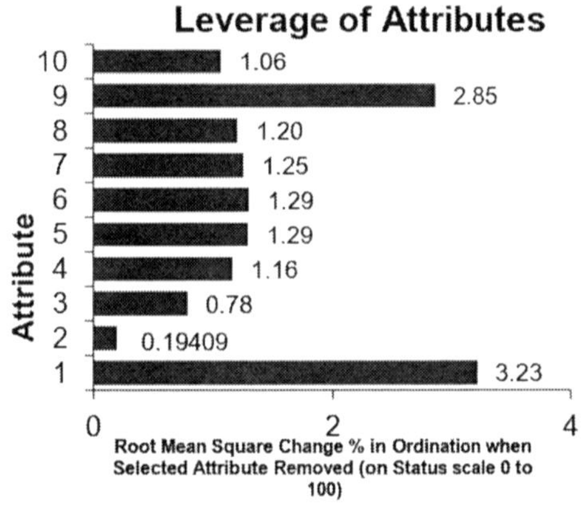

Figure 6. Result of Leverage Analysis on Sensitive Attribute

While other attributes that need to get support whose value is still below the average leverage scale are the attributes:

1. Education level of farmers.
2. How to open farming land, there are still those who do it individually, but the farmers begin to realize to form an oil palm plantation group.
3. Farmers are quite active in social activities, and it is this value that supports the value of sustainability in the social dimension.
4. The active role of farmers in the institutional groups of oil palm plantations, some farmers have joined together.
5. The participation of female workers in oil palm farming activities is partly there, but not in all activities of cultivating oil palm. Women are involved in taking the fallen loose fruit of fresh fruit bunches and bringing them to a location to be transported for market.
6. Institutions and organizations that exist in the village, enough to support the market of oil palm crops. There are Village Unit Cooperatives (KUDs) in the supporting villages palm oil market.
7. Knowledge of the protection of oil palm crops in the village, some farmers learn from the experience of farming over generations.
8. Experience in farming methods obtained from parents, then learning from neighbours or friends who plant oil palm in the vicinity.
9. Farmers' perception of kinship values in giving knowledge about cultivating oil palm plantations. Some farmers say it is very important.
10. Knowledge of riparian area.

3.3 Multidimensional Analysis

Taken together the ecological dimension the economic dimension and social dimension were analysed using Multidimensional Scaling (MDS), depicted in Figures 7 and 8, with a total value of 80.44 (sustainable) community palm oil management.

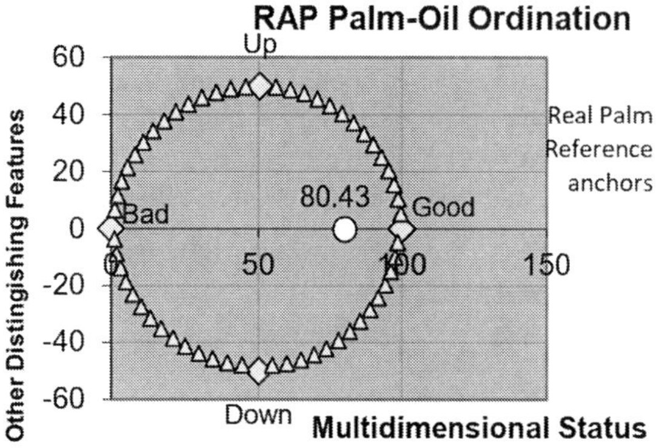

Figure 7. Triangle Value of Sustainability and Index of Smallholder Oil Palm Farm

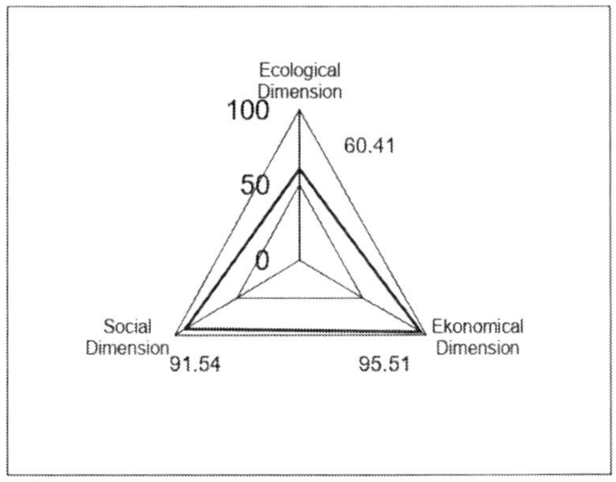

Figure 8. Results of Multi-Dimensional Analysis Integration

The results of the assessment with a continuous status category, this value of total analysis MDS = 80.43 was obtained in the assessment of 30 attributes. Of the three dimensions, namely the dimensions of ecology, social and economy the triangular leverage indicates that the integration resulting from the three dimensions simultaneously has been good, shown in good sustainability status, meaning that the three dimensions still have a have a less point on the deficiency of the MDS value on the ecological dimension. Therefore, to realize sustainable rural economic development, economic, social and ecological aspects must be integrated.

From the economic dimension, palm oil has improved the economy of rural communities. Creating multiplier economic growth in the form of income increase, farmers' income has been able to sustain the welfare of farmers and can improve the level of education of farming families. It has also been able to increase the high kinship for institutions that have a role in economic and social dimensions. The social dimension has shown the attributes of education and the traditions of oil palm cultivation by palm oil growers. Although it is not high enough, the status value is quite sustainable in ecological status, but has also contributed to the environmental sector.

Therefore, other attributes must be triggered to be able to contribute to the ecological sustainability. There is a positive correlation between ecological, economic and social aspects. This, while improving the economic aspect, will be followed by ecological and social aspects, and vice versa. [15]

According to [16] Sustainable palm oil plantation is the activity of plantation sub-sector. Especially palm oil has become a primadonna and a significant contributor to the Indonesian economy. But its management must pay attention to aspects of sustainability from economic, social and ecological aspects. At present, the development of plantations, especially oil palm, often does not consider the environmental aspects. This is evident from the results of the integration that shows the ecological dimension being the weakest dimension.

According to [11], conventional development has succeeded in increasing economic growth but failed in social and environmental aspects. Therefore, sustainable palm oil plantation development is required.

Development of sustainable agriculture, according to [17], is to conserve land, water without damaging the environment, can be technically appropriate, economically feasible and socially acceptable.

Sustainability indicators are tools used to provide information directly or indirectly about a system in the future from various dimensions such as ecology, economics and social aspects. Its use becomes important for consideration in the planning and development of the next system.

In [11], classify sustainability indicators into two types:

1. Condition indicators, i.e. indicators that explain the current state of the system relative to expected conditions.

2. Trend indicators, these are changes in the system based on the time dimension so that it can be used to monitor the tendency that will occur in the system.

Development is a dynamic change towards better conditions. In principle, development is an effort made to achieve a better condition [11]. Development is not just a mere condition or living conditions, but a process that will continue to lead to a better situation than previous conditions.

Development process that utilizes natural resources is often done without regard to environmental sustainability, so that there will be an imbalance or disturbance to the environmental ecosystem. Therefore, during the past decades, experts have established the concept of sustainable development.

Sustainable development is essential for fulfilling human needs and improving the quality of human life. At the same time, development must be based on efficiency and limited use of community resources, natural, human, social as well as economic resources.

This concept is in line with sustainable development globally [18] "Sustainable development is development that embodies current needs without reducing the ability of future generations to realize their needs". Since late 1980s, studies and discussions to formulate a universally sustainable, operational and accepted concept of sustainable development has continued to be in various dimensions and components. They are based on three pillars: economic pillar, social pillar and ecological pillar [18].

With the word sustainability, economic sustainability (profit), the continuity of human social life (people), the sustainability of the natural ecology (the planet), comes the pillar of Triple-P. Sustainable development is development that has the following characteristics:

1. Profitability and economic viability. Farmers can generate profits in a stable production level.
2. Ecological sustainability. Agroecosystem quality is maintained or enhanced, by maintaining ecological balance and conservation of biodiversity. Where ecologically sound agricultural system is a healthy agricultural system and has high resistance to pressure and disturbance.
3. Social justice. Agricultural system that has social character that can guarantee the existence of justice in access and control of land, capital, information and markets for those involved without distinguishing the social, economic, gender, religious or ethnic groups.
4. Humanity and respect of local culture. Respect the existence and treat wisely all kinds of efforts that exist where agricultural development takes place. That is, agricultural development can adapt to ever-changing conditions, such as population growth, new policy challenges and changing market systems.

According to [15] sustainable development is viewed from the aspect of human well-being, a development that has techniques and ways to maintain and sustainably improve welfare in terms of ecological, social, and economic sustainability aspects. The same thing was also expressed by [15] that sustainable development is interrelated between economic, social and ecological systems so that studies on sustainable development and welfare are synergistic studies. The study of human welfare is also a study of sustainable development and vice versa that cannot be separated from each other.

4. Conclusion

The instinct of farmers' farming is quite sustainable, as shown by some farmers who are already practicing sustainable farming, with ecological, economic and social considerations. Proven 30 attributes were analysed with MDS and leverage analysis, showing:

1. Economic dimension of the value of its sustainability status is 95.51 (continuous). Sensitive attribute/lever number 9 is the increase in farmer's income that can improve the welfare of the farmer's family.
2. Social dimension of the value of sustainability status is 91.54 (continuous) attribute sensitive/leverage, attribute number 1 is the level of education of farmers and attribute number 9 is the tradition of farmers in the cultivation of coconut plants, which still retain traditional values.

3. Ecological dimensions for the sustainability status value is 60.41 (quite sustainable). Attribute sensitive/leverage: attribute number 5, i.e. farmers plant other trees among palm oil crops, such as jackfruit, mango, durian, sungkai (Peronema canescens), petai (Parkia spp.) and jengkol (Archidendron pauciflorum).
4. Finally, multidimensional (integrated) status value is 80.43. This means that the status is in the sustainable category.

References

[1] Anonymous Republic of Indonesia, 2018. Data of Oil Palm Plantation in Indonesia, Published by Ministry of Agriculture
[2] Gold M V 1999 *Sustainable Agriculture; Definitions and Terms Compiled* (Baltimore: USDA)
[3] Statistics Bureau, 2017. Data on Plantation Service, Year 2016. Statistics Bureau of Jambi Province
[4] Directorate General of Plantation, 2016. Plantation Statistics Published by Directorate General of Plantation, Republic of Indonesia
[5] Keraf A S 2010 *Krisis dan bencana lingkungan hidup global* (Yogyakarta, Indonesia: Kanisius)
[6] Gatto M, Meike Wollini, Rosyani Asnawi and Matin Qaim 2017 Oil palm boom, contract farming, and rural economic development *World Development* **95** 127-140
[7] Krishna V V, Kubitza C, Pascual U and Qaim M 2017 Land market, property right, and deforestation; Insights from Indonesia *World Development* **99** (November 2017) 335-349
[8] Azis I J, Lidya M N, Arianto A P dan Budy P R 2010 *Pembangunan Berkelanjutan, Peran dan Kontribusi Emil Salim* (Jakarta: Gramedia)
[9] Sugandhy A 2013 Environmental management, spatial planning and sustainable development for human survival *Journal of the Environment (Indonesian Journal of Environment)* **1** (1) 12-17
[10] Gulo 2004 *Metodologi Penelitian* (Jakarta: Gramedia Widiasarana Indonesia)
[11] Frimawaty, Evi, 2012. The Sustainability of Rice Farming and Integrated Beef Cattle Based on Eco Farming. Dissertation of the Postgraduate Environmental Sciences Program, University of Indonesia, Jakarta
[12] Anonymous, 2007. *Indicators of Sustainable Development: Guidelines and Methodologies,* Printed by United Nations, New York
[13] Ruslan, S Sabiham, Sumardjo and Manuwoto 2013 Evaluation of Sustainability of Core-Plasma Oil Palm Management at PT. Plantation Nusantara VII Muara Enim South Sumatra *Ekologia* **13**(1) 33-44
[14] Gallopin, 2003. A System Approach to Sustainability and Sustainable Development and Human Sattlements Devision, Nations Unidas Santiago, Chile
[15] Rosyani, 2008. Sustainability of Indigenous Peoples, Village Communities and Plantation Companies (Land Use and Community Life Study Around Jambi Province Plantation Companies, Dissertation at the Doctoral Level of Environmental Sciences Postgraduate Program at the University of Indonesia
[16] Suandi, DMT Napitupulu, and Yusma Damanyanti, 2014, Linkage Study of Social Capital with Household Food Security in Rural Areas of Jambi Province
[17] FAO, 1989. And Agriculture Organization of The United Nations. 1976. A frame Work For Land Evaluation.
[18] Suryana, Akhmad, 2005. Sustainable Agricultural Development. Mainstay of National Development. Seminar on Sebelas Maret University Sustainable Systems, Solo, February 2005.

The potential of the ecotourism area in *Lubuk Minturun*, Koto Tangah Subdistrict, Padang City

Anelia Siska[1,*], Indang Dewata[1] and Alexander Syam[2]

[1] Environmental Science Study Program, Postgraduate Program, Universitas Negeri Padang, Prof. Dr. Hamka Street, Air Tawar Padang, West Sumatera, Indonesia 25132
[2] High School Teacher Education (STKIP) Pesisir Selatan, Indonesia

*siskaanelia7@gmail.com

Abstract. This study aimed to describe the data potential ecotourism area *Lubuk Minturun* Koto Tangah subdistrict of Padang include an analysis of potential attractiveness, accessibility and infrastructure of ecotourism. This research is a descriptive study with descriptive qualitative done at Village *Lubuk Minturun*. Source of the data required in this study are primary data and secondary data. The data collection technique is by observation, questionnaires, interviews and documentation. Data were analysed using scoring techniques. The results showed that: (1) The potential attractiveness of ecotourism *Lubuk Minturun* Koto Tangah subdistrict with high potential category wuth score 16,3. (2) The potential accessibility of ecotourism *Lubuk Minturun* Koto Tangah subdistrict with high potential category wityh score 6,3. (3) The potential for ecotourism infrastructure *Lubuk Minturun* Koto Tangah subdistrict with high potential category with score 14.6.

1. Introduction

Tourism is part of social, economic, psychological, geographic and cultural phenomena [1]. The social aspect of tourism is a basic human need for while the economic aspect is tourism can increase the income of the local community, local government, small traders [2].

Based on the psychological aspects of tourism, it becomes a means to restore one's moral health. While in terms of geography, tourism is a geographic phenomenon, meaning that special geographical appearance will be an attraction for tourists. The aspect of tourism culture attracts tourists so it is motivated and desires to see and get to know another nation's culture [3].

According to Law No. 10 of 2009 concerning tourism referred to as tourism is a variety of tourism activities supported by various facilities and services provided by the community, businessmen, government and regional government. Efforts to develop the world of Indonesian tourism are supported by Law No. 10 of 2009 concerning tourism which mentions the existence of tourist objects in an area will be very profitable, including increasing Regional Original Income (PAD), increasing people's living standards and expanding employment opportunities given the increasing number of unemployed people, increasing environmental love and preserving the culture local. Therefore the development and implementation of tourism must be pursued in a planned, gradual and sustainable manner after going through careful scrutiny. With the many natural beauty of Indonesia's tourism development, it can for tourists to enjoy the beauty of nature and culture to learn more about being sought for natural tourism and ecotourism [4].

Ecotourism provides an important opportunity for various kinds of living things in it and local culture that develops in the region [5]. Ecotourism activities can increase the income of the surrounding community and for the cost of preserving the area that is used as an ecotourism area [6].

What is meant by tourism is various kinds of tourism activities which are supported by various facilities and services provided by the community, businessmen, government and local government [7].

Ecotourism provides an opportunity for tourists to enjoy the beauty of nature and culture to learn more about the importance of various kinds of living things in it and the local culture that develops in the region [8]. Ecotourism activities can increase the income of the surrounding community and for the cost of preserving the area made ecotourism [9].

Ecotourism potential is ecological, social and cultural potential [10]. This potential will be an attraction for ecotourism that supports environmental conservation and culture in the ecotourism area. The appeal of ecotourism can be in the form of natural attraction in the form of forests, rivers, lakes, beaches and increasing income and welfare of local communities [11]. Areas that have the potential to be used as ecotourism areas are in the city of Padang.

The city of Padang is one of the tourist visit cities in West Sumatra, especially the coastal areas. During this time both local and foreign tourists are only fixated on tourist areas in coastal areas, and forget many areas that are not beaches are also interesting to visit. Many areas that have the potential to become tourist areas are forgotten because tourists are too obsessed with the beaches of Padang City. City development began to be directed towards the east, both the central government, the center of community activities and settlements. Tourism orientation began to move, the non-coastal areas began to be glanced to be developed such as *Lubuk Minturun*.

Lubuk Minturun Village is located about 15 km from the center of Padang City with a distance of about 20 minutes by vehicle. In addition, there is also access to public transportation with the purpose of *Lubuk Minturun*. This area has interesting natural potential, as well as very natural forest conditions. In addition, there is also a *Lubuk Minturun* water stem which is a famous bathing place called Lori Bath, *Lubuk Minturun* is also known as a place to perform Hajj rituals for West Sumatra pilgrims who are located around the cool forest area. Recently found a bathing place called *Ngungun Saok* located in the hilly area of *Lubuk Minturun*, in *Ngungun Saok* there are cliffs and waterfalls that beautify its attraction.

The number of ecotourism objects in *Lubuk Minturun* Village has been preserved by the local community. With the object of ecotourism, it can prosper the community around the object of ecotourism. *Lubuk Minturun* has been designated as an Agrotourism Area by the City Government of Padang, because it is seen from its physical condition suitable for agriculture. In *Lubuk Minturun* there are many natural attractions that have the potential to be ecotourism areas. But *Lubuk Minturun* village has not been designated as ecotourism offering. Ecotourism activities based on environment and culture should be developed in this area, because ecotourism is a green industry that creates quality tourism that allows tourists in small groups to maintain the quality of natural objects and attractions in the form of forests, rivers, lakes, beaches and increase income and welfare of the local community.

Based on the background above, the importance of ecotourism activities in improving the economy of the community and the large potential of ecotourism possessed by this region. The purpose of this study was to determine the potential of the Ecotourism Area of *Lubuk Minturun*, Koto Tangah District, Padang City.

2. Research Method

The type of research used is qualitative with a score approach. this research is more directed at the study of a problem or situation as it is and reveals facts, although sometimes given an interpretation or analysis given [12]. The assessment of tourist attraction objects is carried out using the assessment criteria specified in the Guidelines for Tourism Attraction Assessment. The five main components that become assessments are attractiveness, accessibility, socio-economic environment conditions, accommodation and supporting infrastructure. Each component is described in specific criteria and given a score and weighted. After obtaining the total score, the classification interval is ranked and divided into three classifications namely high, medium and low.The source of information in this study is obtained from primary data and secondary data. Primary data is obtained from direct observation / field observations and interviews, while secondary data can be obtained from library materials, articles, journals, documentaries, company internal data and other supporting documents such as *Lubuk Minturun*.

3. Result

3.1. Description of Lori Ecotourism Object in Lubuk Minturun

Lori is the result of natural formation that flows from the stem of Kandis has been enjoyed since the Meneer era and none of the Netherlands. A cemented inscription from the age at the edge of the *Lubuk Minturun* bath is proof. The prasasti carries four lines of writing that show the inauguration of this bath. The first line is engraved "Verse 8-4-1883", "by" in the second line, "penghoeloe", and the fourth line "Achin". The hallmark of this bathing place is the splash of freshness that is served alongside the lorry, cable car transportation inherited from the Dutch. Therefore, the bath is called Lori Bathing.

Table 1. Potential Attraction of Lori Ecotourism Object

Indicator	Score	Criteria	Information
The uniqueness of natural resources	2	Medium	In the Lori area there is a waterfall with a height of about 10 meters, but the waterfall does not yet have a name and various kinds of flora.
Sensitivity of natural resources	2	Medium	Sensitivity of natural resources in the Lori region is historical value and knowledge.
Variety of activities	3	High	Variations in activities that can be done in the Lori region are enjoying natural beauty, fishing, swimming, camping and education / research.
Prominent natural resources	1	Low	A prominent natural resource is the waterfall.
Location cleanliness	2	Medium	The cleanliness of the Lori ecotourism area is not affected by industry, highways and pollution.
Comfort	3	High	Comfort of the ecotourism area Lori has no disturbing smell, no disturbing traffic, no noise and the air is cool.
Safety	3	High	Lori's ecotourism area has no illegal logging, no theft, no disturbing beliefs and no dangerous diseases.
Total	16	High potential	Regions that have very high attraction to tourism objects based on set parameters.

Source: [10]

Table 2. Potential Infrastructure Facilities of Lori Ecotourism Object

Indicator	Score	Criteria	Information
Criteria for parking spaces, places of worship and toilets	3	High	a. places of worship such as mushalla and mosques are in good condition and complete facilities. b. spacious parking lots can accommodate in a capacity of> 100 vehicles. c. the toilet is in a clean and good condition.
Criteria for accommodation facilities	1	Low	No accommodation facilities available.
Restaurant criteria	3	High	There are restaurants close to tourist areas and

			adequate services.
Criteria for clean water	3	High	The availability of clean water such as PDAM has been evenly obtained by the community around tourist sites.
Criteria for the availability of electricity	3	High	The availability of electricity is evenly obtained by the community around tourist sites.
Criteria for availability of health	3	High	There are health services and are easily accessible in the form of health centers, hospitals and others.
Total	16	High potential	Regions that have very high infrastructure conditions for tourism objects based on the parameters set.

Source: [10]

Table 3. Potential Accessibility Facilities of Lori Ecotourism Object

Indicator	Score	Criteria	Information
Road criteria	2	Medium	Ecotourism areas are easily accessible with road conditions and the level of rough / permanent asphalt road hardness.
Transportation criteria	2	Medium	There are types of transportation such as ojek more than 5 or angkot more than 3 vehicles / hour.
Travel time from the city center	3	High	Travel time from the city center is around 15 minutes.
Total	7	Medium potential	Regions that have moderate accessibility to tourism objects based on the parameters set.

Source: [10]

3.2. Description of Ngungun Saok Ecotourism Object in Lubuk Minturun

Derived from the Minang language, Ngungun means mandanguang (buzzing) while Saok means closed. According to the story of local residents, the origin of this name originated from a rock that covered the flow of water, so there was a buzz. Then the place that has green water clarity is named Ngungun Saok.

Table 4. Potential Attraction of The Ngungun Saok Ecotourism Object

Indicator	Score	Criteria	Information
The uniqueness of natural resources	3	High	Natural resources found in the Ngungun Saok area are waterfalls, caves, flora and fauna.
Sensitivity of natural resources	1	Low	Sensitivity of natural resources in the Ngungun Saok area is the value of knowledge.
Variety of activities	3	High	Variations in activities that can be done at Ngungun Saok are enjoying natural beauty, swimming, fishing, tracking, camping, hiking and research / education.
Prominent natural resources	2	Medium	The most prominent types of natural resources in Ngungun Saok are waterfalls and natural phenomena.
Location cleanliness	3	High	Cleanliness of the area is not affected by industry, busy roads, residential areas and pollution.

Comfort	3	High	Ngungun Saok is very comfortable because there is no disturbing odor, disturbing traffic, noise and the air is cool.
Safety	3	High	The security of the Ngungun Saok area has no illegal logging, there is no theft, disturbing beliefs and infectious diseases
Total	18	High potential	Regions that have very high attraction to tourism objects based on set parameters.

Source: [10]

Table 5. Potential Infrastructure Facilities of Ngungun Saok Ecotourism Object

Indicator	Score	Criteria	Information
Criteria for parking spaces, places of worship and toilets	2	Medium	a. there is no place of worship but can stay in the parking lot cabin. b. the parking lot is quite coordinated c. available but in less clean conditions.
Criteria for accommodation facilities	2	Medium	There are inns, guesthouses with adequate services and facilities.
Restaurant criteria	2	Medium	Stalls are available but only in the form of a coffee shop. With inadequate service.
Criteria for clean water	1	Low	The availability of clean water has been evenly obtained by communities around the ecotourism area.
Criteria for the availability of electricity	3	High	There is no electricity available at tourist sites.
Criteria for availability of health	2	Medium	Health services are available but are still difficult to reach by the community and services are still unsatisfactory.
Total	12	Medium	Regions that have moderate infrastructure conditions for tourism objects based on set parameters.

Source: [10]

Table 6. Potential Accessibility Facilities of Ngungun Saok Ecotourism Object

Indicator	Score	Criteria	Information
Road criteria	1	Low	The ecotourism area can be reached but the road conditions are not good, there are still gravel roads and land.
Transportation criteria	1	Low	There are types of transportation to the ecotourism area, namely ojek with a frequency of <3 vehicles / hour.
Travel time from the city center	1	Low	Travel time from the village center or sub-district to the ecotourism area is> 40 minutes long
Total	3	Low	Regions that have low accessibility to tourism objects based on the parameters set.

Source: [10]

3.3. Description of Miniature Mecca Ecotourism Object in Lubuk Minturun

Nur Zikrillah Mosque is a Mecca Miniature in Lubuk Minturun, the Construction of the Mecca Miniature began on December 13, 2000 and was completed on September 8, 2001. As the name implies, miniature Mecca is not as large and as large as the holy land of Mecca. giant Qur'an monuments like those in Mecca are only smaller in size. In front of the location there is an inscription that mentions the founder of this place. Entering the building turned to the left there is the Nurzikrillah Mosque with a miniature Ka'bah inside it. This mosque besides being used for worship is also used for Hajj rituals. The Nurzikrullah Mosque is opened only at prayer times.

Table 7. Potential Attraction of The Miniature Mecca Ecotourism Object

Indicator	Score	Criteria	Information
The uniqueness of natural resources	2	Medium	The resources contained in the Mecca Miniature are flora and fauna.
Sensitivity of natural resources	2	Medium	Sensitivity to natural resources, namely the value of knowledge and beliefs / religious values.
Variety of activities	2	Medium	Variations in activities that can be done are enjoying beauty, camping and education / research.
Prominent natural resources	1	Low	A prominent type of natural resource is rock.
Location cleanliness	2	Medium	Cleanliness of Mecca Miniature is not affected by industry, waste and pollution.
Comfort	3	High	The Mecca Miniature area is comfortable because there is no annoying odor, noise free and the air is cool.
Safety	3	High	Mecca Miniature Zone Security has no dangerous currents, no illegal logging, no theft, no disturbing beliefs and no infectious / dangerous diseases.
Total	15	Medium	Regions that have very medium attraction to tourism objects based on set parameters.

Source: [10]

Table 8. Potential Infrastructure Facilities of Miniature Mecca Ecotourism Object

Indicator	Score	Criteria	Information
Criteria for parking spaces, places of worship and toilets	3	High	a. places of worship such as mushalla and mosques are in good condition and complete facilities. b. spacious parking lots can accommodate in a capacity of> 100 vehicles. c. the toilet is in a clean and good condition.
Criteria for accommodation facilities	2	Medium	Available lodging, guesthouse with adequate services and facilities.
Restaurant criteria	3	High	There are restaurants close to the ecotourism area and adequate services.
Criteria for clean water	3	High	The availability of clean water such as PDAM has been evenly obtained by the community around tourist sites.
Criteria for the availability of electricity	3	High	The availability of electricity is evenly obtained by the community around tourist sites.
Criteria for availability of	3	High	There are health services and are easily

health			accessible in the form of health centers, hospitals and others.
Total	17	High	Regions that have high infrastructure conditions for tourism objects based on set parameters.

Source: [10]

Table 9. Potential Accessibility Facilities of Miniature Mecca Ecotourism Object

Indicator	Score	Criteria	Information
Road criteria	3	High	Ecotourism is easy to reach, road conditions with hardness and good quality like hotmix asphalt.
Transportation criteria	3	High	Transportation for tourism transportation has been done like pedicab, motorcycle taxi and public transportation with a frequency of more than 10 vehicles / hour.
Travel time from the city center	3	High	Travel time from the city center is around 15 minutes.
Total	9	High	Regions that have High accessibility to tourism objects based on the parameters set.

Source: [10]

4. Discussion

To see the level of tourism potential regarding attractiveness, accessibility, and infrastructure facilities in the Lubuk Minturun Ecotourism Area in Koto Tangah Subdistrict, the overall value is classified based on the ideal score category, which is 3.01 to 4.00 high classification, 2.01 to 4, 00 medium classification and 0.001 up to 2.00 including low classification.

4.1. Potential attraction of the Lubuk Minturun ecotourism area in Koto Tangah District, Padang City

Based on several areas in Lubuk Minturun Village, the results are: On Ecotourism Lori attraction including high classification because there are many variations of activities that can be done, besides that there is a waterfall that can add to the beauty of Lori Baths. The waterfall has a height of about 10 meters, but the Lori waterfall does not have a name, so visitors are not too familiar with the existence of the waterfall. The total score of the Lori ecotourism object is 16, Ngungun Saok is classified as high because there are two waterfall waterfalls which first have a height of about 140 meters and the second waterfall is not too high but has 3 levels. Besides the waterfall, which is the hallmark of Ngungun Saok, namely Batu Tasapik. The total score of Ngungun Saok Tourism Objects is 18, Miniature Mecca attractiveness with a moderate classification because of the lack of prominent natural resources. The total score of Mecca Miniature Ecotourism Objects is 15, so that the average score produced by the three regions spread in Lubuk Minturun village with a score of 49 to determine the results of the attractiveness level of the ecotourism area in Lubuk Minturun Subdistrict.

$$\frac{49}{3} = 16,3$$

That the value of 16.3 is classified as medium potential.

4.2. Infrastructure Facilities for the Ecotourism Area in Lubuk Minturun, Koto Tangah District, Padang City

Based on a number of zones in Lubuk Minturun Subdistrict the results are: On Ecotourism Objects Lori infrastructure includes high with a score of 16, Ngungun Saok prasaranan facilities with a moderate value with a total score of 11, Miniature Mecca with high high infrastructure facilities with a total score of 17 , recording the average score produced by the three regions spread in Lubuk Minturun

Village with a total score of 44 to determine the results of the accessibility conditions of the Ecotourism Area in Lubuk Minturun Village then the costs for the facilities in this study were determined 3.

$$\frac{44}{3} = 14{,}6$$

That the value of 14.6 is classified / categorized as medium potential.

4.3. Accessibility of the Ecotourism Area in Lubuk Minturun, Koto Tangah District, Padang City
Based on some of the areas in Lubuk Minturun Village, the results are: On Ecotourism Objects Accessibility lori includes a moderate classification with a score of 7, Ngungun Saok accessibility with a low classification with a score of 3, Miniature Mecca accessibility with a high classification with a score of 9, so the average score produced by the three regions spread in Lubuk Minturun Village with a score of 19 to determine the results of the accessibility conditions of the Ecotourism Area in Lubuk Minturun Subdistrict, therefore the weight for accessibility in this study was determined 3.

$$\frac{19}{3} = 6{,}3$$

That value 6.3 is classified / categorized as medium potential.

5. Conclusion
Based on the results of the research and discussion of the results of the analysis, it can be concluded as follows:
1. The attractiveness of the object of *Lubuk Minturun* ecotourism in Koto Tangah Subdistrict on object 1, namely Lori Bathing is classified as a high classification. For object 2, *Ngungun Saok* is classified as high. And for object 3, namely the Mecca Miniature is classified as a medium classification. So, the potential appeal of the *Lubuk Minturun* ecotourism area is potentially medium with an average score of 16.3.
2. Accessibility towards the object of *Lubuk Minturun* ecotourism in Koto Tangah Subdistrict on object 1, namely Lori Bathing is classified as a moderate classification. For object 2, *Ngungun Saok* is classified as a low classification. And for object 3, the Mecca Miniature is classified as a high classification. So, the potential for accessibility of the *Lubuk Minturun* ecotourism area is potentially medium with an average score of 6.3.
3. Infrastructure facilities in the object of *Lubuk Minturun* ecotourism in Koto Tangah Subdistrict on object 1, namely Lori Baths are classified as high. For object 2, *Ngungun Saok* is classified as medium coding. And for object 3, that is, the Mecca Miniature belongs to the high category. So, the potential of the *Lubuk Minturun* ecotourism infrastructure facilities is high with an average score of 14.6.

References
[1] Bakarudin 2008 *Perkembangan dan Permasalahan Kepariwisataan* (Padang: UNP Press)
[2] Syam A 2017 Strategi pengembangan objek wisata Pantai Sumedang di Kecamatan Ranah Pesisir Kabupaten Pesisir Selatan *Jurnal Kepemimpinan dan Kepengurusan Sekolah* **2**(2) 191-200
[3] Oka A Yoeti 1996 *Pengantar Ilmu Pariwisata* (Bandung: Angkasa)
[4] Undang-Undang Republik Indonesia Nomor 10 Tahun 1990 tentang Kepariwisataan *(Law of the Republic of Indonesia Number 10 of 1990 concerning Tourism)*
[5] The International Ecotourism Society 2015 Jurnal www.world-ttourism.org.omt/ecotorism 2015.hmtl. accesed on Februari 2016.
[6] Hanif Larantika 2012 Strategi Pengembangan Ekowisata di Jalur Tracking Negeri Suah Hingga Tongkoh Propinsi Sumatera Utara *Tesis* (Bogor: Institut Pertanian Bogor)

[7] Wisantisari Purwani 2005 Penyajian Informasi Pariwisata di Kabupaten Tegal Berbasis SIG *Skripsi* (Semarang: Universitas Negeri Semarang)

[8] Damanik, Janianton and Hemlut F Weber 2006 *Perencanaan Ekowisata dari Teori ke Aplikasi* (Yogyakarta: Penerbit Andi)

[9] Kasuardi, Soekmadi R and Kartodihardjo H 2010 Strategi pengembangan ekowisata di Kabupaten Kepulauan Yapen Provinsi Papua *JMHT* **XVI**(3) 148-154

[10] Romani Siam 2006 Penilaian potensi obyek dan daya tarik wisata alam serta alternatif perencanaannya di Taman Nasional Bukit Dua Belas Provinsi Jambi *Skripsi* (Bogor: Institut Pertanian Bogor)

[11] Latupapua Yosevita 2000 Studi potensi kawasan dan pengembangan ekowisata di Tual Kabupaten Maluku Tenggara *Jurnal Agroforestri* **II**(1) 65-71

[12] Syam A 2016 Kelayakan jalur evakuasi tsunami di Kecamatan Padang Utara Kota Padang *Jurnal Kepemimpinan dan Pengurusan Sekolah* **1**(1) 11-22

IOP Conf. Series: Earth and Environmental Science **314** (2019) 012057 doi:10.1088/1755-1315/314/1/012057

Knowledge system of religious communities in watershed conservation education; case studies in Islamic Boarding Schools in East Java and West Nusa Tenggara, Indonesia

Sukarsono Sukarsono[1,2,*] and Ulfah Utami[3]

[1] Biology Education Department, Faculty of Education, Muhammadiyah University of Malang, Indonesia
[2] Center for Environmental and Population Studies, Muhammadiyah University of Malang, Indonesia
[3] Biology Science Department, Faculty of Science and Technology, State Islamic University Maulana Malik Ibrahim Malang, Indonesia

*sukarsono_umm@yahoo.com

Abstract. Knowledge of conservation education at boarding school residents grew rapidly since 1990. The residents have variety sources of knowledge and insight which affecting the owner's knowledge. Boarding schools citizen's knowledge becomes a key factor to create the basic, instrumental values and the goal values. The successfulness of knowledge transfer in order to build the values, attitudes and behavior are developing by using multiple methods of environmental education, namely: lectures, training, provision of examples and discussions. The resident's conservation knowledge lodges an interrelated system between inputs, processes, outputs and knowledge outcomes. Knowledge system that is built up is more towards emphasizing on the educational component due to the boarding school's background that has role in developing the citizens through education. Knowledge system that is built has a typical religious communities, but in accordance with the conservation objectives. Purpose of conservation education in general is to find happiness in this world and in the Hereafter a continuous basis. The purpose of conservation education is superior which include objectives to gain happiness in this world and in the Hereafter with ongoing basis comparing to the general conservation.

1. Introduction

Damage to watersheds in Indonesia is one of the unresolved issues and tends to be more concerned about. Various governmental regulations acts have been issued in the hope of improving conditions for a better watershed. The biggest impacts of watershed damage visible in various parts of Indonesia are floods and sedimentation of rivers, lakes and reservoirs. The continuing impact of the state is the disruption of economic activity and prosperity. Research by the Office of the Brantas River Basin [1] showed that the capacity of watershed management is not appropriate and knowledge of society and farmers are the main cause of damage Brantas watershed. This condition is exacerbated by high population growth and development patterns that were built on materialism and secular ideology. The ideology in practice has brought into the lives of many human situation alarming, frightening and even to a certain extent has been sparked and spread the spirit of anger and violence [2,3,4].

Research from United Nation (UN) proves that the ideals of sustainable development of the world appear to have failed. This failure was due to the development of the world suffer from distortion, where humans increasingly lost its way, has no identity, only oriented economy, and increasingly shied away from the transcendental relationship [5].

United Nations Educational, Scientific and Cultural Organization (UNESCO) declaration on environment education in the 21st century in the International Conference on Environmental Education, 2007, is stressing the importance of digging and alterative models and vision for the future of sustainability in educational activities for sustainable development [6]. The declaration also stressed the importance of finding the root cause of environmental problems that are getting worse. After three consecutive declarations were executed in 1987 and 1997, UNESCO environmental education conference in 2007 emphasized the importance of social and cultural studies to be broader, consider the context of changes in educational thinking and learning with changes in teaching methods. Besides, considering the range of learning approaches through formal education that is felt slow to make changes within the improvement, thus, the declaration also emphasized the need for immediacy maintenance of education institutions within the non-formal and informal as well as supporting the growth of a new paradigm in education circles. The recommendation has reminded that environmental education must be rooted in the harsh reality that exhausting development is takes place to plunder the planet's resources at unsustainable levels that have caused climate change and bring people to the brink of unimaginable destruction [7]. Environmental education, according to UNESCO is a process that aims to create a global community that has a concern for the environment and related issues in it, and has the knowledge, motivation, and commitment to work, both individually and collectively in the search for an alternative or a solution to the existing environmental problems and to avoid the new problems of life in the environment.

Indonesia is the country with the largest Muslim population. Population census in 2010 showed that of 240 271 522 inhabitants, as many as 85.1% are adherents of Islam [15]. Given the very large number, it is natural that environmental damage is also addressed to Muslims, besides there is no scientific evidence to refute these allegations. On the other hand, a very large number is a great potential to promote the improvement of environmental damage. Study of ethics of the relationship between man and God and nature, have sought to serve as a model for example, to be able to change the view and help people get out of this extraordinary crisis [14]. This needs to be done, given the religious traditions generally do not provide specific guidelines for responding to contemporary issues such as climate change, desertification or deforestation [11]. Need to avoid the tendency to think about the relationship between Islam and the environment using only "the right approach" that is grounded to the awareness of God rather than the fact of consciousness (morality itself), in other words, turning the region and cultural texts. This distinguishes them from environmentalist groups or conservationist grounded in the environment itself rather than text. Therefore, in addition to being a paradigm problem, the solution must also be made by affirmative action by involving all parties [16].

Research done in locations that are considered successful examples of conserving watershed, even the highest earning as environmental savior within the central government. Research was carried out to investigate the knowledge that was acquired and applied to boarding school residents for development of good watershed conservation activities. Research target is religious leaders as well as educators in the field of environment given. Results of this study will hopefully be one of the references in the nature of knowledge development and community education about watershed conservation using religious approach.

2. Theoretical Background

Boarding school is an Islamic educational institution. It's usually has a distinctive pattern of education in both educational materials and delivery methods. The difference is influenced by many factors, but the cultivation of values in order to be strong in attitude of the students is the main aim of education in schools [17]. Many boarding schools are successfully implementing a distinctive educational program, but notes on various matters relating to education were never recorded properly and so difficult to imitate and disseminate, as well as environmental education about the conservation of forests where knowledge and values are inculcated and delivery methods have not been explored and this knowledge is very important to be used by their successors. This situation is greatly feared that would eliminate the good knowledge that has proven to unsuccessfully be transferred to the next generation, given the state of health of the religious scholars and preachers which pioneering in environmental education that has increasingly weak or have to leave the boarding school because of other duties. It is also important to recognize that the knowledge which can be submitted and used [18].

Boarding school is considered as an institution that has more capabilities in the delivery of environmental values through religious approach. The effort to incorporate religious values in conservation activities should be carried out through educational activities, and in the early stages, this process will be easier if done in boarding schools because they have an adequate knowledge of religious authority [8]. The tendency of a paradigm shift towards the inclusion of environmental education of religious aspects is now spread more widely and faster, though still in the form of concepts, opinions and argument. The development of ecological issues and religion shows that this phenomenon includes the idea of the importance of religious values in the solution of environmental problems is a response to dissatisfaction with the approach that had been done. Various ethical thinkers [9,10,11,12,13,14] argued that by looking at how worse and painful state of the environment and its impact on human life, thus man must not only have to change the course of scientific thinking, but also ethical and theological thinking. Moral and religious value system is needed in moving the feelings of the community, including the preservation of the environment for future generations. Religion became a new hope in environmental conservation efforts, given the values and attitudes that make up the concept of human nature which derived from the early practice of religion and ethics every day in the community. Among the academics and social activists in particular, religion is now recognized not only as a set of teachings (value), dogma or something that is normative, but also seen as a case study of interest about how religion is viewed as the object of study for assessment [12]. Based on a cultural perspective, religion is seen at how the divine religions are historically interpreted and practice into everyday social action. So therefore, religion is not something untouchable (untouchable), but something that can be observed and analyzed, because religious behavior must be seen and felt.

The alignments to analyze the behavior of a person or the public on the environment, there are several theories or models [19,21,20,22,23]. The model used as reference in this study proposed by Fietkau and Kessel [19]. The model consists of five variables that influence either directly or indirectly affect the pro-environmental behavior (Fig.1).

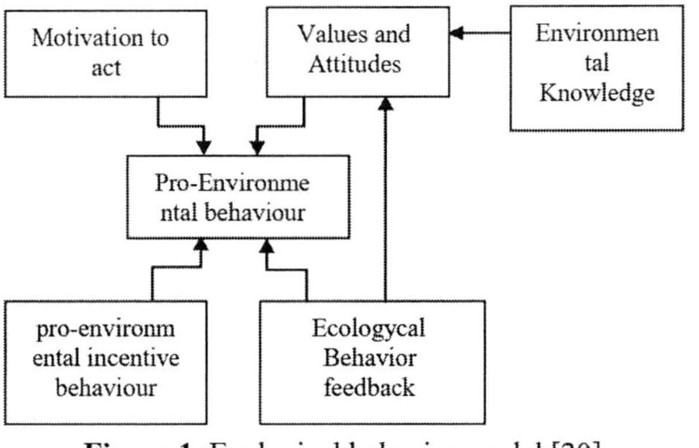

Figure 1. Ecological behavior model [20]

Variable of knowledge (*Wissen*), the main variable model [19] like forming a person's values and attitudes based on shared values and attitudes of a person to act pro-environmentally. Knowledge is an important element for the success and failure of conservation activities. Knowledge is essentially what we all know about a particular object, which includes science, thus science is a part of human knowledge known to others as well as knowledge of art and religion [24] states that knowledge is the result of "know" and happen after people doing the sensing of a particular object. Knowledge is something that is present and manifest in one's soul and mind due to the reaction, contiguity, and relationships with the environment and natural surroundings. Knowledge is the capacity to understand and interpretation of their results of observations and experience [25].

Knowledge can be seen on an individual, group, or organizational level, there is in the form of explicit or tacit, structured or unstructured. Explicit knowledge can be recorded and easily managed (*e.g.* on a computer). While tacit knowledge is generally below our conscious knowledge, even we ourselves cannot realize that we have it. This knowledge includes, among others, insight, intuition, and experience. To him, we need time and personal interaction is quite intense [26,27]. Rediscovery of ecological knowledge at a potential location can be an adaptive management [28] where there is requirement to apply human and ecological [29,30,31] and adaptive strategies for natural management [32] offers prospects for scientists to address the problems that beset conservation biologists and restoration ecologists. Knowledge construction about watershed is created on the community that has proven successful conservation in a long time. The construction of knowledge will be a knowledge structure that consists of variable inputs, processes, outputs and outcomes [33].

3. Methods

The research was conducted from March 2015 to November 2016, at two boarding schools in East Java and West Nusa Tenggara. The two boarding schools are determined as research locations with strong reasons that both have a track record as boarding schools that are very active in carrying out educational activities and conservation of watershed areas. The success of the two *pesantren* is known in Indonesia and even the watershed education and conservation leaders have received environmental awards at the international level.

The determination of informants in both Islamic boarding schools was carried out purposively with the aim to focus the knowledge to be harvested. From each *pesantren*, the main data source was determined by one conservation education actor, namely a senior figure in a *pesantren* who had won an environmental award. While other figures are the helpers of figures who also help conservation education activities with varying amounts from each *pesantren*.

Other data sources are documents owned by informants. Data analysis using Miles and Huberman Model [34] was conducted for source triangulation. While the method of harvesting knowledge is carried out using the stages as suggested by Serrat [35].

Harvesting knowledge is one way to attract and package secret knowledge to help others adapt, personalize, and apply it; building organizational capacity, and preserve institutional memory. The stages of knowledge harvesting are carried out using the method proposed by the Asian Development Bank [35], as follows:

3.1. Focus

Harvesting of knowledge is not done sporadically but rather focuses on information about the knowledge that is owned and used by religious communities in watershed conservation education activities. Thus, information gathering will also only be carried out for leaders or communities who have this knowledge and/or are involved in watershed education and conservation activities. The main consideration of the focus is on the importance of the knowledge that must immediately be known and will be a threat to the environment if not immediately discovered.

Focused data on knowledge explored includes three main things, namely: 1) knowledge of conservation education inputs; 2) knowledge of the conservation education process; and 3) knowledge of conservation education output.

3.2. Find (Finding resource persons/informants)

Planning at this stage is related to logistics operational preparation in the form of tools and research materials and what information will be obtained. Research tools and materials are adjusted to the target informants who will be used as informants, where locations will be interviewed, age, position in the organization, seniority, and others. The data obtained about this informant will affect the choice of the method used.

3.3. Elicit (acquisition of knowledge)

The knowledge acquisition method is done using in-depth interviews, documentation and group discussions. The three methods of obtaining knowledge are a form of triangulation (34). An explanation of the three methods of acquiring knowledge is explained below.

3.3.1. In-depth interviews. In-depth interviews are conducted face-to-face and are the most dominant method of acquiring knowledge. While telephone interviews are only done if there is data that requires further clarification. Interviews are conducted in an unstructured manner. This method is used considering that the information obtained will be more in-depth and able to describe the knowledge that was excavated.

Interview activities are supported by the use of audio and video recordings. Interview activities were carried out anywhere in the area where the informants had the opportunity to be interviewed. Interviews were carried out at the informant's residence, in the garden, and in the pesantren. Interviews in the locations where informants acted further strengthened the atmosphere of the interview because it was supported by evidence of the activities of informants in watershed conservation. Interview guidelines are still made to give direction about knowledge that must be explored and not overlooked. Personal relationships between researchers and informants are attempted to have no problems other than about research.

3.3.2. Documentation. Documents are needed to triangulate the interview data (23). Documents obtained are in the form of books or papers and freelance writings made by the characters themselves. documents also come from good news published in print and electronic media. The document provides an overview of the consistency of thinking with activities in different time paths. Documents are obtained directly from figures or from their assistants.

3.3.3. Group Discussion. Discussion groups are conducted to assist researchers in conducting data validation, examining the knowledge found, analysing using the knowledge of many discussion participants, and deciding whether or not the knowledge obtained by the researcher is correct. The ultimate goal of focus group discussions is to take an agreement that the knowledge found is true and in accordance with the reality in the field.

3.3.4. Organize. The knowledge that has been obtained is then processed using qualitative data analysis steps as suggested by Miles and Huberman (34). The knowledge that has been obtained is examined in terms of its content, consistency, coherence, and systematics. This knowledge is then grouped based on logical considerations into conservation knowledge groups which are input, process, and output.

3.3.5. Package. The knowledge that has been organized (processed) is then packaged into knowledge that can be the knowledge that can be read by teachers and the wider community in the form of publication of research results and information that can be accessed through the internet or delivered in scientific forums and various pieces of training.

3.3.6. Evaluate. Evaluation is carried out by conveying the results of the packaging of knowledge to several parties with an interest in watershed conservation and conservation education. Submission is done through lectures, seminars, discussions, book writing, internet, and other written publications. Evaluation is obtained in the form of feedback on the packaging of knowledge delivered. This feedback is monitored and reviewed at all times and records of repairs are made.

3.3.7. Adapt. Knowledge adjustment is carried out for specific purposes according to needs. So far, the knowledge gained can be adjusted to the need for conservation education in general, not only for the conservation of watershed areas. Adjustment of the latest knowledge is carried out by researchers in conservation education activities in schools through the implementation of conservation-based learning models [36].

4. Results

Research found that knowledge plays an important role and determine the success of watershed conservation. This fact shows that the background of the boarding school residents as educators have a huge impact of conservation strategies undertaken. Results showed that before the clerics have training on environmental basis conducted by NGO for one full year, the activities of the residents have not been done intensively. Environmental activity at that time was only had been practiced by the clerics as a private activity and not as an activity that involve more residents or boarding school residents as a form of environmental concern. These activities were not yet required as a necessity for the residents of the boarding school to help citizens systematically, so it just seems as a hereditary activity but not been taught to students or the general public.

Residents need to know about the knowledge position in order to change the boarding school paradigm thinking about the environment. This knowledge serves to broaden student and people insight about their surroundings. Thus, at an early stage, knowledge serves as an instrument to sensitize students and the community about the issues surrounding them. The next stage, knowledge serves to provide reinforcement or mental empowerment and a belief for students and the community. Based on this knowledge, students and the community have a new value on the environment. Based on the held values, the knowledge serves to shape the students and attitudes of human society to be sympathetic to the environment.

Building of conservation knowledge owned by the boarding school is a unique system that is different from the knowledge system [33]. Uniqueness is found in the knowledge materials on conservation that is more educational in activities. This phenomenon is very reasonable considering implementing conservation activities are the educators in the field of religion and the environment. System consisting of input-process-output and outcome is shown in Fig. 3. The residents of boarding school show community knowledge systems as a value system. It is rather different with several research findings that local knowledge disappearing at a rate that may not allow us even to know what value, if any, such systems had [35,36,27].

4.1. Knowledge System of Watershed Conservation Education

The existence of knowledge conservation at these sites can be distinguished by its function, namely to: 1) establish baseline values and instrumental values; 2) build practical skills of technical support activities; 3) predict and manage the results (output) of conservation; and 4) strengthen beliefs about goodness of the world and the hereafter.

Knowledge of environmental education and watershed conservation form the material for the formation of basic values. This knowledge must be tempered with the knowledge or ability to establish methodological instrumental values. The boarding school is very emphasized the methodologies to develop instrumental values. This requirement is very visible in the teaching lodge which said that "method is more important than the content". Based on these principles, the ability to develop and implement methods should be possessed by every teacher in conservation. Methods will affect the success or failure of any program, including education and conservation. Knowledge and methods is the material to construct instrumental values. Technical knowledge is helping a person or group of people to realize the instrumental values in conservation. In overall, actions of conservation indicate as the behavior of worship and have religious values. Behavior of worship is aimed at achieving goals or terminal values such as goodness in this world and good in the Hereafter. Based on these findings, it can be concluded that knowledge is the critical success factors of conservation. However, knowledge cannot be stand alone as it must be able to establish the values, attitudes and skills in the field of conservation.

The study of knowledge systems has been carried out towards the community in community forests [37]. The study illustrates the need for knowledge (input) to take action on community forest conservation, knowledge about how to take actions (processes) and knowledge about the results obtained (output). This study not only found the three elements of knowledge, but also the knowledge related to the outcomes as shown in Figure 2.

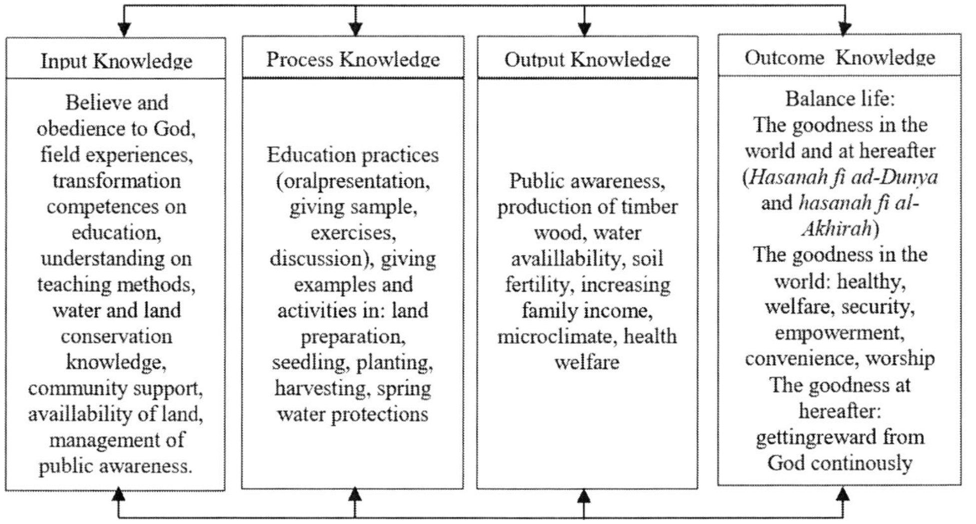

Figure 2. Knowledge of watershed conservation education component system

4.2. Input Knowledge of Conservation Education

Input knowledge of conservation education consists of high confidence and obedience to God, the knowledge component obtained from various literature sources and field experience. The other component is the ability of the material transformation of proselytizing, understanding the methodology or process of education and conservation methods, the ability to give an example, public awareness, support personnel, availability of land and capital. The knowledge used by the lodge comes from the holy book of Qur'an and some books of fiqh. Another source is through sensory experiences and the results of reading and contemplation of natural phenomena. The ability and experience of the people higher up in the environment sector continued to show high ability in concept and implementation in the field. The power of believing of boarding school residents on Lord reward of every good deed, encouraging citizens to act to preserve and be an example to the wider community

Almost the entire watershed conservation knowledge gained by the students and community was originally derived from information submitted by clerics of boarding school. In the following years the knowledge of boarding school residents and community is derived from governmental agencies that are responsible for the environment. The ability of clerics in the acquisition of knowledge about the environment is the key to determine the success of environmental education processes and products. The clerics of boarding school had gained knowledge by themselves about environmental conservation which comes from the observation through the senses of both observation and education or training. The next source of knowledge is from Islamic books that related to the environment. The last source is cleric's intuition which concern about the phenomenon in surrounding environment. The cleric's intuition is developing when dealing with environmental issues as they were demanded for solving any problems using their knowledge. The ability of combining the meaning derived from experience, reading and reflection is then called by the clerics as the ability to transform. Cleric's ability to transform knowledge to conduct environmental education is an important capability that determines the success of conservation.

4.3. Process Knowledge of Conservation Education
Believed by boarding school residents that their knowledge will not transfer properly if it is not done in a way or method either. As one of the processes, methods of conservation education expressed as a very important thing to do. The method is important as the principle of boarding school residents which reads: at *Tariqaatu ahammu min al-maddah, al-mudarrisu ahammu min al-tariqah, ruhu wa al-mudarrisi ahammu min al-mudarris.* Meaning: This method is more important than the material, the teacher is more important than the method, and the soul of the teacher is more important than the teachers themselves. Based on this knowledge, it is believed that the method and the teachers who have a good soul in a process is a requirement for educational and conservational works.

Educational methods are known and applied by the lodge consists of lectures, exercises, giving examples and discussion. Lecture conducted on college students and the general public. This method is performed to improve the knowledge of students and the community about the importance of preservation associated with the commandments of God. Training methods (*riyadhah*) made with reference to the book in the classical Islamic education. Used classics written by Al-Ghazali to infuse personality. Stages of education consist of stages: *taqalluh* (coercion), *ta'awudz* (habituation), *muwadzahah* (chase), and *dhawam* (continuous). Method of deliberation is applied only to the general public. The application of this method is done after the public aware of the importance of conservation. Deliberation conducted to determine the types of activities, planning and preparation of action plans.

Deliberation done well in study groups or in small group meetings for conservation. Residents of boarding school average positioning itself as a facilitator of discussion. This position is believed to be a form of solidarity instill trust between community residents lodge. The presence of the lodge is considered very important to maintain the belief that the activity is a serious activity. Instance method implemented based on the premise that the best educational method is through an example. Method of convincing examples of students and society in a given subject matter, whether delivered in the form of lectures, training or consulting. Giving examples show better results as seen directly.

4.4. Output Knowledge of Conservation Education
The output knowledge related with two main points, namely the awareness and benefits provided by trees. Meanwhile, students and public awareness should be created through a variety of applied educational methods. Awareness will be the basis for conservation activities which independently sustainable. Awareness of students can be seen when they have passed in which they perform their observation activities in where they live. While awareness in the community can be seen from their participation in meetings and actions that produce concrete evidence in the form of forest trees that grow well. Different types of trees planted will provide different benefits in accordance with the characteristics of the tree. Trees will provide certain benefits. Variations in the types of plants that grow will provide a variety of benefits.

Some of the boarding school residents have deliberately planted trees with high variation. Planted tress with a variety of types of behavior which are usually reserved for high school activities. High species variation is expected to provide more knowledge to students and the community. Region with a high variation of the planted trees is used as a medium of education. Knowledge of various types of trees is connected by boarding school residents as significant economic, social and ecological. The ability of trees to provide more benefits to the environment and human well-being referred to as one of nature's balance.

Knowledge of soil types associated with soil characteristics. The type of soil in the study area is referred to as "the land of unloading" or loose soil easily when exposed to rain. Several other properties of the soil is called soil rather white and slippery when wet rain. Based on soil maps of Sumenep Forestry and Plantation, information had obtained that the soils formed from sedimentary rocks belonging to the complex soil Mediterranean of grumusol, regosol and litosol. This soil type according to boarding school residents is difficult to absorb rainwater, including absorbing into the ground. Thus, efforts to save land and water in the watershed is to plant different types of crops that benefit to the economy, water and fertilize soil that hold runoff water carrying soil particle. Some of the boarding schools are still implementing monoculture conservation with economic and conservation goals. Type of monocultures planting by residents lodge no better results when compared with planting using polyculture system. Planting using polyculture system consists of different types of timber plants, fruits and food crops or medicines. Awareness of the public thinking about conservation methods using conservation watershed vegetative is shown in Fig. 3.

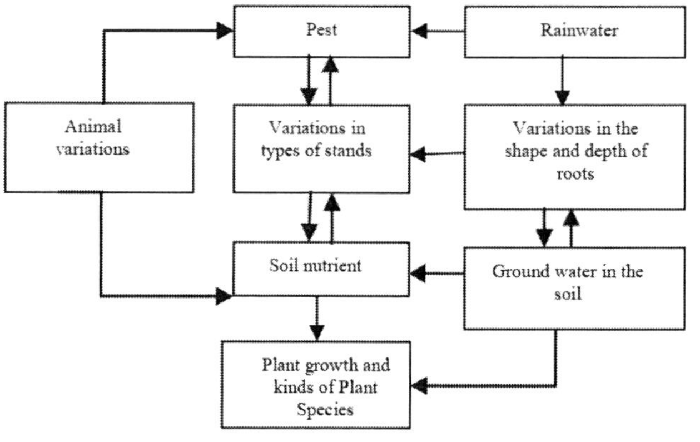

Figure 3. Knowledge about output of watershed conservation "nature balance using vegetative variation".

Planting trees with various types have advantages over monoculture. Some of these benefits by resident's lodge are the emergence of various types of animals to control pests or diseases that would ruin the staple crop. Variations in the type of plant will also produce waste to improve soil nutrients. While the roots of different plant species have different abilities to penetrate the soil. The difference in the ability of roots to penetrate the soil will help water into go through the soil in different variations. Increased nutrients and water in the soil causes plants to grow well. The entry of rain water into the soil causing water availability can be maintained. The complexity of plants and animals in the ecosystem is known by residents to produce more complex ecosystems. The complexity of the ecosystem will maintain a more balanced environment. Knowledge of conservation of the type implied by the knowledge that supports watershed conservation. Some researchers refer to this knowledge as traditional knowledge [27].

4.5. Knowledge Conservation Education Results

Knowledge of the results of conservation education is divided into two forms. The first form is the forms that can be seen and enjoyed as a universal good. Kindness is obtained as a result of the product (output) conservation. The second form is the only good that will be obtained in the afterlife. The second form is a gift or reward for the kindness of God to keep the earth from damage and perform kindness to provide more benefits for humans and the environment. Good in the world that is obtained is safety and environmental health, empowerment and well-being, as well as the convenience of worship. Security quieter environment causes people to worship, protected from environmental threats and even fights between fellow citizens due to limited natural resources. Economic welfare through conservation can be achieved through the method of complex thinking [38][39]. Healthy people are more awake as much clean water that can be used for household needs. Water is also a key ingredient for perfectly conducting religious activities. Empowering communities grew because natural resources are processed and used to enhance prosperity. Happiness world through conservation obtained as prayers were submitted to God.

Knowledge of the population made by boarding school resident's goodness in the hereafter support sustainable conservation. Hereafter believed kindness and consideration are expected to be obtained on a continuous basis even though people have died. Conservation being the key word for the balance of nature as they relate to the goodness in the world and in the Hereafter an ongoing basis. This knowledge provides a new understanding of the phenomenon of human-environment interactions are controlled by scientific thought - religious. Community conservation awareness can be used as knowledge about the sustainability of watershed development [40].

These findings also suggest that knowledge cannot stand alone. In fact, people have knowledge about the values of trust that cannot be proven (the Hereafter). This perspective explains that value-free knowledge. The same knowledge can produce different values in different people. Research shows increasing awareness of ethics requires innovative and policies to maintain ecosystem function. Ethics comes from God and the growth in the community.

5. Conclusions

Conservation education knowledge of boarding school residents has developed well after the training. The main function of knowledge is to build awareness and conservation values in personal and community students. Knowledge is needed to provide more meaning to the phenomena of the environment. Given meaning that better education will provide better results. Knowledge of conservation of boarding school residents includes knowledge about the input, process, output and outcome. This knowledge forms the system of knowledge about the typical citizen of watershed conservation in boarding school residents of An-Nuqayah, Sumenep East Java that can be disseminated to other places.

Acknowledgments

This research will not work well without financial support from the Directorate General of Higher Education, educational figures at the Annuqayah Boarding school and Nurul Haramain Boarding school and all research teams at the University of Muhammadiyah Malang and State Islamic University Malang. For that, we express our deepest gratitude and apologies for all the shortcomings throughout the study.

References

[1] Ajzen I and Fishbein M 1980 *Understanding Attitudes and Predicting Social Behavior* (Englewood Cliffs: Prentice Hall, N.J.)

[2] Al-Qaradhawi Y 2002 *Islam Agama Ramah Lingkungan* (Jakarta: Pustaka Al-kautsar)

[3] Kollmuss A and Agyeman J 2002 Mind the Gap: Why do People Act Environmentally and What are the Barriers to Promote Environmental Behavior? *Environmental Education Research Journal* **8**(3) 239-260

[4] Balai Pengelola DAS Brantas 2011 *Rencana Tindak Lanjut Pengelolaan DAS Brantas* Balai Pengelolaan DAS Brantas Jawa Timur.

[5] Bates D G 2000 *Human Adaptive Strategies: Ecology, Culture, and Politics* (Boston: Allyn and Bacon 2nd Edition) p 238

[6] Bennett G and Jessani N 2011 *The Knowledge Translation Toolkit: Bridging the Know-Do Gap: A Resource for Researchers* (New Delhi, India: SAGE Publications)

[7] Berkes F, Colding J and Folke C 2000 rediscovery of traditional ecological knowledge as adaptive management *Ecological Applications* **10**(5) 1251-1262

[8] Bews J W 1935 *Human Ecology* (London: Oxford University Press) p 312

[9] Biro Pusat Statistik 2010 *Hasil Survey Kependudukan 2010* available at http://www.bpps.org/. retrieved on February 18th 2011

[10] Brodt S B 2001 A systems perspective on the conservation and erosion of indigenous agricultural knowledge in Central India *Human Ecology* **29**(1) 99-120

[11] Burgess J, Harrison C and Filius P 1998 Environmental communication and the cultural politics of environmental citizenship *Environment and Planning A* **30**(8) 1445–1460

[12] Cavalcanti C 2002 Economic thinking, traditional ecological knowledge and ethnoeconomics *Current Sociology* **50**(1) 39-55

[13] Cox P A 2000 Will tribal knowledge survives the millennium? *Science* **287**(5450) 44-45
[14] Davenport T H and Prusak L 1998 *Working Knowledge, How Organizations Manage What They Know* (Boston: Harvard Business School Press)
[15] East E M 1936 Human Ecology *Science* **83**(2152) 305-306
[16] Hines J M, Hungerford H R and Tomera A N 1986/1987 Analysis and synthesis of research on responsible pro-environmental behavior: a meta-analysis *The Journal of Environmental Education* **18**(2) 1-8
[17] ICEE (2007a) The Ahmedaba Declaration 2007: A Call to Action; Education for Life; Life through Education *4th International Conference on Environmental Education United Nation, Ahmedabad, India*
[18] ICEE (2007b) Moving Forward from Ahmedabad; Environmental Education in the 21st Century *4th International Conference on Environmental Education, United Nation, Ahmedabad, India*
[19] Kates R W, Clark W C, Corell R, Hall J M, Jaeger C C, Lowe I, McCarthy J J, Schellnhuber H J, Bolin B, Dickson N M, Faucheux S, Gallopin G C, Grubler A, Huntley B, Jager J, Jodha N S, Kasperson R E, Mabogunje A, Matson P, Mooney H, Moore III B, O'Riordan T and Svedlin U 2001 Sustainability science *Science* **292**(5517) 641-642
[20] Kempton W M, Boster J S and Hartley J A 1996 *Environmental Values in American Culture* (Cambridge, Massachusetts: MIT Press)
[21] Kleijn D, Berendse F, Smit R and Gillissen N 2001 Agri-environment schemes do not effectively protect biodiversity in Dutch agricultural landscapes *Nature* **413**(6857) 723-725
[22] Mangunjaya F M 2010 Developing Environmental Awareness and Conservation through Islamic Teaching *Journal of Islamic Studies* **22**(1) 36-49
[23] Maliki Z 2011 Rekonstruksi Pemikiran Islam untuk Perlindungan Lingkungan *Makalah Seminar Agama dan Lingkungan Pusat Studi Multikulturalisme (PUSAM) Universitas Muhammadiyah Malang*
[24] Muller C H 1974 Human Ecology *Science* **183** 368
[25] *Management* Source: www.library.nhs.uk/ knowledgemanagement, Accessed Oct. 3, 2011.
[26] Nokel S 2009 *Islam, Alam dan Keberlanjutan*. Sources: htt://id.qantara.de/ebcom/show_article.php/_c-769/_nr-1 1/i.html, Accessed Dec. 29, 2010.
[27] Pandey D N 2002 Traditional Knowledge Systems for Biodiversity Conservation. Indian Institute of Forest Management, Bhopal, India. Source: http://www.infinityfoundation.com/mandala/t_es/t_es_p ande_conserve.htm.
[28] Rachman B M 2011 Manusia, Alam dan Lingkungan Hidupnya: Membangun "The Ecologycal Conscience" melalui Pendekatan Filsafat dan Agama *Makalah Seminar Agama dan Lingkungan Pusat Studi Multikulturalisme (PUSAM) Universitas Muhammadiyah Malang*
[29] Rozaki A 2005 Penelitian Agama dalam Perspektif Budaya *Makalah pengantar pada Studium Generale*: '*Penelitian Agama dalam Perspektif Budaya* (Yogyakarta: Fakultas Adab UIN, Sunan Kalijaga)
[30] Rusli 2004 Islam dan lingkungan hidup: Meneropong pemikiran Ziauddin Sardar *Hermeneia, Jurnal Kajian Islam Interdisipliner* **3**(2) 171-190
[31] Schwartz S H 1977 *Normative Influence on Altruism* in L Berkowitz (Ed.) *Advances in Experimental Social Psychology* (New York: Academic Press) **10** pp 221-279
[32] Sunaryo dan Joshi L 2003 *Peranan Pengetahuan Ekologi Lokal dalam Sistem Agroforestri* (Bogor: World Agroforestry Center (ICRAF))
[33] Suriasumantri S Y 1998 *Filsafat Ilmu: Sebuah Pengantar* (Jakarta: Pustaka Harapan)
[34] Tucker M E and Grim J 2009 *Overview of World Religion and Ecology* (US: Yale University)
[35] Serrat O 2010 *Harvesting knowledge* (Washington DC: Asian Development Bank)
[36] Sukarsono Sukarsono 2018 *Conservation Based Learning Model; Teacher Guide* (University of Muhammadiyah Malang)
[37] Awang San fri 2008 Konstruksi Pengetahuan dan Unit Manajemen Hutan Rakyat *Download from:* http://sanafriawang.staff.ugm.ac.id/konstruksi-pengetahuan-dan-unit-manajemen-

hutan-rakyat.html retrieved on 20 January 2018
[38] Tucker M E and Grim J 2010 *World Religion, the Earth Charter and Sustainability* Source: http://fore.research.yale.edu/publicatioons/projects/tuck erec3. pdf. Accessed Feb. 18, 2011.
[39] Witoszek N 2006 Globalization and sustainability: A humanist agenda *Ecotheology* **11**(3) 268-281
[40] Yusdani 2010 Eco-Spritual dalam Berbagai Persfektif *Makalah Workshop Nasional Pusat Studi Lingkungan Universitas Islam Indonesia, Yogyakarta*

IOP Conf. Series: Earth and Environmental Science **314** (2019) 012058 doi:10.1088/1755-1315/314/1/012058

Sand type characteristics analysis and mapping in Bengkulu

Fepy Supriani*, Mukhlis Islam and Yuzuar Afrizal

Department of Civil Engineering, Faculty of Engineering, University of Bengkulu, Bengkulu, Indonesia

*fepy_ilhan@yahoo.co.id

Abstract. Sand can be classified based on the origin of the quarry, namely sea sand, river sand and mountain sand. Sand provides an important role as filler material in building cementation materials such as in mortar and concrete. The characteristics of the sand determine the strength of the building. Characteristics can be in the form of grain size (gradation), sludge/clay content and specific gravity. This study aims to look at the characteristics of sand from different types of sand while looking at the mapping of sand zones based on SKSNI 03 2834 2000. Sand is taken directly from quarry, in five city districts in Bengkulu Province which consist of sea sand, river sand and mountain sand. Determination of sand gradation is based on SNI 03-1975-1990 sieve analysis test. Examination of sludge/clay content and specific gravity based on SNI 6821-2002 and SNI 03-4804-1998 standards. The conclusion obtained is that sand will have different characteristics based on quarries. 81.25% sea sand tends to have uniform granules, with smooth and very fine characteristics and does not enter any zone. River sand has a good gradation and categorized in zones characteristic of smooth, slightly rough and rough. Mountain sand is categorized as fine sand with good gradation. The sludge content of sand from each type of origin of the sand quarry shows that the river sand with a calm flow has a high level of clay content. Clay content levels that are very high on mountain sand are probably ash that was produced by volcanic activity of the mountain. The specific gravity of each type of sand varies, not depending from the origin of the sand.

1. Introduction

Sand/fine aggregate is an important building material and must be used in every building development process. Sand is an aggregate with granules ranging from 0.0625 to 2mm. Natural and artificial sand is one of the main ingredients in forming concrete and mortar. Mortar cementation material is formed from a mixture of sand and cement plus water, while mortar mixture added with coarse aggregate will form concrete. The importance of sand in building materials affects the strength of the building itself.

Bengkulu City as an earthquake-prone area should pay attention to disaster mitigation for a long term, one of which is building in accordance with earthquake resistant standards. One standard that must be possessed is standard in building materials, workmanship and working methods. High quality building materials are obtained from research so that they are in accordance with existing standards.

Bengkulu Province has a long coastline, and also has mountain ranges. This geographical condition is advantageous because it has abundant resources, especially for building materials. The presence of large to small stones can be used as building materials, although massive exploitation of rocks can damage the environment. Sand is one of the building materials whose availability is abundant. Sand can be categorized based on the origin of the quarry, namely sea sand, river sand and mountain sand. The availability of these three types of sand fulfills the need for sand in the city of Bengkulu.

River sand is sand that comes from excavation or mining in rivers. The steep rivers have a swift water flow, so the deposits of the rock particles will vary considerably at a certain distance, usually content only small amount of fine grains and the rocks are clean enough. On gentle slopes, variations in particle size do not change from one place to another, most of the particles are more rounded and dirtier (http://eprints.ums.ac.id/48276/5/BAB%20I.pdf). River sand is obtained directly from the riverbed, which is generally fine grained and rounded due to friction. Adhesion between granules is somewhat lacking due to the round shape of the granules (http://www.forumbebas.com, February 14, 2011). The sand that is mined from the riverbed has been washed directly and the content of the sludge has decreased a lot, the faster the river flows, the less the sludge contained in the sand. The process of river sand mining is also more difficult because it must be sucked and threatened by flooding. River sand has a clean and smooth type of grain with sizes between 0.06 mm - 2 mm [1].

Sea sand is sand taken from the beach. The grain is smooth and round due to friction. This sand is the worst sand because of its salt content. This salt absorbs water from the air and this causes the sand to always get wet and generate swelling when it become a building [2]. Sea sand is one type of fine aggregate material that is quantitatively available along the coast in coastal areas [3]. In general, sea sand has the characteristics of fine and round granules, which have gradations (variations in grain size) that are uniform and have salt chloride (Cl) and sulfur (SO_4) content [4]. The Public Works Department in its regulation in SK-SNI T-15-1991-03 states that it does not recommend sea sand for concrete mixes unless the usage is already approved by experts [5]. Round and fine grains, and uniform gradations will reduce interlocking mechanism between granules.

Mountain sand is obtained directly from the ground or by digging. The form of mountain sand is usually sharp, angular, porous and free of salt content although it usually has to be cleaned from dirt/soil by washing it first [6].

The characteristics of the three types of sand are visually different; some are smooth, rough and very rough. The characteristics of sand can be seen from the value of grain modulus, sludge content, organic content and specific gravity. Grain modulus is a gradation that shows the fineness of the grains that are tested and left in each sieve set. From the sieve analysis test, the data can be plotted into the sand zone graph as in SNI 03-2834-2000 [7]. This study aims to look at the characteristics of sand from the three types of sand origin namely river, mountain and sea sand, as well as the sand zone that will be used for building materials, hence the results of the study could be used as guidelines of sand usage in connection with its technical properties and requirements so that the overusage of one type of sand could be prevented and the envireonmental impact due to overexploitation ot aforementioned type of sand could be prevented.

2. Literature Review

2.1. Impact of sand exploitation on the environment

Excessive and irresponsible mining / exploitation of sand can accelerate environmental damage in a short time which results in deteriorating environmental quality. Although the former mining area can be used for other economic activities, it will still change the balance of environmental land management and cause disruption of the water system in the area that may be far wider than the mining area itself [8].

The impact of over exploitation of mountain sand is already in effect around Merapi mountain area, in which the permit for sand mining is only for river sand, while in practice, mining activity of mountain sand also happen. One of the observed effects is the lowering of ground water table, from average of 4 m to over 4 m [9].

2.2. Standards for sand as fine aggregates

Pamungkas and Hairunnisa stated that fine aggregates are hard mineral grains whose shape is approximately to round, sharp with the size of the grain mostly between 0.07-5 mm [14].

Requirements to produce good cementitious materials, fine aggregates must meet the following requirements:

1. Fine aggregates must consist of sharp and hard grains and continuous gradations. Grains of fine aggregate must be eternal, meaning they are not broken or destroyed by the effects of weather, such as the sun or rain.
2. The composition of the grain size must have a fineness modulus that ranges from 1.50 to 3.80.
3. The sludge / grain content is smaller than 0.07 mm, the sludge content is a maximum of 5%.
4. Organic content test specimen colour is not darker than the standard color after being given 3% sodium hydroxide solution.
5. Grain hardness, when compared with the hardness of the sand grains originating from Bangka quartz sand, gives a number of fraction number not greater than 2.20.

According to Indonesian National Standards (SKSNI-S-04-1989-F: 28) it is stated the requirements of fine sand or fine aggregate as building materials as follows [13]:

1. Fine aggregates must consist of sharp and hard grains with a hardness index <2.2.
2. Eternal properties when tested with a solution of saturated sulfate salt as follows:
   - If it is using natrium sufat maximum decayed parts is 12%.
   - If fine parts of magnesium sulfate are used, up to 10%.
   - It should not contain more than 5% mud and if the sand contains more than 5% mud, the sand must be washed.
   - Sand should not contain too much organic ingredients, which must be proven by color experiments from Abrans-Harder with 3% NaOH saturated solution.
   - The large composition of sand grains has a modulus of fineness between 1.5 to 3.8 and consists of various grains.
   - For high durability concrete, the reaction of sand to alkali must be negative.
   - Sea sand should not be used as fine aggregate for all concrete except with instructions from recognized construction material government agencies.
   - Fine aggregate used for plastering and applied species must meet the requirements of the mortar sand

Aggregate gradations are variations in aggregate distribution and size. The tests were carried out using sieve analysis according to BS 812, ASTM C-33, C136, ASHTO T-27 and SKSNI standards. Grain gradation is divided into 3, namely gap gradation, continuous gradation and uniform gradation.

1. Gap gradation, if one or more grain sizes in a set of sieves is not present, then the gradation will show a horizontal line in the graph. The advantages of gap gradations are: 1) on certain water to cement ratio concrete work will be easier if the sand content is less, 2) for high workability concrete tends to experience segregation, 3) slope gradation does not adversely affect the strength of the concrete
2. Continuous gradation if all aggregate sizes are available and are well distributed. Continuous gradation is often recommended and is often used in concrete mixes. With the availability of all sizes of aggregates there will be good interocking, where aggregates will fill each other in concrete mixes. Continuous gradation is the best gradation
3. Uniform gradation are gradations that have the same grain size. Aggregates with uniform gradations are usually used for lightweight concrete which is a type of concrete without sand. Aggregates with uniform gradations can be used to fill aggregates with poorly aggregated intervals to meet standards

The distribution of sand zones based on the results of sieve analysis will be included in the sand zone on SNI 03-2834-2000 standard [7].

IOP Conf. Series: Earth and Environmental Science **314** (2019) 012058 doi:10.1088/1755-1315/314/1/012058

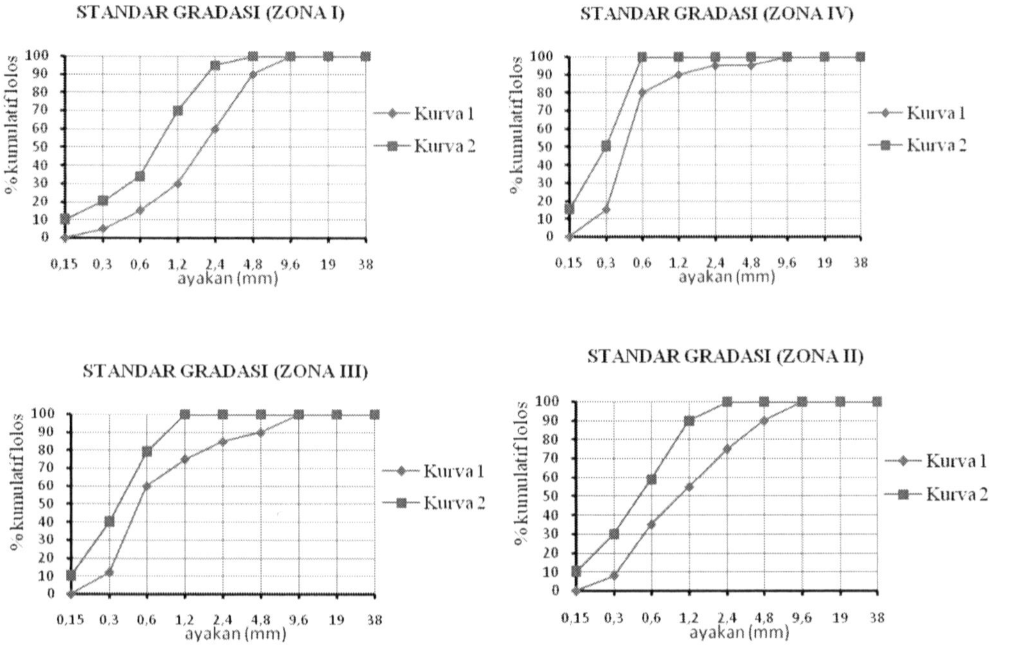

Description: Zone I: Coarse Sand, Zone II: Slightly Coarse sand
Zone III: Slightly Fine Sand, Zone IV: Fine Sand

Figure 1. The distribution of sand zones

2.3. Fineness Modulus

Fineness modulus is an index used to measure the smoothness or roughness of aggregate grains, the greater the MHB value of an aggregate, the greater the aggregate granular sizes. The smoothness or roughness of an aggregate can affect the workability of the mortar. Finenes grain modulus is defined as the number of percent of aggregate items left above a set of sieves, then the value is divided by one hundred. Sand roughness is grouped into 4 zones (SNI 03-2834-2000) [7], namely:

1. Zone 1: Coarse sand, with value of 2.9 - 3.2.
2. Zone 2: Sand is rather rough, with values of 2.6 - 2.9.
3. Zone 3: Sand is rather smooth, with values of 2.2 - 2.6.
4. Zone 4: Fine Sand, with values 1.8 - 2.2.

2.4. Mud Content

Mud is usually mixed in sand. In sufficient quantities it can reduce the strength of concrete, because of its tendency to inhibit the hydration of cement. This mud forms layers that cover the grains of sand, thus preventing the binding of cement. Mud also has properties, which expand in a wet state and shrink at dry time, this property will affect the concrete, namely shrinkage and "creep" in hardened concrete.

The aggregate unit weight is the ratio between the weight of the aggregate and the volume. The unit weight of the aggregate content required in the calculation of the concrete mixture material if the amount of material is measured with a volume size.

3. Research methodology

Research objects are sea sand, river sand and mountain sand from Bengkulu City, Bengkulu Tengah Regency, North Bengkulu Regency, Seluma Regency and Rejang Lebong Regency. After conducting the survey, the type of sand that will be sampled was determined. Sand samples are taken directly from the sand quarry. Testing the characteristics of sand is carried out for the gradation of fineness modulus values, mud content and unit weight. Some sand quarry locations are shown in Figure 2

(a)

(b)

(c)

Figure 2. Sand Quarry Locations (a) river sand, (b) sea sand, (c) mountain sand

3.1. Sieve Analysis (SNI 03-1968-1990 [14])

3.1.1. Purpose and Objectives. This test is intended to determine the distribution of fine aggregate gradation using filters to be used for concrete mix design. Sieve analysis is a determination of the weight held by each sieve expressed in percent.

3.1.2. Tools and Materials

1. Scales with accuracy of 0.5 grams.
2. A set of sieve No.1 ½, No.¾, No.3 / 8, No.4, No.8, No.10, No.30, No.50, No.100 and PAN
3. Oven with a temperature of 110 ± 5°C
4. Sieve shaker machine
5. Pan
6. Fine aggregate

3.1.3. 6.3 Work Procedure

1. Aggregates prepared, weighed and are each divided into 2 aggregate samples 3000 grams each specimen
2. The specimen is inserted into the oven for 24 hours at a temperature of 110 ± 5°C.
3. After being ovened for 24 hours, each aggregate sample is weighed with a weight of 1000 grams for fine aggregate.
4. Sieve set is prepared then arranged according to the sieve number, the lowest one is the smallest sieve.
5. The specimen is inserted into the sieve. Experiments on fine aggregates are carried out with a sieve shaking machine for 15 minutes.

6. The aggregates retained in each filter number are weighed with their respective sieve.
7. Observation data is recorded in the observation table.

3.2. Mud Content Test (SNI 03 – 4141 – 1996 [13])

3.2.1. Purpose and Objective. This test is intended to determine the aggregate mud content which can interfere with cement paste bonding on aggregate which of course can affect the quality of concrete or cementitious material.

3.2.2. Tools and Materials

1. Oven with temperature of 110OC
2. Pan
3. Scales with accuracy of 0.5gr
4. Aggregate samples

3.2.3. Work Procedure

1. Aggregates are taken from the sampling site
2. The weight of the pan is weighed (W1)
3. Aggregate and pan weighed using scales after oven drying for ± 24 hours (W2)
4. Aggregate net weight is calculated using formula W3 = W2-W1
5. Specimens then washed thoroughly.
6. Specimens that have been washed are then dried using oven at 110o C for 24 hours.
7. After being oven dried for 24 hours the specimen was weighed (W4).
8. Calculate the weight of dry specimens using a formula W5 = W4-W1

3.3. Unit Weight Test (SNI 03-4804-1998 [14])

3.3.1. Purpose and Objective. This test is intended to determine the unit weight of the aggregate. The unit weight is the ratio of weight and volume of the aggregate.

3.3.2. Tools & Materials

1. Scales
2. Compacting rod 15 mm in diameter, 60 cm long with rounded edges should be made of steel.
3. 3. 4. A steel container that is quite rigid in the form of a cylinder with holder.

3.3.3. Work Procedure

1. Mold is weighed and then record the weight (W1).
2. Agregate is inserted into the mold carefully hence not to cause grain separation, from a maximum height of 5 cm above the container. Fill it up to the full
3. Flatten the top of a mountainous surface by using a leveling bar.
4. The aggregate and the mold then weighed and recorded (W2).
5. Aggregate weight is calculated using the formula (W3 = W2-W1).

4. Results and Discussion

4.1. Sieve analysis

Sea sand which is the object of research is sea sand obtained from along the coast of Bengkulu City, Bengkulu Tengah Regency (Benteng) and North Bengkulu. After sieve analysis, the results are as shown in Table 1.

Table 1 shows that only 3 types of sea sand (18.75%) entered the zone and had various grains (good gradation). While 13 types of sea sand (81.25%) tend to have uniform and very fine grains. Uniform granules are less preferable because they do not interlock and reduce density if used for building materials.

Table 1. Results of Fineness Modulus testing and zone of Sea Sand

#	Type of Sand	Location	FM	Zone	Characteristics	Ordinat
1	Pasir Laut Pondok Kelapa	Benteng	0.8	not conform to zone IV	Fine	3°41' 748"S 102° 14' 677"E
2	Pasir Laut Air Padang (Kasar)	Bengkulu Utara	1.103	not conform to zone IV	Fine	-3° 30' 41,4"S 102° 00' 06,6"E
3	Pasir Laut Air Padang (Halus)	Bengkulu Utara	1.101	not conform to zone IV	Fine	-3° 30' 41,4"S 102° 00' 06,6"E
4	Pasir Laut Air Lakok	Bengkulu Utara	2.55	Zone II	Slightly coarse	-3° 28' 41,2"S 101° 56' 47,7"E
5	Pasir Laut Serangai	Bengkulu Utara	1.31	not conform to zone IV	Fine	-3° 25' 09,6"S 101° 52' 53,6"E
6	Pasir Laut Selolong	Bengkulu Utara	0.99	not conform to zone IV	Fine	-3° 27' 11,6"S 101° 55' 07,1"E
7	Pasir Laut Selubuk (Kasar)	Bengkulu Utara	2.22	Zone III	Slightly fine	-3° 36' 37,2"S 102° 09' 11,5"E
8	Pasir Laut Selubuk (sedang)	Bengkulu Utara	1.13	not conform to zone IV	Fine	-3° 36' 37,2"S 102° 09' 11,5"E
9	Pasir Laut Selubuk (halus)	Bengkulu Utara	0.94	not conform to zone IV	Fine	-3° 36' 37,2"S 102° 09' 11,5"E
10	Pasir Laut Air Sabu	Bengkulu Utara	2.37	Zone III	Slightly fine	-3° 22' 14,9"S 101° 48' 10,7"E
11	Pasir Laut Air Uray	Bengkulu Utara	1.17	not conform to zone IV	Fine	-3° 24' 49,6"S 101° 52' 17,0"E
12	Pasir Laut Ketahun	Bengkulu Utara	1.33	not conform to zone IV	Fine	-3° 51' 40,9"S 102° 18' 50,5"E
13	Pasir Laut Kota Agung Lais	Bengkulu Utara	0.8	not conform to zone IV	Fine	-3° 33' 44,0"S 102° 05' 44,1"E
14	Pasir Laut Teluk Sepang Bengkulu	Kota Bengkulu	0.87	not conform to zone IV	Fine	03° 54' 04,9"S 102° 17' 06,5"E
15	Pasir Laut Pantai Panjang Bengkulu	Kota Bengkulu	0.86	not conform to zone IV	Fine	3° 46' 215"S 102°15 '894 "E
16	Pasir Laut Pantai Zakat Bengkulu	Kota Bengkulu	0.85	not conform to zone IV	Fine	3° 49' 889"S 102° 16' 988"E

The results of sieve analysis for river sand as shown in Table 2 shows that river sand has a good gradation. Out of 7 types of river sand samples 86% entered into zone I, II and III and categorized as slightly fine, coarse and slightly coarse.

Table 2. Results of Fineness Modulus testing and zone of River Sand

#	Type of Sand	Location	FM	Zone	Characteristics	Ordinat
1	Pasir Rawa Indah Seluma	Seluma	1.96	Almost conform to zone III	Fine	4° 14' 56,7"S 102° 39' 54,4"E
2	Pasir Sungai Talang Rasau	Benteng	2.2	Zone III	Slightly Fine	-3° 30' 19,6"S 102° 04' 02"E
3	Pasir Sungai Limau	Benteng	2.119	Zone III	Slightly Fine	-3° 40' 33,7"S 102° 14' 02,8"E
4	Pasir Sungai Penanding	Benteng	2.097	Zone III	Slightly fine	3° 76' 484"S 102° 49' 276"E
5	Pasir Sungai Kembang Sri	Benteng	2.94	zone II	Slightly coarse	4° 12' 75,3"S 102° 41' 48,8"E
6	Pasir Sungai Lubuk Kebur	Seluma	3.47	zone I	Coarse	4°4,5' 03"S 102° 34' 83,7"E

7	Pasir Sungai Kampai	Seluma	2.628	Zone II	Slightly coarse	4° 9,1' 88"S 102° 41' 59,2"E

Mountain sand is sourced from the City of Curup, Rejang Lebong Regency, which is a mountainous area. The adjacent location produces almost the same characteristics. Rough and fine APJ mountain sand is based on the location of the sand depth. The top of the excavation layers tends to be coarser compared to the bottom layers. The mountain sand of Lubuk Penyamun is also in the same location, the color of which is different based on the depth of the excavation. The top layer tends to be yellowish and the lower layers of the grayish black as shown in figure 3.

Table 3. Results of Fineness Modulus testing and zone of Mountain Sand

#	Type of Sand	Location	FM	Zone	Characteristics	Ordinat
1	Pasir Gunung APJ kasar	Curup	1.998	Zone III	Slightly fine	-3° 29' 54,7"S 102° 30' 19,0"E
2	Pasir Gunung APJ Halus	Curup	1.976	Zone III	Slightly fine	-3° 29' 54,7"S 102° 30' 19,0"E
3	Pasir Gunung lubuk Penyamun kuning	Curup	1.523	almost conform to zone IV	Fine	-3° 49' 80,4"S 102° 50' 18,81"E
4	Pasir Gunung lubuk Penyamun hitam	Curup	1.783	almost conform to zone III	Fine	-3° 49' 80,4"S 102° 50' 18,81"E

Figure 3. Color differences in mountain sand

4.2. Examination of mud content and unit weight

Examination of mud content and unit weight were carried out for 12 sand samples, each representing 4 types of sand origin (Table 4). Maximum mud content according to the standard is below 5%. The mud content obtained from the test results as shown in Table 4 shows that sea sand, except air lakok sand, has a mud content level below 5% (comply to standards). Consequently, the materials do not need to be washed, while Sea sand obtained from Air Lakok has a high mud content level of 9.33%. Kembang Seri river sand has a high level of mud content since it comes from a relatively calm river, hence the sand and mud settle together. The quarry method of Penanding river sand used pumps that directly wash the mined sand so as to reduce the mud content.

Mountain sand has a very high level of mud content, hence in its usage it is recommended to wash it first. However, from some observations for the mountain sand, the mud content contains volcanic ash from the activity of the mountain itself so that it contains beneficial silica as cementation building material mixture.

Table 4. Results of Testing of Mud Content and Sand Unit Weight

#	Sand Type	Unit Weight (kg/m3)	Mud Content (%)	FM	Zone
1	PL. Air Lakok	1880	9,33	2.55	Zone II
2	PL. Selubuk Kasar	1400	0,32	2.22	Zone III
3	PL. Serangai	1380	0,48	1.31	not conform to zone IV
4	PL. Selubuk Halus	1420	0,89	0.94	not conform to zone IV
5	PS. Penanding	1452	2,464	2.097	Zone III
6	PS. Kembang Seri	1184	19,84	2.94	zone II
7	PS. Talang Rasau	1510	1,41	2.2	Zone III
8	PS. Lubuk Kebur	1531	3,176	3.47	zone I
9	PG. APJ Kasar	1420	21,28	1.998	Zone III
10	PG. APJ Halus	1350	21,83	1.976	Zone III
11	PG. Lubuk Penyamun Kuning	1441	20,69	1.523	almost conform to zone IV
12	PG. Lubuk Penyamun Hitam	1329	29,12	1.783	almost conform to zone III

The unit weight of the sand as shown in Table 4 shows the diversity of values. Unit weight is the ratio between the weights of sand to its volume. Sand that has a large unit weight indicates a relatively high density. Solid masses can give strength to sand. Air Lakok sea sand has the highest unit weight, while the mountain sand from Lubuk Penyamun having the lowest unit weight.

For building materials in addition to gradation, mud content and unit weight, several other characteristics also affect the strength of sand such as strength, wear and durability of the sand material itself. Very fine sand and does not enter the zone is usually only used for filling materials / non cementitious materials and filling material between paving blocks. While sand which entering the zone can be used for mortar and concrete building materials.

5. Conclusion

a. Sand based on the quarry location will provide different characteristics.
b. 81.25% sea sand tends to have uniform grain, with fine and very fine characteristics and does not enter any zone. 18.75% sea sand has a good gradation
c. River sand has a good gradation and enters the predefined zone with fine, slightly coarse and coarse characteristics.
d. Mountain sand is categorized as fine sand with good gradation.
e. The mud content level possessed by each type of sand origin shows that river sand with a calm flow has a high level of mud content. Mud content levels that are very high on mountain sand are probably ash that is produced by volcanic activity of the mountain.
f. The unit weight of each type of sand varies, can not be determined from the quarry location nor origin of the sand.
g. All source of sand could be used as construction as long as the said sand characteristincs meet technical requirements among others gradation and mud content. However, the long-term effect of each different sand source should be carefully studied in future researchs.

References

[1] Sjarief R 2010 *Tata Ruang Air* (Yogyakarta: Andi)
[2] Cornelis R, Ramang R and Kandi Y S 2012 Subtitusi agregat halus beton menggunakan kapur alam dan menggunakan pasir laut pada campuran beton *Jurnal Teknik Sipil* **1**(4) 74-86; http://puslit2.petra.ac.id/ejournal/index.php/jurnal-teknik-sipil/article/view/18901
[3] Mangerongkonda D 2007 *Pengaruh penggunaan pasir laut bangka terhadap karakteristik kualitas beton* available at http://library.gunadarma.ac.id/10399066-jurnal.ftsp.pdf
[4] Siregar A H, Satyarno and Tjokrodimuljo 2008 Pemanfaatan pasir pantai Sepempang dan batu pecah asal Ranai sebagai bahan pembuatan beton normal *Forum Teknik Sipil* **18**(1) 739-747; http://puslit2.petra.ac.id/ejournal/index.php/cef/article/view/17400

[5] SKSNI T-15-1991-03 1990 *Tata Cara Perhitungan Struktur Beton Untuk Bangunan Gedung* (Jakarta: Departemen Pekerjaan Umum Republik Indonesia)
[6] http://www.forumbebas.com retrieved on February 14th 2011
[7] SNI 03-2834-2000 2000 *Tata Cara Pembuatan Rencana Campuran Beton Normal* (Jakarta: Departemen Pekerjaan Umum Republik Indonesia)
[8] Suhartini 2006 *Fenomena Penambangan Pasir dan Pembangunan Beserta Dampak Lingkungan Yang Ditimbulkannya* (Yogyakarta: Universitas Negeri Yogyakarta)
[9] https://www.bbc.com/indonesia/majalah/2015/06/150609_majalah_merapi_pasir
[10] Pamungkas B and Hairunnisa S 2007 *Komparasi Mutu Paving block Antara Metode Mekanis dan Konvensional dengan Campuran Endapan Sampah* (Semarang: Universitas Diponegoro)
[11] SK SNI S-04-1989-F 1989 *Spesifikasi Bahan Bangunan Bagian A* (Jakarta: Departemen Pekerjaan Umum Republik Indonesia)
[12] SNI 03-1968-1990 1990 *Metode Pengujian Tentang Analisis Saringan Agregat Halus Dan Kasar* (Jakarta: Departemen Pekerjaan Umum Yayasan Badan Penerbit PU)
[13] SNI 03-4141-1996 1996 *Metode Pengujian Gumpalan Lempung Dan Butir-Butir Mudah Pecah Dalam Agregat* (Jakarta: Departemen Pekerjaan Umum Yayasan Badan Penerbit PU)
[14] SNI 03-4804-1998 1998 *Metode Pengujian Berat Isi dan Rongga Udara Dalam Agregat* (Jakarta: Departemen Pekerjaan Umum Yayasan Badan Penerbit PU)

IOP Conf. Series: Earth and Environmental Science **314** (2019) 012059 doi:10.1088/1755-1315/314/1/012059

Economic growth poverty and degradation of environmental in Balai Gadang Village, Koto Tangah District, Padang City

Supriyadi*, Bustari Muchtar and Abdul Razak

Environmental Science Study Program, Postgraduate Program, Padang State University, Indonesia

*supriyadiparca.sp@gmail.com

Abstract. This study aims to a). Analyze the impact of economic growth and poverty on the carrying capacity of the environment. in Balai Gadang Village, Koto Tangah District, Padang City. b). m analyzing land conversion resulted in environmental degradation in Balai Gadang Village, Koto Tangah District, Padang City. This research method is a descriptive type of research with a qualitative approach. The research location is in lurahan Into Hall Tower to Koto Tangah sub-district, Padang City. Informants in this study came from community leaders, village officials, sub-district officials, and related officials. Data collection techniques in this study, namely interviews, observation and documentation. The analysis technique is in use in this study is inductive data analysis. From the research findings concluded: a). The impact of economic growth on the carrying capacity of the environment is the integration of the environment due to infrastructure that supports facilities and infrastructure to support economic activities built on land that was originally used as agricultural land: b) The impact of poverty on environmental carrying capacity is the business of river dredging for fulfill the demand for sirtukil mining goods as household income. c). There has been an increase in land conversion in Balai Gadang Village. The land converted is paddy fields for residential land and trade d). Land conversion factors in Balai Gadang Village, Koto Tangah District, Padang City are government policies, natural factors, population growth and economic problems. E). The magnitude of settlement and trade expansion in Balai Gadang Village has resulted in environmental degradation, namely 1) environmental degradation to soil productivity; 2) soil degradation; 3). environmental degradation of river widening and shallowness; 4). environmental degradation of air pollution; 5) environmental degradation in the form of air pollution other than sourced from waste also sourced from motor vehicles; 6). environmental degradation of biodiversity.

1. Introduction

The development always raises paradox. One of them is more reduction in the quality and capacity of the environment. Because, all human needs cannot be fulfilled by utilizing resources owned by nature. Therefore, in this case there is an inverse relationship between human needs and natural or environmental resources. That is, the more and varied human needs, the more natural ability to provide them. In on the other hand, in order to carry out their needs, humans carry out economic and industrial efforts that inevitably bring about the side effects of environmental pollution or contamination. In this case, there is a perpendicular relationship between human needs and pollution. The more and varied human needs are met through industrial business, the higher the level of environmental pollution can be ensured. If the trend continues, one day there will be a situation where economic growth cannot be increased again, while the ability and quality of the environment is difficult to be repaired again [1].

According to [2] economies that derive a large portion of their income from natural resources cannot preserve growth by replacing capital accumulation for increasingly bad natural capital. This means that the quality of natural resources (including the environment in it) has a considerable influence. Because the longer the carrying capacity decreases, so that environmental exploitation based on economic interests alone, at one time will cause disruption of ecological balance. This condition is referred to as an "environmental crisis" which is a symptom due to errors in the pattern and method of managing the sources of human life's needs. These symptoms are considered a crisis pressure that endangers human life such as threats to the clarity of air and water sources. Therefore, the growth approach that pays attention to the quality of the environment and the efficient use of natural resources will contribute to the accumulation, investment, economic growth and human welfare. Because, high quality natural capital contributes indirectly to well-being as an essential part of the products of goods and services and a sustainable economy.

According to [3] the concept of carrying capacity of the environment is most easily applied to an agrarian system that is still simple. In the system the human population lives on the agricultural sector in the broadest sense. Basically, the carrying capacity depends on the percentage of land that can be used for agriculture and the number of agricultural products per unit area and time. So, the greater the percentage of land that can be used for agriculture, the greater the carrying capacity of the area. The percentage is determined by the suitability of the land for agriculture, the need for land for other purposes at outside the agricultural sector.

A very important factor in environmental problems is the size of the human population, with rapid population growth, the need for food, fuel, settlements, and other needs and domestic waste also increases rapidly [3]. The more population, the more needs that must be met from nature or the environment. If population growth far exceeds the growth in the availability of natural resources, there will be a massive depletion of natural resources to meet the needs of human life and will lead to higher environmental pollution due to production and consumption processes.

Continued population growth and technological advancements have demanded an increase in the supply of food and energy raw materials. The reality is now, the carrying capacity of the source power or environment is increasingly labile due to the increasingly exploiting of locytes without regard to ecological rules. Excessive use of natural resources without being followed by businesses - businesses that adhere to ecological principles will add the complexity of environmental problems [4].

Land has become one of the elements supporting the continuity of human life. The first activity carried out was the use of land for farming. Land in a broader sense includes those that have been influenced by various fauna and human activities. Every human activity either directly or indirectly is always related to land, such as for agriculture, settlements, transportation, industry or for recreation, so that it can be said that land is a natural resource that is very important for human survival. Land is a physical environment that includes land, climate, relief, hydrology and vegetation, where these factors influence the potential for use [5].

Land conversion is a change in the function of part or all of the land area from its original function (as planned) to another function that has a negative impact on the environment and the potential of the land. Transfer of land functions in terms of changes or adjustments to use is caused by factors that broadly include the need to meet the increasing needs of the population and increasing demands for better quality of life [6]. According to [7] there are seven types of land use changes in carbon stock changes, namely the conversion of natural ecosystems into fields, conversion of natural ecosystems into agricultural land, abandoned fields, abandoned farms, timber production forests, and greening areas.

According to [8] land conversion is the conversion of agricultural land to non-agricultural use or from non-agricultural land to agricultural land. Land conversion is influenced by two main factors, namely:

1. Macro-level factors include industrial change, settlement growth, population growth, government intervention, and economic poverty.
2. Factors in micro level include household income patterns (household economic structure), household welfare (household economic value orientation), and household survival strategies.

Land conversion is closely related to increasing population density. According to [9] with increasing population, the ratio between people and land is getting bigger, even though the utilization of every inch of land is greatly influenced by the level of development of a society's culture. Population growth causes a smaller supply of land.

Economic growth and poverty in Balai Gadang Subdistrict, Koto Tangah District, Padang City have an influence on the carrying capacity of the environment. The transition of productive land to residential land has resulted in low carrying capacity of the environment. Judging from the increase in population and the developments that have taken place, the economic structure of Koto Tangah Sub-district continues to shift, the role of the agricultural sector continues to decline and is replaced by roles in the housing, trade and services sectors. This condition also occurs in Balai Gadang, the transition of agricultural land to non-agricultural land has resulted in a decline in the area of agricultural land followed by farmers desire to grow crops, and shifting eyes, changes in the distribution of economic output are not followed by changes in labor structure work proportionally. There is a tendency for the proportion of workers absorbed in the agricultural sector to decline, the conversion of agricultural land is quite large, is expected to have an impact on changes in welfare of farmers and the environment. The gap between the dominance of output and labor explicitly shows a striking difference in productivity between these two economic sectors.

Besides the problem of income inequality and poverty, along with the stages of development that occur in Balai Gadang is faced with increasingly serious environmental problems. Based on data from the City of [10] there was a change in the use of land where there were increasing uses such as housing, trade, industry, services, mixed gardens, alang-alang bushes, and more. Conversely there are those who experience reductions such as rice fields, vacant land, swamps, forests and rivers. Changes in land use in Balai Gadang Village have various impacts on the environment, both changes in the physical, socio-economic and cultural environment.

Settlement development is carried out horizontally so that more land is spent. In addition, a portion of the community covers the entire surface of the land, either by building a high-rise or covering with cast or giving paving blocks. Community behavior that covers the entire surface of the land will disturb and reduce the infiltration of water into the soil and increase *run-off*. As a result of the increase in run-off, it can result in flooding in lower areas.

Environmental changes from infiltration land into uncontrolled dense land can result in the loss of one of the ecological functions of the soil as absorbent of rainwater. If the land is diminishing infiltration can be increased flow of runoff water *(run off)* that can cause flooding in the rainy USIM. Conversely in the dry season can lead to reduced supply of ground water and an increase in micro air temperature [11].

Degradation environment was due to land conversion, that is fertile for agriculture land many converted to non-agricultural land. The land conversion is carried out to meet the needs of residential land, shops and other developments because population growth is increasing and requires residential land. Displacement of population also leads to land conversion because they need shelter and place of business so that they buy the farm to build houses, shops and others as a place. Thus, the land that was originally intended as an agricultural activity shifted functions as residential land, shops and so on in the form of non-agriculture, so that the quality of land decreased due to environmental degradation [12].

2. Research Method

This research method is a descriptive type of research with a qualitative approach. Bogdan and Taylor [13] define qualitative research as a research procedure that produces descriptive data in the form of written or oral words from people and observable behavior. Meanwhile Moleong defines qualitative research as research that intends to understand the phenomenon of what is experienced by the subject of research such as my behavior, perceptions, motivations, actions, etc. Holistically and by means of descriptions in the form of words and language in a specific context natural and by utilizing various natural methods (2007: 99).

The research location is in district Into Hall Tower to Koto Tangah sub-district, Padang City. Location election based on the consideration that district Into Hall Tower is the largest district of 13 village in the district of Koto Tangah district Padang and Into Hall Tower areas that have large enough land to transfer land functions. Geographically, to Balai Gadang village is at an altitude of $\pm$ 0 - 1600 m above sea level.

The addition of the sample is stopped, when the data is saturated. From various informants, both old and new, did not provide new data anymore. If the selection of a candidate or an infoman really falls on a subject who truly masters the social situation under study (object), then it is an advantage for the researcher because it does not require many more samples, so the researcher is finished quickly. So, the concern for qualitative research is "completion" of information acquisition with a variety of variations that exist, not the number of samples of data sources [14].

Informants in this study came from community leaders, village officials, sub-district officials, and related officials. While secondary data is additional data that is used to perfect primary data [14].

Data collection techniques in this study, namely interviews, observation and documentation. Interviews are meetings of two people to exchange information and ideas through answers, so that meaning can be constructed in a particular topic [14].

According to [14] states that through observation, researchers learn about behavior, and the meaning of the behavior, while [14] states that observation is the basis of science. Observation is systematic observation and recording of the symptoms experienced. Observation is intended to directly see the factual object of research. Observations were carried out at the research location, namely in Balai Gadang Village, Koto Tangah District, Padang City. In the observation the researcher made observations, took pictures, recorded and felt what was in the research location.

According to [15] the documentation method is a way to obtain data or information about various things that have to do with research by looking back at written reports in the form of numbers or information (writing or boards, places and people). In this study, the method used to determine the documentation, land conversion, economic growth rates, the rate of population growth and poverty in district Into Hall Tower in 2006 to 2016.

Mechanical analysis conducted data is descriptive method qualitative analysis, the analysis carefully observes a particular through the collection of facts without doing trial hypothesis [13]. In this analysis method, the results of the exploration from interviews, documentation, observations were concluded to answer the problem formulation that was present in the study. Descriptive analysis is the analysis carefully observe a particular phenomenon through a phenomenon that occurs in the field.

3.Result and Discussion

The impact of economic growth on the carrying capacity of the environment in Balai Gadang Village, Koto Tangah District, Padang City is inseparable from the supporting factors for the ongoing economic activity. The supporting factors are infrastructure, namely shop buildings and transportation means. In addition, the existence of trade facilities was none other than the growing number of settlements that were increasing in the district. With another meaning, where there are settlements there must be found economic activity. So, between settlements, transportation and trade road facilities are interrelated with each other. The existence of these three components of economic activity requires land. The land needed for the existence of these three components in Balai Gadang Village is agricultural land.

Land use for settlements and trade continues to increase. The increase in the rate of increase in settlements and trade was inseparable from economic growth. With the establishment of settlements, trade will also accompany the road of transportation and other infrastructure facilities. In addition, migrants who came to the Balai Gadang Village opened a trading business field instead of an agricultural business field. Likewise, local residents who have capital from the sale of agricultural land also open trade businesses, or shop contracts. Thus, stretching economic activity in Balai Gadang Village is quite dynamic. However, the construction of trade infrastructure and settlements as facilities and infrastructure to support economic growth has an impact on the carrying capacity of the environment. Land used as settlements and trade into hard land cannot absorb water. Likewise, the road that is filled with asphalt affects the quality of the environment. Building waste, such as cement, also affects the quality of the surrounding environment because it can reduce soil productivity. Decreased soil productivity because there is no more water uptake. So, the impact of economic growth on the carrying capacity of the suppressing environment is more on the aspect of infrastructure that supports the facilities and infrastructure to support these economic activities. Facilities and infrastructure that support economic activities are built on land that was originally used as agricultural land and then uses the function as non- agricultural land, so that the environmental carrying capacity becomes damaged or integrated.

While the impact of poverty the carrying capacity of the environment is closely related to the low income and productive resources which ensure sustainable livelihoods. The source of opinion of the residents of the district is generally from the labor market. In the beginning the business fields of the local people were farming. Because, the agricultural land has shifted its function to non-agricultural use, the local population has switched professions as laborers. The transition of agricultural land use to non-agricultural land occurs due to factors of poverty and population growth. The population migration that came in Balai Gadang Village had the consequence of providing land to accommodate the migration. Thus, poverty and population growth are like a welcome dip in land conversion. Thus, the conversion of agricultural land to non-agriculture thus has an impact on changes in people's livelihoods in Balai Gadang Village.

Thus, economic factors encourage the conversion of agricultural land into non-agricultural land. Local people as agricultural land owners sell their land to developers. The reason they sell agricultural land varies, including economic factors, such as for the cost of education and business development itself. In addition, they also set up new buildings for rented houses and shops and kios to be rented or used by themselves.

As a result of the agricultural land being sold, the agricultural business field fell. With the decline in the agricultural business field as a source of household income in the Balai Gadang Village, it shows that agricultural land that will be processed by the community is no longer there because its use has shifted its function to non-agricultural use. To fulfill their daily needs, the farmers switch professions. The business field that he entered as a source of household opinion was as a laborer. In Balai Gadang Village there was an increase in the labor sector, because they did not have professionalism because of the low level of education they had, to work in other informal sectors such as trading, they had no capital and to become motorcycle taxi drivers, they did not have a motorbike.

Poverty affects river flows. Many poor people are mining sirturkil goods. Dredging can lead to silting and widening occur stream flow, water quality declined, and landslides on the cliffs of the river, it can even damage the bridge. Thus, the impact of poverty on environmental carrying capacity is the business activities carried out in the form of river dredging to meet the demand for sirtukil mining good as household income.

There has been an increase in land conversion at Balai Gadang village. The converted land is rice fields. The land that was originally used for agricultural activities, as a source of household income and can absorb labor, shift functions to non-agricultural use, namely settlement and trade.

Land conversion factors in Balai Gadang Village, Koto Tangah District, Padang City are government policies, natural factors, population growth and economic problems. Policy goverment Padang expanding residential areas to the Eastern region of Padang in an effort to improve the opinion of the area through the Building Permit (IMB). This is a form of regional autonomy in order to prepare as much regional income as possible for regional development. In addition, natural factors such as earthquakes or tsunami issues encourage population migration to Balai Gadang Village. Land conversion is not invitable because population growth is increasing, so land is needed for the construction of settlements. Economic problems are also a factor in land conversion in Balai Gadang Village. Because of economic pressure, agricultural landowners are forced to sell their fields to developers or to individuals to be used as residential buildings or shops. The land conversion causes the quality of the carrying capacity of the environment to be low. Therefore, land conversion results in environmental degradation. This is related to the narrowing of rice fields.

The narrowing of the area of rice fields was due to the construction of settlements and trade. Agricultural activities are no longer the primary business sector for household income. The rice field itself is temporarily replaced by a housing complex. Even if there is still paddy fields in the neighborhood. That is, rice fields in Balai Gadang Sub-District of Koto Tangah Subdistrict have become increasingly narrow and on the contrary land of settlements and trade is increasingly widespread. Comparison of settlement and trade expansion in Koto Tangah Subdistrict, Padang City was found that the Balai Gadang Sub-District is the village with the most extensive land as a settlement and trade,

The magnitude of settlement and trade expansion in Balai Gadang Village compared to other district in Kota Tangah Subdistrict, Padang City resulted in environmental degradation. Environmental degradation was found as a result of conversion of agricultural land to residential and trading land in Balai Gadang Village.

Environmental degradation to soil productivity. Environmental pollution due to building activities (such as cement water) has an impact on soil productivity due to reduced absorption of water into the soil, resulting in a decrease in crop yields of farmers. Production rice decreases the more years. Thus, the construction of settlements and trade around rice fields is a factor in decreasing rice production. The land around the rice fields has been polluted because of building waste so that it cannot absorb water, resulting in reduced water supply to the soil.

Environmental degradation of soil. The increase in land that cannot penetrate water will affect the infiltration of ground water, if the infiltration decreases, the volume of groundwater will decrease which will result in drought in the dry season, especially at elevation areas. In addition to the drought, another impact is flooding, because the increase in dense land will cause runoff to *run* high. So that during the rainy season the volume of river water increases to exceed its capacity which ultimately results in flooding. Thus, reducing the flow of river water cannot be separated from the decreasing absorption area which functions as a settlement as a result of the conversion of agricultural land to non-agriculture.

Environmental degradation of river widening and shallowness. The development of settlements and trade which is increasing in the Balai Gadang Village has triggered an increase in sirtukil mining goods. That is, with so much settlement and trade development, the demand for sirtukil mining goods also increased. Thus, dredging can lead to silting and widening occur stream flow, water quality declined, and landslides on the cliffs of the river, it can even damage the bridge.

Environmental degradation of air pollution. The presence of Final Disposal Sites (TPA) of Cold-Water waste has polluted the air environment especially when it rains. The establishment of TPA is a logical consequence of population growth and settlements. The more growing population and settlements, the more the amount of household waste is disposed of into Cold Water Landfill. This causes environmental degradation in the form of air pollution.

Environmental degradation in the form of air pollution other than sourced from waste is also sourced from motor vehicles. Motor vehicles will emit exhaust gas that can pollute the air. vehicle motorized eject some of the gas that can damage the layer of the atmosphere just as carbon dioxide (CO2), carbon monoxide (CO), nitrogen oxides (NOx), sulfur dioxide (SO2), hydrocarbon (HC) and particulate matter (dust, aerosols and tin black (Pb)). Polluted air can cause disruption to human, animal and plant life.

In addition, the narrowing of the area of paddy fields results in environmental degradation of biodiversity. In this case can feed undisturbed habitat of living creatures. Due to the disruption of their habitat, many of them are extinct or migrate to other places.

4. Conclusion

The impact of economic growth on the carrying capacity of the environment is infrastructure that supports facilities and infrastructure to support economic activities that are built on land that was originally used as agricultural land and then shifts its function as non-agricultural land, resulting in damaged or integrated environmental carrying capacity. The impact of poverty on environmental carrying capacity is the business activities carried out in the form of river dredging to meet the demand for sirtukil mining goods as household income.

There has been an increase in land conversion in Balai Gadang Village. Converted land is rice fields for residential land and trade. Land conversion factors in Balai Gadang Village, Koto Tangah Subdistrict, Padang City are government policies, natural factors, population growth and economic problems. The magnitude of settlement and trade expansion in Balai Gadang Village has resulted in environmental degradation, namely 1) environmental degradation to soil productivity; 2) soil degradation; 3) environmental degradation of river widening and shallowness; 4) environmental degradation of air pollution; 5) environmental degradation in the form of air pollution other than sourced from waste also sourced from motor vehicles; 6) environmental degradation of biodiversity.

References

[1] Widodo T 2009 *Ecology Basics* (Yogyakarta: Gadjah Mada University Press)
[2] Vinod 2002 *Environment Resource Nature and Population in Development* (Jakarta: University of Indonesia)
[3] Soemarwoto O 1997*Analysis of Environmental Impacts (*Yogyakarta: Gadjah Mada University Press)
[4] Jumin H B 2002 *Agroecology Something Physiological Approach* (Jakarta: RajawaliPers)
[5] Sarwono 2007 *Evaluation and Land Use Planning* (Yogyakarta: Gajah Mada Univeersity Press)
[6] Utomo S 1992 Development and Control of Land Transfer Functions *Thesis* (Lampung: University of Lampung)
[7] Houghton R A 1991 Tropical Deforestain and Atmospheric Carbondioxide *Climate Change Journal* **19**(3) p 99-118
[8] Sihaloho M 2004 Agricultural Land Conversion and Changes in Agrarian Structure in Mulyoharjo Sub-District, South Bogor District *Thesis* (Bogor: Bogor Agricultural Institute)
[9] Rusli S 2005 *Introduction to Population Sciences* (Jakarta: LP3ES Library)
[10] Padang City BPS 2014 Koto Tangah in Numbers
[11] Salim E 1993 *Development of Environmental Insight* (Jakarta: LP3ES)
[12] Firmansyah 2003 *Cities and Environments: A New Approach to Ecological Insights* (Jakarta: LP3ES)
[13] Moleong L J 2007 *Qualitative Methodology for Research* (Bandung:PT. Teenagers Rosdakarya)
[14] Sugiyono 2008 *Understanding Research Qualitative* (Bandung: CV. Alfabeta)
[15] Arikunto 2000 *Leather Research Methods* (Bandung: CV. Alfabeta)

IOP Conf. Series: Earth and Environmental Science **314** (2019) 012060 doi:10.1088/1755-1315/314/1/012060

The contributions of discipline and environmental knowledge on clean behavior of students in Public Elementary School 19 Kampung Baru Pariaman, West Sumatra

Nurhasan Syah and Sanny Edinov

Postgraduate of Environmental Science, Universitas Negeri Padang, Indonesia

*nurhasan_s@yahoo.com

Abstract. This study aims to obtain a description of the contribution of discipline and environmental knowledge to the clean behavior of students in public elementary school of 19 Kampung Baru, Pariaman. Quantitative approach was used in this research. The sample of this research were the students in grade I-VI at public elementary school and taken by simple random sampling of 81 students. The data were collected by asked to fill in the questionnaires. Results showed that: (1) There was a contribution of discipline (X_1) to students clean behavior (Y) equal to 27.70%, (2) There was a contribution of environmental knowledge (X_2) to students clean behavior (Y) equal to 53.00% And (3) There was a contribution of discipline (X_1) and environmental knowledge (X_2) to the students clean behavior (Y) of 54.20%. The results suggest that the discipline (X_1) and environmental knowledge (X_2), either individually or jointly contributes to students clean behavior (Y). This provides some explanation as to why principals and teachers in school should monitor the students discipline and environmental knowledge. This knowledge can support the clean behavior to the surrounding environment. On the other hand, it is also advisable for students to always improve their own discipline and environmentalknowledge.

1. Introduction

The school aims to prepare the students about the values and competencies that might be needed in the form of knowledge, attitudes and skills. Schools as educational institutions play an important role in train the character early on, during childhood phase. However, the school can also be the place of disease transmission if not appropriately managed. Children in the school age tend to get a various disease. The emergence of different disease that often affects children in the school age associated with clean behavior. Hence, teaching the children about clean behavior in school is an absolute necessity.

A good character must be built in everyone so that they will have responsibility for their behavior. The school has an important role in the development of children personality, for example in their way of thinking, acting and behaving. Clean behavior is one of the awareness that owned by individuals, families and groups to the personal and environmental hygiene. Clean living habits that taught since earlier in life indeed accompanied by a disciplined character that owned by the students. Discipline will help the students to build their behavior, develops their attitude of responsibility that includes increasing awareness which is useful for improving the quality of themselves. Individuals who are disciplined will perform a task with orderly and organized manners and follow the applicable rules.

Apparently, it will give a positive impact because it can make their lives well-organized.

Behavior is a form of real action. The theory states that behavior is a given action or response to a stimulus to achieve a goal or to survive their life [1]. Skinner, a behavioral expert, distinguishes the behavior becomes natural behavior (innate behavior) and operant behavior. Natural behavior is a behavior that brought since the organism was born, in the form of reflexes and instincts, whereas operant behavior is a behavior that is formed through the learning process. In the development of behavior, it is influenced by internal and external factors [2]. The internal factors that affect the development of behavior are the knowledge, intelligence, perception, emotion and motivation, whereas the external factors are both physical and nonphysical include climatic, human, socio-economic and cultural [3].

Roger explains revealed that before people acquire new behavior, there was an on-going process that happens inside the people [3]. Awareness was the first thing that arises in the process, where the person is aware towards the upcoming stimulus, then their interest will focus toward the stimulus. Next, evaluation, the people will evaluate whether the forthcoming stimulus was good for them, then try to do things according to the stimulus and finally adopt the new behavior following the knowledge, awareness and his attitude toward the stimulus. Torndike explains that the stimulus will cause a particular response. The intensity of the relationship between the stimulus and response can be induced through repeated practice. Thus, something that learned later (transfer of training) will influence subsequent behavior [4].

Environmental factors are the most significant factor in determining health status. Therefore, the environmental knowledge is important. Environmental education can increase understanding and awareness of society in the search for solutions and prevention of environmental problems. Some good habits regarding the clean behavior (i.e., taking out the trash to its place, using the toilet and cleaning it after use, etc.) are important to train as early as possible because they will implement this behavior into later in life.

SD Negeri 19, Kampung Baru, which is in Pariaman, a city in West Sumatera. This school is one of the favorite elementary schools which has obtained an A level in the accreditation status. As one of the institutions of formal education, this school is very influential on students' formation. The total of class in this school are 13 class, where the total students in the academic year 2016/2017 as many as 408 people. Green Open Space such as trees, plants, living pharmacy, composter and the trash bin is environmentally friendly infrastructure that is already available in this school. The intensity of using the facilities as well as their environmental knowledge can encourage and improve their behavior to have a cleaner life. The behavior is essentially goal-oriented. Behavior is motivated by a desire to obtain a particular purpose. Some people with a high motivation to act can achieve certain goals that they want.

The Ministry of Health of the Republic of Indonesia explained that the definition of clean behavior is a set of behaviors that practiced based on consciousness as a result of learning that makes a person able to help themselves, which aims to realize a clean conduct public health. Clean behavior is part of a program that has been launched by the government, which is "Perilaku Hidup Bersih dan Sehat (PHBS)" or clean and healthy behaviors.

The clean behavior is a series of various forms of behaviors/ actions towards the waste managements. For example, behavior that pollutes the environments as an irresponsible action or actions to maintain the environment as the responsible one. Clean behavior can be expressed as an act or respond in an environmentally responsible behavior in order to be maintained the environment. Individuals who have the knowledge, skills, positive attitude towards the environment and towards pro- environmental behavior, typically tend to have responsible behavior [5].

People did not realize that they are part of this universe so that the environmental crisis is not yet a common concern. Indeed, humans are part of the environment. Both interact in and ecosystem, discipline does not happen by itself, but it must be generated from the actions of the people. Thus, it is important to train or teach people about the discipline, so they can regulate and control himself in order to do something that is socially acceptable to the environment and avoiding behavior that could damage the environment.

The emergence of self-consciousness in environmental disciplines related to the integration of environmental education that began at early age. Soegeng explains, the discipline can be considered as conditions that created and developed through a series of behaviors that indicate the values of obedience, loyalty and regularity that have become a part of behavior in life. Behavior was created through a process that guided by the family, education and experience. Discipline can be expressed as a conscious behavior of an organism to be obedient, submissive and accountable to the rule of order. In the application of discipline, it needed to make rules and regulations. Discipline is one of the supporting factors in improving the quality of education/school [6].

Environmental knowledge is already integrated in education curriculum that implement Adiwiyata program in their school. Environmental Education is the way to improve the knowledge, awareness and skills of the communities in sustaining the environment. Most of the knowledge gained through the eyes and ears. One's knowledge has different intensity and level [7]. Soekidjo argues that knowledge or cognitive domain is important for the formation of a person's behavior. In the cognitive domain, there are six level of knowledge, namely know, comprehension, application, analysis, synthesis and evaluation [3].

According to Indonesian's Law No. 32 in 2009, The environment is the unity of objects, powers, condition and human including their behavior that will influence the situation itself and other living creatures. Thus, environmental knowledge can be defined as everything that is associated with environment after the direct contact with nature through the instrumentality of the senses and lead to an immediate impression in people's mind.

Hiswari suggests that the level of understanding of the environmental knowledge is the results of an environmental learning process to the students' attitudes toward the environment. Environmental education is expected to be embedded and transformed on the students [8]. Discipline and environmental knowledge of the students are two factors that influence the success of the school in creating and maintaining the environment to keep it clean. The current study is the contribution of discipline and environmental knowledge toward clean behavior of students in the elementary school in Pariaman.

The problem statement from the current study is how much the contribution of discipline and environmental knowledge toward clean behavior of students in Public Elementary School 19 Kampung Baru Pariaman.

2. Methods

This research was conducted with a quantitative approach that belongs to the type of survey research. As a causative research, this study was aimed to reveal their relationship and the contribution of variables that exist. As these variables include the clean behavior of students (Y), discipline (X1) and knowledge of the environment (X2).In this research, as well as interviews to sharpen the results that have been obtained using a questionnaire instrument. The population were students in S Public Elementary School 19 Kampung Baru Pariaman. Proportional random sampling was used to gather the sample of the study and as a results there were 81 students that participated in this study.

To obtain the data from the variables, the researchers used questionnaires in the form of multiple choice where validity and reliability of the tools already met. A correlation and ANOVA were conducted to analyze the data using the SPSS for Windows Version 23.00. interview method was also conducted in order to get the deeper result and these results were reported in a descriptive or qualitative form.

3. Discussions

Based on the results of the normality test, a significance probability of students clean behavior (Y), discipline (X_1) and environmental knowledge (X_2), for 0.200, 0.181 and 0.056, respectively. These results suggesting that the distribution of data is normal (as seen in Table 1).

Table 1. Normality Test Results Summary

Variables	Significance	α (Alpha)	Information
Behavior Clean (Y)	0.200	0.05	Normal
Discipline (X_1)	0.181	0.05	Normal
Environmental Sciences (X_2)	0.056	0.05	Normal

As shown in Table 2, The first hypothesis was analyzed by simple regression correlation analysis. The results showed that the correlation coefficient between discipline (X_1) and students clean behavior (Y) was significantly correlated [$r_{xy} = 0.526$, $p < 0.000$]. Thus, there was a significant correlation between X_1 and Y with 0.277 coefficient determination (27.70% of contribution). These results suggest that the discipline remained significantly related to the variable Y, means that improvement of the students discipline will improve their clean behavior as well.

Table 2. Correlation Test Results Summary The Relationship between Discipline (X_1) with Students Clean Behavior (Y)

Correlation	Correlation Coefficient	Probability Significance	Coefficient of Determination	Relationship X_1 and Y
X_1 and Y	0.526	0.000	0.277	There Relations

The second hypothesis analysis was shown in Table 3. The results showed that the correlation between environmental knowledge (X_2) and students clean behavior (Y) was significantly correlated [$r_{xy} = 0.728$, $p < 0.000$]. Thus, there was a highly significant correlation between variable X_2 and Y with 0.530 (53% of contribution) toward the students clean behavior. This suggests that environmental knowledge remained significantly related to the variable Y. The increasing of environmental knowledge will increase the students clean behavior.

Table 3. Correlation Test Results Summary The Relationship between Environmental Knowledge (X_2) with Students Clean Behavior (Y)

Correlation	Correlation Coefficient	Probability Significance	Coefficient of Determination	Relationship X_2 and Y
X_2 and Y	0.728	0.000	0.530	There Relations

The third hypothesis results were presented in Table 4. The correlation between discipline and environmental knowledge was highly significantly correlated [$r_{xy} = 0.737$, $p < 0.000$]. Therefore, there was a significant correlation between discipline and knowledge of the environment together with a coefficient of determination 0.542 (54.20% of contribution) towards the clean behavior of students.

Table 4. Multiple Correlation Test Results Summary

Correlation	Correlation Coefficient	Probability Significance	Coefficient of Determination	Contributions X_1 and Y
X_1 and X_2 with Y	0.737	0.000	0.542	There Relations

Based on the results of the F test, $F(2,80) = 46.234$, $p < 0.0001$. Thus, the multiple linear regression equations were significant and could be used to express the magnitude of the relationship between independent variables and the dependent variable in this study.

Table 5. Test Results Summary F

Model	Sum of Squares	Df	Mean Square	F	Significance
Regression	8515.196	2	4257.598	46.234	0,000
Residual Total	7182.828	78	92.088		

15698.025	80

The results of interviews revealed that the disciplinary of the students in keeping the environment clean is already good enough. It can be seen from the observations of the cleanliness of the school, as there was not much trash scattered or deliberately discarded by the students are not in place and the cleaning schedule goes according to the plan, Friday cleaning day is held every week by the school. However, to develop the students disciplinary behavior is not easy. Students should be familiar and care with the environment.

The formation of students disciplinary was influenced by the existence of the rules at school. The interviews result suggesting that students are happy to undergo the existing regulations, especially regarding the cleanliness. There were many obstacles that students experience while developing the disciplinary. Thus, the principals may create solutions to make a punishments or penalties for the students who are coming late or break the rules. Rewards and Punishments can be a motivation for the students to implement the disciplinary.

To achieve the implementation of discipline in the students behavior, the awareness of the students is necessary. The support and cooperation from school to facilitate the implementation of current knowledge that owned by the students. Teaching and guiding the students to care about the environment should be apriority. The students expressed their complaint through the interview about the lack of facilities and infrastructure in the school environment also become obstacles to maintain the sanitary condition. For example, the lack of waste management and the lack of toilet hygiene because there was not enough water. In connection with the above findings, the need for particular attention to all parties to improve student disciplinary in order to enhance the students' clean behavior. With the increase of discipline, the students will develop better response on the social environment that will ultimately impact on the comfort and safety of the school environment.

Furthermore, the results of interviews conducted with the students also revealed that the environmental knowledge of the students in maintaining the environment cleanliness was excellent. Students have a good understanding of the environment because the school has been integrating the environmental education in teaching materials. Theories related environmental knowledge that gained since childhood will help in increasing the awareness of the importance of a clean environment because it is also closely related to the health of students in the school. Behavior that is based on understanding will be last longer compared to behavior that is not based on knowledge.

A good knowledge about the environment will improve student awareness in the importance of environmental hygiene so that a clean environment can be maintained. A Cleanliness environment leads to a healthy life of the students. Health, in this case, cannot refer to aspects of treatment but instead refers to the prevention aspect, so it needs to be appropriately addressed so as not harmful to health. Students knowledge in environmental hygiene would be useful in waste management. For example, separate the garbage (organic and inorganic), maintaining the cleanliness of the bathrooms and clean water usage. With the increased knowledge of the environment, the behavior of students will be better for the environment that will ultimately impact on the comfort and safety of the school environment.

The student's behavior should be controlled in order to achieve a clean environment. The role of schools in helping the implementation of the clean behavior of students is crucial so that when the children are in school, teachers need to pay more considerable attention to them, either directly through teaching them in the class or by application.

4. Conclusion

There was a positive and significant contribution to the discipline with the clean behavior of students. Student discipline was one of the factors that influence the clean behavior of students. Based on the results, students have a pretty good control to keep the environment clean. Students are still under the supervision of teachers so that their disciplinary of clean behavior were not optimal yet. There was a significant and positive contribution to clean behavior and environmental knowledge with students. Environmental understanding of student was one of the factors that influence the clean behavior. Based on the results, the students have a good experience in protecting the environment to keep it clean. The integration of environmental knowledge in the teaching material given by the teacher adds confidence the students to behave in a clean environment. Increased discipline and knowledge about the environment, it also can improve the behavior of students towards a clean environment so that the environment can be protected and have a positive impact on health.

References

[1] Purbantara A, Endro pp dan Tri R 2013 Survei Kebersihan Pribadi Siswa Sekolah Dasar Negeri dalam Wilayah Pedesaan dan Perkotaan di Kabupaten Semarang Tahun Ajaran 2012/2013 *Journal of Physical Education, sport, Health and Recession* **2**(6) 368-371 ISSN 2252-6773

[2] Putra G G 2002 The Behaviour of the Community in Cleaning the Settlement Environment to Support the Sapta Program of Enchantment of Tourism a Case Study in Kalibukbuk Village in the Lovina Tourism Area of Singaraja Bali *Thesis* (Universitas Diponegoro)

[3] Notoatmodjo S 2003 *Health Education and Behavior* (Jakarta: Rineka Cipta)

[4] Razali 2007 Contribution of Sanitation Knowledge and Income to Community Behavior in Residential Environmental Management in Kuaro Padang Village Padang City *Thesis* (Universitas Negeri Padang)

[5] Wibowo I 2009 Pola Perilaku Kebersihan: Studi Psikologi Lingkungan tentang Penanggulangan Sampah Perkotaan *Makara, Sosial Humaniora* **13**(1) 37-47

[6] Prasetyo PE dan Mulyadi H 2008 Pengaruh Disiplin Siswa dan Fasilitas Perpustakaan Sekolah Terhadap Prestasi Belajar Siswa Mata Pelajaran Ekonomi *Jurnal Pendidikan Ekonomi* **3**(2) 219-240

[7] Jumadil, Kahar M dan Alimuddin HA 2015 Penerapan Program Adiwiyata pada Aspek Kognitif, afektif dan Psikomotor Tentang Pengelolaan Lingkungan Hidup Sekolah Dasar di Kota Kendari *J. Sains & Teknologi* **15**(2) 195-202 ISSN 1411-4674

[8] Hiswari 2007 Correlation Between Environmental Education and students Attitudes Toward The Environment *Thesis* (Universitas Indonesia)

Validity of integrated calculus module two stay two stray learning model for use in communities independent learning groups

Mishbah Ulhusna[1,*], Sri Diana Putri[1] and Zakirman[2]

[1]Universitas Putra Indonesia YPTK Padang, West Sumatera, Indonesia
[2]Universitas Negeri Padang, West Sumatera, Indonesia

*ulhusna_82@yahoo.com,

Abstract. Calculus is one of the compulsory subjects at the Faculty of Computer Science, UPI YPTK Padang, with the percentage of students passing each semester around 40%. To overcome these problems an independent learning group community was formed at the initiative of students. To support the process of understanding calculus material in the community, an integrated caliber module was designed for Two Stay Two Stray learning models. The conclusion of the study shows that the calculus module developed is valid according to the expert and is ready to be used to support the tutorial process in the independent learning group community of FILKOM UPI YPTK Padang students.

1. Calculus learning in universities

Calculus is one of the topics in mathematics where algorithmic manipulation is learnt as a basic concept in learning [1]. Calculus courses become one of the compulsory subjects in the information system study program Universitas Putra Indonesia YPTK Padang. Calculus becomes a basic subject and has links with several subjects in the next semester. Every student must pass the calculus subject before continue to other subjects such as matrix and vector and algebraic spaces and algorithms.

Teaching experience shows that the concept of calculus learning is not clear enough for students [2]. Most students feel that calculus courses are too difficult to understand and the subjects are most feared at the beginning of each semester. Student learning outcomes show that only 40% of students who pass in calculus courses each semester. This condition is a threat that can reduce students' understanding and success in completing calculus subjects.

Calculus is a field of mathematics studies, plays a key role in shaping students' logical knowledge. But today, as in the past, many students have difficulty with calculus and become dissatisfied because they constantly face conceptual constraints and understanding [3]. Calculus learning at the tertiary level is currently dominated by traditional learning concepts. Learning is teacher-centered and causes passive acceptance of student knowledge and information. Potential problems that arise include: (a) lecturers dominate most of the learning activities and (b) learning acquisition can be very limited because it depends a lot on information by the lecturer. Ideally a good learning process places the lecturer as a manager and instructor of learning. Not only communicating knowledge to students, but also arranging and guiding them to achieve the successful learning [4].

The concept of learning in college has undergone changes from year to year. Active learning has become an important focus today. Research shows that students learn best when they are actively involved, as well as employment opportunities that want graduates who have the ability to work in groups and have developed appropriate teamwork skills [5]. Learning in universities must adopt the concept of team collaboration in lecturing activities, especially in calculus learning which has a high level of learning difficulty according to students.

Innovation in calculus learning can be done with three things, namely learning by uncovering problems, based on experience, and teamwork [6]. Students who experience problems in learning can work in groups based on initiative and without instruction from the teacher / lecturer. This independent work group is a form of student effort in building solid teamwork to solve a problem faced in the lecture system. The limitations of lecture time, the lack of material delivery and the incomplete learning resources can be overcome by the initiative to form group work independently by students.

2. Independent learning group

The formation of a learning group requires several considerations, one of which is the formation of a discussion structure that is not rigid and monotonous so that it is possible for each active member in group work [7]. Learning in group also emphasizes the delivery of learning stimuli and enhances students' conceptual understanding. The activeness of each member is very depending on the path and scheme of a discussion in group learning.

Today there are some obstacles in group learning, and this phenomenon is known as Grouphate. Grouphate phenomenon is an individual's fear and anxiety in interacting with group members [5]. One way to overcome Grouphate is to create a discussion plot and scheme that allows for positive interaction by each group member.

Indications of success in group work are an increase in learning outcomes and increased knowledge [5]. In group learning independently, student knowledge will increase if there is a transfer of knowledge between one student to another. Important stages in group work that must be considered include: providing an understanding that students know their work in group work, know how to work together, arrange time, and delegate tasks.

The independent working group formed to overcome the difficulties of calculus learning at YPTK Padang UPI is a small group discussion category. Group members contain four to five people and are formed based on the wishes of the students themselves. In each caliber class that is taught, at least six to eight independent working groups can be formed outside calculus courses. This group routinely holds meetings twice a week to discuss problems faced by members in calculus subjects and work together in completing college assignments. Small group discussions are interactive, friendly, and bridge the gap between teachers and students [8]. Student communication skills can also be improved in discussions in this independent study group. Learning in small group discussions is felt by students to be more effective than traditional teaching methods.

The combination of lectures and self-study groups on a small scale is the right step in overcoming calculus learning problems. Through observation, it can be understood how effective the discussion is in student self-study groups [9]. The obstacle faced today is that there is no available calculus teaching material to bridge students in discussion activities in independent study groups. For that, we need a teaching material that can guide and direct students to study independently with groups outside the calculus course schedule.

3. Integrated module validation of two stay two stray learning models

The agenda of the independent study group discussion at the YPTK UPI was to discuss the problem of learning calculus courses and repetition of material or understanding of the material to be studied at the next meeting. For this reason, we need a teaching material that can accommodate students' learning independently in groups. Module making by the teacher / lecturer can be a solution in helping students learn in independent groups. Module content can be made more challenging to increase student active participation [10]. Student feedback about teaching practices and feedback on modules is considered to be a reliable variable and is important for measuring module effectiveness [11].

Creating student activity in group learning can be done by collaborating with the learning model in the module breakdown stage. Two Stay Two Stray is one model that is suitable for answering the concept of student activity in working in groups. Two stay two stray is proven to improve understanding, communication skills, increase students' knowledge and activeness in working in groups [12,13,14,15].

To answer the problems in the independent study group at the YPTK UPI, researchers have developed an integrated calculus module for the Two Stay Two Stray learning model. Validity analysis results are shown in table 1.

Table 1. Validity results' summary.

Number	Validator	Rating Result (%)
1	AMA	86
2	NS	75
3	SPS	92
	Average	84,33 (Valid)

Based on the table, it can be concluded that the designed module is valid in the opinion of experts. Modules are said to be valid if the results of the average product evaluation pass the number 80. Modules that are already in the valid category can be used to improve learning outcomes because the testing of these products has been recognized by experts. To test the validity of the module, an initial test was also carried out by distributing modules to four independent study groups/groups of YPTK Padang UPI students. The results of the deployment show that the module has been effective in helping students learn, able to improve students 'understanding and increase students' active participation.

The use of calculus modules that have been developed by researchers can help students to create a warm atmosphere of discussion and enable the transfer of knowledge between students. The basic concept offered by the Two Stay Two Stray model integrated calculus learning module is that each student has a separate role, can be as a guest or remain silent in a group. In group learning, calculus modules designed by researchers can change the pattern of discussion and learning at the college level which initially made passive students become active.

4. Conclusion

The percentage of student failure in calculus subjects can be overcome by group learning. The conditions encountered in group learning in discussion groups formed on the basis of student initiative are that there are no teaching materials that can guide and enhance student activity. Making integrated calculus modules Two Stay Two Stray learning models are the most appropriate solution to assist students in self-study groups. The module developed has been valid according to the expert and has been practically used according to students in independent study groups. Students are helped in understanding the material because the Two Stay Two Stray model syntax allows each student to act as the giver and recipient of information.

References

[1] Michiel D and Jan V M 2008 A Historical Perspective on Teaching and Learning Calculus *Australian Senior Mathematics Journal* **22**(2) 4-14

[2] Abramovich S and Levin L 1994 Spreadsheets in Teching and Learning Topic in Calculus *International Journal of Mathematics in Science and Technology Education* **25** (2) 105-126

[3] Anthony G and Walshaw M 2009 Characteristics of Effective Teaching of Mathematics : A View From the West *Journal of Mathematics Education* **2**(2) 147-164

[4] Rufii R 2015 Developing Modul on Constructivist Learning Strategies to Promote Students

Independence and Performance *International Journal of Education* **7** (1) ISSN 1948-5476
[5] Alison B 2011 Group Work: How to Use Group Effectively *Journal of Effective Teaching* **11**(2) 87-95
[6] Sidabutar R 2016 The Effort to Improve Mathematics Learning Achievement Result of High School Student as Required by Competence Based Curriculum and Lesson Level Based-Curriculum *Journal of Education and Practice* **7**(15) ISSN 2222-1735 (paper) ISSN 2222-288x (online)
[7] Eybe H and Schmidt H J 2004 Group Discussion as a Tool for Investigating Students Concept *Chemistry Education: Research and Practice* **5** (3) 265-280
[8] Annamalai N, Manivel R and Palanisamy R 2015 Small group discussion: Students Perspectives 117.216.56.126
[9] Hadimani C P 2014 Effectiveness of Small Group Discussion Session in Teaching Biochemistry for Undergraduate Medical Students *South East Asian Journal of Medical Education* **8**(1) 77-81
[10] Prendergast M, Spassiani N A and Roche J 2017 Developing a Mathematic Module for Students with Intellectual Dissability in Higher Education *International Journal of Higher Education* **6**(3) 169-177 ISSN 1927-6044 E-ISSN 1927-6052
[11] Deves S and Nasseri D A 2014 Effectiveness of Mathematics Module in Foundation Programme in Majan College *International Journal of Emerging Engineering Research and Technology* **2**(1) 1-7
[12] Wardana L W, Setyani R and Harwida G 2018 The Implementation of Two Stay Two Stray (TSTS) Learning Model and Co-op Co-op for The Improvement of Students Learning Outcome in The Crafts and Enterpreneurship Subject *KnE Social Sciences The First International Research Conference on Economics and Business (IRCEB)* 140-149 DOI: 10.18502/kss.v3i3.1880
[13] Nova R, Asnawi M and Sofyan A G 2016 The Effect of Two Stay Two Stray (TSTS) Technique on Teaching Writing 83
[14] Kurnia N, Degeng I N S and Soetjipto B E 2017 The Implementation of Find Someone Who and Two Stay Two Stray Models to Improve Students' Self-Efficacy and Social Studies Learning Outcomes **7**(3) 66-70 e-ISSN 2320-7380 p-ISSN 2320-737x
[15] Lusiana I A, Setyosari P and Soejipto B E 2017 The Application of Two Stay Two Stray (TSTS) and Fan- N-Pick Learning Models to Improve Students' Motivation and Learning Outcomes on Social studies Subject (A Study on the Fourth-Grade students of SDN Tawun I Ngawi) *International Journal of Academic Research in Progressive Education and Development* **6**(3) 97-108 ISSN 2226-6348

IOP Conf. Series: Earth and Environmental Science **314** (2019) 012062 doi:10.1088/1755-1315/314/1/012062

The influence of infrastructure development and ecological carrying capacity on the economic impact of independent integrated city of *Lunang Silaut*

Damel Van Wanda, Agus Irianto, Sulastri, Erasukma Munaf and Zikri Alhadi*

Evironmental and Development Studies, Universitas Negeri Padang, Indonesia

*zikrialhadi@fis.unp.ac.id

Abstract. Independent integrated city is a city that was built with the aim of increasing the ease of meeting various basic needs that enable the opening of opportunities for socio-economic growth of transmigrate regions and creating centers of business activities that attract investors as an effort to develop the economic activities of transmigrates and surrounding communities. The purpose of the study was to investigate the effect of the independent variables of infrastructure development and ecological carrying capacity on the dependent variable of the Economic Impact of the Independent Integrated City of *Lunang Silaut*. This research is a quantitative research involving 150 respondents of the Independent Integrated City Regional stakeholders *Lunang Silaut* who represent the Government, private parties, non-governmental organizations, and local residents. The sample selection is done in a convenient purposive sampling manner. Data analysis uses multiple regression which is preceded by data requirements test: normality, multicollinearity, autocorrelation, and heteroscedasticity. The research error rate was set at 5%. The results showed that infrastructure development and ecological carrying capacity had a significant effect on the economic impact of Independent Integrated City *Lunang Silaut*, with a Sig. F = 0.000; and the two independent variables can explain the dependent variable of 98.8%, while the rest is determined by other factors outside the research boundary.

1. Introduction

Independent integrated city (KTM) is a transmigration area whose construction and development is designed to be a center of growth, so that it has an urban function through sustainable management of natural resources. Urban functions refer to the transmigration paradigm which include: (a) centers of agribusiness activities that include processing agricultural products into production and / or consumer goods, specialized agroindustry service centers, and superior plant breeding, education and training centers in the agricultural sector , industry and services; and (b) regional trade centers marked by the presence of market financial institutions, wholesale and warehousing markets [1].

The *Lunang Silaut* area is located in the southern part of Pesisir Selatan Regency, West Sumatra Province and administratively covers 3 Sub-districts, namely *Lunang Silaut* District, Basa IV District, Balai Tapan and Pancung Soal District. *Lunang Silaut* area in the regional context is very strategic position to be developed into a new development area because it is located in the border area of West Sumatra Province with Bengkulu Province and Jambi Province. Geographically the *Lunang Silaut* area is located at 2° 05 ', 7' - 2° 28 ', 6 "South latitude and 101° 00' - 101° 12 '3" East longitude, with an area of 56,984 Ha.

The concept of sustainable development began to be formulated in the late 1980s in response to previous development strategies that were more focused on the main objectives of high economic growth, and which proved to have caused degradation of production capacity and environmental quality as a result of excessive exploitation of resources. At the city and local government level, urban development strategies and integrated development plans can prioritize investment decisions and encourage synergy and interaction between several separate urban areas. At the level of the housing environment, plans and layout for the development of roads and public spaces can improve urban quality, social cohesion and inclusion, and protection of local resources.

The success of the KTM *Lunang Silaut* development program cannot only be seen from the perspective of the physical development of city infrastructure, but also needs to be seen from the perspective of its environmental carrying capacity, as well as the perspective of its economic impact on the community. The purpose of the study was to investigate the effect of the independent variables of infrastructure development and ecological carrying capacity on the dependent variable of the Integrated Economic Impact of the Independent City of *Lunang Silaut*, West Sumatra.

2. Literature review

2.1. Independent integrated city

According to [2], a self-contained integrated city (KTM) is a village or area that grows and develops as a collection, processing, distribution and service center of the Transmigrant Development Area (WPT) designed as a direction for structured development of residential unit's transmigration and surrounding villages in one infrastructure network unit and regional economic unit.

The objective of the KTM development is to increase the ease of meeting various basic needs that enable the opening of opportunities for socio-economic growth of transmigrant regions and create centers of business activity that attract investors as an effort to develop the economic activities of transmigrants and surrounding communities. The target of developing the Integrated City of Independent is the availability of social, economic and government facilities to serve the basic / living needs of the transmigrants and the surrounding villages, the infrastructure and facilities to support the business activities of the transmigrants and the surrounding villages and the establishment of business activity centers to grow economic activities in the area transmigration

2.2. Tourism development

Vaughn and Pollard [3], stated that infrastructure in general includes roads, bridges, water and sewage systems, airports, ports, public buildings, and also includes schools, health facilities, prisons, recreation, power generation, security, fire, landfills, and telecommunications.

The prepared infrastructure also needs to be adjusted to the needs of each region, so that it can improve its welfare. Infrastructure needed by developed countries is certainly different from what is needed by developing countries even backward. The same is true for urban and rural areas, or industrial areas with agricultural and coastal regions or islands.

2.3. Ecological carrying capacity

According to the [4], the carrying capacity and capacity of the environment is the ability of the environment to be able to support human life, other living things, and the balance between the two. Thus, the concept of carrying capacity in general can be seen from two sides, namely:

- In terms of availability, by looking at the characteristics of the region, the potential of natural resources in an area.
- In terms of needs, namely by looking at the needs of humans and other living things and the direction of the priority policy of a region.

The carrying capacity and capacity of the environment in spatial planning is intended to make spatial use based on spatial planning not to exceed the limits of environmental capabilities in supporting and accommodating human activities without causing environmental damage. These capabilities include the ability to provide space, the ability to provide natural resources, and the ability to improve environmental quality if there are impacts that disrupt the balance of the ecosystem. Spatial planning that ignores the carrying capacity of the environment will certainly cause problems and degradation of environmental quality such as floods, landslides and droughts, pollution and so on.

Suratmo explained the impact is any change that occurs in the environment due to human activities [5]. The impact of a development project on socio-economic aspects, especially for developing countries, is found in the following components that are defined as indicators of socio-economic, among others: (1) employment, (2) the development of economic structures, namely the emergence of other economic activities due to the project such as shops, stalls, restaurants, transportation, etc., (3) increasing community income, (4) public health, (5) community perception, (6) population growth, and so on.

Therefore, [6] conducted a study which is a study of the development of the Integrated Independent Area model with the One Village One Product (OVOP) approach in an effort to empower the economy of the Transmigration area to become independent by developing superior products with high competitiveness in the domestic and global markets. [7] conducted a study to analyze territorial aspects (suitability of commodity land, superiority, land cover, land characteristics, geology, regional hierarchy), and integrate it for the design of land use and KTM spatial planning.

Accroding to [8] in their research stated that the carrying capacity of resources and the environment is the foundation of sustainable development has often been the concern of researchers. The writing aims to discuss the main concepts of the carrying capacity of resources and the environment and the problems that arise. Accroding to [9] in his research explained that land carrying capacity (LCC) explains whether local land resources are effectively used to support economic activities and / or human populations. Accroding to [10] in his research revealed that electricity infrastructure, labor, and development expenditure have a positive and significant influence on economic growth.

3. Methodology

The research design used is quantitative and causal. Respondents of the study amounted to 150 stakeholders of the Independent Integrated City of Lunang Silaut region representing the Government, private parties, non-governmental organizations, and local residents. Data was collected using a questionnaire instrument designed as a closed question with five choices of Likert scale answers. The infrastructure development questionnaire numbered 10 items, the ecological carrying capacity questionnaire numbered 24 items and the economic impact questionnaire numbered 16 items. Data analysis uses multiple regression which is preceded by data requirements test: normality, multicollinearity, autocorrelation, and heteroscedasticity. The research error rate was set at 5%.

Based on the description of relevant research and research concepts, the research framework can be identified as follows:

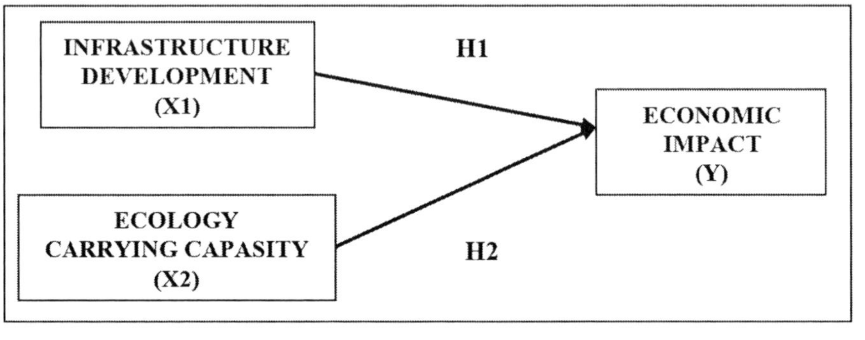

Figure 1. Research Framework

Hypotheses
H1 Infrastructure development (X1) affects the economic impact (Y)
H2 Economic carrying capacity (X2) affects the economic impact (Y)\

4. Results

4.1. Instrument validity and reliability test
Test the validity and reliability of the instrument carried out on 30 respondents with a summary of the results of the instrument validity and reliability testing as follows:

Table 1. Validity and Reability Test

Variable	r-count	r-table	Conclusion	Alpha Cronbach	Conclusion
PI (X1)	0.853-0.891	0.361	Valid	0.973	Reliable
DDE (X2)	0.866-0.927	0.361	Valid	0.990	Reliable
DE (Y)	0.885-0.919	0.361	Valid	0.988	Reliable

The summary of the results of the validity and reliability of the instrument above shows that all research instruments are valid and reliable so that they can be disseminated to collect data.

4.2. Descriptive statistics
The results of research data collection if described statistically are as follows:

Table 2. Descriptive Statistics

	N	Range	Minimum	Maximum	Sum	Mean	Std. Deviation	Variance
PI	150	32	18	50	5100	34.00	5.975	35.705
DDE	150	72	48	120	12600	84.00	14.463	209.181
DE	150	51	29	80	8175	54.50	9.823	96.493
Valid N (listwise)	150							

4.3. Test data requirements
Data collected successfully before being used to test the research hypothesis needs to be tested for data requirements including normality, multicollinearity, autocorrelation, and heteroscedasticity. The data requirements test results show:

- All research variable data are normally distributed
- The research model is free from symptoms of autocorrelation, multicollinearity and heteroscedasticity

4.4. Hypothesis testing
The results of hypothesis testing are as follows:

Table 3. Model Summary

Model	R	R Square	Adjusted R Square	STD. Error of the Estimate
1	.914[a]	.836	.834	4.005

a. Perdictors: (Constant), DDE,PI

Table 4. ANOVA[b]

Model	Sum of Square	df	Mean Square	F	Sig.
1 Regression	12019.218	2	6009.609	374.600	.000[a]
Residual	2358.282	147	16.043		
Total	14377.500	149			

a. Predictors : (Constant), DDE, PI
b. Dependent Variable: DE

Table 5. Coefficients[a]

Model	Unstandardized Coefficients		Standardized Coefficients	t	Sig.
	B	Std. Error	Beta		
1 (Constant)	1.254	1.974		.635	.526
PI	.782	.116	.476	6.757	.000
DDE	.317	.048	.476	6.634	.000

a. Dependent Variable: DE

Hypothesis test results show that:

- Simultaneously infrastructure development and environmental carrying capacity have a significant effect on economic impacts, with sig. F = 0.000 which is smaller than 0.05
- Partially infrastructure development has a significant effect on economic impact with the Sig. t = 0.000 which is smaller than 0.05. Thus H1: proven
- Partially, the ecological carrying capacity has a significant effect on economic impact with the Sig. t = 0.000 which is smaller than 0.05. Thus H2: proven
- Both the independent variables of infrastructure development and the carrying capacity of the economy can explain the dependent variable of the economic impact of 83.6% (0.836)

5. Discussion

The results of the above analysis are in line with Maqin's [10] research which states that infrastructure development has a positive and significant influence on economic growth. The results of the above analysis are also in line with the research of Tian and Wang [8] and Qian et al [9] which state that the carrying capacity of resources and the environment is the foundation of sustainable development, and the carrying capacity of the land explains whether local land resources are effectively used to support economic activities and / or human population.

The new growth theory incorporates infrastructure as an input in influencing aggregate output and is also a possible source of increasing the limits of technological advances obtained from the emergence of externalities in infrastructure development. Infrastructure externalities affect production activities by providing accessibility, convenience and the possibility of more productive production activities. Thus it can be said that public capital and private capital are complementary in production, and infrastructure as part of public capital has an important role in economic growth.

Efforts to reunite economics and environmental have important meaning in the effort to realize sustainable development. The process of integrating both is through the formulation of paradigms and policy directions that rely on partnerships and the participation of development actors in managing resources as optimally as possible. In order for the development to be carried out to grow the economy without causing a lot of environmental damage, the concept of sustainable development is needed, namely the concept of economic growth which continues to maintain the natural resources used, as well as strategies for environmental integration into economic development.

6. Conclusion

Infrastructure development and ecological carrying capacity partially and simultaneously have a significant effect on the economic impact of the Independent Integrated City of *Lunang Silaut*, West Sumatra. It is strongly recommended that economic development in integrated areas such as in *Lunang Silaut* requires the development of measurable infrastructure and prioritizing the carrying capacity of the ecological environment for sustainable development.

References

[1] Depnakertrans (Dinas Tenaga Kerja dan Transmigrasi) 2006 Paradigma Baru Pengembangan Kawasan Transmigrasi Terpadu Menuju Terbentuknya Kota Mandiri Terpadu

[2] Kalsum E and Caesariadi T W 2016 Konsep Permukiman Kota Terpadu Mandiri *Langkau Betang Jurnal Arsitektur* **3**(2) 12-24

[3] Vaughan R J and Pollard R 1983 *Rebuilding America* Council of State Planning Agencies

[4] Kementerian Lingkungan Hidup 2014 Pedoman Penentuan Daya Dukung dan Daya Tampung Lingkungan Hidup (Jakarta)

[5] Suratmo and Gunawan 2004 *Analisis Mengenai Dampak Lingkungan* (Gadjah Mada University Press)

[6] Asriati N 2015 Pengembangan Kawasan Terpadu Mandiri dengan Pendekatan Model One Village One Product (Ovop) Daerah Transmigrasi Rasau Jaya *Jurnal Economia* **11**(1) 72-88

[7] Widiatmaka W A, Sutomo B, Hamzah U and Warsono P 2014 Perancangan Tataguna Lahan dan Tata Ruang Kawasan Perkotaan Berbasis Pertanian: Studi Kasus Kota Terpadu Mandiri Transmigrasi Mesuji, Provinsi Lampung *Prosiding Seminar Nasional Multidisiplin Ilmu, Universitas Budi Luhur, Jakarta. Maret.*

[8] Tian Y N and Wang H Q 2013 Progress of Resources and Environmental Carrying Capacity *Journal of Clean Energy Technologies* **1**(2) 132-135

[9] Qian Y, Tang L, Qiu Q, Xu T, and Liao J 2015 A Comparative Analysis on Assessment of Land Carrying Capacity with Ecological Footprint Analysis and Index System Method *PloS one* **10**(6) e0130315

[10] Maqin A 2014 Pengaruh Kondisi Infrastruktur Terhadap Pertumbuhan Ekonomi Di Jawa Barat *Trikonomika Journal* **10**(1) 10-18

The role of natural science courses to implement the environmental education in elementary school (curriculum 2013)

E Titiek Winanti*, Indiah Kustini and Erina Rahmadiyanti

Faculty of Engineering, Department of Civil Engineering, Universitas Negeri Surabaya, Ketintang, Surabaya, Indonesia.

*titiekwinanti@unesa.ac.id

Abstract. Human behaviour is strongly influenced by prerequisite knowledge, residential, working place, and schools/studying place situation as well as friend or companion's behaviour. Soldiers have different behaviours compared to educators, doctors, farmers, traders, and so forth. To make students become environmentally friendly-human beings, they need to have knowledge on how to protect the environment. Since 1986, the Republic of Indonesia government has included environmental education into school curriculum; and the end is the 2013 curriculum (C-13). There are eight subject courses in elementary school's 2013 curriculum. Of the eight subjects, Natural Science is relevant subjects as a means to implement the environmental education (EE) concepts. Natural Science is taught in grades IV, V, and VI using thematic approaches for 9 x 40 minutes a week. However, the implementation of C-13 in elementary schools is still less optimal because there are still many prerequisite factors to be fulfilled, such as: infrastructure, learning resources, teachers' skills, and assessment systems. In conclusion, Natural Science subjects has a significant role to debrief the environmental education concepts especially in elementary school. Thus, learning model and media of natural science need to be further developed.

1. Introduction

Environmental education (EE) content has been the concern of the Government of the Republic of Indonesia that needs to be included in the school curriculum since 1986. In the beginning, EE content was unclear whether it should be school course or impart in the existing courses. The study conducted by Winanti [1] reported that EE content could be integrated and developed in several subjects, namely: Citizenship Education, Handicrafts, Natural Science, and Social Sciences in the 1994 curriculum. However, in the 2006 curriculum of elementary school, EE content was covered or integrated in seven subjects: Religious Education, Citizenship Education, Indonesian Language, Social Sciences, Cultural arts and skills, as well as Physical and Sports Education, with a total of more than 60 basic competencies [2,3].

In the 2013 curriculum of elementary school, there are eight courses which are divided into two groups. Group A consists of subjects that provide competency on cognitive and affective aspects, namely: religious education and character, Pancasila education, and citizenship education, Indonesian Language, Mathematics, Natural Sciences, and Social Sciences. Group B consists of subjects that emphasize on affective and psychomotor aspects, namely: cultural arts and crafts, physical and physical education and health. Among those eight, only Natural Science and Social Science are given starting in grade IV [4].

The elementary school curriculum employs an integrative thematic learning approach from grade I to VI. Integrative thematic learning is a learning approach that integrates competencies from various subjects into themes. Integration is carried out in two ways, namely: the integration of attitudes, skills and knowledge in learning process, and the integration of various related basic concepts. The theme connects various basic concepts so that students will not learn those concepts partially. Therefore, learning provides full meaning to students as reflected in the various themes available. From a psychological point of view, lower class students (I, II, III) have not been able to think abstractly to understand the content of separate subjects, different from the upper classes (IV, V, and VI) who have begun to be able to think abstractly. Therefore, Natural Science is given in grades IV to VI.

The objectives of the 2013 curriculum include four competencies, namely (1) spiritual attitude competence, (2) social attitudes, (3) knowledge, and (4) skills. These competencies are achieved through the intracurricular, cocurricular, and/or extracurricular learning process. Formulation of the competency of spiritual attitudes, namely: respecting and living the teachings of the religion that adhered to the formulation of social attitude competencies is to show honest, disciplined, responsible, polite, caring, and confident behaviour in interacting with family, friends, teachers, and neighbours. Both of these competencies were achieved through indirect learning, namely exemplary, habituation, and school culture, taking into account the characteristics of subjects and the needs and conditions of students.

The third core competency of knowledge and the fourth core competency of skills include (1) understanding factual knowledge by observing and asking questions based on curiosity about him, God's creatures and activities, and objects found at home, at school and place play; as well as (2) presenting factual knowledge in clear, systematic and logical language, in aesthetically pleasing works, in movements reflecting healthy children, and in actions reflecting a children' faithful behaviour.

In the 2013 curriculum, EE aim is to create people aware with their respective responsibilities to protect and develop the environment. It is expected to make them behave and act according to healthy environmental norms and have the initiative to participate in environmental protection efforts at the local, national and international levels.

The objectives of environmental education are: a) to develop a world population that has awareness and concern for the environment including various related problems and b) to develop a world population that has skills, attitudes, motivation, and commitment to find solutions both individually and collectively to existing environmental problems and prevent new arising problems [5].

2. Method

The 2013 curriculum employs thematic learning approach meaning that each subject is divided into several themes and subthemes. To find the role of Natural Science (NS) subject in Environmental Education (EE), a study of each theme and each subtheme found that EE content were covered in NS subject at grade IV, V, and VI in elementary school (see Table 1 and Table 2).

Table 1. Grouping on Theme and Subtheme of Natural Science at Grade IV

Theme	Sub-theme
1: The beauty of togetherness	1: Diversity of My Nation Culture 2: Togetherness in Diversity 3: Be grateful for diversity 4: Proud of my culture
2 : Always Save Energy	*1 : Kind of Energy sources* *2 : Energy Usage* 3: Motion and Style

3: Cares of Living creatures	*1: Animals and plants in my home environment*
	2: Diversity of sentient beings in my environment
	3: Let's Love the Environment
	4: Living things around us
4: Kind of Occupations	1: Types of Work
	2: Goods and Services
	3: My Parent's Work
	4: Work Around Me `
5: My hero	1: The struggle of the heroes
	2: My hero, my pride
6: Beauty of my country	*1: Diversity of animals and plants*
7. My Dreams	1: Me and my dreams
	2: My dream is great
8: My home	*1: The uniqueness of my home*
9: My healthy and nutritious foods	*1: My healthy and nutritious foods*

Note: Subjects with Environmental Education (EE) content are written italic
Sources: [6]

Table 1 shows that Natural Science (NS) subjects at grade IV has nine themes and 22 subthemes. Four of nine themes (3, 6, 8, and 9) have EE contents in all subthemes, while in theme two, two of three subthemes consist of EE content. Table 1 also shows that five of nine themes (55%) and nine of 22 subthemes (41%) on NS subject at grade IV consists of EE contents.

Similar to grade IV, at grade V, NS subjects has nine themes and 27 subthemes (see Table 2). Six of nine themes (2, 3, 4, 7, 8, and 9) have EE contents in all subthemes, while in theme 1, one of three subthemes have EE content. Table 2 shows that in Natural Science subjects at grade V, seven of nine themes (77%) have Environmental Education (EE) contents, and 19 of 27 subthemes (70%) have Environmental Education (EE) contents as well. Thus, more than half of Natural Science subjects in elementary school level has Environmental Education (EE) basis. If government wants to implement EE further, then the EE contents on Natural Science subjects need to be intensified.

Table 2. Grouping on Theme and Subtheme of Natural Science at Grade V

Theme	Subtheme
	1: The form of objects and their characteristics
1: Objects in the neighborhood	2: Changes in the objects' form
	3: Human and Environment
2 : Events in life	*1: Kind of events in life*
	2: Important events
	3: Human and Natural event
3: Harmony in society	*1: Types of harmony*
	2: Advantages of living in harmony
	3: How to maintain harmony
4: Health is important	*1: The importance of individual and environmental health*
.	*2: Healthy lifestyle*
	3: Healthy environment
5: Proud to be Indonesian	1: My Indonesia, a rich nation
	2: My Indonesia, a cultural nation
	3: My Indonesia, a nation that love peace.

6: Human and animals' organs	1: Human Body 2: Human and Animal Organs 3: The Way of Life for Humans, Animals and Plants
7: History of Indonesian civilization	*1: Islamic Kingdoms in Indonesia* *2: The ruins and heritages of Islamic Kingdom in Indonesia* *3: Preserving the heritages of Islamic kingdom*
8: Ecosystem	*1: Ecosystem components* *2: The relationship between living creatures and ecosystem* *3: Preserving ecosystem*
9: Environment is our friend	*1: Human and Environment 1* *2: Environmental Changing* *3: Environmental Preservation*

Note: Subjects with Environmental Education (EE) content are written italic
Source: [6]

3. Analysis

Study conducted by Titiek Winanti et al. [7] on the implementation assessment of the 2013 curriculum (C-13) in elementary schools in Surabaya indicated that C-13 has been applied at elementary school in Surabaya, although not yet comprehensive. Study also indicated that more than 50% of elementary teachers in Surabaya implemented the C-13. As explained previously, C-13 is good to create human beings to be spiritual meaning that they implement the religion teachings, being social creature, honest, disciplined, responsible, polite, caring, and confident in interacting with family, friends, teachers, and neighbour. However, it should be noted that the current condition of the school is less optimal for running C-13 according to its ideals. Some important things that need priority are: (1) completing the provision of facilities/infrastructure to support learning, (2) providing books and resources other learning is still minimal, (3) the need for training for many teachers who do not understand the various learning tools in C-13, (4) it needs an improvement in the assessment system so teachers will be easily making evaluation.

Regarding the facts, it is important to remember that EE is a cross-curriculum approach to learning that helps individuals or groups of individuals to understand the environment. The main goal is to build a caring attitude and responsibility that will foster the desire and ability to act responsibility in the environment. Environmental Education is not only related to the environment but also with instincts and feelings, attitudes, skills, and relevant social actions [8].

Environmental Education is an effort to prepare humans for their lives as members/parts of the biosphere. This education is to understand, appreciate, cooperate, and support environmental systems comprehensively. Environmental Education is an essential education on philosophical problem solving for continuity, improvement, and guarding. The goal is not only to solve the problem with a narrow focus which worsen the problems as well as correct and restore a status quo, but also to make things better. Because EE is implemented as a process of reorganizing values and for explaining concepts to develop various skills and attitudes to understand and appreciate the interrelationship between humans and their socio-cultural environment as well as biophysical environment [9].

Environmental Education is a way to implement and achieve environmental protection goals. It is not a branch of science or a subject of independent study. The educational process on EE should be carried out according to the principle of lifelong education integrally. EE involves teaching about value and the ability to think clearly about complex environmental problems bearing political, economic, philosophical, and technical characteristics. [10]

The fundamental goal of Environmental Education is a constructive attitude towards the environment in a philosophical and pragmatic sense, which should be the basis for every person in thinking and acting. Environmental Education is an educational process about the relationship between humans and the natural environment and built environment, including human relations with pollution, allocation and management of natural resources, conservation, transportation, technology, urban and rural planning. The resolution of the Belgrade International Conference on Environmental Education (1975) mentioned that there were six environmental education targets, namely, awareness, knowledge, attitude, skills, evaluationability, and participation [11].

4. Conclusion

To answer the challenge given by Environmental Education, it is important to strengthen and develop the learning process, especially in natural science subjects in elementary schools.

For grade IV, themes and subthemes in Natural Science subjects that need to be strengthened are: invitation for energy saving in every activity, awareness of all living creatures, responsibility to maintain the beauty of the country including neighborhood, healthy and nutritious foods consumption.

For grade V, themes and subthemes that need to be added or strengthened are: invitation to foster love for environment by recognizing the nature and potential of any surrounding objects, responsiveness to symptoms and events in life, awareness to create harmony in society, efforts to maintain health, cultural and historical understanding of Indonesian civilization, ecosystem balance to create a friendly environment.

As for grade VI, themes and subthemes that need to be improved are: concrete examples on how to realize ecosystem balance by maintaining forest integrity, no overlogging (either legally or illegally), efforts to prevent hunting of endangered animals and protected animals, chemical fertilizers reduction by changing to organic fertilizers, regulations compliance for wastewater treatment.

There are ways to preserve the population of rare animals and vegetation such as protecting their living place (habitat), breeding, prohibiting wild animals' games, promoting the quality of animals' and vegetations' habitat by creating wildlife sanctuaries, national parks, great forest parks, etc [12]. By strengthening several themes as described as well as added with optimal learning facilities and resources, teachers' training, and assessment system improvement, achieving goals on environmental education are possibly realized.

References

[1] Winanti, Titiek. 2001. *Review of environmental education content in the curriculum of Elementary School Year 1994.* Research Centre Surabaya State University: Surabaya.
[2] Basic Curriculum for Elementary School Year 2006
[3] Winanti, Titiek. 2008. *Review of environmental education content in the curriculum of Elementary School Year 2006.* Research Centre Surabaya State University: Surabaya.
[4] Basic Curriculum for Elementary School Year 2013
[5] The Ministry of Education and Culture of the Republic of Indonesia. 1997. *Guidelines for the implementation of the education of the population and the environment.* Project PKLH year 1997/1998
[6] Teacher and Student Textbook for Elementary School Grade IV and V. 2013. The Ministry of Education and Culture of the Republic of Indonesia.
[7] Winanti, Titiek. 2017. *Evaluation of implementation curriculum year 2013 at Elementary School in Surabaya.* Research Centre Surabaya State University: Surabaya.
[8] Rindawati, 2000. *Understanding of elementary school students in the city of Surabaya against the concept of the environment education.* Research Centre Surabaya State University: Surabaya.
[9] Fien, I (Eds.), 1995. *Teaching for Sustainable World: Environment and Development Education Project for Teacher Education.* AIDAP-AAEE Inc., Griffith University: Brisbane, Australia.
[10] The Law of the Republic of Indonesia No. 20 Year 2003 on Education System in Indonesia
[11] Sutaryono, Retno. 1999. Application of environmental education at the level of secondary school. *National Seminar on Environmental Education for Junior High Schools.* JICA-The Ministry of Environment, Environmental impact Control Agency- Ministry of Forestry and

Plantations-University of Indonesia: Jakarta
[12] The Law of the Republic of Indonesia No. 32 Year 2009 on Environmental Protection and Management.

The effect of contextual teaching and learning approach and motivation of learning on the ability of understanding the mathematics concepts of grade V student

Yuvita Rama Yeni, Hendra Syarifuddin and Riska Ahmad

Basic Education Program Postgraduate, Padang State University, Prof. Dr. Hamka street, Air Tawar Campus, Padang, West Sumatera, Indonesia 25131

*vita17juni1989@gmail.com

Abstract. The purpose of this study was to obtain information about the effect of the *CTL* approach and learning motivation on the ability to understand mathematical concepts in class V students. This research is a quasy type experimental research. The population is all students in class V SDN Gugus I Kecamatan Tanjung Harapan Solok City and the sample is 57 students in grade V SDN 01 Tanjung Paku. This sampling technique uses randomized sampling. Research data is obtained from the ability to understand mathematical concepts and questionnaires. The results of the study showed that: (1) The ability to understand mathematical concepts of students participating in learning with the *Contextual Teaching and Learning* approach was better than using conventional learning; (2) the ability to understand the mathematical concepts of students who have high learning motivation who follow learning with the *Contextual Teaching and Learning* approach is better than using conventional learning; (3) The ability to understand mathematical concepts of students who have low learning motivation who follow learning with the *Contextual Teaching and Learning* approach is better than using conventional learning; (4) There is no significant interaction between the *Contextual Teaching and Learning* approach *and* students 'learning motivation on the ability to understand students' mathematical concepts.

1. Introduction

Given the importance of understanding students' mathematical concepts in learning, it is necessary to have good learning motivation so that students are diligent in facing the task, resilient in the face of difficulties, happy in finding and solving mathematical problems and can achieve the desired learning goals.

While extrinsic motivation is a factor that comes from outside the individual but influences the willingness to learn. Based on this, it can be said that motivation and the ability to understand mathematical concepts is one of the important abilities in mathematics learning [1]

From the statement above, the lack of understanding of mathematical concepts has an impact on students' low motivation to learn mathematics. Which is an indicator of learning motivation are: 1) The existence of encouragement and learning needs; 2) Demonstrate attention and interest in the assignment given; 3) Persevering with the task; 4) Ductile faces difficulties; 5) Successful desires and desires [2].

From these indicators it has not been seen from within the students so that there are still many students who have not issued ideas or ideas because they are still doubtful of their own abilities in the lack of understanding of the concept of matter, and students still experience difficulties in working on questions from the material, so students feel mathematics as a difficult, unpleasant subject, even a frightening specter, and low student learning outcomes are influenced by the lack of learning motivation of students. This can be seen by the presence of some students who are sleepy, daydreaming, looking out of class, lazy to do assignments, and not happy to solve math learning problems. Other factors that influence mathematics learning outcomes are caused by the teacher when learning takes place still using conventional learning models. So, the need to use various models, approaches, and methods in the teaching and learning process that suits the needs of students in the classroom, so that the learning process is not monotonous and boring. In learning, especially mathematics learning the role of the teacher is very large and really challenged to choose a learning approach that can support the achievement of mathematics learning goals. And, the learning process can be said to be quality and effective if the learning process can encourage students to be more active and motivated in learning so as to solve various problems. Therefore, the teacher must strive to present learning that allows students to solve problems in their daily lives, especially those related to mathematics learning. Motivation to learn has an important role in improving learning outcomes for each student, because if the motivation is low then there is no interest in the subject in this subject of mathematics. Learning motivation can be grown from within the students themselves or through encouragement by the teacher. One of the efforts that can be done by the teacher to improve student learning motivation is by choosing the right learning model or approach. The choice of learning model or approach must be adjusted to the material that will be delivered, if the model or approach that is applied is not in accordance with the level of difficulty of the teaching material, it will cause dissonance in the learning process.

One approach to learning that can be done to improve student motivation and ability to understand students' concepts is to use the *Contextual Teaching and Learning (CTL)* approach . Based on several studies found that: 1) the *CTL* approach is appropriate to be used as a learning guide and can develop character and improve learning outcomes (Setyorini; 2014); 2) the use of integer addition and subtraction learning modules with the *CTL* approach can improve concept understanding of fourth grade students [3];

So from the results of the research described above, it can be concluded that the learning process using the *CTL* approach can improve student learning outcomes, this indicates that the *CTL* approach can increase student motivation so that the understanding of students' mathematical concepts also increases.

This is because the *Contextual Teaching and Learning* approach is a conception that helps teachers associate subject content with real-world situations, and motivates students to make connections between knowledge and its application in life [4]

The above definition illustrates that the *CTL* approach is the concept of learning in where the teacher brings the real world into the classroom and encourages students to make connections between the knowledge they have with their application in their daily lives. In the *CTL* approach the task of the teacher is to provide ease of learning to students by providing various means and adequate learning resources. Teachers not only convey learning material in the form of memorization but set the environment and learning strategies. A conducive learning environment is very important and supports the *CTL* approach .

Of the seven components of *CTL* approach *learning* also has several advantages to the ability to understand students 'mathematical concepts and students' learning motivation that is more meaningful learning, meaning students do their own activities related to the material so that students can understand it themselves. Learning is more productive and able to foster understanding of concepts to students because the *CTL* approach requires students to find themselves instead of memorizing subject matter. Increase student motivation to express opinions about the material being studied. Growing curiosity about the material learned by asking the teacher. Develop the ability to work with other

IOP Conf. Series: Earth and Environmental Science **314** (2019) 012064 doi:10.1088/1755-1315/314/1/012064

friends to solve existing problems. Students can make their own conclusions from learning activities. [5].

The above explanation shows that the *CTL* approach can provide opportunities for students to actively learn in understanding and deepening the learning material as a whole. So the researcher will carry out this research with the title: "The Effect of *Contextual Teaching and Learning* Approaches *and Learning* Motivation on the Understanding Ability of Class V Students' Mathematical Concepts at SDN 01 Tanjung Paku, Tanjung Harapan Subdistrict, Solok City".

This research can be a consideration of the use of approaches *CTL* for other material or subjects and it is necessary to review the initial knowledge of students so that they are not considered the same for all conditions. This research is also expected to be used as a variety of learning strategies for teachers to empower the students' ability in understanding concepts in learning.

2. Experimental

The research used in this study is quasi-experiment which aims to investigate the influence of the approach *CTL*. This research was carried out at SDN 01 Tanjung Paku, Solok City in the even semester of 2017/2018 school year which took place on April 2 to 30, 2018 . The population of this study were all students in the school of SDN 01 Tanjung Paku while the sample of this study was the fifth grade students taking two classes as many as 57 people consisting of 29 students in the VA class as the experimental class and 28 students in the VB class as the class control. Sampling by total sampling because it has a homogeneous member or element. The independent variable in this study is *contextual teaching and learning* and conventional learning , and the bound variable is the ability to understand mathematical concepts , while the moderator variable is learning motivation . The research design used in this research is design *Post-Test Only factorial control group 2 x 2* . The instrument used to obtain data in this study is a test of the ability to understand mathematical concepts and learning motivation of students in learning mathematics . The instrument was assessed by a competent validator which included lecturers and teachers at the school and was tested against respondents outside the sample class.

3. Results and Discussion

This study focused on obtaining the average score of the ability to understand students' mathematical concepts . The following is presented the results of the formalities and homogeneity testing data after the research :

Table 1. Tests for Homogeneity Test for Experimental and Control Classes

Class	N	S^2	Count	Ft(α=0,05)	Information
Experiment	29	80,073	2,46	2,54	Homogeneous
Control	28	197,315			

Table 2. Testing for Experiment and Control Class Normality Tests

Class	N	α	L_0	L_t	Information
Exsperiment	29	0,05	0,137	0,1634	Normal
Control	28	0,05	0,146	0,1658	Normal

From tables 1 and 2, both data are homogeneous and normal, said to be homogeneous because $F_{count} < F_{table}$ (2.46 < 2,54 and data dikata ka n normal in the experimental class for $_{observation}$ $L < L_{table}$ (0, 137 < 0.1634) while in the control class there is data $L_{observation} < L_{table}$ (0.146 < 0.1658). As for the acyl h dinalisis posttest learners based on their respective capabilities, an outline of a picture obtained as in the following table:

Table 3. Data on Average Score of Ability to Understand Students' Mathematical Concepts in Experimental and Control Classes

Results	Class	N	Motivation to Learn	$\bar{x}$	X min	X mak	S
Postest	Exsperiment	7	Low	74,917	60,3	85,3	8,93
		22	High	79,035	65,7	95,1	8,93
		29	Total	78,04	60,3	95,1	8,95
	Control	10	Low	67,235	40,7	87,9	16,89
		18	High	71,535	45,7	85,7	12,47
		28	Total	69,9	40,7	87,9	14,05

From Table 3 shows the average results of the ability to understand mathematical concepts in the experimental class students is higher (78.04) than the control class (69.9) after being given treatment. This happens because students in the experimental class are able to associate the old knowledge with the newly acquired after being given treatment. Whereas in the control class students are still less able to associate the old knowledge with the new ones obtained after being given treatment. Calculation of variance and standard deviation aims to determine the diversity of a data group. Per standard deviation count from the above bell also reveals that the variance or diversity of the experimental group is different from the control class at the posttest. In experiments a little variant and control class have more variants.

4. Hypothesis testing

4.1. The first hypothesis

Testing the first hypothesis is done by using the t test because the data is normally distributed and has a homogeneous variance. The combined variance of the two samples is 11.73 for the real level α = 0.05 and 55, so that the $t_{calculated}$ is 2.53. Being right t_{table} obtained was 1,673. Because t_{count} greater than t_{table}, then Ho is rejected and H_{1} is accepted. So it can be concluded that the ability to understand mathematical concepts of students who use the *CTL* approach is higher than students who use conventional learning.

4.2. The second hypothesis

The second hypothesis testing is done using t test because the data is normally distributed and has a homogeneous variance. the combined variance of the two samples is 10.66 for the real level α = 0.05 and 38, so the resulting t_{count} is 2.199. While the t_{table} obtained is 1.69. Because t_{count} greater than t_{table} then Ho is rejected and H_{1} is accepted. So it can be concluded that the ability to understand mathematical concepts of students who have high learning motivation using the *CTL* approach is higher than students who have high learning motivation who use conventional learning.

4.3. The third hypothesis

The third hypothesis testing is done using t test because the data is normally distributed and has a homogeneous variance. the combined variance of the two samples is 14.25 for the real level α = 0.05 and 15, so that the $t_{calculated}$ is 1.80. While the t_{table} obtained is 1.75. Because t_{count} greater than t_{table}, then Ho is rejected and H_{1} is accepted. So it can be concluded that the ability to understand mathematical concepts of students who have low learning motivation using the *CTL* approach is higher than students who have low learning motivation who use conventional learning.

4.4. The fourth hypothesis

To find out how the interaction of students between experimental classes and the control class. Based on the summary of the results of the two path ANAVA calculations above can be stated that: 1) The alternative hypothesis (Ha) states that there are differences in the effect of the learning approach between groups with CTL approaches and conventionally accepted, because the calculation results show that Fcount = 6.63> Ftable = 2.78. Complete calculation results can be seen in Appendix 11. 2) Alternative hypothesis (Hb) which states that there are differences in the influence of high and low motivation in the results of the ability to understand students' mathematical concepts, because the calculation results show that Fcount = 2.88> Ftable = 2.78 . The results of the complete calculation can be seen in Appendix 11. The research hypothesis proves that there is no influence of interaction between learning approaches and learning motivation on the results of students' ability to understand mathematical concepts in mathematics learning.

Based on the fourth hypothesis test results obtained there is no interaction between approaches *CTL* with *learning* motivation towards the ability to understand mathematical concepts . The absence of these interactions can be seen from the average test results of the ability to understand mathematical concepts through approaches *CTL* and conventional learning. If approach *CTL* and conventional learning are applied, then the average test results of the ability to understand the mathematical concepts of students taught by approach *CTL* have better learning motivation compared to students taught with conventional learning. When treated with an approach *CTL* , the average obtained tends to be higher than conventional learning. This can be seen in Figure 1 below:

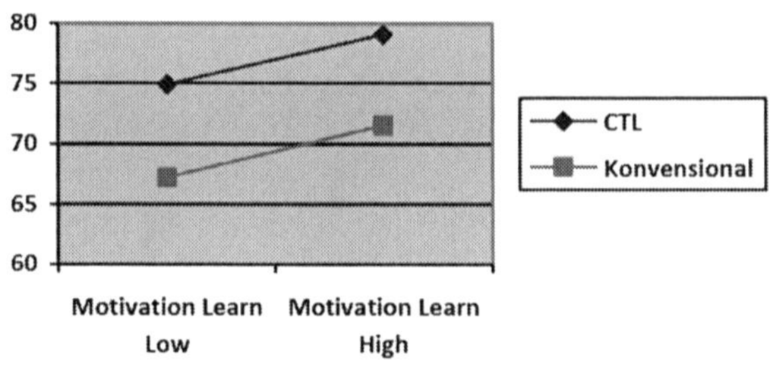

Figure 1. Interaction Graph between Contextual Teaching and Learning Approach and Motivation of Learning on the Ability of Understanding the Mathematics Concepts

Based on Figure 3 , it can be seen that point A is the students with low learning motivation found in the experimental class with an average of 7 4.92 , while point B is the students with high learning motivation who are in the experimental class with an average of 79.04 , Point C is students with low learning motivation in the control class with an average of 67.24 , while point D is students with high learning motivation in the control class with an average of 7 1.54 . So the AB line with the CD line does not intersect. Explains that the presence or absence of interaction can be predicted from the graph of the independent variable profile. If the profiles of the first and second independent variables do not intersect, there is likely to be no interaction between the two variables. Because lines are parallel AB and CD, factors have no interaction.

Social interaction that occurs in groups can improve concept comprehension skills because children who feel less intelligent will be helped and children who feel smart will try to explain to their less intelligent friends so that their understanding becomes more profound. One of the factors that most influence the ability to understand students' concepts is the *CTL* learning approach . approach *CTL is* carried out on 7 components . In the first component is constructivism, in which the teacher develops thinking students will learn more by working on their own, finding themselves, and constructing their own new knowledge and skills . Both found, where the learning process was built in the search process, the process of moving from observation became understanding for students . All three ask, which teacher develops students' curiosity through interactive dialogue through question and answer by all elements involved in the learning community. Fourth learning society, where k raft learning community is learning the findings resulting from the collaboration with others. Teachers in Contextual Learning (*CTL*) always carry out learning in groups whose members are heterogeneous students who are good at teaching the weak, who already know to tell the unknown, and so on. Fifth modeling, in which d nature of vocational learning and specific knowledge, necessary to have a model that can be replicated by students. The model in this case can be various ways, how to throw or kick a ball in sports, how to recite in a foreign language, or the teacher gives an example of how to do something. Sixth reflection, which is an effort to see, organize, analyze, clarify, and things that have been learned. Practice in the classroom at the end of each lesson, namely the way the teacher leaves time to provide opportunities for students to reflect: students 'direct statements about what happens after learning, notes or journals in the student book, students' impressions and suggestions on day learning that, discussion, and work. The seven authentic assessments, namely the achievement of students is not enough just with the test, the learning outcomes are measured by authentic assessments that can provide true and accurate information about what students really know and can do or about the quality of educational programs. At the end of the lesson the teacher will give reinforcement so that the child does not misunderstand. This matter is inversely proportional to conventional learning where the teacher gives the material directly from the beginning of learning until the end of the lesson. Based on the above description of these two lessons there are differences in the process of knowledge formation by the teacher. This difference is considered to support the results of research that shows that the approach *CTL* effective for the ability to understand students' concepts rather than conventional learning. [6] The Impact of Motivation on Student's Academic Achievement and Learning Outcomes in Mathematics among Secondary School Students in Nigeria.

"Both classes are given a test in the form of a test of the ability to understand mathematical concepts" [7]. In this activity, students get the opportunity to empower their understanding of the concept . According to [8], "the intended understanding is students' understanding of the qualitative basis everywhere the facts are interrelated with their ability to use that knowledge in new situations" . Understanding is a process of thinking, and learning, said to be because the direction of understanding needs to be followed by learning and thinking. Understanding is a process, action, and way of understanding. Understanding according to [9] is encompassing the ability to capture meaningful meanings of material learned. Furthermore, Bloom's taxonomy states that understanding is the lowest level in the cognitive aspects related to mastery or understanding of something. [10].

5. Summary

The *CTL* approach can influence the ability to understand students' mathematical concepts. First, understanding the mathematical concepts of students who follow learning with the *Contextual Teaching and Learning* approach is better than using conventional learning. Second, understanding the mathematical concept of students who have learning motivation who follow learning with the *Contextual Teaching and Learning* approach is better than the learning motivation of students who use conventional learning. Third, there is no significant interaction between the *Contextual Teaching and Learning* approach *and* students 'learning motivation on the ability to understand students' mathematical concepts.

6. Suggestion

For guru expected to approach *Contextual Teaching and Learning* in the learning process. Because it is proven to improve the ability to understand the mathematical concepts of students in problem solving.

This research only examines initial knowledge and critical thinking skills. For that reason, it is suggested to further researchers to examine other aspects, such as creativity, intelligence and others.

For other researchers who are interested in conducting research on the influence of the *Contextual Teaching and Learning* approach, *in* order to be able to examine the influence of the learning approach on the ability to understand concepts and other subjects.

References

[1] Sari I N 2015 Experimental Cooperative Learning Model of TAI Type Assisted by Adobe Flash Learning in the Main Material of Quadrats Viewed from the Learning Motivation of

Students of State Junior Hight Schools in Sukaharja Regency in the 2013/2014 Academic Year *Journal of Electronic Mathematics Learning* **3**(1) :113-123

[2] Lestari K E2015 *Penelitian Pendidikan Matematika* (Bandung : Refika Aditama)

[3] Dyah T W and Raddin N S 2017The Use of Addition and Reduction of Round Numbers Learning Module with CTL (Contextual Teaching and Learning) Approach to Improve Understanding of Concepts of Fourth Grade Elementary School Students". *Journal of Education* **2**(1) : 8-13 e-ISSN: 2527-6891

[4] Trianto 2015 *Designing Innovative, Progressive, and Contextual Learning Models* 9 (Jakarta: PT Kharisma Putra Utama)

[5] Ahmad R 2016 Information Service Effectiveness by Using *Contextual Teaching and Learning* Approaches in Reducing Students' Attitudes towards Hedonism Lifestyle *Indonesian Journal of Educational Research Print ISSN: 2477-8524-ISSN Online: 2502-8103* **2**(1) 1-11

[6] Adedeji T 2007The Impact of Motivation on Student's Academic Achievement and Learning Outcomes in Mathematics among Secondary Students in Nigeria *Eurasia Journal of Mathematics, Science and Technology Education* 2007 **3** (2) 149-156 ISSN: 1305-8223

[7] Syarifuddin H 2018 Effectiveness of Contextual Learning with REACT Strategies on Understanding Mathematical Concepts *Journal of Mathematics Education* **7**(1) pp. 72-77

[8] Trianto 2009 *Constructivist-Oriented Innovative Learning Models* (Jakarta: Achievement of Publisher Library)

[9] Winkel W S 1996 *Psikologi Pendidikan* (Jakarta : Grasindo)

[10] Suherman E 2003 *Contemporary Mathematics Learning Strategy* (Bandung: Indonesian Education University)

[11] Susanto A 2013 *Learning and Learning Theory in Primary Schools* (Jakarta: Kencana)

[12] Tilaar A L F2014 Effect of Cooperative Learning Type of Team Assisted Individualization (TAI) and The Performance Assessment of Learning of Achievement to Linear Program Course *International Journal of Science and Engineering Investigations* **3**(24) No: 25-29

[13] Yannuansa N 2015 The Influence of Cooperative Learning Model Type TAI with Contextual Approach on the Subject of Building a Flat Side in Terms of the Learning Style of the Eighth Grade Students of Kediri Middle School *Journal of Electronic Mathematics Learning* **3**(3): 294-304

IOP Conf. Series: Earth and Environmental Science **314** (2019) 012065 doi:10.1088/1755-1315/314/1/012065

Model of socio-economic recovery of farmers in erupted areas of mount Sinabung in Karo Regency

Hotden Leonardo Nainggolan[1,*], Albina Ginting[1], Jongkers Tampubolon[1], Johndikson Aritonang[1] and Jef Rudianto Saragih[2]

[1]Faculty of Agriculture, University of HKBP Nommensen. Jl. Sutomo No. 4A Medan North Sumatra, Indonesia.
[2]Faculty of Agriculture, Simalungun University, Pematang Siantar, North Sumatra, Indonesia.

*hotdenleonardo76@gmail.com

Abstract. This study aims to find a model of socio-economic recovery of farmers in erupted areas of Mt. Sinabung in Karo Regency. The study was conducted in Payung and Namanteran Districts with a sample of 30 respondents determined purphosically, with descriptive research methods. The result of research; a) Social recovery of farmers carried out by the government through community group development, customs are still minimal, b) Training on adaptation/ mitigation technology related to farming for the process of socio-economic recovery has not been optimal, c) There has been no use of new varieties that are tolerant of diversity that match power support land, d) There is no access to cooperation with partners for the agricultural product market. In accordance with the results of the study it is recommended; a) Efforts are needed to restore the socio-economic conditions of farmers, through empowerment to improve the ability of farmers, b) A holistic approach is needed to touch all aspects of socio-economic farmers, c) The government must provide free space for farmers to increase their socio-economic strata through activities that productive nature, d) The government must carry out a specific identification process to determine the slow process of recovery in the socio-economic area of farmers affected by eruption.

1. Introduction

Indonesia is a tropical country and has many volcanoes. This causes natural resources in Indonesia to become fertile. But the existence of the volcano can cause disasters if it erupts. Mount Sinabung is an active volcano located in Karo Regency and is one of the proud tourist attractions of North Sumatra Province. Recorded since 2010 Mount Sinabung has been erupting until now. The first eruption place on 29 August 2010 at 00.08 am, followed by a second on 30 August at 6:30 am. At approximately 4:45 am. local time on September 3, Mount Sinabung erupted again for the third time. This was the latest activity and stronger than the previous two eruptions [1].

On December 21, 2013, the Government of Indonesia issued a level 4 alert, requiring residents living within a 3 km radius of Mount Sinabung to evacuate. The level 4 alert remained in place until January 31 when nearly 14,000 people returned to their houses as volcanic activity had decreased. On September 15, 2013, Mount Sinabung experienced a violent eruption of smoke and ash, causing mass evacuations of villages surrounding the volcano. Local government officials reported that the volcano threatened more than 29 villages [2].

Between 2013 and 2014 the alert for a major event was increased with no significant activity. On 2 June 2015 the alert was again increased, and as of 26 June 2015 at least 10,000 people have been evacuated, fearing a major eruption. A pyroclastic flow in May 2016 killed 7 people [3]. Following is a short overview of activities to date by other agencies to the Sinabung volcano eruption: a) In response to the Sinabung volcano eruption, the Indonesian government through *Badan Nasional Penanggulangan Bencana* (BNPB/ National Disaster Management Agency) has allocated IDR 15 billion (CHF 1.69 million or USD 1.67 million) to deal with the internally displaced people camps in Karo Regency. This fund will be used for food and drugs as needed and also to ensure security in all affected areas. In addition to this, BNPB has distributed 1,000 blankets, 1,000 sleeping mats, and 500 family tents, b) The local government has allocated 50 tonnes of rice, 14,000 cans of fish, one water truck, 1,000 bottles of ketchup, 240 kg of instant noodles, 500 blankets, 50 family tents, 200 sleeping mats [1].

The eruption of Mount Sinabung caused a decrease in harvested area and production of food commodities such as; paddy rice, corn commodities and horticulture. Since 2012-2014 corn land decreased by an average of 14.5% and the average production decline was 12.4%. Tomato land area decreased by an average of 42% followed by an average production decline of 44.3%, as well as citrus production fell by an average of 32.4%, causing farmers' income to decline. The agricultural sector has a strategic role for the economy of Karo Regency. This area is an important commodity production center in North Sumatra, which acts as a supplier to other regions and farmers' sources of income.

The socio-economic conditions of the people affected by the eruption of Mount Sinabung are seen from four aspects, namely (i) social relations that reflect the socio cultural environment can affect the farm household system, (ii) infra structure that reflects the physical environment, (iii) farming activity and (iv) income generating through off farm employment.

Other socio-economic impacts faced by farmers after the eruption of Mount Sinabung were: a) farmers had difficulty developing their farming due to limited production factors such as; seeds, fertilizers and pesticides at a very expensive price, which greatly affects the productivity of farming; b) limited farm labor with very expensive wages, because many people have moved to a place of residence; b) the social ties of the community in the research location become tenuous and generally focus more on taking care of their respective families; c) local community social institutions that have previously developed well, but due to the eruption of Mount Sinabung it has not been taken care of and some have dispersed, because some people have been displaced and even moved to different areas, c) previous non-governmental organizations develop and play a role in helping farmers in various ways; like counseling no longer exists.

Furthermore [4] conveyed; the worst socio-economic conditions faced by the community after the eruption of Mount Sinabung were; a) the marketing network of agricultural products does not work well which causes the level of farmers' income to decrease; b) facility distributors and production factors become far from farm center locations so that the price of production inputs that farmers have to pay becomes more expensive, so that farming costs become higher, c) poverty rates increase due to declining farmer income, d) local social support systems that does not work.

The worst condition occurs in the aspect of infrastructure with a score of 39.41, thus disrupting public services in the field of education (score 25.88), health services (score 31.18) and access to clean water (score 35.29). Disruptions in socio-economic conditions that affect farming include five elements, namely (i) availability of production facilities, (ii) availability of wage labor, (iii) access to business capital credit, (iv) production price information and (v) marketing of farming product [5].

[4] Convey financial damage as a result of Sinabung eruption which can be seen from farming analysis. If before the eruption of carrot, cabbage, tomato and chili farming shows good business feasibility (the R/ C ratio between 1.43 and 4.79), but after eruption only cabbage farming is economically feasible to cultivate (R/ C ratio= 2.46) while tomato farming, carrots and chili are run in a loss condition with the R/ C ratio below 1.

Eruption disasters not only have a physical impact but have a socio-economic impact, such as; stagnation in economic growth, weakening social relations, rising poverty rates, loss of people's livelihoods [6] even disasters destroy infrastructure systems, social facilities, and other economic support facilities. The events of the Mount Sinabung eruption have a negative impact on the economy of the region and even the socio-economic conditions of the community, therefore it needs remedial action to restore economic conditions through increasing crop productivity and social recovery of local communities. This is in line with [6,7] to reduce the negative impacts of disasters and encourage sustainable life, recovery efforts are needed. The recovery phase provides an opportunity to rebuild the affected physical, social and economic systems.

[8] disaster recovery is grouped above; recovery of economic resources, individual resources, social and political resources. Individual resources that are affected can be recovered through: improving health, physical ability, psychiatry, education/ knowledge/ skills. So that people take advantage of opportunities, access resources and face difficulties after a disaster. Thus this study aims to identify the policy of social recovery of farmers affected by the eruption of Mount Sinabung in Karo Regency. Through this research formulated a collaborative model for socio-economic recovery of farmers. To achieve these objectives, it is necessary to understand the community to restore their socio-economic conditions due to the impact of the eruption, so that the programs implemented in the future are more effective, efficient and sustainable.

2. Research Metodhology

This research was conducted in the Mount Sinabung area that is Payung and Naman Teran District, Karo Regency. The location of the study was determined purphosive, where the farmers of both districts were affected by the eruption of Mount Sinabung that occurred since 2010 ago. The population of this study were farmers affected by the eruption in the Mount Sinabung area located in Naman Teran District and Payung District which amounted to 950 families. Farmers interviewed were 30 respondents who were determined purphosically from Payung Village in Payung District with a total of 15 respondents and Naman Village in Naman Teran District as many as 15 respondents who were determined purphosically.

Data analysis method used is descriptive analysis. Data analysis is done after the data is obtained from relevant parties; through interviews, focus group discussions and public hearings. This analysis is part of statistics to describe or describe data sets or observations. Descriptive analysis is used to determine the socio-economic recovery activities of farmers affected by the eruption of Mount Sinabung in Karo Regency.

3. Result and Discussion

3.1. Social recovery of farmers affected by the eruption of Mount Sinabung through aspect of social development.

[9,10] the resources needed by disaster victims in recovery depend on the use of their social networks. [11] shows strong ties between community members in the village of Tamil Madu, South India, helping communities form informal mechanisms to access logistical needs and financial assistance from outsiders. The role of government and actors outside of other community systems is important for the post-disaster recovery process [11].

In line with the research [6] that the socio-economic recovery process of the community must be fully restored, it is necessary for the role of government as a planner to synergize recovery strategies. In some cases, post-disaster recovery efforts often focus more on physical recovery such as; housing, infrastructure, while the improvement of the social conditions of affected communities is often overlooked [12,13]. This is because the physical impact can be seen in real terms, the level of loss can be calculated, and can be recovered in a relatively short period of time, while the socio-economic impact is difficult to measure, but can cause social and economic deterioration in the long term [14]; [15,6].

The role of planning is needed in regulating, synergizing recovery efforts and strategies by considering the timeframe needed and synergizing it with other development plans and effective use of resources [6]. In accordance with the results of interviews with the respondent farmers in the study location, that the government's role in recovering is not maximal, even some programs do not exist, as presented in Table 1.

Table 1 shows that social development in the context of restoring the social life of farmers is still very minimal, only 40% of respondents said that there was a government program to conduct inter-village community group/ communication coaching. While 37% of respondents conveyed farmer group development programs conducted by the government and the private sector to improve the communication between farmers in one village or between villages.

Table 1. The role of the government and private parties for the social economic recovery of farmers affected by Mount Sinabung eruption through aspect of social development.

Aspects of social development	respondents' opinions about the program of socio-economic recovery (existing conditions)/ (%)			
	not existing	existing/ non-implemented	existing / implemented	limited to socialization
Group development / communication coaching between and within the village	37%	7%	40%	13%
Kinship building between farmers	33%	20%	37%	7%
Guidance and development of cultural and cultural activities	27%	20%	47%	3%
Guidance and development of church/ mosque activities	20%	0%	60%	17%
Formation and development of village level community organizations (eg. karang taruna).	17%	7%	70%	7%

Source: Primary data, processed 2018.

There were 47% of respondents who delivered the development program of cultural/ customs activities carried out by the government had begun to run but had not been maximized. And 60% of respondents said that the Church, Mosque activity development program carried out by the government and the private sector had gone well. Then 70% of respondents conveyed the government program in the framework of forming and fostering community organizations that have started to run well.

The results of this study are in line with [10,16] who convey community development and empowerment will have an impact on increasing networking and ultimately affecting the social and economic activities of the community after a disaster and social capital will encourage the creation of collective action to utilize resources in the recovery process. Social resources can be linked to social capital [8,10]. Social capital acts as an informal mechanism that helps disaster victims to share resources, even reducing dependence on external assistance [16] so that reduced social capital in the community will limit the speed and effectiveness of the recovery process.

3.2. Economic recovery of farmers affected by Mount Sinabung eruption through farming aspects and disaster mitigation technology.

To support the post-disaster recovery process, the role and support of the government is needed especially through the distribution of resources [17]. In 2015, the Government of Karo Regency received a fund allocation IDR.190,674,100,000 through a central government grant for post-disaster rehabilitation and reconstruction activities, but the results have not been maximized and the recovery process is still ongoing.

[18] that economic recovery is only part of the broader community recovery effort, though a crucial aspect. Economic recovery provides basic necessities of life, jobs to sustain the economy in terms of income and tax revenues, and reinstatement of wealth through detailed property values. At the same time, economic recovery is not likely to be effective, or at all, unless the physical and sociopolitical fabric is repaired: a) civil orders must be restored, b) infrastructure must be repaired, c) capital markets

must be accessible, d) workers must be available and labor markets functioning, e) government services need to be available.

Based on the results of the study through interviews with respondents in the research location, that the role of the government in carrying out economic recovery based on farming aspects and disaster mitigation technology is not yet running as in Table 2.

Table 2. The role of the government in the social economic recovery of farmers affected by Mount Sinabung eruption through farming aspects and disaster mitigation technology.

Farming aspects and disaster mitigation technology	respondents' opinions about the program of socio-economic recovery (existing conditions)/ (%)			
	not existing	existing/ non-implemented	existing / implemented	limited to socialization
Calendar and planting schedule	60%	23%	10%	7%
New superior varieties / tolerant diversity, dry resistance	63%	20%	13%	3%
Technology of water resources processing / rain harvest technology / identification of water potential / rain prediction technology	87%	0%	13%	0%
management of land resources/ application of fertilization technology	83%	0%	10%	7%
irrigation channel development planning	73%	17%	0%	10%
Spatial evaluation (plant type and land carrying capacity)	93%	0%	0%	7%
Development of systems information and disaster early warning (earthquake / eruption).	67%	10%	17%	7%
Understanding climate change (application of adaptation and mitigation technologies)	93%	0%	0%	7%

Source: Primary data, processed 2018.

Table 2 shows the government program in the context of recovery of socio-economic farmers in the affected areas of the Sinabung eruption in Karo Regency has not been carried out properly. The results of the interviews showed that 93% of respondents said there were no training/ counseling programs on understanding climate change including the application of adaptation and mitigation technologies and their relation to the development of farming in the context of socio-economic recovery of the community.

Then 93% of respondents also said that the spatial evaluation of the use of plant species that was adjusted to the carrying capacity of the land also did not exist. This is supported by the opinion of 63% of respondents who convey new superior varieties that are tolerant with no diversity and drought.

[19] reveals there are three basic components to household recovery. These are housing recovery, economic recovery and psychological recovery. All three of these components require resources to recover, but house- holds must invest time to obtain these resources. This includes time to find and purchase alternate shelter, clothing, food, furniture and appliances to support daily living [20]. Time is also needed to file insurance claims, apply for loans and grants, and search for jobs. The time required for these tasks is increased by multiple trips to obtain required documentation and understaffing of providers [21] finally, victims need skill and self confidence to cope with the disaster assistance bureaucracy.

Economic recovery. Some households 'economic recovery takes place quickly, but others' takes much longer. For example, the percentage of households reporting complete economic recovery after the whittier earthquake was 50% at the end of the first year but 21% reported little of no recovery even at the end of 4 years [22]. Economic recovery was positively related to household income and negatively related to structural damage, household size, and the total number of moves [22]. In some cases, this is due to the loss of permanent jobs that are replaced only by temporary jobs in temporary shelter management, debris cleanup, and construction – or are not replaced at all [20].

3.3. Economic recovery of farmers affected by the eruption of Mount Sinabung through aspects of local wisdom and traditions of the local community.

The eruption of Mount Sinabung that occurred some time ago certainly had a tremendous impact, not only on physical impacts, but on the economic impact, so a recovery program was needed for the condition of the people affected by the eruption so that the role of the community itself was needed to carry out the recovery process.

The results of interviews with respondents showed that the economic recovery program of the communities affected by the Sinabung volcano eruption by regulating cropping patterns and community understanding of social and economic erosion impacts had not yet been implemented as in Table 3.

Table 3. The role of government/ society in the economic recovery of farmers affected by mount Sinabung eruption through aspects of local wisdom and tradition.

Aspects of local wisdom and tradition	Respondents' opinions about the program of socio-economic recovery (existing conditions)/ (%)			
	not existing	existing/ non-implemented	existing / implemented	limited to the program
Determination of cropping patterns / types of plants based on local wisdom / local community beliefs / related to disasters	40%	10%	47%	3%
Understanding disaster / climate change based on ancestral traditions	43%	10%	40%	7%

Source: Primary data, processed 2018.

Table 3 shows that the economic recovery of farmers through farming development has not paid attention to cropping patterns, and 47% of respondents said that the determination of cropping patterns that are in accordance with local wisdom has not yet been implemented. Then 43% of respondents said there were no programs implemented by the government and the private sector regarding public understanding of disaster. The results of this study indicate that the role of local wisdom has not been utilized optimally in the process of socio-economic recovery of farmers in areas affected by the eruption of Mount Sinabung.

In general, the recovery process carried out by related parties [23] is still focused on physical recovery and basic needs, as stated by arly recovery focused on the basic functions and services of the community and the recovery of vital facilities and infrastructure. Including recovery of religious activities, revitalization of arts and culture. Socio-economic recovery of farmers can be done through; revitalization of farmer groups, gardens and livestock, diversification/ alternative programs for agricultural businesses, provision of fast harvesting seeds, business capital assistance for traders and small and medium industries. In connection with this, Table 3 shows that only 43% of respondents understand that local wisdom in accordance with the tradition of heredity (ancestors) plays a role in the process of socio-economic recovery due to climate change or due to an eruption disaster.

3.4. Socio-economic recovery of farmers affected by the eruption fo Mount Sinabung through market aspect, institutional and capital resources.

[24] implied that economic recovery in post-disaster in Yogyakarta has performed well because of many factors ranging from local government roles, people's perception of disasters, monetary policies conducted by central bank in regard with loan of the victims, and others. Right decisions made by local and central government have been implemented well with support of people impacted and not impacted by the disasters. The recovery discussed also indicated the importance of development

quality to ensure standard of education, economic life and social activities which were proven helpful in the awakening in the aftermath of disasters.

The results of interviews with respondents in the locations affected by the Mount Sinabung eruption showed that the socio-economic recovery program of the communities affected by eruption disasters based on market aspects, institutions and sources of capital had not gone well as in Table 4.

Table 4. The role of government and private sector in the social economic recovery of farmers affected by Mount Sinabung eruption through market aspects, institutional and capital resources.

Market aspects, institutional and capital resources	Respondents' opinions about the program of socio-economic recovery (existing conditions)/ (%)			
	not existing	existing/ non-implemented	existing / implemented	limited to socialization
Information on agricultural production prices	87%	0%	1%	7%
agricultural market information	87%	10%	3%	0%
development and formation of farmer groups	20%	0%	19%	17%
access to collaboration with partners (third parties) for domestic, industrial and export markets)	83%	13%	3%	0%
farmer counseling and agricultural labor training	27%	20%	40%	13%
Providing capital assistance and capital access (government and private banks, etc.)	53%	30%	13%	3%

Source: Primary data, processed 2018.

Table 4 shows that the economic recovery program of farmers in the affected areas of Mount Sinabung eruption has not yet taken place, where 87% of the respondents said that the production price information and market information on agricultural products did not exist. Likewise, with access to cooperation with partners, 83% of respondents expressed access to cooperation with partners (third parties) for market purposes, both domestic and export not yet available.

The partnership model is needed by farmers to help develop their farms. The efforts made are with financial policies such as; property compensation purchases, job opening, skills training, tax relief, asset purchases, low interest credit offers and others. For example, many victims of the Aceh Tsunami disaster received training from donor agencies and the government to develop alternative livelihoods or improve previously owned skills, such as entrepreneurship skills, farming techniques, and aquaculture techniques [25].

[24] the important role showed by government in several levels during the recovery along with continuous support from people implied that different cultures resulting different attitude toward help, support and gift during and after disasters. This is notewor thy since Indonesia has many disaster risks all over its wide area and understanding inhabitants' point of view can be the key for mitigating disaster and its recovery in the aftermath. There are of course some limits in this paper that should be addressed in the further research such as additional data on macroeconomic indicators related to economic recoveries covering wider aspects of victims' life; enrichment in modes of analysis; inclusion of others natural disasters effected areas in Indonesia; comparison with other cases of different cultures.

3.5. Model of recovery of socio-economic farmers in areas affected by the eruption of Mount Sinabung in Karo Regency.

The socio-economic recovery program of farmers implies to increase the independence of farmers so that all their potential is empowered to improve social life (the distribution of power to manage their farms) and economic power (redistribution of economic resources through productive activities). In the paradigm of social and economic development of farmers, farm managers must be able to become planning, implementing and beneficiaries. This strategy is carried out so that the deterioration conditions as a result of a disaster can immediately recover, thus; farmers' readiness to face globalization and the utilization of available opportunities [26] can be better.

Based on the results of the research through interviews with respondents, several very important aspects were identified to restore the socio-economic conditions of the farmers. Important aspects that need to be considered to support the socio-economic recovery of farmers include; technological innovation activities, counseling activities, training, mentoring and strengthening of local economic institutions. And based on the results of data analysis as in Tables 1, 2, 3, 4 and referring to the results of previous studies can be formulated models of socio-economic recovery of farmers affected by the eruption of Mount Sinabung in Karo Regency in Figure 1.

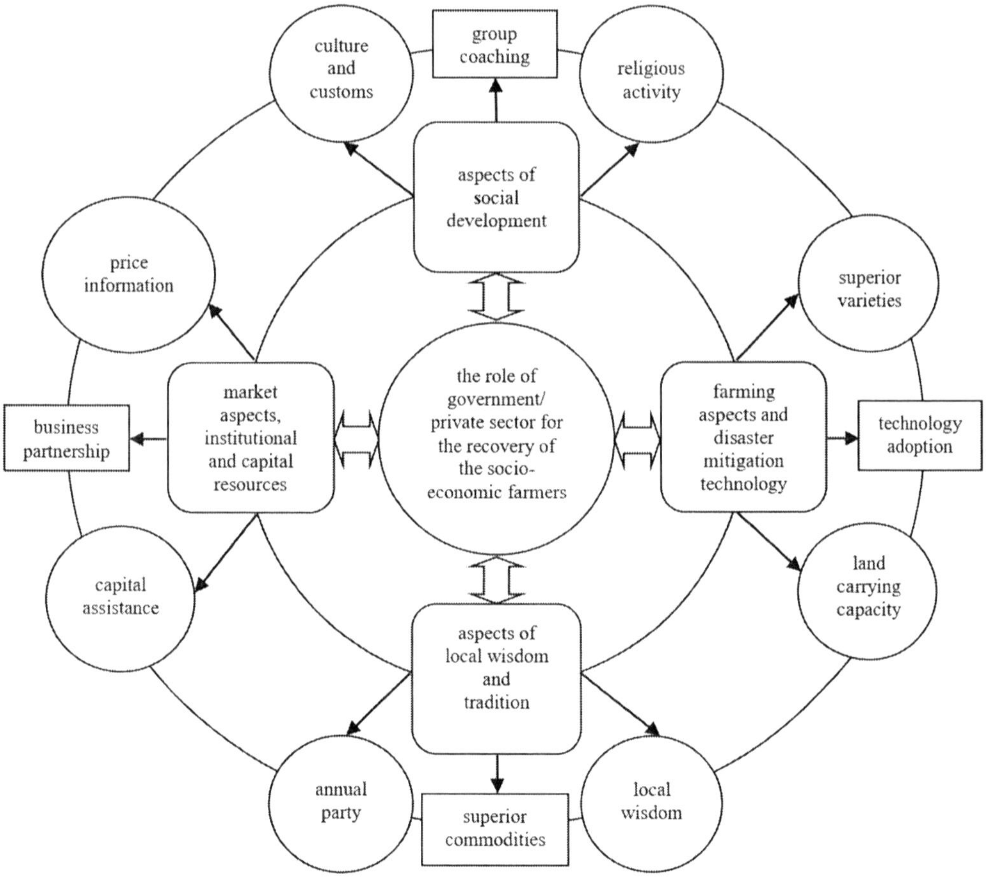

Figure 1. Model of socio-economic recovery of farmers in erupted area of Mount Sinabung in Karo Regency.

Figure 1 shows the model of socio-economic recovery of farmers affected by eruptions in the Mount Sinabung area. The recovery process must involve the government including and the private sector, with the main focus being; 1) development of social aspects which include; training of farmer groups and communities, increasing religious activities and fiber culture of local customs, involving the community to maintain traditional values and local wisdom, 2) developing farming aspects based on disaster mitigation technology, taking into account; nursery technology that is susceptible to climate change, the application of technology in the field of farming and water resources technologies such as; rain harvest technology, as well as paying attention to land carrying capacity, 3) development of social and cultural aspects, including traditional values that play a role in developing local potentials in the community, such as the work culture of the year which serves to improve kinship between communities and farmers, 4) development of market aspects and partnerships as sources of capital. It is very important to build partnerships with related parties to accommodate the production of community agricultural products at competitive prices, as well as clear market information for farmers.

The results of this study are supported by the findings of [27] who stated that the objectives of the farmers' socio-economic recovery program were welfare and to make people can survive, which can be seen in terms of increasing income, consumption levels and purchasing power of farmers. And when viewed from economic anthropology, welfare is seen as inter-cultural relativity, because the concept of sociology views welfare from the aspect of social stratification in the form of opening up access to the lower socio-economic strata in order to increase. Thus, the model of socio-economic recovery of farmers in Karo Regency aims to empower the community's economy towards self-reliance so as to improve farmers' adaptive behavior in the face of future disasters.

4. Conclusions and Suggestions

4.1. Conclusions

Based on the results of the research and the discussion carried out, it can be concluded; a) social recovery of farmers carried out by the government through guidance of farmer groups and communities as well as cultural activities and customs is still minimal, b) Social recovery of farmers conducted by the government through training and counseling programs on the implementation of adaptation and disaster mitigation technologies related to farming development has not been implemented, c) The use of plant species that are adapted to the carrying capacity of the land and new superior varieties that are tolerant with diversity and drought do not yet exist, d) Farmers 'economic recovery program through information on production prices and information on agricultural markets is not yet running, e) Farmers' economic recovery program through access to cooperation with partners (third parties) for market purposes, both domestic and export does not yet exist, while this partnership is needed by farmers to help develop their farms.

4.2. Suggestions

Based on the conclusions the suggestions are; a) integrated and integrated efforts are needed to restore the socio-economic conditions of farmers, through economic empowerment in the framework of developing the ability of farmers in areas affected by the eruption of Mount Sinabung, b) a holistic handling effort is needed to touch all aspects of socio-economic farmers, and not a form of empowerment which is sloganistic and only touches on the structural aspect of empowerment, c) The government must provide free space for farmers to improve their social and cultural economic strata, through activities that are productive and on target, d) The government must carry out a specific identification process to find out the root causes of the problem of the slow recovery process of socio-economic farmers in the area affected by the eruption of Mount Sinabung.

Acknowledgement

The authors would like to thank The Ministry of Research, Technology and Higher Education of Republic Indonesia for providing grant to fund the institution's national strategic research in 2018.

Reference

[1] International Federation of Red Cross and Red Crescent Societies 2010 Indonesia: Sinabung Volcano Eruption DREF operation http://www.ifrc.org/ docs/appeals/ 10/MDRID005 dref.pdf (accessed Oktober 6 2018)

[2] Lutheran World Relief (LWR) 2014 Indonesia – Mt. Sinabung Volcanic Eruption February 3 2014 No 1 Lutheran World Relief (LWR) Sustainble Development Lasting Promise. https://reliefweb.int/sites/reliefweb.int/files/resources/LWR%20Indonesia%20Mt.%20Sinabung% 20 Eruption%20 Situation%20Report%201.pdf (accessed Oktober 6 2018)

[3] Muzambiq Said, Syafriadi Wijaksana, Budi Satria, Umar Rosaidi 2017 Characteristic of Sinabung volcano deformation of 2011-2012 estimated based on GPS data *Australian Journal of Basic and Applied Sciences* **11**(9) pp 59-71 ISSN 1991-8178 EISSN 2309-8414 Journal home page: www.ajbasweb.com
[4] Tampubolon Jongkers, Nainggolan H L, Ginting A, Aritonang J 2018 Mount Sinabung eruption impact on local economy and smallholder farming in Karo Regency North Sumatra 2017 4th International Conference on Environmental Systems Research (ICESR 2017). *IOP Publishing IOP Conf. Series Earth and Environmental Science* **178** 012039 Doi 10.1088/1755-1315/178/1/012039
[5] Tampubolon Jongkers, Ginting Albina, Nainggolan Hotden L, Aritonang Johndikson 2017 Farm management responses to natural disaster a case study mount Sinabung eruption in North Sumatra Indonesia *International Journal of Agriculture and Biosciences* P-ISSN 2305-6622 E-ISSN 2306-3599 www.ijagbio.com editor@ijagbio.com
[6] Olshansky R and Chang S 2009 Planning for disaster recovery emerging research needs and challenge progress in planning **72** 200-209
[7] Tobin G A 1999 Sustainability and community resilience the holy grail of hazard planning *Environmental hazards* **1**(1) 13 – 25
[8] Morrow B H 1999 Identifying and community vulnerability disasters **23**(1) 1-18
[9] Aldrich D P 2010 The power of people social capital's role in recovert from the 1995 *Kobe Earthquake Nat Hazards* **56** 595-611
[10] Nakagawa Y and Shaw R 2004 Social capital a missing link to disaster recovery *International Journal of Mass Emergencies and Disasters* **22**(1) 5-34
[11] Aldrich D P 2011 The externalities of strong social capital post-tsunami recovery in Southeast India *Journal of Civil Society* **7**(1) 81-99
[12] Alexander B, Chan-Halbrendt C and Salim W 2006 Sustainable livelihood consideration for disaster risk management implications for implementation of the government of Indonesia Tsunami recovery plan disaster prevention and management **15**(1) 31-50
[13] Chhotray V and Few R 2012 Post-disaster recovery and ongoing vulnerability ten years after thesuper-cyclone of 1999 in Orissa India: *Global environmental change* **22** 695-702
[14] Gordon R 2004 The social system as site of disaster impact and resource for recovery *The Australian Journal of Emergency Management* **19**(4) 16-22
[15] Lindell Michael K and Prater C S 2003 Assessing community impacts of natural *Disasters Natural Hazards Review* **4** 176 – 185
[16] Gupta M and Sharma A 2006 Compounded loss: The post tsunami recovery experience of Indian Island communities *Disaster prevention and management* **15(1)** 67-78
[17] Tobin G A and Whiteford L M 2002 Community resilience and volcano hazard: The eruption of Tungurahua and evacuation of the faldas in Ecuador Disasters **26**(1) 28-48
[18] Chang Stephanie E and Rose Adam Z 2012 Towards a theory of economic recovery from disasters *International Journal of Mass Emergencies and Disasters* August 2012 **32**(2) pp 171–181
[19] Lindell Michale K 2013 Recovery and reconstruction after disaster Encyclopedia of Natural Hazards Bobrowsky P T ed) http://www.springer.com/978-90-481-8699-0
[20] Yelvington K A 1997 Coping in a temporary way The tent cities In Peacock W G Morrow B H and Gladwin H (eds.) *Hurricane Andrew Ethnicity Gender and the Sociology of Disaster* London: Routledge pp 92–115
[21] Morrow B H 1997 Stretching the bonds: The families of Andrew in Peacock W G Morrow B H and Gladwin H (eds.) *Hurricane Andrew Ethnicity Gender and the Sociology of Disaster* London Routledge pp 141–170
[22] Bolin R C 1993. Household and community recovery after earthquakes Boulder CO: University of Colorado Institute of Behavioral Science
[23] Arain F 2015. Knowledge-based approach for sustainable disaster management: Empowering

emergency response management team *Procedia Engineering*; **11**:232-239

[24] Andriansyah Yuli 2014 Post-disaster economic recovery in Yogyakarta Indonesia: *Conference Paper* November 2014 DOI 10.1201/b18146-62 https:// www. researchgate. net/publication/ 268509307

[25] Thorburn C 2009 Livelihood recovery in the wake of the tsunami in Aceh *Bulletin of Indonesian Economic Studies* **45**(1) 85-105

[26] Wilkinson E, Lovell E, Carby B, Barclay J, Robertson RE 2016 The dilemmas risk-sensitive development on a small volcanic island Resources **5**(21):1-20.DOI: 10.3390/ resources5020021

[27] Hariyono Eko and Liliasari Solaiman 2018 Disaster mitigation model of eruption based on local wisdom in Indonesia. Additional information is available at the end of the chapter http://dx.doi.org/10.5772/intechopen.79217

Environmental pollution monitoring using a Web-based GIS in Surakarta

Alif Noor Anna[1,2,*], Rudiyanto[1,2] and Vidya Nahdhiyatul Fikriyah[2]

[1] Centre of Environmental Studies, Universitas Muhammadiyah Surakarta, Indonesia
[2] Faculty of Geography, Universitas Muhammadiyah Surakarta, Indonesia

*alif_noor@ums.ac.id

Abstract. The case of environmental pollution is usually not all exposed onmedia. As such, environmental pollution monitoring needs to be done, hence the government can take legal measures against it. Monitoring environmental pollution will be more effective when involving community. The development of geospatial technologies such as Web-based GIS is helpful for monitoring pollution inenvironment. This is due to the Web-based GIS has several advantages including easy tooperate, easily accessible and able to present data (spatial and non-spatial) in real time. The purpose of this research is to develop a web-based GIS for monitoring environmental pollution. The software was developed using Rapid Application Development (RAD). Software development using technology based on opensource (CodeIgniter, Bootstrap, OpenLayers, and GML). The results show that the utilization of web-based GIS technology for environmental pollution monitoring is easily accessible and operated by user. Furthermore, the trial results show that the utilization of developed web-based GIS technology can operate properly and present data interactively.

1. Introduction

The intensity of environmental pollution incidence has been increasing from year to year in Indonesia, including in Surakarta, such as the cases of pollution in groundwater [1] and in Bengawan Solo river [2]. All environmental pollution occurs due to economic activities (industry) that are less environmentally friendly as well as the negligence of business actors in managing waste from industrial products. There are many more environmental pollution incidents going on, but not all of them are exposed on media which leads to the slow response from the government. Therefore, there is a need for an effort to make the incidence of environmental pollution can be exposed and immediately responded by the law enforcement authority. One of the efforts is by making a web-based GIS information system of environmental pollution monitoring.

The technology of web-based Geographic Information Systems (GIS) is utilized in the application development of monitoring environmental pollution due to its easily accessible and simple to operate by users from various backgrounds of profession and education. The utilization of web-based GIS technology will anticipate the shortage of GIS-based desktop, i.e. in terms of ease of access and usage. Lately, the utilization of web-based GIS technology for various purposes has been growing rapidly. A wide range of development has been done by some researchers such as utilization of web-based GIS for renewable energy after the reconstruction and rehabilitation process in Aceh Province [3], a web-based GIS development for groundwater management in Karanganyar Regency, [4] the utilization of web-based GIS for land cover monitoring in Mekong Delta, Vietnam [5], and application development of Web GIS to estimate land price [6].

Considering the importance of immediate action from the governmental authority in response to pollution incidence as well as the importance of involving community for a more effective environmental monitoring, in this study, we developed a web-based GIS application that can accommodate both parties. The development of web-based GIS application for monitoring environmental pollution used an opensource-based software. The software used to develop this application included the CodeIgniter Framework, Bootstrap Framework, OpenLayers, and Geography Markup Language (GML). Meanwhile, some previous studies have developed a web-based GIS application with the Server Folders [7], PostgreSQL, PostGIS, and Bootstrap Framework [8]. Web-based GIS application development by using opensource software is profitable because the software is easy to download (accessible), easy to use, and free. The development of this web-based GIS application for environmental pollution monitoring involved several stakeholders including the local government Surakarta City, NGOs, and the community. The purpose of this research is to develop a web-based GIS application for environmental pollution monitoring in the city of Surakarta.

2. Literature

2.1. Environmental Pollution

According to Indonesian Law Number 32 of 2009, the environmental pollution is the entrance or the inclusion of living things, energy, and matter, or other components into the environment by human activities until it goes beyond the environmental quality standard that has been set. Contamination occurs due to the presence of a contaminant source which is any activity that disposes of pollutants. Contaminants usually consist of hazardous and non-hazardous materials. Hazardous and toxic materials, abbreviated as B3 (hazardous and toxic materials), are substances, energy, and other components which because of their characteristics, amount, or concentration, and, either directly or indirectly, can contaminate or damage the environment, and harm to the environment, health, survival of human beings, and other living beings. These contaminants can be solid, liquid, gas, or particle suspended in certain levels into the environment, either through the air, water, or land which eventually will go to human [9].

2.2. Web GIS Application

Web GIS is one of the application developments of Geographic Information Systems (GIS) that represented in the form of web services. It is a combination of GIS technology and the web. There are several methods and software that can be used to develop a web-based GIS technology, including Quantum GIS, Map Server, ArcGIS Online, OpenLayers, and GML (Geography Mark-up Language).

Web GIS is a tool or a product of digital mapping that based on internet network as its medium of communication. Functions of Web GIS as the communication media are to show, deploy, integrate, provide spatial and non-spatial information in various forms (such as digital maps, texts, diagrams), and run the main functions, namely analysis and queries which are still associated with GIS technology through internet network [10].

2.3. Bootstrap Framework

Bootstrap is one of CSS framework that is helpful to design a responsive admin page and the front page of a website. Utilization of the bootstrap framework has been rapidly growing lately due to its open source characteristics and cleaner, lighter, and more responsive template compared to the desktop, tablets, and mobile version. The bootstrap framework also supports some file types such as HTML, CSS, and JavaScript. These advantages of bootstrap make website developers prefer this framework. Utilization of the bootstrap framework to design a responsive website is advantageous because the display of the website will automatically adjust with the type of device you use to access the web [11].

2.4. CodeIgniter Framework

CodeIgniter is a framework for the development of a PHP-based application. It is software to build an open source web application that has an MVC model (model, view, and controller). The advantages of this framework include (a) the construction of the application can be done quickly and easily, (b) easy maintenance, (c) a complete facility available such as pagination, validation, multiple database, ORM, scaffolding, session setting, error handling, etc., so no need to create from scratch and (d) users have more freedom in developing applications. Currently, the latest version of this framework is 3.1.7 and the stable version is 4.0 [12].

2.5. OpenLayers

OpenLayers is a JavaScript-based client application for displaying map data in web browsers and does not depend on the used web server. OpenLayers implements a JavaScript API that is used to build a web-based GIS application. It is similar to Google Maps and MSN Virtual Earth API, but with an important difference. It is free software that was developed for and by the community of Open Source software.

2.6. Geography Markup Language (GML)

GML or Geography Markup Language is one of the computer programming languages like XML that based on spatial or geographic reference. Spatial data that can be included is in the form of point, line, or polygon data.

3. Research Methods

3.1. System development method

The method to develop the web-based GIS application for environmental pollution monitoring is Rapid Application Development (RAD). RAD is a software development method that was invented to compress the time to design and to implement information systems, so it can produce a very short development cycle. Visually, the plot of application development with the RAD method can be seen in Figure 1.

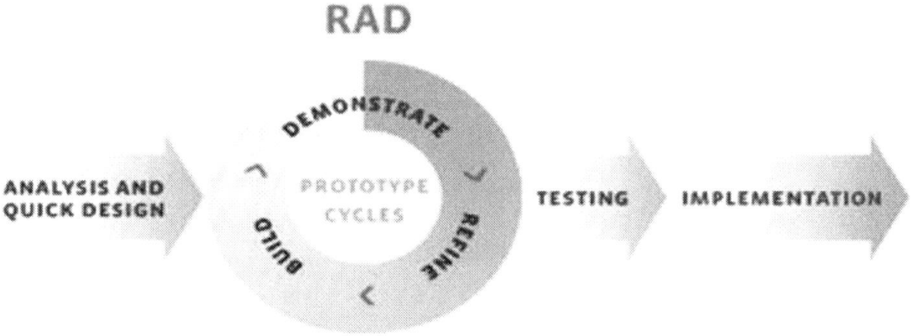

Figure 1. Development of Information System of Environmental Pollution Monitoring with RAD Method [13]

RAD is a system development method for designing applications in a short time. There are several stages or application development cycles with the RAD method. The stages are as follows:

1. Requirement Planning: Users and analysts conduct meetings to identify the objectives of the system and information needs to achieve the goals. At this stage, the most important thing is the involvement of both parties.
2. System Design: At this stage, the activity of the users involved determines the achievement of objectives. The process of design, repair, and improvement are carried out during this stage if there is still a mismatch between the user and the analyst. A user can immediately comment

when there is a discrepancy in the design, and be actively involved in designing the system by referring to the documentation of user needs made in the previous stage. The output of this stage is software specifications that include system organization in general, data structures and others.

3. Implementation: This is the stage where the programmer develops the design of the program that has been approved by both the user and the analyst. Before being applied to an organization, a program testing isdone at first to check whether there is an error or not. At this stage, the user usually gives a response to the system that has been made and gives approval for it.

3.2. The use of software in system design

The software used to compose information system of environmental pollution monitoring in this study was opensource software. The software used to develop the program and to display spatial and non-spatial model were as follows:

1. Quantum GIS Valmiera, for spatial data processing
2. The software that is running on the server (server-side), such as MySQL spatial, serves as a database system that stores both the spatial data as well as the non-spatial data; Apache and PHP, are software that acts as a web server.
3. The software that is running on the client (client-side), such as Internet Browser (Microsoft Internet Explorer, Mozilla Firefox, Opera, and others), are used to access the application
4. OpenLayers 2.2, for displaying the layers of spatial data on the system
5. GML, for making web-based spatial data (polygon, polyline, point)

3.3. Research data

The data used in this study include primary and secondary data. Primary data obtained through field surveys, while secondary data obtained from the Geospatial Information Agency. The type of data is spatial and non-spatial data. The primary spatial data is the coordinate locations of environmental pollution; while the secondary spatial data is the map of Surakarta City (scale 1:50,000). The non-spatial primary data includes description and documentation of contamination sites. Detailed information on the used data can be seen in Table 1.

Table 1. Research Data

No	Data	Data Type	Structure of data	Source
1	Administration map of Surakarta, scale 1:50.000	Secondary data	Spatial	Geospatial Information Agency
2	Coordinates of pollution sites	Primary data	Spatial	Survey
3	Documentation of pollution	Primary data	Non spatial	Survey
4	Description of pollution sites	Primary data	Non spatial	Survey

Source: Author, 2018

The Surakarta administration map data as shown in Table 1 includes regional administrative data and regional boundary data. Coordinate data for pollution sources are obtained from users who report pollution events in the field accompanied by documentary evidence and a brief narrative of pollution events. Data of pollution reports that enters the system will be validated and checked in the field, so that the environmental pollution complaint is 100% valid before entering the follow-up stage.

3.4. Stages of the Research

The implementation of this research is divided into 3 stages, namely: the preparation of system procedures, system testing, and system implementation. The order about the stages of implementation of the research in detail can be seen in Figure 2.

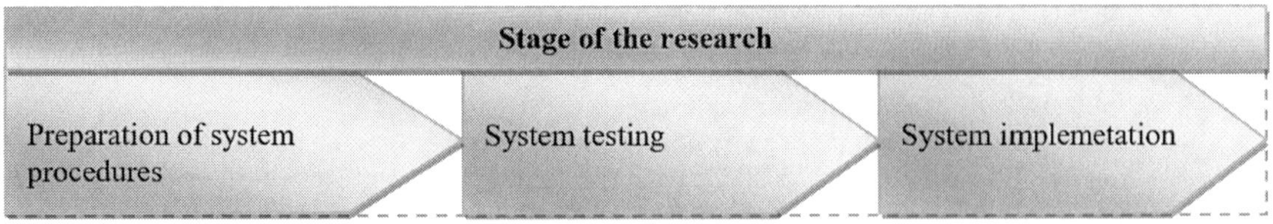

Figure 2. Stages of the Research
Source: Author, 2018

3.4.1. Preparation of System Procedures. The system to be built is expected to be implemented in environmental pollution monitoring activities. The information contained in the system includes coordinates and location of pollution, documentation of pollution events, and chronological descriptions of the events. In order to the system procedures to be in accordance with the needs of users, the system is designed by holding FGD (Focus Group Discussion) activities together with stakeholders such as the Department of Environment in Surakarta (DLH), NGOs, and community.

3.4.2. System Development. There are several steps that must be done in developing an environmental pollution information system. The stages include:

1. Inventory of equipment and information needs related to the monitoring of environmental pollution (user requirements).
2. System Analysis is the system design stage based on the analysis of system requirements in the previous stage. At this stage, a system design that will be used to support environmental pollution monitoring is produced.
3. Spatial database design is the stage of making a spatial database scheme that is in accordance with the analysis that has been made.
4. Designing the system interface. At this stage, an interface design is made according to user needs.
5. Workshop on system design with users.The system analysis and the resulting design need to be adjusted to users. If all the designs are approved by users, the next stage will be carried out. However, if there still need a change in the design, adjustments will be made with the user in this workshop session.
6. The process of coding in applications. At this stage, the design that has been made is then implemented in programming scripts to produce the application as it has been designed.

3.4.3. System Testing

1. A system operational manual making.
2. System socialization that has been made.
3. System testing, which involves users
4. Evaluation of the system with users. After the user tries the application, then an evaluation session is carried out for the improvement in the application.

3.4.4. System Implementation

1. Stage of server and infrastructure preparation.
2. Socialization stage to stakeholders.
3. Stage of system utilization

4. Research Results

4.1. System procedure preparation

The preparation of system procedures in this study was carried out through a Forum Discussion Group (FGD) with Department of Environment of Surakarta City. The system procedures compiled include system procedures for admin, field officers, and users. The details are as follows:

1. System procedure for admin
   a. Admin log in to the system
   b. Admin looks and list of user requests
   c. Admin checks the completeness of registration documents and activates the registration of users and field officers. At this stage, the username and password of the user and field officer will be sent by the system via email and phone number
   d. The admin selects a list of complaints about environmental pollution from users and continues those information to field officers
   e. The admin receives and prints the verification report from the field officer
   f. The admin reviews the results of the report and gives a follow-up decision based on the verification report from the field officer
   g. Admin provides notification of follow-up results from complaints of environmental pollution to users
   h. Admin closes complaints of environmental pollution
2. System procedure for field officers
   a. Field officers log in to the system
   b. Field officers see a list of complaints that must be surveyed or checked
   c. Field officers verify complaints to the field
   d. Field officers report the results of field checks and print reports
   e. Field officers close the list of complaints that have been surveyed
3. System procedures for users
   a. Users register online
   b. Users fill out the registration form and upload the documents or files needed during registration (copy of ID card, and user's photograph and his ID card)
   c. Users log in to the system
   d. Users report pollution events with the documentation of evidence in the field
   e. The user prints the report
   f. Users receive follow-up notifications from the admin regarding their complaints

4.2. System architecture

System architecture in this research is divided into a user architecture, a web server, and database. The architecture of the system is presented in Figure 3. Spatial database stored in the database can be visualized on OpenLayers (JavaScript framework) after changed into GML format using PHP [13].

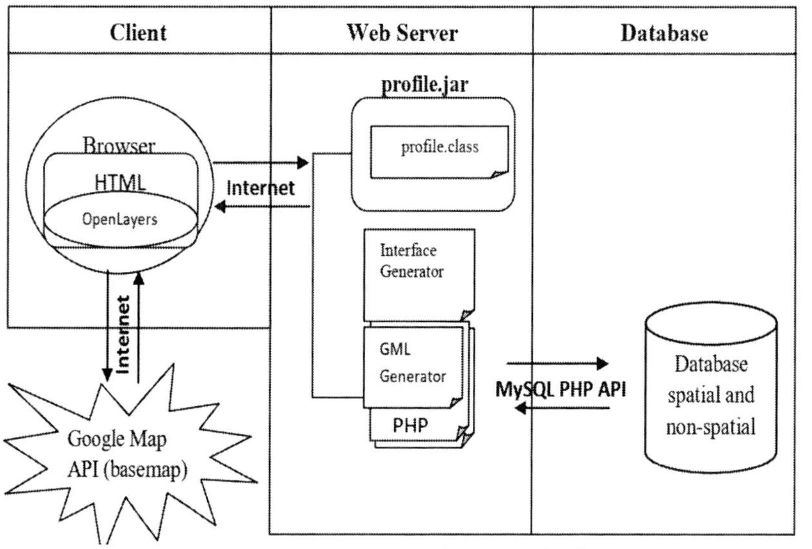

Figure 3. System Architecture [13]

Based on Figure 3, it is known that the system architecture consists of a database, web servers, and clients. The database in this system consists of spatial and non-spatial data. The spatial database includes the administrative map data, pollution coordinates, and Surakarta administrative boundaries, while non-spatial data includes data on pollution events (documentation and description). The web server consists of various PHP and GML files. On the client page, there is a web browser, HTML file, OpenLayers and also Google Map API.

4.3. Application design

The design of the main page (interface) includes spatial and non-spatial representations, menu, navigation tools, and geographic analysis. It is generally divided into 8 main parts, namely: (1) header, footer, and application title, (2) main menu, (3) legend, (4) map navigation menu (map tools), (5) layer control tab and data tracking, (6) control layer, (7) space for map (map space), and (8) panel for showing the coordinates of pointer. The application design can be seen in Figure 4, Figure 5, Figure 6, and Figure 7.

Figure 4. Application Home Page
Source: Author, 2018

Figure 4 shows the front page of the information system application. There are several menus that can be accessed by all users including the menu of river maps, downloads, web links, contacts, memberships, and logins. The river map menu contains spatial information about the point of pollution, documentation of pollution, and its description. The download menu shows data or file information that can be downloaded by users. While the link menu contains information, links related to the information systems. Further, the contact menu provides information of address and contact of the system maker and manager. The membership menu contains information and registration forms for new members or users. The login menu contains information on user access to the system.

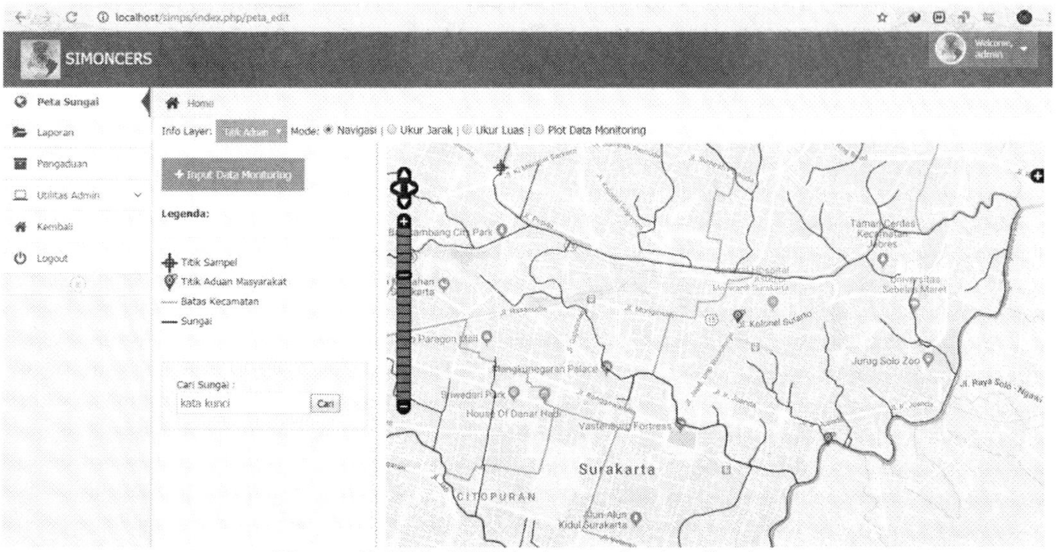

Figure 5. Web GIS Display on the Application
Source: Author, 2018

Figure 5 shows the GIS Web view on information systems. This page contains some information, such as view maps with the basic Google MAP API, legend maps, wind direction, zoom in and zoom out maps, data search, map navigation, tools for measuring distance and area, plot map coordinates, and the forms to input environmental pollution data.

Figure 6. Form Page of Environmental Pollution Complaint on User Application
Source: Author, 2018

Figure 6 presents the page of the complaint form for environmental pollution from users. Reports of environmental pollution via Android phones and tablets will automatically record the coordinates of the location of the pollution. If reporting is done through a notebook or personal computer, then the coordinate data can be searched through the plot coordinate menu which is available in the system. Next, the user uploads a photo of the pollution event and describes the chronology of the event. After all forms are filled, user can report the pollution incident by pressing the save menu on the system.

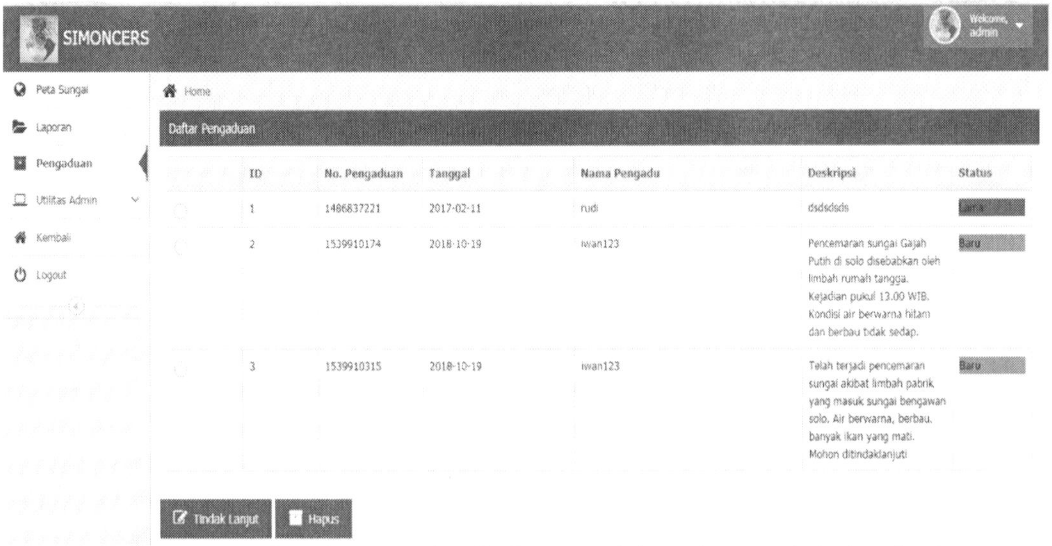

Figure 7. Page of Environmental Pollution Complaints List on Admin Application
Source: Author, 2018

Figure 7 shows a list of reports in the system admin page or web administrator. On this page, the admin can select complaints from users. There are two menu options in response to the complaints of environmental pollution from users: follow up or delete from the system. Valid pollution reports will be followed up, while the invalid reports (not in accordance with the field conditions) will be deleted from the system. The follow-up response regarding those reports is in the form of environmental policies from the authorized agency to handle the pollution.

4.4. Application prototype
At this stage, an application prototype has been built which will later be evaluated together with stakeholders. Stakeholders who will be involved in the evaluation of the system are the community, the local government of Surakarta, business actors, and NGOs. This application prototype is available and can be accessed through http://simps.kreasibundanhk.com.

5. Discussion
The use of GIS in this study combined with the decision-making system is in accordance with the Spatial Decision Support System (SDSS) criteria [14]. In this case, GIS has also been adjusted to the needs of users in order to facilitate the use by lay people. The research on the development of web-based GIS applications for environmental pollution monitoring has a fairly good progress compared to previous studies in the same study focus. Based on aspects of technology utilization reflected in the system architecture, this study shows an improvement that has been adjusted to technological developments compared to the architectures developed by other previous studies, such as the use of web-based GIS to map renewable energy after the reconstruction and rehabilitation process in Aceh Province [3], web-based GIS development for groundwater management in Karanganyar [4], use of web-based GIS for land cover monitoring in Mekong Delta, Vietnam [5], and the use of Web GIS applications to estimate land price [6]. The architecture of this system theoretically has better performance and accessibility assessed in several aspects. First, by relying on JavaScript framework, the web-based GIS of environmental pollution monitoring information system will be more independent because it does not depend on third-party applications. Second, the use of the OpenLayers framework is considered more in line with the Open Geospatial Consortium (OGC) standard as a guarantor institution for quality standards for spatial data, thus enabling better data sharing. Third, the use of Google Map API as a provider of spatial data is considered reliable and has a good and automatic level of data updates.

6. Conclusion

Based on the results and the software testing that has been done, it can be concluded that problems related to environmental pollution are the domain that involves several stakeholders so that synergistic collaboration between stakeholders is required. Furthermore, the applications developed in this study can run properly and be accessed by the admin, users, and field officers.

References

[1] Anna Alif Noor 2004 Pemintakatan daerah yang rawan pencemaran airtanah sebagai dasar perencanaan penyediaan air bersih di daerah Surakarta *Jurnal Penelitian Forum Geografi* **18** 115–133 doi:10.23917/forgeo.v18i2.612

[2] Hermawan P and Yoshanti G 2016 *Unfolding the problem of batik waste pollution in Jenes River, Surakarta, using critical system heuristics and drama-theoretic dilemma analysis* (Tokyo: Springer) pp 93–108 doi:10.1007/978-4-431-55273-4_6

[3] Nizamuddin, Hizir, Ardiansyah, D Pertiwi and P Handayani 2017 Development of Web GIS for information of renewable energy in Aceh Province after rehabilitation and reconstruction process *IOP Conf. Series: Earth and Environmental Science* **56**(2017) 012016 doi:10.1088/1755-1315/56/1/012016

[4] Jumadi dan Priyana Yuli 2015 Development and evaluation of Web GIS Application for groundwater management of Karanganyar Regency *Jurnal Penelitian Forum Geografi* **29**(2) 165-174 https://doi.org/10.23917/forgeo.v29i2.1482

[5] Le Van Trung and Dao Minh Tam 2018 Web GIS solution for monitoring the forest-cover in the Mekong Delta, Vietnam *Journal of Geographic Information System* **10**(5) 491-502 doi: 10.4236/jgis.2018.105026

[6] Yongjun Yang, Yaqin Sun, Songnian Li, Shaoliang Zhang, Kuoyin Wang, Huping Hou and Shishuo Xu 2015 A GIS-based web approach for serving land price information *ISPRS Int. J. Geo-Inf.* **4** 2078-2093 doi:10.3390/ijgi4042078

[7] Puyam S Singh, Dibyajyoti Chutia and Singuluri Sudhakar 2012 Development of a Web-based GIS application for spatial natural resources information system using effective open source software and standards *Journal of Geographic Information System* **4** 261-266 http://dx.doi.org/10.4236/jgis.2012.43031

[8] G Vaccaa, D R Fiorinoa and D Pilia 2017 A Web GIS for the knowledge and conservation of the historical buildings in Sardinia (Italy) *The International Archives of the Photogrammetry, Remote Sensing and Spatial Information Sciences* Volume **XLII-4**/W2, 2017FOSS4G-Europe 2017 – Academic Track, 18–22 July 2017, Marne La Vallée, France

[9] Arya Wardana Wisnu 2001 *Dampak Pencemaran Lingkungan* (Yogyakarta: Penerbit Andi)

[10] Prahasta Eddy 2007 *Sistem Informasi Geografis: Membangun Aplikasi Web-Based GIS Dengan MapServer* (Bandung: Informatika)

[11] Pancal Priyanka 2017 Responsive web design and web development using bootstrap frond-end framework *International Journal of Advance Research in Science and Engineering* **06**(07) July 2017

[12] About Codeigniter 2018 https://codeigniter.com [online]

[13] Priyana Yuli, Jumadi, Sigit, Agus Anggoro dan Rudiyanto 2015 Pengembangan aplikasi SIG berbasis web untuk mendukung kolaborasi dan pengambilan keputusan dalam pengelolaan airtanah di Kabupaten Karanganyar *Prosiding Seminar Nasional Aplikasi Teknologi Prasarana Wilayah (ATPW) Surabaya* 11 Juni 2015 (Surabaya: Institut Teknologi Surabaya)

[14] Sugumaran R and DeGroote J 2010 *Spatial Decision Support Systems Principles and Practices* (Boca Raton, Florida: CRC Press)

Mapping of potential green city attributes in Batu District, Batu City

Sri Utami Azis, Kartika Eka Sari* and Hanita Nirvana

Faculty of Engineering, Universitas Brawijaya, Malang, Indonesia

*kartikaekasari46@gmail.com

Abstract. Batu District is located in Batu City defined as BWK I which is designated as central government development, modern trades and services development, tourism development, supporting services of tourism accommodation and education area. As the center of city development, the area / development that occurs in Batu District will have an impact on the increase in temperature. This research aims to identify the Batu District based on green city attributes such as green open space, green waste, green water, green transportation, green energy, green building, green planning and design, and green community that can be developed in Batu District. The method of analysis in this research are using descriptive analysis to determine the attributes of green city that can be found in Batu District. Based on the description of the seven attributes of Batu City, it is known that Batu City has 5 green city attributes that are being developed, namely Green open space of 32.317 m^2, Green Waste (separate waste bins and 5 garbage bank communities), Green Water (29 9 units of recharge wells, 4 communal *waste water treatment plant* (WWTP) and 1 individual ~~WWTP.~~ Green Energy (30 units of biogas) and Green Community (5 green city observers). As for the development of Green Building and Green Transportation attributes have not been done in Batu City.

1. Introduction

Green has become a new urban development model in the Americas, Asia, Europe, Australia [1]. Green city concept emerged as a refinement of the concept of garden city and the urban forest. The concept of garden city is considered less effective in dealing with increasingly complex issues of environmental issues, because the concept of green city at the beginning only emphasized on how to green the city with urban forests and city parks.

According to the Spatial Planning, the Green City is one approach to the concept of sustainable city planning, green city is also known as ecological city, which means the balance between city development and development with environmental sustainability.

Green city according to Detail City's Spatial Planning, Batu is a city development which one of them is supported by green concept supported by structured green open space network system, efficient energy with minimal pollution, and can reduce global warming by reducing buildings with glass materials.

Green City according to the 2011 Implementation Guide of *Green City Development Program* (P2KH) has 8 (eight) attributes that must be fulfilled: Green Planning and Design, Green Open Space, Green Community, Green Water, Green Waste, Green Energy Green Transportation. One of the city that can be developed to the green city is Batu City, especially located in Batu District.

Batu District is located in Batu City defined as Urban Area, which is designated as central government development, modern trades and services development, tourism development, supporting services of tourism accommodation and education area, which has development goals in the form of an acttractive, comfortable, and eco-friendly city tourism center as part of its comparative advantages (RDTRK Batu). As a district that is developed as a city center, Batu District is not separated from the rampant development, whether the construction of settlements, offices, and artificial tourist attractions, and these affects the carrying capacity of the environment.

The existing development issues in Sub District Batu the areas / developments of the city that have an impact on the increase in temperature, which affects the hotter temperatures. It also causes discomfort, decline in biodiversity due to land use changes caused or influenced either directly or indirectly, by conversion of forest land into agricultural areas and built areas. With the existence of these developments, the concept of environmental arrangement of Batu District leads to the efforts of green city establishment with the intention of restoring the city's microclimate to be cooler than it is today, one of the concepts that will be developed is using green transportation where green transport is one of the green city attributes (RDTRK Batu).

2. Literature Review

Green City according to [2] is a development which one of them is supported by green concept supported by structured green open space network system, efficient energy with minimal pollution, and can reduce global warming by reducing buildings with glass materials. According to [3] definition of green city is an eco-friendly city by effectively and efficiently utilizing water and energy resources, reducing waste, implementing an integrated transportation system, ensuring environmental health, synergizing natural and artificial environments based on urban planning and design in favor of sustainable development principles. A city that prioritizes the balance of the living ecosystem with the built environment so as to create comfort for the population of the city that live in it as well as for the city's visitors. In this study, green city related theories are used to identify the attributes of green cities in the Batu District, Batu City

2.1. Green City's attributes

The green city attributes that can be used to determine the attributes of a green city according to the 2011 Implementation Guide P2KH can be seen from several attributes that have been mentioned by experts which are eight city attributes that can be regarded as green city, attributes that must be fulfilled according to [3] Development Program (P2KH) are:

- Green Open Space (Availability of Green Open Space)
- Green Waste (Waste management under the 3R Principle)
- Green Water (Effective water management)
- Green Building (Energy-efficient buildings or green buildings)
- Green Transportation (The implementation of a sustainable transport system)
- Green Planning and Design (Urban planning and design that is environmentally friendly)
- Green Energy (Efficient energy consumption)
- Green Community (Ehancement of community role as a green community)

2.2. Green Open Space (availability of Green Open Space)

Green Open Space is an increase in the quantity and quality of Green Open Space according to the characteristics of the city / district with a target of at least 30% of all urban areas as planned in the RTRW. Activities related to these attributes include the construction of green city parks, urban forests, green corridors in urban areas to increase the area of urban green space [4].

The Green Open Space is a more elongated / open-ended and / or more open-ended, plant-growing, plant-growing area [5]. Based on the Regulation of the [6], the function of green open space is also divided into two: intrinsic (main) and extrinsic (additional) functions. According to [5] About Spatial

Planning, availability of green open space are 30% of city area, 20% of private green space, and 10% of public green open space [7].

2.3. Green Waste (waste management under the 3R Principle)

Green waste is the application of waste and urban waste management by applying the concept of zero waste, 3R principle, ie reducing waste (reduce), increasing the value of waste (reuse), and developing recycling process (recycle) [4].

Green Waste or commonly called Zero Waste is an integrated waste processing so no garbage is wasted. Waste is a very crucial issue. Disposal sites are becoming increasingly difficult to find while garbage continues to be produced. The mixed waste disposal destroys and reduces the value of the material which may be utilized again. Organic materials can contaminate the materials that may still be recycled. One alternative waste management that can be the best solution is the management and processing of waste at the source in an integrated based Zero Waste (RDTR Batu).

The concept of Zero Waste essentially prevents the disposal of household garbage out of the house so that each individual must process the garbage themselves. Includes the process of reducing the volume of waste generation and handling waste as closely as possible from the source with approach through legal aspects (rules), organizational aspects (institutional), operational technical aspects, financing aspects (retribution) and aspects of active role of the community.

2.4. Green Water (effective water management)

Green water is the improvement of water resource utilization and management efficiency, conservation of water resources, and coverage of clean water access. Examples of the application of green water are the zero run-off concept in urban park / private green space, reuse of used water, rainwater harvesting, increased water absorption to the soil, the manufacture of surface water management systems in urban areas, and improving the quality of water-related disaster land [4].

The Green Water concept is more focused on the efficiency of water resource utilization for survival by maximizing water absorption, reducing water runoff, and water use efficiency. The application of this approach is in the form of procurement of biopori to increase water catchment area, procurement of lake as water retention, and selection of vegetation that can absorb water in large quantity so that ground water reserves can be well preserved [8].

2.5. Green Building (energy-efficient buildings or green buildings)

Green building is an application of building requirements with measurable performance in energy, water and other resources savings according to function and classification. The construction of a green building is a development that is both conceptually and environmentally responsible from the selection of the place to the design, materials and construction, operational, maintenance, renovation and utilization [4].

Green building is a combination of engineering designs, building materials and technologies that minimize negative impacts on the environment and human health [9,10]. This is achieved by a better site planning, design, building materials, construction, maintenance, and waste disposal.

2.6. Green Transportation (the implementation of a sustainable transport system)

Green Transportation is a development of sustainable transportation system, through the development of public transportation, pedestrian paths, and bicycle paths, as well as integration between modes [4].

Green Transportation is an environmental transportation device [11]. The concept of green transportation is a concept intended to make transportation modes more environmentally friendly. Green transportation is one of the eight components of green city where the development of sustainable transportation system through green transportation aims to encourage people to use the environmentally friendly public transportation. Green Transport refers to low-impact transport facilities in the environment, including non-motorized transport of walking and cycling, use of green vehicles, carsharing, and attempting to build or protect a fuel-efficient urban transport system and space so that can create a healthy lifestyle.

Criteria for green transportation are the use of environmentally friendly fuels, green fuel types, the use of environmentally friendly technologies (Hybrid), mass transit, bicycle lane facilities, pedestrian facilities, emission levels, and Smart Transportation Management System (STMS) The variables of green transportation implementation are the use of environmentally friendly fuel, the amount of emitted emissions, the type of fuel, the use of environmentally friendly technology, mass transportation, pedestrian facilities, and bicycle lane facilities [12].

2.7. Green planning and design (urban planning and design that is environmentally friendly)

Green planning and design is an improved quality of urban spatial and urban design plans that are more adaptive to the natural physical (biophysical) environment of the region, as well as seeking adaptation and mitigation to climate change issues. Activities related to these attributes include the preparation of a green city masterplan, and the preparation of detailed plans such as the RDTR and RTBL [4].

2.8. Green energy (efficient energy consumption)

Green energy is the use of energy that efficient and environmentally friendly, such as the reduction in the use of non-renewable energy, or the use of alternative renewable energy (sunlight, water flow, geothermal, sea tides). Examples of green energy are the use of solar and / or electric power for the public street lighting [4,13].

Green energy is energy generated from environmentally friendly sources, or has little negative impact on the ecosystem of the environment. This concept evolved because of the tremendous negative impact of fossil energy use. The goal of green energy is to find alternative energy sources other than fossil energy. Energy sources that can minimize negative impacts on the environment. Already many alternative energy has been developed, ranging from the utilization of wind energy, solar, water, tides, and others. In addition to reducing the negative impact on the environment, this energy can also be used as an alternative energy for the future as a substitute for fossil energy [14].

2.9. Green community (enhancement of community role as a green community)

Green community is an active participatory enhancement of society and private institutions in the realization of a sustainable city vision. Examples of green community activities are the preparation of green community maps, the socialization of green city programs to communities, the inclusion of educational institutions through school programs and green campus [4].

Green community according to City Planning and Dvelopment Department BAPPEDA Banda Aceh is a strategy of involving various stakeholders from government, business and society in green city development. The Green Community aims to establish stakeholder participation in green city development and build communities that have environmentally-friendly characteristics and habits, including in the habit of disposing of waste and the active participation of communities in government green city programs. Green community indicators according to [3] the 2011 Green City Urban Development Program Guidelines are as community sensitivity, community initiatives, and partnerships.

3. Methodology

The method used in this research is descriptive analysis. This research gives model and concept of Batu District development with theme of green city based on criteria that suitable with existing condition of Batu District. Collecting data in this study conducted by collecting primary data by direct observation in the study area, interviews with government Batu related research such as Regional Development Planning Agency, Office of Public Works Highways and Irrigation, Public Works Department of Human Settlements and Spatial Planning, Department of Tourism, Office of the Environment, and the Batu District Office and secondary data collection from the existing literature.

IOP Conf. Series: Earth and Environmental Science **314** (2019) 012067 doi:10.1088/1755-1315/314/1/012067

The scope of the area in this study was conducted in Batu District, Batu City. The area of Batu District is 4.545,81 km2 or 22.83% of the total Batu City area.

Here are the administration's boundaries of the research area, covering:

West : Malang Regency (Kabupaten Malang);
South : Blitar and Malang Regency (Kabupaten Blitar dan Malang);
East : Junrejo District;
North : Bumiaji District

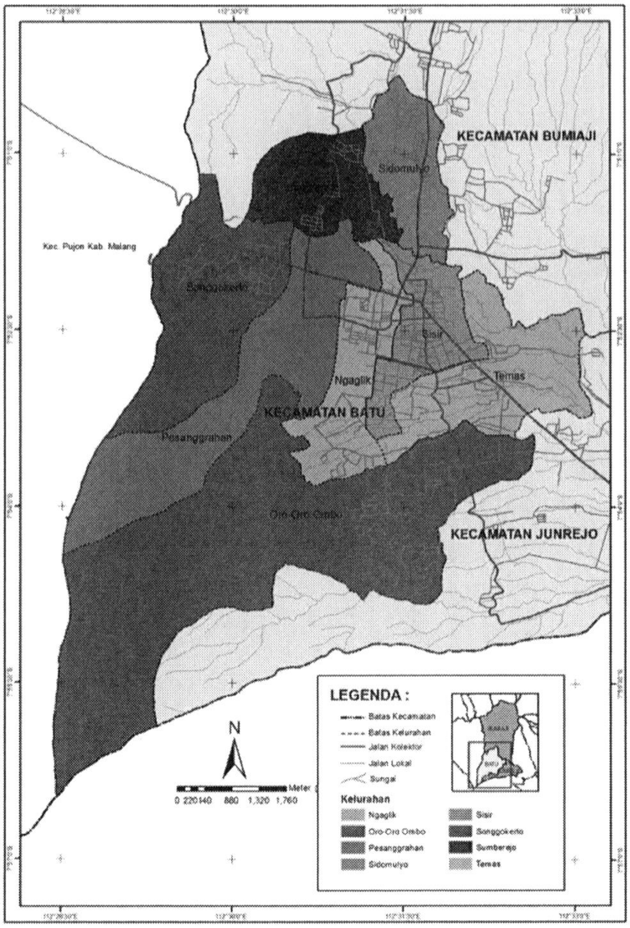

Figure 1. Location map of research area

4. Result and Discussion

4.1. Profile of Green City in Batu District

4.1.1. Green Open Space (availability of Green Open Space). *Green Open Space Under Law No. 26 of 2007 is an elongated / lane and / or clumped area whose use is more open, where crops grow, whether naturally grown, or intentionally planted*

Based on the existing condition, Batu District has green open space in the form of city park that is in the form of Batu Plaza (Alun-alun Kota Batu), Jalan Sultan Agung park, Jalan Kasiman park, Jalan Imam Bonjol park, Jalan Diran park, Jalan Gajah Mada park, and also Batu District has green open space in the form of city forest located in Oro-oro Ombo Village.

Table 1. Area of Green Open Space in the Batu District

Type of Green Open Space	Area (m^2)	Location
Bondas City Forest	22.051 m^2	Sisir Sub-district
Temas City Forest	7.826 m^2	Temas Sub-district
Jl. Sultan Agung Park	12.405,93 m^2	Sisir Sub-district
Jl Kasiman Park	630 m^2	Sisir Sub-district
Jl. Imam Bonjol Park	317,64 m^2	Sisir Sub-district
Jl. Diran Park	193 m^2	Sisir Sub-district
Jl. Gajah Mada Park	264 m^2	Sisir Sub-district
Adipura Park	25,8 m^2	Sisir Sub-district
Batu Plaza (Alun-alun)	3.963,4 m^2	Sisir Sub-district
Tugu Apel Park	214 m^2	Sisir Sub-district
Ganesha Park	90,3 m^2	Sisir Sub-district
Jl. Indragiri Park	971 m^2	Sisir Sub-district
Jl. Dewi Sartika Green Path Park	534,75 m^2	Temas Sub-district
Jl. Mawar Park	168 m^2	Pesanggrahan Village

Source: Primary survey, 2017

4.1.2. Green Waste (waste management under the 3R Principle).

Green Waste or commonly called Zero Waste is an integrated waste processing so no garbage is wasted. The concept of Zero Waste essentially prevents the disposal of household garbage out of the house so that each individual must process the garbage themselves. Zero waste concept consists of reduce, reuse, and recycle. In addition, green waste can be done by increasing the added value of waste by the presence or absence of creative economic programs from waste in the community, for example such as garbage bank.

Based on the existing condition, Batu District already implementing the waste management with the concept of green waste. The procurement of garbage facilities that have been equipped with the separation of waste types are along protocol roads such as Batu Plaza (Alun-alun Kota Batu), Jalan Dipenogoro, Jalan Panglima Sudirman, Jalan Gajah Mada, Jalan Brantas, Jalan Imam Bonjol, and Jalan Sultan Agung. Batu District also has five garbage bank groups which are built by Environmental Office of Batu City.

4.1.3. Green Water (effective water management).

The Green Water concept was introduced to conserve and manage the existing water, thus supporting the efforts to create a green city. Green water can be applied by maximizing water absorption, reducing water runoff, and streamlining water usage.

Based on the existing condition, Batu District in reducing water runoff that aims to control rain water to be able to seep back into the soil that is by making a recharge well. The absorption wells in Batu District are 29 units spread over several villages. The following table shows location and the number of absorption wells in Batu District.

Table 2. Location of catchment wells in Batu District

No.	Location of Catchment Wells	Unit
1.	Sisir Cemetery	2
2.	SMAN 01 Batu	4
3	SDN 01 Oro-0ro Ombo	2
4.	SDN 02 Oro-0ro Ombo	2
5.	SDN 01 Temas	2
6.	SDN 02 Temas	2
7.	SDN Ngaglik 02	2
8.	Lahor Agung Housing	4

No.	Location of Catchment Wells	Unit
10.	SDN Pesanggrahan 02	1
11.	SMPN 01 Batu	1
12.	SDN Sisir 05	1
13.	SMKN 03	1
14.	Dusun Toyomerto, Pesanggrahan Village	1
15.	Jl. Lahor, Pesanggrahan Village	2
16.	Jl. Ngukir, Temas Sub-district	2
Total		**29**

Source: Primary survey, 2017

In addition, Batu District has WWTP for water quality fulfillment. WWTP is a communal WWTP and medical WWTP scattered in several sub-district. The following table shows the location of WWTP in Batu District.

Table 3. Location of WWTP in Batu District, Batu City

No	Type	Location	Administrator
1.	communal WWTP	RT 05 RW 08 Pesanggrahan Village	Mr. Suropati
2.	communal WWTP	Sisir Sub-district RT 08 RW 11	RW
3.	communal WWTP	Baranrejo, Oro-oro Ombo Village	Oro-oro Ombo Village community
4.	communal WWTP	Pesanggrahan Village	Pesanggrahan Village community
5.	Medical WWTP	Sisir Sub-district	Puskesmas Sisir

Source: Primary survey, 2017

4.1.4. Green Building (energy-efficient buildings or green buildings).

Green Building is a process of designing buildings and infrastructure in such a way as to minimize the use of building materials to reduce the negative impact on ecology. Based on the existing condition of Batu Sub District has not fulfilled the concept of green building.

Table 4. Area of KDB and KDH in Batu District, Batu City

Type of Land Use	KDB	KDH
House Settlement	60-80%	5-10%
Commercial Facilities	60-80%	5-20%
Office Facilities	60-70%	10-20%
Public and Social Facilities	70%	10-20%

Source: RDTRK Batu, 2014

4.1.5. Green Transportation (the implementation of a sustainable transport system).

According P2KH green transportation is to develop public transport that connects the service centers and settlements, a use of environmentally friendly vehicles that are intermodal, congestion reduction through the implementation of policies aimed to reduce congestion at peak hours. Based on the existing condition, Batu District already has pedestrian facility in Oro-oro Ombo Village up to Batu Plaza (Alun-alun Kota Batu). Unavailability of bike lane facility Batu District make people prefer to use private vehicles such as motorcycles and cars to reach their destination. Public transportation that can be used in Batu District is angkot. The angkot route in Batu District has been able to reach activities centers such as schools, shopping areas and city plaza. In addition, the angkot route is equipped with the presence of stops scattered in Batu District.

4.1.6. Green Planning and Design (urban planning and design that is environmentally friendly).
Green planning and design is the concept of urban planning that applies a green city planning approach. Implementation of green city planning approach can be seen from the existence of documents such as the existence of city planning documents.

4.1.7. Green Energy (efficient energy consumption).
Green energy is energy generated from sources that are environmentally friendly or have little negative impact on the ecosystem of the environment. Based on the existing condition, Batu District has not implemented the two indicators in green energy. One Batu District does not use solar energy as a replacement for street lights, and also there are no absence of a house or building with energy-efficient design by considering good lighting and ventilation that can lead to the use of lights in the day.

Table 5. Distribution of Biogas (Livestock Waste) in Batu District, Batu City

No.	Location	Unit
1.	Songgoriti Village, Songgokerto Sub-district	1
2.	Dresel Village, Oro-oro Ombo Village	13
3.	Toyomerto Village, Pesanggrahan Village	8
4.	Gondorejo Vilalge, Oro-oro Ombo Village	1
5.	Tambuh Village, Songgokerto Sub-district	1
6.	Srebet Barat Village, Pesanggrahan Village	1
	Total	30

Source: Primary survey, 2017

4.1.8. Green community (enhancement of community role as a green community).
Green community is a strategy of involving various stakeholders from government, business and society in green city development. Based on the existing condition, Batu District has a community that is concern to the implementation of green city, the community is Ijo Royo-Royo, FKMPL (Forum Masyarakat Masyarakat Peduli Lingkungan), Alam Sejahtera, Earth Hour, Eco Ijo, and also Forum Kota Hijau (FKH). Batu District also has a program of one of the green communities, Earth Hour with activities such as distributing tree seedlings to the community, socializing green lifestyles to students in school, socialization through radio and also in Car Free Day Batu City, as well as clean action and unplug the nail on tree along with Batu City Environmental Office. But there is still no implementation of green city program in Batu District which cooperate with private / CSR.

Table 6. Green Community di Batu District, Batu City

No.	Name of Green Community	Chairman	Address
1.	Ijo Royo - Royo	Kuswardiyoko	Sisir Sub-district
2.	FKMPL (Forum Komunikasi Masyarakat Peduli Lingkungan)	Kariadi	Temas Sub-district
3.	Pamali Manggala Bala Bhakti	Muji Wiyanto	Sidomulyo Village
4.	Earth Hour	M. Alfien Zuliansyah	Jalan Imam Bonjol Atas no. 65
5.	Eco Ijo	Stefan Krisna Priawan	Jl. Palem Raja 18, Sidomulyo Village

Source: Primary survey, 2017

5. Conclusion
Based on the analysis result, it can be concluded that Batu District has seven green city attributes spread in seven urban villages namely green open space, green waste, green water, green transportation, green energy, green planning and design, and green community. Green city attributes

that have not been found in Batu Sub-District are green building. Each of the green city attributes that already exist in Batu District still do not meet the indicators of the 2011 Implementation Guideline for Green City Development Program (P2KH), therefore it needs an increase in every green city attribute in Batu District.

References

[1] Nugroho dan Syaodih 2013 Strategi Peningkatan Kualitas Empat Atribut Green City di Kecamatan Bandung Wetan Kota Bandung *Jurnal Perencanaan Wilayah Kota Universitas Islam Bandung* **13**(2) 1-9
[2] RDTR (Rencana Detail Tata Ruang) Kecamatan Batu Tahun 2014 – 2034
[3] Kementerian Pekerjaan Umum 2011 *Panduan Pelaksanaan Program Pengembangan Kota Hijau (P2KH)* (Jakarta)
[4] Kementerian Pekerjaan Umum 2017 *Panduan Pelaksanaan Program Pengembangan Kota Hijau (P2KH)* (Jakarta)
[5] Undang-Undang Republik Indonesia 2007 Undang-Undang Republik Indonesia Nomor 26 tahun 2007 Tentang Penataan Ruang (Jakarta)
[6] Peraturan Menteri Pekerjaan Umum 2008 Peraturan Menteri Pekerjaan Umum Nomor 5 Tahun 2008 Tentang Pedoman Penyediaan dan Pemanfaatan Ruang Terbuka Hijau di Kawasan Perkotaan (Jakarta)
[7] Hermon D, Putra A and Oktorie O 2018 Suitability Evaluation of Space Utilization Based on Enviromental Sustainability at The Coastal Area of Bungus Bay in Padang City *International Journal* **14**(41) 193-202
[8] Sagala A R, Prasetyo A, Syakur D A, Amania N R, radnawati D, Syahadat R M dan Putra P T 2017 Perencanaan Taman Kota Sebagai Salah Satu Atribut Kota Hijau di Kecamatan Gedebage, Bandung *Jurnal Arsitektur, Bangunan, & Lingkungan* **6**(3) 85-90
[9] Siagian 2005 Bahan Bangunan yang ramah lingkungan *e-USU* (Medan : Fakultas Teknik Universitas Sumatera Utara)
[10] Putra A and Mutmainah H 2016 The Mapping of Temporary Evacuation Site (TES) and Tsunami Evacuation Route in North Pagai Island, Mentawai Islands Regency-Indonesia. *IOP Conference Series: Earth and Environmental Science*. **47**(1) 012020
[11] Andriani D M dan Yuliastuti N 2013 Penilaian Sistem Transportasi yang Mengarah Pada Green Transportasi di Kota Surakarta *Jurnal Pembangunan Wilayah & Kota* **9**(2) 183-193
[12] Tjokronegoro dan Pradono 2014 Penilaian Kesesuaian Penerapan Green Transportation di Kota Bogor *Jurnal Perencanaan Wilayah dan Kota B SAPPK* **3**(2)
[13] Hermon D 2017 *Climate Change Mitigation* (Jakarta: Rajawali Pers Radjagrafindo)
[14] Hermon D 2014 Impacts of Land Cover Change on Climate Trend in Padang Indonesia *Indonesian Journal of Geography* **46**(2) 138-142

Analysis of land use change using Landsat image from 2006 - 2016 in Solok Regency

Dasrizal[1,2,*], E Barlian[1], Farida[2], A Rezki[2], A Z P Ulnie[2] and Y S Syafruddin[2]

[1]Universitas Negeri Padang, Prof. Dr. Hamka Street, Air Tawar Padang, West Sumatra, Indonesia 25131

[2]STKIP PGRI Sumatera Barat, Gunung Pangilun Street, Padang, West Sumatra Indonesia 25173

*dasrizal204@gmail.com

Abstract. Land use change is an important issue in the regional planning and development. This research uses a remote sensing approach of Landsat images and *Geographic Information System* (GIS) analysis to detect land use change in Solok Regency by time series of 2006, 2011, and 2016. Land use changes were interpreted through Landsat TM satellite images of 2006 and 2011 and Landsat 8 OLI 2016. GIS is used in analysing land use classifications. Land use classification found in Solok Regency is classified into 9 classes which dominated by primary and secondary forests followed by rice fields, mixed gardens, crop field, water bodies, settlement, shrubs and plantations. Based on the results of research, there is a continue increasing of agricultural and settlement expansion. There also expansion of mixed garden against secondary forests, and then there is a significant expansion of settlement land against the rice field. These results indicate that there has been a change of land conversion from non-cultivation to cultivation, and the development of residential areas.

1. Introduction

Land use is defined as human intervention to the land in order to meet their needs, while land cover changes refer to vegetation changes. Land use changes can generally be observed using spatial data from land use maps from the same point in different years. Remote sensing data such as satellite imagery, radar, and aerial photographs are useful in observing changes in land use. Land use change includes shifting land use to different land use (conversion) or diversifying to existing land use. In general, land use changes will change: 1). river flow characteristics, 2) number of surface streams, 3). hydrological properties of the area [1].

Land use is linked to human activities in certain areas of land. Land use information can be recognized directly by using the appropriate remote sensing. The phenomenon of land conversion is part of the transformation of the national economic structure. Economic growth and populations centering in urban areas require wider space out of town for various economic activities and settlements. As a result, the mostly marginal areas of wetland farming are converting to non-agricultural land with varying transition rates between periods and regions [2]. Land conversion often has interrelated problems, so it is not independent and cannot be solved by partial approaches but requires integrative approaches. The problems are: 1) Efficiency of allocation and distribution of resources from the economic point of view,

2) their relationship with equity and resource control issues, and 3) their relation to degradation process and damage of natural resources and environment.

The development of horizontal moving land uses spatially toward areas that are easy to cultivate. Land use also moves vertically in order to raise its quality [3]. The trigger is the growing population that increases the need for land in order to raise income for the fulfillment of needs and raise the standard of living. Community life in Solok Regency is dominated by people who rely on agrarian based life. The development of spatial land use in Solok Regency benefits from physical factors with sloping areas. Population growth and settlement in the slopes resulted the utilization of space to change. Various phenomenon of land use change has occurred from time to time. Changes in land use that occur in line with the increasing number of people who directly affected the need for land.

2. Method

This research used remote sensing image interpretation method, *Geographic Information System* (GIS) approach, field survey and spatial based descriptive analysis. Remote sensing image classification is a complex process which involves many steps, including the determination of a land cover classification system, collection of data sources, selection of a classification algorithm, extraction of thematic information, and accuracy assessment [4,5]. The interpretation of remote sensing imagery is done to find out information on actual land use types on previous years based on digital values recorded on remote sensing data. Field survey was conducted to find out the interpretation of land use and environmental conditions. The research location was conducted in Solok Regency, West Sumatera Province. The materials used in this study were Landsat 7 TM satellite imagery acquisition in 2006, 2011 and Landsat 8 OLI in 2016 obtained from USGS to obtain information on land use change and digital topographic map of Solok Regency.

Land use information is required to obtain information and classification processes for the development of land use structures and patterns in the study area. In the process of classification carried out a combination of supervised and unsupervised classifications. The supervised classification is performed after segmentation process. Segmentation is the process of dividing region based on pixel algorithm which then followed by giving the sample based on object and survey result so that segmentation can be used as classification boundary. This process uses training sample such as pattern approach, hue, color, and association with nearest neighbor method and assisted with manual editing [5,7]

Land use change was obtained by comparing land use occurring in actual year to previous observation year. From the results of the analysis will be obtained area, direction and trends of development and change of each type of land use within the period of two points of observations. Field checks are also required as a way to improve accuracy in the classification process. In this research, some geometric correction is done to improve the accuracy of coordinate position and image enhancement to improve visual quality of image data [2,7]

3. Results and discussion

3.1. Remote Sensing Image Analysis and Geographic Information System (GIS)

Landsat image of Solok Regency on 2006, 2011 and 2016 is limited by the administrative boundaries of Solok Regency located on the UTM grid of 47 South zone. Image processing is intended to extract the information contained both spatial and descriptive information. Image recovery process is done through geometric correction caused by position shift to reference coordinate system using ground control point data, the process is called "resampling", while radiometric correction has not done because it has been corrected by the image supplier. Radiometric correction is performed for errors caused by recording time or errors resulting from the sunlight traveling from an object to the camera recorder through the atmospheric medium.

The sharpening of the contrast is done by modifying the image values of each band, in order to obtain clearer information. The image reinforcement process is done by creating a composite image composite for bands 5, 4, 3 for Landsat 7 TM and band 6, 5, 4 for Landsat 8 OLI. Bands are selected based on the most representative colour discrimination to show the best shape of the object. Improved image contrast (image stretching) is done through histogram alignment (histogram equalization). The largest frequency value of the histogram that has the highest peak can be selected according to the combined combination of its image. Furthermore, from the resulting composite image can be interpreted visually on the monitor screen, so that class division can be detected by the process of classification.

The image classification process uses maximum likelihood technique (maximum likeness). The results of classified imagery in the form of land use consisting of nine types of use. The interpretation of the object for each class according to the characteristics and reflectance graph using the signature comparison module. The results of interpretation of classified imagery is divided into 9 classes as primary forest, secondary forest, paddy field, settlement, mixed plantation, crop field, water bodies, bushes and plantations.

Based on the interpretation result above then ground check and field verification was done in order to get the information according to the field condition [8]. Ground check is the process of matching the result of image classification that has been interpreted with land cover conditions in the field, while field verification is a step to get certainty of objects classified based on secondary data and observation result.

3.2. Land use in 2006, 2011 and 2016

Based on image classification of land use in 2006, 2011 and 2016 can be identified 9 types of land use as primary forest, secondary forest, paddy field, settlement, mixed plantations, crop fields, water bodies, bushes (bushes), and plantations. The following table will explain the area of each land use type (table 1).

Table 1. Land use in 2006, 2011 and 2016 of Solok Regency

No	Land use	Year					
		2006		2011		2016	
		Area (ha)	%	Area (ha)	%	Area (ha)	%
1	Primary Forest	162,883	48.3	162,653	48.23	159,544	47.31
2	Secondary Forest	17,207	5.1	16,088	4.77	14,679	4.35
3	Paddy Field	45,361	13.5	45,666	13.54	45,591	13.52
4	Settlement	3,709	1.1	3,770	1.12	3,963	1.18
5	Mixed Plantations	70,805	21.0	71,789	21.29	76,478	22.68
6	Crop Fields	27,489	8.2	27,489	8.15	27,489	8.15
7	Water Bodies	6,690	2.0	6,690	1.98	6,690	1.98
8	Bushes	2,454	0.7	2,454	0.73	2,165	0.64
9	Plantations	620	0.2	620	0.18	620	0.18
	Total	337,220	100	337,220	100	337,220	100

Based on the table 1 shows that the preliminary data in 2006 was Primary Forest area of 162,883 Ha (48.30%), then Mixed Plantations with an area of 70,805 ha (21%). Five years later in 2011 there was a change where most land use was Primary Forest covering 162,653 Ha (48.23%), then Mixed Plantations with 71,789 ha (21.29%). Then in 2016, the Primary Forest area of 159.544 Ha (47.31%), then Mixed Plantations with an area of 76.478 ha (22.68%). The distribution of each land use type forms a typical pattern of district land use types (rural) where the highest percentage of land use is forest and agricultural land.

IOP Conf. Series: Earth and Environmental Science **314** (2019) 012068 doi:10.1088/1755-1315/314/1/012068

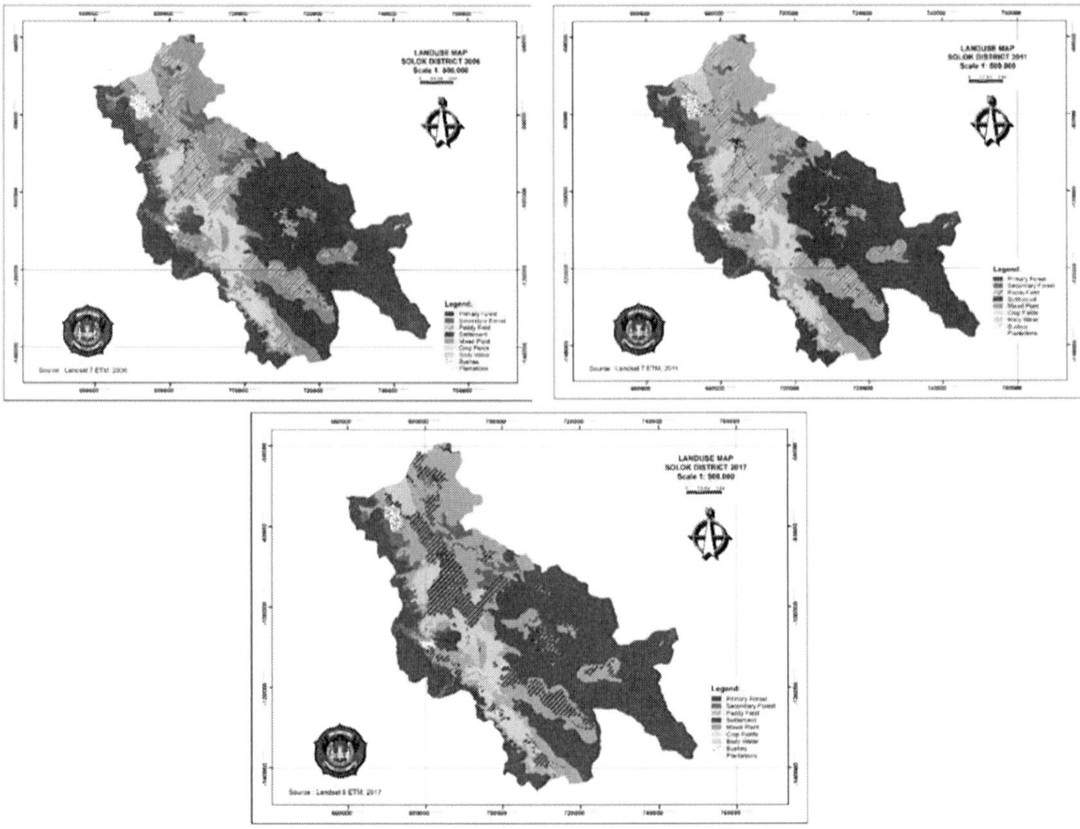

Figure 1. Land use distribution map of Solok Regency in 2006, 2011 and 2016

3.3. Land use change in 2006 – 2011

The widest land use change is the Secondary Forest change to Mixed Plantations with an area of 747 ha or an area of 64.33% of the total land area change. The total area of land that has been changed is 1.161 ha, and the total unused land is 336.847 ha. The most widespread land use change took place on Secondary Forest land use that turned into Mixed Plantations, followed by a change from Primary Forest to Paddy Field.

Table 2. The distribution of land use change in 2006 - 2011

	Land use	Area of Year 2011 (ha)									Total in 2006
		PF	SF	PdF	S	MP	CF	WB	Bs	Pl	
Aree of year 2006 (ha)	Primary Forest (PF)	162,65		218	13						162,88
	Secondary Forest (SF)		16,08			747					16,83
	Paddy Field (PdF)			45,66	48						45,71
	Settlement (S)				3,77						3,77
	Mixed Plantations (MP)			135		71,78					71,92
	Crop Fields (CF)						27,48				27,48
	Water Bodies (WB)							6,69			6,69
	Bushes (Bs)								2,45		2,45
	Plantations (Pl)									620	620
	Total in 2011	162,65	16,088	46,019	3,832	72,535	27,489	6,690	2,454	620	338,38

3.4. Land use change in 2011– 2016

The widest land use change is the change of Primary Forest to Mixed Plantations with a width of 3.109 ha or an area of 61.21% of the total land area change. The total area of land that has been changed is 5.078 ha, and the total unused land is 334,798 ha. The most extensive land use change occurred on the use of Primary Forest land that changed to Mixed Plantations, followed by a change from Secondary Forest to Mixed Plantations.

Table 3. The distribution of land use change in 2011 - 2016

	Land use	Area of year 2016 (ha)									Total in 2011
		PF	SF	PdF	S	MP	CF	WB	Bs	Pl	
Aree of year 2011 (ha)	Primary Forest (PF)	159,54				3109					162,65
	Secondary Forest (SF)		14,67			1409					16,08
	Paddy Field (PdF)			45,59	147	4					45,74
	Settlement (S)				3,96						3,96
	Mixed Plantations (MP)			76	45	76,47					76,59
	Crop Fields (CF)						27,48				27,48
	Water Bodies (WB)							6,69			6,69
	Bushes (Bs)					289			2,16		2,45
	Plantations (Pl)									620	620
	Total in 2016	159,54	14,67	45,66	4,15	81,28	27,48	6,69	2,15	620	338,38

4. Conclusion

There are 9 types of land use available in Solok District. The land uses are Primary Forest, Secondary Forest, Paddy Field, Settlement, Mixed Plantations, Crop Fields, Water Bodies, Bushes, and Plantations. The most widespread land use in Solok District is Primary Forest land use, amounting to 48% of the total area of Solok district. In 2006 to 2011, the most widespread land use change took place on Secondary Forest land use which later became Mixed Plantations. In 2011 to 2016, the most widespread use of land occurred in the use of Primary Forest type of land which later became Mixed Plantations.

Acknowledgments

We would like to send our gratitude to Ministry of Research and Higher Education of Republic of Indonesia for providing us the research funding through grand scheme of Applied Product Research on 2017/2018. Our highly appreciation for the Geography Education Department of STKIP PGRI Sumatera Barat specially for the facility to use their GIS/RS laboratory equipment during the Landsat image analysis for this research.

References

[1] Turner B L, Meyer W B and Skole D L 1994 Global land-use/land-cover change: towards an integrated study *AMBIO* **23**(1) 91-5
[2] Hermon D 2014 Impacts of land cover change on climate trend in Padang Indonesia *Indonesian Journal of Geography* **46**(2) 138-42
[3] Agaton M, Setiawan Y and Effendi H 2016 Land use/land cover change detection in an urban watershed : a case study of upper Citarum Watershed, West Java Province, Indonesia *Procedia Environ. Sci.* **33** (2016) 654-60 doi:10.1016/j.proenv.2016.03.120
[4] Kim C 2016 Land use classification and land use change analysis using satellite images in Lombok Island, Indonesia *Forest Sci. Technol.* **12**(4) 1-9 doi:10.1080/21580103.2016.1147498
[5] Putra A, Wisha U J and Kusumah G 2017 Spatial analysis of the river line and land cover changes in Kampar River Estuary: the influence of Bono Tidal Bore Phenomenon *Forum Geografi* **31**(2) 220-31
[6] Irawan S and Sirait J 2017 Perubahan kerapatan vegetasi menggunakan citra landsat 8 di kota Batam

berbasis web *J. Kelaut.* **10** 174–84
[7] Hermon D 2016 Estimate of changes in carbon stocks based on land cover changes in the Leuser Ecosystem Area (LEA) Indonesia *Forum Geografi* **29**(2) 187-96
[8] Hermon D 2016 The changes of carbon stocks and CO_2 emission as the result of land cover change for tin mining and settlement in Belitung Island Indonesia *Journal of Geography and Earth Sciences* **4**(1) 17-30 doi:10.15640/jges.v4n1a2
[9] Zhu Z, Fu Y, Woodcock C E, Olofsson P, Vogelmann J E, Holden C, et al 2016 Including land cover change in analysis of greenness trends using all available Landsat 5, 7, and 8 images: A case study from Guangzhou, China (2000–2014) *Remote Sens. Environ.* **185** 243–57 doi:10.1016/j.rse.2016.03.036
[10] Butt A, Shabbir R, Ahmad S S and Aziz N 2015 Land use change mapping and analysis using Remote Sensing and GIS: A case study of Simly watershed, Islamabad, Pakistan *Egypt. J. Remote. Sens. Sp. Sci.* **18**(2) 251–9 doi:10.1016/j.ejrs.2015.07.003

IOP Conf. Series: Earth and Environmental Science **314** (2019) 012069 doi:10.1088/1755-1315/314/1/012069

Estimations of limestone resources using three dimension block kriging method, a case study: limestone sediment at PT Semen Padang

M Gusman[1,*], B Muchtar[1], N Syah[1], M D Akbar[2] and A V Deni[2]

[1] Department of Environmental Science, Universitas Negeri Padang, Prof. Dr. Hamka Street, West Air Tawar, Padang-West Sumatera 25131 Indonesia

[2] Department of Mining Engineering, Universitas Negeri Padang, Prof. Dr. Hamka Street, West Air Tawar, Padang-West Sumatera 25131 Indonesia

*mgusman1974@gmail.com

Abstract. Making of block model was performed by geostatistic approach using kriging method to estimate and modeling of limestone source quality at Karang Putih Hill. Modeling and counting of source were based on three-dimension block model concept which was divided into block units with grid size $100\times100\times15$ m which was appropriate with the size of mining block at location of the study. Variable used for limestone source estimation were CaO, $SiO_2$and H_2O levels, according to conditions of raw material quality which has been specified.The result from variogram analysis showed that spatial characteristic of CaO, SiO_2 and H_2O levels in limestone had nugget effect percentage of 47,36%, 30,92%, dan 33,3% with range 145 - 230 m and range for vertical direction 21 - 40 m. Percentage (%) error of limestone estimation with confidence level 95% for CaO level were 54,74% $\pm$ 0,04%, SiO_2: 0,72% $\pm$ 0,04% and H_2O: 0,36% $\pm$ 0,007% respectively. Limestone resource quantity estimation at Karang Putih Hill up to elevation 214 meters above sea level which fulfilled the standard quality of CaO level $\geq$ 48 %, SiO_2 level < 10%, and H_2O level < 6 % was 158,01 million ton, while tonnase in the form of 799,94 thousand ton of silica and 1,17 million ton of basalt.

1. Introduction

Limestone is one of the main and dominant raw materials in the process of making cement which must be used in the right quality to produce a good cement product. PT. Semen Padang uses a standard reference from ISO 9002 to control the quality of the limestone to be mined in the limestone quarry of Bukit Karang Putih. With the discovery such as basalt and silica intrusions be side of limestone in the mining site that are not known for their distribution. This condition has severely hampered of mining operations because the mining boundary targets and pit plans are not appropriate and limestone production targets cannot be absolutely complete.

To controlled the quality and planning of pit designs, the relationship of spatial variables is very important to see. In this journal, we will measure the quality of limestone in Bukit Karang Putih, PT. Semen Padang. The parameters that used to estimates from the variables in limestone deposits are CaO, SiO_2, and H_2O and use a geostatistical method called kriging. Kriging method works to estimate the average levels of CaO, SiO_2, and H_2O in a several blocks that have not been known the grade of limestone as well as horizontally and vertical. And it will produce a sedimentary model in the form of

3 (three) dimensional blocks. block Model was made based on the kriging method that will calculate the volume and tonnage of limestone so that it will generate a resource of limestone based on the quality requirements.

The aim of Modeling and estimating of the limestone resources is to describe the geometry and grade of limestone sedimentation and make design of process of the mining and it applications that will be carried out. In the mining industry assessment the quality and quantity of the resources is a very important part that is carried out to determine mine planning, mine life and others.

2. Research Methodology

The location of this research is in Bukit Karang Putih PT. Semen Padang, were located in Indarung Village, LubukKilangan District, Padang City, West Sumatra Province. Mapping area can be seen on the Figure 1 bellow:

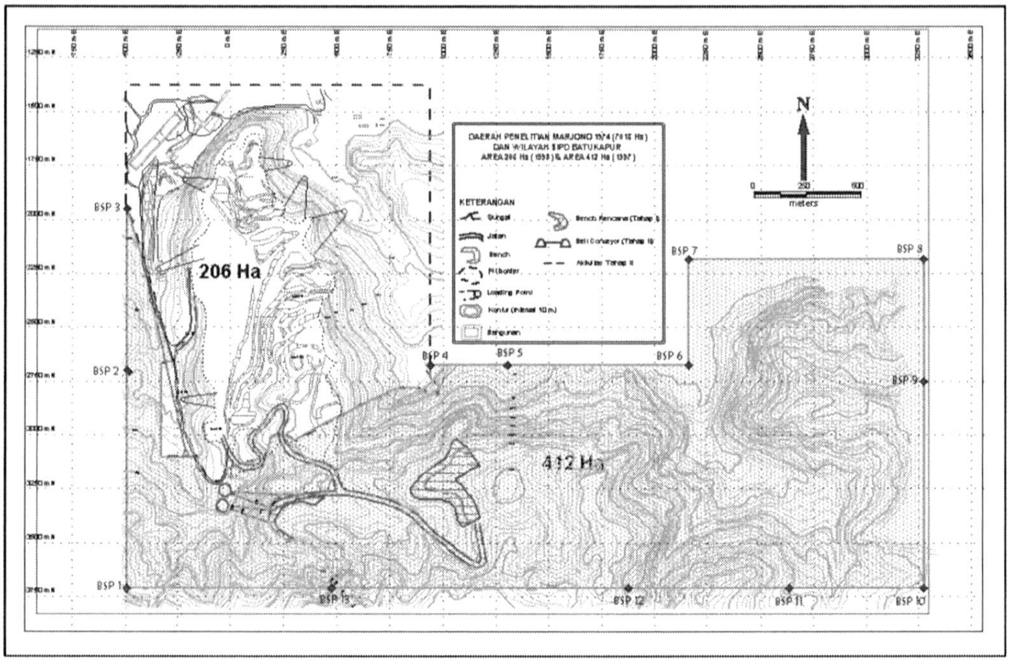

Figure 1. Area of Mining Business License PT. Semen Padang

Carrying out of this research, several work steps were prepared to facilitate and detailed the activities, bellow is the following steps:

2.1. Study of literature

Study literature in this research area was carried out by collecting various references on génesis from limestone deposits and studying previous research reports with the aim of finding out the research area in general.

2.2. Data collecting

The data obtained in theformofsecondary data which is supporting data in the subsequent data processing. Secondary data obtained are: drill log data, assay data, drill point distribution map, topographic map of the study area. The drilling data which were used comes from the exploration results of drilling at Bukit Karang Putih conducted by PT. Gamma Epsilon in 1997 as many as 13 (thirteen) core drill holes BH-01, BH-02, BH-03, BH-04, BH-05, BH-06, BH-08, BH-09, BH-10, BH-11, BH-12, BH-13, BH-14 with an average depth of 100 m . Limestone quality variables tested in the laboratory are assay values of all drill cores for CaO, SiO_2 and H_2O levels (in the form of percentage weight). Quality requirements for raw materials provided by PT. Semen Padang can be seen in Table 1.

Table 1. Quality Requirements for Material bI PT. Semen Padang

Raw material	Composition of Chemical Quality Standards			
	SiO_2(%)	Al_2O_3(%)	CaO(%)	H_2O(%)
Limestone	<10	-	$\geq$ 48	<6
Silica	$\geq$ 65	-	-	< 10
Clay	-	$\geq$ 27	-	< 37

2.3. Data processing

Based on the results of the analysis of secondary data that has been collected, the drill hole data is recapitulated as a database which will be used for further data processing, where the initial data processing is doing geological modeling and continued using descriptive statistical methods. In addition, calculations such as variogram studies, determination of estimation parameters, estimation by kriging method were then followed by modeling in 3 (three) dimensions of limestone deposits. The data base processing process uses the SGeMS program (Stanford Geostatistical Earth Modeling Software) version 2.0 where in this program will produce an estimated level in the block model.

2.3.1. Data processing is done by statistical analysis. Statistical analysis is used to see the tendency of the distribution pattern of a data set. Based on this statistical analysis can be explained the relationship/ correlation and trend of the data so that it can be determined the assessment method that is suitable with the pattern of data distribution [2]

2.3.2. Import Data and Data Interpolation. Import data is the data base entry process assay (drill hole, drill hole coordinates, and the assay data CaO, SiO2, and H2O) into SGeMS program. The result of import data is the output file in the form of a three-dimensional (3D) drill hole distribution display. Data interpolation of CaO, SiO_2, and H_2O levels was carried out on the block model that had been made based on the existing content data for every 5 (five) meters of drill hole depth interval. Interpolation is carried out on each grid which has not known the value or the grid that has a grade value from the drill assay results , in various directions in three dimensions. In this case the geostatistical method is used in the form of block kriging with attention to the distribution of the surrounding data that are interconnected with a certain radius obtained from the variogram analysis. The variable made to get the kriging parameter in this study is the three-dimensional variogram, because it takes into account spatial aspects in terms of x, y, and z [3].

2.3.3. Experimental Variogram and Fitting Variogram. Variogram can be used to analyze the level of similarity or variability between each data level CaO, SiO_2, and H_2O. To determine the appropriate variogram, it is necessary to make an experimental variogram. The experimental variogram was made based on each level data in limestones so that there would be 6 (six) experimental variograms. The process of making an experimental variogram is done with the SGeMS program, which requires first selection of variogram parameters. The selection of these parameters is done by observing the data patterns and sample data used, because the selection of good parameters will produce a good experimental variogram so that it can facilitate the variogram fitting process to produce variogram models. There are 2 (two) methods commonly used in fitting variogram, namely: visual method and least square method. Geostatistics experts use more visual (manual) methods for variogram fittings because the results are satisfactory [4].

2.3.4. Geological Modeling. Geological modeling is the activity of making a geological model of a deposit of quarry material, in this case limestone deposits, based on data taken in the field to obtain a representative model. From the resulting model can be calculated resources. Limestone resource

IOP Conf. Series: Earth and Environmental Science **314** (2019) 012069 doi:10.1088/1755-1315/314/1/012069

modeling Bukit Karang Putih, PT. Semen Padang is done using a program mining. Resource modeling and estimation is an activity that is the basis for mine planning. Resource modeling and estimation must be carried out before mining activities begin [5].The data needed as input for the basic modeling material are as follows [6]:

1. Topographic map of Bukit Karang Putih.
2. Coordinate data (X, Y, Z) position of each core drilling point.
3. Lithological data from core drilling results.

Core drilling data becomes input for making drillhole files. The drillhole file contains information [6]:

1. Collar data, contains information on coordinates (X, Y, Z) surface position of each core drilling point, total depth, drill ID ,
2. Survey data, contains bearing, dip and drill ID information.
3. Sample data, contains information of penetration (from, to) and data content and drill ID
4. Geological data, contains information containing information, drill ID, Depth (from, to) and lithology.

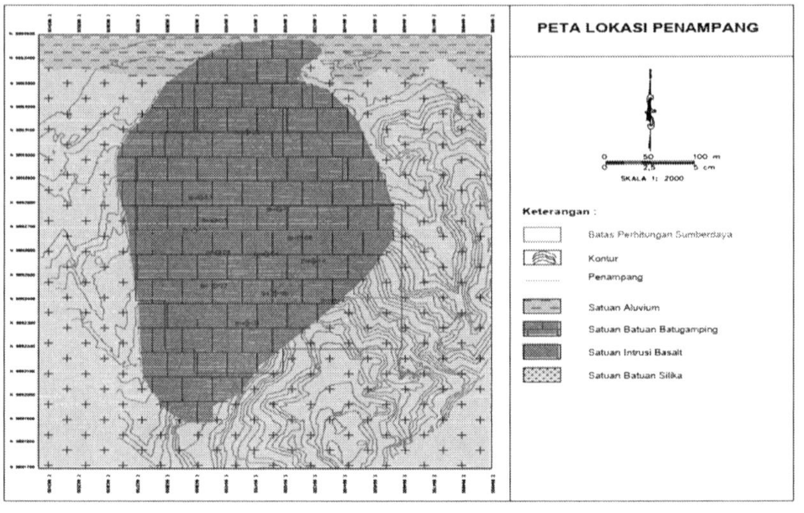

Figure 2. Cross Section Directions West-East [6]

2.3.5. Estimation with Kriging. Value estimation of the smaller block model units (grid) is done by geostatistical method in the form of kriging, where the type of kriging method used in this case is ordinary block kriging. In this study estimates were made of limestone as ore, while silica and basalt were considered as waste. The process of calculation by the kriging method to get the estimated values of CaO, SiO_2, and H_2O useds SGeMS program [7].

2.3.6. Estimation with Kriging. The block model that has been estimated with kriging contains information about the values of CaO, SiO_2, and H_2O levels estimation results. This information is very important for the next process, namely estimating limestone resources in the study area. Estimated levels using the kriging method with certain cut-off values in accordance with limestone quality requirements of PT. Semen Padang. Before calculating resources, the model must be corrected to the topographic value of the calculation area, the elevation of the bottom borehole, the silica and basalt boundary because the existing model is a solid block [8].

The correction of the model cannot be done with the SGeMS program, so it must be done with a separate program. Where, the correction material is the output value of the model, which in the value of the correction results will be modeled again to represent the distribution of quality and limestone resources resulting in 3 (three) dimensional kriging blocks [9].

IOP Conf. Series: Earth and Environmental Science **314** (2019) 012069 doi:10.1088/1755-1315/314/1/012069

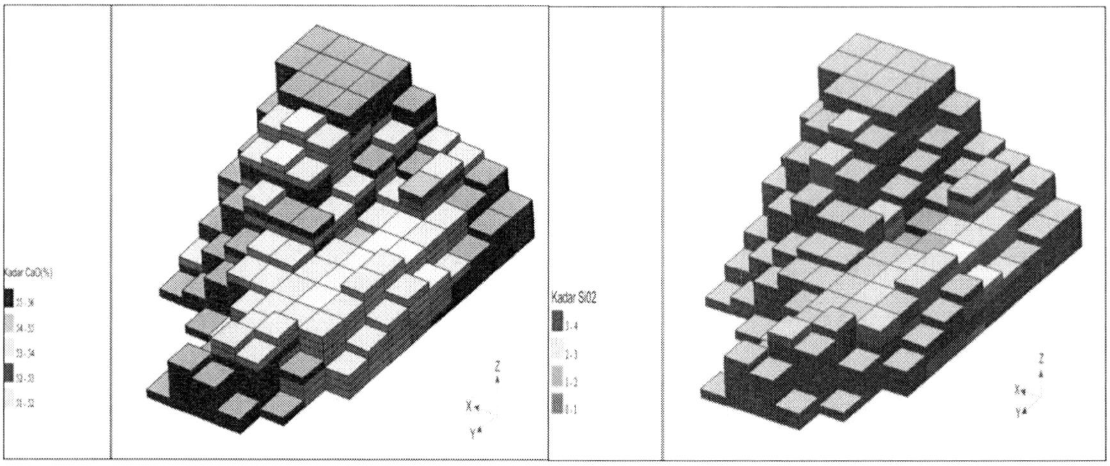

Figure 3. Block estimation model for CaO levels and SiO_2 levels in limestone with kriging which has been corrected to the topography and bottom borehole.

2.3.7. Calculation of Resources. Limestone resource calculation is done using a grid model. Where each grid is 100 x 100 x 15 m corresponds to the smallest units in the block model system. In this study the distribution of limestone quality and estimation is adjusted to the quality requirements of raw materials at PT. Semen Padang. Resource calculation is done based on the results of estimation of levels in limestone by block 3 (three) dimensional kriging method as shown in Figure 4.

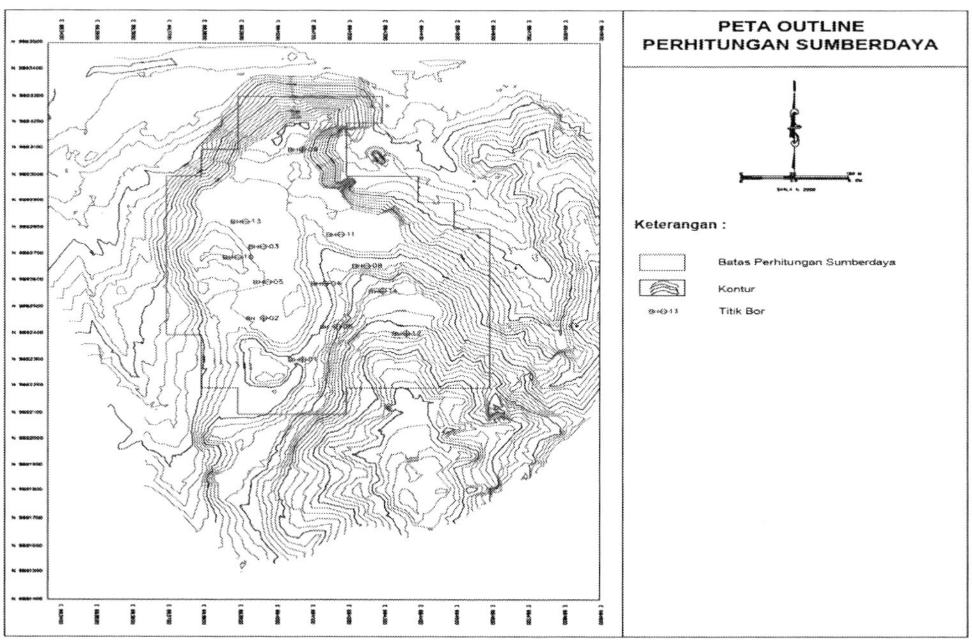

Figure 4 . Map outline of resource calculation

Resource modeling is carried out based on the block modeling framework. With the dimensions of each block adjusted to the dimensions of mining. The principle of volume calculation is the volume of one cell block multiplied by the number of existing block cells , which are defined as follows[10]:

$$\text{Tonnage} = \text{Vol.}_{\text{one block}} \text{ Number of Blocks density } (\rho) \quad (1)$$

$$= \sum_{i=1}^{n} x_i y_i z_i \times \text{density}(\rho)$$

$$Tonnage_{\,limestones} = {}_{total}\,tonnage - (Geological\ Losses\ {}_{Total}\,tonnage) \quad (2)$$

The tonnage of Bukit Karang Putih limestone resources is calculated by the parameters:

1. Density (ρ) limestone used is 2.6 tons /m^3.
2. Geological losses in limestone deposits are 10%.
3. Limestone resource estimates are carried out up to level 214.
4. In modeling the silica rock boundary has been localized, and the basaltic rock intrusion boundary. The area within the boundaries in the estimation is not counted as limestone resources at the study site.

At this stage, data analysis is carried out both qualitatively and quantitatively. Qualitative analysis is carried out based on a limestone sediment model made from secondary data. Whereas quantitative analysis is carried out based on statistical analysis, variogram studies, and parameters that are determined in the process of estimation of limestone deposits in the study area.

3. Result and Discussion

3.1. Database Assay

The assay database is the initial data that contains information about the results of drilling and sampling activities, which can later be used in various activities, such as resource calculations. This assay database can be obtained from exploration drilling, test pit , or channel sampling activities. In this study the assay database used was obtained from exploration drilling activities. Database used in this study include the coordinates of the drilling location, the depth of drilling, the content of the elements consisting of levels of CaO, SiO_2 , and H_2O.

The basic information of the assay database for this study was obtained from exploration drilling activities with depths ranging from 84 meters to 100 m, while the level analysis of samples obtained from drilling was carried out on average every 5 (five) meters of the sample depth.

This assay database was made for the purpose of importing data into the SGeMS (Stanford Geostatistical Earth Modeling Software) program, which is software or programs for processing data in geostatistics using the kriging method on a geometry support in the form of 3 (three) dimensional blocks which are models from processed data. The assay database used in this study is divided into 2 (two) parts, namely: Drill hole data , which contains the position / coordinate data of drill holes in the form of Northing , Easting , and elevation. Level data that contains information about the levels at each specific depth interval according to assay or analysis of the levels carried out.

3.2. Preparation of Database Assay

The assay database is an important data or information that forms the basis of the process of modeling and estimating resources obtained from the results of the sampling activities and the results of the level analysis of the sample. In this study, the data collected to obtain information about the characteristics of the data and the population of subsurface limestone deposits were obtained based on drilling activities by PT. Semen Padang.

Database assay for this study consists of 2 (two) parts, namely: drill hole data in the form of coordinate and elevation data, as well as assay yield data on average every 5 (five) m depth interval of drill hole samples obtained. Drill hole coordinate data is very important because it will determine the data distribution in the calculation area or block, where the data of this drill hole will be used as the initial data to estimate the CaO, SiO_2 and H_2O values in the model blocks that have been made. The spatial characteristics of this initial data population can be known from the results of the variogram analysis of these data. Assay spacing carried out every 5 m for limestone deposits is not representative enough to produce preliminary data in resource modeling and estimation activities, because it will get little or limited initial data although generally limestone deposits have a spread with relatively homogeneous continuity laterally, but constraints the existing data contained in the Bukit Karang Putih is only limited to 13 (thirteen) drill holes.

3.3. Distribution and Drill Hole Intervals

Regular distribution and drill hole intervals will be crucial for getting good initial data in resource modeling and calculation. In general, for limestone deposits there is no convention regarding drilling intervals to state the classification of resources and deposits of limestone, it is different for coal deposits that have been clearly formulated regarding the classification of resources in SNI (Indonesian National Standard). However, based on the reality in the field and exploration activities regarding the previous limestone sediments stated that the average drilling interval of 100 m was representative enough to do the modeling and calculation of resources.

Judging from the data in the form of a base map and drill hole distribution map for exploration of limestone deposits in the Bukit Karang Putih research area, PT. Semen Padang, it can be seen that the distribution of data is not good enough, where drilling is done with a grid that is not regular. Likewise, the assay data space should be done every 1 m interval so that the estimation of limestone quality distribution is better. However, with the increasingly tight data assay samples taken, the costs to be incurred will be even greater.

3.4. Variogram Analysis

Variograms for levels in limestones are made in various directions, namely horizontal direction and vertical direction with omnidirectional. The goal is to determine the continuity of data in three dimensions and get a representative estimation parameter for subsequent data processing, namely estimation using 3-dimensional block kriging method.

3.4.1. Variogram in Limestone. The variogram model used for the data levels of CaO, SiO_2 and H_2O is a spherical model where the data behaves linearly near the starting point stating that the data has moderate / moderate continuity. In addition, the CaO variogram in these limestones has a percentage Nugget effect is 47.36%, SiO_2 : 30.92% and H_2O: 33.33%, and for CaO levels in the vertical direction is 15.62%, SiO_2 : 45%, and H_2O: 35.29%. The nugget effect data values of CaO, SiO_2 and H_2O in various directions are included in the medium nugget effect, except for CaO levels for vertical direction included in the classification of low nugget effect. However, in general, the levels of limestone deposits are relatively uniform, and the tendency of erratic data is almost non-existent. Total value sill obtained for CaO content data on limestone is 3.80%, SiO_2 : 4.85% and for H_2O: 0.12% data, while for vertical direction on total CaO data sill of 3.20%, SiO_2 : 3.37% and H_2O: 0.08%. Persentase Nugget effect 25-50 % is included in the medium nugget effect[11].

However, the range (a) value is different for the data of CaO, SiO_2 and H_2O levels, namely the range variogram value for the data of CaO, SiO_2 and H_2O levels in limestone of 150 m, 230 m, and 145 and the variogram range value for vertical direction for levels of CaO, SiO_2 and H_2O for 40 m, 40 m, 21 m. Range values state the extent to which a data relates or has a correlation with other data. The level of SiO_2 has a relatively low variability or high continuity from the data of CaO and H_2O levels in limestones which can be seen with a large range value.

Based on the available data, it shows that the drilling spacing distance conducted at the Bukit Karang Putih research site is an average of 100 m where this distance is smaller than the range obtained from the variogram analysis. To get the data that has a spatial correlation and the same characteristics, based on the CaO level the drilling should be carried out at a distance of 150 m, SiO_2 230 m, while the H_2O 145 m, but the overall data content, then the drilling space should be performed with a distance 145 m follow kontiniutas levels of H_2O which is the lowest, while the vertical direction seen from the data content of the entire sample taking should be done at a distance of 21 m follow kontiniutas H_2O which is the lowest.

3.5. Kriging estimates

The estimation process with the kriging method is part of the procedure for modeling and estimating limestone sediment resources. Kriging will estimate each value in the block model units (grids) that have been made using the initial data, namely the assay results in the drill hole. The estimation grid is 100 x 100 x 15 m. This dimension corresponds to the size of the mining block and high level at PT. Semen Padang. Although on some grids there are CaO, SiO_2 and H_2O levels from the assay results, but kriging will re-estimate each level of CaO, SiO_2 and H_2O in the grid. Because it has been explained in the previous chapter that the data collected from the drill hole does not necessarily represent the actual values of CaO, SiO_2 and H_2O from the grid / block represented by the drill hole.The estimation with this kriging method uses the following parameters:

- The maximum data used is 15 data, where the data used is in the search area which is generally the closest data to the data that will be estimated at a certain distance.
- The area of the search area is in the form of " Ellipsoid 3D " which is stated by the axis parameter, namely: r max = 300 meters, r med = 300 meters, and r min = 50 meters, and the angular parameters are α, β, and θ which are all 0 angles made of 0 °.

The ellipse field in the search area will be located in the horizontal plane and there is no subduction because all the angles are made 0 °, this is based on the characteristics of limestone deposits which have a high homogeneity and lateral dispersal, in contrast to deposits in the form of veins that have a specific orientation.In the figure below shows a level map of the estimation results with kriging on the levels of CaO and SiO_2 at the elevation of basalt and silica intrusions at the study site.

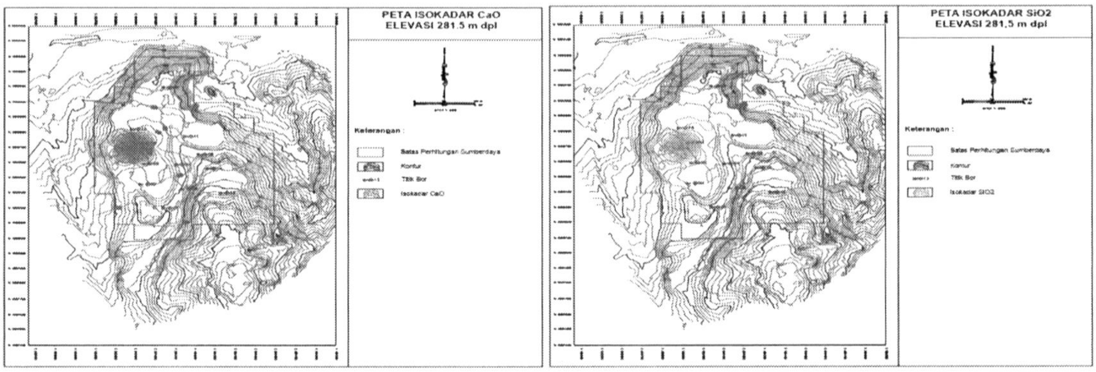

Figure 5. Map of Isograde: (a) CaO and (b) SiO_2 at an elevation of 281.5 meters above sea level.

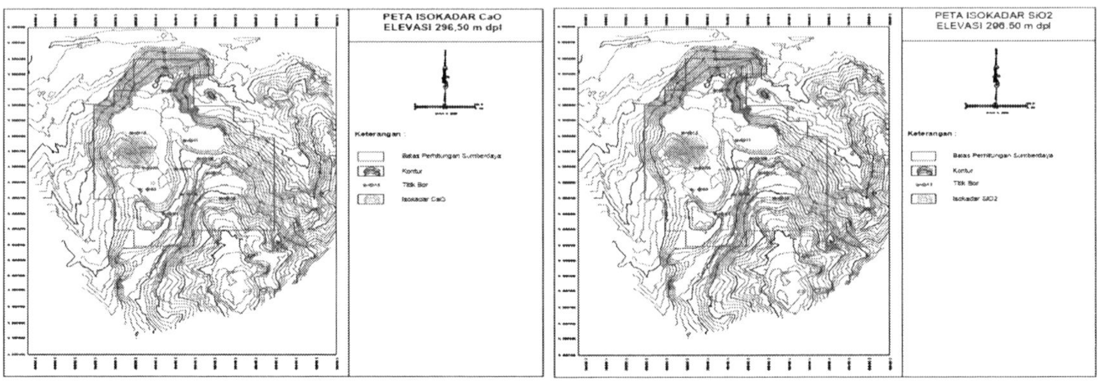

Figure 6. Map of Isograde: (a) CaO and (b) SiO_2 at an elevation of 296.50 meters above sea level

The presence of basalt intrusion will inhibit the mining activity process, therefore it is necessary to know the position and elevation of its density. From Figure 5 through 6 above it can be seen that the basalt intrusion is at the elevation of 281.5 m and 296.5 m, and this can be seen from the decreasing levels of CaO and increasing SiO_2 levels to direction of intrusion. Likewise, the presence of silica in the study location is at an elevation of 386.5 m, and 401.5 m. Breakthrough rocks found in the research location are igneous rocks with basaltic composition consisting of mafic minerals. Limestone as a side rock will experience quality changes towards intrusion with reduced levels of CaO and increased SiO_2 levels.

4. Presentation and Tabulation of Resource Calculations

Presentation and tabulation of the results of resource calculation is a very important step in the assessment and modeling of a quarry and to display a model of resources that are considered potential value and calculate the number of deposits based on the model that has been made taking into account certain aspects. The resource calculation block model system contains the smallest block units that function as geometry support for estimated values of CaO, SiO_2 and H_2O. The size of the block model units is 100 x 100 x 15 m as described earlier. The size of the block model units considers the dimensions of the mining block in the study area.

Table 2. Tabulation of Limestone Resources at Bukit Karang Putih PT. Semen Padang Up to Elevation 214 Meters Above Sea Level

Category	Total
Resources (*insitu*) Limestone (Ton)	175,566,800
Geological Losses 10% (Tons)	17,566,800
Resources (Tons)	158,010,120

Tabulation of results of limestone sediment resources by using 3 (three) dimensional block kriging methods are shown in Tables 2 and 3 where limestone sediment resources are expressed by tonnage and the calculation is based on the size of the mining block. The calculation of the amount of limestone sediment resources in this study has considered aspects of the cut off grade value and limestone density value of 2.68 tons / m^3.

Table 3. Tabulation of Silica and Basalt Tonnage Bukit Karang PutihPT. Semen PadangUp to Elevation 214 Meters Above Sea Level

Category	Total
Silica Tonnase (Ton)	799,949.30
Tonnase Basalt (Tons)	1,179,045.40

References

[1] Remy N, Boucher A and Jianbing W 2009 *Applied Geostatistics with SgeMS* (London: Cambridge University Press) pp 50-61
[2] Anonymous 2007 *Reports and Data from the Mining Department of PT. Semen Padang*
[3] Jean P Chilès and Piere D 2012 *Geostatistics Modeling Spatial Uncertainty* (Canada: John Wiley & Sons, Inc) pp 147-238
[4] Lee J Bain and Max E 1992 *Introduction to Probability and Mathematical Statistics 2nd ed* (USA: DUXBURY, Brooks/Cole) pp 358-384
[5] Heriawan M N and Koike K 2008 Uncertainty assessment of coal tonnage by spatial modelling of seam structures and coal qualities *International Journal of Coal Geology* **76**(2008) 217–226
[6] Heriawan M N, Rivoirard J and Darijanto T 2004 Grade estimation and geometric modeling of a lateritic nickel deposit using ordinary block kriging *Jurnal Teknologi Mineral* **11**(1) 41-52

[7] David M 1977 *Geostatical Ore Reserve Estimation 3rd ed* (Holand: Elsevier Scientific Publishing Company Amsterdam)
[8] Hustrulid and Mark K 1995 *Open Pit Mine Panning & Design Fundamentals* (Rotterdam, Netherland: A.A. Balkema)
[9] Annels A E 1991 *Mineral Deposit Evaluation: A Practical Approach* (London: Chapman & Hall)
[10] Journel G A and Hujibregts J 1978 *Mining Geostatistic* (London: Academic Press)
[11] Clark I 1979 *Practical Geostatistics* (London: Applied Science Publisher)
[12] Darijanto T 1999 Geostatistik Diktat Kuliah Departemen Teknik Pertambangan ITB Bandung
[13] Remy N 2004 *Geostatistical Earth Modeling Software User's Manual* (London: Cambridge University Press)
[14] Journel G A and Hujibregts J 1978 *Mining Geostatistic* (London: Academic Press)
[15] Stuart S 1976 *Recent Advances in Mining and Processing of Low-Grade and Submarginal Mineral Deposits* (New York: Pergamon Press Inc) p 100

IOP Conf. Series: Earth and Environmental Science **314** (2019) 012070 doi:10.1088/1755-1315/314/1/012070

Dynamic model of land use change in landslide hazard zones in Tanah Datar District, West Sumatra

Iswandi Umar[1,2,*], Indang Dewata[2] and Eri Barlian[2]

[1]Department Geography, Faculty of Social Sciences, Padang State University. Airtawar Padang-Indonesia
[2]Pascasarjana, Padang State University. Airtawar Padang-Indonesia

*iswandi_u@yahoo.com

Abstract. Population growth has an impact on land resource needs, pressure on land use will have an impact on environmental degradation. As an effort to reduce pressure, policy efforts in land use are needed. This study aims to develop dynamic models in land use and develop land use policy direction.The method used in dynamic modeling uses system analysis and land change policy direction using ISM analysis. In dynamic modeling, land use change uses three scenarios, namely: optimistic scenario, muder scenario, and pessimistic scenario. In determining the direction of the policy involving all stakeholders as many as 15 experts. The results of dynamic model analysis show that forest area changes from time to time of 1.6 percent per year. Changes in forest areas will have an impact on increasing environmental disasters. As an effort to save the environment, there is a need for law enforcement and strict sanctions against perpetrators of forest area destruction.

1. Introduction

Landslides are a natural phenomenon that is influenced by many factors, uncontrolled land use can cause landslides. In addition, the high intensity of rainfall causes the soil cavity to be filled with water, and forms a sloping area [1,2,3]. Today, landslides often occur in various parts of Indonesia, and landslides have caused property losses and loss of life. Landslide disasters have caused the destruction of settlements and other facilities and infrastructure, so that it is necessary to restructure the settlements of the population into areas without landslides [4,5,6,7]. During recent decades, the study of Land Use Change has become a prominent research topic, as change in land use has been recognized as one of the most important factors of environmental modification in the world [8]. The land use refers to objects that represent human activities that results in the production of goods and services for society [9].

Uncontrolled population growth affects land use. Built area will increase, but the water catchment area will be even faster. Unbalanced and supported by the amount of water absorbed in the soil will encourage landslides during the rainy season [10,3].The dynamics of land use change for settlements are influenced by human movements in building settlements and the movement of regional functions, such as education, industry, trade, etc [11]. Furthermore, the rapid development will lead to changes in land use patterns, where built space increasingly dominates and urges natural spaces to change function [12,13,3,14]. In addition, [15] and [16] explained that changes in land use patterns resulted in fluctuations in the carrying capacity of land resources, resulting in landslides. Land use change has been recognized throughout the worldwide as one of the most important factors influencing the occurrence of landslides hazard around the world [17]. [18] and [19] explain that landslides are essentially caused by the inability of the soil to withstand loads above it because the soil has undergone degradation of soil properties.

Tanah Datar District has experienced an increase in landslides from both frequency and area affected in the period 2000-2017 [17]. According to [21], landslides can cause serious casualties dan severe damage to property. Furthermore,[22], explained the death toll caused by landslides worldwide is large. Disaster mitigation that can be done to minimize the impact of risk by determining the area of landslide hazard. Tanah Datar District based on physical characteristics has areas that are very prone to landslides, including: a) morphologically, around 55 percent of the area is relatively steep (<27%); b) around 70 percent of the Limapuluh Kota District area primary forest area is converted into a secondary forest area; c) increasing rainfall intensity in the upper watershed area; and d) the development of the developed area. The growth of world population is currently increasing every year and requires a shorter period of time. The faster the growth of the world's population has an impact on increasing the need for land, especially for the development of residential areas. Limitations of the earth's surface space to support land needs for settlements have an impact on the use of land that is not in accordance with its designation. Tanah Datar District has a population growth rate of around 1.3 percent per year. High population growth increases the need for land for residential areas. While the physical characteristics of the Tanah Datar District area have many limiting factors to be used for residential areas, namely: a) relatively steep morphology, b) around 70 percent are primary forest areas, and c) are hazard to landslides and volcanic eruptions .

2. Research Methods

2.1. Location and Time of Research

This research is found in Tanah Datar District. Geographically the location of the study is in longitude $100^{0}18$'E – $100^{0}50$'E and latitude $0^{0}18$'S-$0^{0}40$'S. Furthermore this study was conducted in April - October 2018. Figure 1 is presented in the study location.

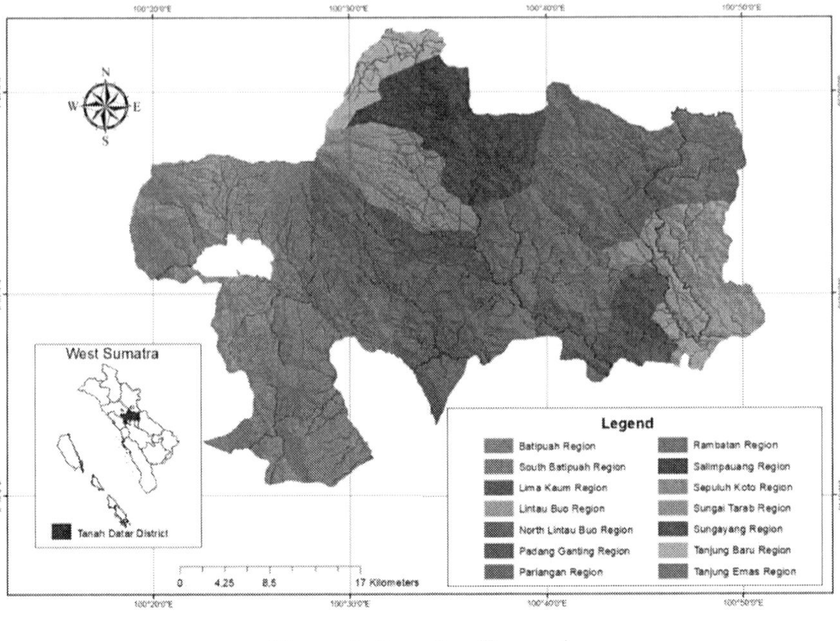

Figure 1. Location Research

2.2. Data Type

The type of data used in this study is categorized into two types, namely primary data and secondary data. Primary data is generated from measurements and collection from the field, while secondary data is obtained from official documents, information and records from various relevant agencies. According to [13,3,14] stated that some map needs in the dynamic model of land use in landslide hazard zone research are land use sourced from Landsat 7+ ETM imagery in 2010 and Landsat 8+ ETM imagery in 2017.

2.3. Analysis Technique

When dynamic systems there is often a need to use assumption because; data may not exist for variables and we don't have empirical data about the future [23]. In the study of dynamic models of land use in the landslide hazard area a comparative analysis of land use was carried out with different interpretations of the time. The results of interpretation of changes in land use are analyzed using dynamic system approach. In the dynamic system approach there are several stages, including: needs analysis, problem identification, problem formulation, system identification, validation, model supplementation, and model evaluation.

To determine the direction of the policy using the Interpretative Structural Modeling (ISM) method. ISM is an established qualitative tool which can be applied in various disciplines [24]. The ISM method is effective enough to structure complex issues because it can be used to define and clarify issues, assess impacts and identify relationships between policies. Broadly speaking, the stages of the ISM method are as follows:

a) Decomposition of each element into several sub elements.
b) Determination of the contextual relationship between sub-elements in each element that shows pairwise comparisons there / no contextual relationships are used by expert opinions.
c) Structural Self Interaction Matrix (SSIM) using symbols V, A, X and O.
d) Making the Reachability Matrix (RM) table, replacing the symbols V, A, X and O with numbers 1 or 0.
e) Perform calculations based on transivity rules where the SSIM matrix is corrected until a closed matrix occurs.
f) Perform sub element level on each element according to vertical or horizontal levels. Preparation of the Power Dependence Driver (DPD) matrix for each sub element. Element classification is divided into four, including:
   1) Quadrant I: Autonomous consists of sub elements that have driver power (DP) value $\leq$ 0.5 X and dependence value (D) $\leq$ 0.5 X. Where X is the number of sub elements in each element. Sub elements in quadrant I are generally not related to the system.
   2) Quadrant II: Dependent consists of sub elements which have a power driver (DP) value $\leq$ 0.5 X and dependence value (D) $\geq$ 0.5 X. Where X is the number of sub elements in each element. Sub elements in quadrant II are sub elements that depend on elements in quadrant III.
   3) Quadrant III: Link (Linkage) consists of sub elements that have driver power (DP) value $\geq$ 0.5 X and dependence value (D) $\geq$ 0.5 X. Where X is the number of sub elements in each element. Sub elements that enter the third quadrant need to be studied carefully, because every action on one sub element will affect the other sub elements in quadrants II and IV.
   4) Quadrant IV: Independent consists of sub elements that have a power driver (DP) value $\geq$ 0.5 X and dependence value (D) $\leq$ 0.5 X. Where X is the number of sub elements in each element.

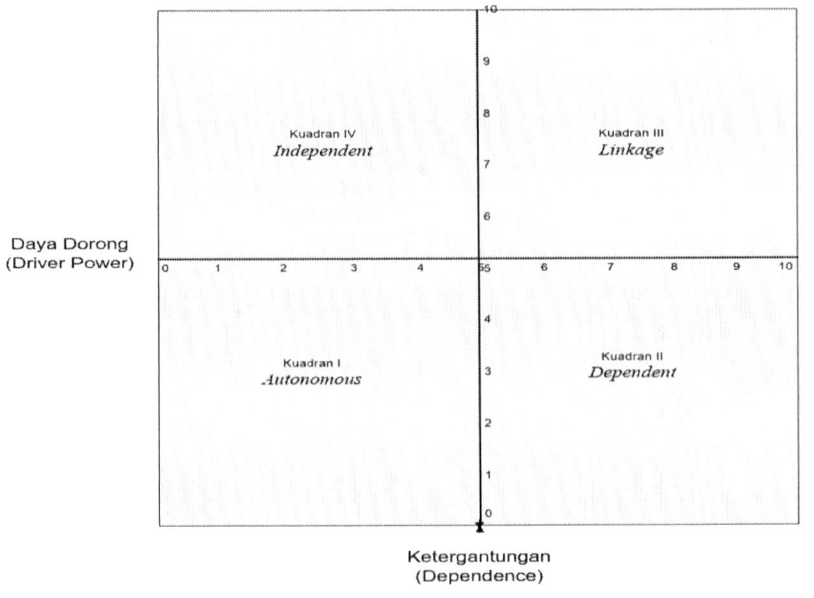

Figure 2. Driver power and dependence matrix

3. Research Result

Tanah Datar District has around 45 percent of the landslide hazard area. High population growth (1,6 percent) causes the use of land that is not suitable to be used or developed for settlements [3]. The study area has the characteristics of the area around 70 percent of the morphology of the hills. Based on the geology of the research area, it is an area of Bukit Barisan and Semangko faults.

Table 1. Land use change for the period 2010-2017 in Tanah Datar District

Land Use	Forest	Farm	Intercropping	Sattlement	Paddy	Swamp	Clearing	Total (ha)
Forest	**75,864**	1	0	0	0	0	0	75,865
Farm	0	**12,678**	7,717	1,507	9,004	0	62	30,968
Intercropping	0	10,201	**589**	273	902	21	8	11,994
Settlement	0	0	0	**5,371**	0	0	0	5,371
Paddy	0	1,022	966	294	**6,406**	0	0	8,688
Swamp	13	71	0	11	111	**182**	0	388
Clearing	0	0	0	0	0	0	**0**	0
Total (ha)	75,877	23,973	9,272	7,456	16,423	203	70	

Source : Analysis (2018)

In Tanah Datar District the period of 2010-2018 has experienced changes in land use.In 2010 around 70 percent were forest areas, but this region experienced a reduction of 0.4 percent/year. Although the changes have no significant effect on landslides, this condition will be a threat to the future.

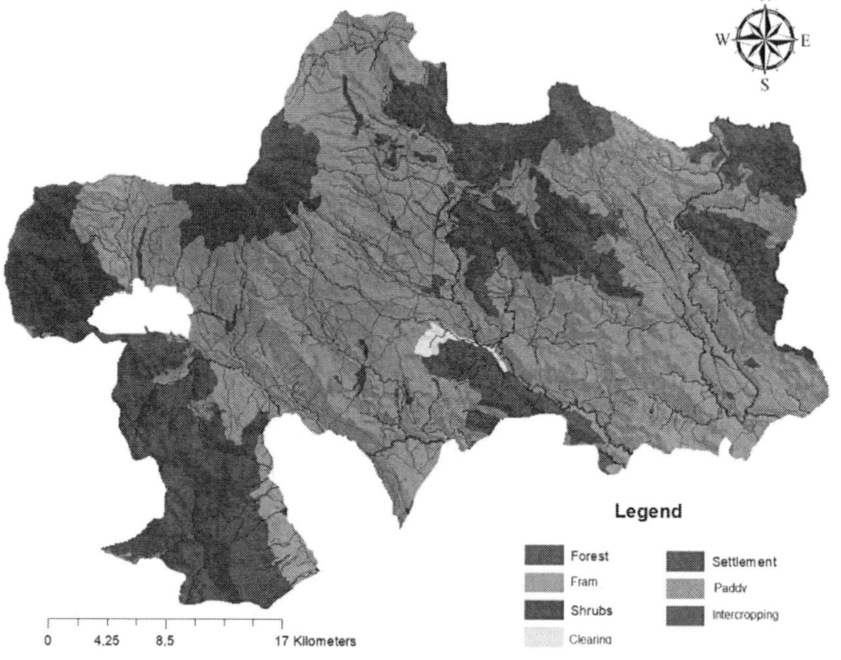

Figure 3. Land use in Tanah Datar District (2010)

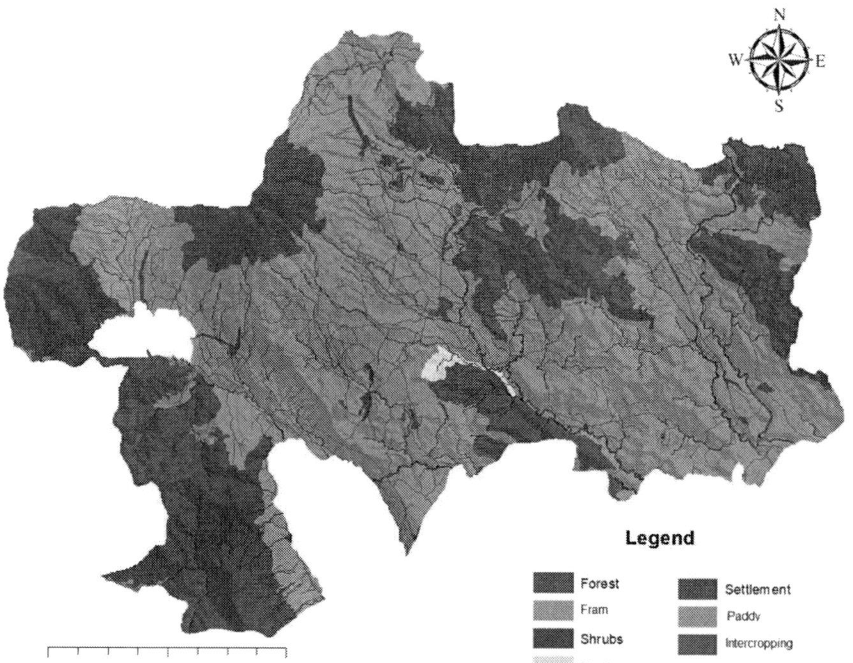

Figure 4. Land use in Tanah Datar District (2017)

The results of problem identification have four causal relationships, including: population growth, land use change, land needs, and policies. Population growth in the study area increased about 1 percent/year, and this problem caused land change about 1,4 percent/ year.

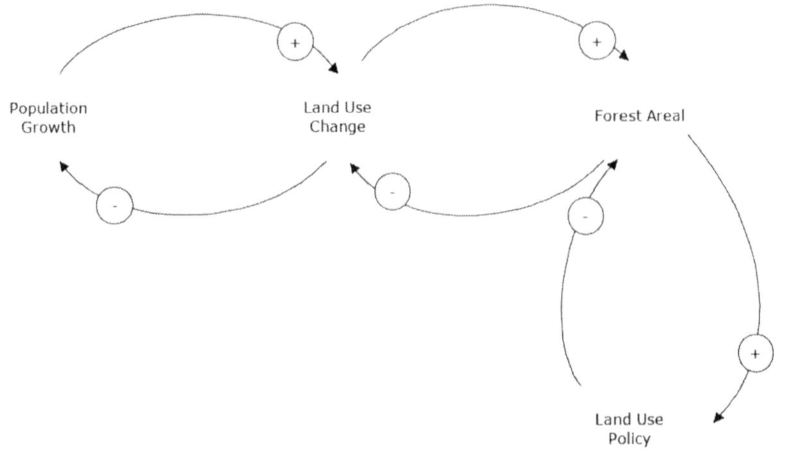

Figure 5. Causal loop dynamic model land use change in Tanah Datar District

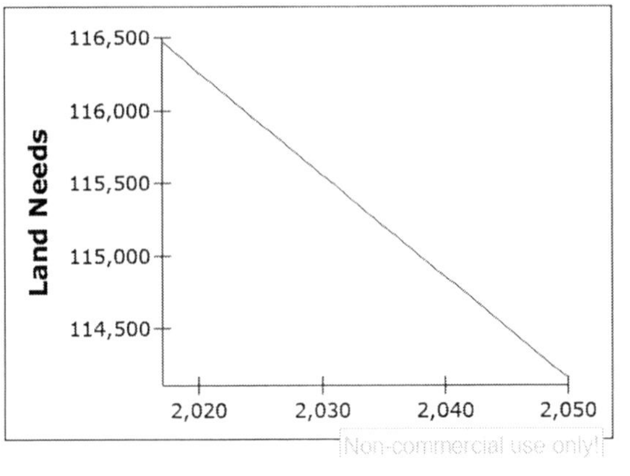

Figure 6. Graph land needs impac population growth in Tanah Datar District

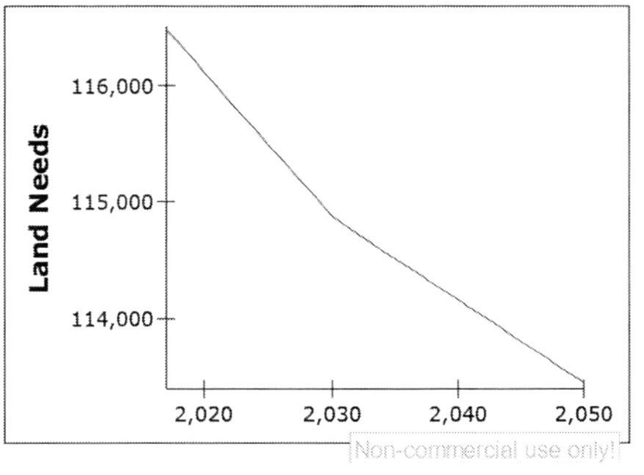

Figure 7. Graph of land needs for government policy scenarios in reducing population growth

The results of an analysis of 10 sub elements for the direction of spatial policy in Tanah Datar Regency indicate that two elements belong to the independent category (quadrant IV), and eight sub-elements are in quadrant III. Figure 8 presents that sub-elements of E7 consistency in the implementation of spatial plans and sub elements of E8 law enforcement in spatial violations have a high thrust (drive power). In addition, both sub elements have a low dependency value on other sub elements.

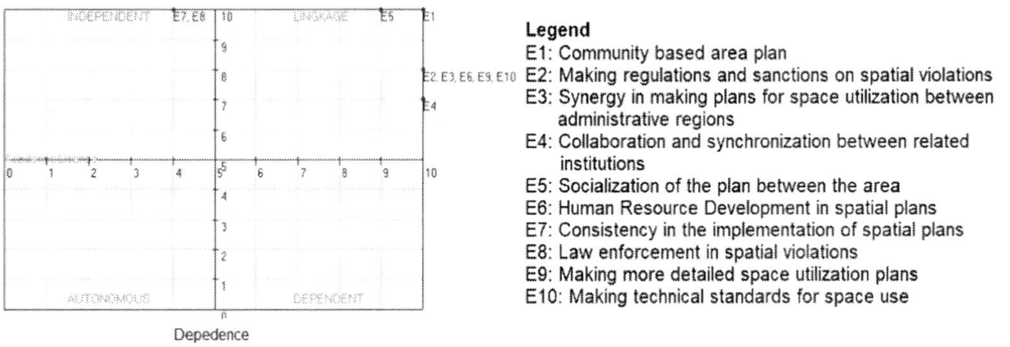

Figure 8. Quadrant drive power and dependence

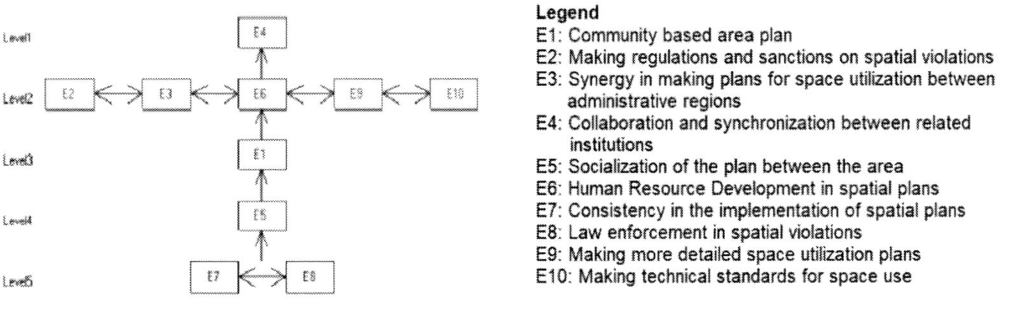

Figure 9. Land use policy structure in Tanah Datar District

Figure 8 presents the implementation structure of the Tanah Datar District space utilization plan, there are five levels. At the highest level (level 5) there are two sub elements are sub elements of E7 consistency in the implementation of spatial plans and sub elements of E8 law enforcement in spatial violations. As a sub element of the summit, the two sub elements become the main policy in the direction of the implementation of spatial planning in Tanah Datar District.

4. Conclusion

The population growth of around 1.6 percent in Tanah Tanah District has an impact on land use. Settlement areas have increased every year, thus reducing water catchment areas.

References

[1] Utoyo B S, E Anwar, I M Sandy, R S Saefulhakim dan H Santoso 2001 Analisis Keterkaitan antara Pertumbuhan Wilayah dengan Pola Perubahan Struktur Penggunaan Lahan *Forum Pascsarjana* **24**: 159-162
[2] Canuti P, N Casagli and R Fanti 2003 Landslide Hazard for Archaeological Heritage: The Case of Tharros in Italy *Landslides News* **14** (15): 40-43
[3] Umar I, Dewata I., Barlian E dan Hermon D 2017 Zonasi Rawan Longsor Pada Kawasan Permukiman Di Kabupaten Tanah Datar *Prosiding Seminar Nasional DAS PKLH UNRI* (Pekanbaru: Universitas Negeri Riau)
[4] Virdin J W 2001 Understanding the Synergies between Climate Change and Desertification UNDP
[5] Syahrin A 2003 *Pengaturan Hukum dan Kebijakan Pembangunan Perumahan dan Permukiman Berkelanjutan* (Pustaka Bangsa Press)
[6] Suryani R L dan A Marisa 2005 *Aspek-Aspek yang Mempengaruhi Masalah Permukiman di Perkotaan* (Medan : Program Studi Arsitektur. Fakultas Teknik USU)
[7] Martono D N, Surlan dan B T Sukmana 2005 Aplikasi Data Penginderaan Jauh untuk Mendukung Perencanaan Tata Ruang di Indonesia. http://io.ppi.jepang.org/article
[8] Xiao J, Shen Y, Ge J, Tateishi R, Tang C and Liang Y 2006 Evaluating urban expansion and land use change in Shijiazhuang, China *by using GIS and remote sensing. Landscape and Urban Planning* **75**(1e2), 69e80
[9] Mendoza M E, Granados E L, Geneletti D, Pérez-Salicrup D R and Salinas V 2011 Analysing land cover and land use change processes at watershed level: a multitemporal study in the Lake Cuitzeo Watershed, Mexico (1975–2003) *Applied Geography* **31**(1): 237-250
[10] Sitorus S R P 2005 Pengembangan Lahan Berpenutupan Tetap sebagai Kontrol terhadap Faktor Resiko Erosi dan Bencana Longsor. Makalah. Lokakarya Penataan Ruang sebagai Wahana untuk Meminimalkan Potensi Kejadian Bencana Longsor (Jakarta 7 Maret 2006)
[11] Kaur E, Palang H dan Soovali H 2004 Landscape in Change Opposing Attitudes in Saaremaa Estonia *Landscape and Urban Planing* **67**(2): 109-120.
[12] Pribadi D O, D Shiddiq dan M Ermyanila 2006 Model Perubahan Tutupan Lahan dan Faktor-Faktor yang Mempengaruhinya *Jurnal Teknologi Lingkungan Pusat Pengkajian dan Penerapan Teknologi Lingkungan* **7**: 35-51
[13] Umar I 2016 Mitigasi Bencana Banjir pada Kawasan Permukiman Di Kota Padang, Provinsi Sumatera Barat *Disertasi (*Bogor: Sekolah Pascasarjana IPB)
[14] Umar I dan Dewata I 2017 *Pendekatan Sistem* (Jakarta: Rajawali Press)
[15] Harun U R 1992 Dinamika Penggunaan Sumberdaya Lahan di Jawa Barat 1970-1990. *Jurnal PWK* **3**: 48-53
[16] Kustiawan I 1997 Permasalahan Konversi Lahan Pertanian dan Implikasinya terhadap Penataan Ruang Wilayah. Studi Kasus: Wilayah Pantura Jawa Barat *Jurnal PWK* **8**: 49-60
[17] Glade T 2003 Landslide occurrence as a response to land use change: a review of evidence from New Zealand. *Catena* **51**(3-4) : 297-314
[18] Buol S W, F D Hole and R J Cracken 1980 Soil Genesis and Classification. Second Edition (Amess : The Iowa State University Press)
[19] Djakapermana R D 2010 *Pengembangan Wilayah Melalui Pendekatan Kesisteman* (Bogor (ID): IPB Pr)

[20] [BPBD] Badan Penanggulangan Bencana Daerah Kabupaten Tanah Datar 2017 *Data Statistik Bencana Alam Kabupaten Tanah Datar*

[21] Huang Y and Zhao L 2018 Review on landslide susceptibility mapping using support vector machines *CATENA* **165** : 520-529

[22] Flentje P, Chowdhury R 2016 Resilience and sustainability in the management of landslides. In: Proceedings of the Institution of Civil Engineers-Engineering Sustainability Thomas Telford Ltd pp. 1–12.

[23] Mai T and Smith C 2018 Scenario-based planning for tourism development using system dynamic modelling: A case study of Cat Ba Island, Vietnam. *Tourism Management* **68**: 336-354

[24] Abuzeinab A, Arif M, and Qadri M A 2017 Barriers to MNEs green business models in the UK construction sector: An ISM analysis. *Journal of cleaner production* **160** :27-37

Transmission of dengue hemorrhagic fever and climate variability in Jakarta

Haryoto Kusnoputranto[1,*], Margareta Maria Sintorini[2], Suyud Warno Utomo[1], Nurusysyarifah Aliyyah[3], Epi Ria Kristina Sinaga[1] and Okky Assetya Pratiwi[4]

[1] Department of Environmental Health, Faculty of Public Health, Universitas Indonesia
[2] Environmental Engineering, Faculty of Landscape Architecture and Environmental Technology, Trisakti University, Indonesia
[3] Public Health Study Program, Institute of Health Indonesia
[4] Environmental Health Study Program, Institute of Health Indonesia

*haryoto_k@yahoo.com

Abstract. Dengue hemorrhagic fever (DHF) has become an endemic in major cities in Indonesia. The climate change and poor level of awareness and knowledge of the community in Indonesia have caused an increase in the DHF cases. Outbreaks on January until April 2015, the morbidity rate reached 50.75. In 1996, the Intergovernmental Panel on Climate Change (IPCC) predicted threefold increase in the dengue hemorrhagic fever incidents in Indonesia by 2070, if the environment and community conditions do not improve. This study aims to provide a dynamic model to predict the dynamics of the DHF incidents based on climate variability in the Special Capital Region of Jakarta. The design of this study was ecologic study with hypothesis test, modelling, simulation, and intervention. In collecting the data, interview with respondents and measurement of climate variables were conducted. Interviews with respondents include the level of knowledge, attitudes, and behaviour (PSP) of the community. Measurements of climate factors include rainfall, temperature, humidity and CO_2 level in the ambient environment. The DHF system dynamics model simulation shows that the significant effect on the decline of Breeding Places and the decrease of DHF cases were achieved by increasing the participation of the community to actively control water places that are potential for mosquito breeding places.

1. Introduction

The increasing population has caused higher cases of communal disease. The DHF cases in Indonesia first appeared in Surabaya in 1968. Based on the number of cases, Indonesia ranked second after Thailand. The Ministry of Health reported that the number of DHF continued to increase from 0.05 in 1968 to 35.19 in 1998 when it was recognized, there was a relationship between environmental changes and an increase in cases of disease. This morbidity rate continues to increase every year. It became 43.31 in 2015; 39.80 in 2014 with a total of 100,347 cases and it reached 50.75 in 2015. In 2014, 71,668 cases with 641 of them died were recorded in 34 provinces in Indonesia.

The number of environmental factors that are related to incidence of DHF which tend to recurred encourage researchers to investigate environmental factors can be used as indicators in predicting the occurrence of DHF cases and DHF transmission system in Jakarta, as well as predicting the incidence of DHF through intervention models on system dynamics.

This study provides a dynamics model of DHF transmission in relation to climate variability that will contribute to improvements in development on a local and even national scale. The dynamics model of DHF transmission is expected to be an innovation to reduce the incidence of DHF, especially in Jakarta.

2. Methods

The study design was an ecological study with a hypothesis test. The modeling and simulation were carried out to identify factors that are relevant to future DHF cases. The added value of the study is the association between environmental factors in ecology such as rainfall, temperature and humidity as well as basic indicators of air quality, namely CO2, with vectors of DHF.

The study related to the dynamics model in providing the early warning for the emergence of DHF cases has been started since 2006. The patterns which were found in this study are suggested to be applied in a broader area, namely in Jakarta, Bogor, Depok, Tangerang and Bekasi (Jabodetabek).

2.1. Samples

The samples in this study were air quality sample, vector sample, case sample, house sample, and respondent sample. Air quality sample was taken from the research location. Vector sample was adult *Aedes* mosquitoes caught from the respondents' house. Case samples were DHF experienced by respondents in the last two years. House sample was determined based on WHO standards for larvae and mosquito surveys according to the area and population: "Cluster Design Sampling" [1]. The sample size for PSP study was calculated based on hypothesis test for two proportions by using following equation:

$$n = \frac{\left[Z_{1-\frac{\alpha}{2}} \sqrt{2P(1-P)} + Z_{1-\beta} \sqrt{P_1(1-P_1) + P_2(1-P_2)} \right]^2}{(P_1 - P_2)^2} \quad (1)$$

The minimum number of samples calculated based on the prevalence of DHF in the preliminary study was 14%, while the expected prevalence of control variables that had been found was 4%. The calculation used 5% confidence level and 90% test strength. The distribution of household sample was carried out by dividing the sample proportionally according to the number of cities in the Special Capital Region of Jakarta, namely Central Jakarta, East Jakarta, South Jakarta, West Jakarta, and North Jakarta. In each region, a total of 40 houses were taken with individual analysis units.

2.2. Data Collection

The study was conducted within three years starting in 2018. The following data were collected from the Special Capital Region of Jakarta:

- The environmental data: rainfall, temperature, humidity, vector, and CO_2 level in the ambient air through direct measurements at the sampling point
- The community data to find out the knowledge, attitudes and behavior of the community about DHF

Rainfall measurements were carried out every rainy-day using rain gauge. Temperature and humidity were measured using thermo hygrometer, and CO_2 level was measured using RAC sampler. The vector data collection of adult *Aedes* mosquitoes was carried out by purposive sampling in several households in each region.

Community data collection was carried out through interviews using a questionnaire to determine social, economic and education status of the community, as well as knowledge, attitudes and behavior of the community about DHF, prevention and eradication of *Aedes* vectors.

2.3. Data Analysis
To obtain the concept of DHF control in the future, a system dynamics model was used. The following stages were conducted to obtain the research outcome: (1) validating the environmental components (2) demonstrate the model in the relevant environment, (3) reporting the activity comprehensive. In this study, the case of DHF is considered as one of the components of the causal system in a scenario simulated through intervention.

2.4. System Modeling
The simulation was carried out through model construction, model development, and validation of the simulation results to determine the compatibility between the simulation results and established simulation mechanism. The simulation results were used to understand the process behavior and predict the future trends [2]. Based on the understanding of the systemic events, the following steps were carried out to build the modeling concept in Figure 2:

- Identifying the process that produces the actual event.
- Identifying the desired event.
- Identifying the gap between the actual state and the desired situation
- Analyzing the policies.

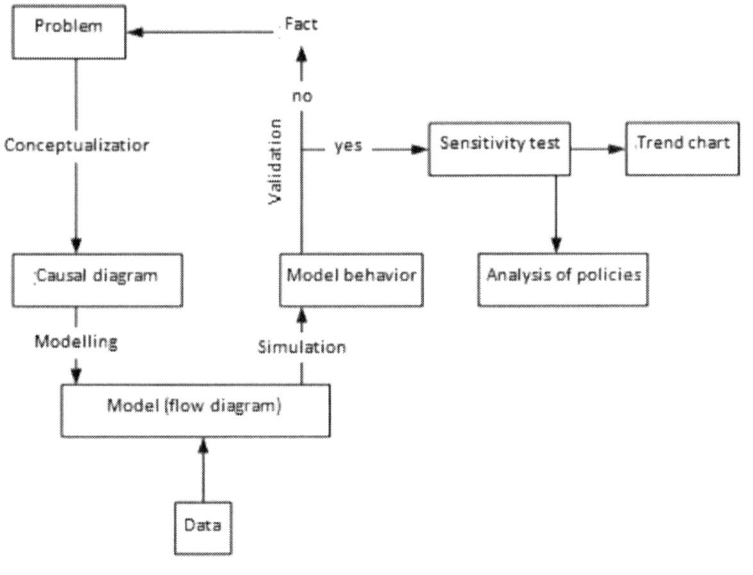

Figure 1. The Modelling Concept

3. Results and Discussion
The first year of the study has been conducted in the Special Capital Region of Jakarta, namely in the Central Jakarta (Harapan Mulya Urban Village, Kemayoran Sub-district), East Jakarta (Malaka Jaya Urban Village, Duren Sawit Sub-district), South Jakarta (Lenteng Agung Urban Village, Jagakarsa Sub-district), West Jakarta (Tomang Urban Village, Grogol Petamburan Sub-district), and North Jakarta (East Kelapa Gading Urban Village, Kelapa Gading Sub-district). The collection of environmental data and community data about DHF were carried out in the five regions from July to September 2018.

3.1. System Dynamics Model
3.1.1. DHF Case Prediction based on Model Simulation. Based on Figure 3, it is known that climate variability patterns affect the incidence of DHF because the life of the disease vectors and the dengue

virus as an agent is very dependent on the environmental conditions. The temperature, the humidity, the chemical composition in the air and of mosquitoes breeding places, the rainfall, the wind speed and other environmental factors are the limiting factors of its life. Therefore, the environmental system associated with the incidence of DHF has 4 subsystems, namely: climate subsystem, *Aedes* mosquito, human and DHF cases. These four subsystems are interconnected and they influence each other.

The climate subsystem (blue) is a series of climate factors related to global climate change that triggers an increase in global earth temperature. As a result, this global warming phenomenon can affect living things. The most dominant insect causing DHF is *Aedes aegypti*.

CH_4, N_2O, CO_2 and CFC-11 are gases that cause the global climate change or the so-called greenhouse effect if it is at a concentration exceeding normal concentration. The increase in global temperatures will increase sea surface temperatures. It will later trigger the El Nino phenomenon. According to the World Meteorological Organization (WMO), there is a relationship between the occurrence of El Nino and the incidence of DHF in Indonesia [3,4]. This cycle forms the climate subsystem in the model.

The second subsystem is the *Aedes* mosquitoes, from eggs to adult mosquitoes, which is indicated by a positive arrow. The connecting factor of the climate subsystem and mosquito subsystem is the breeding places for mosquitoes. The breeding places are highly affected by rainfall. When rainfall is high, the containers are easily filled with water and this relationship is indicated by a positive arrow. Another connecting factor of the climate subsystem is the ambient temperature which then affects the Extrinsic Incubation Period (EIP). The EIP is influenced by environmental temperature, humidity, level of viremia in humans, and viral strains [5]. The increase in temperature will shorten the EIP and increase the transmission. The temperatures that increase to 34°C will affect the temperature of the water in breeding places which also accelerates the hatching of eggs into larvae [6,7].

The vector subsystem and the disease subsystem are connected by the Landing Rate and the role of dengue virus factors. The presence of DHF is determined by the contact between mosquitoes and humans. It is assumed that the larger the mosquito population the higher the Landing Rate, so the connecting arrow becomes positive. The larger the *Aedes* mosquito population the larger the infective *Aedes* population that carries the Dengue virus. This relationship is also indicated by a positive arrow.

After the Aedes mosquito bites a human, the virus replicates in the human body. The more viruses incubated into humans the more humans become infectious so that the number of DHF cases in the community increases. This link is indicated by a positive arrow.

Human activity is indicated by the intensity of petroleum fuels usage. The impact of fuel utilization is the increase of CO_2 emissions; the relationship is indicated by a positive arrow. The Landing Rate factor is also related to the human subsystem through the individual activeness factor. The more inactive a person is, the easier for mosquitoes to approach, especially during peak hours.

In this case, intervention in increasing the knowledge, attitudes and behaviour (PSP) of the community is required. From the analysis, it is known that the factors that are significantly related to the occurrence of DHF in the community are knowledge, attitudes and behaviour factors. This program provides community awareness education. This relationship is indicated by a positive arrow for PSP. Figure 3 shows the diagram stock flow scheme which is the model development stage, where all variables are included as factors that influence the incidence of DHF.

3.2. Knowledge Attitude and Behaviour

The research on the first year had been conducted in the Special Capital Region of Jakarta (DKI Jakarta) area, in the Central Jakarta (Harapan Mulya Urban Village, Kemayoran Subdistrict), East Jakarta (Malaka Jaya Urban Village, Duren Sawit Subdistrict), South Jakarta (Lenteng Agung Urban Village, Jagakarsa Subdistrict), West Jakarta (Tomang Urban Village, Grogol Petamburan District), and North Jakarta (East Kelapa Gading Urban Village, Kelapa Gading Subdistrict). Environmental and community data collection about dengue fever was performed in those five areas starting in July and continued until September 2018. The data on resident knowledge about Dengue Fever in DKI Jakarta is as follows.

Table 1. Distribution of Resident Knowledge in DKI Jakarta about Dengue Fever in Higher Education Applied Research (PTUPT) in 2018

Variable	Frequency	Percentage (%)
Knowledge about Dengue Disease		
Know	199	99.5
Does not Know	1	0.5
Dengue Fever (DBD) Cause		
Virus/Germ	2	1.0
Insect Bite	183	91.5
Poorly Processed/Unsanitary Food/Drink	2	1.0
Does not Know	6	3.0
Other	7	3.5
Infectious Disease Dengue Fever		
Yes	110	55.0
No	90	45.0
Dengue Fever Transmission		
Other than Mosquito	9	4.5
Mosquito Bite	99	49.5
Missing	92	46.0
Dengue Fever Symptoms		
Fever 2-7 days		
Yes	191	95.5
No	9	4.5
Headache		
Yes	30	15.0
No	170	85.0
Bleeding: red spots on the skin		
Yes	138	69.0
No	62	31.0
Nosebleed, blood feces, blood vomiting, etc.		
Yes	51	25.5
No	149	74.5
Total	200	100

Referring to the data on table 1, DKI Jakarta Residents who have sufficient knowledge about Dengue Fever (DBD) was 99.5%. If there are any Dengue Fever patients, more than 50% of DKI Jakarta's residents have known the signs and symptoms of individuals who are infected by the disease. However, 91.5% of residents still consider that insect bites are the main cause of Dengue Fever and not the virus.

Table 2. Distribution of Resident Knowledge in DKI Jakarta about Dengue Fever in Higher Education Applied Research (PTUPT) in 2018

Variable	Frequency	Percentage (%)
Characteristics/Behaviour of Infected *Aedes Aegypti* Mosquito		
White Spots/Dapple		
Yes	90	45.0
No	110	55.0
Biting during the day		
Yes	17	8.5
No	183	91.5

Found in the settlement		
Yes	4	2.0
No	196	98.0
Breeding in the clear water		
Yes	18	9.0
No	182	91.0
Total	200	100

The knowledge of DKI Jakarta resident about the characteristics of Infected *Aedes Aegypti* Mosquito can be seen in the table 2. Based on the data, 45% of the population know the Infected *Aedes Aegypti* Mosquito characteristics by the white spots or striped on its body. In addition to the knowledge, the resident's PSP (Attitude and Behaviour Knowledge) can be assessed from their attitude on dengue fever disease, ranging from prevention to willingness to make the prevention effort [8,9]. Former research shows that knowledge has a significant relationship with community attitude and behaviour regarding DHF control [10,11].

Based on the data in table 3, it is found that almost all DKI Jakarta residents (99%) think that the dengue disease prevention must be done as soon as possible and those who are responsible for the disease control are all parties, and both government and society must be involved (39.5 %). Around 91.5% of DKI Jakarta residents are willing to follow the efforts to prevent dengue disease because 100% of the residents agree if the routine dengue fever prevention activities would be held.

Table 3. Distribution of Resident Knowledge in DKI Jakarta about Dengue Fever in Higher Education Applied Research (PTUPT) in 2018

Variable	Frequency	Percentage (%)
Immediate Prevention		
Yes	198	99.0
No	1	0.5
Does not Know	1	0.5
Prevention responsibility		
Government	15	7.5
Dengue patients and families	35	17.5
Communities	56	28.0
All parties	79	39.5
Others	15	7.5
Routine prevention		
Agree	200	100
Not Agree	0	0.0
Willingness to follow the prevention effort		
Willing	183	91.5
Not Willing	17	8.5
Total	200	100

DKI Jakarta residents' attitude about the dengue fever prevention (table 4) is very good, such as the need to drain the bathtub (99.5%), agree to 3M (drain, close and bury) plus and the government effort (98.5%), not hanging clothes at home (85.5%), and the need to monitor mosquito larvae (98.5%).

Table 4. Distribution of Resident Knowledge in DKI Jakarta about Dengue Fever Prevention in Higher Education Applied Research (PTUPT) in 2018

Variable	Frequency	Percentage (%)
Prevention Form		
Bathtub drain		
Not necessary	1	0.5
Necessary	199	99.5
3M plus and the government efforts		
Agree	197	98.5
Not Agree	3	1.5
Hanged Clothes		
Allowed	29	14.5
Not Allowed	171	85.5
Efforts to monitor mosquito larvae		
Necessary	197	98.5
Not necessary	3	1.5
Insecticide Fogging		
Effective	106	53.0
Ineffective	94	47.0
Dengue Fever Prevention		
Pay attention to the personal health and 3M plus	94	47.0
Only pay attention to the personal health	40	20.0
Only 3M plus	61	30.5
Does not Know	3	1.5
Other	2	1.0
Total	200	100

DKI Jakarta residents' awareness in preventing Dengue Fever can be seen from the behaviour about 3M plus (table 5), in which 91% of the residents drain and clean the bathtub/ TPA (Final Disposal Site) at home, with a minimum frequency of once a week (59.5%). In addition, DKI Jakarta residents are also regularly clean and bury the used goods (69.5%) and close the window, ventilation, or door with mosquito net (67.5%).

Table 5. Distribution of Resident Knowledge in DKI Jakarta about 3M Plus of Dengue Fever in Higher Education Applied Research (PTUPT) in 2018

Variable	Frequency	Percentage (%)
Drain and clean the bathtub/TPA at house		
Yes	182	91.0
No	18	9.0
Frequency of drain and clean the bathtub/TPA at home		
Once a week	119	59.5
Once every two weeks	27	13.5
Once every three weeks	19	9.5
Once a month	17	8.5
No	18	9.0
Clean up/Bury the used goods		
Regularly	139	69.5
Seldom	43	21.5
Never	16	8.0
Missing	2	1.0

The use of abate		
Yes	65	32.5
No	135	67.5
Close the Window/Ventilation/door with mosquito net		
Yes	135	67.5
No	65	32.5
The habit of storing clothes		
Hung in the room	27	13.5
Kept in the dirty clothes places	173	86.5
Garbage Disposal Method		
Transported/collected by officers	196	98.0
Burned/buried	3	1.5
Other	1	0.5
Total	200	100

3.3. Environmental Measurement Result

The environmental data measurement was performed simultaneously in the five research areas starting from July 2018 to September 2018. The data measured including temperature and environment. The environmental data from the environmental temperature and humidity measurement result in the research areas are presented in table 6 and table 7. Based on the data of table 6, the highest average environmental temperature in North Jakarta was 31.37 °C and the highest environmental humidity in West Jakarta was 62.06%.

Table 6. Environmental Temperature (°C) in DKI Jakarta July-September 2018

Sampling number-	West Jakarta	East Jakarta	North Jakarta	South Jakarta	Central Jakarta	Average Temperature
1	31.03	30.33	31.68	32.33	30.39	31.15
2	30.85	31.87	31.73	30.93	31.7	31.42
3	32.46	31.65	31.69	31.27	30.35	31.48
4	31.57	30.89	31.23	31.63	30.71	31.21
5	29.45	30.77	30.87	30.85	30.49	30.49
6	30.41	31.35	31.55	30.26	30.25	30.76
7	31.27	31.54	31.75	31.14	30.77	31.29
8	30.59	31.47	30.48	30.36	30.42	30.66
Average Temperature	30.93	31.23	31.37	31.1	30.63	31.05

Table 7. Environmental Humidity (%) in DKI Jakarta July-September 2018

Sampling number-	West Jakarta	East Jakarta	North Jakarta	South Jakarta	Central Jakarta	Average Humidity
1	55.16	56.34	60.35	60.8	45.41	55.61
2	62.63	57.88	55.42	56.58	58.65	58.23
3	54.63	59.82	57.54	58.72	59.95	58.13
4	62.83	60.44	60.25	59.45	63.8	61.35
5	69.61	65.25	62.43	60.36	65.27	64.58
6	68.1	63.37	57.85	58.87	51.86	60.01
7	62.4	61.55	60.52	62.77	55.34	60.52
8	61.08	60.73	61.68	63.25	58.72	61.09
Average Humidity	62.06	60.67	59.51	60.1	57.38	59.94

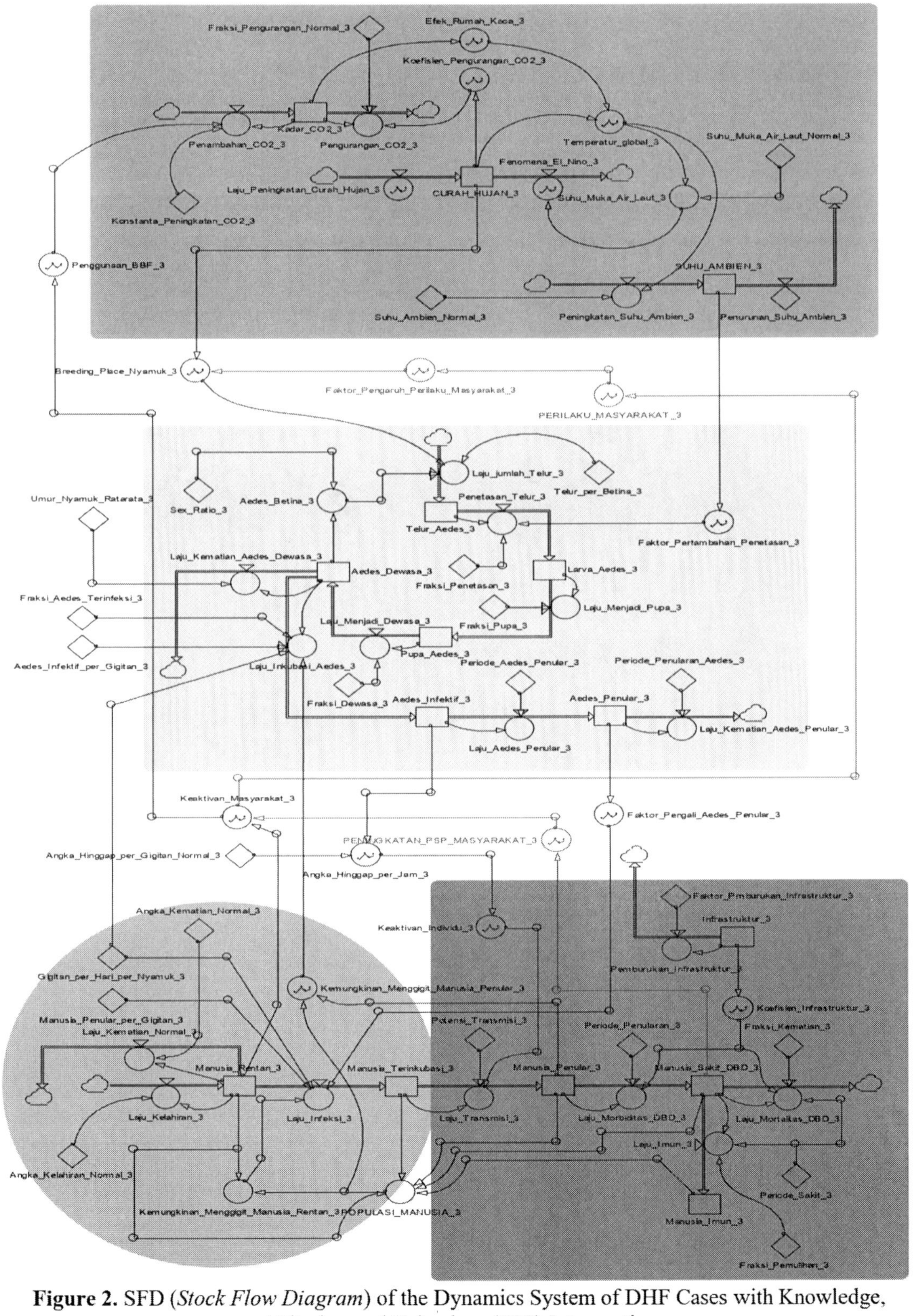

Figure 2. SFD (*Stock Flow Diagram*) of the Dynamics System of DHF Cases with Knowledge, Attitudes, and Behaviour (PSP) Intervention

Figure 3 shows the Dynamics System model of DHF cases with Community Role Improvement program intervention, which was built from the model contained in the previous figure. This model is the basis for programs simulation that are considered to be able to control the increase of DHF cases.

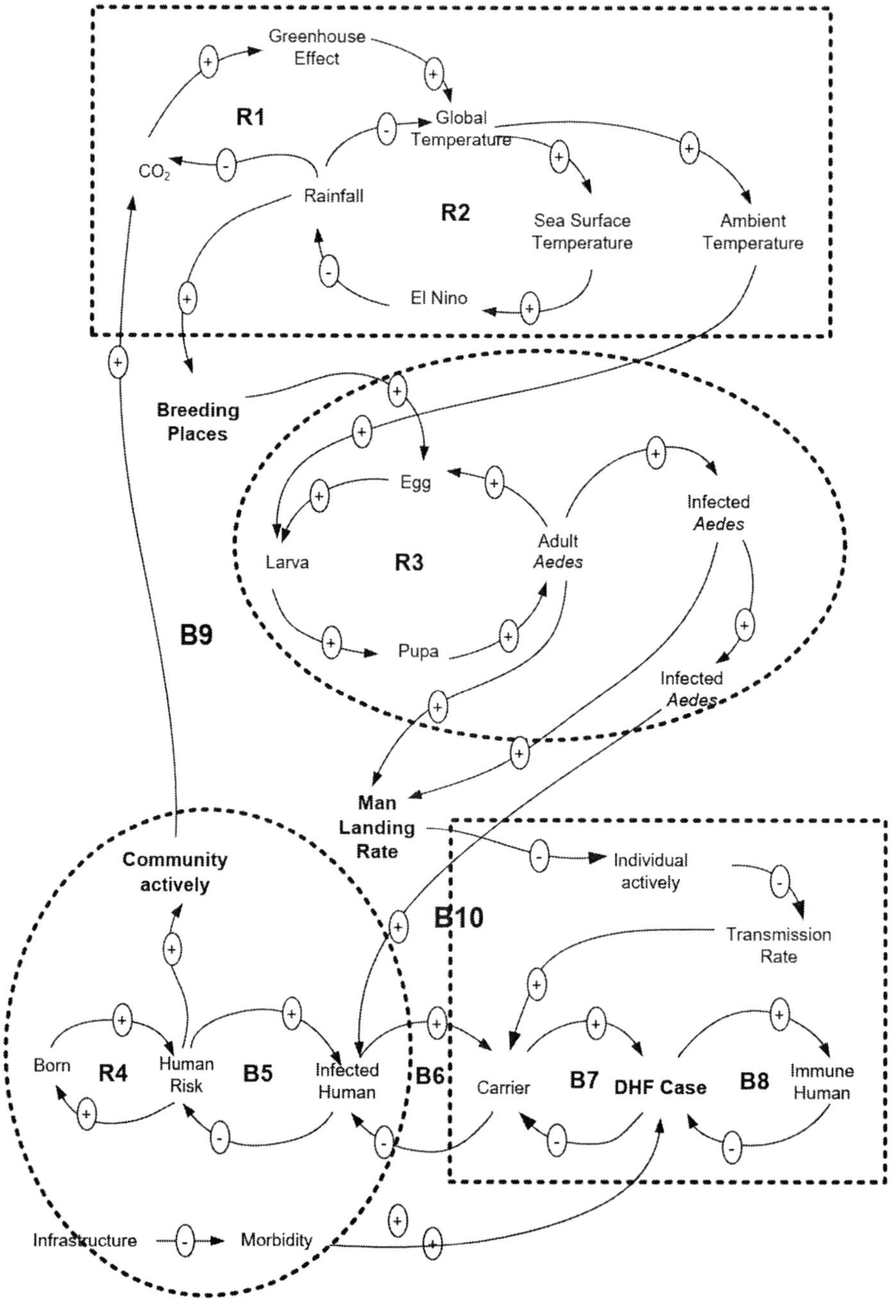

Figure 3. Dynamics System Model of DHF Cases with Community Role Improvement (PSP) intervention

Figure 4 below shows that naturally if the adult *Aedes* mosquito population increases, the infective *Aedes* also increases, which means that the infectious mosquito population increases. The following figure 5 shows that if there is no disease control intervention, the humans who have been incubated with dengue virus will persist (blue line), and the number of vulnerable humans continues to increase (red line).

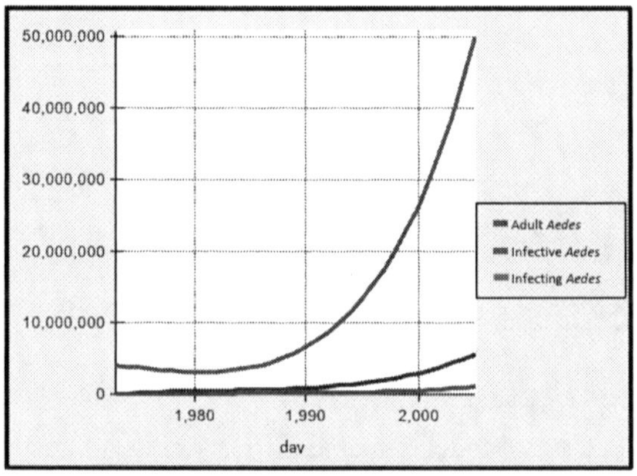

Figure 4. Results of Natural DHF Cases Model Simulation

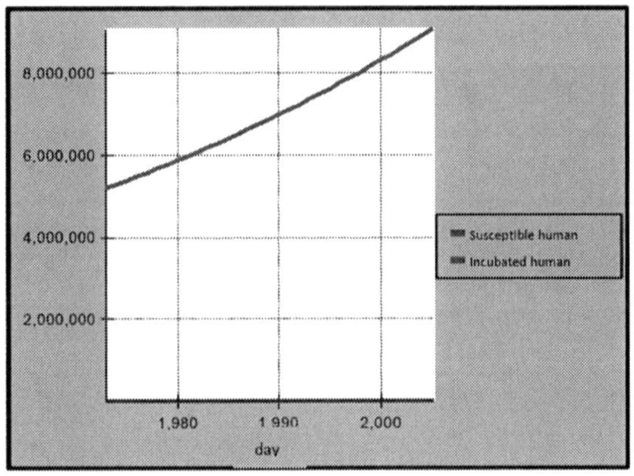

Figure 5. Results of DHF Cases Model Simulation without Intervention on the Number of DHF patients

Referring to the charts above, it is necessary to conduct a program to control mosquito-borne diseases. The most effective program to reduce infective mosquito population is to increase the community role in controlling breeding places for mosquitoes to lay their eggs [1].

The simulation results are shown in Figure 7. The tendency of the chart seems different before and after the intervention. The red line indicates the state of no intervention, while the blue line indicates the condition after the PSP improvement intervention. The results of interventions on breeding places show an evident decrease compared to before the intervention. Although breeding places were still high at the beginning of the program intervention, it declined rapidly after the intervention. In the case chart, the incidence of DHF is lower compared to before the intervention.

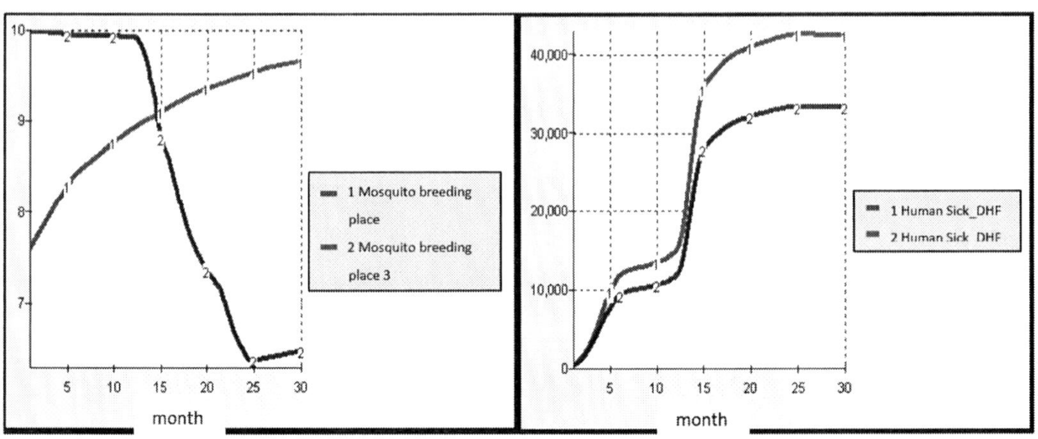

Description: Red→before intervention; Blue→after intervention
Figure on the left→TPN Chart; Figure on the right → DBD Case Chart

Figure 6. Results of DHF Cases Model Simulation with Knowledge, Attitudes and Behaviour Improvement on Breeding Places and DHF Cases Intervention

IOP Conf. Series: Earth and Environmental Science **314** (2019) 012071 doi:10.1088/1755-1315/314/1/012071

Figure 7 shows the results of simulation model, if a disease control program is carried out, namely by improving the community role, the adult *Aedes* population can be reduced so that it continues to decline (green line).

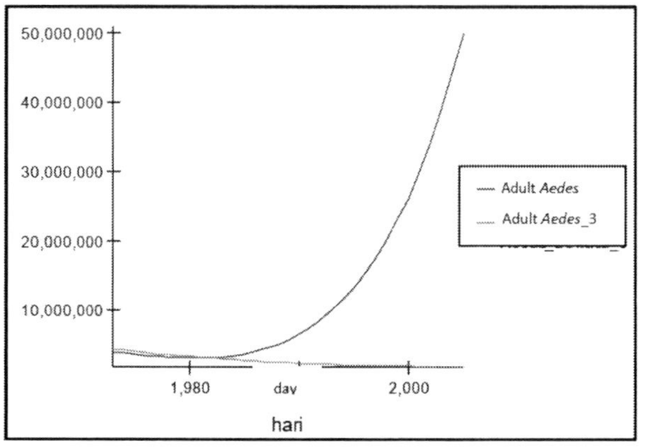

Figure 7. Results of DHF Cases Model Simulation with Knowledge, Attitudes and Behaviour Improvement on Adult Aedes Population Intervention

4. Conclusion and Suggestions

4.1. Conclusion

The system dynamics model simulation of DHF shows that the significant effect on the decline of Breeding Places and the decrease in DHF cases were achieved by increasing the participation of the community to actively control water places that are potential to be mosquito breeding places.

The simulation in this study has produced a model that can be used for other infectious diseases model. Through the system dynamics modelling, the transmission model of the disease can be known in details, so the most effective interventions can be determined to control the infectious disease cases.

4.2. Suggestions

The intervention program can be improved by adding the role of the Larvae Monitor. They can encourage the community to carry out the Eradication of Mosquito Nests independently and voluntarily as well as improving knowledge, attitudes and behaviour regarding DHF, including recognizing the initial symptoms so that they can receive treatment before it is too late.

The DHF cases can be reduced by improving environmental hygiene including puddles in parks, flower pots, ponds, bird and other pets' cages, and paying attention to waste disposal sites that have the potential to become mosquito breeding places, especially during the rainy season.

References

[1] Health Department 2003 *Prevention and Control of Dengue Fever and Dengue Hemorrhagic Fever, Complete Guide to Translation from WHO Regional Publication No. 29, "Prevention and Control of Dengue and Hemorrhagic Fever"* WHO and Health Department Cooperation, Jakarta.

[2] Muhammad E and Soesilo B 2001 *Analysis of dynamic system, living environment, social, economy, management* (Jakarta: UMJ Press)

[3] Gagnon A, Bush A and Smoyer-Tomic K 2001 Dengue epidemics and the el nino southern oscillation *Climate Research* **19** 35-43

[4] Kovats S, Hajat S, Worall E and Haines A 2003 *El Nino and Health* http://image.thelancet.com/extrax/02art5336web.pdf

[5] Gubler J and Kuno G 2001 *Dengue and dengue hemorrhagic fever* (New York, USA: CABI Publishing)
[6] Tun-Lin W, Burkot T R and Kay B H 2000 Effects of temperature and larval diet on development rates and survival of the dengue vector Aedes aegypti in north Queensland, Australia *Medical and Veterinary Entomology* **14**(1) 31-37
[7] Eastin MD, Delmelle E, Casas I, Wexler J and Self C 2014 Intra-and interseasonal autoregressive prediction of dengue outbreaks using local weather and regional climate for a tropical environment in Colombia *The American journal of tropical medicine and hygiene* **91**(3) 598-610
[8] Hafeez F, Akram W, Suhail A and Arshad M 2012 Knowledge and attitude of the public towards dengue control in urban and rural areas of Punjab *Pakistan Journal of Zoology* **44**(1) 15-21
[9] Bloom Benjamin S 1956 *Taxonomy of Educational Objectives: The Classification of Educational Goals* (New York: David McKay)
[10] Hairi F, Ong C H, Suhaimi A, Tsung T W, bin Anis Ahmad M A, Sundaraj C and Soe M M 2003 A knowledge, attitude and practice (KAP) study on dengue among selected rural communities in the Kuala Kangsar District *Asia Pacific Journal Public Health* **15**(1) 37-43
[11] Santoso S and Budiyanto A 2008 Knowledge, attitude and practice relationship of the community towards dengue hemorrhagic fever (DHF) in Palembang City South Sumatra Province *Indonesian Journal of Health Ecology (Jurnal Ekologi Kesehatan)* **7**(2) Agt 2008 732-739

Critical land mapping for the development of biomass-based energy in East Lombok Regency, Indonesia

B H Narendra[1, 6,*], Widiatmaka[2], C Kusmana[3], L Karlinasari[4] and Machfud[5]

[1] Study Program of Natural Resource and Environmental Management, IPB University, Indonesia
[2] Department of Soil Science and Land Resources, Faculty of Agriculture, IPB University, Indonesia
[3] Department of Silviculture Faculty of Forestry, IPB University, Indonesia
[4] Department of Agro-Industrial Technology, Faculty of Agricultural Technology, IPB University, Indonesia
[5] Department of Forest Products, Faculty of Forestry, IPB University, Indonesia
[6] Research, Development and Innovation Agency, Ministry of Environment and Forestry, Indonesia

*narendra17511@gmail.com

Abstract. Utilization of critical land as woody biomass plantations need to be supported by information on the potential of critical land. The objective of this research was to delineate a critical land using remote sensing and GIS techniques combined with Analytical Hierarchy Process method. The parameters used were land use/cover, slope, rainfall erosivity, tolerable soil erosion, soil erodibility, land management, and land productivity. The Analytical Hierarchy Process method was used to provide the weight of each parameter. Land use/cover map was interpreted from SPOT-7 satellite imageries acquired in 2017. The other parameter maps were built based on reference maps, secondary and field data collection as well as soil sample analysis. The resulted map showed that 32.7% area of East Lombok Regency was categorized as a critical area. Based on the land status, the critical land area located in the State Forest was smaller (42%) than on the Area for Other Utilization (58%). Utilization of critical land as woody biomass plantation should be prioritized on the shrub-covered land using community forest scheme, while for the area of State Forest, the plantation can be only planted in the production forest area applying industrial plantation forest or community plantation forest scheme.

1. Introduction

Increasing energy demand is commonly accompanied by higher pollution and greenhouse gas emissions [1]. Dominantly use of fossil energy sources tends to increase the CO_2 content in the air and leads to global warming and climate change that are increasingly felt in this century [2]. To overcome this global issue, coupled with the prediction of fossil fuel reserve depletion, the use of renewable energy is encouraged including energy from woody biomass. Biomass plantation is very potential to be developed because providing multiple benefits such as bioenergy production, carbon sequestration and regaining ecosystem services [3]. Biomass energy is predicted to be a driver of growth in renewable energy because biomass can be developed almost anywhere, and biomass energy can be produced on demand [4].

East Lombok Regency currently has the largest consumption of wood-based bioenergy in the form of firewood compared to other districts. This firewood is used for household consumption, small scale industries, and the largest proportion is used for Virginia tobacco industries [5]. Currently, the use of energy wood is dominantly used as firewood and in the future potential to be developed as a substitute fuel in power plants gradually replacing fossil fuels.

To supply woody biomass for fulfilling the increasing energy demand, more land resources are required to expand the woody plantation. East Lombok as an agrarian area with high population density, has limited land area to be cultivated as a bioenergy source. The arable land will be prioritized as agricultural's areas to produce more food to feed the growing population. Therefore, Indonesia Government proposed to utilize critical land as expandable areas for biomass plantation development. The use of the critical area as woody plantation can avoid the conflict between food and energy production. The international trend is tending to scale up the optimization use of critical land as fast-growing woody plantation specifically as a source of power plant fuel. Many of them are developed on the local and small scale to anticipate the fluctuations of economic appropriateness level of produced energy [6,7,8,9,10,11].

Currently, the information of critical land on a medium or detail scale for the development of biomass plantation has not been mapped. The government has released critical land map at the national level that comes from small-scale map inputs so that it is necessary to classify critical land based on the quality and productivity of the land. Several methods in delineating critical land are still a trend that continues to be developed by researchers. The use of remote sensing and GIS techniques is the most extensive method in classifying a kind of degraded land and desertification. The important parameters involved in this method are continued to be developed. Some adjustments were made between the affecting parameters and the specific characteristics of the observed ecosystem observed [12]. The level of importance among the parameters can be easily characterized using the AHP method, a robust and flexible decision-making method in finding solutions such as in the case of erosion level determination [13,14,15]. Therefore, the objective of this research was to delineate a critical land area using remote sensing and GIS techniques based on AHP weighted parameters.

2. Materials and Methods

2.1. Research area

East Lombok is a regency in Nusa Tenggara Barat province located at the east part of Lombok Island. As shown in Fig.1, astronomically the position is between 116° - 117° east longitudes and 8°- 9° south latitudes. The land occupies about 161 thousand km^2 (33.9% area of Lombok Island), laying from mountainous area with 3,726-meter altitude up to coastal area, and 60% of the area has 2 – 15% slope.

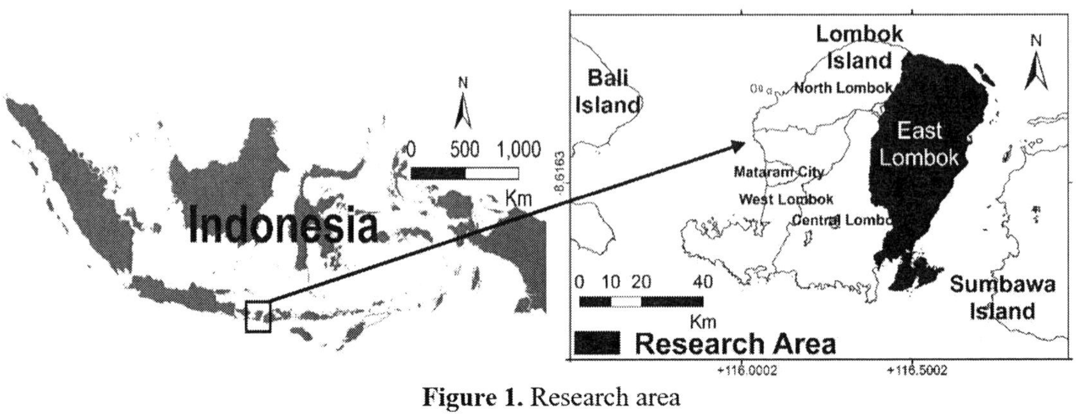

Figure 1. Research area

Most of the land is utilized as agricultural land. East Lombok consists of 20 districts, and it has the largest population compared to the other regencies in the provinces. In 2017, the population number was about 1.2 million, with a density of 737 inhabitants/km^2, and 0.8% population growth. East Lombok has a tropical climate with 31 – 33°C average maximum temperature and 22 – 25 °C average minimum temperature. In 2017, the annual rainfall 1580 mm with 109 rainy days [16]. Based on soil type, East Lombok has five soil ordo with 14 soil types in the great-group category. The main soil ordo namely Andisol, Inceptisol, Entisol, Alfisol, and Aridisol. Inceptisol has the largest area with 56,791.7 ha or 55.4% of all area, followed by Alfisol with 19,779.9 ha or 19,3% [17].

2.2. Data

In determination of critical land level, firstly the object area (East Lombok Regency) was divided into three specific function areas: (a) protected forest and conservation forest, (b) production forest area, and protected area outside the state forest area, and (c) agriculture cultivation area, and area for other utilization. The division was based on the guidance of critical land mapping from Ministry of Forestry that was changed in 2018. East Lombok Regency was divided into three categories based on the map of forest area status, combined with the map of *Official Spatial Land Use Plan* (OSLUP). Determination of critical land for each specific area was based on a set of parameter maps namely land use/cover, rainfall erosivity (R), slope (S), tolerable soil erosion (T), soil erodibility (K), land management, and land productivity. Those parameters were determined through expert discussion to eliminate the weaknesses of previous guidance [18].

Land use/cover map was delineated from the interpretation of SPOT-7 satellite imageries acquired in 2017. The interpretation was done using on-screen digitizing in ArcGIS software. The pan-sharpened mode was used with a spatial of resolution 1.5 m, and the result accuracy was assessed by field check and verified using available QuickBird imageries from 2015.

Rainfall erosivity map indicates the effect of rainfall intensity on soil erosion [19]. It was built based on monthly rainfall data, a number of rainy days and maximum daily precipitation in a certain month. Ten years rainfall data from 16 rain gauge stations were obtained from Agricultural and Livestock Services of East Lombok Regency. The erosivity was calculated using Bols [20] as in equations (1) and (2).

$$\boldsymbol{R = \sum_{m=1}^{12}(Rm)} \tag{1}$$

$$\boldsymbol{Rm = 6.119 \times (Rain)m^{1.21} \times (Days)m^{-0.47} \times (Max)m^{0.53}} \tag{2}$$

Where Rm is average monthly rainfall erosivity; (Rain)m is average monthly rainfall (cm); (Days)m is a number of rainy days in a certain month; (Max)m is average maximum daily rainfall in a certain month (cm). Distribution of erosivity in the study area was mapped using the interpolation technique in the ArcGIS software.

The slope map was derived from 8 m resolution DEM acquired from the Geospatial Information Agency, Indonesia. The analysis was done using the 3D analyst tool in ArcGIS software, and the result was classified into five class: < 8, 8-15, 16-25, 26-40, and > 40%. Soil characteristic maps at scale 1:50,000 used in this research were obtained from Agricultural Land Resource Agency. Based on those maps, East Lombok regency was divided into 75 land mapping units. To complete physical soil characteristics needed for generating critical land parameter maps, soil samples were taken from the field using soil core sampler tools. Soil properties involve five fractions of soil texture, bulk density, permeability, and organic matter content were analyzed in the laboratory of Soil Research Institute, Ministry of Agriculture using standard laboratory methods [21].

Tolerable soil erosion (T) indicates the maximum allowable erosion rate that soil fertility and productivity still can be maintained [22,23]. T was calculated based on [24] equation (3).

$$T = \left(\frac{De - Dmin}{S} + F\right) \times BD \quad (3)$$

Where,
De = soil effective depth (cm) x soil depth factor;
Dmin = minimum soil depth for plant root (cm)
S = soil sustainability (300 years)
F = rate of soil formation (1 – 1.5 cm/year depending on the soil type)
BD = bulk density (gr/cm^3)

Soil erodibility (K) reflected soil susceptibility to the erosion is one of the important factors in determining soil erosion using USLE and RUSLE model [25]. The value varies from 0 to 1; lower K value indicates less prone to soil erosion [26]. Equation (4) was used to determine K value based on soil texture (% of silt, sand, and very fine sand), soil organic matter content, soil structure, and permeability [19,27].

$$K = \frac{\{2.713\,(M^{1.14})(10^{-4})(12 - a) + 3.25(b - 2) + 2.5(c - 3)\}}{100} \quad (4)$$

Where,
M = (very fine sand % + silt %) x (100 – clay %)
a = organic matter content (%)
b = soil structure code: (1) very structured or particulate; (2) fairly structured; (3) slightly structured, and (4) solid
c = soil permeability code: (1) rapid; (2) moderate to rapid; (3) moderate; (4) moderate to slow; (5) slow, and (6) very slow

Land management map was built based on secondary data acquired from Dodokan Moyosari Watershed Management Institute. In this parameter, land management in each district was scored based on completeness of land boundary markers, activities to secure the land, extension activities, and application of soil and water conservation. This data was validated based on field observation and interview with the related stakeholders.

Land productivity map is one of the parameters used to assess the critical land in an agricultural cultivation area. In this parameter, land productivity score was carried out based on a comparison between the average annual production of rice and maze with optimal production that can be reached. These data sources were acquired from Agricultural and Livestock Services [16].

2.3. Analysis

Each parameter for a certain specific function area was weighted using the *analytical hierarchy process* (AHP). The AHP, developed by the mathematician Thomas L. Saaty, was conducted by fulfilling a set of the questionnaire by eleven experts. The questionnaire consists of pairwise comparisons between the parameters, and the experts valued using a scale of 1 to 9 [28,29].

Scoring for each parameter was done based on parameter values resulted in the previous steps. The scoring used value 1 (for the worst condition) up to 5 (for the best condition). GIS analysis to delineate the critical land was done by overlaying all prepared thematic maps according to each area status. Total values for each land mapping unit were resulted from overlay operation in ArcGIS software or known as weighted index overlay [26,30]. This operation would result in the total value range from 100 to 500 and the critical land was categorized into five levels: severe critical (100 – 180), critical (181 – 260), moderately critical (261 – 340), potentially critical (341 – 420), and uncritical (421 – 500).

3. Results and Discussion

Land use/cover map resulted from remote sensing analysis was shown in figure 2. The map indicated that East Lombok area was dominated by agricultural land (44%), followed by forested land (25%). In East Lombok, agricultural land can be found in the form of dryland agriculture, shrub - mixed dryland agriculture, and paddy field as the largest portion. Based on tabular data in table 1, East Lombok also has unutilized land covers included shrub, grassland, and barren land with 14% of total area. On the other hand, plantation forest has the smallest area compared to the others. In East Lombok, there is only one private company manages plantation forest producing energy wood. Other energy woods are produced by the community in the form of the community forest. Utilization of unproductive land cover as plantation forest will help this regency to fulfill its biomass energy from sustainable sources.

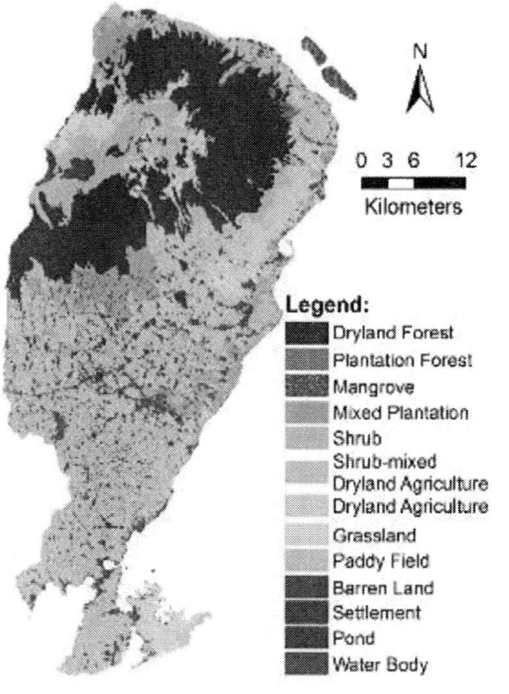

Figure 2. Land use/cover map

Table 1. Area of each land use/cover

Land use/cover	Area (ha)	%
Dryland Forest	39,062	24.3
Plantation Forest	426	0.3
Mangrove	1,680	1.0
Mixed Plantation	15,539	9.7
Dryland Agriculture	11,020	6.9
Shrub-mixed Dryland Agriculture	15,949	9.9
Paddy Field	43,647	27.1
Shrub	14,681	9.1
Grassland	6,439	4.0
Barren Land	1,482	0.9
Settlement	9,021	5.6

Pond	599	0.4
Water Body	1,270	0.8
Total	160,814	100

Source: the result of analysis

Land use/cover condition often can be used as a simple mark of critical land. Based on the weight of parameter resulted from AHP in table 2, land use/cover was the highest weight especially in the protected and conservation forest area. Deforestation affecting the reducing of forest covers was mainly due to the opening of forest for dryland farming and the massive of illegal logging. Forest conversion becomes agricultural land was also caused by higher pressure to the productive agricultural land which was often converted to build up area [29].

Table 2 Weighting factor for each parameter

Parameter	Weight value		
	Protected and conservation forest	Production forest area and Protected area outside the state forest	Agriculture cultivation area and area for other utilization
Land use/cover	0.254	0.184	-
Slope	0.192	0.278	0.183
Rainfall erosivity	0.206	0.157	0.162
Tolerable soil erosion	0.094	0.107	0.108
Soil erodibility	0.122	0.137	0.148
Land management	0.134	0.137	0.128
Land productivity	-	-	0.272

Source: the result of analysis

The other parameter maps built to classify critical land were listed in figure 3. Based on the slope map in figure 3a, topographic characteristic of East Lombok area was dominated by the steep slope in the north parts and slightly flat slope in the south parts. Table 2 indicates that the slope parameter has a big role in determining critical land. It has the highest weight in the protected area outside the state forest area, and production forest area. Rainfall erosivity map in the figure 3b showed that estimated erosivity value ranges from 711 to 2687. The west part of East Lombok regency with relatively higher annual rainfall has the highest rainfall erosivity value.

Map of soil erodibility as shown in figure 3c was generated based on soil texture, organic matter content, permeability, and soil structure. These values were retrieved from soil characteristics stated in the soil map, completed trough soil sample analysis, and the result was listed in table 3. Using equation (4), calculation of K value was range between 0.15 and 0.38. The lower K value indicates low permeability value and also affected by soil texture.

Table 3. Soil sample characteristics

Variable of analysis	Value range	
	Minimum	Maximum
Organic C (%)	1.0	2.7
Silt (%)	14.4	53.3
Clay (%)	27.3	51.4
Sand (%)	13.7	46.2
Very fine sand (%)	1.7	4.7
Bulk density (gr/cm^3)	1.1	1.4
Permeability (cm/hour)	1.8	58.4

Source: the result of soil sample analysis

Tolerable soil erosion (T) in figure 3d indicated that area in East Lombok mainly has low T value (<25 ton/ha/yr). The area with low T values needs more attention because only a minor erosion will affect serious consequences and will aggravate critical land level, especially in hilly areas with a steep slope and shallow solum [31].

Land management map in figure 3e showed that north area of East Lombok relatively has better land management compared to other areas. North area is dominated by state forest area that has better completeness of land boundary markers, higher extension activities, and more applied of soil and water conservation. For the land productivity in the cultivation area, as shown in figure 3f, the center part of East Lombok has higher productivity. This area is dominated by Typic Eutrudepts soil type with good characteristics [17] coupled with good water supply.

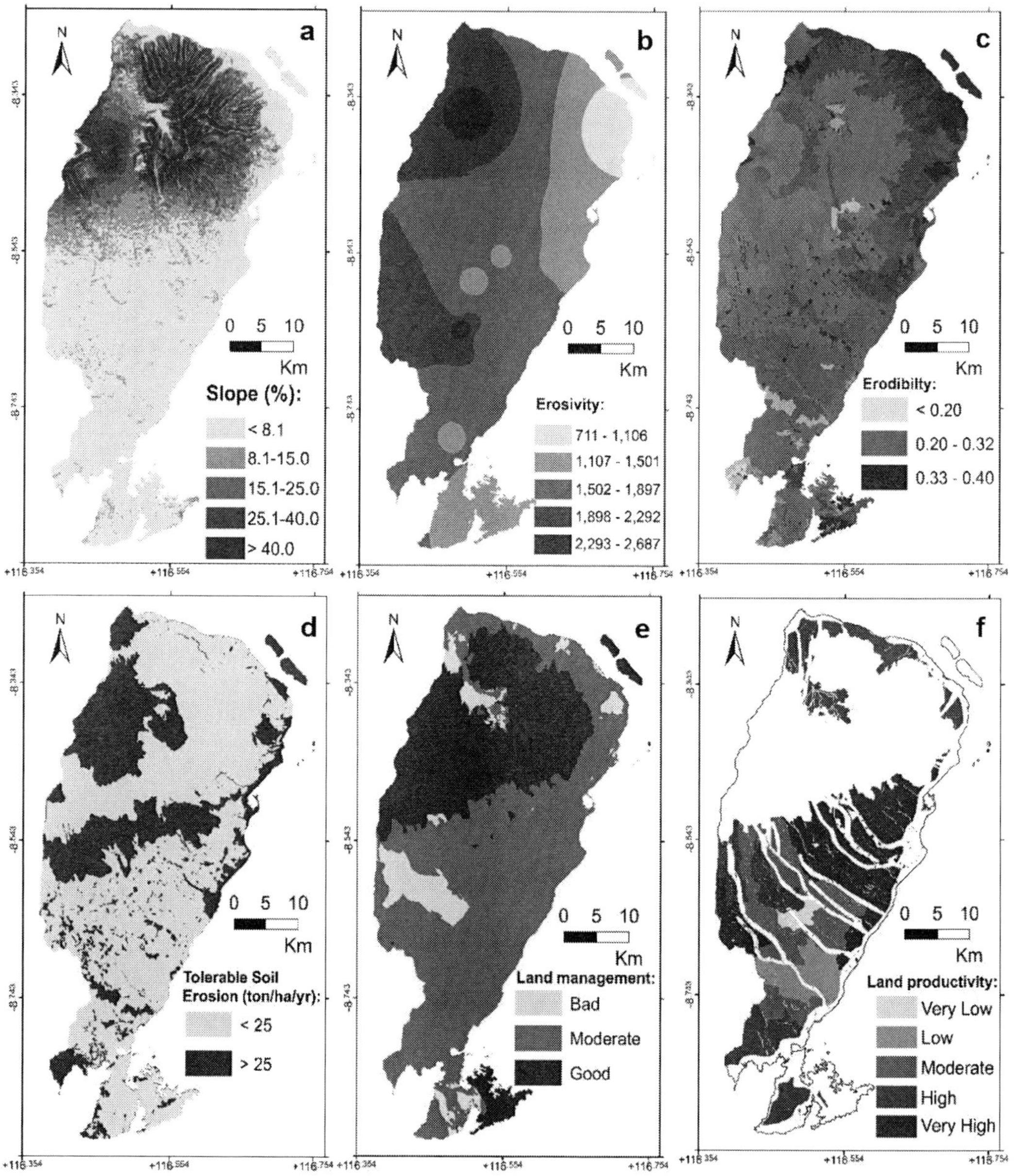

Figure 3. Map of slope (a), rainfall erosovity (b), soil erodibility (c), tolerable soil erosion (d), land management (e), and land productivity (f)

GIS analysis has been carried out to delineate the critical land level. Distribution and level of degraded land in East Lombok is presented in figure 4.

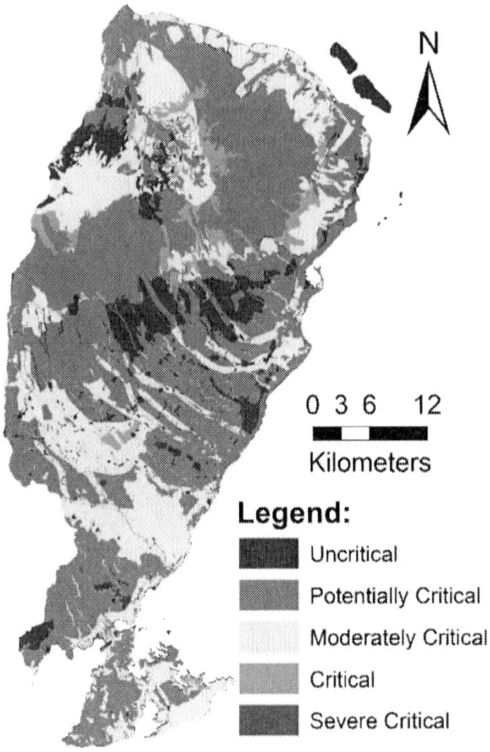

Figure 4 Distribution of critical land levels

Critical land in East Lombok consists of three levels i.e. moderately critical, critical, and severe critical. Those critical level also can be divided based on the ecological status of the area during scoring and weighting process. Finally, those critical land area can be categorized based on the status of forest area as stated in table 4.

Table 4. Area of critical land (in hectare)

Critical level	Protected and conservation forest (State Forest Area)	Protected area outside the state forest area and production forest area		Agriculture cultivation area, and Area for Other Utilization (Non-Forest Area)	Total	%
		State Forest Area	Non-Forest Area			
Moderately Critical	13,036	3,635	14,023	15,110	45,804	87.04
Critical	4,589	776	880	560	6,805	12.93
Severe Critical	4		11		14	0.03
Total	17,629	4,411	14,914	15,670	52,623	32.70[a]
%	33.50	8.38	28.34	29.78	100	

[a] percentage from the total area of regency
Source: the result of analysis

The map resulted from this research showed that 32.7% or 52,623 ha area of East Lombok Regency was categorized as critical land. Although 32.7% is a quite large portion area in a regency, need a more analytical procedure to assess available critical land for producing woody biomass. Based on the area calculation of critical land map intersected with land use/cover map and land status, indicated that the critical land area was distributed on several land use/cover types. The distribution was dominantly located in agricultural land (50.3%) and followed by shrub (22.4%), while, based on the land status, the critical land located in the State Forest Area was smaller (42%) than on the area for other utilization (58%).

In the development of woody biomass plantation, it should be considered to prioritize the utilization of degraded land covered by shrub or other unutilized lands such as grassland and barren land. On the Area for Other Utilization, the biomass plantation can be built using community forest scheme, while for the State Forest Area, the plantation can be only planted in the production forest area using industrial plantation forest or community plantation forest schemes. Although critical land is not an ideal land for farming, optimized management practices can support biomass productivity. The management includes the use of suitable species [31], optimized fertilizer inputs, harvesting time management, and available water handling. Utilization of water resource is not only the concern in biomass production but also a critical issue for environmental sustainability [7].

4. Conclusion

The map resulted from this research showed that East Lombok Regency has 32.7% area of critical lands. Those critical land are dominantly located in agricultural land (50.3%) and shrub (22.4%), while based on the land status, they are located in the area of State Forest (42%) and Area for Other Utilization (58%). Utilization of critical land as woody biomass plantation should be prioritized on the shrub-covered land using community forest scheme, while for the area of State Forest, the plantation can be only planted in the production forest area applying industrial plantation forest or community plantation forest scheme. Further research will assess suitable land and species for the biomass plantation development.

5. Acknowledgment

The authors would like to acknowledge The Directorate General of Strengthening for Research and Development, Ministry of Research, Technology, and Higher Education for the funding support of this research through Operational Support for State Universities (*Bantuan Operasional Perguruan Tinggi Negeri - BOPTN*), Bogor Agricultural University, Indonesia.

References

[1] Abhilash PC, Singh B, Srivastava P, Schaeffer A and Singh N 2013 Remeditaion of Londane by *Jatropha curcas L :* Utilization of Multipurpose Species for Rhizoremediation *Biomass Bioenergy* **51**: 189–193

[2] Höök M and Tang X 2013 Depletion of Fossil Fuels and Anthropogenic Climate Change : A Review *Energy Policy* **52**: 797–809

[3] Edrisi SA and Abhilash PC 2016 Exploring Marginal and Degraded Lands for Biomass and Bioenergy Production an Indian Scenario *Renewable and Sustainable Energy Reviews* **54**: 1537–1551

[4] Clair L and Fordham P 2009 Biomass: An Emerging Fuel for Power Generation (Florida, US) *Power Engineering Articles* pp 1-8

[5] Bae JS, Kim C, Kim YS, Latifah S, Afifi M, Fisher LA, Lee SM, Kim IA, Kang J, Kim R and Kim JS 2014 *Opportunities for Implementing REDD+ to Enhance Sustainable Forest Management and Improve Livelihoods in Lombok, NTB, Indonesia.* Working Paper. (Bogor, Indonesia: CIFOR) p. 151

[6] Saha M and Eckelman MJ 2018 Geospatial Assessment of Regional Scale Bioenergy Production Potential on Marginal and Degraded Land *Resources, Conservation & Recycling* **128**: 90–97

[7] Mehmood MA, Ibrahim M, Rashid U, Nawaz M, Ali S, Hussain A and Gull M 2017 Biomass Production for Bioenergy Using Marginal Lands *Sustainable Production and Consumption* **9**: 3-21

[8] Niblick B, Monnell JD, Zhao X and Landis AE 2013 Using Geographic Information System to Assess Potential Biofuel Crop Production on Urban Marginal Lands *Appl.Energy* **103**: 234–242

[9] Saha M and Eckelman MJ 2015 Geospatial Assessment of Potential Bioenergy crop Production on Urban Margninal Land *Appl. Energy* **159**: 540–47

[10] Smith SL, Thelen KD and MacDonald SJ 2013 Yield and Quality Analyses of Bioenergy Crops Grown on a Regulatory Brownfield *Biomass Bioenergy* **49**: 123–130

[11] Xue S, Lewandowski I, Wang XY and Yi ZL 2016 Assessment of the Production Potential Miscanthus on Marginal Land in China Renew *Sustain Energy Rev* **54**: 932–943

[12] Qi S and Cai Y 2007 Mapping and Assessment of Degraded Land in the Heihe River Basin Arid Nothwestern China *Sensors* **7**(25): 65-78

[13] Wu Q and Wang M 2007 A Framework for Risk Assessment on Soil Erosion by Water Using an Intergrated and Systematic Approach *J. Hydrol* **337**: 11-21

[14] Rahman MR, Shi ZH and Chongfa C 2009 Soil Erosion hazard Evaluation an Integrated Use of Remote Sensing, GIS an Statistical Approaches with Biophyscal Parameters Towards Management Strategies *Ecol. Model* **220** 1724-3174

[15] Ni JR, Li XX and Borthwick AGL 2008 Soil Erosion assessment Based on Minimum Polygons in the Yellow River basin China *Geomorphology* **93**: 233-252

[16] Statistics of East Lombok regency 2018 *East Lombok Regency in Figures 2018* (Selong, Indonesia: Statistical Agency of East Lombok Regency) p. 396

[17] Widiatmaka, Ambarwulan W, Munajati SL, Munibah K, Murtilaksono K, Tambunan RP, Nugroho YA, Santoso PBK, Suprajaka dan Nurwadjedi 2013 Perencanaan Spasial Peningkatan Produksi Kedelai Berbasis Kesesuaian Lahan di Kab.Lombok Timur Prov. Nusa Tenggara Barat *Globe* **15**(2): 161-169

[18] Indrihastuti D, Murtilaksono K and Tjahjono B 2016 Analisis Lahan Kritis dan Arahan Rehabilitasi lahan dalam Pengembangan Wilayah Kabupaten Kendal Jawa Tengah *Tata Loka* **18**(4): 222-239

[19] Wischmeier WH and Smith DD 1978 *Predicting Rainfall Erosion Losses – A Guide to Conservation planning* Agricultural Handbook (Washington: United States Department of Agriculture) p 58

[20] Bols PL 1978 *The iso-erodent maps of Java and Madura*: Report of Belgian Technical Assistance Project ATA 105 (Bogor: Soil Research Institut)

[21] Tan KH 2009 *Environmental Soil Science* (New York: CRC Press) p 600

[22] Roose E 1996 *Land Husbandry-Components and Strategy* (Rome: 70 FAO Soils Bulletin) p 380

[23] Morgan RPC 2005 *Soil Erosion and Conservation* (Oxford: Blackwell Publishing) p 304

[24] Wood SR and FJ Dent 1983 *LECS: A Land Evaluation Computer System Methodology, Manual 5: Methodology* (Bogor, Indonesia: Centre for Soil Research, Ministry of Agriculture) pp.221

[25] Al Rammahi AHJ and Khassaf SI 2018 Estimation of Soil Erodibility Factor in Ruste Equation for Euphrates River Watershed Using GIS *International Journal of GEOMATE* **14**(46): 164-169 ISSN: 2186-2990

[26] Ganasri BP and Ramesh H 2016 Assessment of Soil Erosion by RUSLE Model Using Remote Sensing and GIS-A Case Study of Nethravathi Basin *Geoscience Frontiers* **7**(6): 953-961

[27] Hardjowigeno S 2003 *Ilmu Tanah* (Jakarta: Akademika Pr) p 286

[28] Saaty TL 2008 Decision Making with the Analytic Hierarchy Proces *Int. J. Service Sci.* **1**(1): 83-98

[29] Hermon D 2017 *Climate Change Mitigation* (Jakarta: Rajawali Pers - Radja Grafindo)

[30] Hermon D, Putra A and Oktorie O 2018 Sustaibility Evaluation of Space Utilization Based on Environmental Sustainability at the Coastal Area of Bungus Bay in Padang City, Indonesia *International Journal of Geomate*. **14**(41): 193-202

[31] Chaidar AN, Soekarno I, Wiyono A and Nugroho J 2017 Spatial Analysis of Erosion and Land Criticality of the Upstream Citarum Watershed *International Journal of GEOMATE* **13**: 133-40

Mapping of mercury pollution in GIS-based small-scale mining (case study: Cineam Tasikmalaya District)

Andi Nur Rachman[1,*], Indra Mahdi[2], Agi Nurhidayah[1], Anto Purwanto[3] and Cecep Muhamad SR[1]

[1] Department of Informatics, Siliwangi University, Indonesia
[2] Department of Civil Engineering, Siliwangi University, Indonesia
[3] Department of Public Health, Siliwangi University, Indonesia

*andy.rachman@unsil.ac.id

Abstract. The use of mercury in Small Scale Gold Mining (ASGM) in Cineam District, Tasikmalaya Regency is one of the most important problems in the environment. The long-term accumulation of mercury use results in uncontrolled and invisible mercury pollution. This threatens the health of the environment and has a very bad impact on humans in the long run. Geographical Information System is one way to map and record mercury pollution in Cineam District. The creation of a web-based GIS application can automatically display the mercury pollution distribution radius in the area around the Cineam District ASGM Tasikmalaya District.

1. Introduction

Small-Scale Gold Mining (ASGM) in Indonesia or commonly referred to as Gold Small Scale Mining (GSM) are numerous and difficult to detect their existence. In 2010, around 280 tons of illegal mercury were imported into Indonesia for use in ASGM. This figure has doubled in 2011 [1]. In the process, traditional gold processing (amalgamation) produces mining industry waste (tailings) [2]. Tailings are mercury wastes which increase every time along with mining activities. Mercury is a chemical widely used by large-scale gold miners and Small-Scale Gold miners (ASGM). The majority of researchers agree that mercury use provides greater losses than the results obtained from ASGM [3]. In accordance with regulations the threshold for safe concentration of mercury (Hg) is only allowed 0.001 ppm (President of Republic of Indonesia No. 28 of 2001 and Decree of the State Minister of Environment No. 112 of 2013) [4].

According to the United Nations Environment Program (UNEP) Global Mercury Assessment. Who found that mercury is found in fish worldwide at levels that harm humans and animals [5]. In humans, hair is generally accepted as a means of estimating the body's methyl mercury load, which is most likely derived from fish consumption [2,6], however mercury is useful for industrial purposes such as making batteries, cellulose and plastics, pharmaceuticals, paints, and laboratory and electronic equipment. In water, mercury can spread on the surface, or settle on sediments.

The Cinema Area of Tasikmalaya Regency has many ASGM and the number is difficult to know, because it operates in a hidden and far from monitoring from the local government. The ASGM has been in operation for a long time and has been gradually decreasing natural resource mining in the form of gold. Excavation of land and logging of forests and the use of mercury in a negative impact, not only on the environment, but also on humans around the mining area. This is often not recognized by the general public, because mercury does not directly show its impact but slowly but surely. Nonetheless, the existence of ASGM is difficult to control because ASGM is a source of income for the local community which has been carried out for a long time.

Information technology has been widely used for the benefit of humans, one of the technologies often carried out in the interest of knowing environmental changes. Is a Geographic Information System (GIS), of course it will be useful if it is used for mapping pollution due to the use of mercury around ASGM to provide a clear and comprehensive picture of the distribution of mercury map-based pollution with web technology. The method of making web-based GIS applications intended for mercury pollution mapping uses the V-Mode method with the approach of the object oriented process (OOP) paradigm.

2. Theoretical Basis

2.1. Mercury

Mercury pollution is one of the most important problems in the environment [2]. Mercury at certain concentrations can cause death and health problems that cannot recover in a short time [7].

2.2. Mapping

Mapping is grouping a collection of regions that are related to some geographical location of the region which includes the highlands, mountains, resources and potential of the population that influence the socio-cultural which has special characteristics in the use of appropriate scale [4].

Another definition of mapping is a step that must be done in making maps. The initial steps taken in making data, followed by processing data, and presenting in the form of maps [8].

So, from the two definitions above and adapted to this research, mapping is the process of collecting data to be used as a first step in map making, by describing the spread of certain natural conditions in a spatial manner, moving the real situation into a base map, expressed by using a map scale.

2.3. Geographic Information System

Geographic Information Systems (GIS) always develop, increase, and vary. Geographical Information Systems (GIS) is also a field of study of science and technology that is relatively new, used by various fields of disciplines, and is developing rapidly. Geographical Information Systems (GIS) are organized collections of computer hardware, software, geographic data and personnel that are designed efficiently to obtain, store, update, manipulate, analyze and display all forms of information that are geographic in reference.

Geographical Information System (GIS) is a set of tools that allows processing of spatial data into information in assisting policy making about the earth. The Geographical Information System (GIS) is composed of the concepts of several layers and relations [9].

2.4. Google map

Google Maps is an online map application service provided by Google for free. The Google Maps map service is officially accessible through the site http://maps.google.com. On the site, geographical information can be seen on almost all surfaces on the earth except the north and south poles. This service is made very interactive, because in it the map can be shifted according to the user's wishes, change the zoom level, and change the appearance of the map type.

Google Maps has many facilities that can be used for example location search by entering keywords, the intended keywords such as the name of the place, city, or road, other facilities namely calculating travel routes from one place to another [1].

3. Methodology

System development methods used during the research using the Rational Unified Process (RUP) method. RUP uses the object oriented concept, with activities that focus on developing models using the Unified Model Language (UML). Through the picture below it can be seen that RUP has, namely:

1. The first dimension is drawn horizontally. This dimension represents the dynamic aspects of software development. This aspect is described in the development phase or phase. Each phase will have a major milestone that marks the end of the next phase. Each phase can stand from one iteration. This dimension consists of Inception, Elaboration, Construction, and Transition.
2. The second dimension is described vertically. These dimensions represent the static aspects of the software development process which are grouped into several disciplines. The software development process described in several disciplines consists of four important elements, namely who is doing, what, how and when. This dimension consists of Business Modeling, Requirement, Analysis and Design, Implementation, Test, Deployment, Configuration and Change Management, Project Management, Environment.

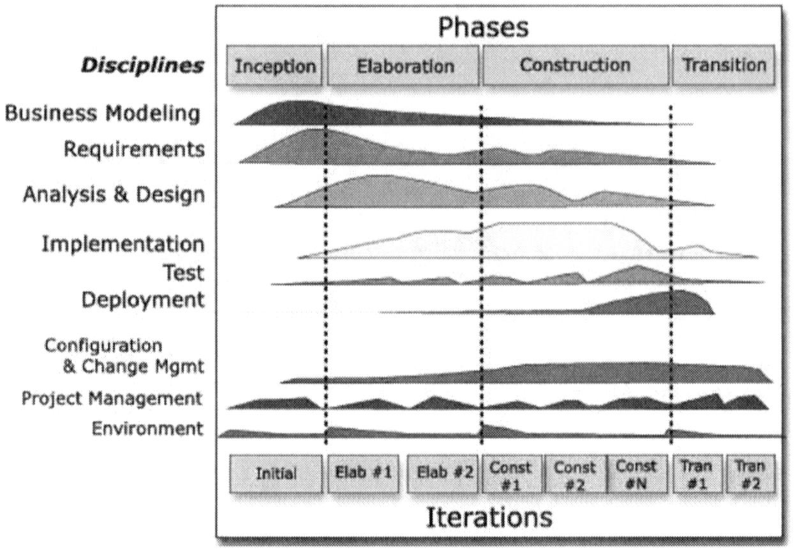

Figure 1. Rational Unified Process Architecture
Source: Pressman [9]

The application creation phase uses the RUP method as follows:
1. Inception, namely determining the scope of the project, creating a 'Business Case', answering the question 'what is done can create a' good business sense 'so that the project can be continued.
2. Elaboration, that is analyzing various requirements and risks then setting the base line.
3. Construction, that is doing a series of iterations, in each iteration will involve the following processes: analysis of design, implementation and testing.
4. Transition, namely making what has been modeled into a finished product, in this phase beta and performance testing are carried out, making additional documentation such as; training, user guides and sales kits, plan product launches to the user community.

3.1. 5W + 1H method
The research method used for mapping the distribution of pollution around PESK uses the 5W + 1H approach. The 5W + 1H method is used to formulate problems and solutions related to pollution in the ASGM operational area. The problem approach mechanism uses the 5W + 1H method as shown in Figure 3.2 below:

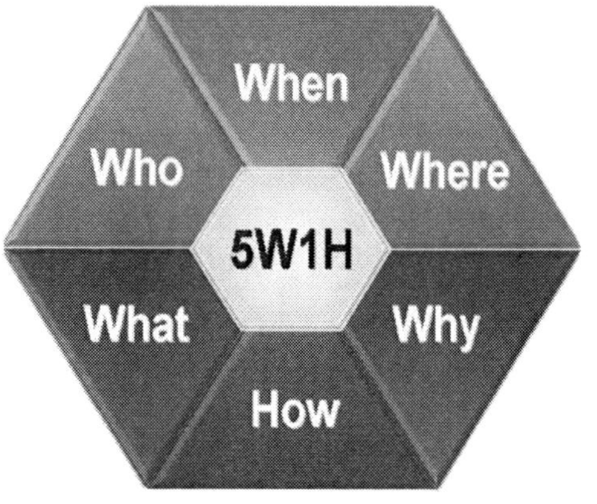

Figure 2. 5W1H Method

Step	*Action*
Where	Identifying the location, namely in the area around the Cineam District ASGM of Tasikmalaya District
What	• The material analyzed from this study is the distribution of pollution of heavy metals, namely mercury in the area around ASGM. • The technology that will be applied is web technology • The scientific field needed in this study is the science of Public Health or the Environment and information technology
Why	• Mercury is a substance that is harmful to ecosystems, especially humans, because in the long run it can cause fetal disability and human genetic disorders. • Mercury causes negative effects that are not felt by humans • Mercury distribution in the environment cannot be seen in plain view, so GIS technology helps in knowing the distribution of mercury pollution. • Mapping of mercury pollution in the area around ASGM is important to help the community, government and researchers know the extent of the distribution of environmental pollution as a result of GIS-based ASGM. • Ministry regulations regarding the environment and pollution from heavy metals in the mining environment.
Who	• Identify who is the user of the pollution mapping application, Who is the miner. • Analysis based on 5M (Man / Manpower, Machine, Material, Method, Mother Nature (work environment).
When	• Identify how long the ASGM has been operating and when problems have occurred to the community around ASGM. • The extent of distribution of mercury pollution in water and sediment.
How	• Taking water and soil samples (sediments) along with spot coordinates for sampling is done in stages • Phase I is carried out by the closest part (radius <50 m) from ASGM. Phase II surveys are taken samples with a radius of 100 m from ASGM and a 200m radius sample is carried out in stage III. • Analyze mercury levels in water and sediment using the irradiation process in the laboratory • Coordinate data from samples and lab results from samples are converted into spatial data to be displayed on GIS • Designing and Making a GIS application using a web-based RUP method • Dissemination to the public about mercury hazards to communities around ASGM.

4. Results And Discussion

4.1. Software Implementation

This section shows the results of interface design that has been implemented into a program created using PHP with the database used is MySQL.

1. Main page

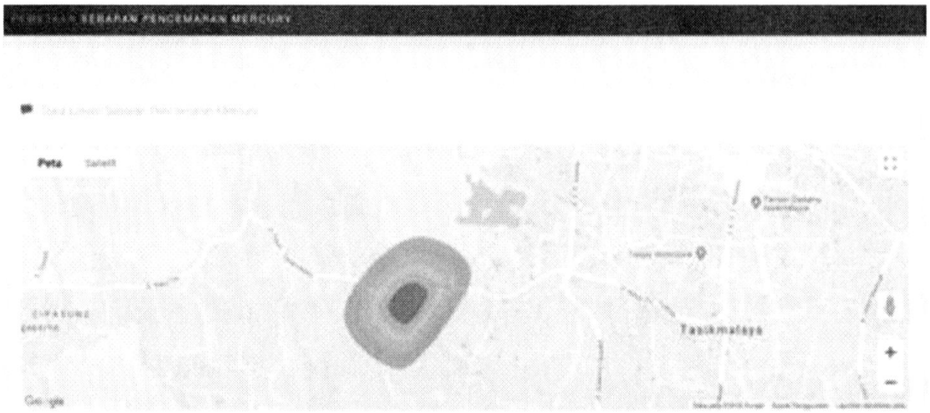

Figure 3. Main Page

On this page you can see the point and area of mining and the spread of mercury in certain locations.

2. Data Input Page

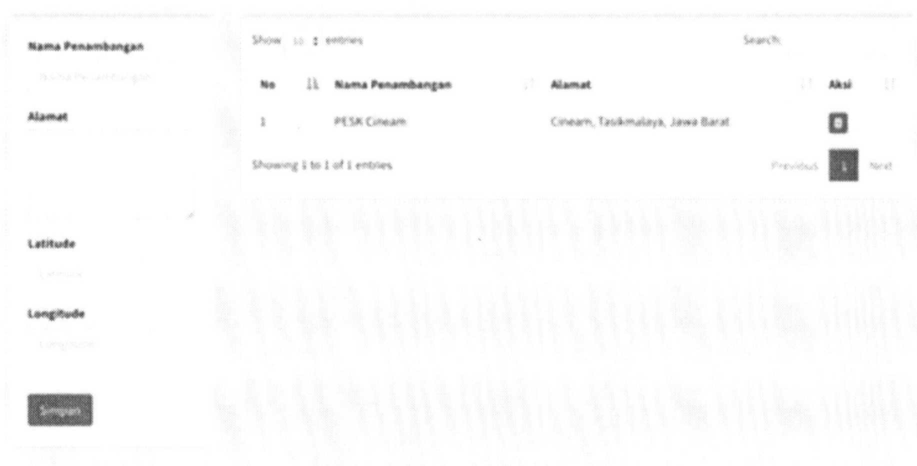

Figure 4. Data Input Page

On this page you can enter the location data points that have been obtained, so that the location point of PEST and the distribution of pollution will also appear on the main page.

3. ASGM Location Page

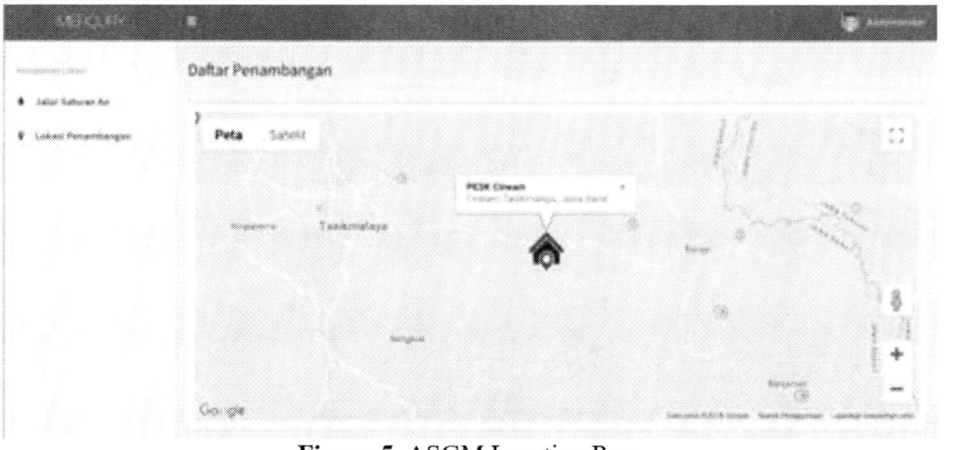

Figure 5. ASGM Location Page

On this page you can see the location points for Cineam sub-district of Tasikmalaya District.

4.2. Advantages and disadvantages

The advantages of this application are as follows:

1. In this application it can show the distribution of ASGM location points, mark the point of pollution and store it in the database.
2. In this application data will appear in the form of markers with layer level in accordance with the level of pollution in the field.

Deficiency :

1. Display the map and location point must be online.
2. The application only covers points located in Cineam District, Tasikmalaya Regency.

5. Conclusions And Suggestions

5.1. Conclusion

Based on the results of the research that has been done can be concluded as follows:

1. Geographical information system mapping the distribution of mercury pollution in Cineam Subdistrict, Tasikmalaya Regency can be used as one of the information media regarding location points and locations in Tasikmalaya Regency.
2. Geographical information system mapping the distribution of mercury pollution can display information about the location and point of pollution at the layer level according to the level of pollution in the field.

5.2. Suggestion

In this study, of course there are many shortcomings found, so suggestions can be given if they can be used and support as a reference for the next system development. Geographical information system mapping the distribution of mercury pollution can still be developed in the future. There are some drawbacks to the features found in the geographic information system mapping the distribution of mercury pollution, which include, in the absence of the nearest road route displayed in the geographical information system mapping the distribution of mercury pollution, the study only covers mercury pollution in Tasikmalaya.

References

[1] Ismawati Y 2011 *Interviews with importers and ASGM workers by Yuyun Ismawati*

[2] Michalska A and Michalski K 2009 *Implementation GIS for the modeling of the mercury pollution decomposition in soils around power stations* In: D Kereković (ed.) *Time, GIS & Future* Hrvatski Informatički Zbor – GIS Forum, University of Silesia, Zagreb, 198-203

[3] Ismawati Y, Jindrich P and Joe D 2013 *Titik Rawan Merkuri di Indonesia* (Indonesia: BaliFokus)

[4] Ismawati Y 2010 *Policy Brief: ASGM in Indonesia* (Denpasar: BaliFokus)

[5] UNEP 2002 *Global Mercury Assessment* (Geneva, Switzerland: UNEP) 258

[6] Harada M, S Nakachi, T Cheu, H Hamada, Y Ono, T Tsuda, K Yanagida, T Kizaki and H Ohno 1999 Monitoring of mercury pollution in Tanzania: relation between head hair mercury and health *Science of The Total Environment* **227**(2–3) 249-256

[7] Bernhardsen Tor 2002 *Geographic Information Systems: An Introduction, 3rd Edition* (Wiley)

[8] Chor Pang Lo and Albert K W Yeung 2002 *Concepts and Techniques of Geographic Information Systems* (Prentice Hall)

[9] Kang-Tsung Chang 2002 *Introduction to geographic information systems, Volume 1* (McGraw-Hill)

Examining the effectiveness of prodira policy on improving human development index at Province of Gorontalo

Arwildayanto*

Management of Education Department, Universitas Negeri Gorontalo, Indonesia

*arwildayanto@ung.ac.id

Abstract. The purpose of this study was to determine the effectiveness of the implementation of an education program for the local people policy - *Program Pendidikan untuk Rakyat* (PRODIRA) in the improvement of Human Development Index (HDI) in the Province of Gorontalo. Through PRODIRA the local government has already issued the regional budget (APBD), which the impact of the budget allocation will be measured by the HDI. The research approach employed a qualitative method of semi-structured interview, observation, and focus group discussion on key informants, which involved in compilation and implementation of the PRODIRA in Gorontalo Province. The study result showed that the PRODIRA provided acceleration on educational equity for the population aged 16-18 years, which enrolled at the level of secondary education. PRODIRA has also significantly impact in increasing the Human Development Index of Gorontalo Province during 2012-2016, that it proved by the increasing the literacy rate from 95% to 99.81%, the average number of school years from 8 years to 12.88 years. The result also revealed that the PRODIRA is able to reduce levies in some schools and it increase acceleration of all education services to school age 16-18 years through accelerated national education service. The research recommended that the central government needs to encourage local governments to innovate in developing the PRODIRA in order to achieve the target of to ards the nation of Indonesia is independent and highly competitive in 2025.

1. Introduction

Education is an effort that can accelerate the development of human resources in carrying out the tasks assigned to him. Realizing that, the Indonesian government is committed to providing educational services, as stipulated in the 1945 Constitution of the Third Amendment of article 31 paragraphs 2, which every citizen is obliged to follow basic education and the state finances it [1]. The derivation is reinforced by Law Number 20 Year 2003 on National Education System, in article 34 states that the Government and regional governments guarantee the implementation of compulsory education at the minimum level of basic education without charging [2]. The mandate of this law provides consequences for the central and local governments to allocate funds for free basic education to improve the quality of human resources in minimum standards. Free basic education policy so that every citizen of Indonesia has opportunity to complete the minimum education at the level of basic education in preparation his life free from illiteracy, as well as preparation to continue on higher education level.

The government's commitment, as evidenced by the development and improvement of the national education budget is moving up. This needs to be appreciated as a form of government's promise to pay attention to the quality of education. The increasing of budget allocation for education is seen as a strategic move to achieve quality education in line with Government Regulation No. 19/2005 providing the legal framework and vision of education standards and quality, which require clear strategic steps to make it happen [3].

The central government's commitment to free basic education is the foundation for conducting basic education as a whole no primary school-age children lack access to schools. In addition, provincial, district and municipal governments are given discretion to provide free education at a higher level towards free education at the secondary and higher education levels. As the statistical data, the 2015 budget of the central government prepares the financing of a 12-year compulsory education program. The 12 year compulsory education requires commitment and support of all parties, namely the center, local government, community and parents [4].

To respond to the increase in compulsory education programs held by the central government, the Province of Gorontalo has launched a Free Education Program policy "in the form of an education program for the people (PRODIRA) which is the flagship program of Gorontalo provincial government since 2012 until now [5]. This policy is based on several considerations, among others, that human resources is a very decisive asset for the retreat Gorontalo. Therefore, the development of human resources has become one of the priorities, and become the flagship program of Gorontalo Provincial Government.

Political will the provincial government of Gorontalo make educational programs for the people (PRODIRA) listed in the financing of regional income and expenditure budget (APBD) is inspired from the idea that education functions to develop ability an and shape the character and civilization of dignified nation in order to educate the nation 3 of RI Law No. 20 Year 2003 on National Education System). Local governments make pro-people policies, reducing the burden of the people in financing education. The political will of the government is in line with efforts to improve the human development index (HDI) of Gorontalo province.

Contextually, the problem of low quality of education so far was due to the low source of education costs. This is evident from the 2004 Human Development Index report that placed Indonesia in 111th position under Malaysia 58, Thailand 76 and even Philippines 83 [6]. Hadiyanto explained specific circumstances at the primary level of education, seen from the results of a study conducted by the International Educational Achievement (IEA), which shows the reading ability of Indonesian Elementary School students are in the 38th out of 39 countries studied [7]. Gorontalo Provincial Government's policy as a form of commitment of provincial government of Gorontalo more accelerate the success of free education funding at high school/vocational school level by accommodating the burden of existing costs in the community and parents are given financial incentives from the budget of regional income and expenditure (APBD) Education program for the people (PRODIRA) which was launched by Gorontalo Governor Rusli Habibie and Vice Governor Idris Rahim [8].

The main objective of PRODIRA is to help parents not to burden with school lives that are operational and personnel. Arfan Arsyad explained that PRODIRA is the work program of Gorontalo Provincial Government for the period 2012-2017 which is budgeted through the government budget, Gorontalo Provincial Government to encourage accelerated increase of human development index (IPM) in Gorontalo province [9].

Buhler, Patricia and Alpha define effectiveness as a measure of the targeted output level, or a measure of success for achieving output [10]. Robbins states effectiveness as a degree to which an organization or institution can realize its purpose [11]. Effectiveness becomes an essentially causal concept, in which the meaning to end relationship, and the cause-effect relationship. Panggulu argues that the success of the organization, program is positively dependent on the effectiveness achieved. Therefore, effectiveness is quite important for the work unit, both government and private institutions in order to be able to provide the best service to its stakeholders, in the affairs of public services, in accordance with what has been planned and right on the target [12], as well as in managing PRODIRA policy is an integral part of the free education program mandated expressly in the 1945 Amendment III Law of Article 31 paragraph 2 which stated, every citizen is obliged to follow Basic Education and the

State to finance it. The derivation of Law Number 20 Year 2003 Article 34 affirms that the Government and Local Government guarantee the implementation of compulsory education at the minimum level of Basic Education without collecting fees. To achieve these objectives, the state shall provide free quality education services in accordance with the interests and talents of its citizens irrespective of social, racial, ethnic, religious, and gender status, efforts to carry out the mandate [13].

Puluhulawa and Puluhulawa stated that the policy of free education program is one of the central government policy from elementary level to junior high school, which regulated as compulsory education (WAJAR) 9 years through school operational fund (BOS). In addition, it is also one the central government's efforts in implementing the constitution on the right of the citizen in obtaining education. However, the free education program that implemented widely in Indonesia has not yet reached the education of Senior High School (SMA) or equivalent so that it becomes a problem for the people who cannot afford [14].

Based on Law Number 32 Year 2004 regarding Regional Government, article 13 states that it has decentralized some obligatory matters of provinces and districts/municipalities, among others, concerning education [15]. The decentralization policy of education affairs to the government provides the discretion of managing education based on local regulations (PERDA) respectively. One of Gorontalo Local Government's policies is Free Education can be exceeded from the national policy of 9-year compulsory education. Gorontalo Province has the initiative to implement education program policies for the people from primary to secondary school level or equivalent, regulated by Local Regulation No. 7 of 2012, on the implementation of the PRODIRA policy, in the general provision it is explained that the Free Education Service (FES) is an education service that complimentary participants students, parents or guardians from the levy of operational costs of education units at secondary education level [16]. Gorontalo Provincial Government is committed to implement free education program for people who send their sons and daughters in high school level or equal. Local government policy on free education is one of the flagship programs of Gorontalo Governor. The free education program has been implemented since January 2012 is expected to provide educational services for the underprivileged/less favoured to complete education up to senior high school level or equivalent.

FES program up to senior high school level is also regulated by Gorontalo Governor Regulation No. 08/2012 on Guidance of Grant Fund Management for School Operational Fund (BOS) of Senior High School, Senior High Vocational School (SMK) and Senior High Islamic School (MA) in the Framework of Free Education Program of Gorontalo Province [17]. Subsequently changed to the Education Program for the People (PRODIRA) as outlined in Gorontalo Provincial Regional Regulation No. 2 of 2012 on Regional Development Plans of Gorontalo Province in 2012-2017 which makes education for the people and free health become the leading program [18].

The conception of PRODIRA has been fully explained in the Regional Regulation No. 7 of 2012 on the Implementation of Education for the People (PRODIRA) in Chapter I General provisions of Article 1 states that PRODIRA is the implementation of educational programs that complimentary learners, parents or guardians from the levy of operating expenses on the unit secondary education except for International Standard Schools (RSBI), International Standard Schools (SBI), and community-managed education units, as well as providing quality improvement costs at primary education and incentives for PAUD educators [19].

The function of the implementation of the policy of PRODIRA by the government of Gorontalo province can be observed in the PRODIRA guideline (Juknis PRODIRA), namely: a) as an effort of expanding and equalizing the opportunity to obtain free, affordable, quality and justice for all citizens, b) as supporting national education programs Universal secondary education held jointly by the Regional Government, and the Education Unit [20].

The objective of the PRODIRA Policy is in accordance with the Decree of Head of Province District Education Office (Dikbudpora) No.188.4/Dikbudpora/ 172.a/Dikmen/ I/2014, among others: a) increase the participation rate of early childhood education, primary and secondary education, b) improve services at all levels of education for the quality and relevance of education, c) Free of charge for all students of SD/MI/SMP/MTs, and SMA/SMA/MA/ SMK, both state and private school to the operational cost of education unit. From PRODIRA goal it is clear that PRODIRA policies are in line with efforts to increase the Human Development Index (HDI) of Gorontalo [21].

The Human Development Index (HDI) is a comparative measure of life expectancy, literacy, education and living standards for all countries in the world (Bureau of Statistics and UNDP [22]. HDI is used to classify whether a country or region is in an advanced, developed or underdeveloped category and also to measure the influence of economic policy on quality of life [23]. HDI is an important indicator to measure success in building the quality of human life or community/population [24]. IPM explains how people can access development outcomes in obtaining income, health, education, and so on. The HDI introduced by UNDP in 1990 underwent a change in the method of calculation in 2010. Where the Indonesian Statistic Bureau (BPS) is adopted a new methodology change of HDI calculation in 2014 and back casting since 2010 [25].

HDI is formed by three basic dimensions, namely longevity and healthy life (a long and healthy life), knowledge (knowledge), and decent standard of living (decent standard of living). Longevity and healthy living are illustrated by Life Expectancy at Birth (AHH), the number of years expected by newborns to live, assuming that the pattern of death rates by age at birth is the same throughout the age of the infant. Knowledge is measured through the average indicator of Age/Old School and School Expectation. Average School Length (RLS) is the average length (years) of population aged 25 years and over in formal education. School Long Expectancy (HLS) is defined as the length of (year) formal school that is expected to be felt by the child at a certain age in the future.[26]

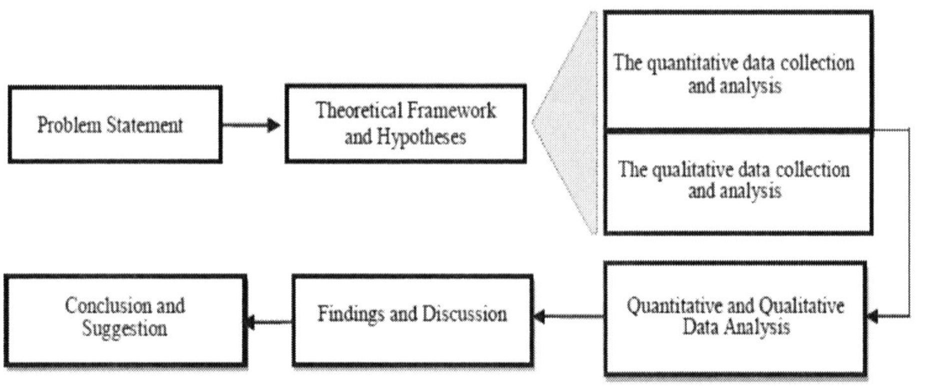

Figure 1. Research methods

HDI is an indicator used to view development progress in the long term, including in seeing the effectiveness of PRODIRa Program implemented by the local government of Gorontalo Province. To see the progress of human development in Gorontalo Province, there are two aspects to note, namely the speed and status of achievement of PRODIRA Program. In Law no. 33 of 2004 on regional financial balances states that the government, both central and local, should allocate 20 percent of the budget for education outside of salaries and official fees. Thus, the educational budget referred to here is included in the development budget category because it does not include routine budgets in the form of salaries and others [27]

2. Methods

The study used a combined method of quantitative and qualitative methods (mixed methods). Creswell states that mixed method is an approach to inquiry that combines or qualitative quantitative forms of research [28], as shown in figure 1.

The purpose of combining methods together, or sequentially in a study, obtained a comprehensive data, valid, reliable and objective. In this research used combination research method (mixed) with design or concurrent embedded model (unbalanced mixture). According to Sugiyono, the combination model or concurrent embedded method is a research method that combines both quantitative and qualitative research by mixing [30]. Where both quantitative and qualitative methods are used are not balanced. Because this study is the first year, the dominant method used is in quantitative form in qualitative support. The target population is distributed according to the quantitative method established as the primary method in the first year. The survey was conducted at 144 SMA/SMK/MA. The sample calculation is proportionally based on the representation of each regency and city region as Gorontalo Province, as much as 20% of the existing SMA.SMK/MA population. Obtained as many as 29 school units. Respondents who fill the instrument consist of principals, treasurers, teachers, and parents (school committee) so the number of respondents of each school's study there are 4 people x 29 schools as many as 116 people.

3. Results and Discussions

From 116 respondents who gave answers obtained information about the policy of PRODIRA according to Principal, Teachers, Treasurer and Parents Students of SMA/SMK/MA in Gorontalo Province able to increase Human Development Index (IPM) Gorontalo Province.

Table 1. Prodira's policy in improving human development index (HDI) of Gorontalo Province

NO	Assessing items	Category
1	Production policy increased the participation rate of rough (APK) of SMA / SMK / MA	76.90 Good
2	Prodira's policies increase the Net Enrollment Rate (APM) of SMA / SMK / MA	77.76 Good
3.	Prodira's policies increase School Enrollment Rate (APS) SMA / SMK / MA	79.31 Good
4.	Policy Prodira increase the Graduation rate (AL) SMA/ SMK / MA each year	75,52 Good
5.	Prodira policy lowers dropout rates (DO) students SMA / SMK / MA	82.07 Good
6	Policies Prodira completed illiteracy level SMA / SMK/ MA	80,52 Good
7	The prodira policy completes the 12-year compulsory education of	86.56 Good

Table 1 showed that the contribution of Prodira's financing policy is in the "Good" category in completing the 12-year compulsory education. In general, policy of PRODIRA supports the increase of Human Development Index by Average = 625,89/8 = 78.24% (good). This is also confirmed by the interview with one of the principals of SMA Negeri 1 Telaga Biru Gorontalo Regency, PRODIRA which was implemented by Governor Rusli Habibie is a commitment in the effort to build the quality of education, when compared with the previous years are still below. After this PRODIRA is rolled out it looks increasing especially from the number of Human Development Index Gorontalo reach 71,76. This PRODIRA policy has been growing public trust to school, in term of transparency for instance [W/E/A/1/11/2016] dalam Warni T. Sumar [31]

The increasing trend of HDI in Gorontalo Province can be seen from the report of Gorontalo Central Bureau of Statistics 2017, as as shown in figure 2.

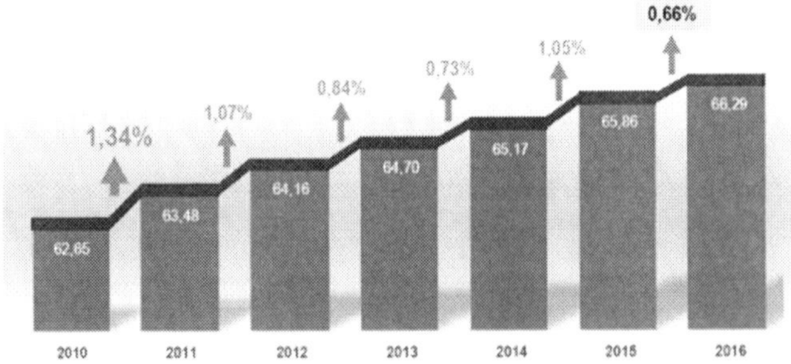

Figure 2. Trend of Increase in HDI of Gorontalo Province from 2010-2016 (Source: Report BPS Gorontalo Province [32]

The graph showed that, the HDI of Gorontalo Province continues to progress during the period 2010 to 2016 HDI Gorontalo Province increased from 62.65 in 2010 to 66.29 in 2016. During that period, HDI in Gorontalo Province grew on average by 0, 73% per year. In terms of rate increase has not showed significant figures but in terms of commitment to promote education sector in improving education is felt and evident from the flagship program of the provincial government of Gorontalo. As stated by the Provincial Secretary of Gorontalo that in 2017 the budget for APBD in the education sector rose to 28%. Since 2012, the provincial government of Gorontalo is also prepared budget for PRODIRA in order to accelerate the development of human resources. The total amount of PRODIRA funds for SMA/SMK/MA depends on the total of BOS funds. Before the BOS funds for SMA/SMK/MA were rolled out by the central government, education funding in Gorontalo province was almost assisted by PRODIRA funds from 2011 to present 2016 [33]. The amount of PRODIRA funds varies every year following BOS funds and BOS funds that are rolled out by the central government.

Table 2. Prodira allocation budget from the year 2011-2016 [34]

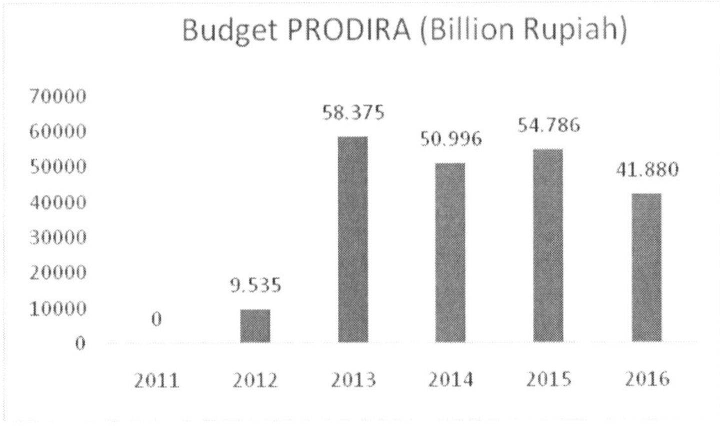

PRODIRA budget allocation (see table 2) is contributed to the increase of Human Development Index of Gorontalo, according to the report of Gorontalo Provincial Central Statistics Agency which stated that HDI is supported from the knowledge aspect, consisting of two indicators namely Average School Length (RLS) and School Expectation (HLS), as shown in figure 3.

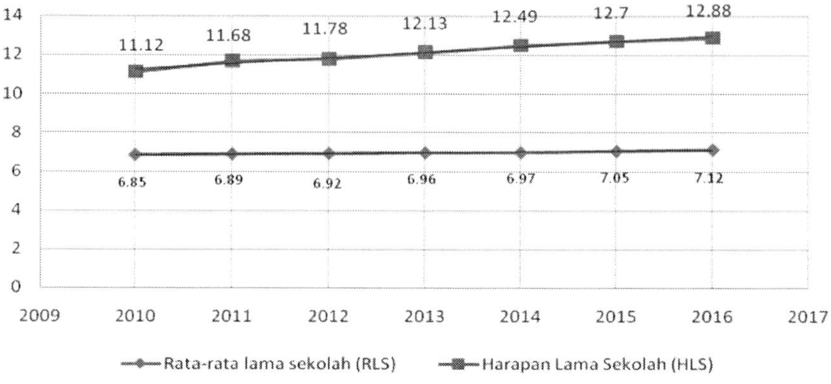

Figure 3. Trend of RLS and HLS of Gorontalo Province from 2010-2016 (Source: Report BPS Gorontalo Province [35]

In accordance with the research report Warni T. Sumar, explained that the implementation of PRODIRA in selected Senior High School in Gorontalo Province is seen from several indicators, such as education condition, 75% support, education index 63% support, Gorontalo provincial superior program 72% support [31]. This is also stated by the Head of Dikbudpora Gorontalo Province Weny Liputo explaining, Contribution Prodira in increasing HDI Gorontalo seen from aspect of average of school length have significant increase which year 2011 about 8,2 year. Then, Central Bureau of Statistics (BPS) data in 2015, the average school rate in Gorontalo is estimated to be 12.88, it means that it continues to increase, and is already above the national rate of 9.2 [37]. Similar things are supported by research Meylina Nikensari and Kuncara, concluded that the level of local government spending on the education sector has a significant influence on HDI, where every change occurs On regional government spending on the education sector will be followed by changes in HDI. From t test, from the results of his research calculations concluded partially the variables of local government spending on education sector have an effect on the HDI (T count 3.023> t table 1, 674116), while the local government spending on the health sector has no effect on the HDI (t hitung0,412 <t table 1.674116) [38]. This is proven BPS of Gorontalo Province acknowledged the increase of RLS and HLS positive impact of the government's development in education several years ago. Seen also his contribution to the population aged 15 years and over can read and write has reached 98.44%. Residents aged 16-18 years as many as 7 out of 10 people still in school. This is of course from the support of PRODIRA funds prepared by Gorontalo Provincial Government [39].

The same thing was also explained by Ekosiswoyo, Kardoyo, and Raharjo that the economic benefits of educational investment (rate of return) such as the PRODIRA policy, were higher than physical investment with an average ratio of 15.3% and 9.1% [40]. So the investment of Gorontalo provincial government through PRODIRA fund policy is the right choice to make investment in the field of education very profitable Gorontalo generation forward more and productive in work.

Dwi Atmanti and Hastarini explained that investment policy through PRODIRA by Gorontalo Provincial Government is also in line with the basic assumption of Human Capital theory which states that one can increase their income through education improvement. Each additional one year of school means, improving work ability, productivity and income levels, as well as delaying receipt of income for one year in education [41].

The PRODIRA policy which frees up the operational costs of the students up to the SMA/SMK/MA/LB level certainly contributes greatly in increasing the public interest to continue the level of SMA/SMK/MA/LB even to universities. Since government policies such as this certainly change the public mindset of the importance of government stimulus providing education financing through PRODIRA policy is a long-term macro investment, the benefits of this investment can only be felt after decades [42].

Psacharopoulos, George., This is in line with the World Bank's recommendations since 1960 that define four criteria for human resource development investment, including 1) the need for skilled labor in vocational and technological fields; 2) the expansion of primary and secondary education, Rate of return) with respect to lower costs. 4) justice and equity which shows the importance of the distribution of educational opportunities and other forms of human resource development, both geographically, socially and economically [43]. Thus, Gorontalo Provincial Government is already on the right track to put education as a superior program.

It is clear that the policy of PRODIRA at SMA/SMK/MA level contributes to the improvement of Human Development Index in Gorontalo Province. Moreover, if followed by increased awareness of the community participate in the financing of the acceleration that can be achieved will exceed if only sourced from the central government and local governments only. It requires the participation of people who have a strong economy to support local government efforts and programs to innovate in developing the PRODIRA in order to achieve the target of Towards the nation of Indonesia is independent and highly competitive in [44] and the Indonesian Education Outlook 2025: toward a sustainable world class quality level [45]. It is expected that there is a breakthrough to boost the economic potential of the community to help finance education for the underprivileged while improving the expected quality .

The success of the program underscores the importance of strong political will and commitment that Gorontalo provincial government has shown in launching the initiative and sneering that budget allocation is available to finance its implementation.

4. Conclusions

Based on the research findings and theoretical description above are concluded as follows:

1. The success of education development is in need of financing from various sources, especially from central and local government. The stimulus of education financing through PRODIRA policy by Gorontalo Provincial Government is able to accelerate the achievement of better human development index (HDI) target. This can be seen from the school's expectation number (HLS) growing above the national rate.
2. To maximize the policy of PRODIRA this certainly need support all education stakeholders. Including ensuring this PRODIRA policy can be included in Gorontalo Province *Rencana Pembangunan Jangka Menengah Daerah* (RPJMD) 2017-2022 for the smoothness of its budgeting.

With the commitment of all parties believe Gorontalo Province can advance regional development beyond national development through acceleration of education development by implementing the policy of PRODIRA more effective and efficient.

Acknowledgements

The author would like to thanks the Directorate General of Research and Development Reinforcement, Ministry of Technology Research and Higher Education of the Republic of Indonesia for the research grant, and Gorontalo Provincial Youth and Sports Culture Education Office for allowing author access to data used in this research.

References.

[1] Undang-undang-Dasar Tahun 1945 Lembaran Negara Republik Indonesia Nomor 14 (Jakarta)

[2] Tim Penyusun Undang-undang No 20 Tahun 2003 tentang Sistem Pendidikan Nasional (Jakarta: Departemen Pendidikan Nasional)

[3] Indonesia, Presiden Republik 2005 Peraturan Pemerintah Republik Indonesia Nomor 19 Tahun 2005 Tentang Standar Nasional Pendidikan (Departemen Pendidikan Nasional Republik Indonesia)

[4] Mohamad R 2014 Persepsi Masyarakat Terhadap Kebijakan Pendidikan Gratis Studi Kasus di Kecamatan Sumalata Kabupaten Gorontalo Utara *Diss* (Gorontalo:Universitas Negeri

Gorontalo)
[5] Yusuf M 2015 Peran Dinas Pendidikan Kabupaten Boalemo Dalam Menjalankan Tugas Pembantuan (Studi Kasus Implementasi Pendidikan Gratis Bagi Masyarakt *Diss* (UNG)
[6] Widayati A 2008 Penelitian Tindakan Kelas *Jurnal Pendidikan Akuntansi Indonesia* **6**:1
[7] Hadiyanto 2014 Mencari Sosok Desentralisasi Manajemen Pendidikan di Indonesia (Jakarta: PT. Rineka Cipta) 204; p.5
[8] Gorontalo Post 2012 Program Pendidikan untuk Rakyat, Gorontalo Post, Senin 20 Februari 2012 Gorontalo
[9] Arfan A 2016 Influence of Knowledge of Management, Principals' Attitude, And Effectiveness of Prodira Towards School Performance In Gorontalo Provincial *Jurnal Ilmiah Education Management* **7**(1) pp. 1186-1204
[10] Buhler P 1998 *Alpha Teach Yourself; Managemen Skill in 24 Hour* (Jakarta; Prenada)
[11] Robbins S P and Timothy A J 2006 *Perilaku organisasi Edisi Bahasa Indonesian* (Jakarta: PT. Indeks Kelompok Gramedia)
[12] Panggulu Y T 2013 Efektivitas Kebijakan Retribusi Pada Dinas Pengelolaan Pasar Kebersihan Dan Pertamanan Di Kabupaten Kepulauan Talaud *JURNAL ACTA DIURNA* **2**(4) 7-18
[13] Tim Penyusun, 2003 Undang-undang No. 20 Tahun 2003 tentang Sistem Pendidikan Nasional. (Jakarta: Departemen Pendidikan Nasional)
[14] Jusdin P dan Moh R P 2013 *Implementasi Kebijakan Pendidikan Gratis (Studi Kasus Di Provinsi Gorontalo)* (UNG Press, Gorontalo)
[15] Undang-Undang Republik Indonesia Tahun 2004 Pemerintahan Daerah. 15 Oktober 2004 Lembaran Negara Republik Indonesia Tahun (2004)
[16] Peraturan Daerah Provinsi Gorontalo Nomor 7 Tahun 2012 tentang Penyelenggaraan Pendidikan Untuk Rakyat Pemerintah Provinsi Gorontalo
[17] Peraturan Gubernur Gorontalo Nomor 08 Tahun 2012 tentang Pedoman Pengelolaan Dana Hibah Untuk Bantuan Operasional Sekolah (BOS) Tingkat SMA, SMK dan MA dalam Rangka Program Pendidikan Gratis Provinsi Gorontalo, Pemerintah Provinsi Gorontalo
[18] Peraturan Daerah Provinsi Gorontalo Nomor 2 Tahun 2012 tentant Rencana Pembangunan Jangka Menengah Provinsi Gorontalo Tahun 2012-2017, Pemerintah Provinsi Gorontalo
[19] Peraturan Daerah Provinsi Gorontalo Nomor 7 Tahun 2012 tentang Penyelenggaraan Pendidikan Untuk Rakyat, Pemerintah Provinsi Gorontalo
[20] Surat Keputusan Kepala Dinas Pendidikan Kebudayaan Pemuda dan Olahraga Provinsi Gorontalo Nomor 188.4/Dikbudpora/172.a/ Dikmen/ I/2014 tentang Petunjuk Teknis Penyelenggaraan Program Pendidikan untuk Rakyat Tahun Anggaran 2014. Gorontalo; Dinas Pendidikan Kebudayaan Pemuda dan Olaharaga Provinsi Gorontalo (Dikbudpora Gorontalo Press)
[21] Dinas Pendidikan Kebudayaan Pemuda dan Olahraga 2012 *Juknis Program Pendidikan Gratis, Dikpora Provinsi Gorontalo* (Gorontalo) p. 17
[22] Badan Pusat Statistik Provinsi Gorontalo 2017 *Provinsi Gorontalo dalam Angka (Gorontalo Provinve In Figures 2016* (BPS Gorontalo) p. 87
[23] Anand S and Amartya S 2000 The income component of the human development index *Journal of human development* **1**(1) 83-106
[24] Sagar A D and Adil N 1998 The human development index: a critical review *Ecological economics* **25**(3) 249-264
[25] Badan Pusat Statistik Provinsi Gorontalo 2017 *Indeks Pembangunan Manusia Provinsi Gorontalo 2010-2016* (BPS Gorontalo)
[26] Angka Melek Huruf, Rata Rata Lama Sekolah, Pengeluaran Perkapita Dan Jumlah Penduduk Terhadap Tingkat Kemiskinan Di Jawa Tengah *Diss* (Universitas Diponegoro)
[27] Undang-Undang tahun 2004 tentang perimbangan keuangan pusat dan daerah Lembaran Negara Pemerintah Republik Indonesia (33)
[28] Creswell J W 2013 *Research design: Qualitative, Quantitative, and Mixed Methods Approaches*

(Sage publications)

[29] Sugiyono 2016 *Metode Penelitian Kombinas (mixed Methods)* (Alfabeta: Bandung)

[30] Sugiyono 2016 *Metode Penelitian Kombinas (mixed Methods)* (Alfabeta: Bandung) p.537

[31] Warni T S 2017 Evaluasi Implementasi Program Pendidikan untuk Rakyat (Prodira) di SMA Se Provinsi Gorontalo *Disertasi* p 202

[32] Badan Pusat Statistik Provinsi Gorontalo 2017 Ikhtisan Bulanan Data Strategis Provinsi Gorontalo Edisi 30 Juli 2017 (BPS Gorontalo)

[33] Badan Perencanaan Pembangunan Provinsi Gorontalo 2017 Profil Provinsi Gto Tahun 2016, UPTB Data dan Analisis Pembangunan Bappeda Provinsi Gorontalo, dipublikasikan tanggal 8 Februari 2017 Sumber: https://issuu.com/fachrymusa/docs/buku_profil_2016_final__revisi_iv, 2017; pp. 2-5

[34] Badan Perencanaan Pembangunan Provinsi Gorontalo 2017 Profil Provinsi Gto Tahun 2016, UPTB Data dan Analisis Pembangunan Bappeda Provinsi Gorontalo dipublikasikan tanggal 8 Februari 2017 Sumber:https://issuu.com/fachrymusa/docs/buku_profil_2016_final__revisi_iv, 2017; pp. 2-6

[35] Badan Pusat Statistik Provinsi Gorontalo 2016 *Indeks Pembangunan Manusia Provinsi Gorontalo 2010-2016* (BPS Gorontalo) p 7

[36] Gorontalo Post 2016 Angka Rata-Rata Lama Sekolah di Gorontalo Melebihi Nasional, Gorontalo Post; Gorontalo p. 2

[37] Astri, Meylina, Sri I N, and Harya K 2013 Pengaruh Pengeluaran Pemerintah Daerah Pada Sektor Pendidikan Dan Kesehata Terhadap Indeks Pembangunan Manusia Di Indonesia *Jurnal Pendidikan Ekonomi dan Bisnis (JPEB)* **1**(1) 77-102.

[38] Badan Pusat Statistik Provinsi Gorontalo 2016 *Indikator Pendidikan Gorontalo 2016* (Grafika Karya, Gorontalo) p. 16

[39] Rasdi E, Kardoyo and Tri J R 2008 Strategi Akselerasi Pencapaian Ipm Bidang Pendidikan Untuk Mendukung Keberhasilan Pembangunan Jangka Menengah Kota Semarang *Riptek* **2**(2): 1 - 6

[40] Dwi A H 2005 Investasi sumber daya manusia Melalui pendidikan *Jurnal Dinamika Pembangunan (JDP)* **2**(1): 30-39

[41] Astri M, Nikensari S I and Kuncara H 2013 Pengaruh Pengeluaran Pemerintah Daerah Pada Sektor Pendidikan dan Kesehata Terhadap Indeks Pembangunan Manusia di Indonesia. *Jurnal Pendidikan Ekonomi dan Bisnis (JPEB)* **1**(1) :77-102

[42] Psacharopoulos G 1994 Returns to investment in education: A global update *World development* **22**(9) : 1325-1343

[43] Ali M 2009 *Pendidikan untukPpembangunan nasional: Menuju Bangsa Indonesia yang Mandiri dan berdaya Saing Tinggi* (Grasindo) p. 1

[44] Suryadi A 2012 Outlook (2025) *Pembangunan Pendidikan Indonesia: Menuju Kualitas Yang Berdaya Saing Secara Global (The Indonesian Education Outlook 2025: Toward A Sustainable World Class Quality Level)* (Jakarta: Badan Penelitian dan Pengembangan Kementerian Pendidikan dan Kebudayaan)

[45] Arwildayanto 2018 Assessing The Contribution of The People's Education Program (Prodira) Policy Toward Senior High School Education Performance in Gorontalo Province *JKAP (Jurnal Kebijakan dan Administrasi Publik)* **22**(1) 19-31. doi. org/ 10.22146/jkap.29976

IOP Conf. Series: Earth and Environmental Science **314** (2019) 012075 doi:10.1088/1755-1315/314/1/012075

Improving the capacity of the alertness disaster group in managing the environment and disaster risk

Nasfryzal Carlo* and Eva Rita

Civil Department, Faculty of Civil and Planning, Bung Hatta University, Padang, Indonesia 25135

*carlo@bunghatta.ac.id

Abstract. Disaster Alertness Group (DAG) is a community group formed in the disaster-prone areas. DAG establishment is preceded by a briefing knowledge of threats and risks of disaster that may occur in the neighbourhood. By having the knowledge, DAG and the community identify the potential threats of disasters and plan the appropriate activities or actions to reduce the impact of the disasters in the future. Furthermore, the action conducted jointly activities based on the scale priority. All activities have been done in Nagari Salayo (Salayo village), District of Solok, West Sumatra by using Participatory Rural Appraisal (PRA) method. As a result, the capacity of DAG has been improved in managing the environment from threat disaster (hazards) in Nagari Salayo. Some actions have been done based on the discussions of the citizens who are members of DAG. There are some alternative plans to be done in order to reduce the disaster risks, such as tree planting, installing signals for evacuation on evacuation tracks, improving the irrigation and drainage canals, and facilitating a briefing knowledge of Bank Sampah or "Waste Bank" to the Disaster Alertness Group.

1. Introduction

According to [1], the government has established the Nagari Salayo as one of the disaster resilient villages since 2013 along with the Nagari Batu Bajanjang [2]. This establishment is the manifestation of Disaster- Resilient Village Program in West Sumatra. The resilient village is one of way to reduce disaster risk based on the community [3]. This community-based risk reduction is any effort to reduce disaster risk and vulnerability in the community to increase the preparedness capacity planned and conducted by the community as the main actors.

Nagari Salayo according to [4] is one of 8 villages found in Kubung Sub-district, Solok Regency. The position of Nagari Salayo is at the centre of Kubung Sub-district with a distance of 3 km from the centre of Solok City, 21 km from Arosuka, the capital of Solok Regency, and 60 km from Padang, the capital city of West Sumatra. The region of Nagari Salayo area is divided into 4 jorongs, namely Jorong Batu Palano, Jorong Galanggang Tangah, Jorong Sawah Suduik, and Jorong Lurah Nan Tigo. The area of the Nagari Salayo is 21.44 km square at an altitude of 390-550 meters from the sea surface and the precipitation of 2,141 mm per year and an average of rain days of 145.1 days per year. Nagari Salayo has the borders as follow: in the north is bordered with Solok city, in the south is bordered with Nagari Gantung Ciri and part of Nagari Koto Baru, in the west with Nagari Koto Hilalang and Bukit Barisan; in the east with Nagari Kotobaru and Nagari Panyakalan. Nagari Salayo is passed by Batang Lembang River which is the confluence of the Batang Sumani River from the direction of Nagari

Gantung Ciri and Batang Lawas River from the direction of Nagari Koto Baru. Map of Nagari Salayo as shown in figure 1.

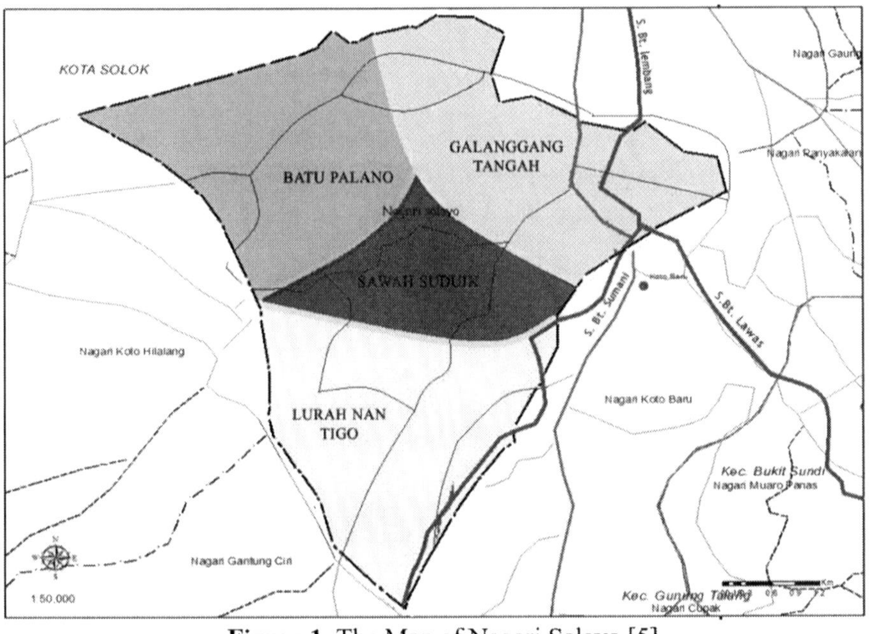

Figure 1. The Map of Nagari Salayo [5]

Furthermore [1], explained that the purpose of the resilient village/nagari is based on the Regulation of Disaster Management No. 1 year 2012 about the General Guideline on Disater-Resilient Village is to (1) protect the community living in disater-prone areas from negative impacts, (2) to increase the participation of the community, especially the vulnerable groups in managing resources in order to reduce disaster risk, (3) increase the capacity of community institution in managing the resourses and nurturing local wisdom for disaster risk reduction, (4) increase the government capacity in giving the resource and technical support for disaster risk reduction, (5) increase the cooperation between the stakeholders in disaster risk reduction, the local government, private sector, universities, Non Govermental Organizations, social organization and other groups who have concerns.

A village/nagari must meet 20 indicators to actieve a resilient village/nagari [3]: (1) the availability of policy/ regulations in the village about disater management or disaster risk reduction, (2) availability of disaster management plan, community action plan, and/or contingency plan , (3) the establishment of Disater Risk Reduction (DRR) forums, (4) the establistment of disaster management volunter (disaster alertness groups) in the nagari, (5) there is corporation between actors and regions, (6) the availability of emergency alertness budget, (7) the availability of DRR budget, (8) training is conducted for village administrative/authority, (9) the availability of training for voluntary teams, (10) the availability of training for village inhabitants, (10) the involment or participation of the inhabitants, (12) the involvement of women in volunteer teams, (13) the availability of maps and risk analysis, (14) the availability of map and evacuation track and refugee sites, (15) the availability of early warning systems, (16) physical structural mitigation is conduction, (17) the availability of economic resilience patterns to reduce vulnerability of the community, (18) the availability of medical protection for vulnerable groups, (19) the availability of natural resource management for DRR, and (20) the availability of protection to the communities major productive assests.

To achieve the above targets, 2 established Disaster Alertness Groups (DAG) in 2 jorong in Nagari Salayo have been actively involved in the resilient village program since 2014. However, in 2 other jorongs have been established yet [5]. Therefore, through the activity of KKN-PPM (student comunity service program to empower the community), the learning process and the establishment of DAG will be conducted to finally realize the 4 DAG in Nagari Salayo, wich in turn make Nagari Salayo as a disaster-resilient village and manage the environment well.

2. Method

The method used in increasing the capacity Disaster Alertness Groups (DAG) is PRA (Participatory Rural Appraisal). At first the community in four jorongs in Nagari Salayo who care their environment is given knowledge about disasters and effects of earthquake, floods and landslides, as well as whirlwinds. Through the PRA method, the community is actively involved in studying, analyzing, and planning the action plans for disaster risk reduction by using resources and be able to integrate the sustainable development policies [6]. The community discussion in DAG results in alternative action plan. Then, the action plan is conducted together based on the priority scale in a limited time.

3. Result and Discussion

3.1. Disaster Alertness Group (DAG)

One indicator of the resilient village is the establishment of disaster management volunteers in the Nagari (Disaster Alertness Group) [3]. In 2014 two DAG were established in two jorongs (Gelanggang Tangah and Batu Pelano). Both DAG have been active in activity of disaster risk reduction and environment management activities in the area of Solok Regency, West Sumatra. However, the other 2 jorongs (Sawah Suduik and Lurah Nan Tigo) have not got any volunteer organization so that there is an imbalance in overcoming disasters and the environment related to disaster resilient villages. Therefore, the starting step is to establish DAG and the results of community discussion in each jorong. As the results of community discussion, 2 new DAG have been established in Jorong Sawah Suduik and Lurah Nan Tigo so that the 4 jorongs have their own DAG. For 2 established DAG (Jorong Gelanggang Tangah and Batu Palano), regeneration is done by some administrators. The administration of the four DAG has been established by the local government on August 13, 2016. It shows that the local government has a concern and be serious to make Nagari Salayo a disaster-resilient village [1]. The concern of Solok regency government is not limited only to Nagari Salayo. Six nagaris in the subdistrict of Lembang Jaya with their respective DAG were established by the regent in 2014 [7].

3.2. The Improvement of Capacity of DAG

Another indicator to achieve a disaster-resilient nagari is the improvement of DAG [2]. It is done by giving the learning on the threat and effect of earthquake, floods, landslides, whirlwind climate change. Based on the knowledge, each DAG is asked to identify the disaster potential threatening them in each area. The results of the identification show that in beside the earthquake threatening the community of Nagari Salayo, flood is a major threat in the region of Jorong Gelanggang Tangah. Almost every year floods occurs as well as of the overflow of Batang Lembang River [1,5]. Jorong Sawah Suduik receives the overflow flood from Jorong Gelanggang Tangah. This flood is worsened by a pile of rubbish in the drainage channels and Batang Lembang River passing Nagari Salayo.

In the Jorong Nan Tigo, the major threat is critical land and drought caused by the effects of climate change. This effect has damaged the farming cycle of local inhabitants. Consequently, plant diseases by rats, pigs and plant hoppers often occur. Another threat is the high potential for landslides on the local people's house location on the hillsides.

In the area of Jorong Batu Palano, there are threats of landslides and flash floods, whirlwinds and disturbance in some several irrigation canals passing Jorong Batu Palano. There is an isolated area in the region of Kayu Manang in Jorong Batu Palano. In this area, the major threat is critical land and landslides as well as the low level of education at the age of compulsory education. At the moment, there are only four local inhabitants studying at the Junior High School [5]. Generally, local people do not graduate from Elementary School. The low level of education will also be a threat on their environment management because of their low capacity to anticipate the situation and conditions of the environment.

3.3. The Action Plan and the Implementation

Based on the threats and identification disasters potential, DAG conducted the community discussions by the help of student facilitators who are the participant of KKN-PPM (student community service program to empower the community) to identify the alternative action plan to reduce disaster risk as well as to manage their environment. First the alternative action plan identified in Jorong Sawah Suduik are: (1) to determining the evacuation path/track, (2) to identify the gathering site to install the evacuation signals on the existing evacuation tracks, (3) to clean the drainage canals from rubbish, and (4) to manage the rubbish with waste bank.

Second, in Jorong Nan Tigo, the action plans are: (1) to plant the trees on the critical lands, (2) to conduct agricultural extension on plant diseases by plant hoppers and pigs and how to overcome them, (3) to spread the information about climate change and how to overcome the effects.

Third, in Jorong Batu Palano the action plan are: (1) to clean the existing evacuation track (2) to design, make and install evacuation signals and gathering sites/spots, (3) to plant the appropriate trees on the critical lands potential to landslides and flash floods especially in Kayu Manang, (4) to spread information on how important education is the age of compulsory education, (5) to fix the irrigation canals and broken drainage.

Finally, in Jorong Gelanggang Tangah the action plan are: (1) to design, make and install the evacuation signals and gathering spots/sites at the appropriate location, (2) to work together with the community on the gutters and the potential area to disturb the water flow, (3) to spread the information and do the coaching to establish and operate waste banks to the members and administrators of DAG, (4) to make the signals of water levels as the early detection to do the evacuation.

The action plans for the respective jorong are done based on the priority scale during the KKN-PPM program. The actions that can be done are (1) to clean the evacuation tracks, to determine the evacuation tracks and gathering spots, (2) to make and install evacuation signals in Jorong Sawah Suduik, Batu Palano, and Gelanggang Tangah; (3) to plant trees in the critical land areas in Jorong Batu Palano and Lurah Nan Tigo; (4) to coach the DAG with waste bank in Jorong Sawah Suduik and Gelanggang Tangah. By the operational of Waste Bank, the waste could be a blessing [8]. As wel as to help DAG to get the capital cost in managing their environment.

The identified and conducted action plans in Nagari Salayo is the manisfestation to be a disaster resilient nagari in Salayo gradually [1] as required to be disaster-resilient nagari [3]. Some action plans which are not yet done are (1) to install the signals for early flood detection in Jorong Gelanggang Tangah, and (2) to conduct the agricultural extension on plant diseases and how to overcome them. For this purpose, DAG is hoped to continuously bring the agricultural experts to help the people of Jorong Lurah Nan Tigo to overcome the threat on their agriculture land.

4. Conclusion

The establishment and affirmation of DAG in the 4 jorongs in Nagari Salayo is the concern of the community and the local government in managing the environment to realize a disaster resilient nagari. The learning on the potential and threat of disaster has contributed to improve the capacity of DAG in managing the environment and disaster threat in Nagari Salayo. The Disater Alert Group (DAG) have succeeded to identify the potential and threats of disaster in respective jorong. Floods, landslides, whirlwinds and climate change effects are the threats for local community. The action plans that can be done are planting the trees, installing the evacuation signals on evacuation tracks, improving irrigation canals and drainage, coaching on Waste Bank to the DAG. Two action plans which are not done yet, and supposedly to continue by DAG are installation the signals for early flood detection in Jorong Gelanggang Tangah and the agriculture extension in Jorong Lurah Nan Tigo.

References

[1] Carlo, N., Yanti, Y. Septe, E. 2017. Cummunity Empowerment of Nagari Salayo Disaster Vulnerable to Become Disaster Resilient Nagari. Ethos (Journal of Research and Community Service): 130-135 https://ejournal.unisba.ac.id/index.php/ethos/article/view/2257

[2] Carlo, N., Budianto, W., Rita, E. 2016. Community Preparedness of Nagari Batu Bajanjang Towards Disaster Resilient. Journal of Environment and Development 02 (01) 2016, pp 324-331. http://www.bkpsl.org/jurnal/jurnal-lingkungan-dan-pembangunan
[3] Anon, 2012. Regulation number 1 of 2012 Concerning Disaster Management General Guidelines for Village Resilience. BNPB.
[4] Witrianto, 2010. Nagari Salayo Subdistrict Kubung Solok Regency http://witrianto.blogdetik.com/2010/12/10/tata-ruang-trTradisional-nagari-selayo/ [March 12, 2015].
[5] Carlo, N. Yanti, Y. Septe, E. 2016. Final Report on the Grant of KKN-PPM, Empowerment of Vulnerable Disaster Communities Nagari Salayo, Kubung Subdistrict, Solok Regency to Realize Resilient Nagari Disaster.
[6] Bemson, C. Twigg, T., and Rossetto, T., 2007. Tools for Mainstreaming Disaster Risk Reduction: Guidance Notes for Development Organizations. The International Federation of Red Cross and Red Crescent Societies / the ProVention Consortium.
[7] Carlo, N. Septe, E., Yanti, Y. 2015. Participation of Community at Nagari Batu Bajanjang Towards a Disaster-Resilient Nagari. Journal of Environment and Development 01 (01) 2015. pp 68-76. http://www.bkpsl.org/jurnal/jurnal-lingkungan-dan-pembangunan
[8] Carlo, N. 2011. Garbage Bank, Who Cares? Article in Padang Ekpres Daily. Monday, March 11, 2011

Knowledge of mother about the household environment againt acute respiratory infection in Padang Pasir: A literature study

Linda Handayuni[1,2,*], Azyyati Ridha Alfian[2], Ali Amran[1] and Abdul Razak[1]

[1] Department of Environmental Science, Postgraduate Program, Universitas Negeri Padang, Prof. Dr. Hamka Street, West Air Tawar, Padang-West Sumatra 25131 Indonesia
[2] STIKES Dharma Landbouw Padang, Jhoni Anwar Street No. 29 Ulak Karang Padang-West Sumatera 25135 Indonesia

*lindahandayuni@yahoo.co.id

Abstract. Acute Respiratory Infections (ARI) is one cause of death in infants and toddlers. The incidence of ARI at West Sumatera in 2016 was 154,962 cases and increased by 2017 as many as 169,031 cases. The highest incidence of ARI was found in Padang Pasir for age <1 year 568 people and age 1-4 years 1648 people. The high number of ARI cases in children under five is due to lack of knowledge of the mother about the household environment to prevent ARI of toddlers. Sample was 86 mothers. The study was conducted by literature study technique. The result showed that 30,2% mother knowledge about ARI still low. Activities to improve knowledge of mother toddlers need to be intensified again by way of counseling about the prevention of ARI through Posyandu cadres and health personnel so as to reduce the incidence of ARI.

1. Introduction

ARI is one of the causes of death in infants and toddlers in developing and developed countries and has been able to. The results of the International Conference on ARIs in Canberra, Australia, in July 1997 found that four million children under five in developing countries die each year from ARI. In Indonesia, ARI is the number one child killer. Setiyorini [1] states that although the ARI eradication program has specifically begun in 1984, with the aim of trying to reduce morbidity and mortality, especially in infants and toddlers caused by ARI, the morbidity and mortality rates are still high. In West Sumatra Province, at least 17,622 children under five in various cities were infected with ARI. The toddler was partly suffering from a non-pneumonia type of ARI and was relatively small with a pneumonia (severe) ARI and was referred to a hospital for intensive care until healed. While the cases in the community health centre that mostly affect toddlers are non-pneumonia ARI with symptoms of cold cough accompanied by symptoms of shortness of breath (fast) and a feeling of withdrawal of the lower chest wall during breathing [2].

Table 1. Case finding of ARI at Padang City Health Centre January - December 2017 period

No	Primary Health Care	< 1 years old	1-4 years old
1	Padang Pasir	568	1684
2	Air Tawar	306	787
3	Alai	343	1221
4	Lapai	411	1050
5	UlakKarang	280	973
6	Nanggalo	461	1245
7	LubukBuaya	646	2712
8	Air dingin	1253	4897
9	Belimbing	526	1880
10	Kuranji	283	800
11	Lb. Kilangan	259	487
12	Penggambiran	425	770
13	LbBegalung	445	1327
14	Rawang	565	936
15	Andalas	None	
16	Seberang Padang	553	1577
17	Pemancungan	236	1166
18	Pauh	570	1848
19	Bungus	203	682
20	Ambacang	327	943
	Total	8660	27985

From research conducted by Setiyorini [1], toddlers with mothers who have low knowledge of ARI will have a 0.3 times greater risk of developing Non-Pneumonia ARI than mothers who have high knowledge of ARI. According to Suryana [3], the low quality of health of family members including children is due to the low knowledge of a mother regarding health. The results of statistical tests on the study showed that maternal knowledge and attitudes affect the incidence of ARI in infants / toddlers.

Suryana [3] showed that out of 157 toddlers who died of ARI, there were 30 toddlers (30%) who were not taken to health care at the time of illness, this was due to a lack of knowledge, attitudes, level of education and low economic status in these communities that can cause death in toddlers. From the initial survey conducted at Padang Pasir Health Center, of the 7 mothers who were asked about their understanding of ARI, 5 mothers said they did not know about the causes and prevention of the ARI, while 2 people could only find the ARI shortness of breath. Knowledge of mothers who have children under five can affect the high incidence of ARI in Padang Padang Health Center.

2. Study of literature

ARI is an infectious disease in the upper and lower respiratory tract, which is caused by the entry of germs or microorganisms (bacteria and viruses) into the respiratory tract which can last up to 14 days.[4] The cause of ARI consists of 300 types of bacteria, viruses and rickettsia. Bacteria that causes ARI include genera Streptococcus, Staphylococcus, Pneumococcus, Haemophilus, Bordetella and Corynebacterium. Viruses that causes ARI include Mycovirus, Adenovirus, Coronavirus, Picornavirus, Mycoplasma, and ARI Classification divided by age groups, namely:

a. Age group 2 months - 5 years
   - Severe pneumonia, when accompanied by shortness of breath, the chest pulls down the lower chest wall when the child inhales.

- Pneumonia is not severe, if accompanied by rapid breathing, the limit of rapid breathing is ages 1- 5 years = 40 times per minute or more.
- Not pneumonia (common cold cough), if there is no pull of the chest wall and not fast breathing.

b. Group age <2 months
- Severe pneumonia, if accompanied by a sign of strong pull of the lower wall or rapid breathing. Fast breathing limits at this age are 60 times per minute or more.
- Not pneumonia (common cold cough), if not found a sign of strong pull in the lower chest wall or rapid breathing. [5]

Transmission of ARI disease can occur from sufferers of ARI, also called reservoirs of disease that are transmitted to other people through direct contact or through objects that have been contaminated with diseases, including air, airborne transmission means the mode of transmission that occurs without contact with patients or objects contaminated. [5,6] Prevention of ARI is very important to be done in order to reduce the incidence of ARI. These precautions include: [6]

a. Improvement and improvement of nutrition. Children who experience malnutrition will have a lower body resistance, especially children under 5 years old who are vulnerable to health and nutrition.
b. Environmental improvement and sanitation. Environmental factors play an important role in determining the process of interaction between causes and hosts in the process of occurrence of disease. An unhealthy living environment is at risk of an increased incidence of ARI. The most important thing in preventing ARI is:
   - Ventilation. This ventilation is very important to facilitate the circulation and replacement of fresh air from outside with air inside the house. We recommend ventilation openings in the direction of the wind.
   - Lighting and adequate lighting during the day. A healthy house should be facing the sunrise, so that in the morning the sun's rays go straight into the house. With the sun can kill germs that are in the house.
   - Personal health care

Personal health is very important to be maintained because individual health will also affect the health of others, especially in families. ARI prevention is carried out through efforts to improve health such as: [7]

a. Complete immunization in an effort to provide immunity to toddlers against ARI attacks.
b. Increasing the distribution of health service coverage and quality.
c. Avoid places where there are coughs and colds
d. Prevent children from dealing with ARI patients.

Child home care with ARI include: [8]

a. Feeding children to avoid weight loss which can lead to malnutrition.
b. Adding fluids to avoid dehydration.
c. If there is a fever, immediately apply a warm compress.
d. Encourage mothers to give children adequate rest.
e. Take the child to a health facility, if the ARI does not improve or other serious symptoms occur.

Knowledge or cognitive is a domain that is very important for the formation of one's actions. Someone who has good knowledge is able to think more critically in understanding everything. Because a knowledge can also be formed from experience, information obtained from non-formal education such as reading books, newspapers and magazines. Activities like this can make a person have high knowledge. High knowledge produces high actions, and vice versa. But it does not rule out the possibility of low knowledge resulting in high actions, and vice versa. [9,10]

3. Research Methods

This research is a quantitative research with descriptive research design. Descriptive research is research that aims to explain or describe a situation, event, object whether people, or anything related to variables that can be explained either by numbers or words. This study provides an overview of the level of knowledge of mothers of toddlers on the cleanliness of the home environment which greatly affects the incidence of ARI in infants. The type of data used in this study uses secondary data where secondary data is data obtained by researchers from existing data. Secondary data used in the study are journals or other existing or published research results about the level of maternal knowledge about the home environment on the incidence of ARI in infants. Data were collected using literature study techniques. Literature study is looking for theoretical references that are relevant to the case or problem found. Literature study was carried out by analysing as many as 5 (five) fruits from 2002-2005 related to ARI in infants and toddlers. The results of the analysis of the 5 (five) journals obtained 86 research samples of mothers of children under five. Based on the number of samples obtained, then an analysis was carried out so that the level of maternal knowledge of the incidence of high or low ARI was known. Data analysis used univariate test analysis. Univariate test analysis is an analysis carried out for one variable so that a description of the variable is obtained. The results of the univariate analysis in this study produced a table of frequency distribution regarding the level of knowledge of mothers of toddlers on the incidence of ARI so that it can be known how much percentage of mothers' level of knowledge of infants with high and low ARI incidence. mother's knowledge of children with ARI.

4. Results and Discussion

The results of this study indicate that out of 86 respondents found at 30.2% have low knowledge about prevention of ARI. This ignorance can be seen from the answers of respondents who were mostly unable to answer correctly, including: question No. 1 relates to the definition of ARI as much as 59%. Question No. 2 relates to the class of ARI as much as 35%. Question No. 6 related to the cause of ARI as much as 74%. Based on the distribution of respondents' answers, it was clearly seen that the knowledge of mothers of toddlers about understanding, class, causes, from ARI was still lacking. This shows the low intelligence of respondents especially regarding ARI. This may be due to a lack of information relating to ARI. Thus, providing counselling can improve respondents' understanding, especially about ARI. In this study it is seen that knowledge is the main form of intervention in changing health behaviour factors to reduce the occurrence of ARI. Limited level of knowledge, this is the possibility that causes ARI.

Table 2. Distribution of Frequency of Respondents Based on Knowledge of Toddlers Against ARI Prevention Efforts in Padang Pasir Health Center in 2018

Educational Level	f	(%)
High	60	69,8
Low	26	30,2
Total	86	100,0

Table 3. Distribution of Frequency of Respondents Based on Preventive Efforts of Mothers Against ARI at the Padang Health Centre Year 2018

Prevention Program	f	%
Good	53	61,6
Less	33	38,4
Total	86	100,0

The results of this study indicate that out of 86 respondents found that 38.4% of prevention efforts for ARI were lacking. This ignorance in prevention efforts can be seen from the answers of respondents who mostly cannot answer correctly, among others: question No. 10 related to prevention of ARI as much as 85%. The results of previous research conducted by Setyorini [1] on the relationship of predisposing factors of mothers who have children under five on ARI with the incidence of acute respiratory tract infections (ARI) in Air Tawar Health Center in 2008 found out of 31 mothers who had a toddler some mothers 61.3% negative attitude. The results of research conducted by M. Irfan [5] in Nanggalo Padang found as many as 52.8% of mothers under five under five behaved negatively towards the incidence of ARI.

Based on the results of the research and the above theory, it is clear that efforts to prevent ARI by mothers under five are not good for prevention of ARI, this shows the misperception of mothers in under five in interpreting healthy and sick lives according to their life experiences or values passed down by previous generations. hence prevention of ARI disease which is often reported to occur due to a lack of maternal attitudes which depend on the perception of the ARI. That is, if the ARI is perceived as a non-serious disease and does not threaten its life, the prevention of ARI is not too serious. Conversely, if the mother of a toddler perceives that ARI is a health problem that needs to be watched out, automatically the mother of a toddler will react seriously to this disease by developing preventive behaviours.

5. Conclusion

Considering that there are still many respondents who have low knowledge and efforts to prevent diseases that are less supportive of ARI, it is necessary to intensify activities that might improve knowledge and various prevention efforts, so that in the end it can reduce the incidence of ARI, namely in a way:

a. Conduct information about prevention of ARI through Posyandu cadres.
b. Mapping toddlers with ARI by Posyandu cadres with routine monitoring.
c. Increase cross-sector cooperation (PKK) because to reduce the incidence of ARI, it needs support from the community, especially mothers of children under five.

References

[1] Setyorini 2009 *Pengaruh Status Imunisasi DPT, BBLR, Paparan Asap Rokok dan Tingkat Pengetahuan Ibu Terhadap ISPA Non Pneumonia* (Surabaya: FKM UNAIR)
[2] Dinas Kesehatan Kota Padang 2010 *Profil Kesehatan Kota Padang* (Indonesia: Padang)
[3] Suryana A 2005 *Berbagai Masalah Kesehatan Anak dan Balita* (Jakarta: Khilma)
[4] Departemen Kesehatan RI 2002 *Pedoman Penatalaksanaan ISPA* (Jakarta)
[5] M Irfan 2005 *Hubungan Faktor Presisposisi dengan kejadian ISPA pada balita di Puskesmas Andalas Padang* (Padang: Universitas Andalas)
[6] Ike Silvia 2006 *Faktor-faktor yang berhubungan dengan kejadian ISPA pada balita di Puskesmas Sukamenati Pasaman Barat*
[7] Umar AF 2002 *Pemberantasan Penyakit Menular dan Penyehatan Lingkungan* (Jakarta)
[8] WHO 2003 *Penanganan ISPA Pada Anak di Rumah Sakit Negara Berkembang* (Jakarta: EGC)
[9] Notoatmodjo S 2003 *Ilmu Kesehatan Masyarakat* (Jakarta: Rineka Cipta)
[10] Notoatmodjo S 2003 *Pendidikan dan Perilaku Kesehatan* (Jakarta: Rineka Cipta)

The role of agriculture, forestry and fishery sector in the development of Malinau District (location quotient and shift share approach)

D Kartikawati*, Darsono and M T Sundari

Department of Agribusiness, Faculty of Agriculture, Universitas Sebelas Maret Surakarta, Indonesia

*dinakartikawati15@gmail.com

Abstract. This study aims to reveal the role, the role changes in the future, and the causative factors of these changes in the sector and each subsector of Agriculture, Forestry and Fishery in Malinau District. The basic method of this study was descriptive method. The analytical tool used were Location Quotient (LQ), Dynamic Location Quotient (DLQ) and Shift Share. This study was located in Malinau District, North Kalimantan Province. The types of data used were secondary data in the form of Gross Regional Domestic Product (GRDP) on the basis of constant prices in 2010 of Malinau and North Kalimantanin 2012-2016. The results shows that the Agriculture, Forestry and Fishery sector in Malinau District isa non-basis sector with an average LQ <1 of 0.76. The basis subsector in Malinau District is the Forestry and Logging subsector with an LQ> 1 value of 136,269.17. Based on a combined analysis of LQ and DLQ, the sectors of Agriculture, Forestry and Fishery have not undergone a role change where the sector remains a non-basis sector at present and in the future. While the subsectors of Agriculture, Animal Husbandry, Agricultural Services and Hunting and the Fisherybecome the basis subsector and the Forestry and Loggingsubsector becomes the non-basis subsector in the future.

1. Introduction

National economic development aims to create a fair and prosperous society. One of the economic sectors which become the main priority in national economic development is the Agriculture, Forestry and Fishery sector. The agricultural sector is the biggest support of Indonesia's economy, where almost all of Indonesia's economic activities are centered on the agricultural sector. This can be seen in the formation of GDP, foreign exchange earnings, employment, food supply, and supply of industrial raw materials [1]. Malinau District is one of the districts located in the youngest province in Indonesia, namely North Kalimantan. Based on the data from Central Bureau of Statistics (BPS) of Malinau District in 2016, the economic sector that provides the largest contribution is the Mining and Excavation sector with the main coal output, which accounts for around 52% of the GRDP [2]. This condition reflects that Malinau District in its regional economy relies heavily on the economic sector with non-renewable resources. Therefore, Malinau District must develop the economic sector with the use of long-term or sustainable resources, namely the Agriculture, Forestry and Fishery sector. It is important for Malinau District to know the role, the role changes, and the causative factors of the changes in the roles of the sectors and each subsector of Agriculture, Forestry and Fishery in the development of their regions. This is in line with one of the Malinau District's missions, namely increasing the role of the Agriculture, Forestry and Fishery sectors in the regional economy. Through this research Malinau District is expected to be able to take policies in the Agriculture, Forestry and Fishery sectors as well as an evaluation of the policies that have been carried out, so that in the future Malinau District has a leading sector with long-term use of resources, while preserving the environment.

2. Research Method

This study used secondary data in the form of time series data, namely the Gross Regional Domestic Product (GRDP) of Malinau District and North Kalimantan Province based on the 2010 constant price throughout the year of 2012 to 2016, and other sources relevant to this study. These data were obtained from the Central Bureau of Statistics (BPS) of the Malinau District and North Kalimantan Province. The location of the study was chosen deliberately / purposively namely Malinau District with the consideration that Malinau District had one mission to increase the role of the Agriculture, Forestry and Fisheriy sectors in the regional economy, in addition to this is because Malinau District is the border district of Indonesia - Malaysia. The analytical tool used in this study were:

2.1. Location Quotient (LQ)

The LQ analysis is a comparison of the magnitude of the role of the sector / industry in an area towards the magnitude of the role of the sector / industry nationally. The LQ analysis in this study was used to determine the position of the sectors and the respective subsectors of Agriculture, Forestry and Fishery in Malinau District [3]. The formula of LQ is:

$$LQ = \frac{Xr/RVr}{Xn/RVn} \tag{1}$$

Information:

Xr = GRDP of the Agriculture, Forestry and Fishery sector or the GRDP of the subsectors in Malinau District in a given year

RVr = Total GRDP or total GRDP of Agriculture, Forestry and Fishery sector in Malinau District

Xn = GRDP of the Agriculture, Forestry and Fishery sector or the GRDP of the subsectors in North Kalimantan in a given year

RVn = Total GRDP or total GRDP of Agriculture, Forestry and Fishery sector in North Kalimantan

From the Location Quotient (LQ) analysis, there are two criteria that will be generated:

$LQ > 1$, means that the sector/subsector is the basis sector, which means it is able to meet the needs of local communities and the surplus is exported outside the region.

$LQ<1$, means that the sector/subsector is the non-basis sector, which means it is unable to meet the needs of local communities and the deficits imported from outside the region.

2.2. Dynamic Location Quotient (DLQ)

Dynamic Location Quotient (DLQ) analysis is carried out by introducing the growth rate with the assumption that each sector value added and GRDP has an average growth rate per year on its own over a period of time (0) and year (t) [4]. The DLQ analysis in this study was used to find out whether the sectors and the respective subsectors of Agriculture, Forestry and Fishery in Malinau District experienced a change of role in the future. The formula for DLQ analysis is as follows) [4]:

$$DLQij = \left[\frac{(1+gij)/(1+gj)}{(1+Gi)/(1+G)}\right]^t \quad (2)$$

Information:

DLQij = Potential index of Agriculture, Forestry and Fishery sector / subsector

gij = Average growth rate of Agriculture, Forestry and Fishery subsector in Malinau District

Gj = Average growth rate of total GRDP/GDRP of Agriculture, Forestry and Fishery sector in Malinau District

Gi = Average growth rate of GDRP of Agriculture, Forestry and Fishery sector/subsector in Malinau District

G = Average growth rate of total GDRP/GDRP of Agriculture, Forestry and Fishery sector in North Kalimantan

t = the research period

From the Dynamic Location Quotient (DLQ) analysis, there are two criteria that will be generated:

DLQ > 1, means that the sector / subsector of Agriculture, Forestry and Fishery in Malinau District is estimated to be the basis sector / subsector in the future.

DLQ < 1, means that the sector / subsector of Agriculture, Forestry and Fishery in Malinau Districtis estimated to be non-basis sector / subsector in the future.

2.3. Combined Analysis of Location Quotient (LQ) and Dynamic Location Quotient (DLQ)

A combination of Location Quotient Analysis (LQ) and Dynamic Location Quotient (DLQ) analysis is used to determine how changes in roles occur in the sector and subsector of Agriculture, Forestry and Fishery in Malinau District) [5]. There are four criteria resulting from this combination of analysis, namely:

LQ > 1 and DLQ > 1, means that Agriculture, Forestry and Fishery sector/subsector remains basis both in the present and in the future.

LQ > 1 and DLQ < 1, means that the sector/subsector of Agriculture, Forestry and Fishery will has changed position from basis to non-basis in the future.

LQ < 1 and DLQ > 1, means that the sector/subsector of Agriculture, Forestry and Fishery will has changed position from non-basis to basis in the future.

LQ < 1 and DLQ <1,means that Agriculture, Forestry and Fishery sector/subsector remains non-basis both in the present and in the future.

2.4. Shift Share

Shift share analysis in this study is used to determine the factors that cause role changes that occur in the sector and each subsectorof Agriculture, Forestry and Fishery in Malinau District. The formula for Shift Share analysis is as follows) [6]:

$$TSS = SSS + LSS$$
$$SSS = \sum(gn - gin)Xino + \sum(Gi - G)Xino \quad (3)$$
$$LSS = \sum(gin - Gi)Xino$$

Information:

TSS = Total Shift Share

SSS = Structural Shift Share

LSS = Locational Shift Share

gn = Average growth rate of total GRDP / GRDP of Agriculture, Forestry and Fishery sector in Malinau District

gin = Average growth rate of GRDP of Agriculture, Forestry and Fishery sector / subsector in Malinau District

Gi = Average growth rate of GRDP of Agriculture, Forestry and Fishery sector / subsector in North Kalimantan Province
G = Average growth rate of total GRDP / GRDP of Agriculture, Forestry and Fishery sector in North Kalimantan Province

From this analysis, there are two criteria that will be generated:
SSS >LSS, means that the most determining factor on the occurrence of changes in Agriculture, Forestry and Fishery sector and subsector in Malinau Districts the factor of its economic structure.
SSS <LSS,means that the most determining factor for the change of sector and subsector of Agriculture, Forestry and Fishery in Malinau District is the location factor.

3. Result and Discussion

3.1. Location Quotient (LQ)

The results of the LQ analysis for the sector and respective subsectors of Agriculture, Forestry and Fishery in Malinau District can be seen in Table 1.

Table 1. LQ Value of Sector and Subsector of Agriculture, Animal Husbandry, Hunting and Agricultural Services Sector and Subsector in Malinau District 2012-2016

Sector/Subsector	Year					Average of LQ	Info.
	2012	2013	2014	2015	2016		
Agriculture, Forestry and Fishery Sector	0.91	0.81	0.73	0.69	0.69	0.76	Non Basis
1. Agriculture, Animal Husbandry, Hunting and Agricultural Services	0.83	0.87	0.90	0.93	0.93	0.89	Non Basis
2. Forestry and Logging	**2.14**	**2.20**	**2.31**	**2.42**	**2.40**	**2.29**	**Basis**
3. Fishery	0.06	0.06	0.07	0.07	0.07	0.06	Non Basis

Source: Secondary Data Analysis

Based on the Location Quotient (LQ) analysis in the sectors of Agriculture, Forestry and Fishery throughout 2012-2016, the LQ value tends to decrease with an average LQ <1 value of 0.76. This means that the sector of Agriculture, Forestry and Fishery is non-basis sector which means that this sector is unable to meet the needs of local communities so that these needs are obtained from outside the region. Several factors that cause the Agriculture, Forestry and Fishery sector to become a non-basis sector are due to geographical conditions that are different from other regions, infrastructure that is not evenly distributed [7] so that it has an impact on the high prices of agricultural inputs to farmers. The high price of inputs affects the output prices of agricultural products which tend to be high so that it is difficult to compete with agricultural products from other regions [8]. Based on the LQ analysis of each of the Agriculture, Forestry and Fishery subsectors, the Agriculture, Animal Husbandry, Hunting and Agricultural Services subsector has an LQ value that tends to increase throughout 2012-2016, but the resulting average value of LQ <1 is 0.89. At present, Malinau District is improving food security through rice field generating programs, by providing assistance to farmers in the form of seeds, and irrigation facilities and infrastructure in the form of ponds. However, this is considered less than optimal considering that there are still many obstacles faced by farmers, namely pests and diseases that attack crops, especially rice, and the high price of inputs to farmers [9]. The Forestry and Logging Subsector has an LQ value> 1, which is 2.29. This shows that the Forestry and Logging subsector is the basis subsector so that it is able to meet the needs of local communities and the surplus is exported to other regions. Malinau District has abundant potential in this subsector, of which 52% of forests in North Kalimantan are in Malinau District. In 2016 the total forest area in Malinau District reached 3.6

million Ha, of which a total area of 1.5 million Ha could be utilized by both the private sector and the government.

3.2. Dynamic Location Quotient(DLQ)
The results of DLQ analysis for the sector and subsector of Agriculture, Forestry and Fishery in Malinau District can be seen in Table 2.

Table 2. Value of DLQ of Sector and Subsector of Agriculture, Animal Husbandry, Hunting and Agricultural Services in Malinau District

Sector/Subsector	DLQ	Information
Agriculture, Forestry and Fishery Sector	0.00	Non Basis
1. Agriculture, Animal Husbandry, Hunting and Agricultural Services	**237.26**	**Basis**
2. Forestry and Logging	0.00	Non Basis
3. Fishery	**172,873.04**	**Basis**

Source: Secondary Data Analysis

Based on the results of the Dynamic Location Quotient (DLQ) analysis, the Agriculture, Forestry and Fishery sector is estimated to remain a non-basis sector in the future, with the value of DLQ <1 which is 0.00. Based on DLQ analysis in each subsector of Agriculture, Forestry and Fishery in Malinau District, the subsectors of Agriculture, Animal Husbandry, Hunting and Agricultural Services and Fishery are expected to be the basis subsector in the future with DLQ> 1 values of 237.26 and 172,873.04. The Forestry and Logging Subsector is expected to become a non-basis subsector in the future with the value of DLQ <1 which is 0.00.

3.3. Combined Analysis of Location Quotient (LQ) and Dynamic Location Quotient (DLQ)
The combination of LQ and DLQ analysis in the sectors and respective subsectors of Agriculture, Forestry and Fishery in Malinau District can be seen in Table 3.

Table 3. Combined Analysis of LQ and DLQ Sector and Subsector of Agriculture, Forestry and Fishery in Malinau District

Sector/Subsector	Average of LQ	DLQ	Information
Agriculture, Forestry and Fishery Sector	0.76	0.00	Remains Non-Basis
1. Agriculture, Animal Husbandry, Hunting and Agricultural Services	0.89	236.26	Non Basis→ Basis
2. Forestry and Logging	**2.29**	**0.00**	Basis → Non Basis
3. Fishery	0.06	172,873.04	Non Basis → Basis

Source: Secondary Data Analysis

Based on a combination of Location Quotient (LQ) and Dynamic Location Quotient (DLQ) analysis, the Agriculture, Forestry and Fishery sector is expected to remain a non-basis sector in the future.While the subsector of Agriculture, Animal Husbandry, Hunting and Agricultural Services and the Fishery subsector experienced a change of position, namely from the non-basis subsector to the basis subsector in the future. There is no change in position / role in the Agriculture, Forestry and Fishery sector in general where it remains a basis sector due to the uneven infrastructure that can be used by the community. Therefore, the local government of Malinau District still has to focus on building infrastructure, especially in the border area in the future. Infrastructure development is expected to facilitate the accessibility of the community, especially the border community, one of which is in obtaining affordable agricultural inputs and marketing agricultural products . The existence of this infrastructure is expected to foster economic activities of the people in this sector. In each subsector of Agriculture, Forestry and Fishery in Malinau District, there are two analysis criteria for combined LQ and DLQ generated, namely the subsectors that change roles from non-basis subsectors to basis and subsectors that experience a change in role from basis to non-basis. The Agriculture, Animal Husbandry, Hunting and Agricultural Services subsector and Fishery subsector experience a change from the role of the non-basis subsector at present to be the basis subsector in the future.This condition shows that the current Malinau District government effort in developing these subsectors can succeed in the future or can make these two subsectors become the regional superior subsector. One example in supporting the central government program is to create food security, the local government prepares 207 ha of paddy fields to be planted and the construction of irrigation facilities in the form of ponds. In addition, in changing the traditional pattern of society, which tends to store their crops in the granary rather than being sold, the local government through Perusda (Regional Company) buys rice grain owned by farmers and then milled, packed in such a way, and marketed by Perusda. This can not only improve the regional economy, but also improve the economy of the farming community. In the Fishery subsector, the local government policy in supporting this subsector is the establishment of the Minapolitan area in the North Malinau District, as well as the construction of facilities and infrastructure to the region. Forestry and Logging Subsectors experience a change in the role of the basis subsector, becoming a non-basis subsector in the future. Although the Forestry and Logging subsector has a significant contribution to the regional economy, the regional government cannot continue to depend on this subsector because in the future the forest area that can be utilized will certainly decrease along with the increase in development [10], in addition Malinau District as a conservation district based on the Malinau District Regulation No. 4 of 2007 which refers to Law No. 32 of 2004 concerning Regional Government [11], it is necessary to preserve existing forests.

3.4. Shift Share

The results of the Shift Share analysis with the Total Shift Share formula in the sector and each subsector of Agriculture, Forestry and Fishery in Malinau District can be seen in Table 4.

Table 4. Results of Shift Share Analysis on the Agriculture, Forestry and Fishery Subsector in Malinau District

Subsector	SSS	LSS	Information
1.Agriculture, Animal Husbandry, Hunting and Agricultural Services	**-2,212,736.51**	-3,107,467.59	Economic structure
2.Forestry and Logging	**-2,428,854.75**	-2,891,349.35	Economic structure
3.Fishery	-3,038,892.40	**-2,281,311.70**	Location

Source: Secondary Data Analysis

In the previous combination of LQ and DLQ analysis, it is known that the Agriculture, Forestry and Fisherysector does not experience a change in role in the future, therefore the Shift Share analysis in this study is only carried out in the agriculture, forestry and fisherysubsectors that have changed roles. Based on the Shift Share analysis, the subsector of Agriculture, Animal Husbandry, Hunting and Agricultural Services which has undergone a change in role from the non-basis subsector is currently the basis subsector in the future due to the economic structure of Malinau District. This is because there are regional government policies that encourage the growth of primary activities in addition to Mining and Excavation. The Forestry and Logging Subsector also experienced a change in the role of the basis subsector at present to a non-basis subsector in the future due to the economic structure of Malinau District. Factors of economic structure in this subsector are influenced by the government's policy of preserving forests, considering that Malinau District is a conservation district. Fishery subsectors experience a change in the role of the non-basis subsector at present being the basis subsector in the future due to location factors. The high public demand for freshwater fish encourages the Malinau District government to develop freshwater aquaculture in the Minapolitan area located in North Malinau Subdistrict, this is because the area is a tidal river area so that the water availability is guaranteed [12]. In addition, there are also several assistances from the regional government, for example the creation of ponds, the help of fish seeds, and the construction of facilities and infrastructure in the Minapolitan area to facilitate the accessibility of farmers in order to encourage the growth of this subsector.

4. Conclusions and Recommendations

4.1. Conclusions

Based on the results and discussion above, the conclusions that can be drawn in this study are:

1. The Agriculture and Forestry Sector in Malinau District is not a basis sector during 2012 - 2016.
2. The basis subsector in Malinau District throughout 2012 - 2016 is the subsector of Forestry and Logging.
3. The Agriculture, Forestry and Fishery Sector in Malinau District remains a non-basis sector at present and in the future.
4. The subsector that is predicted to be the basis subsector in Malinau District in the future is the subsector of Agriculture, Animal Husbandry, Hunting and Agricultural Services, and the Fishery Subsector.
5. The Agriculture, Animal Husbandry, Hunting and Agricultural Services subsector and the Forestry and Logging subsector will experience a role change in the future due to factors of economic structure while Fishery subsector will change due to location factors.

4.2. Recommendations

Suggestions that can be given based on this research is highly needed for Malinau District government to side in developing an economic sector that utilizes renewable resources, namely the Agriculture, Forestry and Fishery sectors, given the largest contribution to GRDP is the Mining and Excavation sector. This is because Malinau District has enormous potential resources in the Agriculture, Forestry and Fishery sectors which have not been utilized optimally and at the same time can preserve the environment because Malinau District is a conservation district.

Reference

[1] Zulkifli S 2017 *Pembangunan Pertanian Dalam Pusaran Kearifan Lokal* (Makassar: CV SAH Media)
[2] BPS 2017 *PDRB Kabupaten Malinau 2012 - 2016 (GRDP of Malinau Regency 2012 – 2016)* (Malinau: Central Bureau of Statistics (BPS))
[3] Bendavid-Val Avrom 1991 *Regional and Local Economic Analysis For Practitioners Fourth Ed.* (Westport, Connecticut: Praeger)
[4] Kurniawan B 2016 *El-JIZYA, Jurnal Ekonomi Islam (Islamic Economics Journal)* **4**(1) 1-26
[5] Widodo T 2006 *Perencanaan Pembangunan: Aplikasi Komputer* UPP STIM YKPN Yogyakarta
[6] Wibowo H, Darsono and Nurjayanti E D 2016 *MEDIAGRO* **12**(1) 17-26
[7] Darma D C, Haryadi and Umiyati E 2015 *Jurnal Perspektif Pembiayaan dan Pembangunan Daerah* **2**(4) 173-180
[8] Setiawan E, Hartoyo S and Sinaga B M 2016 *Jurnal Agro Ekonomi* **34**(2) 81-104
[9] Rizal M and Tarmisol 2015 *Prosiding Seminar Nasional Masyarakat Biodiversitas Indonesia* **1**(6) 1502-1507
[10] Balkis S, Imang N and Mursidah 2018 *Internasional Journal of Humanities and Social Science* **8**(7) 116-128
[11] Malinau District Regulation Number 4 of 2007: *Concerning Malinau District as a Conservation District*
[12] Djumlani Y 2015 *Jurnal Administrative Reform* **13**(1) 222-233

IOP Conf. Series: Earth and Environmental Science **314** (2019) 012078 doi:10.1088/1755-1315/314/1/012078

The relationship between organization's structure, leader behavior and personality with citizenship behavior on managing environment

Rina Lusianil*, Made Putrawan and Rukaesih Achmad

State University of Jakarta, Indonesia

*rina_lusiani@yahoo.com

Abstract. This research is aim at determining relationship between organization's structure, leader behavior and personality with citizenship behavior on managing the environment at Cilegon, Banten Industrial Park. The method used was survey with a correlational study by involving 150 sample. There were four instruments which measured citizenship behavior on managing the environment (rel. 0.94), organization's structure (rel. 0.93), leader behavior (rel. 0.93), and personality (rel. 0.97).Data has been analyzed by two-way ANOVA. Research results revealed that there is positive and significant correlation among those variables, even though it has been controlled by second-order correlation, it is still significant among them. Therefore, it could be concluded that if citizenship behavior on managing the environment would be improved, then organization's structure, leader behavior and personality should be taken into consideration

1. Introduction

Behavior in managing the environment is based on the response of environmental problems to the management such as how management actions in the process of using natural resources efficiently and effectively. The company itself is often faced with many priorities to do, on one hand for business competition and on the other hand for attention to the natural environment. Behavior in managing the environment is needed in order to improve employee behavior regarding environmental impacts that can contribute positively and negatively in achieving organizational or company goals.

The previous research was conducted by [1], over the past few decades, India has seen the rise of corporate social responsibility (CSR) among the corporate. In the middle of economic and social upheaval and from the immense pressure from environmentally conscious consumers, government and environment activists, Indian organizations are increasingly rectifying their CSR strategies in order to express their responsible nature. Different studies have been conducted by CSR and environmental sustainability whereby showing different kinds of relationship human behavior and CSR practices.

Furthermore, research by [2] is successful in implementingthe practice environment (ie, implementation of sustainability policies, pollution prevention measures, industrial ecological promotion, environment-training programs) execution of active environmental commitment from top managers. It is based on an institution or organization that cares about the environment around it is highly dependent on the extent to which leaders have commitment and leadership in positioning themselves as top managers in the implementation of management functions, so that it can improve the function of organizational structure.

IOP Conf. Series: Earth and Environmental Science **314** (2019) 012078 doi:10.1088/1755-1315/314/1/012078

According to [3], although environmental leadership is generally regarded as a prerequisite for environmental improvement in greening institutions, few studies focus on management systems and the ability to influence sustainability commitments. Thus, in improving employee's environmental performance, leaders need to be oriented toward sustainable environmental awareness. Integrative model of organizational behavior, according to [4], leaders and employees as individuals can be assessed from individual achievements (individual outcomes) of job performance or behavior that contributes good or bad and organizational commitment performance is strong or weak individual desires in the organization or institution.

There are several factors affecting individual achievement (individual outcomes) within an organization. Individual characteristic factors are personality and group mechanisms, one of which is leadership behaviors. Based on perceptions, an environmentally based company should be able to improve wise behavior in managing the environment and minimize negative impacts on the environment. Where the influence of positive leadership and personality is what can lead to a positive environmental performance.

Based on some of the International Journal mentioned above, the novelty in this research is on behavior in managing the environment which is related to organizational structure, leader behavior and personality. Researchers are interested in conducting research on the relationship between organizational structure, leader behavior and personality with citizenship behavior on managing the environment. Where the influence of positive leadership and personality is what can lead to positive environmental performance.

[5] suggests that, one's behavior can be maintained or maintaned for longperiods of time. [6] argue that organizing is grouping activities, assigning activities an providing the authority necessary to carry out the activities. In simple terms according to [7], behavior in managing the environment based on organizational citizenship behavior can be interpreted as an attitude or organizational behavior that is aligned and serves for the benefit of the organization.

Some experts defined organizational citizenship behavior defines OCB as individual behavior that is discretionary, not directly or explicitly recognized by the formal reward system, and that in the aggregate promotes the effective functioning of the organization. The personality foundation for these organization citizenship behaviors reflects the employee's predispositional traits to the cooperative, helpful, caring, and conscientious. The attitudinal foundation indicates that employee engage in OCBs in order to reciprocate the actions of their organizations.

Further explaination, [7] stated that the behavior of organizational personnel is in some form, but can be classified as, OCBs can take many forms, but the major ones could be summarized as: 1) altruism (e.g., helping out when a coworker is not feeling well), 2) conscientiousness (e.g., staying late to finish a project), 3) civic virtue (e.g., volunteering for a community program to represent the firm), 4) sportsmanship (e.g., sharing failure of a team project taht would have been success full by following the member's advice), 5) courtesy (e.g., being understanding and empathetic even when provoked.

This suggestion implies that for every member of the organization must build an attitude of self-interest, build a high awareness, social, honest about everything, and courtesy. Some of these attitudes woke up would certainly raise the life of the organization. From some of the description and analysis of the concept above, the synthesis of wise behavior in managing the environment is an action done by someone who contributes in managing the environment for organizational goals through dimension helping, sportsmanship, loyalty, organizational compliance, individual initiative, civic virtue, and self-development.

According to [4], organizational structure is formally dictates how jobs and tasks are divided and coordinated between individuals and groups within the company. One way of getting a feel for an organization's structure is by looking at an organizational chart. An organizational chart is a drawing that represents every job in the organization and the formal reporting relationships between those jobs. It helps organizational members and outsiders understand and comprehend how work is structures within the company.

While [8] define the organization's structure as the formal pattern of activities and interrelationships among the various subunits of the organization. Viewing an organization chart on a piece of paper or framed on a wall, one sees only a configuration of positions, job duties, and lines of authority among the parts of an organization. Organizational structures can be more complex than. Based on the above description, the synthesis of the organization's structure is the activity of employees in the division of tasks between individuals and groups within the company include the dimensions of work specialization, control range, command line, and formalities, delegates (centralization and decentralization) to preserve the environment in achieving green industry.

According to [9], leadership is in terms of an influence process that occurs naturally within a social system and is diffused among the members. Leadership is the process of influencing a person or group of people in a group or an organization. According to [4], leadership is defined as use power and influence to direct the activities of followers toward goal achievement. Leadership is the use of power and influence directing subordinate activities in achieving goals.

According to [10], leadership is the ability to influence a group of people in achieving a vision or set of goals. In the Trait theories of leadership, one's leadership is seen in one's quality and character. Whereas in behavioral theory of leadership, subordinates see the leader from the aspect of initiating structure is how far the leader explains the plan and the role of the subordinate and the attention (consideration), that is how far the leader builds friendship relationship with the staff to trust each other (mutual trust), attention to the ideas of subordinates (respect for employee's). From some research, 66% of staff is motivated by the attention of the leadership.Thus, the synthesis leader behavior is the action of a leader who is assessed by employees in managing the environment to achieve corporate goals that have a positive impact through the dimensions of consideration and initiating structure.

According to [5] in [11], personality is the complex organization of cognitions, affects and behaviors that gives direction and pattern (coherence) to the person life Personality consists of both structures and processes and reflects both nature (genes) and nurture experience. Personality is a complex cognitive organization that influences behavior to give direction and pattern of one's life, personality consists of structure and process, both natural (gen) and experience. Thus, personality is a complex knowledge that can affect behavior on a particular purpose.

[12] in The Big Five personality model explains that there are five factors underlying human personality that is conscientiousness, agreeableness, neuroticism, openness, and extraversion. From some of explanations above, the synthesis of personality is one's traits in responding, reacting, and interacting relatively stable with others and its environment includes conscientiousness, agreeableness, neuroticism, openness, extraversion to achieve organizational goals. Based on the description, the researcher is interested to conduct research on the relationship between organization's structure, leader behavior and personality with behavior in managing environment, correlational study on employees 'in Krakatau Steel Company Cilegon, Banten year 2016.

2. Method

The purpose of the research is to find a relationship between organization's structure, leader behavior, and personality together with employees' citizenship behavior on managing the environment. A correlational study was used by selecting randomly 150 employees' as sample. There were four measurable instruments that are employees' citizenship behavior on managing the environment (rel. 0.94), organization's structure (rel. 0.93), leader behavior (rel. 0.93), and personality (rel. 0.97). Data werecalculated by descriptive and inferential statistics. Then, it was analyzed by simple and multiple regression, followed by finding out its second-order correlation and verified by t-test.

3. Results and Discussion

3.1. Result

The relationship between organization's structurewith citizenship behavior on managing the environment could be seen in the regression model with regression coefficients b = 0.44 and a = 77.96.

Table 1. Anova table for regression model of $\hat{Y} = 77.96 + 0.44X1$

Source of Variances	Degrees of Freedom (DF)	Sum of Square (SS)	Means Square (MS)	F_{cal}	F_{table}	
Total	149				0.05	0.01
Coefficient (a)	1					
Regression (b/a)	1	7353.043	7353.043	35.01 **	3.91	6.81
Residual	148	31091.791	210.080			
Deviation from Linearity	55	12799.482	232.718	1.18^{ns}	1.45	1.69
Error	93	18292.308	196.691			

**ρ : 0.01 (very significant)^{ns} : nonsignificant

Based on table above, it can be seen that, the resulting regression equation was significant with a regression model $\hat{Y} = 77.96 + 0.44X_1$. Therefore, there was a positive relationship between citizenship behavior on managing the environment with organization's structure which was very significant. Next, a large correlation coefficient between organization's structure with citizenship behavior on managing the environment was found, as shown in the following table 2.

Table 2. Partial correlation (RY 1.23)

Sample (n)	Correlation Coefficient (zero)	Second order correlation	t_{cal}	t_{table}	
				0.05	*0.01*
150	0.437	0.246	5.91^{**}	1.97	2.61

**ρ <0.01

There was highly significant correlation between the organization's structure with citizenship behavior on managing the environment. The variation of citizenship behavior on managing the environment is determined by 6.1% of the organization's structure variation through the regression model $\hat{Y} = 77.96 + 0.44X_1$.

The relationship between leader behavior with citizenship behavior on managing the environment could be seen in the regression equation which generates the direction of regression coefficients b = 0.62 and constants = 58.68 to determine the degree of significance (level of significance). The F was subsequently tested, as shown in the following table 3.

Table 3. Anova Table of regression model $\hat{Y} = 58.68 + 0.62X_2$

Source of Variances	Degrees of Freedom (DF)	Sum of Square (SS)	Means Square (MS)	F_{cal}	F_{table}	
Total	149				0.05	0.01
Coefficient (a)	1					
Regression (b/a)	1	9662.144	9662.144	49.68**	3.91	6.81
Residual	148	28782.689	194.478			
Deviation from Linearity	42	9168.540	218.299	1.18^{ns}	1.49	1.76
Error	106	19614.149	185.039			

**ρ <0.01 ^{ns} : nonsignificant

Based on table above, it can be seen that the resulting regression equation was significant with a regression model $\hat{Y} = 58.68 + 0.62X_2$. Therefore there was a positive relationship between citizenship behavior on managing the environment with leader behavior which was very significant. Next a large correlation coefficient between leader behavior with citizenship behavior on managing the environment was found, as shown in the following table 4.

Table 4. Partial correlation (RY2.13)

Sample (n)	Correlation Coefficient (zero)	Second order correlation	t_{cal}	t_{table}	
				0.05	*0.01*
150	0.501	0.373	7.05^{**}	1.97	2.61

** p<0.01

There was highly significant correlation between leader behavior with citizenship behavior on managing the environment. The variation of citizenship behavior on managing the environment is determined by 13.9% of the leader behavior variation through the regression model of $\hat{Y} = 58.68 + 0.62X_2$.

The shape of the relationship between personality with citizenship behavior on managing the environment could be seen in the regression equation which generates the direction of regression coefficients b = 0.29 and a = 107.11.

Table 5. Anova table for regressin model of $\hat{Y} = 107.11 + 0.29X_3$

Source of Variances	Degrees of Freedom (DF)	Sum of Square (SS)	Means Square (MS)	F_{cal}	F_{table}	
Total	149				0.05	0.01
Coefficient (a)	1					
Regression (b/a)	1	3670.516	3670.516	15.62**	3.91	6.81
Residual	148	34774.317	234.962			
Deviation from Linearity	49	14724.739	300.505	1.48^{ns}	1.49	1.82
Error	99	20049.578	202.521			

**ρ <0.01 ^{ns} : nonsignificant

According to the table above, it showed that, the regression equation was significant with a regression model $\hat{Y} = 107.109 + 0.292X_3$. Therefore, there was a positive relationship between citizenship behavior on managing the environment with personality which was very significant. Correlation coefficient between personality with citizenship behavior on managing the environment was found, as shown in the following table 6.

Table 6. Partial correlation (RY3.12)

Sample (n)	Correlation Coefficient (zero)	Second order correlation	t_{cal}	t_{table}	
				0.05	0.01
150	.309	0.184	3.95^{**}	1.97	2.61

** p< 0.01

There is highly significant correlation between personality with citizenship behavior on managing the environment. The variation of citizenship behavior on managing the environment is determined by 3.4% of the personality variation through the regression model $\hat{Y} = 107.11 + 0.29X_3$.

The shape of the relationship between organization's structure, leader behavior, personality with citizenship behavior on managing the environment could be seen in the regression equation which produced a directions regression coefficient = $b_1 = 0.24$; $b_2 = 0.45$; $b_3 = 0.15$; and constant = 35.78. To determine the degree of significance (level of significance), the F was tested, as shown in the following table 7.

Table 7. Anova table for multiple regression of $\hat{Y} = 35.78 + 0.24X1 + 0.45X2 + 0.15X3$

Source of Variances	Degrees of Freedom (DF)	Sum of Square (SS)	Means Square (MS)	F_{cal}	F_{table}	
Total Direduct	149				0.05	0.01
Regression	3	12886.763	4295.588	24.53**	2.67	3.92
Residual	146	25558.071	175.055			

**ρ <0.01

Based on table above, the resulting regression equation was significant with a regression model $\hat{Y} = 35.78 + 0.24X_1 + 0.45X_2 + 0.15X_3$. Therefore, there was a positive relationship between citizenship behavior on managing the environment with organization's structure, leader behavior, and personality which was very significant. A large correlation coefficient between organization's structure, leader behavior, and personality with citizenship behavior on managing the environment could be seen the following table 8.

Table 8. Multiple correlation (RY 123)

Sample (n)	Multiple Correlation Coefficient	Multiple Determination Coefficient.	F_{cal}	F_{table}	
				0.05	0.01
150	.579	.335	24.55**	2.67	3.92

** p<0.01

It indicated that around 33.5% of the citizenship behavior on managing the environment variation determined all together by the variations of the organization's structure, leader behavior, and personality through the linier regression model, as follows: $\hat{Y} = 35.78 + 0.24X_1 + 0.45X_2 + 0.15X_3$. Based on table above, the value of the correlation coefficient between organization's structure, leader behavior, and personality with citizenship behavior on managing the environment of $ry_{123} = 0.579$. The higher the organization's structure, leader behavior, and personality, the higher was citizenship behavior on managing the environment.

3.2. Discussion

The first hypothesis proposed that organization's structure was positively related to citizenship behavior on managing the environment. The success of citizenship behavior on managing the environment a company was determined by its organization's structure of the company. The organization's structure in this study includes the dimensions of work specialization, span of control, chain of command, formalization, delegation (centralization and decentralization) to preserve the environment.

The basic theoretical reference used in explaining citizenship behavior on managing the environment such as organization's structure the theory Integrative Model of Organizational Behavior from [12]. In addition, in line with the results of research conducted by [1]that over the last few decades, India has socialized corporate social responsibility (CSR) in various companies. In the economic and social upheaval and from the tremendous pressure of environmentally conscious consumers, the Government, environmentalists, and Indian organizations are increasingly improving the CSR strategy by putting emphasis on corporate and employee responsibilities. The results show the differences in human behavior with CSR practices. Research suggests how organizations tend to be positive and wise to environmental management by meeting needs but without compromising the needs of future generations.

In addition, according to [13], facing the situation of product or service demand and competition to gain customer trust, then the pro-environment company must be able to solve the environmental problems, it will have a positive ability in affecting the performance employees, and ultimately for the environmental performance of the company. Based on results above, then it is obtained a positive relationship between organization's structure with citizenship behavior on managing the environment.

Another hypothesis test result depicted that the leader behavior was positive and highly significant with the citizenship behavior on managing the environment. The success of citizenship behavior on managing the environment of a company is determined by the leader behavior of a company. Leader behavior in this research includes the dimensions of helping, sportsmanship, loyalty, organizational compliance, individual initiative, civic virtue, and self-development. The basic theoretical reference used in describing citizenship behavior on managing the environmentis leader behavior using the umbrella theory of Responsible Behavioral Behavior toward the Environment that has been adapted by Blaikie & Ward (1993) from Model Hinnes et al (1986/1987), [14].

According to [9], leadership is the ability to influence a group of people in achieving a vision or set of goals. In the theory of leadership character (Trait theories of leadership), one's leadership is seen in one's quality and character. Whereas in behavior theories of leadership theory, subordinates see the leader from the aspect of initiating the plan (ie the extent to which the leader explains the plan and the role of the subordinate and attention (consideration). It is the extent to which leaders build friendships with staff to each other trust (mutual trust), attention to the ideas of subordinates (respect for employees). Based on other research, 66% of staff are motivated by the attention of the leaders.

According to [15], in history, some leadership researchers emphasize the importance of situations in the leadership process, in which situations are not based on the person's traits or skills, but rather a situational perspective in handling a problem in times of economic crisis, social unrest, problems such as natural disasters and so forth. One attempt to include situational factors in leadership theory is the multiple influence model of Hunt and Osborn. This model is differentiated that micro variable (task factor) with makro variabel (external environmental factors) in the situation in an organization or institution.

Research conducted by [2]is the successful implementation of the practice environment (ie, implementation of sustainability policies, pollution prevention measures, industrial ecological promotion, environmental training programs) top manager. It is based on an institution or organization that cares about the environment around that highly dependent on the extent to which leaders have commitment and leadership in positioning themselves as top managers in the implementation of management functions. Thus, it can improve the function of organizational structure.

Based on explanation above, an institution or organization that cares about the environment around it is very dependent on the extent to which leaders. Leaders are committed and leadership in positioning themselves as top managers within a company supported by behavior in managing the environment. Thus, there is a very significant positive relationship between leader behavior and citizenship behavior on managing the environment.

The rest of other hypothesis verification revealed that personality was positive and highly significant with citizenship behavior on managing the environment. The success of citizenship behavior on managing the environment of a company was determined by employees' personality. Big-five personality model consist of conscientiousness, agreeableness, neuroticism, openness and extraversion. [16] several studies have revealed that positive personality in employees indicated good job performance as well, because the positive environmental impacts in the company facilitate the involvement of the personality work and the resulting resources.

It is important for employees to demonstrate proactive behavior and optimize their own work environment. Furthermore, research conducted by [17] showed that the five factors of human personality in adults, showing a strong relationship between one another. These findings reinforce previous research data on a five-factor model in personality research. Based on the above explanation there is a positive and highly significant correlation between personality and citizenship behavior on managing the environment. Comparing to research conducted by [13], it supported this finding which people pro-ecological behavior could be best predicted by their personality, especially environmental personality.The results of the fourth hypothesis test show that the organization's structure, leader behavior, and personality are together positively and highly significant with citizenship behavior on managing the environment. The success of citizenship behavior on managing the environment is determined by the organization's structure, leader behavior, and the personality of its employees.

Research conducted by [18] is the impact of integration of intervention implementation of the intention more effective to promote individual behavior. In this case, the implementation of a strong and weak intention also determines one's behavior. [15]stated that some studies have revealed positive personality in employees showing good job performance anyway. The positive corporate environmental impacts facilitate job engagement from personality outcomes. Therefore the management can affect the work of employees and resources generated. It is important for employees to demonstrate proactive behavior and optimize the work environment of employees.

Research conducted by [19] is about the development of environmental leadership that can holistically face the existing environmental problems. Environmental leaders promote the sustainability of environmental issues to protect the natural environment in decision-making and action processes. Thus, it can be concluded that in this study the transformational-style leadership variables affect the environmental performance of the organization. Based explanation above, there is a positive and very significant relationship between the organization's structure, leader behavior, and personality together with the citizenship behavior on managing the environment.

The pure contribution of each variable is known by controlling other independent variables. The results of the analysis can show the rank of the relationship as shown in the following table 9.

Table 9. Second-order correlation among variables

Partial Correlation	Second-Order Correlation Coefficient	Strength in contribution
$r_{y1.23}$	0.246	Second
$r_{y2.13}$	0.373	First
$r_{y3.12}$	0.184	Third

Based on table above, the first relationship strength rank is behavior leader of 0.373; organization's structure equal to 0.246; and the third personality of 0.184. These findings would be expected to add a treasure that enriches scientific knowledge, especially about citizenship behavior on managing the environment.

4. Conclusions

Based on these findings it could be concluded that when employees citizenship behavior on managing the environment would be positively improved, then factors such as organization's structure, leader behavior and personality could not be neglected. Therefore, these findings could be beneficial in contributing the theoretical basic if employees' citizenship behavior for managing the environment would be measured, by considering the role of leader behavior, organization's structureand especially personality.

References

[1] Mukherjee R dan Indranath G 2014 Going Green: The New Strategy for Environmental Sustainability and CSR in India *The International Journal of Business & Management* **2** (2): 109-113 ISSN 2321-8916
[2] Boiral O, Charles B and Olen G 2014 Environmental Leadership and Counsciousness Development: A Case Study Among Canadian SMEs *Journal Bus Ethics* **123**: 363-383
[3] Robertson J L and Julian B 2013 Greening Organization Through LeadersInfluence on Employees Pro-environmental Behaviors *Journal of Organizational Behavior* http://doi.org/10.1002/job.1820
[4] Colquitt, Jason A., Jeffery A L dan Michael J W 2011 *Organizational Behavior: Improving Performance and Commitment in the Workplace* (New York: McGraw Hill)
[5] Pervin L A and John P O 1997 *Personality Theory and Research* (USA: John Willey and Sons, Inc)
[6] Byars L, Llloyd dan Rue L W 2006 *Human Resource Management 8 edition* MCGraw-Hill, Irwin)
[7] Luthans F 2008 *Organizational Behavior Eleventh Edition* (Singapore: McGraw-Hill International Edition)
[8] Ivancevich J M 2011 *OrganizationalBehavior& Management* (New York: McGraw-Hill Education)
[9] Yukl G 2010 *Leadership in Organizations* (New Jersey: Pearson)
[10] Robbins S P and Timothy A J 2013 *Organizational Behavior* (USA: Pearson)
[11] Rhodewalt F (Ed) 2008 *Personality and Social Bahavior* (New York: Taylor & Francis Group)
[12] Colquitt J A, Jeffery A L dan Michael J W 2009 *Organizational Behavior Improving Performance in The Work Place* (New York: McGraw-Hill Inc)
[13] Chen Y, Guiyao T Jiafe Jin, Ji lin and Pascal Paile 2015 Linking Market Orientation and Environmental Performance: The Influence of Environmental Strategy, Employee's Environmental Involvement, Environmental Product Quality *Journal of Business Ethics*
[14] Putrawan I M 2017 Predicting Responsible Environmental Behevaior (REB) based on Personality, NEP and Naturalistic Intelligence *Advanced Science Letters* **23** (9) 5586-5593
[15] Newstrom J W 2007 *Organizational Behavior* (New York: McGraw-Hill)
[16] Bakker A B, Maria T and Daantje D 2012 Proactive Personality and Job Performance: The Role of Job Crafting and Work Engagement *Journal of Human Relations* https://doi.org/10.1177%2F0018726712453471
[17] McCrae R R. dan Paul T C, 1987 Jr Validation of the Five-Factor Model of Personality A cross Instrumens and Observers. National Institute on Aging and National Institudes of Health *Journal of Personality and Social Psychology* **52**(1): 81-90
[18] Ransom-Flint, Terry 2006 *ImplementationIntentions, Personality, and ExerciseBehavior* (The Ohio State University)
[19] Mino T and Keisuke H 2013 *Environmental Leadership Capacity Building in Higher Education Experience and Lessons from Asian Program for Incubation of Environmental Leaders* (Springer: Tokyo)

Environmentally friendly photo road modeling in Tasikmalaya city to improve urban transportation framework

Indra Mahdi[1,*], Nina Herlina[1], Agi Nurhidayah[2], Anto Purwanto[3] and Andi Nur Rachman[4]

[1]Department of Civil Engenering, University Siliwangi, Indonesia
[2]Department of Informatics, University Siliwangi, Indonesia
[3]Department Public Health Faculty, University Siliwangi, Indonesia
[4]Department of Informatics University Siliwangi, Indonesia

*indramahdi@unsil.ac.id

Abstract. The development of a region can be seen from population growth and a high average growth rate, which has implications for socio-economic activities, infrastructure and other physical buildings that are increasingly complex. To accommodate all these activities, space is needed as a place of activity. The road infrastructure sector is one of the arteries in the region's economic growth, so the accuracy of its provision through the amount of investment is very important. This will not cause a bottleneck in the flow of goods / services and people at regional, national and even international levels which causes lower economic and social costs. Even so, in Tasikmalaya City, the road problems at this time have shown serious problems, especially on the main roads, traffic jams often occur, not only due to the increasing number of vehicles, but due to the non-functioning of the sidewalk as a place to walk because some sidewalks are used to sell street vendors, while walking is a tool for internal movement of the city, the only tool to meet the needs of face-to-face interactions that exist in commercial and cultural activities in urban life. This study will examine whether the sidewalk functions in accordance with existing regulations, and whether the sidewalk design has fulfilled the requirements of safety, fun, comfort, and has an appeal. This study uses a qualitative descriptive research method. Data obtained through direct observation in the field. Observations were made on the location of the main road that many pedestrians traversed such as the Jl. Martadinata, Jl. Doctor Sukardjo, and Jl. H. Mustopa arrived at Padayungan intersection. The analytical method uses descriptive qualitative with the planning model. Stages of analysis are drainage channel analysis, analysis of the physical condition of the sidewalk, analysis of the quality of safety and comfort of the sidewalk as a pedestrian path. The results of the research are in the form of directional and environmentally friendly sidewalk designs in urban areas, which can be used to complement the shortcomings in variables and to correct variables that have not been fulfilled according to the perception and behavior of pedestrian users in the City of Tasikmalaya.

1. Introduction

The development of a region can be seen from population growth and a high average growth rate, which has implications for socio-economic activities, infrastructure and other physical buildings that are increasingly complex. To accommodate all these activities, space is needed as a place of activity. Especially with the dynamic nature of regional development that requires the provision of wider space to accommodate it. The road infrastructure sector is one of the arteries in the region's economic growth, so the accuracy of its provision through the amount of investment is very important. With regard to economic development, road investment has a broad influence both for road users and for the region as a whole. For this reason, proper policies are needed in the implementation of roads so that they can support regional development and economic growth. One of the successes of city development is the availability of good transportation facilities and infrastructure in the area (Minister of Public Works 03/2014). In addition to playing a role in supporting the smooth socio-economic activities, it will also support physical development in the area concerned. Based on Law Number 38 of 2004 concerning Roads, that Regional Governments have an obligation to carry out road operations. Strategic issues faced in the implementation of roads, especially national roads and / or urban roads include the inadequacy of the primary road network system and / or collectors in serving continuous traffic flow and or urban traffic flow [1].

This has caused a bottleneck in the flow of goods / services and people at the regional, national and even international levels, which causes higher economic and social costs. Even so, in Tasikmalaya City, the road problems at this time have shown serious problems, especially on the main roads, traffic jams often occur, not only due to the increasing number of vehicles, but due to the non-functioning of the sidewalk as a place to walk because some sidewalks are used to sell street vendors, while walking is a tool for internal movement of the city, the only tool to meet the needs of face-to-face interactions that exist in commercial and cultural activities in urban life. Walking is a means of connecting between other modes of transportation. Judging from the speed, the walking mode has many advantages, namely low speed so it is advantageous because it can observe the surrounding environment [2]. Pedestrians are an important form of transportation in urban areas. Pedestrians are quite essential activities of the transport system and must get the right place. Traveling by public transport always begins and ends on foot, if pedestrian facilities are not provided properly, the community will be less interested in using public transportation [3], so that the existence of several sidewalks on the main road in Tasikmalaya City need to get attention, meaning whether the existence of the sidewalk has met the standards of its width, or still needs to be improved. Aside from being a pedestrian morning walkway, sidewalks can also be used as places / locations to put up billboards to promote goods / services products, so that they can be a source of Regional Original Income from advertisement taxes or advertising fees. According to Unterman, the sidewalk must fulfill several conditions, namely safety, pleasant conditions, comfort and attractiveness.

In the downtown area, a good and appropriate pedestrian pathway is needed so that people are more interested in choosing to walk compared to using a private vehicle. Perception and behavior of pedestrian path users is needed to be able to find out the needs of the community so that the pedestrian arrangement is in line with expectations and can attract the public to use pedestrian paths without neglecting the applicable pedestrian arrangement regulations. Therefore, research on the arrangement of lanes in the downtown area uses 4 variables. This is used to determine a design that is good and appropriate according to the perception and behavior of pedestrian path users. Directives and designs are used to complement the shortcomings in variables and to correct variables that have not been fulfilled according to the perception and behavior of pedestrian users at the research location.

2. Methodology

This study uses a qualitative descriptive research method. Data obtained through direct observation in the field. Observations were made on the location of the main road that many pedestrians traversed such as the Jl. Martadinata, Jl. Doctor Sukardjo, and Jl. H. Mustopa arrived at Padayungan intersection.

The analytical method uses descriptive qualitative with the planning model. Stages of analysis are drainage channel analysis, analysis of the physical condition of the sidewalk, analysis of the quality of safety and comfort of the sidewalk as a pedestrian path.

Broadly speaking, the implementation of the sidewalk masterplan work in Tasikmalaya City is carried out in 5 (five) stages of work. The five stages of the implementation of interrelated work are:

- The first stage is more oriented to the preparation and preparation of the work program where after this stage a preliminary report has been made.
- The second stage is the stage of data collection and data compilation and initial identification of problems
- The third stage is the identification stage of the basic concept of site selection planning the development of sidewalks in the City of Tasikmalaya After this stage a product will be produced in the form of a report between.
- The fourth stage is the location selection stage where there will be an analysis based on the complete criteria for the selection of the location of the sidewalk development planning in the City of Tasikmalaya.
- The fifth stage is the last stage of the work of preparing this master plan is to provide conclusions on all the work that has been done and provide recommendations so that traditional market operations can run well. After this stage a product will be produced in the form of a final report

3. Contruction

3.1. Construction of sidewalks to prioritize pedestrian interests in order to provide comfort and safety for user.

- The use of sidewalks for other purposes that interfere with the sidewalk function needs to be disciplined by the application of strict sanctions
- If the sidewalks built on the sewers must be guaranteed the smooth flow of water in the drains so as not to cause flooding
- Construction of sidewalks to be adjusted to modern standards so that it is comfortable for users and road-use vehicles.
- Construction of sidewalks should not be cut off, especially to the edge of a sewer or river. If the disconnection must be given a warning sign.
- Construction of sidewalks to synchronize with the construction of sewers, roads, electricity networks, telephones, drinking water and reforestation trees.
- The height of the sidewalk height from the road surface is not too low but also not too high because it will reduce the road capacity. The height of the recommended road pavement is 150 mm.
- Aian Slackness on road access to allow disabled people who use wheelchairs to be able to use the sidewalk easily and easily.
- Trajectories that can be passed by blind disabled people.
- Width in accordance with the number of pedestrians who use sidewalks

3.2. Drainage Arrangement and Road Utility in the Research Area

3.2.1. Drainage. Drinase structuring concepts and utilities on roads and the environment will be integrated with drainage channels and utilities. Based on SNI 03-1733-2004 Procedures for Urban Housing Planning in Urban Areas, road sections consist of the following

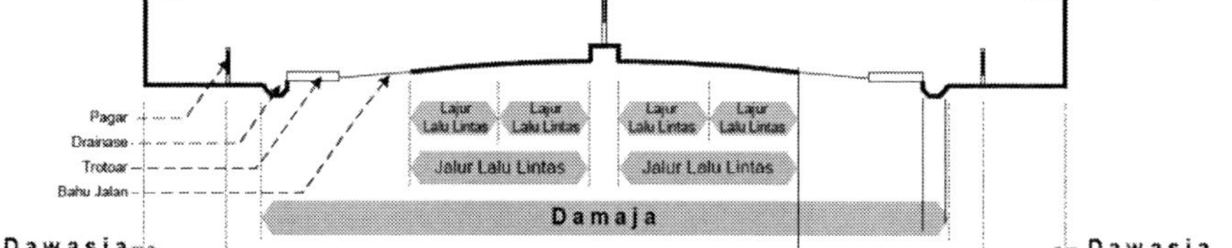

Figure 1. Road Parts

IOP Conf. Series: Earth and Environmental Science **314** (2019) 012079 doi:10.1088/1755-1315/314/1/012079

Model of urban drainage channels using materials from precast holes with holes with the concept of Green Water, where drainage using this model is expected rainwater can seep into the soil.

The ZRO (Zero Run-Off) concept for drainage is the use of large capacity rainwater harvesing, supporting environmental-based drainage, and can reduce stagnant water in the settlements. The ZRO concept is the integration between the intensity of rain, storage, absorption, benefits or consumption, and the flow (ITRMA) of the remaining water runoff out of the area up to 0% (zero run off).

Figure 2. Illustration of Closed Semi Drainage Design

3.2.2. Utility. Urban utilities in the study area are based on observations in the field, not yet integrated such as PLN Cable Channels, Telephone Cables, PDAM Fiber and Pipe Fiber Cables, all of which are planted but are in their respective lanes.

For urban utility planning, there needs to be integration so as to facilitate monitoring of care and beauty of the koat, then special routes must be made which together with drainage

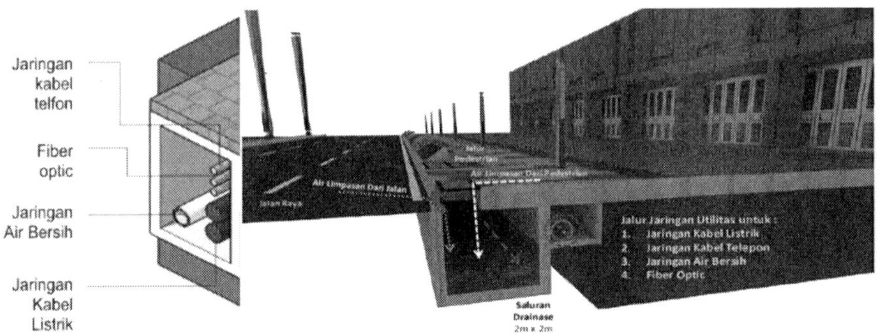

Figure 3. Illustration of Utility System Details

3.3. Sidewalk Arrangement and Completeness

3.3.1. Sidewalk Material. The sidewalks built as far as possible are placed on the inside of open drainage channels or above closed drainage channels by using the use of materials that are environmentally friendly, so that rainwater can seep into the soil, use materials with local equipment that is in place and easily found, so it does not need to bring material from outside the area except manufacturing materials that cannot be made in that location such as the examples and motives below

Figure 4. Sidewak motif Illustration

The use of materials for sidewalks is various but strong and environmentally friendly materials such as Pavingblock containing Titanium Dioxide (TiO2) are selected. Construction of sidewalks with pavingblock is done with 2 materials, namely pavingblock material with TiO2 content and without TiO2 content. The color and surface of paving blocks with and without TiO2 content is no different.

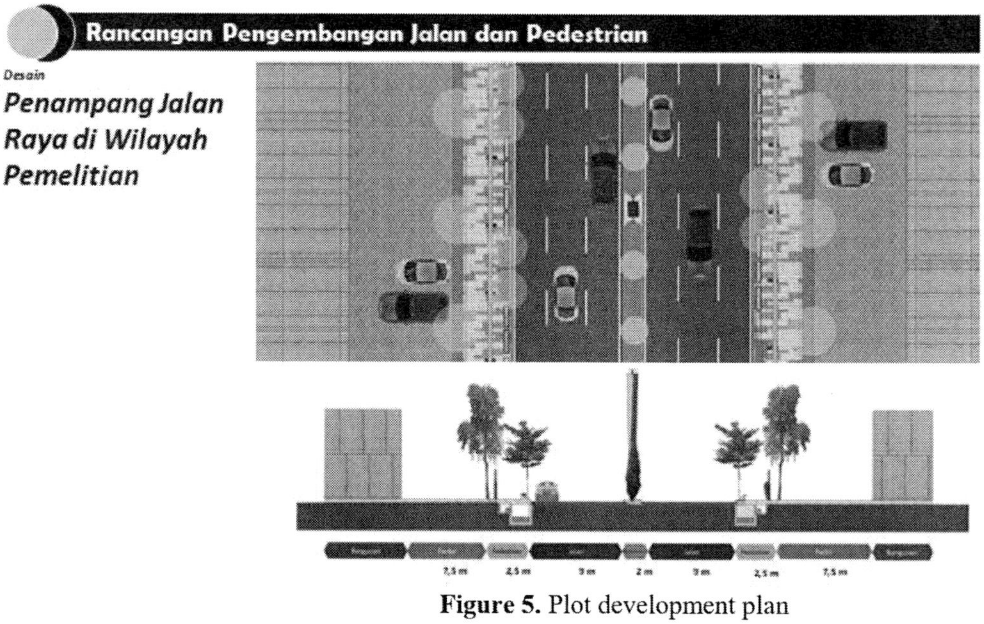

Figure 5. Plot development plan

3.3.2. Sidewalk Motives and Design. In designing the sidewalk motif in the research area, it is necessary to consider and bring up a model with local wisdom, because Tasikamlaya City is a handicraft industry especially for webbing, so for planning sidewalk motifs and designs it should utilize attractive woven motifs and special materials for diffable.

Figure 6. Sidewalk Motif Design [1]

3.3.3. Sidewalk lighting motifs and designs.

- Installation of lighting lamps Public roads aim to accommodate safe movements for pedestrians and vehicles. Where movement, road users can be helped by their orientation by providing the right hierarchy of lighting effects. The lighting hierarchy can be seen from the difference in distance, height and color of the light used. Lighting must also be functionally suitable and on a scale that is suitable for both pedestrians and vehicle lanes. For pedestrian lighting, lamps with relatively low altitudes can be used to provide a human scale and illuminate the lower canopy of roadside trees. The lighting properties for the pedestrian walk should not be uniform throughout the road, otherwise the vehicle path must be uniform overall. The average street lighting has a height of 6-15.2 meters, while for pedestrian paths, the vertical lighting distribution must reach 2 meters so that the vision towards other pedestrians remains clear. Street lighting facilities should meet the following requirements:
- a. Placed on the left side of the traffic lane according to the direction of traffic or on the island of traffic;
- b. Distance of street lighting poles is at least 0.60 meters from the edge of the traffic lane
- c. The lowest part of the street lighting is at least 5.00 meters from the road surface

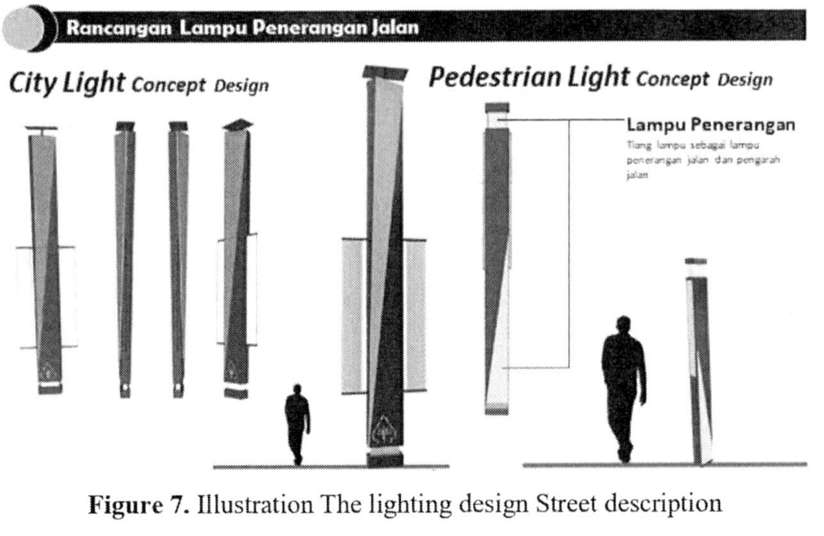

Figure 7. Illustration The lighting design Street description

Figure 8. Details of the lighting system

Pedestrian Information Facility
Pedestrian lighting is designed with the following components.

- Solar Cell, The use of solarcell as street light energy
- Lighting Lamps, Lampposts as street lighting lamps and road directors
- Digital Advertising Board, The existence of placement of advertising media and digital information on street lights as controlling billboards or advertising
- Digital Direction Board / Road, There are placement of digital road signposts at the bottom of digital billboards as a guide for pedestrians.
- Emergency Button, There are buttons, mic, and speakers that are directly connected to the nearest security to anticipate an emergency, such as an accident, crime, or fire.
- Wi-Fi Router and charger, There are internet facilities and charging power for pedestrians.

3.3.4. Street Furniture. Street furniture/Street furniture is a facility that is placed along the road as a complement or support for pedestrian paths. The placement is adjusted to the location of the area used by the pedestrian path. The material used must use materials that are easy to get around, strong against the weather, easy to maintain, easy to repair, strong and safe for road users and the environment around this road complement facilities must fulfill its functions such as:

1. The function of safety and comfort for pedestrians.
2. Complementary function as a seat,
3. Aesthetic functions can be seen from the shape, texture and color.

Some considerations that must be considered in planning road furniture such as:

1. Uniformity of shape and size of signs;
2. Sign design; colors, shape sizes, and material textures that meet the standards will attract the attention of road users.
3. The location of street furniture is placed in a location that is easily visible and not obstructed in front of it.
4. Maintenance of signs; needed to keep functioning properly.

Figure 9. Model Street Furniture [4]

3.3.5. Signage.

3.3.5.1 Signage

- Signage in the form of a building name can be installed on the facade / facade of the building, at an altitude between +2.20 to +5.00 (between the top of the ground floor to the middle of the second floor).
- The design of the marker must be integrated with the architecture of the building.
- Road Signage can refer to existing regulations

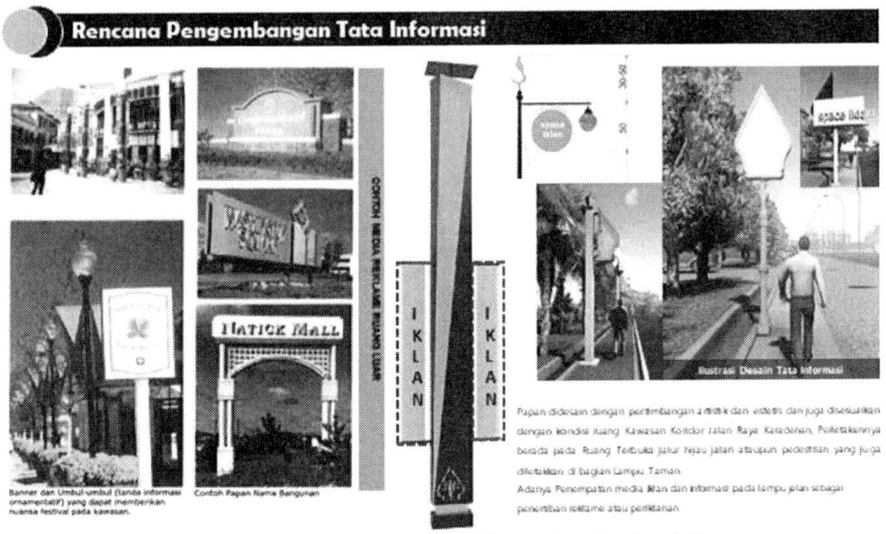

Figure 10. Illustration of Information Design [4]

3.3.5.2 Advertising

The installation of billboards must pay attention to the safety and security factors, the aim to be achieved is to strive for a balance, linkage and integration of all types of information governance elements and road faces in terms of function, aesthetics and social, both in the region and the surrounding region.

- The form of billboards and information boards must be integrated with the architect of the building.
- Advertising is only permitted in certain locations.
- Advertisements attached to buildings are allowed at certain locations.

The installation of billboards / billboards is an integral part of the development of the region. Because of that, it is necessary to match the planning of the billboard with the arrangement of the area. The billboard is no longer seen as an addition, but must be part of the arrangement of the area by taking into account applicable regional regulations. Included in the free area of billboards in the planning area are:

- Settlements and conservation areas on the banks of rivers
- The side of the plot facing the knot
- Utility paths and pedestrian lines
- The installation point of the billboard / billboard in the planning area can be placed in an open space formed on the road nodes

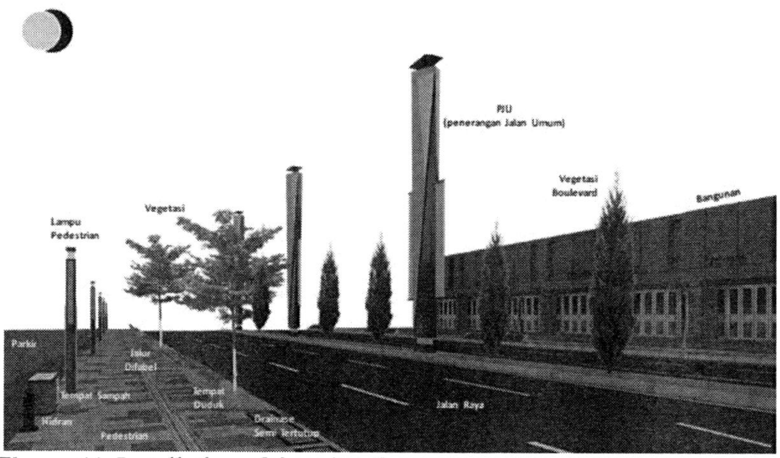

Figure 11. Installation of the advertisement on street lighting [4]

3.3.6. Waste management. Based on Government Regulation No. 81 of 2012, public facilities, especially sidewalks, are required to provide trash cans in the form of TPS 3R

Figure 12. Trash Shelter [4]

3.3.7. Implementation of green road open space facilities.

- The minimum sidewalk width is 2 (two) meters so that the ecological functions that arise from the arrangement of pedestrian paths and green lanes can improve pedestrian comfort. Aesthetic functions between the sidewalks and greening of the road where the two elements need to be arranged on the sidewalk, on the side of the road, which is adjusted to the width of the land from the road shoulder to the channel threshold or Rumija boundary. The height and size of the plant must be adjusted so that the area has aesthetic value. Likewise, the color of the plant is adjusted and uniformed, but not flashy, not monotonous, giving an interesting impression.
- Aesthetic types of plants in urban areas that can be considered are:
- a. Beautiful,
- b. Titled easily formed,
- c. Beautiful leafy,
- d. Beautiful flowering, and
- e. A distinctive scented or fragrant aroma

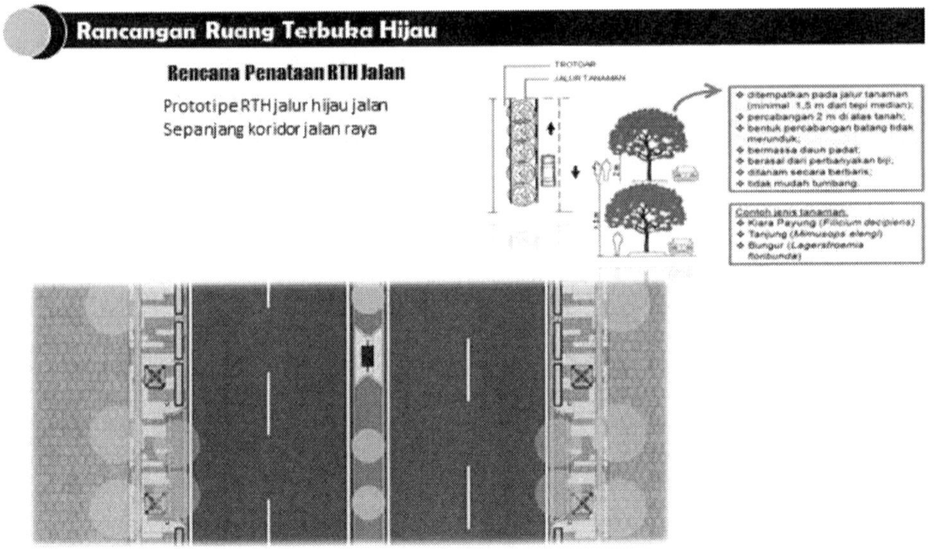

Figure 13. Pavement and Plant Pathways to Support Traffic Comfort and Smoothness

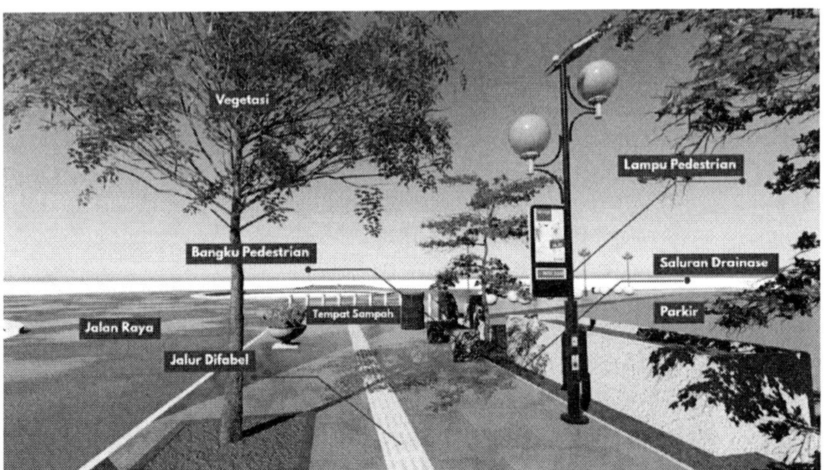

Figure 14. Models of Integrated Road and Sidewalk Development Plans [4]

3.3.8. Circulation and parking. Parking of motorized vehicles consists of parking inside the yard or inside the parcel (off-street) and parking inside the area owned by the road (on-street). For the parking area, the office is directed to develop Green Parking. Parking areas are built without dominant pavement so that rainwater can still be absorbed into the soil.

Figure 15. Illustration of environmentally friendly parking land [4]

References

[1] Haryadi B S 2010 *Architecture, Environment and Behavior* (Yogyakarta: Gajah Mada University Press)
[2] Anggriani N 2009 *Pedestrian Ways in City Design* (Klaten : Humanities Foundation)
[3] Guidelines for Planning Walkways on Public Roads No.032 / T / BM / 1999 1999 Attachment No.10 General Decree of Bina Marga No. 76 / KPTS / Db / 1999 Public Works Department
[4] Guidelines for Provision and Utilization of Urban Pedestrian Space and Infrastructure and Facilities 2000 Directorate General of National Spatial Planning Public Works Department.
[5] Asmoro D 1990 *Sidewalk Planning Instructions* (Jakarta: Directorate General of Highways)
[6] Public Works Department 1990 No.007 / T / BNKT / 1990 Guidelines for Sidewalk Planning Department of Public Works Directorate General of Highways and Directorate of Urban Road Development.
[7] Directorate of National Spatial Planning. Standardization Technical Committee for Construction Materials 2011 Guidelines for Provision and Utilization of Pedestrian Space Infrastructure and Facilities in Cities
[8] Tanan N 2011 *Pedestrian facilities* (Jakarta: Center for Research on Roads and Bridges)
[9] United States 2000 *Highway Capacity Manual (HCM)* (Washington D.C: Transportation Research Board)
[10] Unterman Richard K 1984 *Accomodating the Pedestrian (Advent Towns and Neighborhoods for Walking and Bicycling)* (New York: Van Nostrand Reinhold Company)
[11] Waldock R 2011 Planning *and designing for Pedestrians: Guidelines. Perth* (Western Australia: Department of Transport)
[12] Yang, Ching Chow 2005 The Refined Canoe's Model And It's Application *Journal Total Quality Management* **16** (10):1127-1137

IOP Conf. Series: Earth and Environmental Science **314** (2019) 012080 doi:10.1088/1755-1315/314/1/012080

Child-friendly Kampoong: quality of play value criteria for children's identity and play place in Malang, Indonesia

Agung Murti Nugroho*

Department of Architecture, Engineering Faculty, University of Brawijaya Malang, Indonesia

*sasimurti@yahoo.co.id

Abstract. The paper discusses child-friendly kampoong. A kampoong with a complex and diverse environment is a dense residential reality in urban areas that gives the impression of slums and unorganized. The kampoong actually contains cultural values of shared life that must be maintained and preserved especially for children. The paper aims to study the criteria for developing the quality of play value for children identity and play place in Malang, Indonesia. The reserch method about the social identity and children play place using narative method with a qualitative approach to obtain the general condition of child-friendly kampoong. The method used in determining the quality of play value criteria using Analysis Hierarchy Process through interviews with experts. Discussions about child-friendly kampoong included child-friendly city, the character of children's friendly play value space; children's friendly kampoong as social identity; children play place and quality of play value space. The results indicate the main criteria for quality of play value according to the experts is safe spaces with value of 0.141. Using several broader perspectives, it is hoped that this paper can foster awareness about safe, comfortable and enjoyable playgrounds in urban dense environments.

1. Child-Friendly Kampoong as The Implementation of Children Friendly City

As an effort to accelerate the fulfillment of children's rights, several cities in Indonesia have the initiative to form a child-friendly kampoong program. Child-friendly kampoong is held in the form of a pilot project. This pilot project is an activity that makes a community group in an area as a model for other regions, and runs according to the program objectives that have been established in the rules. The implementation of child-friendly cities in the form of child-friendly kampoongs contains rules that have been legally legitimate, so that all components involved together take action in achieving their goals. Application of child-friendly kampoongs through a series of activities. The kampoong area is analyzed as a unit of the smallest unit after the family to be able to accommodate a variety of child-friendly indicators which means ensuring the condition of the child and their rights in living life. Thus, child-friendly kampoongs can be defined as a place to provide interaction space so that the community is easier in socializing and building awareness about the rights of children. Child-friendly kampoong is a program unit carried out by citizens who are members of the kampoong association in the form of efforts to fulfill children's civil rights to provide opportunities to grow and develop based on realistic conditions towards the kampoong that are able to provide comfort, livable, and deserving of development on the basis of health, education and protection law based on independent initiative. This program is implemented in an integrated manner with the activities of the regional and neighborhood units as a fulfillment of basic living needs.

Child-friendly kampoongs are a tangible manifestation of increased awareness that guarantees the fulfillment of children at the kampoong level and ensures efforts to pay attention to the needs of aspirations, attention, and appreciation for children without discrimination. Child-friendly kampoongs are expected to be physically and non-physically feasible in meeting the needs and rights of children. The development of child-friendly kampoongs is expected to unite the commitment and resources of the kampoong, the community and the business world in the kampoong to be able to respect, guarantee and fulfill the rights of children, protect children from acts of violence, exploitation, harassment and discrimination and hear children's opinions consciously planned, comprehensive and sustainable.

The concept of child-friendly city policies as stated by Corsi [1], that there are two supporting models of child-friendly city policies, first, model with an orientation of education, cognition and normative, and second, a promotion model of social participation. The concept of environmental security as formulated by Tranter and Sharpe [2] that the danger of traffic causes parents to pay attention to their children. The concept of fulfilling basic rights was conveyed by Wilks [3] that the fulfillment of children's basic rights to obtain all basic services and security as well as protection from exploitation efforts is also a concern in realizing a child-friendly city.

Criteria for play value include aspects of location and size; safe spaces; easy access to entrances; circulation path; signage; seating; fence; playground equipments; lighting facilities; trees and plants; garden; environmental sustainability; sand and water playgrounds. The location of play spaces should be in areas that are easily accessible to children and have visual views from all directions [4]. Criteria for a safe space include adequate lighting facilities, emergency telepohones and the fence precisely surrounds [5]. Criteria for appropriate signage are in the form of important information about the space and provide direction of traffic inside the open spaces [6]. Seating arrangements can support or hinder social interaction [5]. According to Shackell [4] the following design considerations must be considered when giving a fence in a child-friendly playing space, which is barrier, protector, and aesthetic. Playground equipment are multipurpose and support the creativity development and coordination [4]. Lighting can be used as follows: security; protection against crime and aesthetic [7]. Trees create a variety of play activities such as climbing, hiding and searching, exploring, discovering, imaginative playing, gathering and stimulate senses of children [4,6,8]. Criteria for the garden is the best way to allow children for interacting with each other and with nature. They learn about ecological cycles, how to preserve environment and foster cooperation between children. Trees in the garden are chosen according to their roots, water requirements, endurance and growing behavior [8]. Sand area in child-friendly spaces must be close to a pathway that is easily accessible to children with disabilities and located near or under a tree for sun protection and for wind protection [4,6]. Water playground can be integrated into child-friendly playing spaces through various shapes such as fountains, spray pools and water tables [5].

2. The Character of Child-Friendly Play Value Space

A city if the greater the flow of urbanization will result in a more populated population. Gradually it began to erode its calm and comfort. Replacing with the hustle and bustle of community activities, job competition was in the midst of expensive needs. This situation makes a city unfriendly again, especially for children. Many parents or communities forget to pay attention to their children. Parents only fulfill the need for clothing, food and shelter. Even though they also need more love and attention. Besides that, many of their playgrounds have been lost, switching to residential functions. Children now play more at home with their gadgets or access the internet at the side of the road, watch television all day, and play themselves with games on the computer. The child-friendly kampoong program is a form of city government effort to empower people to care for children's growth. Encouraging people to be aware of the importance of preparing the next generation better. A good successor generation certainly needs a conducive environment as a growth and development medium. The added value of a child-friendly kampoong is to make children feel more comfortable. This means that the child must be comfortable in his own kampoong through planting moral values in daily life. For example, for adults who smoke are provided in certain places. In addition, children are also involved in various activities, such as the waste bank program. Thus, the children in this kampoong are more focused on their activities. Not only that, the kampoong is child-friendly, children are also involved in various activities, such as the establishment of reading gardens, sustainable food house areas, children's playgrounds, arts stage, healthy toddlers, and fond of learning, garbage banks, playgrounds, nursery gardens kampoong, skills for school dropouts.

The kampoong dominates the allotment of land in cities in Indonesia (around 70 percent), the kampoong becomes the foundation of housing 70 to 85 percent of the city population. Meanwhile, housing provision through formal channels by the private sector and the government is only able to provide around 15 percent of the total needs of homes in urban areas. Data and facts show that self-help residents are the largest supplier of housing. The kampoong follows the compact city principle with mixed uses. The kampoong is a kind of mini-urban collage that allows them to continue to develop the principles of diversity, tolerance and solidarity. One room is used for various needs, such as streets in kampoongs used for passing vehicles, playing, hanging out, and earning a living. State protection against children's rights to play is a protection of human rights so there is no reason for the state not to enforce children's rights to play. Playing is a direct and spontaneous activity for fun purposes. Therefore, every child wants to always play because playing with children feels comfortable, happy and not depressed. The playing function not only can improve language development, discipline, moral development, creativity and physical development. Thus, children's play activities are a way for children to do activities that contain elements of learning and are carried out with pleasure and relaxation without any pressure on the child. The element of learning in children's play activities will be beneficial for physical and motoric development, psychic development (emotion, attitude, intelligence, perception) and the media for children to develop their social relationships.

Play is considered very important for physical and psychological development so that all children can be given time and opportunity to play and are also encouraged to play without regard to their family's socioeconomic system [9]. During play children will develop various social skills so that it is possible to enjoy group membership in the children's community. Hurlock [9] explained further about children's play patterns that are classified into playing activities in the early childhood and playing activities in the late childhood. Playing activities in early childhood are often called the toy stage because in this period almost all games use toys. His interest in playing with toys began to decrease and when he reached school age.

Hurlock [9] states that playing in childhood is a serious activity which is an important part in the development of the first years of childhood. In this perspective the playing activities of early childhood differed from one location to another. For example, the pattern of playing American children is different from the playing patterns of Asian and African children. Likewise, not all playing patterns from time to time will be as popular. Playing activities at the end of childhood are different from playing activities at the beginning of children. In the early days of children, playing activities tend to be individualized while in the late childhood, the play activities prioritize playing together or groups and prioritizing popular playing activities. Playing activities at the end of the children tend to play more constructively, explore, collect, sports and games that contain elements of entertainment. The character of children's friendly play value space provides an overview of the importance criteria such as children must feel safe and want to play the area; improve health and well-being; social meeting space; various types of paths that support a variety of different activities; designed for playing purposes; encourage interaction between children; provide a sense of security, closure and support for activities; supporting body muscle development, social interaction and fantasy play; provide a safe and aesthetic atmosphere; stimulate behavior to explore, discover, and also encourage fantasy and imaginative play; increase social interaction, develop fine motor skills and stimulate sensory; a very good media for creative play and social interaction; multisensory characters include sounds and textures that make children interested and relaxed.

3. Child-Friendly Kampoong Method

Development of a child-friendly city policy cannot be separated from the basic rights of children who become an important locus in fulfilling children's rights. The rights of the child in question are things that must be fulfilled by parents, community and government to children. Thus, children will avoid discrimination, have special protection, and can participate in activities. This is important for achieving the fulfillment of children's rights fundamentally. In the kampoong area, children's rights are based on 5 clusters that have been established by the government and translated into 61 child rights indicators. The cluster can be explained as follows: civil rights and freedom; family environmental rights and alternative care; the right to basic health and welfare; the right to education, use of leisure and cultural arts activities; the right to get special protection. As an implementation program, child-friendly kampoongs cannot be separated from various problems. The reality that occurs in the field shows that the historical value and plurality of social, economic, and cultural conditions experienced by the community in a kampoong have provided space for its citizens to construct and give subjective meaning to the current existence of child-friendly kampoongs [10]. The reserch method about the social identity and childern play place in child-friendly kampoongs used is narative method with a qualitative approach to obtain the general condition in Malang city. The reserch method about the quality of play value place in child-friendly kampoongs used is descriptive method with a quantitative approach to obtain the results of analysis hierarchy process (AHP) to experts related to child-friendly campoong. Questionnaire design of AHP was focused on quality factors of child-friendly play value space in the city campoong. Criteria used in AHP analysis consist of thirteen aspects as follows: distance and dangerous road (A), safe space (B), entrance (C), pathway lane (D), signage (E), seating (F), fence (G), playground equipment (H), toilet (I), garden (J), vegetation (K), sand playground (L) and water playground (M).

4. Result and Discussion

4.1 Child Friendly Kampoong as Social Identity

Social change is something that is sure to happen in every society. Even though social change is fairly slow, society will not stagnate and will continue to undergo social reality. Social change becomes a series of events that bring people to a new historical dimension. And includes aspects of acculturation, assimilation and enculturation in the culture that is carried out. Social change brings people to realize the conditions they judge are no longer relevant in social life. One of them concerns identity issues. The existence of identity is important, because identity is a characteristic that distinguishes humans from others. In the root of philosophy, according to Aini [11] the formation of identity is divided into three approaches, namely: primordialism, constructivism and instrumentalism. The primordialism approach explains that identity is something that is obtained naturally (given) which is formed through the process of hereditary socialization. Whereas the constructivism approach explains that identity is a complex social process through cultural ties built in society. An instrumentalist approach, identity as something constructed for the sake of the elite and for the sake of power.

Identity becomes a form of affirming the existence of an individual and his group. According to Kinasih [12] identity becomes a dimension of necessity inherent in human relations because the existence of a person becomes part of an ethnic group, religion, tradition and language in a particular cultural system. Humans individually or in groups will place themselves in the corridor of identity in a cultural context. With the existence of identity, the individual will be recognized for its existence as well as its existence in the social space. Jeefrey Weeks quoted by Kinasih [12] explains the importance of identity for an individual

Identity is about the similarities of a number of people and about what distinguishes you from others. As the most basic thing. Identity will give someone a sense of personal location, a stable core for individualism. Sense of belonging in this context will provide a sense of security for individuals. Security will provide stability in the social system that is being carried out by individuals in society. Through identity, individuals outside the community will provide an assessment that consciously shapes selfhood for the individual. As revealed by Barker [13] that social identity is the hope or opinion of others about selfhood. However, it is necessary to realize that the identity of a time can change shape as social change returns. Because its nature is not taken for granted, identity can be used inconsistently. Adapted to the needs of the individuals and groups concerned. This adjustment occurs as a step in forming a positive identity. This is a meaningful identification step, an effort to identify the identity that is already attached to get a better view and assessment than others [12].

The social construction built by the residents of a kampoong regarding the child-friendly kampoong program, was born through a simultaneous process that took place dialectically, namely: objectivation, internalization and externalization. As a government program, the reality of child-friendly kampoongs is present as a manifestation of policies that have a legally binding set of rules. So far, if based on reality, the implementation of child-friendly kampoongs has a gap when judged in the frame of social construction. These differences can be grouped into several things, namely: context, issues, agents, strategies and results. In context, by the government, the area of a kampoong is assumed to be an urban sprawl with a negative stigma that is inherently considered not responsive to children so that the child-friendly kampoong program becomes a program that is believed to be able to change this. But on the other hand, the kampoongrs constructed a child-friendly kampoong as a form of adaptation to social change in the region. Although it was realized that the child-friendly kampoong program had not bind the subjective awareness of citizens.

Child-friendly kampoongs have given meaning to the components of society which consist of: children, parents, administrators as cadres and government. Simultaneously, the reality of child-friendly kampoongs is an effort to form a positive image of a kampoong. Positive images that are present are believed to have an impact on changing the views of outsiders towards the territory of a kampoong. So that people can visit the kampoong area and can tell about the social changes that have now taken place. For children, the reality of child-friendly kampoongs has been interpreted as an appreciation of the voices and aspirations conveyed by children. Modern instrumental views are still held by most kampoongrs. So that it still places children as objects, as well as individuals who must submit and obey. The absence of respect for votes has an impact on the lack of involvement of children in decision making at the institutional level. Based on environmental aspects, child-friendly kampoongs are interpreted collectively by parents as an area that can provide a comfortable and safe environment for children to do activities. A safe environment is perceived as a place that is far from danger, while a comfortable environment is interpreted as a place that gives children a sense of comfort when playing and utilizing leisure time.

4.2 Children Play Place

Facilities and infrastructure for children's play should be educative, namely: the development of children's personality attitudes and abilities, talents, mental and physical abilities to achieve their optimal potential; development of respect for human rights and fundamental freedoms; developing respect for parents, cultural identity, language and its own values, national values where children live, where children come from, and civilizations that differ from their own civilization; child preparation for a responsible life; and developing respect and love for the environment. Playground facilities in the open space area according to Kusumo [14] can be divided into three categories, namely: play lot, playground and play field. In kampoongs that are densely populated and access to outside as well as entering residential areas connected with footpaths and alleys, which illustrate that there is insufficient public space available. Population dense settlements cause the distance between individuals to be physically very close so that there is no private space, let alone public space. Space conditions like this affect the interaction between citizens. Interaction among citizens is very high and has an impact on the emergence of pro-social behavior such as the emergence of mutual help behavior among citizens. Social interaction in space - children's play time naturally creates a type of children's play that utilizes spatial situations and game repetition. It can be seen on a daily basis that there are children who play along the aisle with repeated games all the time. Although children have many opportunities to play but children are less satisfied playing because in addition to influencing the mobility of the people passing by on the kampoong streets or aisles, less involvement of children plays in large numbers and

IOP Conf. Series: Earth and Environmental Science **314** (2019) 012080 doi:10.1088/1755-1315/314/1/012080

the games played by children are very limited. The phenomenon that happens a lot is the very lack of children's playground in the kampoong so they play around the house or play on the road. The game also has no traditional games, except running around, playing ball in the alley which sometimes disturbs people who go outside in the alley, are playing a bicycle so they go to the big road which is certainly dangerous

Game tools used by children what they can to play like bricks, bamboo / sticks, pipes, and so on. Sometimes due to land conditions that there are not many children playing in public spaces government facilities such as schools. Child games that are repeated all the time are running without rules or sometimes they also play running with certain rules that are created by themselves. Other games that children often play are playing and playing cards. The game of children in the urban kampoong generally does not discriminate gender, because men and women play together. This situation is advantageous to foster a pro-gender attitude since childhood and build togetherness. Child games that involve a group of children can help train children's social development. Social development means that the ability to behave in accordance with social demands. Good social development is needed so that children in urban kampoongs are able to socialize or make social adjustments.

Several factors that influence the social adjustment of children in the kampoong of the city as follows. First, children in urban kampoongs have full opportunities to socialize because every day children have time to play and socialize. Playing while socializing is important for children to learn to live in a community. If most of the time they are only used alone or activities play alone then children will lose the opportunity to socialize and learn to live in a community. Second, children who play while socializing are able to talk socially, and ultimately make them socially acceptable. Children like this easily associate and place language in relationships, even though in general they use the local language. Third, children will learn socialization if they have motivation. Motivation largely depends on the satisfaction provided by social activities given to children. Although children's play satisfaction is limited by the conditions of place and space but does not reduce the motivation of children to socialize. This can be seen from the frequency of children in urban kampoongs in groups.

Children's games are limited because the land can still develop motor skills such as running movements that are performed when playing in principle is physical movement. Motor development means the development of physical movement control through coordinated central requirements, and muscle movements. If a child has no environmental or physical disturbances or mental barriers that interfere with normal motor development, children will be ready to adjust to playing with peers who strongly support motoric development. Good motorbike provides an opportunity to learn many things including social skills, physical security that will give birth to a sense of psychological security and will ultimately lead to self-confidence that will affect children's behavior. The limited place to play and less varied play activities can still develop children's emotions.

Emotional development has a very important role in life. The emotional development of children in the city kampoong is influenced by two factors, namely: first, the maturity factor as a form of intellectual development. This factor is obtained from playing ball and cards by children. Ball games are not only physical and cooperative, but also help children's intellectual maturity. While playing cards contributes to children's emotional control efforts. Second is a factor of learning experience. Playing is learning and learning will determine which potential reactions children will use to express anger or pleasure. Availability of facilities and infrastructure or space and playground for children is an indicator that must be conditioned gradually to be achieved and fulfillment is improved. The right of children to get a decent place for recreation and play is very important for children to develop their basic potential such as emotion, intelligence, creativity, motoric, social relations, and others. One of the activities of children in playing in a narrow alley is running. Even though the playground is very limited, children can still run along the alley. Even though children run without rules, they are still useful for training their psychomotor development.

Climbing games are one of the basic rock-climbing sports. This game can train children's motor development because one of the functions of the game is to train children's motor skills. Climbing is a form of motor skills, but it is unfortunate that a child climbs an electric pole that can endanger his life because he does not pay attention to the elements of safety. Children play using small poles embedded in the alley. The child does not give the name of the game, but the game is played very often by children in the kampoong. Although children play with game tools as they are but this game can train collaboration as one element of children's psycho-social development. Play ball in a very narrow alley. Children cannot run free to chase and dribble and fool opponents and kick the ball so that it is difficult for children to develop skills to play football well. Playing ball has this technique and technique trained in an adequate field, not in a narrow field so that children have reliable football playing skills. Playing ball on a narrow field is solely a function of health sports not for achievement sports. No wallpaper can be used for drawing. Children try to practice their creativity even though the walls are littered. Behind the children training drawing skills on the wall is reflected in the low economic ability of the family to buy drawing paper. On the other hand, littering the walls of the wall reflects the slums of a city kampoong.

4.3 Child Friendly Kampoong: Quality of Play Value Space

Based on the weighting results of all experts, it can be seen that the highest value of the expert's average geometry is the B criteria, which is in the form of security with a value of 0.141 (Figure 1). This is because there are three experts who argue that security is a priority in the development of child-friendly play value. The security has been explained, can be in the form of security of playground equipment, security in interacting with each other, security in absorbing information and security in consuming food available in the open space. This is in accordance with a research conducted by Munoz [15] which states that playing space of children must avoid danger without reducing the child's motivation to be more courageous through providing safe facilities.

The criteria for safe space are the highest because of the need to maintain child safety in using all elements of the playing place. It aims to minimize the dangers that occur in children, therefore a child-friendly kampoong concept is created. The criteria for safe space are also stated by Parson [8] that the creation of safe open spaces refers to the health and well-being of all children protected from all danger conditions.

The lowest criteria value based on the weighting calculation of all experts is the entrance or criteria E with a value of 0.018. This is because the seating is a very important thing, but less significant since there is enough one gate entrance without having many entrances for a child-friendly play value. The importance of one gate entrance system is also used to maintain safety of children in playing in kampoong open space. This evaluation is in harmony with the statement by Parson [8] said that the seating to child-friendly open spaces mainly includes aspects of functionality and waiting zones.

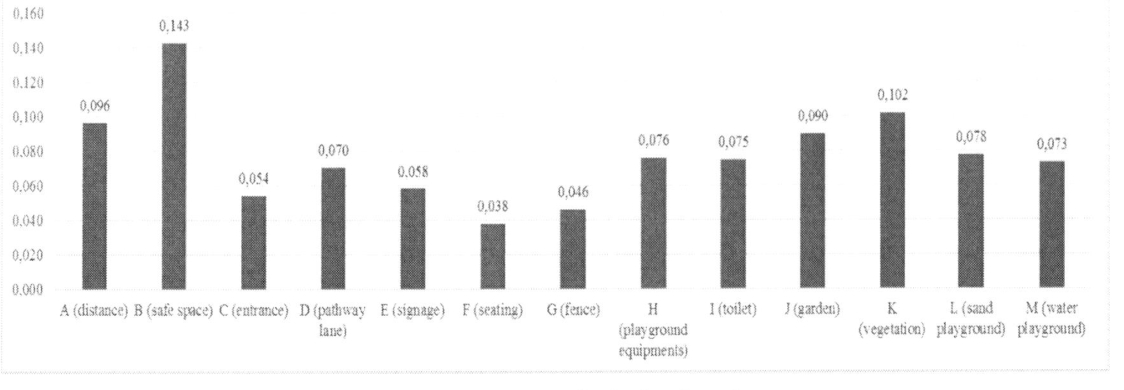

Figure 1. AHP Calculation Results

5. Conclusion

It is fitting that child-friendly city policies must be known by the community. So that the participation of the community in realizing a children's playground can be continuously improved. This will encourage the government not to ignore the function of children's playgrounds. So that it can be avoided in a city there are many parks but not for children's playgrounds, but for the beauty of the city. The right of children to play is sometimes overlooked by the level of income of parents of children in urban kampoongs who are generally informal sector workers. So that the government is obliged to provide a space or playground area for children who live in densely populated areas such as the urban kampoongs. The availability of space or area for children's playgrounds that are both play lot, play ground or play field in urban areas that are very limited of play value space. The most influential criteria of play value space for child-friendly kampoong in Malang city are determined by the weighting results from all experts in AHP. The highest value from the average geometry of the experts is safe space with a value of 0.141. Three experts argue that security is a top priority in child-friendly spaces in the shape of playground security, security in interacting with others, security in absorbing information, and security in consuming food available in the park. These criteria are considered as the main priority to maintain safety of children in using all playing space elements and to minimize the dangers that occur.

Acknowledgements

This research was supported by Center of Tourism Study, University of Brawijaya, Malang, Indonesia

References

[1] M Corsi 2002 The Child Friendly City Initiative in Italy *Environment and Urbanization* **14**(2) 169-179
[2] P Tranter and E Pawson 2001 Children Access to Local Environments: a case- study of Christchurch, New Zealand *Local Environment* **6**(1) 27-48
[3] J Wilks 2010 Child-Friendly Cities: a place for active citizenship in geographical and environmental education *International Research in Geographical and Environmental Education* **19**(1) 25-38
[4] A. Shackell, "Design for Play: A guide to creating successful play spaces". England, pp.1-130, 2008
[5] C C Marcus and C Francis 1998 *People Places: Design Guidelines for Urban Open Spaces, 2nd ed* (New York: Wiley) pp 5-293
[6] S. Herrington, "CHILD: Seven Cs An informational guide to young children's outdoor play spaces", pp. 1-59, 2006
[7] M L Christiansen 1977 *Park Planning handbook* (New York: Wiley) pp 1-219
[8] A. Parsons, "Young Children and Nature: Outdoor Play and Development, Experiences Fostering Environmental Consciousness, And the Implications on Playground Design". Virginia, pp 4-74, 2011
[9] E B Hurlock 1980 *Psikologi Perkembangan: Suatu Pendekatan Sepanjang Rentang Kehidupan* (Jakarta: Gramedia)
[10] P L Berger and T Luckmann 1990 *The Social Construction of Reality: A Treatise in The Sociology of Knowledge* (Harmondswirth: Penguin Books Ltd.)
[11] Widayanti C G S 2009 Fenomena Bullying di sekolah negeri Semarang: Sebuah study kualitatif *Jurnal Psikologi Undip* **5**(2) Desember 2009
[12] A W Kinasih 2005 *Identitas Etnis Tionghoa di Kota Solo* (Yogyakarta: Laboratorium Jurusan Ilmu Pemerintahan Universitas Gadjah Mada)
[13] V Barker 2009 Older adolescents' motivations for use of SNS: The influence of gender, group identity and collective self-esteem *Cyberpsychology & Behavior* **12** 209-213
[14] W. S. Kusumo, 2010, Perubahan Pemanfaatan Ruang Bermain Anak di Perumahan Griya Dukuh Asri Salatiga, Tesis tidak diterbitkan, Program Studi Magister Teknik Pembangunan

Wilayah dan Kota Univesitas Diponegoro
[15] S Munoz 2009 *Children in the Outdoors* (Scotland: SDRC) pp 5-24

Model of sustainable development of smallholders in Riau Province

Nurhamlin[1,*], Aslim Rasyad[2], Zulkarnain[3], and Suwondo[4]

[1] Department of Sociology, Faculty of Political and Social Sciences, University of Riau 28293 Pekanbaru, Indonesia
[2] Faculty of Agriculture, University of Riau 28293 Pekanbaru, Indonesia
[3] Faculty of Economics and Business, University of Riau 28293 Pekanbaru, Indonesia
[4] Faculty of Teachers Training and Education, University of Riau 28293 Pekanbaru, Indonesia

*ocuhamlin@gmail.com

Abstract. Natural rubber has countless significant roles in human life. Rubber plants have the ability to maintain natural functions (they serve as carbon producers, conservationists of land and water, and habitats for wildlife) and they also produce many added values in terms of social and economy for rubber farmers or smallholders as well as become sources of their living and income. The price of natural rubber in the international market has significantly dropped in the last decade; leaving many rubber farmers or smallholders to live in poverty. Many programs to elevate the prosperity of rubber farmers or smallholders have been created and applied by the government of Indonesia but none of the programs has been deemed successful according to the analysis results of Multi-Dimensional Scaling (MDS), the sustainable ecological index (52.95%), technological sustainability (62.35%), economical sustainability (60.18%), social sustainability (53.62%), and institutional sustainability (20.73%). The institutional management of rubber farmers is highly expected to increase the development of rubber farmers so that each and every one of them may have high competitiveness which in turn will sustainably improve their prosperity in the future.

1. Introduction

Many programs with various strategies and schemes aimed to improve the quality of life and prosperity of rubber farmers have been applied, and one widely known scheme is the core estate and smallholder (Perkebunan Inti Rakyat or PIR), which has been in effect since 1977. This scheme in its development has been applied in the transmigration areas known as Trans Core Estate and Smallholders (Indonesian: PIR Tran), and solely related to new land clearings, meanwhile another scheme or strategy known as the available plantation land rejuvenation is known as PPKR (Indonesian: Peremajaan Kebun Karet Rakyat) or rubber plantation revitalization. The Plantation Revitalization Program is the Indonesian government's effort to accelerate the development of smallholders' rubber plantations through expansion, rejuvenation, and rehabilitation of plantation plants supported by banking credit investment and interest subsidy given by the government by involving corporations in plantation business sector as partners in the plantation development, processing, and marketing. Additionally, a new scheme called UPPB or Lump Processing and Marketing Unit was developed to create a number of good smallholders and increase the number of qualified smallholders and rubber quality. This scheme is a business entity or a business unit formed by two or more groups of smallholders as the training venue of farming, processing, and temporary storing techniques, and smallholder's lump marketing. Despite all these efforts, the prosperity of the rubber farmers has not significantly improved. A concrete solution to improve the smallholders' prosperity must be found to create rubber business sustainability.

Based on the problem background stated above, an integrative and collaborative approach of ecological, technological, economic, social and institutional factor must be created toward plantation ecosystem of smallholders' plantation in Riau Province. The problems that have to be solved through this research are as follows:

- What are the ecological, technological, economic, social and institutional characteristics of smallholders' rubber plantation sustainability in Riau Province?
- What are significant attributes of the ecological, technological, economic, social and institutional dimension that become the factors of sustainable development arousal of smallholders in Riau Province?

The results of this research are hoped to bring about a model and strategy for empowering smallholders in order to accelerate the sustainable economic development of in villages in Riau.The research findings are going to be beneficial to all agricultural entrepreneurs as well as the government as decision makers in regards to the business of plantation development. It is highly expected that there is progress and improvement which increase the smallholders' income to improve their prosperity.

2. Literature Review

Sustainable development is a concept of development that has been widey applied in many countries in the world. It is aimed to give maximum solutions of all different interests for development. This sustainable development concept was first introduced by the World Commission on Environment and Development (WCED) in 1987 in their report entitled "Our Common Future" [1] . It is a simple yet complex concept causing the definition of sustainability to be very multidimensional and multi-interpretational, including the implementation of such concept in sustainable rubber plantation. [2] stated that there are eight strategic factors that influence the development of sustainable rubber business, those factors are: availability of technology, empowering resources, farmers training, policy support, plantation size, farmers' skills, farmers' institutionality, production and productivity. The first four strategic factors are categorized in the agribusiness system as input as they have strong influences on other factros, yet their dependency on the other factors is not as strong. Meanwhile the other four factors are connecting factors in that they have strong influences on the other factors as they have strong dependency on the other factors. Many studies have been conducted to analyze the relationship between smallholder development and their family income as reported by [3, 4, 5], Studies on the impact of the rubber plantation development on the farmers' social and economy were carried out by [6, 7, 8, 9].

Results of previous studies concluded that the majority of smallholders in various regions in Indonesia including Riau Province are poor (pre-prosperous). Nurhamlin [10] stated that the income of smallholders in Kampar, Riau Province is just Rp2.500.000 per month with average number of family members of five per household, if that is compared to the minimum wage of Riau Province in 2016 which is Rp2.266.722,53, it can be said that household income of the smallholders in Riau is categorized as low income. Results of studies conducted in other regions also showed the same condition as stated by [11] that the agricultural development in Riau Province has not yet reached poor rubber farmers. [12] who conducted a research in West Kutai Regency, Provinsi East Kalimantan Province also concluded that the average income of the rubber farmers there is only Rp14.909.608,70 a year or Rp1,187,609.50 per month or lower than the minimum wage of Riau Province in 2017. This is confirmed by the report released by the Riau Agriculture Office in 2013, that the income level of rubber farmers in 2013 was just Rp14.251.314,00 a year or Rp1.187.609,50 per month. Research result by [13] concluded that the average income of rubber tappers in Pangkal Baru village, Tempunak

Sub-District, Sintang Regency, West Kalimantan Province in a month is Rp2.800.000 with the average size of the plantation of one hectare classified as medium level.

According to [13], factors affecting the income of the rubber tappers are buyers, climate and wheather, and rubber quality. Efforts made by the rubber tappers to increase their income are by working outside the activities of rubber tapping and through intensification and extensification of agriculture such as maintenance and vertilization of rubber plants and also adding fields of tapping elsewhere. Accrodint to [14] in Sannia B. *et al.* stated that income is a form of reward for maintenance service which utilizes land, labor, and capital owned in an agricutural activity. Farmers' prosperity will improve if their income is bigger than the costs that they reduce balanced by high quality production and good rubber prices. Influence of price and productivity that fluctuate causes fluctuations in farmers' income. Price and productivity are two factors of uncertainty in an agricultural activity. According to the data issued by [15], the price of rubber at the factory door in Riau is only Rp13.000 for a kilogram of dried rubber or KKK (Indonesian: kilogram karet kering) of 100%; the price of a kilogram of rubber is inadvertenly only around Rp5.000 to Rp6.000 per kilogram among the farmers. Due to the low price, many farmers ditched their plantations and changed professions. If this continues to occur, in a long run the production of smallholders' rubber will significantly drop and will impact the continuity of rubber processing factories in Riau, which in effect will decrease the amount of non-oil and gas export to regions like Riau especially in the sub-sector of agriculture. With that consideration, a research is deemed necessary in order to find a development model for smallholders so that their prosperity can improve and rubber plantations can be sustainable.

3. Research Methods

This research uses a combination of quantitative and qualitative research which is done by utilizing the survey method with questionnaire instrument and structured interview guidelines. The research location is purposively determined based on the total sizes of the plantation areas and number of smallholder households. With that in mind, three regencies (Kuantan Singingi Regency, Kampar Regency, and Pelalawan Regency) are set up as the locations for this research.Data required is primary and secondary data. The secondary data is collected from related institutions as well as private parties and rubber associations. Meanwhile the primary data is obtained by conducting interviews with the experts and distributing questionnaires to the rubber farmer households. To ascertain the sustainability level of smallholder plantations, an analysis of *Multi-Dimensional Scaling* (MDS) was done with the help of *RapEst software* [16, 17, 18].

The result of the determination of the sustainability from each dimension is then converted to a *kite diagram* to see the sustainability *trade-off* from the management of the rubber plantations. The index value of the sustainability in each dimension is depicted in Figure 1.

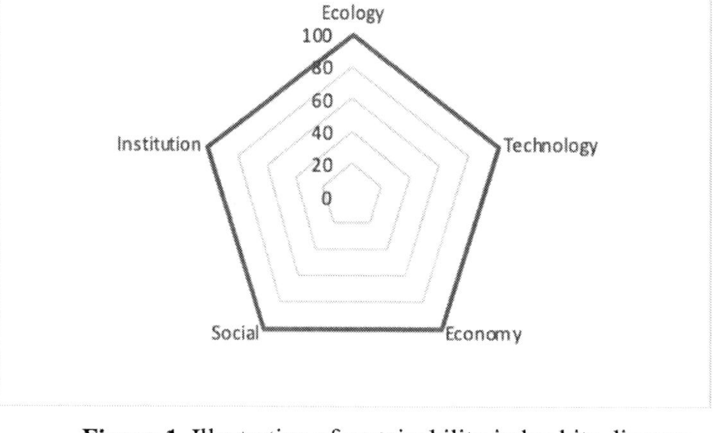

Figure 1. Illustration of sustainability index kite diagram

IOP Conf. Series: Earth and Environmental Science **314** (2019) 012081 doi:10.1088/1755-1315/314/1/012081

To learn which attribute is the most crucial in each dimension a prospective analysis is done. This prospektive analysis aims to determine the rankings of inquisitive attributes so that key attributes or *driving variables* in the management of rubber plantations are achieved in the research locations. The prospective of the *output* analysis comes up with four quadrants which are the ranks of the inquisitive attributes as seen in Figure 2 [19,20,21].

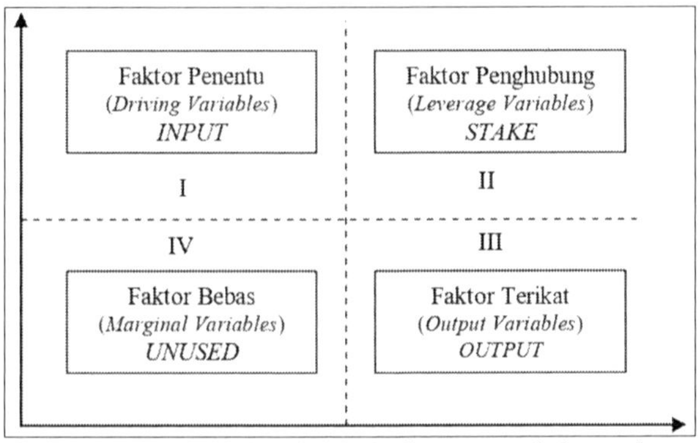

Figure 2. Graph of variable influence and dependency

4. Result and Discussion

4.1. Research results

4.1.1. Ecological Dimension. The result of the ecological parameter measurement on the smalholders' plantations shows that there are seven attributes which predictably affect the sustainability of the ecological dimension, and those attributes are: (1) rubber seed type; (2) number of rubber tapping days/rubber tapping frequency; (3) rainfalls; (4) land administrative status; (5) land area; (6) land fertility; (7) land type. The result of the *Multi-Dimensional Scaling* (MDS) analysis of the ecological dimension can be seen in Figure 3.

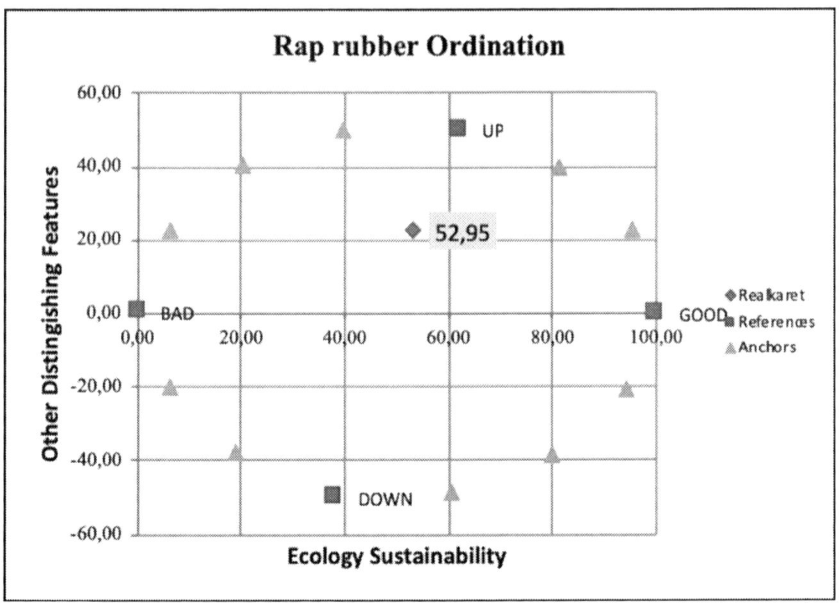

Figure 3. Index of ecology dimension sustainability of rubber plantations

The MDS analysis of the ecology dimension attributes shows an index score of 52.95 which fits in enough sustainability. Role of each attribute in the ecology dimension is then analyzed with the *leverage* analysis which is aimed to look for an attribute which is sensitive in contributing to the sustainability of the ecology dimension. The result of the *leverage* analysis is gained from the *Root Mean Square* (RMS) value of each attribute. The result of the *leverage* analysis the ecology dimension can be seen in Figure 4.

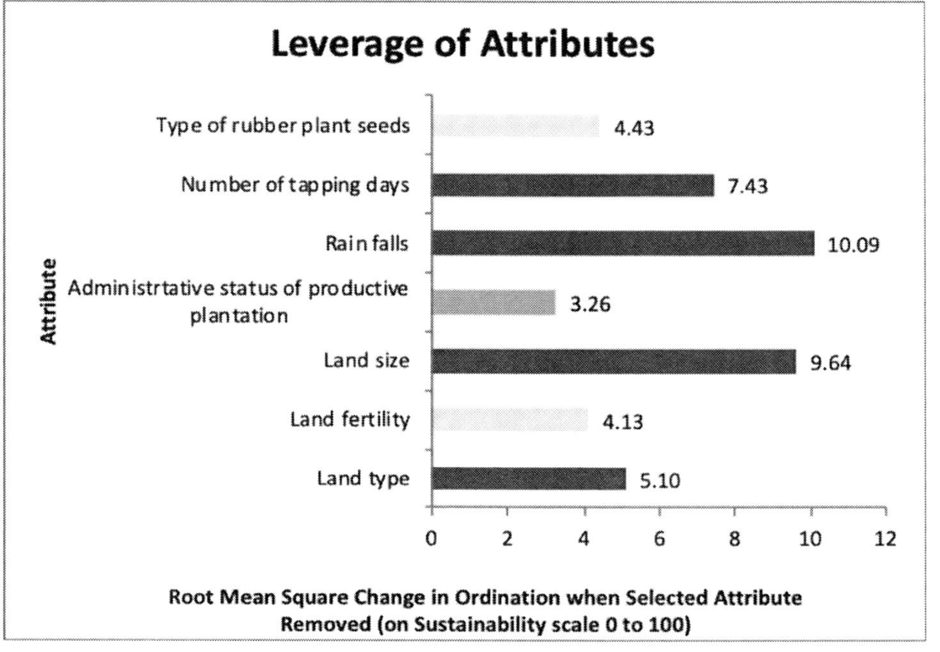

Figure 4. Role of each attribute influencing the sustainainability of ecology dimension of smallholders plantations

The determination of the sensitive attributes which influence the sustainability of the ecology dimension uses a combination of *leverage* analysis and Pareto analysis Kusbimanto [22]. The Pareto analysis was conducted by sorting/arranging the RMS values of the *leverage*analysis of the highest to the lowest value and then weighting them in percentage and then accumulating the till the maximum cumulative value limit of 75%. The percentage of the RMS value to determine the sensitive attributes in the ecology dimension can be seen in Table 1.

Table 1. Percentage of RMS Value to Determine Sensitive Attributes in Ecology Dimension

No	Attributes	RMS Value	Percentage
1	Rainfalls	10.09	22.32
2	Land Area	10.04	22.14
3	Tapping frequency	7.43	16.97
4	Land type	5.10	11.36
	Total percentage		72.80
5	Rubber seed type	4.43	10.37
6	Land fertility	4.13	9.27
7	Adm status of land deed	3.26	7.55
	Total	**21.28**	**100.00**

There are four sensitive attributes in the ecology dimension which affect the sustainability of the smallholders' plantations. They are derived from the total cumulative percentage of the RMS which is 72.80%. If another attribute is added, it shows a value over from the requirement (93.07%) and that is the maximum cumulative value limit of 75%. Those sensitive attributes are (1) rainfalls; (2) land area; (3) tapping frequency, and (4) land type. These four attributes of the ecology dimiension that are considered for the next step in the efforts to develop the sustainable development model for the rubber farmers in Riau Province.

4.1.2. Technology Dimension. The result of the technology parameter measurement mentions that there are 14 attributes that are forecast to be influential on the sustainability of the technology dimension, and those attributes are: (1) fertilization frequency of productive plantations; (2) care frequency of productive plantations; (3) pestiside frequency/pest control frequency of productive plantations; (4) behaviour of farmers regarding clean lump; (5) knowledge of clean lump; (6) knowledge of coagulant; (7) knowledge of tapping technique; (8) knowledge of care/maintenance technique of productive plantations; (9) spacing/planting time of productive plantations; (10) knowledge of spacing/planting time; (11) seed source; (12) knowledge of quality seeds; (13) terrace making, and (14) knowledge of farming. The result of the Multi Dimensional Scaling (MDS) analysis of the techology dimension can be seen in Figure 5.

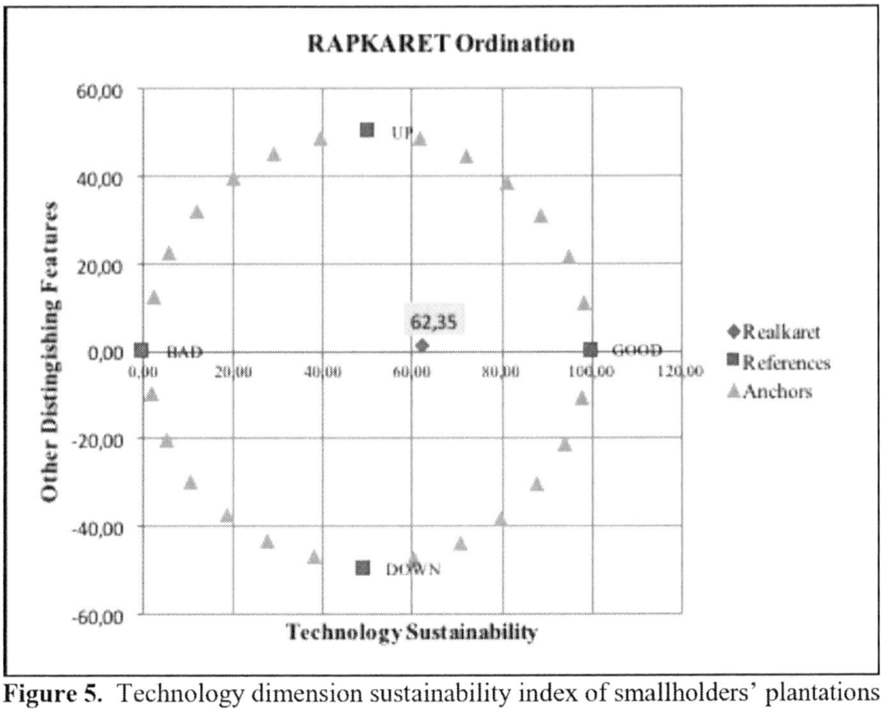

Figure 5. Technology dimension sustainability index of smallholders' plantations

The MDS analysis of the technology dimension attributes shows an index value of 62.35 and that is classified as *pretty sustainable*. The role of each attribute was then analyzed with the *leverage* analysis which was done to find any sensitive attributes contributing to the sustainability of the dimension. The result of the *leverage* analysis was gained from the *Root Mean Square* (RMS) value in each attribute which can be seen in Figure 6.

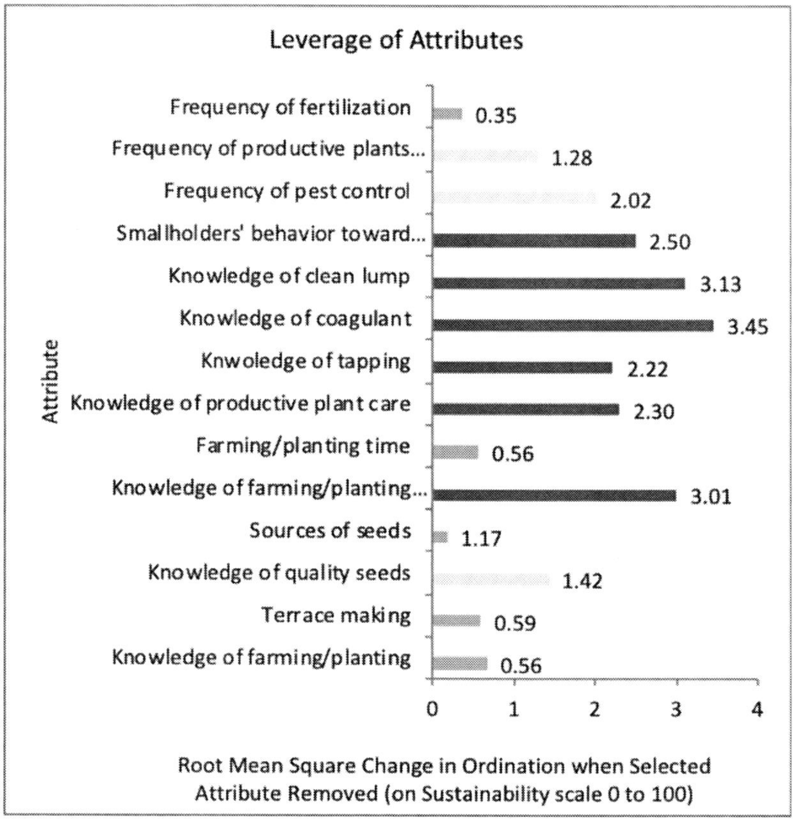

Figure 6. Role of each attribute influencing smallholders' plantation sustainability in technology dimension

The determination of the sensitive attributes that influence the sustainability of the technology dimension uses a combination of *leverage* and Pareto analysis [22] The Pareto analysis was conducted by sorting the RMS value, a result of *leverage* analysis, from the biggest to the lowest, and then weighting in percentage was done and they were then accumulated to the maximum cumulative value limit of 75%. The percentage of the RMS value used to determine the sensitive attributes in the ecology dimension can be seen in Table 2. Six sensitive attributes were found in the technology dimension influencing the sustainability of the smallholders' plantations. Those attributes were retrieved from the total percentage of the RMS cumulative which is 70.20%. If another attribute is added, it will be over limit (78.74%), overthe maximum cumulative value limit of 75%. The six sensitive attributes are (1) knowledge of coagulant; (2) knowledge of clean lump; (3) knowledge of spacing/planting time; (4) behavior of farmers regarding clean lump; (5) knowledge of productive plantation care, and (6) knowledge of tapping technique. These six attributes were the consideration for the next step in creating the model.

Table 2. The RMS value percentage to determine sensitive attributes in technology dimension

No	Technology Attributes	RMS Value	Percentage
1	Knowledge of solidification material	3.45	14.58
2	Knowledge of clean lump	3.13	13.23
3	Knowledge of spacing/planting time	3.01	12.72
4	Smallholders' behavior regarding clean lump	2.50	10.57

No	Technology Attributes	RMS Value	Percentage
5	Knowledge of care productive plants	2.30	9.72
6	Knowledge oftapping technique	2.22	9.38
	Total		**70.20**
7	Frequency of pest control of productive plantations	2.02	8.54
8	Knowledge of quality seeds	1.42	6.00
9	Frequency of care of productive plantations	1.28	5.41
10	Knowledge of gardening/farming	0.66	2.79
11	Terrace making	0.59	2.49
12	Spacing/planting time of productive plantations	0.56	2.37
13	Frequency of fertilization of productive plantations	0.35	1.48
14	Source of seeds	0.17	0.72
	Total	**23.66**	**100.00**

4.1.3. Economy Dimension. The result of the calculation of the economy parameters shows that there are 12 attributes that are assumed to be able to influence the sustainability of the economy dimension, and they are: (1) perception of rubber price; (2) perception of nine basic needs adequacy; (3) access to price information; (4) access to capital; (5) access to transportation; (6) access to clean water; (7) obstacles in selling agricultural products; (8) access to market; (9) *marketable right*; (10) debt; (11) saving, and (12) household income. The result of the *Multi-Dimensional Scaling* (MDS) analysis in the economy dimension can be seen in Figure 7.

The MDS analysis of the economy dimension attributes shows an index value of 60.08 which is classified as *good* (*pretty sustainable*). The role of each attribute in this dimension was then analyzed with the *leverage* analysis which was done to search for any sensitive attributes contributing to the the ecology dimension's sustainability. The result of the *leverage* analysis was taken from the *Root Mean Square* (RMS) value in each attribute. The result of the *leverage* analysis in the ecology dimension can be seen in Figure 8. The determination of the sensitive attributes influencing the sustainability of the economy dimension uses a combination of the *leverage* and Pareto analysis (Kusbimanto *et al.* [24] The latter analysis was performed by sorting the RMS values as the results of the *leverage* analysis from the biggest to the smallest and then were weighted in percentage and accumulated to the maximum accumulative value limit of 75%. The percentage of the RMS value to determine the sensitive attributes in the economy dimension can be seen in Table 3.

There are six sensitive attributes found in the economy dimension influlencing the sustainability of the smallholders' plantations. They were gained from the total cumulative percentage of the RMS which is 67.29%. If another attribute is added, it will be over limit (75.59%), which is the maximum accumulative value limit of 75%. Those six sensitive attributes are (1) access to clean water; (2) access to transportation; (3) debt; (4) obstacles in marketing agricultural products; (5) perception of nine basic needs adequacy, and (6) access to rubber price information. They were considered for the next step in creating the model.

IOP Conf. Series: Earth and Environmental Science **314** (2019) 012081 doi:10.1088/1755-1315/314/1/012081

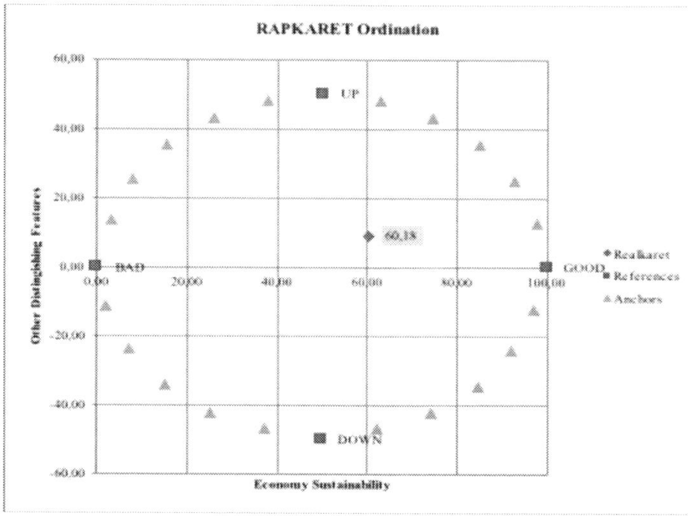

Figure 7. Smallholders' Plantations Sustainability Index

Table 3. Percentage of RMS value to determine sensitive attributesin economy dimension

No	Economy Attributes	RMS Value	Percentage
1	Access to clean water	3.70	15.49
2	Access to transportation	3.46	14.49
3	Debt	2.81	11.77
4	Obstacles in marketing agricultural products	2.06	8.63
5	Perception of nine basic needs adequacy	2.05	8.58
6	Access to rubber price information	1.99	8.33
	Total		**67.29**
7	Access to market	1.98	8.29
8	Access to capital	1.86	7.79
9	Household income	1.79	7.50
10	Marketable rights	1.47	6.16
11	Perception of rubber price	0.63	264
12	Savings	0.08	0.34
	Total	**23.88**	**100.00**

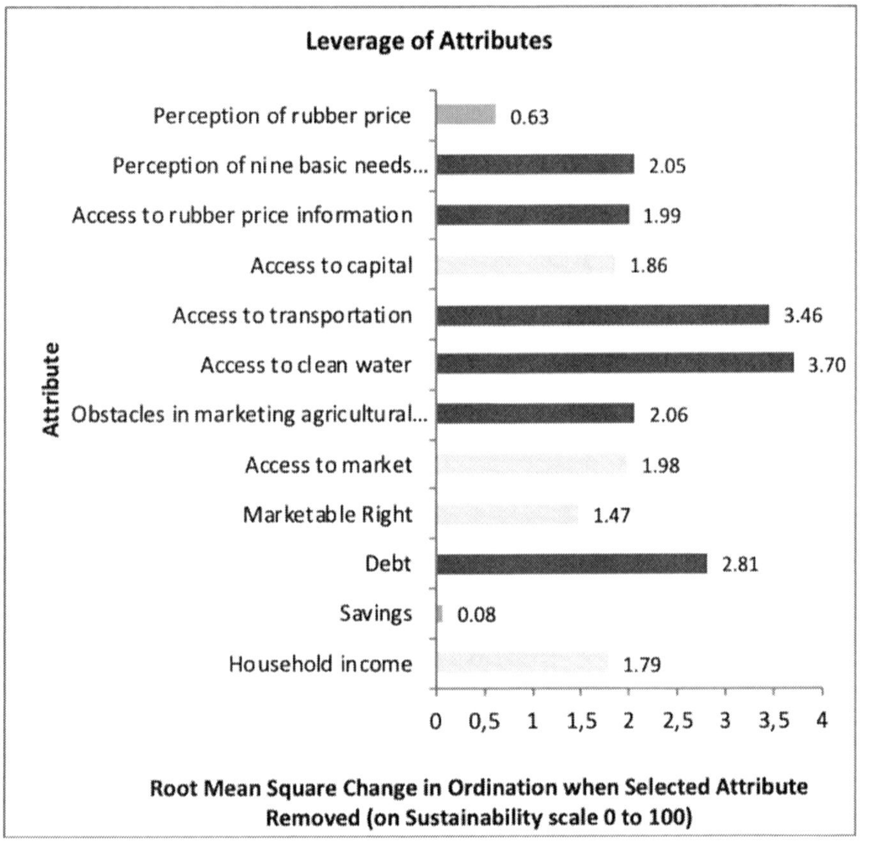

Figure 8. Role of each attribute influencing sustainability of smallholders plantations in economy dimension

4.1.4. Social Dimension. The result of the social parameter calculation results in 10 attributes that are thought of being influential on the sustainability of the social dimension, and those attributes are: (1) social relationship pattern in society; (2) relationship pattern of farmers and proprietors; (3) employment rate; (4) local wisdom and knowledge; (5) population density; (6) rate of social conflict; (7) economical facilities/infrastructures; (8) educational facilities; (9) social facilities; (10) society participaton rate. The *Multi-Dimensional Scaling* (MDS) analysis result in the economy dimension can be seen in Figure 9.

The MDS analysis of the attributes in the social dimension shows an index value of 53.62 and it is classified as (pretty sustainable). The role of each attribute in the economy dimension was then analyzed with the *leverage* analysis which was aimed to find out all the sensitive attributes contributing to the sustainability of the ecology dimension. The result of the *leverage* analysis was gained from the *Root Mean Square* (RMS) value in each attribute. The result of the *leverage* analysis in the ecology dimension can be seen in Figure 10.

The determination of the sensitive attributes influencing the sustainability of the social dimension uses a combination of the *leverage* analysis and the Pareto analysis [22] The Pareto analysis was done by sorting the RMS value of the *leverage*analysis result from the highest to the lowest and then weighting was done in percentage and then accumulated until the maximum cumulative value limit of 75%. The percentage of the RMS value to determine the sensitive attributes in the social dimension can be seen in Table 4.

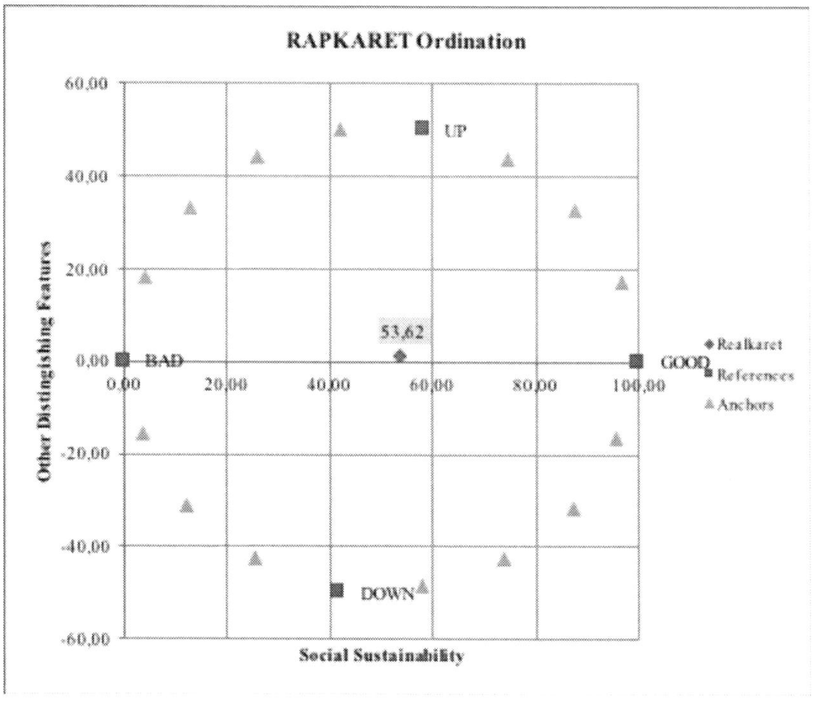

Figure 9. Smallholders' plantation sustainability index in social dimension

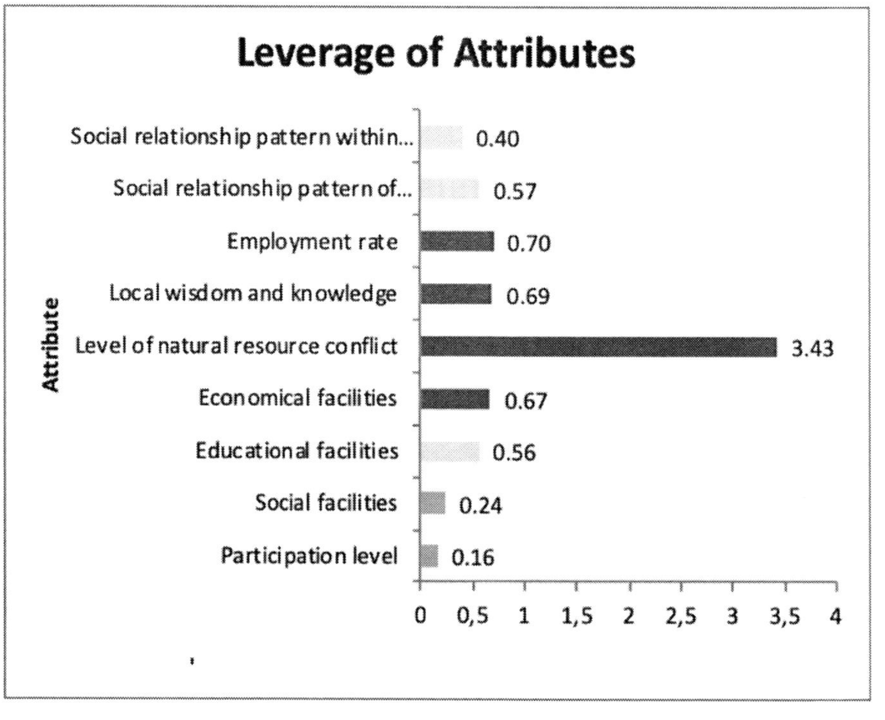

Figure 10. Role of each attribute influencing smallholders plantation sustainability in social dimension

Table 4. Percentage of RMS value to determine sensitive attributes in social dimension

No	Atribut Sosial	Nilai RMS	Percentage
1	Conflict rate	3.43	46.23
2	Employment rate	0.70	9.43
3	Local wisdom and knowledge	0.69	9.30
4	Economic facilities	0.67	9.03
	Total		**73.99**
5	Social relationship of farmers and proprietors	0.57	7.68
6	Health facilities	0.56	7.55
7	Social relation within society	0.40	5.39
8	Social facilities	0.24	3.23
9	Participation rate	0.16	2.16
	Total	**7.42**	**100.00**

There are four sensitive attributes found in the social dimension influencing the sustainability of the smallholder plantations. These four sensitive attributes were found from the total percentage of the RMS cumulative which is 73.99%. If another attribute is added, it will be over the limit (82.67%) as the maximum cumulative value limit is 75%. Those sensitive attributes are (1) conflict rate in society; (2) employment rate; (3) local wisdom and knowledge, and (4) economic facilities. These social dimension attributes were then considered for the next step in creating the model.

4.1.5. Institutional Dimension. The result of the institutional parameter calculation produced 8 attributes which are presumed to be influential on the sustainability of the institutional dimension, and those 8 attributes are: (1) role of plant breeding institution; (2) role of government institution in empowering the smallholders; (3) role of the *tauke* or proprietors; (4) role of auction market; (5) role of UPPB; (6) role of agro-cooperative; (7) role of farmers groups; and (8) regulation completeness. The result of the *Multi-Dimensional Scaling* (MDS) analysis in the economy dimension can be seen in Figure 11.

The MDS analysis of the institutional dimension attributes shows an index value of 20.73 which is classified as *bad* (*unsustainable*). The role of each attribute in the institutional dimension was then analyzed by using the *leverage* analysis to find out which of the attributes are sensitive and contribute to the sustainability of the ecology dimension. The result of the *leverage* analysis was gained from the *Root Mean Square* (RMS) value in each attribute. The result of the *leverage* analysis in the institutional dimension can be seen in Figure 12.

The determination of the sensitive attributes influencing the sustainability of the institutional dimension uses a combination of the *leverage* analysis and the Pareto analysis. The Pareto analysis was done by sorting the RMS value of the *leverage* analysis result from the highest to the lowest value and then weighting was done in percentage and it was accumulated until the maximum cumulative value limit of 75%. The percentage of the RMS value to determine the sensitive attributes in institutional dimension can be seen in Table 5.

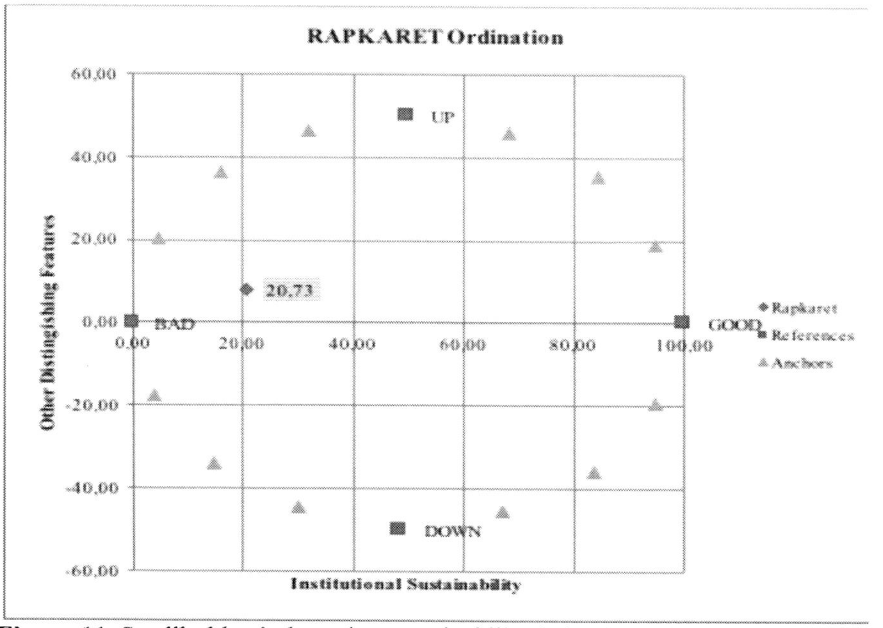

Figure 11. Smallholders' plantation sustainability index in institutional dimension

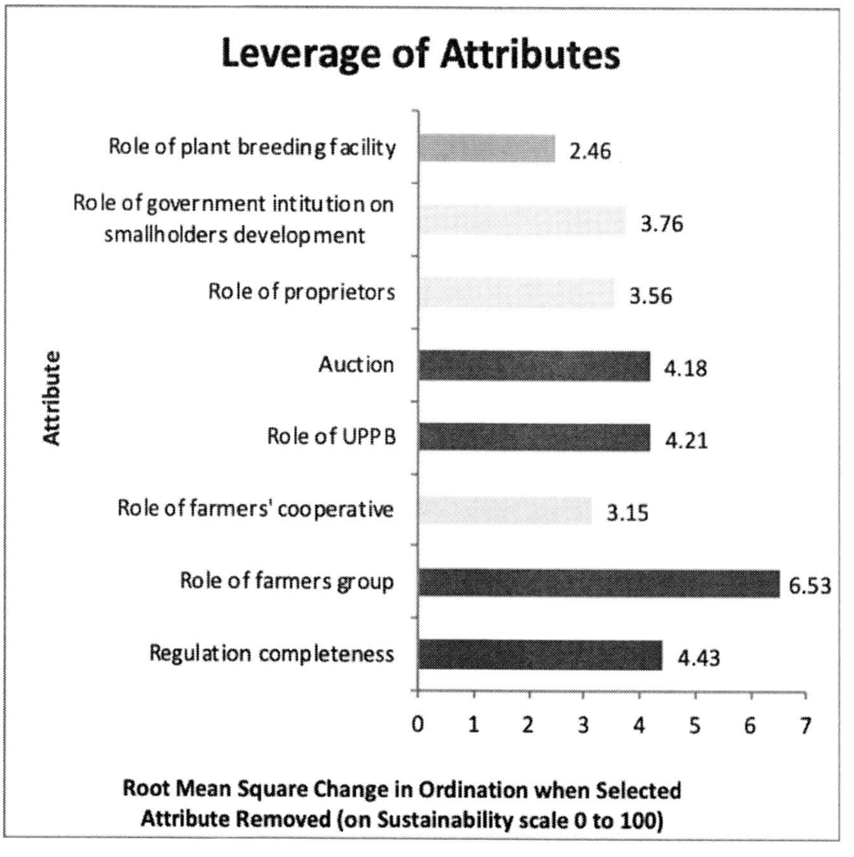

Figure 12. Role of each attribute influencing smallholders plantation sustainability in institutional dimension

Table 5. Percentage of RMS Value To Determine Sensitive Attributes In Institutional Dimension

No	Institutional Attributes	RMS Value	Percentage
1	Role of farmer group	6.53	20.23
2	Regulation completeness	4.43	13.72
3	Role of UPPB	4.21	13.04
4	Auction	4.18	12.95
5	Role of government instritutions on smallholders' development	3.76	11.65
	Total		**71.59**
6	Role of proprietors	3.56	11.03
7	Role of farmers cooperative	3.15	9.76
8	Role of plant breeding facility	2.46	7.62
	Total	**32.28**	**100.00**

In the institutional dimension five sensitive attributes were found influencing the smallholders' plantation sustainability. They were taken from the RMS total percentage cumulative which is 71.59%. if another attribute is added, it will be over limit (82,63%), which is over the maximum cumulative value limit of 75%. Those five attributes are (1) role of farmer group; (2) completeness of regulations; (3) role of UPPB; (4) auction/bid market, and (5) role of government institution in smallholder development. These attributes were considered for the next step to create the model.

4.1.6. Level of Smallholders Plantations Sustainability. The partial analysis of the smallholders' plantation sustainability for each dimension where five dimensions have sustainability value in between 52.95 and 60.35 except for the institutional dimension which is only 20.73. Meanwhile the average stress value is 15% and the Determination Coefficient value (R^2) can be seen in Table 6.

Table 6. Sustainability value, pressure value, and determination coefficient value (R^2)

Dimension	**Sustainability Value (%)**	**Pressure Value (%)**	**Determination Coefficient Value (R^2)**
Ecology	52.95	13.73	94.60
Technology	62.35	13.10	95.47
Economy	60.18	13.64	95.30
Social	53.62	15.40	94.55
Institutional	20.73	13.75	95.25
Smallholders' rubber	53.73	12.90	95.77

Source: MDS Process

Table 6 shows that the sustainability index of the smallholders' rubber plantations in multi dimensions is categorized as **pretty sustainable** with a value of 53.73 or> 50%. The pressure value on the model is just 12.90% or <20%. Meanwhile the determination coefficient value (R^2) reaches 95.77% which means that only 4.23% that can not be explained by the model. As a whole, the result of the MDS analysis on five dimensions of smallholders' rubber plantations can be seen in Figure 13.

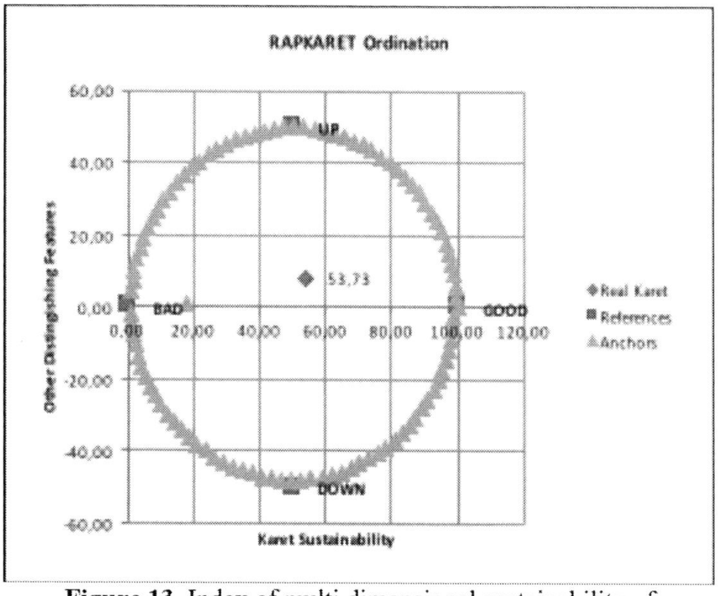

Figure 13. Index of multi dimensional sustainability of smallholders' rubber plantations

Even though the sustainability index of the smallholders' rubber plantations is categorized as *pretty sustainable*, by taking the five dimensions into account, the sustainability value of the institution dimension is considered *less sustainable* as seen in the *kite diagram* in Figure 14.

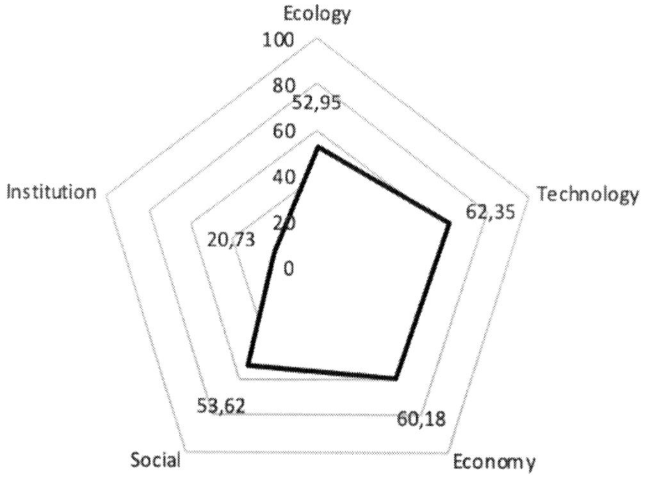

Figure 14. Five-Dimension Analysis of Smallholders' Rubber Sustainability

4.2. *Discussion*

4.2.1. Key Attributes of Sustainability of Smallholders Rubber Plantations in Riau. The key attributes that influence the sustainability of smallholders' plantations in Riau are considered from the development of the farmers based on the *leverage* analysis as seen in Table 7.

Table 7. Recap of key attributes influencing sustainability of smallholder rubber plantation in Riau

No	Dimension	Attributes	*Leverage*	*MDS*
1	Ecology	Rainfalls (HH	10.09	52.95
		Land size (LLK)	10.04	
		Tapping frequency (JHS)	7.43	
		Land type (JL)	5.10	
2	Technology	Knowledge ofbahan pembeku (PP)	3.45	62.35
		Knowledge ofclean smallholder's lump (PBB)	3.13	
		Knowledge ofspacing/planting time (PJT)	3.01	
		Behavior of farmers regarding clean smallholder's lump (PPBB)	2.50	
		Knowledge of PRODUCTIVE PLANTS maintenance (PPTM)	2.30	
		Knowledge oftapping techniques (PTP)	2.22	
3	Economy	Access to clean water (AAB)	3.70	60.18
		Accessto transportation (AT)	3.46	
		Debt (Ht)	2.81	
		Obstacles in Marketing Rubber (HPK)	2.06	
		Perception of food kecukupan pangan (PKP)	2.05	
		Access to rubber price information (AIHK)	1.99	
4	Social	Level of resource conflict (TKS)	3.43	53.62
		Level of labor recruitment (PTK)	0.70	
		Local wisdom and knowledge (KPL)	0.69	
		Economic facilities (FE)	0.67	
5	Institutional	Role of farmers groups (PKT)	6.53	20.73
		Regulation completeness (KA)	4.43	
		Role of UPPB (PUPPB)	4.21	
		Auction market (PL)	4.18	
		Role of government toward rubber farmers (PIP)	3.76	

There are 25 attributes identified as influential attributes on the ecology, technology, economy, social, and institutional dimension and those 25 attributes clearly affect the sustainability of the smallholders' rubber plantation in Riau. The next procedure was to determine the key attributes out of those 25 attributes which will be put into consideration in deciding the steps to create a development model for for farmers to achieve sustainable rubber business in Riau. The perpective analysis method was used to determine the key attributes.

4.2.2. Perspective Analysis. The Prospective Analysis (AP) was then conducted which was aimed to determine the positioning of the driving attributes so that the key attributes of the *driving variables* in carrying out the development of smallholders in the framework of the sustainable rubber plantation management could be found. The perspective analysis *output* was gained in the form of four quadrants which are the positioning of the driving attributes as seen in Figure 15.

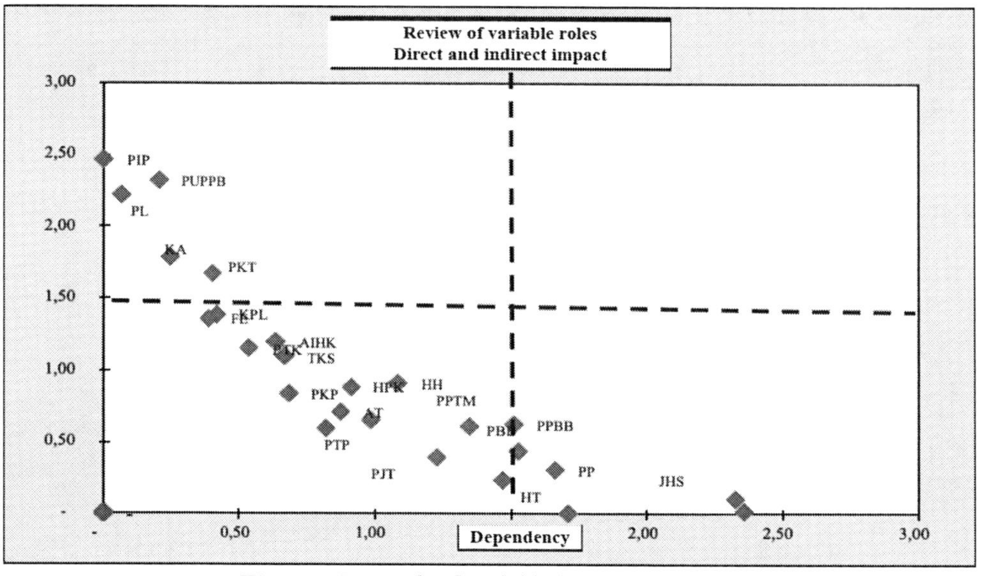

Figure 15. Graph of variable influence and dependency

Based on the participative prospective analysis, there is an indication that the five factors below determine the sustainability of the smallholder development and their correct order can be seen in Figure 15.

- Improvement of the role of the government
- Improvement of the role of UPPB
- Improvement of the role of Rubber Auction Market
- Improvement of regulation completeness
- Development of agricultural groups

5. Conclusions

From the analysis result and discussion of the previous parts, it can be concluded that:

- Based on the Multidimensional Scaling (MDS) method as a whole the smallholders' plantations are categorized as **pretty sustainable** with an index value of 53.73. Of the five dimensions, the institutional dimension is the only unsustainable dimension with an index value of 20.73 and the technology dimension is the most sustainable one. Therefore, to create sustainable smallholders' plantations good institutional management is a must.
- Based on the *leverage* analysis, there are twenty five driving attributes of the fifty attributes that influence the sustainability of the smallholders plantations, and based on the participative prospective analysis, of the twenty five driving attributes there are five which are categorized as *driving variable*s i.e. improvement of the role of the government institution, improvement of UPPB, improvement of the rubber bid/auction market, improvement of the completeness of the regulations and development of the farmer groups.

Acknowledgements

This research is part of the writer's dissertation in the Environmental Sciences PhD Program at University of Riau and guided by Prof. Dr. Aslim Rasyad, MSc, as the promotor, by Prof. Dr. Zulkarnain, MM, as the co-promotor, and by Dr. Suwondo as the co-promotor. I am especially indebted to them for their valuable assistance for me to able to complete this research.

References

[1] Komisi Brundtland 1987 *Our Common Future* (Inggris : Oxford University Press) pp 383

[2] Damanik S 2012 Pengembangan Karet (*Hevea brasiliensis*) Berkelanjutan di Indonesia *J Perpektif* **11**(1): 91-102

[3] Goswami S N, Challa O 2007 Economic Analysis of Smallholder Rubber Plantation in West Garo Hills District of Megahalay *Indian Journal of Agricultural Economics* **62**(4): 649

[4] Septianita 2009 Faktor-Faktor yang Mempengaruhi Petani karet Melakukan Peremajaan Karet di Kabupaten Ogan Komering Ulu *Jurnal Agrobisnis* **1**(1): 130-136

[5] Syahza A 2014 Strategi Percepatan Pembangunan Ekonomi Melalui Penataan Kelembagaan dan Indutri Karet Alam (Pekanbaru Perprinas MP3EI 2011-2015)

[6] Myria A 2002 Kajian Strategi Pengembangan Perkebunan Karet Rakyat sebagai Komoditi Unggulan Kasus Kabupaten Kapuas Provinsi Kalimantan Tengah *Thesis* (Bogor: Program Pascasarjana Institut Pertanian Bogor)

[7] Liu W, Hu H, Ma Y and Li H 2006 Environmental and Socioeconomic Impacts of Increasing Rubber Plantations in Menglun Township, Southwest China *Mountain Research and Development* **26**(3): 245-253

[8] Haryoto B S 2008 Kebijakan Pemerintah Daerah untuk Pemberdayaan Petani karet Rakyat kasus Kecamatan Pangean, Kabupaten Singingi Provinsi Riau *Tesis* (Malang: Program Pascasarjana Universitas Brawijaya)

[9] Sadikin, Idan R I 2010 *Dampak Pembangunan Perkebunan Karet rakyat terhadap Kehidupan Petani di Riau* (Bogor: Pusat Penelitian dan Pengembangan Sosial Ekonomi Pertanian)

[10] Nurhamlin 2012 Kesejahteraan Petani Karet terhadap Pendapatan Petani Karet di Desa Padang Muntung Kecamatan Kampar *Prosiding* Seminar Internasional Lingkungan

[11] Husinsyah 2006 Kontribusi Pendapatan Petani karet terhadap Pendapatan Petani *Jurnal Sosial Ekonom* **3**(1): 9-20

[12] Kurniawan A 2012 Analisis Pendapatan Petani Karet Lateks di Desa Pangkal Baru Kecamatan Tempunak Kabupaten Sintang Provinsi Kalimantan Barat

[13] Hemanto 2005 Pendapatan Usaha Tani Ubi Kayu dan Efisiensi Pemasaran http://repository.usu.ac.id

[14] Gakpindo Riau 2015 *Hasil Kongres Gapkindo ke XVII Tahun 2015* (Jakarta)

[15] Pitcher T J 1999 Rapfish a Rapid Apprasial Technique for Fisheries and its Application to the Code of Conduct for Responsible Fisheries *FAO Fisheries Circular* No.FIRM/C: No 947 : 47

[16] Liling Y, Daud M, Baharuddin N and Betta P 2016 Analysis of Private Forest Resources Management in Sociocultural and Institutional Aspects a Case Study of Tana Toraja and North Toraja Regencies Indonesia *International Journal of Science Basic and applied Research (IJSBAR)* p 349-358

[17] Tesfamichael D and T J Pitcher 2006 Multidisciplinary Evaluation of the Sustainability of Red Sea Fisheries Using Rapfish *Vancouver Fisheries Research* **78** 227-235

[18] Hardjomidjojo H 2002 *Metodeanalisis Prospektif* (Bogor: Jurusan Teknologi Industri Pertanian Fakultas Teknologi Pertanian IPB)

[19] Wibowo Y 2010 Analisis Prospektif Strategi Pengembangan Daya Saing Perusahaan Daerah Perkebunan *AGROINTEK* **4**(2) : 104-113

[20] Yusuf M, Achmad F, Cecep K and Mukhlis K 2016 Nalisis Faktor Penentu dalam Pengelolaan Berkelanjutan Estuaria Das Tallo (Driven factors Analysis on Sustainable Management of Tallo Watershed Estuaries) *Analisis Kebijakan* **13**(1): 41-51

[21] Kusbianto I W, Santun R P, Machfud L F, Poemomosidhi P and Mohammad Y 2013 Analisis Keberlanjutan Pengembangan Prasarana Transportasi Perkotaan di Metropolitan Mamminassata Provinsi Sulawesi Srlatan *Jurnal Jalan Jembatan* **3**(1): 1-15

Conflict among stakeholders in karst area management of Pati Regency

Hartuti Purnaweni[1,2,*], Kismartini[1,2], Arif Budy Pratama[3] and Catur Wulandari[3]

[1]Public Administration Department, Faculty of Social and Political Sciences, Universitas Diponegoro, Indonesia
[2]Master and Doctoral Program of Environmental Science, School of Post Graduate, Universitas Diponegoro, Indonesia
[3]Public Administration Department, Faculty of Social and Political Sciences, Universitas Tidar, Indonesia

*hartutipurnaweni@gmail.com

Abstract. Environmental emntal problem leading to conflict among stakeholders occurred in Pati Regency, Central Java Province, when local community of *Samin* people and mining company were involved in dispute concerning the situation in the North Kendeng karst mountain area. The mountain has been known for limestone richness as a result of natural processes through geological space and time. This research applied a descriptive-qualitative method, focused on the aspect of conflict between local people and PT. Semen Gresik, Tbk., and between the local people and Pati Regency Government. Research data were collected by observation, in-depth interviews, and analyses of secondary sources. The qualitative data were analysed by a descriptive-qualitative approach, more familiarly known as qualitative-descriptive analysis. There is conflict of interest between those agree and disagree with the karst exploitation for cement factory, which has been them known nationally. Opposition strategy by the local people around the karst area in Pati Regency by opposing the cement industry cooptation was similar to the strategy applied in other conflicts. The "shifting strategy" of the cement conflict in Pati is difficult to vanish because it has been strongly rooted in *Saminism*. Therefore, better mediation must be well practiced.

1. Introduction

Environmental problem leading to conflict among stakeholders occurred in Pati Regency, Central Java Province, when local community and mining company were involved in dispute concerning the situation in the North Kendeng Mountain. The mountain has been known for limestone richness as a result of natural processes through geological space and time. The limestone is potentially utilized for cement and other uses. The product of earth dynamics that continues to occur from the ancient era to the present day have resulted in unique natural phenomenon. North Kendeng Mountain is definitely the major site on which the local people have been depending on, because it is an important area of ground water storage for potable water and rice field irrigation [1,2]. Therefore, strong social problem arises as the karst landscape becomes a potential target of cement producers using the karst as the raw material when Central Java's government wants to make the province as an investment-friendly province [3]. Tension becomes apparent when the natural resources are wrongly exploited or when the government forgets to take into account the environmental impacts of the investment projects [4, 3].

Tension between local people and cement company has been taking place since 2006. PT Semen Gresik, Tbk., the cement producer, was attempting to exploit the karst area in North Kendeng Mountain, which are in Sukolilo and Kayen Districts, Pati Regency [5,3]. The company's mining plan caused a conflict, which threatened the tranquillity among the local inhabitants because of intimidation, black campaign towards key figures, and criminalisation against the local farmers.

The conflict began with irresponsible act of suspected individual and *makelar tanah* (land broker) by making effort alledgedly on behalf of the company's interest. It was eventually found that some did it for their own sake, placing the cement company as a scapegoat. Another factor of the conflict was the one sided policy of the government, which favored the corporate. Such policy was definitely against the will of the grassroot actors. The intervening factor that causes a social movement is government policy, which stands on the opposite direction of popular interest [6] adds that conflict occurs because the government issue policies that do not prioritise the popular interest.

2. Literature Review

Conflict in a general meaning is a perceived divergence of interest, a belief that aspirations of the conflicting parties lead to a deadlock. This definition deals with either micro- or macro-space, in which a conflict always relates to actors with different interests. The fundamental issues of a conflict that become the conflict sources include difference in aspiration and perception, which develop as the difference in interest. The interest entails what an individual really wants, which develops as such that it conflicts with other's interest. Such feeling tends to be the centered in the mind and action of an individual, creating the core of attitudes, objectives, and intentions [7].

Jurgen Habermas contends that conflict is an inherent object within a community system. Power, or authority, closely relates to social system, with its dominance nature and becomes something to compete. An authority structure group is capable of making and directing policies to dominate, where the interest in dominating and conquering particular class structures below it occur, creating misunderstanding of instrumental communication [8].

The above concept of conflict reveals that critical conflict opposes functional conflict, which strengthens dominant authority as the center for social system. The concept of domination comprises the following conditions: (a) unequal authority/power, (b) dependence, and (c) asence or disobedience. According to Althusser, domination forms can be divided into RSA (Repressive State Apparatus) and ISA (Ideological State Apparatus).

The conflict that occurred in Pati Regency concerning the emergence cement company affected local people and government. This conflict among stakeholders began in 2006 and still continues up to now. The mining conflict in Pati Regency was coincident with other areas, where aspirations and complaints converted into poorly accumulated perceptions [9]. Accordingly, values of mutual agreement in terms of concensus did not exist as the conflict grew into an open encounter, even physically.

3. Method

This research applied a descriptive-qualitative method using a subject of Policy on Environmental Management in Central Java. It focused on the following aspects: (1) Causes of conflict among stakeholders concerning on North Kendeng karst area, in Pati Regency, Central Java Province; (2) Conflict and social opposition over the karst area; and (3) Government response to the conflict.

Research data were collected by observation, in-depth interviews, and analyses of secondary sources. The qualitative data were analysed by a descriptive-qualitative approach, more familiarly known as qualitative-descriptive analysis [10]. Figure 1 illustrates the research scheme:

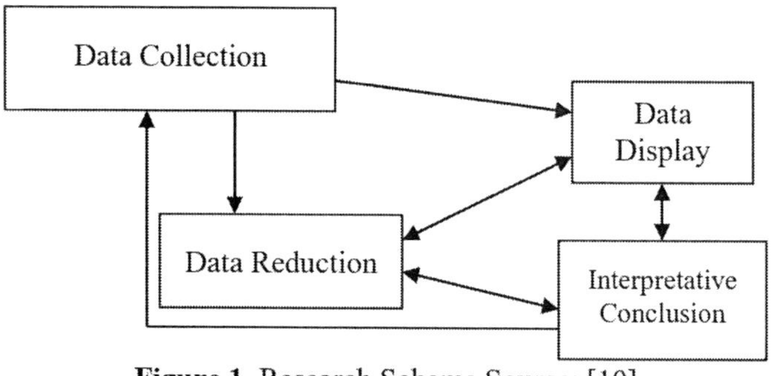

Figure 1. Research Scheme Source: [10]

4. Result and Discussion

4.1. Causes of conflict leading to social oppositions

Conflict that occurred between stakeholders (local people, mining company, and government) can be described into the following themes:

4.1.1. Long-standing conflict. Mining conflict that involved local people or local communities, miners, mining company, and the government has been a long-standing problem in some parts of Indonesia, including in Pati Regency [11,12,13,14,15]. The miners are referred to as big, legal investors or small, illegal mining. The conflict between the local people and the cement company in Pati began when lands, as stated by the local people, unfairly traded by the corporation. This conflict continued by the growing actors of interest. Since then, the local people sufferred from many bad experiences due to unfriendly deeds of irresponsible group of individuals who acted as if they were on behalf of the mining company. These misconduct behaviours sacrificed the popular interest and marginalised their rights and existence [16].

4.1.2. Land conflict. The termination of land property added the conflict, in which one might found two important cases. *Firstly,* the mining corporation annexed the land had been taking place long before PT. Semen Gresik came on the stage. The cases were still in dispute with no sign of problem-solving. Farmers' land possession became diminished due to the road extension project to the mining site. The miners tricked the farmers by not informing them about the site extension and land exploitation. The local people appealed such case to the village government, but the government did not give any meaningful decision. *Secondly,* land trade was full of fraud, in which the trading activity was dominated by land brokers. The brokers take benefit from the trading process. They came in different levels, e.g. 1st-tier, 2nd-tier, and 3rd-tier brokers, causing the loss of profit at the expense of the land owners.

4.1.3. Emergence of PT. Semen Gresik Tbk., a new conflict. The emergence of a large scale cement industry in the shape of PT. Semen Gresik Tbk. had given birth to negative perceptions of most local people with past experience of injustice concerning the mining business. A new opportunity had opened for the locals to express their unsatisfactory attitudes towards the miners. Similar to the past experience, PT. Semen Gresik Tbk. also posed threats to the local people. The company, however, also became the prolonging opportunity to irresponsible individuals who used to take benefits at the expense of the local people, such as land trade and brokerage. In other words, those individuals irresponsibly acted fraudulent manners as the brokers towards the land owners, but with the interest of PT. Semen Gresik Tbk. (it was possible, however, that some individuals in the company also conspired in the fraud). According to the local people, the new conflict with PT. Semen Gresik Tbk., was caused by agents bringing new culture with them without having capacity to become an agent. Therefore, anything that happened during the process often violated the norms and code of conduct.

4.1.4. Change in policy. Since having had a license to develop facilities and infrastructures, the cement company had made use of land owned by the Ministry of Forestry under the terms of land function or other reasons, which was claimed to be a result of mutual understanding between the company and the authority. The authority encouraged economic policies without considering the genuine needs of the localities, favoring the company, i.e. PT. Semen Gresik Tbk., at the expense of the local people. Deals between the government and the company were not disclosed to the other stakeholder, i.e. the affected people. Meanwhile, the local people were too busy (and even too satified) with good political promises given by the authorities during electoral process. The conflict with PT. Semen Gresik, Tbk. entailed problems surrounding intention, interest, and necessity that must be in an equal proportion. What people want and need must be put into the priority because they will take the immediate impact. In other words, the company must fulfill the people's interest. The policy on the establishment of the cement industry on the hundreds of hectares of land owned by the Ministry of Forestry has brought about significant effects on the agricultural life of the local people who are mostly farmers. Therefore, such situation lead the local people to make movements [17], both by internal or external actors, which leads to social opposition. The local people who are worried about the environmental impact of limestone mining, encouraged with the emergence of democracy values. Besides, as farmers for generations since their ancestors, the industrial concept is in contradictory with their present daily life. There are also external factors such as support and comments from various religious leaders and scholars from various background, make the situation more complex. Table 1 illustrates a brief description of the ongoing conflicts in question and their causes.

Table 1. Causes of Mining Conflict

No	Cause of conflict	Description
1	Past conflict	The conflict had existed long before the new cement company had been established, beginning with the dispossession of land property. The conflict occurred between the actors of interest and the local people. The conflict of land property dispossession included land trade and trade fraud. Furthermore, village chief election contributed to extend the conflict even worse due to misconducted behaviors and intervention of externalities.
2	New conflict with cement company	The conflict began with violation against humanity, emergence of critical community members, corporate-bound government policies by allowing land exploration, especially those owned by the Ministry of Forestry, where the local people made their livings.
3	Other causes of conflict that triggered social opposition	The conflict that triggered the social opposition continued to grow as time went by, i.e. change in policy, environmental impact, human rights, and democracy, as well as unpreparedness of the local people to accept new cultures, difference in perception among religious leaders and scholars.

4.2. Conflict and Social Opposition

Conflict among the stakeholders (PT. Semen Gresik Tbk., the government, and the local people) in the karst area management triggered a social opposition in Pati Regency against PT. Semen Gresik Tbk., and the government.

4.2.1. Institutional Strategy. Institutional strategy explains from the start about a process of organisational establishment that functions as an aspiration medium for the people, and explains the other organisations/institutions that also support the social opposition against cement industry in Pati (i.e., local people *vs.* PT. Semen Gresik, Tbk.). Institutions are established coexisting with community groups in delivering their identities, such as what have been performed by local community in several districts in Pati Regency, i.e. Sukolilo, Tambakromo, and Kayen. They established unions to develop as a social movement [17]. Institutional strategy in the local term in Pati Regency was called *seduluran*, a term derived from the Samin people, which means kinship without social class and interest. It has purpose of creating the mutual feeling, construct, and value. *Seduluran* is similar to kinship or friendship, which values highly the members, consistently and committed to togetherness. *Para dulur* is equal to *dulur*, a primordial nature which is take for granted. The fundamental objectives of the establishment of an organisation in the cement conflict in Pati were opposing any mining activity that exploits the karst area at small or large scales, preserving and maintaining nature power over the mining efforts, recovering the loss of the rights on land, and struggling in justice and humanity in an agricultural economy. Local people in three districts of Pati Regency who stood against the establishment of cement industry by exploiting the karst area joined together in an institutional network namely Community Care Network for Kendeng Mountain (*Jaringan Masyarakat Peduli Pegunungan Kendeng,* JMPPK) of Pati Regency. The JMPPK collaborated with JMPPK Blora, Rembang, and Grobogan. The successful effort of the local people to oppose the cement industry in Pati was a result of action contribution by a local community group namely *Sedulur Sikep* which are also known as *Samin* community. Members of this group were proactive during the protests and demonstrations by coordinating one to another in Rembang, Pati, Blora, and Grobogan regencies. They worked in concert towards mutual objective. The involvement of community leaders from *Sedulur Sikep* in the social movements under the JMPPK was proven in some events, such as casting legs with concrete mold, long march from Pati to the Governor Office of Central Java in Semarang, and other events.

4.2.2. External Support Strategy. The role of external institutions in helping sustain the social movement opposing the karst area exploitation in Pati by the establishment of the cement industry had a signficant effect. It was due to the fact that all had the similar ideas and principles taken for granted. The role of external institutions was proven in practice by strengthening material and immaterial aspects. The externalities helped sustain the social movements, which often found the deadlock if they were performed by solely internal actors. The establishment of PT. Semen Gresik, Tbk. in three districts of Pati Regency were opposed by community groups (internally or externally), mass media, scholars, religious leaders, community leaders, environmentalists and environmental activists, and NGOs, such as Walhi (Indonesian Environmental Association) or JATAM (Mining Advocacy Network). Equally important in the conflict among stakeholders in the karst area management was the rise of social vulnerability due to women participation in the social opposition movements against the cement company and the government.

4.3. Government response to the conflict

Governments, either at regency, provincial, or national levels, became the most prominent actors during the conflict in cement industry in Pati Regency. They proved inconsistent in the decision-making process and implementation.

Difference in perception between regency and provincial governments encouraged the national government to analyse and examine the cement conflict in Pati, Central Java. The conflict extends and to Rembang Regency, and get stronger when another cement investor wanted to invest in Rembang Regency, next to Pati. The unrest forced the President to establish a special team namely Strategic Environmental Assessment (SEA) or *Kajian Lingkungan Hidup Strategis* (KLHS). This assessment was important to put any problem under the dispute on the right track, since the aim of SEA is sustainable development [18].

The KLHS eventually decided the future status of the mining company to be established in Kendeng Mountain area, including in Pati. The first assessment had been resulted in a discussion that mining industry would disturb the karst sustainablity in Watuputih Mountain. This assessment should play an important role in determining the action taken by the President in responding to the future development of the cement industry in Pati, Central Java, in particular, and Java Island, in general. This is due to the fact that the ecological condition of the island has been more vulnerable.

5. Conclusion

The conflict concerning the cement industry in Pati is closely related to the conflict of the karst management in North Kendeng Mountain between the local people and PT. Semen Gresik Tbk., who intended to establish the mining site in Kendeng area. The local people resistance prevented this intention and pushed PT. Semen Gresik, Tbk., out of Pati region.

The cement conflict in Pati is a build up conflict beginning with old instruments, i.e. small scale mining businesses. As complaints accumulated, it eventually crystalised and culminated in the terms of struggle, opposition, and resistance towards the idea of the current mining site planning by big cement business.

Opposition strategy by the local people around the karst area in Pati Regency by opposing the cement industry cooptation was similar to the strategy applied in other conflicts. The "shifting strategy" of the cement conflict in Pati is difficult to vanish because it has been strongly rooted in *Saminism*, that witholds a principle of "*resistance to the death*". Therefore, better mediation must be well practiced.

6. Acknowledgments

The authors highly appreciate the research fund from the Indonesian Ministry of Research and Higher Education (*Kemenristekdikti*) under PKPT (University Cooperation Program) scheme for Universitas Tidar and Universitas Diponegoro, in 2017-2018.

References

[1] Hadi S P, Purnaweni H, Kismartini 2018 Inequality in Public Decision Making Process: Case Studies of Cement Factories in Central Java, Indonesia, *E3S Web of Conference* **73**

[2] Mojo E, Hadi SP, Purnaweni H, 2015 The Role of Karst as Water Supplier for Community Case Study: Sukolilo Karst Pati, Central Java, Indonesia, *Int. J of Applied Envir Sci*, **10** (5): 1793-1801

[3] Mojo E, Hadi SP, Purnaweni H, 2017 Sedulur Sikep's Environmental Wisdom Rejected the Plan of Cement Industries in Karst Kendeng Mountains Sukolilo, Pati, Central Java, *Advanced Science Letters* **23** (3): 2504-2506

[4] Awang S A 2005 *Negara, Masyarakat dan Deforestasi, Kontruksi Sosial atas Pengetahuan dan Perlawanan Petani terhadap Kebijakan Pemerintah, Unpublished Dissertation* (Yogyakarta: Universitas Gadjah Mada (UGM))

[5] Solihin M, Kurnia N 2017 Pemaknaan Konflik Pabrik Semen Kendeng dalam *Framing* Media Berita Online *kompas.com* dan *suara merdeka.com*, *J Ilmu Komunikasi* **15**(1): 15-27

[6] Ngadisah 2003 *Konflik Pembangunan dan Gerakan Sosial Politik di Papua* (Yogyakarta: Pustaka Pelajar)

[7] Raven and Rubin 1983 *Protest: Studies for Collective Behaviour and Social Movements* (New

Brunswick: Transaction Book)
[8] Dahrendrof 1986 *Konflik-Konflik dalam Masyarakat Industri, Sebuah Analisa Kritik, (translated by Alimandan)* (Jakarta: CV Rajawali)
[9] Gurr T R 1970 *Why Men Rebel* (Princestone NJ: Princeston UP)
[10] Miles M B, Huberman M 2016 *Qualitative Data Analysis (ed 3)* (California: SAGE)
[11] Manalu D 2006 Gerakan Sosial dan Kebijakan Publik: Studi terhadap Gerakan Perlawanan Masyarakat dalam Mengubah Kebijakan Pemerintah Mengenai PT. Inti Indorayon Utama, di Porsea Sumatera Utara *Published Theses* (Yogyakarta: UGM)
[12] Yurida T D 2006 Konflik antara Masyarakat Lokal dengan Perusahaan Tambang Asing (Studi Kasus Teluk Buyat dan PT. Newmont Minahasa Raya *Published Theses* (Yogyakarta: UGM)
[13] Zeitlin I M, in Dwicipta and Ardianto H T (ed) 2015 *Dwicipta and Ardianto HT (ed Rembang Melawan; Membongkar Fantasi Pertambangan Semen di Pegunungan Kendeng (*Yogyakarta: Literasi Press)
[14] Wulandari C, Pratama A, Kismartini dan Purnaweni H 2018 Policy conflict and change in Kendeng karst: A Chronological Narrative of Advocacy Coalition within Local People, Presented *at Int Conf of Indonesian Assoc of Public Administration (IAPA)* (Palangkaraya)
[15] Novri S, 2014 *Konflik Pertanahan dalam Kebijakan Pertambangan Pasir Besi di Kabupaten Kulon Progo Daerah Istimewa Yogyakarta. Published Dissertation* (Yogyakarta: UGM)
[16] Sidiq P 2018 Konflik Pertambangan di Kabupaten Rembang. Unpublished Dissertation Semarang: Post Graduate Program (Semarang :Universitas Negeri Semarang)
[17] Carthy J D and Zald M 1977 Resource Mobilization and Social Movements: A Partial Theory. *American J of Sociology* **82** (6): 1212-1241
[18] Brontowijono W, Ribut L, Donan W, 2012 KLHS untuk Pembangunan Daerah yang Berkelanjutan, *J Sains dan Teknologi Lingk* **4**(1): 43-54 ISSN: 2085-1227
[19] Prabawani B, Hadi S P, Purnaweni H, 2018 A Power Distance in the Sustainable Utilization of Kendeng Mountain, *presented at ICISPE 2018* (Semarang: FISIP Universitas Diponegoro)

Determining the public understanding on the public policy based on cosmopoliteness and innovativeness

Rahmatulloh*, **I Made Putrawan and Yufiarti**

[1]Universitas Negeri Jakarta, Indonesia

*rahmat.algharamy@gmail.com

Abstract. The aim of this research is to determine the relationship between Cosmopoliteness, and Innovativeness with the public understanding on public policy in Jakarta, Republik Indonesia. A survey method was applied with a correlational study. The respondents were the heads of families and 120 samples were chosen randomly. There were four instruments which measured the public comprehension (rel..922), Cosmopoliteness (rel. .912), and Innovativeness (rel. .935). Data were analyzed by two-way ANOVA. The result of this research was revealed that there was positive and significant correlation among those variables. Therefore, it is concluded that to increase the public's understanding on public policy, reinforcement in aspects of Cosmopoliteness and Innovativeness need to be improved.

1. Introduction

The aim of the research is to find out how public understand the aspects of cosmopoliteness and innovativeness policy in Jakarta.

The level of public understanding of public policy as a component in the social system is influenced by many environmental factors and depends on the extent to which a person is able to adapt themselves in the face of changes in their environment [1].

There is a space for change through cosmopolitanization and innovativeness in individuals in society to adapt themselves in a changing environment. As cosmopolitan is characterized by the capacity of individuals to open dialogue with diversity, it certainly can have the breadth of social communication. Innovative individuals are characterized by ownership of attitudes to adopt new things to become more advanced than others because of their ability to follow change, accept new ideas, and be critical in understanding public policy.

According to people who have a higher level of openness to experience and learn new things will be more innovative than those who are reluctant to develops their potentials [2].

The process of someone adopting a new thing or innovation involves diffusion activities, which are defined as a process through which new innovation is communicated through a variety of specific channels. State that, the process of transmissing new information to the society can be done through online or virtual communication [3].

In the study of [4] it is found that information and exchange are the core of the public domain for the idea of citizenship and become a factor in the development of cosmopolitanism.

Recent developments in information technology have enabled individuals to receive access to information anywhere as a discursive activity in the public sphere. Because of that, the spread of information, nowadays has formed individuals or modern societies accustomed to information literacy.

The advancement of information and technology that drives public literacy, in the public sphere, is widely used by the state or government institutions, including the government innovations to improve the quality of democracy and efficiency in public services. Affirm that government policy involves massive publicity to create a broad perception, that the higher the public understanding on government policies, the greater the support will be gained [5]. Similarly according to the E.U study, [6] inform that: "Understanding can affect concerns and support for policies, as the struggle to frame the issue indicates".

Similarly, [7] says that the development of information literacy as a policy of countries to advocate the importance of the information is intended to build awareness and create connections. Because of this, [8] in their study provide recommendations on the importance of information literacy for the recognition of knowledge access to the public.

Interprets cosmopolitan as a predisposition that is specifically displayed, namely a set of attitudes, values, behavior, practices to be cosmopolitan, including: (1) cultural openness; (2) global pro-sociality; and (3) respect for diversity [9].

Cosmopoliteness according to [10] can also be applied to interpersonal communication, which is about the study of values, communication behavior and political participation including measures of "discussing diversity" measured on a scale of indicators how often respondents engage in busy discussions where others have many different points of view from they are on three types of topics, race relations, religious beliefs and general (radical) political views.

Distinguishes some of the following Cosmopoliteness aspects: (1) Diversity of Interest: (2) Cosmopolitan Identification; (3) Appreciation of Different Cultures; (4) Tolerance of Different Cultures; (5) Knowledge of Different Cultures; (6) Knowledge of Current Events and International Affairs; (7) Cultural Diversity of media content to which one is expossed; (8) Diversity of Interpersonal Communication Network [10].

Furthermore regarding to innovativeness, [11] states that: "Innovative is the degree to which individuals or other units of adoption are relatively earlier in adopting new ideas than the other members of a system". In his study, innovativeness is defined as the level of individual adoption of new ideas from other members in a social system. It relates to their understanding on public policies according to several indicators, including: (1) the level of activity to follow changes, (2) the acceptence toward new ideas to be implemented, (3) criticism toward the existing system, and (4) the courage to accept risks on new ideas, strategies or models.

Regarding to understanding, [12] argues that, "a comprehension is defined as "Here we are using the term comprehension" to include those objectives, behaviors, or responses which represent an understanding of the literary message contained in a communication". Activities or behaviors measured in comprehension are translation, interpretation, and extrapolation.

According to [13]: "Public policy also may be viewed as whatever governments choose to do or not to do". Similarly [14] states that public policy is: "whatever governments choose to do or not to do". The government acts can have just a great impact on society" [14].

Thus, the understanding of citizens as individuals towards public policy is related to their literacy ability in responding to every information spread. The globalization factor with the advancement of information and technology also spur citizens to accept things that are foreign or from outside enabling the occurrence of cosmopolitanization. In addition, global changes also relate to a number of individuals to receive new experiences that are considered better resulted in allowing individuals to adopt innovation amid the current changes.

The various thought literature and findings above, regarding the cosmopoliteness variable, and innovativeness at different individual levels it seems interesting to study, especially for urban communities in Jakarta (Indonesia), related to the level of literacy or understanding of existing public policies.

Thus, each of the variables, empirically can be further investigated together, namely the extent to which the variables are related to other variables as a new contribution in this study.

2. Methods

The aim of this research is to find out whether there is a relationship between Cosmopoliteness, and Innovativeness with the Community's understanding on public policy in Jakarta (Indonesia). This research was conducted in July 2016 until April 2017. This research applied correlational approach.The population of this study was the residents who lived in Jakarta region. The respondents were 120 heads of families.

In this research, there are 4 (four) variables involved: firstly, measuring the public's understanding on public policy as something that is done or not done by the government in three dimensions, namely translation, interpretation, and extrapolation (conclusion or calculation) with a reliability coefficient of 0.9225.

Secondly, measuring Cosmopoliteness with six indicators, including respecting differences in interests; identifying yourself in a broad association; appreciating cultural differences; having knowledge or insight into contemporary life and international relations; familiarizing yourself with the diversity of cultural media content; and building a different interpersonal communication network, through an instrument with a reliability coefficient of 0.912.

Thirdly, measuring innovativeness with four indicators, namely following changes; receiving new ideas to implement; criticizing the existing systems; and accepting the risk of a new idea, strategy or model that becomes the choice or decision through an instrument with a reliability coefficient of 0.935. To analyze the data collected, descriptive and inferential data analysis techniques were used.

3. Results and Discussion

3.1. Results

This research found that, firstly; there is a relationship between Cosmopoliteness (X_1) and the public understanding on public policy (Y). In accordance with the calculation of the significance and linearity of the Community Understanding on Public Policy (Y) of Cosmopoliteness (X_1), the regression equation is $\hat{Y}$= 0.244 + 4.900. From the analysis of variance (ANOVA), it is found that the score of the regression model is 31,489. Meanwhile at the real level α is 0.05, F_{table} is 3.89. Because F_{cal} is 31,489 > F_{table} is 3,89, the regression coefficient is significant. The result of the linearity test shows that F score is 0.167. Meanwhile at the real level, α is 0.05 and F_{table} is 1.60. Because F_{cal} is 0.167 < F_{table} is 1.60, the regression is linear.

The score of the regression equation model $\hat{Y}$ is 0.244 + 4.900 in which X_1 is linear (0.167 < 1.60). This results suggest that there is a tendency for the score of public understanding on public policy to change into 4.900 at a constant of 0.244.

The correlation coefficient between cosmopoliteness and public understanding on public policy is 0.240. The second order correlation is 0.219. The test results in table I below:

Table 1. Test of coefficient correlation ($RY_{1.2}$)

Sample (n)	Correlation Coeficient (zero)	Second order correlation	T_{cal}	T_{table}	
				0.05	*0.01*
120	0,240	0,219	2,686	1,980	2,617

The result of the t-test shows the value of t_{cal}> t_{table} is 0.05 or 2.686 > 1.980. This means that the correlation coefficient between cosmopoliteness and public understanding on public policy is significant. From these results it can be said that there is a significant positive relationship between cosmopoliteness and public understanding on public policy. This means that the higher the cosmopoliteness, the greater the public understanding on public policy. The contribution of the relationship between cosmopoliteness and public understanding on public policy on the second order correlation is $(0.219)^2$ x 100% or 4.80%.

The results of the partial correlation coefficient describes the relationship between cosmopoliteness and public understanding on public policy, if the innovativeness controlled the result is the coefficient $ry_{1.2} = 0.219$ and $t_{cal} = 2.413 > t_{table} = 1.980$ (0.05).

Secondly, there is a relationship between Innovativeness (X_2) and the public understanding on public policy (Y). The result of the significance and linearity of the public understanding on public policy (Y) on innovativeness (X_2) shows that the regression equation $\hat{Y}$ is 0.275 + 8.472. The result of variance (ANOVA) analysis on this model is 34.287. Meanwhile at the real level, the value of α is 0.05, F_{table} is 3.89. Because F_{cal} is 34.287 > and F_{table} is 3.89, the regression coefficient is significant. Furtehrmore, the result of linearity test shows that F value is 0.800. Meanwhile at the real level, α is 0.05, and F_{table} is 1.60. Because $F_{cal} = 0.800 < F_{table} = 1.60$, the regression is linear.

The regression equation model $\hat{Y} = 0.275 + 8.472\ X_2$ is linear to (0.800 < 1,60).This means that there is a tendency for the score of public understanding on public policy to change into 8.472 at a constant of 0.275.

The correlation coefficient between innovativeness and public understanding of public policy is 0.200. The second order correlation is 0.171. The test results in table II below.

Table 2. Test of coefficient correlation ($RY_{2.1}$)

Sample (n)	Correlation Coeficient (zero)	Second order correlation	T_{cal}	T_{table}	
				0.05	0.01
120	0,200	0,171	2,223	1,980	2,617

The results of the t for $t_{cal} > t_{table}$ is 0.05 or 2.222 > 1.980. This means that the correlation coefficient between innovativeness and public understanding of public policy is significant. From these results it can be said that there is a significant positive relationship between innovativeness and public understanding of public policy. This means that the higher the innovativeness, the greater the public understanding on public policy will be.

The contribution of the relationship between innovativeness and public understanding on public policy on the second order correlation is $(0.171)^2$ x 100% indicating that 2.92% of the variation in public understanding on public policy is determined by innovativeness.

The results of the partial correlation coefficient describes the relationship between innovativeness and public understanding on public policy, if cosmopoliteness controlled by the result, the coefficient of $ry_{2.1}$is 0.171 and $t_{cal} = 1.895 < t_{table}$ is 1.980 (0.05). This result indicates that the variables of innovativeness and public understanding on public policy do not show a significant relationship. However, there is a pure contribution which is the partial correlation coefficient of 0.171.

The third finding is a relationship between Cosmopoliteness (X_1) and Innovativeness (X_2) Together with Community Understanding on Public Policy (Y). The relationship between cosmopoliteness and innovativeness together with people's understanding on public policy result in the equation $\hat{Y} = 5.249 + 0.143\ X_1 + 0.117\ X_2$.The results of the test and multiple regression linearity are presented in table III below.

Table 3. Anova for multiple regression $Y = 5.249 + 0.143X_1 + 0.117X_2$

Varians Source	Dk	JK	RJK	F_{cal}	F_{table}	
					0.05	*0.01*
Regression	228.470	2	114.235	5.484	3.07	4.78
Residual	2437.397	117	20.832			
Total	2665.867	119				

Based on the above table III, it can be seen that there is a very significant relationship between cosmopoliteness and innovativeness together with public understanding of public policy ($p < 0.05$).

The plural correlation coefficient between cosmopoliteness and innovativeness together with the public's understanding of public policy ($Ry._{12}$) of 0. 293, is presented in table IV below.

Table 4. Multiple correlation test (RY_{12})

Sample (n)	Correlation Coeficient (zero)	*Second order correlation*	F_{cal}	F_{table} *0.05*	F_{table} *0.01*
120	0.293	0.086	5.484	3.07	120

Based on Table IV, it can be said that the better cosmopoliteness, innovativeness and individual modernity, the better the public comprehension on public policy. The contribution of the relationship between cosmopoliteness and innovativeness together with public understanding on public policy on the second order correlation is $(0.086)^2$ x 100% indicating that 0.47% variation in public understanding on public policy is determined by cosmopoliteness and innovativeness.

The contribution of each variable can be known by controlling other independent variables. The results of the analysis above show the relationship ratings in table V below.

Table 5. Partial correaltion rank

Correlation	Correlation Coeficient	Rank
$r_{y1.2}$	0,219	First
$r_{y2.1}$	0,171	Second

The explanation of table V above is, *first*: the contribution of pure cosmopoliteness to people's understanding if innovativeness is controlled, then the partial correlation coefficient is 0.219. *Second*; contributions are purely innovativeness to people's understanding if cosmopoliteness is controlled, then the partial correlation coefficient is 0.171.

3.2. Discussion

From the findings above, the cosmopolitan itself is related to the public understanding on public policy which is interpreted as a government strategy in dealing with urban problems and population. Through globalization, every time people may receive information from various media and interact with others from out of their local custom. This fact has been emphasized by [11] stating that people who communicate with networks from external sources (cosmopolitan) outside their social systems will learn the new information to expect the development of public policy happens.

Citizens who are literate from outside, either through media coverage and social interaction networks to discuss public affairs issues or about changes that occur in their environment contribute to the level of their understanding of the policy.

It turns out that information literacy obtained by making use of information and communication technology, accompanied by its tolerance for different views, openness to things considered foreign contributes to the shaping of knowledge and enriching its thinking as citizens regarding a number of public policies that have been or are being implemented in Jakarta. This is in line with the study conducted by [8] showing information literacy for the recognition of knowledge access to the public in reducing gaps and the need to be developed in all areas of life is important. Similarly, [7] states that information literacy is important to build awareness and create connections, and as part of human rights which is very important in achieving the success of personal, social, and economic citizens.

Innovativeness has been proven to be related to people's understanding on public policy. Innovativeness is related to behavioral change from acceptance of a thing that has novelty, either in the form of adopting ideas, methods, or habits in the changes faced [2]. Various things that lead to the change there are adoptions of innovation that occur because it is a relatively new thing from the individual's activities in his social life.

As found in individuals who adopt innovation that is marked attitudes or behavior to accept the risk of change in following policies that are in line with relatively new progress. Online application system regarding to administrative practices in services that support the public interest, such as for the fulfillment of aspects of transparency, accountability, and efficiency is an example.

As new innovations in public services, the results of public policy are understood by the community as a breakthrough in progress in accordance with technological developments, also due to serving the interests of city residents who have a busy level of activity, and of course to be more easily accessible.

Through innovativeness there is an increase in people's understanding of public policies that occur in their environment because of their adaptation to the latest developments in urban development. Citizens have the awareness to take the decision to adopt innovation as the demands of a changing era, as well as an opportunity to keep up with the fast and open development progress. The awareness of innovation adoption is also based on the risk of experience in following daily developments in public services in the capital.

There is a relationship between cosmopoliteness and innovativeness together with the public understanding on public policy. Support for progress and experience on innovation practices that occur in various countries, both obtained dialogically, directly and indirectly through information media, encourages citizen initiatives to respond to changes in public policy as well as empathy for the progress of the modern world.

From this research, it is revealed that citizens' understanding on public policy is supported by its cosmopolitanity and ability to accept innovation. Thus, the cosmopolitan experience and ability to adopt a person's innovation relate to his ability to understand the progress of the world of public services in adapting to the progress and demands of the global world.

4. Conclusion

Based on the findings of the study, it can be concluded that in order to improve someone understanding on public policy, it is necessary to pay attention to cosmopoliteness and innovativeness aspects. It means that if cosmopoliteness and innovativeness is increasing, the understanding on public policy in the social environment is also increasing.

Considering that cosmopoliteness and innovativeness bring the lives of individuals to get used to new things, creativity, openness to experiences something come from outside, and critical thinking in adapting the changes that occur regarding to the affairs of the public interest are needed. They might be gained through a number of interaction, experiences or strong use of information received. These then would affect the level of public understanding on public policy in social environment

References

[1] Rambo A T 1983 *Conceptual Approaches to Human Ecology* (Honolulu, Hawaii: Research Report East-West Environment and Policy Institute)

[2] Ali I 2018 Personality traits, individual innovativeness andsatisfaction with life *Journal of Innovation & Knowledge* **4**:38-46

[3] Tolba A H and Maha M 2011 Individual and cultural factors affecting diffusion of innovation *Journal of International Business and Cultural Studies* **5**:1-16

[4] Olsson M R 2014 Information Practices in Contemporary Cosmopolitan Civil Society *Cosmopolitan Civil Societies Journal* **6**(2): 87-88

[5] Hyung P E and Lee J W 2015 A study on policy literacy and public attitudes toward government innovation focusing on Government 3.0 in South Korea *Journal of Open Innovation: Technology, Market, and Complexity* **1**(23): 1-13

[6] Weber E U and Paul C S 2011 Public Understanding of Climate Change in the United States *American Psychological Association* 0003-066X/11/$12.00 **66** (4) 315–328

[7] Bradley C 2013 Information Literacy Policy Development in Canada: Is It Time? Partnership: the Canadian *Journal of Library and Information Practice and Research* **8** (2) :1-28

[8] Pinto, Maria., Francisco-Javier Garcia-Marco, Gloria Ponjuán, and Dora Sales 2017 Information

literacy policies and planning in Ibero-America: Perspectives from an international digital survey *Journal of Librarianship and Information Science* **1**(16) p 12
[9] Leung, Angela K.-Y., Kelly Koh, Kim-Pong Tam 2015 Being environmentally responsible: Cosmopolitan orientation predicts pro-environmental behaviors *Journal of Environmental Psychology* journalhomepage:www.elsevier.com/locate/jep. 2015. P 80-81
[10] Jeffres, Leo W, Cheryl Campanella Bracken 2002 *Cosmopoliteness, Cultivation and Media Uses* (Ohio Departmen of Communication Cleveland State University)
[11] Rogers E M 1983 *Diffusion Of Innovations* (Nw York: The Free Press)
[12] Bloom B S 1956 *Taxonomy of Educational Objectives* (Printed in the USA)
[13] Anderson J E 2003 *Public policy making: An introduction (*Boston: Houghton Mifflin Company)
[14] Dye T R 2013 *Understanding Public Policy*. (14th ed United States of America; Pearson Education)

UNP Extinct Metric for *Bilih* Fish (*Mystacoleucus padangensis Bleeker*)

Abdul Razak[1,3*], Ganefri[2], Eri Barlian[3] and Indang Dewata[3]

[1]Biology Department, FMIPA, Universitas Negeri Padang, Indonesia
[2]Department of Electrical Engineering, Faculty of Engineering, Universitas Negeri Padang, Indonesia
[3]Environmental Science Study Program, Postgraduate Program, Universitas Negeri Padang, Indonesia

*ar710322@gmail.com

Abstract. This paper write about UNP Extinc Metric. It is new approach and methods how to calculate extinc condition for wild animals especially bilih fish from Lake Singkarak. We introduce new criteria and ombine to seven categories from IUCN. For bilih fish we can calculate number extinc 49 and the meaning is endangered We conclude after that result we can conserve bilih fsh soon as soon for our life in biosfer.

1. Introduction

Indonesia is one of the countries rich in freshwater fish species. Indonesia is in 3rd place. As the country with the most freshwater fish species in the world, the total species reaches 1155 species. For the number of endemic freshwater fish species, Indonesia has a total of 440 endemic freshwater fish species, in 4th place, after Brazil (1716 species), China (888) and the United States (593 species). Pressure on the habitat of freshwater fish and sea water fish is a threat and poses a risk of extinction.

High human needs cause habitat damage. Damage to habitats such as lake forests, rivers and coral reef ecosystems causes freshwater fish and marine fish to lose food sources, habitats and roaming spaces to reproduce [1]. Wild fish whose habitat is disturbed causes the population to decline. The decline in population due to disruption of the reproductive process or a place for reproduction is damaged so that it is difficult to breed.

Human needs continue to increase along with the increase in human population. This triggers pressure on the environment, especially fish resources. This occurs in rivers as freshwater ecosystems and in the sea of saltwater ecosystems which results in a decrease in aquatic biota populations including fish. This situation causes some species of fish to become rare and endangered.

Furthermore, in terms of humans, human behavior to meet their needs often leads to conflict and over-exploitation which threatens the sustainability of freshwater fish and marine fish. Excessive exploitation (over fishing) does not only cover aspects of trade, hunting and overfishing, but also triggers extinction of the species. This is also related to the many types of fishing and bycatch where protected dead fish are caught accidentally, for example due to fishing practices that are not environmentally friendly.

The above conditions are experienced by Bilih fish in Singkarak lake, Bada lake Maninjau fish, Napoleon fish that live in coral reefs, Padang flying fish (flyingfox) and Garing fish. The level or level of disturbance to wild animals such as fish are not the same as land animals. This happens because of differences in habitat and animal habit and different human interests.

The method or method of the IUCN is often late without quantitative analysis, which is why there must be a preventive way so that animal extinction can be anticipated as early as possible. The EXTINCT METRIC UNP (EM UNP) offers 10 criteria that will be ensured through research funded by all member states that ratify the IUCN.

This article offers a solution on how to measure and analyze the level of risk of extinction of fish or other wild animals discussed on other occasions. After studying the causes and consequences of endangered fish and endangered fish in Indonesia in general, West Sumatra in particular. In addition, the author has also examined the ecological aspects of the eye and the chemical content of selected fish in 2015 and 2016 and has been published in proceedings [5] and journals indexed by Scopus [6]. The results of the study of literature and research are the rationale for developing tools or tools that can analyze the risk of extinction of wild animals including freshwater fish and marine fish.

2. UNP Extinct Metric Criteria

From the analysis of literature studies and research that has been done, there are several criteria that become indicators that become the basis of UNP EXTINCT METRIC (UNP EM). UNPEM is composed of several criteria that trigger the threat of extinction of freshwater fish and marine fish. The criteria are:

2.1. Over exploitation is symbolized by a factor 1 denoted = F1

This criterion describes fish resources that are exploited with utilization rates that exceed the threshold, captured by fishing gear with many types. On the other hand, economically, like fish, the demand level is high because it is nutritious and tastes good.

2.2. Uninterrupted /Damaged habitat is symbolized by Factor 2 denoted = F2

Habitats that are damaged can be as a result of human behavior or natural causes such as global warming.

2.3. Hard/not yet can be cultivated symbolized by factor 3 denoter =F3

This situation is a natural cause of genetic factors and environmental factors that cause fish to be difficult or not yet able to be cultivated because the key factors for successful cultivation have not been found.

2.4. Production/edndemic hard is symbolized by Factor 4 denoted = F4

This situation is a natural cause of genetic factors

2.5. Habitat has not been protected = F5 is symbolized by Factor 5 denoted = F5

This situation is due to the lack of government and community attention to the natural habitat of wildlife including fish.

2.6. Deforestation = F6.

This situation is due to the lack of attention from the government and the community towards the forest which is a natural habitat for wildlife including fish, as well as the security key of the ecosystem and affecting the ecosystem as a whole

2.7. Climate change = F7.

This situation is a natural cause triggered by rising earth temperatures due to activities and activities of volcanic eruptions such as Mount Sinabung that have an impact on the community and natural habitat of wildlife including fish.

2.8. Pollution = F8.

Damaged habitats can be as a result of human activities such as pollution or pollution.

2.9. Invasive species = F9
This situation can be as a result of human activities that include foreign species that have the effect of invading other species in an ecosystem. A case in point is the glass fish that invaded the Bilih Toba fish.

2.10. Expert opinion = F10
Expert opinion is a strengthening aspect so that the data obtained can be supported and validated by experts.

3. Scoring
Giving a score with a range of 1 to 5. This range is made so that the score range is stable or equal 2. Each factor starting from factors 1 to 10 is also considered equally important so that linearity is generated because each factor contributes the same

SCORE FLOWS = 1 - 5.
1. = SMALL
3. = MEDIUM
5. = BIG

If the total score is above 30 then the condition is threatened and requires in-depth research action for conservation. This is also adjusted for the interpretation of the final score adjusted to the category of the seven red list of extinctions including: low risk (least concern), near threatened (vulnerable), vulnerable (vulnerable), critical (endangered), critical or endangered (critically endangered), extinct in the wild (extinxt in the Wild), to extinct [4]. Furthermore, the 7 criteria are given the following this below:

- The least concern is 10-19
- Near threatened (weight threatened) 20 = 29
- Rentan (vulnerable) weighs 30-39
- Genting (endangered) weights 40-49
- Critical endangered weight 50
- Extinct in the wild (extinc in wild) weights above 51-55
- Extinct (extinc) weight 56-60

Besides that, it is adjusted to the scope of environmental science indicators. These indicators are in accordance with the scope of the viewpoint of Environmental Sciences. This is important to be part or as a discipline. The scope:

- Social environment : Created human, unfortunated, beautiful for people, economic debit water decreased by human skill, shrinking forests.
- Natural environment : habitat damaged due to human
- Made environment : it cannot be cultured , protected with prohibited

4. Implementation of extinct metric UNP (EM UNP)
For example, we implement the bilih fish (*Mystacoleucus padangensis Bleeker*).

$$\text{UNP EXTINCT METRIC (UNP EM)} = F1 + F2 + F3 + F4 + F5 + F6 + F7 + F8 + F9 + 2F10$$

Figure 1. Bilih Fish [6]

Example calculation :

(UNP EM) = F1(5) + F2 (3) + F3(5)+ F4(5)+ F5 (5)+ F6 (3)+ F7(3)+ F8(5) + F9(5)+2 F10(5)
= 49

The results of this calculation indicate that the Bilih fish has reached the endangered stage. Because this picky fish has reached a critical or dangerous level but has not been included in the IUCN list. This was revealed in Suwelo's study [5]. These results indicate that bilih fish is an endemic freshwater fish resource and is endangered. This is an indication that the Bilih fish must be immediately conserved so as not to experience extinction.

5. Conclusion

Based on the analysis and discussion above, the EM UNP can be used for quantitative analysis of the level of extinction of selected fish. These results indicate that pickled fish must be conserved immediately so that they do not become extinct.

References

[1] WWF Indonesia 2018 http://wwf.indonesia.com

[2] Kartamihardja E S, Hedianto D A dan Umar C 2015 *Strategi pemulihan ikan Bilih dan pengendalian ikan Kaca (*Parambassis siamensis) di *Danau Toba, Sumatera Utara* (Balai Penelitian Pemulihan dan Konservasi Sumberdaya Ikan Jl. Cilalawi, No.1 Jatiluhur, Jawa Barat)

[3] Suyastri C 2012 Mengukur Efektivitas CITES Dalam Menangani Perdagangan Satwa Liar Dengan Menggunakan Identifikasi Legalisasi Artikel CITES *Jurnal Transnasional* **4**(1) 790-808.

[4] https://sains.kompas.com/read/2018/03/21/170200223/bagaimana-memastikan- suatu-hewan-terancam-punah-atau-tidak-.

[5] Suwelo I S 2014 Spesies Langka yang Perlu Di Lindungi Oleh Undang-Undang *Tesis* (Bogor : Magister IPB Bogor)

[6] Razak A 2015 *Adaptasi Ekologi Mata ikan Bilih (M.padangensisi Bleeker)* (Laporan Penelitian Dosen PPs UNP Tidak dipublikasikan)

IOP Conf. Series: Earth and Environmental Science **314** (2019) 012085 doi:10.1088/1755-1315/314/1/012085

Revitalization of cloves cultivation in Central Java, Indonesia

E W Riptanti [1,2,*], A Qonita[1] and R Uchyani[1,2]

[1] Department of Agribusiness, Faculty of Agriculture, Universitas Sebelas Maret, Jl. Ir. Sutami 36 A Surakarta, Indonesia
[2] Study Center and Mentoring of Cooperation and Micro, Small and Medium Enterprises (CMSMEs), Universitas Sebelas Maret, Jl. Ir. Sutami 36 A Surakarta, Indonesia

*erlynawida@staff.uns.ac.id

Abstract. Clove (Syzygium aromaticum) is an Indonesian indigenous plant, which belongs to the most valuable spices in the world. Central Java has been one of the clove production centers in Indonesia, but in the last six years, the area and production tend to decline due to poor management. In order to revive the glorious existence of the clove, the act of revitalization in the clove cultivation becomes crucial, given the role of clove as one of the world's important commodities. This research uses purposive sampling in some regions, namely Wonogiri Regency and Pemalang Regency. The respondents selected as subjects of this research are 30 clove farmers chosen from each regency in purposively using snowball-sampling technique. Apart from the clove farmers, the information is gained from the Field Extension Officer, farmer groups, clove traders, and any representatives from the related agencies. Revitalization is carried out by increasing the capacity building, the institution, and the cultivation activity of the clove farmers. The key of success in this effort is the good collaboration between the farmers, farmer groups, and the stakeholder involved.

1. Introduction

Clove (Syzygium aromaticum) is one of the most valuable spices that has been used for centuries as food preservative and for many medicinal purposes. Clove is native of Indonesia [1] but nowadays is cultured in several parts of the world including Brazil in the state of Bahia [2]. Clove tree is cultivated in the islands of Tanzania (Pemba and Zanzibar), Madagascar, Comoros Islands, and Sri Lanka, but the major producer is Indonesia [3,4]. This plant represents one of the richest sources of phenolic compounds such as eugenol, eugenol acetate and Gallic acid and possess great potential for pharmaceutical, cosmetic, food and agricultural applications [2].

Clove is a unique and strategic commodity for national economy. It is unique because Indonesia plays role as both the producer and the consumer. It is strategic because this commodity plays the essential role in absorbing the labors, both directly and indirectly. In direct way, it absorbs the labors in the whole cultivation activities. In indirect way, it absorbs labors in other activities related to cigarette industries and other clove processing industries [5].

Central Java has been one of the clove production centers in Indonesia, but in the last six years, the area of production tends to decline, as shown in the figure 1 [6,7]. Clove commodity becomes an interesting topic of research, given the fact that it has been the competitive commodity of farmers in the mountainous area and becomes the export mainstay in Indonesia. The volume of clove export in Central

Java in the recent years can be seen in the figure 2 [8]. This commodity has also experienced price downturn, forcing most farmers to cut down the clove trees and replace them with other plants, while the others abandoned the plants.

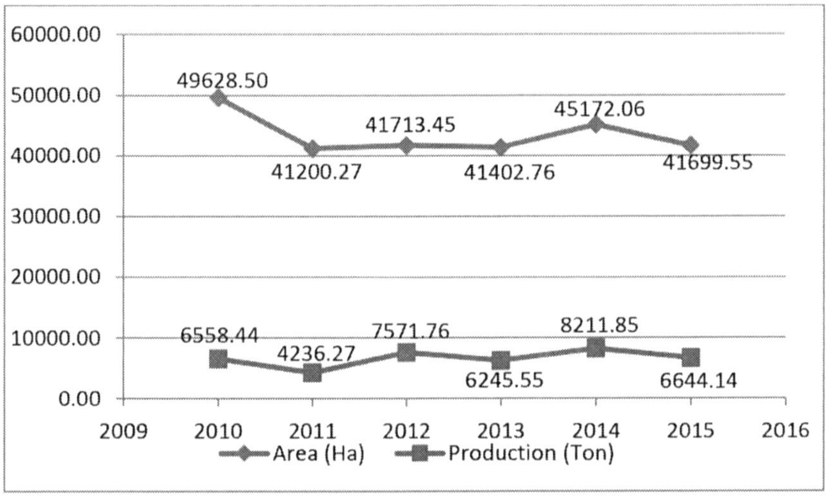

Figure 1. Area and production in 2010-2015
Note: Data for 2016 and 2017 are not found

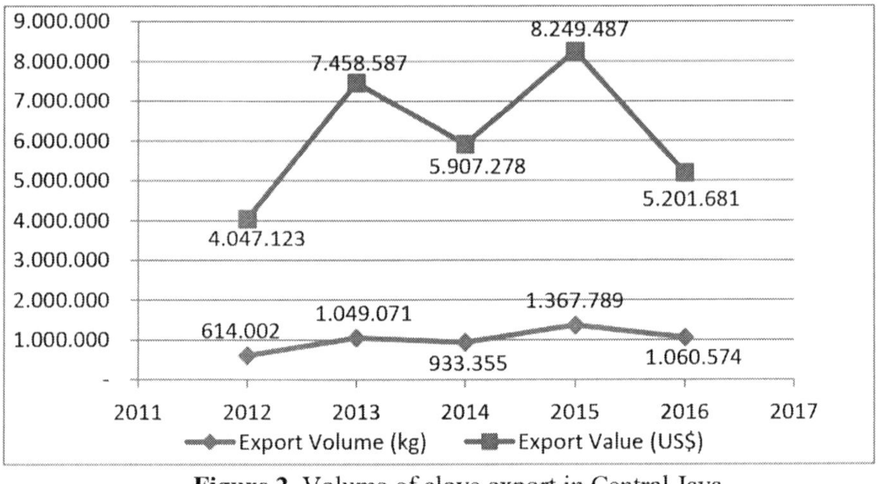

Figure 2. Volume of clove export in Central Java

The clove cultivation areas in Central Java amounted to 42,348 in 2015 and increased to 42,468 ha in 2016 (estimated number). Approximately 97 percent of the areas were people's plantations. The total production was only around 58.9 percent of the total clove cultivation areas. The productivity of cloves in Central Java was only 265 kg/ha in 2015, far lower than the productivity of cloves in the clove production centres in Indonesia. This indicates that the potential of clove has not been optimized, and therefore, there is an opportunity to improve the productivity of cloves in Central Java.

The main problem faced by clove farmers in Indonesia especially in Central Java is the remarkable triennial and quadrennial fluctuation depending on the total year rainfall, the exposure to low and high temperatures, the length of the dry season and other ecological and physiological parameters [9,10]. Problems, the worst of which are tree diseases, ageing tree populations and occasional neglect of trees [11,12]. Besides, the high-quality standard is applied in both domestic industry and the global market [13]. Some cloves are not cultivated but are still wild collected. The need for renewable sources of industrial products as well as the need to protect plant biodiversity creates an opportunity for farmers to produce such crops [14].

The change of strategic environment in the forms of economic globalization-trade liberalization, urbanization and market segmentation phenomena, consumers preference change, and disruptive innovation phenomena insist the change of mindset on the operation of the business organization of agricultural and food industry [15]. In macro sense, the disastrous spiral of price stress syndrome will lead to the collapse of national clove plantations. The national production plummets, trade deficit widens and Indonesia will again depend on imported cloves [16]. In recent years, clove price shows significant increase, but it has not gained proper respond from the farmers, and leads to a decline instead.

The opportunity of clove commodity for both domestic and international market is very promising, indicated by the development of clove processing industry. On the other hand, the potential of clove cultivation in Central Java has not been optimized. The farmers are reluctant to do an intensive cultivation. For this reason, this research article aims to revitalize the clove cultivation in Central Java.

2. Methods

The design of this study uses explanatory research to gain an overview of clove cultivation in Central Java. Explanatory research aims at finding the explanation of the observed phenomena, problems, or behavior [17]. The methodological approach used in this research belongs to the cross-sectional study, which is a research, conducted only at a certain time without doing a treatment on the object so as to produce a portrait of the situation at a certain time.

Data used in this research include primary and secondary data. Primary data involve the farmers' characteristic condition and the management of clove cultivation. The data were collected by doing in-depth interview using questionnaires with the clove farmers, Field Extension Officer, and the District Forestry and Plantation Services. The secondary data were gained from the Central Bureau of Statistics, Provincial Trade and Industry Service, and other institutions related.

The research was conducted in Central Java Province (Figure 3) and the research locations selected using purposive method were Wonogiri Regency as the largest clove cultivation area and Pemalang Regency as the area with the highest level of productivity. From each regency, the sub-district with the largest production was selected, namely Karang Tengah and Moga Sub-district.

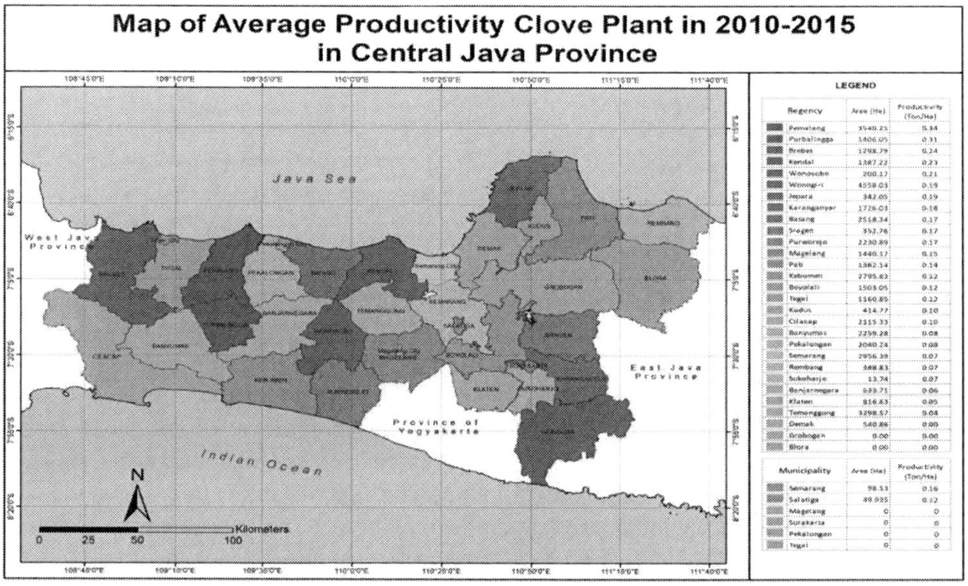

Figure 3. Distribution of clove production areas in Central Java

The sample were selected in purposive way using snowball-sampling technique in which from each sub-district 30 respondents were drawn and the total number of respondents were 60 farmers. The information was also gained from the head of the farmer group, Field Extension Officer, collector trader, and representatives from the District Forestry and Plantation Services.

Data analysis applied in this research is the descriptive analysis, which carries out the research finding as the portrait of the actual situation. The activities included in the qualitative data analysis were conducted interactively and continuously until the adequate data were completed and saturated. Activities in the data analysis include data reduction, data display, and conclusion drawing/verification [18].

3. Result

Cloves in Central Java are planted in the front yards, fields, as well as the forest borders. By doing so, farmers in Southeast Asia have developed a great variety of land use systems to be classified as either home gardens or forest gardens, which are important sources of supplementary subsistence and/or income. In contrast to forest gardens, home gardens are confined to the area immediately surrounding the home [19,20,21]. Cloves in the research location are planted using "*tumpang sari*" (polyculture), technique with the perennial plants such as kinds of fruits and coconuts [22] and with cassava or other food plants if they are planted on fields [23]. Clove plants are cultivated as secondary crops, instead of as main crops. This plant is mostly planted as a land separator.

The result of observation in the farming level shows that the cultivation has not been optimized, due to several reasons: 1) farmers are reluctant to do treatments because they prefer seasonal plants; 2) without any treatments, clove is still able to bloom every year and there are variations in production from year to year [9] ; 3) farmers experience trauma when the plants were properly treated but the price dropped (Clove Marketing Buffer Agency). In general, the characteristics of clove farming can be seen in the Table 1. Variety of cloves planted by farmers are mostly Zanzibar because the production is higher, compared to Siputih.

Table 1. Characteristic of Clove Farming

No	Description	Annotation
1	Area of production (m^2)	3,882
2	Total number of plants (trunk)	38
3	Age of plant (year)	Less than 10 years = 20% 10 to 25 years = 30% Above 25 years = 50%
4	Range of plantation (m)	Irregular Planted on the edge of the land as a barrier
5	Intensification on clove cultivation	No specific treatment
6	Average amount of production (kg) per stalk	7 – 69
7	Price of wet cloves (IDR)/kg	20,000 – 33,000
8	Price of dry cloves (IDR/ kg	75,000 – 120,000

Source of data: primary data analysis (2018)

Table 1 shows that the business of clove plants is relatively wide, but is still in need of known to be fluctuating, characterized by the plenty harvest in a year and then followed by very small harvest in another year. The different amount between the plenty harvest and small harvest can reach up to 60%. Periods of harvesting are varied between 2-4 years, depending on the age, genotype, and plant maintenance in the field. Table 1 displays that the production is relatively low, far from the targeted potential production; the production of Zanzibar and Siputih varieties reached 161.8 kg/tree/year and 93.1 kg/tree/year [24]. The low production was also due to the traditional harvesting method in which pickers climb the trees, lash branches together and use ropes or hooked sticks to reach the cloves on distant branches. He also indicated the difficulties of controlling branch breaking; in particular, the picking period is very short and if overseers were too strict pickers were likely to leave and the crop could be lost [25]. Low production is also caused by the lack of care on clove plants.

Most clove plants cultivated by farmers are more than 25 years old, and therefore, their productivity begins to decline. In fact, old plants are no longer productive and some of them are dead. However, rejuvenation has not been carried out. Improvement has to be made by setting a good cropping pattern in order to maximize farmers' income.

4. Discussion

Community management of natural resources fosters a greater sense of responsibility and better stewardship of the land. These community-based management approaches also have better integrate indigenous knowledge and traditions in management decisions. Further, indigenous people's efforts to develop their own management plans are important not only in promoting ecological and economic sustainability, but also in promoting cultural sustainability, which is essential for the survival of the indigenous medical knowledge, values and in traditions [26].

The development of agribusiness is a great way to improve the economy and competitiveness of the region. To build and accelerate the development of agribusiness requires determination of agribusiness service centers as well as measures of operational strategies to suit local circumstances. The determination of growth centers through the analysis of comparative advantage plays an important role in the agribusiness development [27]. The suitable area, climate, and human resource in the mountainous land in Central Java [28] contribute as the basic capital of the comparative advancement of the cloves. Therefore, this comparative advancement needs to be managed at optimum level by revitalizing the clove cultivation.

Revitalizing the cloves cultivation aims at raising the consciousness of the cloves farmers to put back the clove as the essential commodity as a source of livelihood which can compete in the global market, by means of refreshing the vitality, utilizing the skill and enhancing the quality of the commodity. Clove plant production and productivity can be potentially optimized through revitalization of clove cultivation. Revitalization is realized through:

a. Reviving clove farmers' awareness about the importance of cultivating the cloves by referring to the Good Agriculture Practices (GAPs) [29,30,31] and Sustainable Agricultural Intensification (SAI) [32,33]. This effort is fundamental to be taken considering traumatic past experiences when the clove price collapsed below the production operational cost. If the farmers consider this matter, then there will be a desire to work for better production, the product will be accepted by the global market, and the farming business shall continue to exist [34].

b. Increasing the capacity of human resources, particularly the clove farmers, by training and empowerment. Clove cultivation management comprises the optimization of ecological factor maintenance by upgrading the suitability level of the soil from actual to potential, the maintenance of clove production factors from the genetic aspect by selecting the advanced variety, intensification of clove cultivation in accordance with GAPs and SAI, access to the marketing and financial institutions, and optimization of social factor of local people by strengthening the clove farmers institution [35,36].

   The role of the clove seeds suppliers is very crucial in providing certified seeds with the guaranteed quality. The availability of local clove genetic resources is highly fundamental for the utilization through the population and parent trees selection program, which aims to produce qualified parent trees [24].

c. Strengthening the cloves farmer group institution, including the institutional policy, institutional structure, and personnel and operational skill [37,38]. This will lead to the increase of bargaining position of farmers in both input purchase and output sales, and access to the marketing and financial institution. Furthermore, strengthening the institution can be realized by improving and enhancing the role of clove association groups, science and new technology adopted by farmers.

d. Replanting the plants

Clove plants aged over 25 years old should be replaced with new plants due to their decreasing level of productivity. Replanting can also be applied to dead plants caused by diseases. Replanting has been crucial to the maintenance of the industry, which has also been strengthened by government assistance [12]. This step is very necessary to be taken when the condition of the plants is terrible. The success of replanting will depend upon the availability of good quality seedlings and the adoption of methods of tree establishment, which are appropriate for the climate and farming systems to grow Zanzibar or Siputih. Zanzibar is less influenced over external factors and present trends, suggesting that increasing Indonesian production will result in a smaller world clove market in which there will be increasing competition between the main producers [12].

The availability of the certified seeds is the key of success of this replanting program. In the research location, sufficient certified seeds must be available, to note that this type of seed is rarely found.

e. Improving the harvesting technique to produce qualified cloves, which meets the standard
This technique improvement is realized by utilizing high triangle ladders to ease the farmers in harvesting and prevent them from breaking the branches or the flowers. This method is considered to be more effective than harvesting by climbing or using long pole. Cutting the top of the last leaves using scissors will increase new shoots. Clove flowers are plucked when they are still buds in the age of six months since the emergence of the flowers from the flower base [39].

Revitalization will succeed when there is a good collaboration among many parties namely the clove farmers, Field Extension Officer, farmer group, and the related institution. Supporting policies by the government are in forms of incentive replanting [40] input subsidy policy, facilitation of capital and marketing access, and the improvement of infrastructures which ease the farmers in revitalizing the cloves.

5. Conclusion

Guarantee upon the sustainability of clove production is precisely vital in providing important commodity for the global market. Revitalizing the cultivation is also necessary to guarantee the availability of cloves as the fulfilment of raw material in both domestic industry and foreign market. Revitalization is implemented by raising the awareness of the farmers about the importance of clove cultivation by referring to GAPs and SAI. Furthermore, the enhancement of capacity building, the strengthening of institution, the clove replanting program, and the improvement of harvesting technique are very necessary to be taken in order to guarantee the quality and the quantity of the production.

Acknowledgements
We would like to express our sincere gratitude to Universitas Sebelas Maret for funding this research. Our sincere appreciation is also extended to all parties who have provided information to complete this article.

Conflicts of Interest. The authors declare there is no conflict of interest

References

[1] Dufournet R 1968 Le giroflier et sa culture à Madagascar *Bulletin de Madagacascar, Nr 262* Tananarive Madagascar
[2] Rojas D F C, Fernandes C R and Oliveira W P 2014 Clove (Syzygium aromaticum): a precious spice *Asian Pacific Journal of Tropical Biomedicine* **4**(2) 90-96
[3] Leela N K and Sapna V P 2008 *Clove* In: Parthasarathy V A, Chempakam B, Zachariah T J (Eds) *Chemistry of spices* pp 146-164 CAB International. Cambridge, USA.

[4] Barichello R and Patunru A 2009 Agriculture in Indonesia: Lagging performance and difficult choices *Choices* **24**(2) 37

[5] Rumajit G A J 2007 Kajian ekonomi keterkaitan antara perkembangan industry cengkeh dan industri rokok kretek nasional *Disertasi* (Bogor Indonesia: Institut Pertanian Bogor)

[6] Badan Pusat Statistik 2013 *Jawa Tengah dalam Angka Tahun 2013* (Jawa Tengah, Indonesia: Badan Pusat Statistik)

[7] Badan Pusat Statistik 2016 *Jawa Tengah dalam Angka Tahun 2016* (Jawa Tengah, Indonesia: Badan Pusat Statistik)

[8] Dinas Perindustrian dan Perdagangan Jawa Tengah 2017 *Realisasi ekspor – impor Jawa Tengah periode 2012 -2017 (Oktober)* Bidang Perdagangan Luar Negeri. Dinas Perindustrian dan Perdagangan Jawa Tengah Indonesia

[9] Martin P J and Dabek A J 1988 The role of agronomic factors in the juvenile decline condition of clove trees in Zanzibar *Tropical Pest Management* **34**(3) 271-277

[10] Baietto M 2014 Bud fall induction in clove (Syzygium Aromaticum) *Academic Research International* **5**(4) 23-29

[11] Martin P J, Butler D R and Dabek A J 1988 Causes of irregular clove production in the islands of Zanzibar and Pemba *Experimental Agriculture* **24**(1) 105-114

[12] Martin P J 1991 The Zanzibar clove industry *Economic Botany* **45**(4) 450-459

[13] Godoy R and Bennett C 1990 The quality of smallholder cloves in Maluku: The local response to domestic demand for a high-quality product *Bulletin of Indonesian Economic Studies* **26**(2) 59 – 78

[14] Lubbe A and Verpoorte R 2011 Cultivation of medicinal and aromatic plants for specialty industrial materials *Industrial Crops and Products* **34** 785 – 801

[15] Saptana 2016 Pertanian Indonesia dalam menghadapi disruptive innovation di bidang pangan *Agrimedia* **21**(2) 3-12

[16] Simatupang P 2003 Opsi kebijakan memulihkan anjlok harga cengkeh *Analisis Kebijakan Pertanian* **1**(4) 297-305

[17] Neuman W Lawrence 1994 *Social research methods: qualitative and quantitative approach* (Boston: Allyn and Bacon)

[18] Patton M Q 2001 *Qualitative Research and Evaluation Methods* (Thousand Oaks, CA: Sage Publications)

[19] Wiersum K F 1982 Tree gardening and taungya on Java: Examples of agroforestry techniques in the humid tropics *Agroforestry Systems* **1**(1) 53–70

[20] Michon G 1983 *Village forest gardens in West Java* In: Huxley PA (ed) Plant Research and Agroforestry pp 13–24 ICRAF Nairobi Kenya

[21] Soemarwoto O and Conway G R 1991 The Javanese home garden *J. Farm. Syst. Res. Ext.* **2** 95–118

[22] Kaya M, Kammesheidt L and Weidelt H J 2002 The forest garden system of Saparua island, Central Maluku, Indonesia, and its role in maintaining tree species diversity *Agroforestry Systems* **54**(3) 225–234

[23] Michon G and Mary F 1994 Conversion of traditional village gardens and new economic strategies of rural households in the area of Bogor, Indonesia *Agroforestry Systems* **25**(1) 31-58

[24] Randriani E and Syafaruddin 2011 Keragaan Pohon Cengkeh Terpilih Tipe Zanzibar dan Siputih Palabuhanratu *Buletin RISTRI* **2**(3) 405 – 410

[25] Tidbury G E 1949 *The clove tree* (London UK: Crosby Lock- wood)

[26] Pahlen M C and Grinspoon E 2002 Promoting traditional uses of medicinal plants as efforts to achieve cultural and ecological sustainability *Journal of Sustainable Forestry* **15**(1) 81 – 93

[27] Darmansyah A, Rochana S H, Sutardi A and Zuraida U 2014 The new growth centres and strategy for building and accelerating agribusiness development in Cirebon Regency, Indonesia *Procedia - Social and Behavioral Sciences* **115** 296 – 304

[28] Riptanti E W, Qonita R A and Fajarningsih R U 2018 The competitiveness of medicinal plants in Central Java Indonesia ICSAE *IOP Conf. Series: Earth and Environmental Science* **142** (2018) 012018

[29] Hobb J E 2003 Incentives for the adoption of good agricultural practices (GAPs) *Paper for the FAO consultation on Good Agricultural Practices November 10-12 Rome Italia*

[30] Cruz A G, Cenci S A and Maia M A 2006 Good agricultural practices in a Brazilian produce plant *Food Control* **17**(10) 781-788

[31] Amekawa Y 2009 Reflections on the growing influence of good agricultural practices in the global south *Journal of Agricultural and Environmental Ethics* **22**(6) 531–557

[32] Maertens M, Zeller M and Birner R 2006 Sustainable agricultural intensification in forest frontier areas *Agricultural Economics* **34**(2) 197 – 206

[33] Ortiz R and Alfaro D 2014 Sustainable agricultural intensification – A perspective from Latin America and the Caribbean *World Sustainability Forum 2014 Conference Proceedings Paper* November 1 – 30 https://pub.epsilon.slu.se/12818/ (accessed on 22 June 2018).

[34] Thankamani C K, Sivaraman K, Kandiannan K and Peter K V 1994 Agronomy of tree spices (clove, nutmeg, cinnamon and allspice) - a review *Journal of Spices & Aromatic Crops* **3**(2) 105-123

[35] Bingen J, Serrano A and Howard J 2003 Linking farmers to markets: different approaches to human capital development *Food Policy* **28**(2003) 405-419

[36] Sajeev M V and Singha A K 2010 Capacity building through KVKs: Training needs analysis of farmers of Arunachal Pradesh Indian *Research Journal of Extension and Education* **10**(1) 83 – 90

[37] Merrill-Sands D M, Biggs S D, Bingen R J, Ewell P T, McAllister J L and Poats S V 1991 Institutional considerations in strengthening on-farm client-oriented research in national agricultural research systems: lessons from a nine-country study *Experimental Agriculture* **27**(4) 343-37

[38] Keleman A 2010 Institutional support and in situ conservation in Mexico: biases against small-scale maize farmers in post-NAFTA agricultural policy *Agriculture and Human Values* **27**(1) 13–28

[39] Suwarto Oktavianty Y 2010 *Budidaya 12 tanaman perkebunan unggulan* (Jakarta, Indonesia: Penebar Swadaya)

[40] Newman S M 1985 A survey of interculture practices and research in Sri Lanka *Agroforestry Systems* **3**(1) 25-36

IOP Conf. Series: Earth and Environmental Science **314** (2019) 012086 doi:10.1088/1755-1315/314/1/012086

The role of gender as a model of climate change adaptation in fisherman settlement communities

M Riviwanto[*] and A Basuki

Health Polytechnic Ministry of Padang, Siteba Street, Padang-West Sumatera 25146 Indonesia

*muchsinr@yahoo.com

Abstract. Women are the largest proportion of the world's poor, including children and young women. Women are very vulnerable to climate change. But until now gender considerations are still not explicitly stated in various climate change policies. In the marine and fisheries sector, extreme climate change makes many fishermen unable to go to sea and fish declines. The unequal division of gender roles causes the level of vulnerability to climate change to differ between women and men. This study aims to analyze the vulnerability and gender roles of fishing communities due to climate change in West Sumatra. A positivistic-deductive research approach. Indicators / parameters used by the index to measure vulnerability in 2 coastal areas, namely Bungus Beach and Pariaman Beach. Secondary data obtained from the Geophysical Meteorology Agency station. Primary data through interviews with questionnaire tools on 60 respondents. The vulnerability model (V) is a function of exposure (E), sensitivity (S), and adaptive capacity (AC), which are then expressed mathematically. The results showed a vulnerability index of 305 with a vulnerable classification. Efforts to increase adaptive capacity and reduce women's vulnerability to climate change through a comfortable place to live, better access to climate change information and livelihood options. Gender-based climate change adaptation models for coastal areas are increasing women's political role, increasing women's skills capacity through field schools Climate and enhancement productive work.

1. Introduction

Climate change will have different impacts in each country, region, generation, class, occupation, gender, age and income. From 1975 - 2006, most natural disasters occurred on the Asian continent. Of the vulnerable groups, including 3.4 million people from groups of poor people, children, indigenous peoples, farmers and fishermen. Women are the largest proportion of the world's poor, including children and adolescent girls, very vulnerable to climate change [1].

Climate change has the most severe impact on women from the lowest social groups. In each disaster it turns out that female victims are larger than men with a ratio of 4: 1. The results of the analysis of disasters that occurred in 141 countries proved that differences in the number of victims due to natural disasters were closely related to women's economic and social rights [2] .

Gender is an important issue because of the inequality factor and the specific needs of women in disaster situations [3] In a hazard perspective, the spatial dimension is very important to identify risky societies, but must be linked to understanding socio-economic differentiation, linkages and dynamics within the region. But until now gender considerations are still not explicitly stated in various climate change policies [3] .

In the maritime and fisheries sector, after being hit by various destructive efforts in coastal areas, extreme weather changes made many fishermen unable to go to sea. The results of aquaculture and fishing are decreasing. Previously, the number of days they went to sea was 240-300 days a year, now it is only 160-180 days. As a result, the income of traditional fishermen has also reduced to around 50 to 70 percent. That is, the income of fishermen's family is reduced. And usually, the woman redeemed the deficiency.

Natural signs that are usually used by fishermen determine the harvest season. Now you can't be a guide anymore. On the coast of Indramayu, people used to know about the signs of nature. Until 2003, fishermen were still able to estimate the west and east winds. The west season was marked by abundant catches of Rajungan and Sotong. But, for the past three years, since 2005, the public has been unable to predict the weather changes [4]. Desmawan's research due to climate change is the presence of floods such as in the coastal areas of Central Java with the impact of Damage to residential buildings because in addition to flooding the floor and yard surfaces, Rob flooding is corrosion and damaging to buildings [5].

West Sumatra Province has a coastline of approximately 375 km, in the form of lowland as part of the archipelago of the bow face. The mapping location along the coast belongs to the Province of West Sumatra, and covers two cities and four districts, namely Padang City, Pariaman City, Pesisir Selatan Region, Agam Region, west Pasaman Region, and Padang Pariaman Region [6].

On this occasion, a gender perspective will look at the level of vulnerability of fishing communities and the role of gender that is created in efforts to deal with climate change in West Sumatra Province. The level of vulnerability and community adaptation patterns were analyzed using gender analysis that emphasized gender relations and roles created in fishermen's households. The unequal division of gender roles causes the level of vulnerability to climate change to differ between women and men. This study aims to analyze the vulnerability and strategy of gender roles of fishing communities due to climate change in West Sumatra. So that research is beneficial for the government as information that can influence decision making in determining development policies, especially development in the fisheries and health sectors.

2. Research Methods

The Research was conducted with a quantitative approach that was positivistic-deductive. Starting from the concept and operationalized into indicators / parameters in the form of indexes to measure vulnerabilities. Quantitative research methods identify variables that influence the index of vulnerability. In a variety of quantitative research, this study classified research explanations to describe the gender disparities in adaptation to environmental changes and find adaptive strategies. The population is all family heads of people living in coastal areas in Padang city, West Sumatra Province. The number of coastal areas of the city as many as 7 regions, researchers took 2 coastal areas namely Bungus beach and Pariaman beach with 234 families. The sample is a portion of the population with a probability proportional to size (PPS) sampling. Based on the calculation of the sample, the number of samples for the city of Padang was 30 respondents and the Pariaman beach was 30 respondents.

The data used were primary and secondary data. The collection of secondary data was obtained from the climatology Geophysical Meteorology Station. Primary data through interviews with questionnaire tools. Processing and analyzing data using a computer. The sequence of analysis carried out in this study is as follows: a. direct data collection to analyze the criteria and indicators used in determining the level of community vulnerability to climate change. The vulnerability index model constructed in this study is a static model of coastal area environmental vulnerability index. Turner (2013) uses the concept of environmental vulnerability. Where vulnerability (V) is an function overlay of exposure (E), sensitivity (S), and adaptive capacity (AC) [7]. Then expressed mathematically by Metzger et al. (2006) as in equation (1). Equation (1) is further expressed in the form of mathematical equations by Hamzah et al. (in press) and also has similarities developed by UNU-EHS (2006) to [8]:

$$V = (E \times S) / AC \quad (1)$$

The EVI report for an area is arranged on one page. The information available on the report includes the overall EVI score in points, with the percent of data calculated and the overall vulnerability classification. The classification, shown below, quickly identifies whether the environment of an area of vulnerability is very high overall [9].

Table 1. Classification of Vulnerability Indices an Area

Classification	Total scores
- Extremely vulnerable	375 +
- Highly vunerability	315 +
- Vulnerability	265 +
- at risk	215 +
- resilient	< 215

3. Result and Disscusion

3.1. General Description

The research location for the city of Padang is at the Bungus district. Bungus district has a coastline length of 21,050 meters and a bay length of 5,418 meters, volume 223,255,052.2 m^3, has a rounded surface shape and a surface area of 1383.86 Ha located south of Teluk Bayur and has a strategic position overlooking the Indian Ocean. This bay is included in Bungus district and is one of the coastal sub-districts in the southern region of Padang City. Astronomically this sub-district is in the position 01^{0}01'21 " - 01^{0}05'02 " South Latitude (SL) and 100°21'58 " - 100^{0}26'36 " East Longitude (EL) and located on the west coast of Sumatra Island. Bungus district is at an average height of around 0-5 m above sea level for coastal areas, and <850 m for hilly areas. The temperature ranges from 22.5°C - 31.5°C and rainfall is 314.47 mm / month.

Bungus beach is affected by the condition of the Indian Ocean. Dominant incoming wave directions from the West and Northwest, where waves come from the Indian Ocean towards the Bay of Bungus. Significant wave height, which is an average of 1/3 of the measured average wave height, shows the range of wave height ranges between 0.02 m - 0.2 m, with the highest wave amplitude of 0.189 m and the average wave amplitude (mean) 0.0975 m. The dominant tidal current direction at high tide is westward, while at low tide the dominant tide moves towards the bay to the southeast. Bungusin male population is more than the female population.

Geographically, Pariaman City is located on the West Coast of Sumatra Island and faces directly to Indonesia Ocean. The astronomical position of Pariaman City is between 00°33' 00" -00° 40 '43" South Latitude and 100° 04' 46" – 100° 10' 55" East Longitude, with an area of about 73.36 Km^2, and coastline length of 12.00 Km. The land area of this area is only 0.17 percent of the land area of West Sumatra Province. Pariaman City is also identical with the coastal city where all Sub-districts in Pariaman City have an area bordering the beach except East Pariaman District. The average height of the region is in the range of 0-15 meters above sea level.

Pariaman Beach are strongly influenced by the strong Indian Ocean waves reaching the coast. The abrasion process (coastal erosion) occurs predominantly along the coast, while the process of land erosion is also intensive in the upstream area, marked by the high supply of sediment carried by the river flow to the sea. Wind direction comes from the west with the percentage of wind incidence as much as 8.49%. While the minimum percentage is achieved by winds from the south with an incidence percentage of 0.04%. wave conditions in these waters are relatively normal with altitudes ranging from 0.1-1 m. The most common wave is the westward wave with the percentage of occurrence of 8.49%, the highest percentage is 8.47% with an altitude between 0.1-1 m.

3.2. Climate Change

Based on the Sicincin BMKG analysis of the monthly temperature of the Padang City during 2017 January and December shows the temperature variations of these two regions. The average temperature in the city of Padang is higher than in the Pariaman City and the increase in temperature is higher in May in both regions.

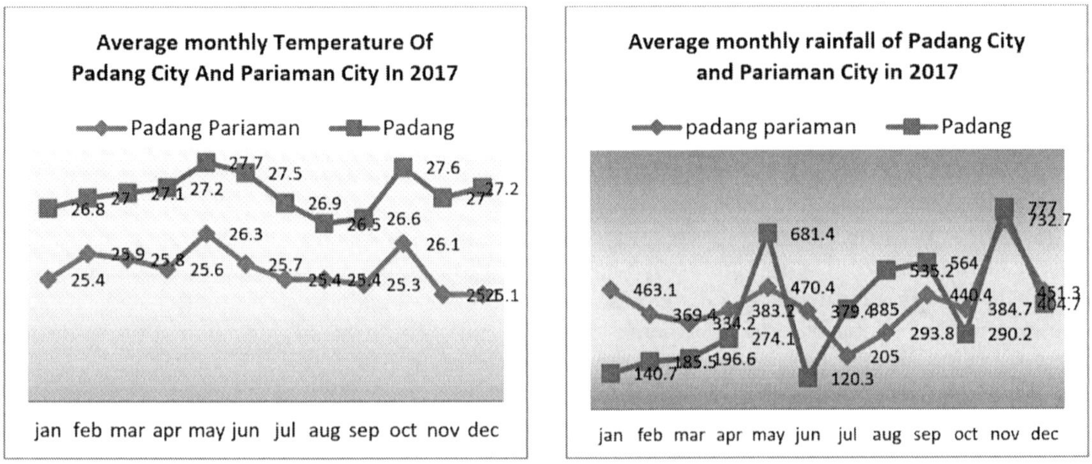

Figure 1. Average temperature and monthly rainfall of Padang City and Pariaman City in 2017
Source: BMKG Sicincin Station

Based on temperature trend data during the 2008-2017 period at Sicincin BMKG Station, which represents the coastal area of Padang city, the average annual temperature is around 27, 43°C with a minimum temperature of 27.1 C and a maximum temperature of 27.7 C. Pariaman City obtained an annual average temperature of around 26.51°C with a minimum temperature of 25.6 C and a maximum temperature of 27.0 C. Results of analysis of rainfall trends annual shows that Padang City has an average annual rainfall of around 364.67 mm with a minimum rainfall of 271.9 C and a maximum temperature of 471.9mm. Pariaman City obtained an average annual rainfall of around 392.97 mm with a minimum rainfall of 361.7mm and a maximum temperature of 435 mm. from both regions there was a decrease in rainfall.These dynamics indicate atmospheric disturbances, minimal disturbances on local and regional scales that affect rainfall in the Padang City and Pariaman city in 2017.

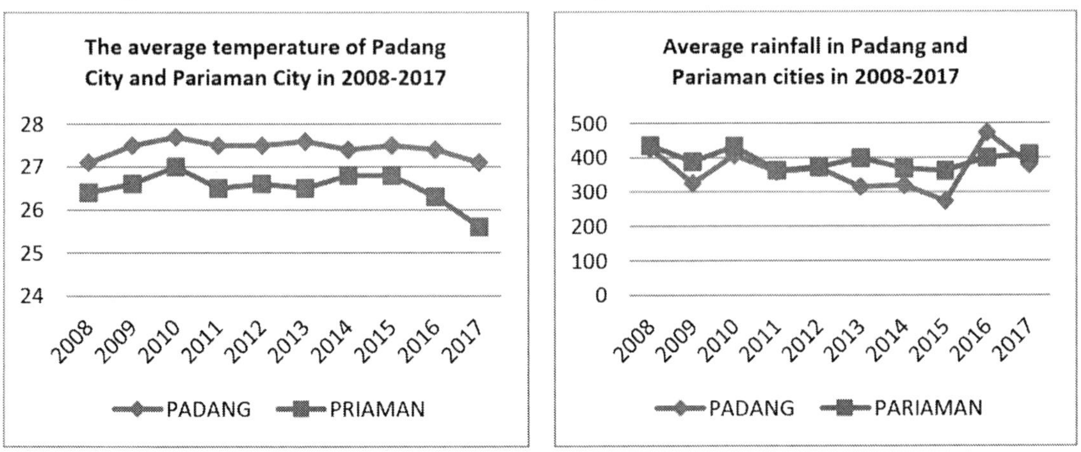

Figure 2. Average temperature and rainfall of Padang City and Pariaman City in 2008- 2017
Source: BMKG Sicincin station

IOP Conf. Series: Earth and Environmental Science **314** (2019) 012086 doi:10.1088/1755-1315/314/1/012086

3.3. Fisherman Profile

3.3.1. Productive Activities of Education and Income. Profile profiles of respondents conducted daily include productive activities, reproductive and social activities to be examined in this study. The average time for the fishermen's productive activities in one day is 8.35 hours, while for the activity is 6.7 days a month and the average time to go to sea is 7.87 hours a day. The level of formal education of respondents varies from not completing primary school to graduating from first grade. The average level of education of fishermen's family is still dominated by the low education level, namely graduating from elementary school where the husband is 36.7% and wife is 23.3%.

The level of education is very important in the management of a business, both formal and non-formal education. Formal education is a very valuable capital to get a decent economic life. Education is also very influential on the pattern of life in each individual, both ways of thinking, ways of making decisions and how to behave. The low level of education of respondents was due to the very small income level of their parents, so that respondents had to help their parents to take fish at sea. On the other hand, there are still some parents who have the perception that girls are not required to be highly educated on the grounds that women are more important to work on domestic activities such as taking care of children, taking care of their homes and husbands. The level of education is an influential factor for understanding climate change and its impact on health [10].

The level of income of fishermen in research locations is generally low, where 43.3% and wives 56.7%. Fisherman's wife's income is generally still below the regional minimum wage. The level of income of fishermen is a very important factor in family life and well-being, one's income will affect the level of family welfare. The higher the level of income, the higher the level of life can be expected [11].

3.3.2. Family Status of Fishermen. In accordance with the abundance of natural resources support in the fisheries and marine sector in West Sumatra most of the community's main occupation of their husbands in the family is as fishermen and this work has been passed down from their parents. Husband respondent category is greater with the status of fisherman ship owner, that is 46.7%, but few have status as sea skipper / captain / helmsman, which is 21.7%. Likewise, with wife respondents, only a few wives worked in the fisheries sector with categories as fishermen by 1.7%, as fish traders 10% and as fish processors by 15%. This happens because many wives do not work and only do homework or wives who work outside the fishery sector such as trading.

3.3.3. Decision Making in Productive, Reproductive and Social Activities. Decision making on productive activities in fishing households carried out by husbands, wives and husbands and wives such as preparing for fishing, fishing, drying and storing fish, selling fish and labor work. In Figure 3, this activity is more dominated by husbands (58%). Reproductive activities such as cooking, washing, bathing children, taking school children and integrated health post are dominated by wives (80%). Social gathering activities, community service, death and marriage and health cadres are dominated by wives 36%.

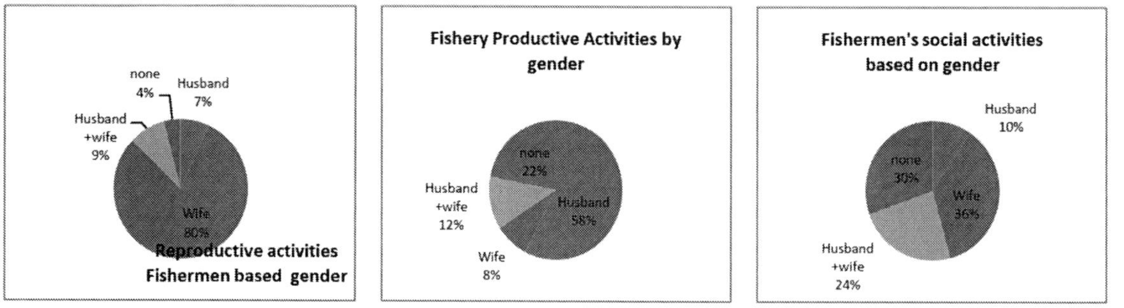

Figure 3. Decision making in productive, reproductive and social

Activities In the productive activities of fisheries business management both in preparing the needs for fishing, fishing, drying and selling the catch of the husband's role is very dominant in decision making. In general, fishermen's wives are not much involved in decision making. This shows that in productive activities the role of the wife is still low. Thus, there is an indication of a gender gap in decision making on productive activities.

In social activities, there is a tendency for institutional separation for women and men. Formal social institutions such as PKK and posyandu activities are a forum for women, while the activities of fishermen groups such as community service are a forum for men, while non-formal institutions such as meeting are a forum for women. Besides that, the influence of boat ownership and fishing gear is a gap for women in the research location, where fishermen's wives cannot do much in their management, so the full decision is in the hands of the husband, for the activities of educating schoolchildren the decision making is done together [12].

3.4. Knowledge of Fishermen About Climate Change

3.4.1. Acceptance of Information on Climate Change. Climate change information received by fishermen from various sources of information. It was seen that fishermen had received information about the change of seasons by 63.3%. Changes in sea level by 45%, Changes in intensity / rainfall 63.3%, Change / shift in direction and current strength by 45%, Changes in the west wind season and east wind season 45%, Changes in the rainy season and dry season 55%, Increased extreme weather events of 45% and ice melting were 28.3%.

Table 2. Climate Change Information Received by Respondents

Information Climate change	%
Change in wave season: strong, calm and transition	63.3
Change in sea level	45.0
Change in intensity / bulk rain	63.3
Changes in direction and current strength	45.0
Changes in the seasons of the west wind and east wind season	45.0
Changes in the rainy and dry seasons	55.0
Increases in extreme weather events	45.0
Ice melting poles	28.3

3.4.2. Acceptance of information the main causes on climate change. Receipt of information the main causes of climate change by fishermen are industry/factories by 23.3%, logging/burning of forests/mangroves by 25%, Damage to coral reefs by 18.3%, development / coastal reclamation by 23.3%, solid sea transportation 21.7%, annual cycle of 65% and ordinary natural occurrence of 48.3%.

Table 3. Information on The Causes of Climate Change Received by Respondents

Information on The Causes of Climate Change	%
Industry / factory	23.3
Logging / burning of forests / mangroves	25.0
Coral reef damage	18.3
Development / coastal reclamation	23.3
Dense sea transportation	21.7
Mining / offshore drilling	18.3
Annual cycle	65.0
Ordinary natural events	48.3

3.4.3. Changes in Weather Conditions Felt by Fishermen. Climate change information felt by fishermen states that what is most felt is an increase in sea water temperature (73.3%), erratic current direction, fish in the catchment area are decreasing (71.7%), and uncertain sea wave season (71.7%).

Table 4. Changes in Weather Conditions Perceived by Respondents

Changes in perceived weather conditions	%
1. Strong winds occur more frequently	71.7
2. Seasons waves (strong waves, calm, transition)	71.7
3. current direction	73.3
4. Fish in the area decreases	71.7
5. Catchment Sea level decreases	68.3
6. Changes in seasons west and east wind season shift	70.0
7. Flow getting stronger	66.7
8. Temperature / sea water temperature increases	73.3
9. Flood	15.0
10. Flood rob	16.7

3.4.4. Impact Felt by Fishermen Due to Climate Change. Perceived impact due to climate change fishermen had felt a few years ago. Generally, fishermen stated that the perceived impact of climate change was to disrupt time to go to sea (73.3%), disrupt fishing areas (70.0%), and disrupt fisheries production (70.0%).

Table 5. Impacts of Climate Change On

Climate change impacts on fishermen activities	%
Disrupt time when fishing	73.3
Disturb capture areas	70.0
Affect species / fishing gear	68.3
Fisheries Disrupt fisheries production	70.0
Activities Disturb aquaculture activities	23.3

3.4.5. Fishermen Efforts Towards Climate Change. Generally, the effort of fishermen's adaptation in anticipating the perceived impacts of climate change is to maintain the source of clean water (75.5%), not throw garbage into the sea (71.7%) and make changes in fishing time (70.0%).

Table 6. Attempts to Adapt Fishermen to Climate Change

Efforts to adapt fishermen to climate change	%
Change time and time of capture	70.0
Change the type of fish caught	61.7
Form a group of mothers	21.7
Looking for side jobs	41.7
Not burning garbage / opening land	36.7
Not cutting trees carelessly	63.3
Doing crops resistant to drought	16.7
Removing places for breeding mosquitoes	36.7
Do not dispose of trash into the sea and rivers	71.7
Maintaining clean water sources	75.5

Based on knowledge of climate change related to understanding climate what is felt, the causes, impacts, and efforts made, will be useful to improve resilience and health and health due to the effects of climate change [10].

3.5. Coastal Area Vulnerability Index Vulnerability

Consists of three dimensions or components namely exposure, sensitivity and adaptive capacity. Adaptive capacity has an element in resource readiness that influences the resilience of a system and adaptation actions that can be taken to overcome and reduce vulnerability. Sensitivity reflects the response of a system to changes or events that occur such as an increase in temperature and low rainfall and the rate of change caused by these changes. Vulnerability in coastal areas will be seen in Figure 4.

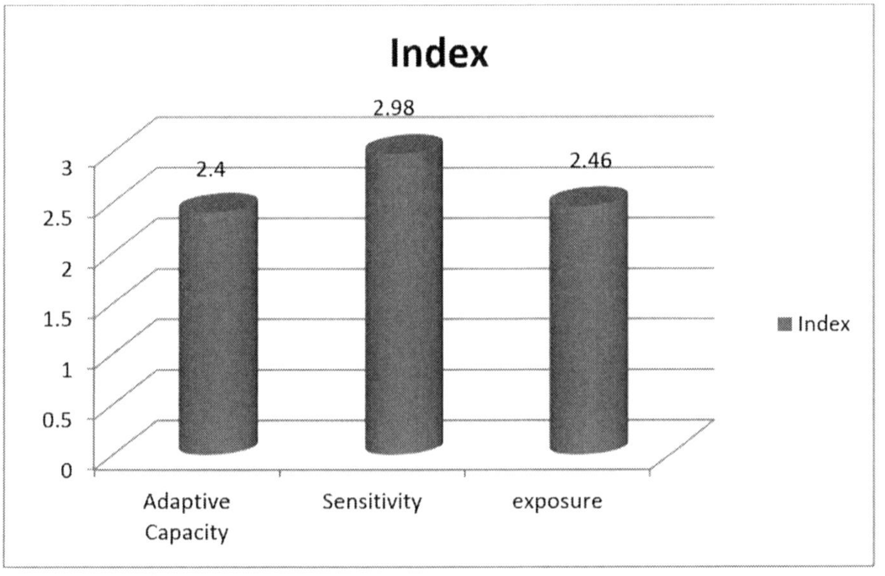

Figure 4. Vulnerability index change in fishing community

Figure 4 shows the index of adaptation ability of coastal communities depending on the value of weight and score. The indicators of adaptability are: indicators of the number of family dependents, income, education level, Climate Information and adaptation efforts 2,4 (low class).Sensitivity index in the coastal area is 2.98 (classification is caused by the weight value and score of each indicator of different sensitivity. Indicators of sensitivity are: length of activity, length of day of going to sea, inequality of domestic decision making, inequality of environmental health responsibilities, inequality of service request responsibility HealthIndex Coastal area exposure is 2.46. This change in outcrop shows that coastal areas are vulnerable to climate change with indicators of rainfall, population density, disease, perceived changes in conditions and perceived impact. Figure 4 above shows the level of vulnerability of the community to climate change beach depending on size of index exposure, sensitivity and adaptive capacity, the vulnerability index is a positive function of exposure and sensitivity and negative functions of adaptability. Thus obtained coastal vulnerability index of 305 (classification vulnerable).

3.6. Model gender-based adaptation of coastal communities

The role of husband and wife in fishing settlements as a form of adaptation in the face of climate change is grouped into productive roles, productive roles and social roles. The reproductive (domestic) role is the role or task related to reproduction, and household management. Productive (Public) is the role of a breadwinner or other role performed outside the home to make money. The social role is a role in its relationship with other community members. Changing roles is a response to the impacts of climate change. Changes in roles cause women's adaptive capacity or vulnerability to climate change to increase [13].

McCright (2010) argues that women are knowledgeable in climate change rather than men, but women have a greater concentration on climate change than men. Women face special barriers from their capacity to adapt to the impacts of current and future climate change [14], so a strategy is needed specifically to mainstream women in climate change adaptation. This strategy is needed to ensure efforts to increase adaptive capacity and reduce women's vulnerability to climate change. Adaptation needs and priorities are classified into 3 (three) aspects, namely:

1. a comfortable place to live, better
2. access to climate change information and its impact, health services, and access to transportation,
3. other livelihood options through increased technical knowledge and availability of fisheries resources and local counseling and assistance and knowledge of climate change at sea

Based on the description above, the strategy that can be taken as a gender-based climate change adaptation model for coastal areas in the research location includes:

1. Increasing the political role of women in terms of leadership, community organizations, and disaster management. This political role has begun to look good for women in Bungus, the city of Padang, with the election of women as village heads.
2. Enhancing women's skills can be in the form of climate education through the field climate school, ornamental fish cultivation, fish salting and others.
3. Increased productive work is a type of work that makes money . Husbands not only earn income as fishermen, but also have the ability to market catches. And when you don't go to sea your husband must find a side job. Also when a husband does not go to sea, he can also help his wife for reproductive roles such as taking children to school.
4. Equalization of reproductive work is work that guarantees the management and reproduction of the workforce (which includes taking care of domestic work and childbearing. This reproductive work usually does not make money. Although the husband does not do the reproductive role in full, but household activities such as home management such as the provision of clean water, waste management, riol routine household hygiene and family nutrition fulfillment.

The community needs to connect climate vulnerability to current problems and identify future challenges that are important to themselves, their families and livelihoods [15]

4. Conclusion

The vulnerability index is positive function of outcrops and sensitivity and negative function of adaptation ability. Then obtained coastal area vulnerability index of 305 (vulnerable classification) Efforts to increase adaptive capacity and reduce women's vulnerability to climate change through privacy to stay, better access to climate change information and other livelihood options. Gender-based climate change adaptation models for coastal areas are enhancing women's political roles, enhancing women's skills capacity through Climate Field Schools and improving productive work.

Acknowledgment

Authors would like to thank the Director of the Padang Health Ministry Health Polytechnic for both moral and material.

References

[1] Brody A, Demetriades J and Esplen E 2008 *Gender and climate change : mapping the linkages A scoping study on knowledge and gaps* (Bridge, Institute of Development Studies, University of Sussex) vol. 44 June 2008 pp 1–23
[2] DTE 2012 *Climate Justice and Sustainable Livelihood Volume II* (Jakarta, Indonesia)
[3] UNDP 2009 *Gender and Climate Change, First Edit.* (Mexico: United Nation Development Programme)

[4] Khalid P, Maimunah K S and Dewi 2011 Gender Justice in Climate justice *Civil Society Forum* Jakarta 2011

[5] Desmawan B T dan Sukamdi S 2012 Adaptasi Masyarakat Kawasan Pesisir Terhadap Banjir Rob di Kecamatan Sayung, Kabupaten Demak, Jawa Tengah *Jurnal Bumi Indonesia* **1**(1) 1–9 available at http://lib.geo.ugm.ac.id/ojs/index.php/jbi/article/view/38

[6] Yudhicara 2008 Kaitan antara Karakteristik Pantai Provinsi Sumatera Barat dengan Potensi Kerawanan Tsunami *Jurnal Geologi Indonesia* **3**(2) 95–106

[7] Turner B L et al 2003 A framework for vulnerability analysis in sustainability science *PNAS* **100**(14) 8074–8079

[8] Villagrán de Léon, Juan Carlos 2006 *Vulnerability: a conceptual and methodological review* (Bonn, Germany: UNU-EHS)

[9] Kaly U L, Pratt C R and Mitchell J 2004 *The Demonstration Environmental Vulnerability Index (EVI) 2004, SOPAC Technical Report 384*

[10] Kabir I et al 2016 Knowledge and perception about climate change and human health : findings from a baseline survey among vulnerable communities in Bangladesh *BMC Public Health* **16** (266); https://doi.org/10.1186/s12889-016-2930-3

[11] Azizi A, Hikmah dan Pranowo S A 2012 Peran gender dalam pengambilan keputusan rumah tangga nelayan di Kota Semarang Utara, Provinsi Jawa Tengah *Jurnal Sosial Ekonomi Kelautan dan Perikanan* **7**(1) 113–125; http://dx.doi.org/10.15578/jsekp.v7i1.5740

[12] Rochmayanto Y and Kurniasih P 2013 The role of gender on climate change adaptation in the mountainous ecosystem at Solok District, West Sumatera) *Jurnal Analisis Kebijakan dan Kehutanan* **10**(3) 203–213 available at http://ejournal.forda-mof.org/ejournal-litbang/index.php/JAKK/article/view/328/0

[13] McCright A M 2010 The effects of gender on climate change knowledge and concern in the American public *Popul. Environ.* **32(**1) 66–87; https://doi.org/10.1007/s11111-010-0113-1

[14] Bridge, *Gender and climate change : mapping the linkages A scoping study on knowledge and gaps June 2008*, vol. 44, no. June. UK, 2008.

[15] Haque M A, Yamamoto S S, Malik A A and Sauerborn R 2012 Households' perception of climate change and human health risks: A community perspective *Environ. Health* **11(**1); https://doi.org/10.1186/1476-069X-11-1

Policy implementation evaluation about quality management and pollution control of water in Regency of Bekasi

Arief Saefudin[1,*], Nadiroh[1] and Rukaesih Achmad[2]

[1]Environmental Education Postgraduate Program, State University of Jakarta, Indonesia
[2]Lecturer Environtmental Education Postgraduate Program, Trisakti Insurance Business School Jakarta, Indonesia

*asaefudin@unj.ac.id

Abstract. This study aims to determine the implementation of public policy about Water Quality Management and Water Pollution Control in the Regency of Bekasi of Indonesia in order to minimize water pollution. Based on previous research can be synthesized that: Pollution of river water occurs caused by the disposal of waste containing organic or inorganic substances to water sources that cause various diseases and silting the river and have an impact on the environment. The method used in this Policy Research is content analysis. Data of the research results in the analysis by using Merilee S. Grindle model. The results show from of policy content : Objectives of policy that has not adhered. Context of implementation: Inter-institutional coordination in the realization of policy has not gone well. Implementation strategy used: Maximizing policy support from Ministry of Environment, Overcoming the lack of human resources in the field of environment, Maximizing infrastructure and financial facilities.

1. Introduction

Water resources both groundwater and water that are on the surface of the soil should be maintained and kept not polluted so that in turn can cause disasters that affect the lives of living things that are on this earth. It needs to be wise to reduce and improve if water resources are polluted, if effective measures to reduce and improve water pollution and its sources are not found, it is estimated by the UN that 2.7 billion people will face water shortages by 2025 [1]. Similarly, the increased negative effects on food safety due to water and soil pollution make more people at risk of carcinogenic diseases, which contribute to the "cancer village" [2]. Furthermore, the use of chemicals continues to increase in both the number and range of different substances that cause river pollution in Liberia, it warns the need for increased public awareness to protect ecosystems and health from risks associated with chemical pollution problems [3]. This requires a review of river and coastal management policies as well as in Japan that suffered severe water pollution during economic growth in the 1960s. With population and industrial and coastal concentrations producing large amounts of pollutants such as chemicals and so on into the sea and this causes health hazards, also damaging the fisheries and livelihoods environment [4]. Similarly, past pollution incidents in China due to industrial water pollution, in environmental governance are meaningful to ensure prevention and mitigation of future pollution environments [5]. The same is true of the history of water pollution in Switzer land due to the sediment of Big Lake Perialpine and Lucerne, where the peak of water pollution by toxic metals due to the release of industrial wastewater was achieved in the mid-20th century. Subsequently after the 20th century both lakes showed a decrease in metal pollution after implementation of the wastewater treatment plant [6].

Control of rural non-sourced water pollution in Hubei Province of China which is a farming area shows that water pollution has continued to increase over the last ten years by chemical fertilizers, poultry and livestock, aquaculture, and rural life, these pollutants flow into drinking waterways [7]. Similarly, the quality of Marreco river water, Brazil, based on the analysis of major components and toxicological tests showed strong evidence of increased levels of K and Ca associated with anthropogenic activity installed in watersheds [8]. In line with the above, there is also pollution of river water in the Danube River in Serbia, so it is necessary to clarify the ecological and identification of para meters that affect the water quality variables, which emphasizes the importance in realizing the principles of water management [9]. Similarly, the assessment of the status of water quality for the Selangor River in Malaysia where the condition of water degradation in the river will continue in the coming years because the pollutant load from poultry, municipal wastewater, and industrial waste water is less actively handled causing problems water quality used for some purposes [10]. Similarly, in West Bengal, India, where research results show irrigation water contaminated with arsenic so that rice, wheat, vegetables and other cultivation consumed by humans in Nadi District, India will be exposed to arsenic [11]. Similarly, mercury flows affecting Water Quality in the EU where know ledge of the mainstream of priority substances and production systems and consumption structures in society causes this flow to be a prerequisite for any attempt to predict and understand the fate of their environment and minimize future environ mental burden [12]. As well as river condition in the Regency of Bekasi, currently not escape from pollution problem where the quality of river water in nine watersheds in the Regency of Bekasi is seriously polluted [13].

Based on previous research, it can be synthesized that: Pollution of water occurs caused by the disposal of wastes containing organic or inorganic substances to water sources that cause various diseases and silting the river and have an impact on the environment. The results of this research provide in-depth information on the causes, the contents of pollutants, while the last five years research has not found research on public policy in an effort to minimize water pollution, it means increasing the accuracy sequence steps to solve the problem by researching the implementation of public policy on water quality management and water pollution control needs to be done, so in this study, the focus of research on the implementation of policy of Regional Regulation No. 9 year 2005 about Water Quality Management and Water Pollution Control in the Regency of Bekasi using Merilee S. Grindle model. Where according to this model the activity of policy implementation is influenced by Policy Content and context Implementation.

2. Method

The purpose of this study is to know the Implementation of Regional Regulation Policy No. 9 the year 2005 about Water Quality Management and Water Pollution Control in order to minimize pollution of river water in the Regency of Bekasi. The method used in this Policy Research is content analysis. Data of the research results was analyzed by using Merilee S. Grindle model [14].

3. Results and Discussion

3.1. Results

The results of the study are based on three perspectives, namely: (1) The content of the policy influenced by factors such as the interests that effect, the benefits generated, the degree of change, the location of decision-making, implementing policies and available resources have not been able to minimize water pollution. This led to the implementation of the policy have not been realized and can be indicated based on BPLHD document on 2016, that the Cikarang watershed is still polluted.

IOP Conf. Series: Earth and Environmental Science **314** (2019) 012087 doi:10.1088/1755-1315/314/1/012087

Based on the parameters of COD and BOD the results of the survey of sampling in water bodies showed the value of COD lowest monitoring point 67.22 mg / L, highest 115.52 mg / L, for the value of BOD lowest monitoring point 24.48 mg / L, highest 62,58 mg / L. This indicates that the Cikarang River is above the normal limit required by Government Regulation No. 82/2001 on Water Quality Management and Water Pollution Control, Class III is COD 50 mg / L, BOD 6 mg / L where the intended Water may be used for the cultivation of freshwater fish, farms, water for irrigating gardens, and / or other designations which require the same water quality as that purpose. Thus, water from the Cikarang River Basin (DAS) cannot be used for the people along the river. (2) The policy environment is the factors affecting the policy implementers such as the interests, powers, and strategies of the actors involved, the Characteristics of the ruling institutions and regimes, the Compliance and the Response of Executives, have not met the expectations of the policy objectives of ensuring water according to water quality standard. The existence of the interests of the actors involved so that the policy cannot be implemented, then the characteristic of the ruling institution that the policy is a political product made by the legislative and executive institution (head of the region) and implemented by the administrative body responsible to the chief executive while the compliance and the responsiveness of the policy implementer in accordance with the main duties and functions set forth in the Regent Regulation No. 38 of 2009 About the Organization and Work Procedures of the Environmental Control Agency of the Regency of Bekasi, although the lack of human resources in the field of environment (3) The performance of the institution is the implementation strategy used in realizing the policy using SWOT analysis.

3.2. Discussion

Factors causing unrealized implementation of the policy are viewed from the perspective of Policy Content, as follows:

First,the interests that affect the policy. The result of the research indicates that the interest will influence the policy target, where the society and industry residing in a region or industrial zone are prohibited to dispose waste into river body unless they have permission from the local government, but the community does not all understand about the local regulation. 9 of 2005 except for large industries in the region and industrial zones that must comply with the regulation. In fact, the river in Bekasi Regency is still polluted and this condition is not in line with the expectations of Article 14, Article 15 About the Liquid Waste Disposal Permits and Article 24 About disposal of liquid waste. Local Regulation is a form of public policy aimed at regulating both community and institutional policy objectives, as follows: Public policy is a set of inter related decisions regarding the selection of targets and the benefits to achieve them, where the goals are community and institutions [15]. The other opinion that public policy is the study of what the government is doing, why the government took the action, and what the consequences of that action [16]. The substantive elements of public policy and procedural steps taken by the decision-making authority during the policy cycle affect the perceptions of policy and legitimacy held by stake holders and the public. In the case of substantive, the content of the policy should align with the dominant attitude of the affected policy community and ideally the wider society [17]. Thus, it can be stated that the regulation on water quality management and water pollution control in Articles 14, 15 and 24 has not affected the policy objectives, especially the community, the general public has not yet learned that disposing of domestic waste into the river body violates local regulations.

Second, Benefits resulting from the policy. The results of research show that the benefits generated from the policy cannot be felt directly by the community that is clean water, and the industry to obtain the ease of licensing services and transparent in terms of financing, but the reality is still far from their desire, so this is not in line with expectations. Article 4 About Goals and Objectives Policies and Article 25 About Waste Water Disposal. According to the above Articles it is implied that local legislation implies that the policy should have benefits for the policy objective, especially the community, as it follows that the design of public policy and management must be closely connected, so that public policy is implemented to achieve the desired outcomes [18], and the Conformity between public will and public policy is of particular importance in representative democracy, and whet her the public gets the desired policy [19]. Thus, it can be stated that the regulation on water quality management and water pollution control in Article 4 and Article 25 has not been beneficial to the policy targets, especially the people who have not been able to utilize clean water from the river since the river water has been contaminated.

Third, the resulting degree of policy. The result of the research shows that the absence of degree of change resulting from the policy with river indication until now is still polluted and has not seen any efforts to recover, so the community considers that the polluted river is assumed to be allowed to dispose garbage or household waste and this has an impact on pollution of river water. This condition is certainly not in accordance with the expectations of Article 21 About the obligation of persons or institutions. Regional Regulation is a form of public policy aimed at regulating the policy objectives of both society and institution and how its effect on society, such as the following understanding that policy evaluation is objective, systematic, and empirical examination of the effect of the policy on the target in terms of goals to be achieved, after policy implemented [20. In another sense, policy implementation reflects a complex process of change in which government decisions are converted into programs, procedures, regulations or practices aimed at social improvement [21], and potential influences that have implications for designing and implementing public policy whereby private institutions can encourage effective, voluntary environ mental management. in this sense, should involve the potential of private institutions to encourage the expected environmental change [22]. Thus, it can be stated that the law on water quality management and water pollution control in Article 21 has not been able to change the behavior of policy targets, where up to now the river's water condition is still contaminated.

Fourth, the position of the policy maker. The result of the research shows that the position of the policy maker is the regent based on the approval of the Regional Representative Council of Bekasi Regency. The Regent is the head of the region directly elected by the people of Bekasi Regency, with the direct election of the people should also support the implementation of policies that have been issued by the local government, but in terms of policies on water quality management and water pollution control is still far from expectations, where pollution on the river is still on going, this means not in accordance with the hope of Article17 About Handling of Water Pollution.

Regional Regulations are a form of public policy aimed at regulating and complying with both public and institutional policy objectives, as follows: Public policy is an action, aims and a government statement, and an explanation they provide for what happened (or did not happen). That sense means that if the government chooses to do something or not do something, it has its purpose and purpose [23]. In another sense that to implement policies and achieve policy objectives effectively, adequate public support is an important prerequisite; most will fail, and the government may also experience setbacks in other policy areas in the absence of public support [24. Thus, it can be stated that the regulation on water quality management and water pollution control especially in Article 17 has not received support from the policy objectives.

Fifth, implementing the policy. The result of the research shows that the right regional apparatus unit as the implementer of the policy is the BPLH of the Regency of Bekasi, which should be able to implement the policy of Regional Regulation no. 9 Year 2005 on Water Quality Management and Water Pollution Control, although it has implemented according to the mandate of the local regulation, but see the reality until now the results are not optimal, this is due to the limitations possessed by the Regency of Bekasi BPLHD such as human resources who understand the limited field of environment and the lack financial problems in supporting the preparation of planning, implementation, monitoring and evaluation, so that this condition is not in accordance with the expectations in Article 5 About authority in water quality management, Article 12 About authority in water pollution control and Article 27 About Supervision.

Regional Regulation is a form of public policy aimed at regulating the policy objectives of both society and institution and implemented by the implementing agency policy. In this sense, the Public Policy is a series of instructions from decision makers to policy implementers explaining the objectives and ways of achieving that goal [25] and another understanding that the Public Policy addresses the question of how issues and the issues are constructed and defined, and how they are placed in the policy agenda and other political agenda [26] that in the public policy process, joint learning between policy actors is important in shaping how the process is rolling and the type possible policy outcomes [27]. as the task of implementation is to establish a linkage that makes it easy for policy purposes to be realized as a result of a government activity (Grindle, S Merilee 1980). Thus, it can be stated that the regulation on water quality management and water pollution control in Article 5, Article 12, and Article 27 result has not been optimal to overcome the pollution of river water.

Sixth, sources available in policy sup port. The results showed that the resources available in support of policies, especially related to finance are still minimal. The budget managed in the BPLHD of the Regency of Bekasi is limited to about 3 percent of the Regency of Bekasi's Budgets [28] so that the planned program has not been fully realized. The wide range of supervision and limited supervisory officers is one of the weaknesses of supervision that lead to violations of local regulations so that if pollution occurs, it is the environmental polluters that must be responsible, but in reality, the river is always in polluted condition while the polluters are difficult to ask accountability. This is certainly not in line with expectations. Article 16 About the Emergency Countermeasures, and Article 26 About coaching.

Local Regulation is a form of public policy aimed at regulating the policy objectives of both public and institutions, of course, the need for resource support, as one of the opinion that, the implementation of policy is the stage where the policy has been taken by administrative units that mobilize financial and human resources [29]. Another opinion is that the state budget is one of the most important public policy instruments because it sets the priorities of government policy and has the potential to determine win and lose, which can be interpreted to succeed or fail [30]. Thus, it can be stated that the regulation on water quality management and water pollution control in Article 16 and Article 26 has not been realized to minimize the contamination of river water in Bekasi Regency due to the lack of support of resources, especially financial resources.

Factors causing unrealized implementation of the policy are seen from the perspective of the Implementation Context. In addition to policy content, the context or environment of a policy implementation is an important factor that determines the success of policy implementation. There are 3 indicators according to Grindle that are the context in the implementation of the policy, as follows:

First, he powers, interests, and strategies of actors involved in the policy. The results showed that the power, interests, and strategies of actors involved in policy are very influential in the successful implementation of a policy. The closure of the waste management plant ever conducted by BPLHD in an industrial area in Jababeka has an impact on the laying off industrial workers, the economics of the community in disturbed areas, the pressure of trade unions and employers and politicians who overshadow the employment field, so that the implementation of policies relating to the application of sanctions the result is not optimal. This is certainly not in accordance with the expectations of Article 30 about Criminal provisions. Regional Regulation is a form of public policy aimed at regulating both individual and group policy goals. Policies should be understood as processes involving various elements that do not necessarily move in an integrated way but can thrive in various steps or even opposite directions [31]. In regions, contested policies such as migrant integration, organizational set ting patterns often have the high-level character, leading to different policy frameworks and thus complicating government in multi-level arrangements, so coordination is required between levels to enable policies to be implemented [32]. By constructing a conceptual map that takes into account the constellation of ideas, institutional arrangements, and interests involved in dealing with policy issues, so that there should be no tug of interest in the field in order to have no barriers to implementation [33]. Thus, it can be stated that the regulation on water quality management and water pollution control in Article 30 regarding criminal sanction to violators of Regional Regulation No, 9 of 2005, due to the pressure and tug of interest of actors involved in the implementation phase of the policy cause not yet can be implemented maximally.

Second, characteristics of institutions and authorities in policy. The result of the research shows that the Regency of Bekasi Environmental Management Agency (BPLH) as the Regional De vice Work Unit (SKPD) is the implementer of regional regulation no. 9 of 2005 on Water Quality Management and Water Pollution Control in the Regency of Bekasi . Regional regulation is a political product created by the executive body with the approval of the legislature, but in its implementation by the administrative body is responsible to the regent as the executive body. Communities reporting pollution, if slowly responded by policy implementers, then they channel the report to the legislature or DPRD and then legislatively coordinate with the executive or regent to be followed up by the implementer of the policy. So this process takes a long time to solve environmental problems quickly and this is not in line with the expectations of Article 17 About Handling of water pollution. Regional Regulation is a form of public policy aimed at regulating both community and institutional policy objectives and implemented by policy implementers, as it understands that the Policy cannot work without implementation, all or one of which includes implementation. There must be a starting point. If no action is started, implementation cannot be performed. There should be an ending point. Implementation cannot succeed or fail without purpose that decides [34]. in another sense that policymakers may encourage or initiate some cumulative, small-scale experiments that are translated into additional policy changes so that problems can be resolved quickly [35]. Thus, it can be stated that the regulation on water quality management and water pollution control in article 17, although the policy implementers are working, the results are not yet optimal.

Third, compliance and responsiveness of policy implementers. The results show that the level of compliance and responsiveness of policy implementers is still not optimal results, this is due to the policy objectives that are not yet compliant with local regulations, where still the disposal of liquid waste into the river body and the absence of real action against the violation, this is certainly not in accordance with the expectation of Article 28 About Super vision and Article 2 About General Provisions.

Regional Regulation is a form of public policy aimed at regulating both public and institutional policy goals. The following opinion is that one way to analyze implementation problems is to begin by thinking about what 'perfect administration' is like in the way economists apply the perfect competition model. A perfect administration can be defined as the condition in which the 'external' element of re source availability and political acceptance is combined with 'administration' to produce the perfect policy implementation [36], as well as that the Science of socialization and implementation is an important help build field skills. Meaning that policy implementers should be able to disseminate policies to policy goals for successful implementation [37]. Thus, it can be stated that the regulation on water quality management and water pollution control in Article 2 and Article 28 even though the implementer of the policy has worked, but the result is not yet optimal.

4. Coclusion

Based on the results of research and discussion can be concluded as follows:

1. Factors that have not realized the implementation of policy of Regulation No. 9 of 2005 on Water Quality Management and Water Pollution Control in Bekasi Regency in order to minimize River Water Pollution vie wed from the perspective of policy con tent, namely: (1) Objectives of policy that has not been complied, (2) Human and financial resources are still limited.
2. Factors that have not realized the implementation of policy of Regulation No. 9 of 2005 on Water Quality Management and Water Pollution Control in Bekasi Regency in order to minimize River Water Pollution from the perspective of implementation context, namely: (1) Coordination between institutions in realizing the policy not going well. (2) Control of industry and environment is still weak.
3. Implementation strategies used in the realization of policies are: (1) Maximizing policy support from the Ministry of the Environment; (2) Overcoming the lack of human resources in the field of environment; (3) Maximizing infrastructure and financial facilities; (4) Addressing non-compliance of policy objectives.

References

[1] Satinder A dan Hristovski K Novel Solutions to Water Pollution https://global.oup.com/academic/product/novel-solutions-to-water-pollution-9780841227545?cc=id&lang=en&#, 2013

[2] Lu Y 2015 Impacts of soil and water pollution on food safety and health risks in China *Journal Environment International* **77**: 5-15

[3] Kuznetsov A N and Yu A Fedorov 2014 Oil Com ponents in The Mouth Area of The Don R. And in The Sea of Ozov: Results of many year studies *Journal Water Quality and Protection: Environmental Aspects* **41**(1): 55-64

[4] Tomita A 2015 Review of coastal management policy in Japan *journal of Coastal Conservation* **19**(4): 393-404

[5] Miao X, Yanhong T, Christina W Y Wong and Hongyu Z 2015 The Latent Causal Chain of Industrial Water Pollution in China *Journal Environmental Pollution* **196**: 473-477

[6] Thevenon F and John P 2012 Water Pollution History of Switzerland Recorded by Sediments of the Large and Deep Perialpine Lakes Lucerne and Geneva *Journal Water, Air and Soil Pollution* **223**(9): 6157–6169

[7] Wu M, Xianqiang T, Qingyun L, Wenjung Y, Feng J, Minzhe T and Miklas S 2013 Review of Ecological Engineering Solutions for Rural Non Point Source Water Pollution Control in Hubei Province, China *Journal Water, Air and Soil Pollution* **224**:1561

[8] Palacio, Moreno, Soraya 2016 Assesment of Anthropogenic Impact on The The Water Quality of Marreco River, Brazil, Based on Principal Component Analysis and Toxicological Assays *Journal Water, Air and Soil Pollution* **227**: 307

[9] Takic and Ljiljana 2017 The Assesment of The Danube River Water Pollution in Serbia *Journal Water, Air and Soil Pollution* **228**: 380

[10] Fulazzaky and Mohamad A 2010 Assessment of Water Quality Status for the Selangor River in Malaysia *Water, Air, and Soil Pollution* **205**:63

[11] Bhattacharya P 2010 Arsenic Contamination in Rice, Wheat, Pulses, and Vegetables: A Study in an Arsenic Affected Area of West Bengal, India *Journal Water, Air, & Soil Pollution* **213**(1-4): 3-13

[12] Sundseth and Kyrre 2011Substance Flow Analysis of Mercury Affecting Water Qua lity in the European Union *Water, Air, & Soil Pollution* **223**(1): 429-442

[13] Rencana Strategis Badan Pengendalian Lingkungan Hidup Kabupaten Bekasi Tahun 2012-2017 2010

[14] Grindle M S 1980 *Politics and Policy Implementation in the Third World* (New Jersey: Princeton University Press)

[15] Hill M 2005 *The Public Policy Process* (England: Pearson Longman)

[16] Dye T R 1975 *Understanding Public Policy* (Englewood Cliffs, New Jersey)

[17] Wallner J 2008 Legitimacy and Public Policy: Seeing Beyond Effectiveness, Efficiency, and Performance *Policy Studies Journal* **36**(3): 421-443

[18] Mintrom, Michael and Jonathan L 2015 Creating Public Value: Tightening Connections Between Policy Design and Public Management *Policy Studies Journal* **45**(1): 170-190

[19] Wlezien C 2016 Public Opinion and Policy Representation: On Conceptu alization, Measurement, and Interpretation *Policy Studies Journal* **45**(4): 561-582

[20] Parson W 2006 *Public Policy: Pengantar Teori dan Praktik Analisis Kebijakan* (Prenada Media Group)

[21] DeGroff A dan Margaret C 2009 Policy Implementation: Implication for Evaluation *Journal Policy Implementation* (124): 47-60

[22] Khanna M, Cody J, Patricia K and David E E 2007 Motivations for Voluntary Environmental Management *Policy Studies Journal* **35**(4): 751-772

[23] Wilson in Kevin B. Smith and Christopher W Larimer 2009 *The Public Policy Theory Primer* (Central Avenue Boulder, CO: Westview Press)

IOP Conf. Series: Earth and Environmental Science **314** (2019) 012087 doi:10.1088/1755-1315/314/1/012087

[24] Wan C 2017 A Review on Political Factors Influencing Public Support for Urban Environmental Policy *Journal Environmental Science and Policy* **75**: 70-80

[25] Winarno B 2012 *Kebijakan Publik, Teori dan Proses* (Yogyakarta: Media Pressindo)

[26] Parson W 2006 *Public Policy: Pengantar Teori dan Praktik Analisis Kebijakan* (Prenada Media Group)

[27] Heikkila, Tanya and Andrea K. G 2017 Building a Conceptual Approach to Collective Learning: Lessons for Public Policy Scholars *Policy Studies Journal* **41**(3): 484-512

[28] Regional Statistics of Bekasi Regency in 2017

[29] Dunn, N W 2000 *Pengantar Analisis Kebijakan Publik* (Gadjah Mada: Universitas Press)

[30] Langer, Ines, Ana and Iñaki Sagarzazu 2017 Are All Policy Decisions Equal? Explaining the Variation in Media Coverage of the UK Budget *Policy Studies Journal* **45**(2): 337-358

[31] Candel, Jeroen J L dan Robbert B 2016 Toward A Proccessual Understanding of Policy Integration *Journal Policy Sciences* **49**(3): 211-231

[32] Scholten P W A 2012 Agenda Dynamic and The Multi-Level Governance of Intrsctable Policy Controversies: The Case of Migrant Integration Policies in The Netherlands *Journal Policy Science* **46**(3): 217-236

[33] May P J dan Ashley E J 2013 Policy Regime Perspectives: Policies, Politics, and Governing *Policy Studies Journal* **41**(3): 426-452

[34] Pressman J L and Aaron W 1984 *Implementation: How Great Expectation in Washington are Dashed in Oakland* (California: University of California Press)

[35] Mei C and Zhilin L 2014 Experiment-Based Policy Making or Conscious Policy Design? The Case of Urban Housing Reform in China *Journal Policy Sciences* **47**(3): 321-337

[36] Hill M and Peter H 2002 *Implementing Public Policy: Governance Theory and in Practice* (London: Sage Publications Ltd)

[37] Meissner I H 2013 The U.S. training institute for dissemination and implementation research in health *Implementation Science* **8**: 12

Evaluating the implementation of Konawe regional regulation number 5/2007 on public mining management

Yuslan*, Nadiroh and Mukhneri Mukhtar

Department of Environmental Management, State University of Jakarta, Indonesia

*yuslan_pl12s3@mahasiswa.unj.ac.id

Abstract. This research aims at determining the implementation of Konawe regional regulation number 5/2007 on public mining management, with the focus on a C-Class mining activities in Tuoy Village. Tuoy Village, Unaaha District, is one of the center of C group mining activities in Konawe Regency, Southeast Sulawesi Province. It is recommended to the Regional Government and all elements involved in mining management to take concrete steps in preventing further damage to the environment in the futureby maximizing the implementation. To find out the implementation of Konawe Regional Regulation Number 5/2007 on Public Mining Management the writers use Merilee S. Grindle model of approach. The level of success of a public policy according to Grindle is determined by the level of policy implementation itself, which consists of the Content of Policy and the Context of Policy.

1. Introduction

Konawe District is, located in Southeast Sulawesi Province, a potential mining.The Regional Government of Konawe Regency has issued a Regional Regulation of Konawe Regency Number 5/2007 on Public Mining Management. This regulation then became the reference of management procedures for business actors engaged in mining. Unfortunately, because mining activities continue to invite the spotlight of the surrounding community due to environmental destruction, unauthorized mining, besides damaging the environment, also endangers the lives of miners. This is due to the limited knowledge of miners and also due to lack of supervision from the relevant agencies. This kind of condition happens in Konawe Regency, spesifically in the location of sand mining in Tuoy Village. To keep the mining activity going well without ignoring the environmental damage factor, then through the Konawe Regional RegencyNumber 5/2007, the provision of General Mining Management is stipulated. The regulation is intended to be a patron in mining activities, especially in the region. The existence of an indication of non-conformity to the form of implementation of the regulation requires a proof that is supported by scientific rules so that it can be accepted as a scientific process. To find out more about the existence of regional regulation, the researchers conducted a research on implementation of Konawe Regional Regulation Number 5/2007 on Public Mining Management.

2. Research Method

Basically, the general purpose of the research is to find out the implementation of Konawe Regional Regulation Number 5/2007 on Public Mining Management by using Merilee S. Grindle model approach. The level of success of a public policy according to Grindle is determined by the level of policy implementation itself, which consists of the Content of Policy and the Context of Policy. The

Dimension of Content of Policy consists of: Interest, Type of Benefits, Extent of Change Envision, Site of Decision Making, Implementer program, Resources Committed. While the Context of Policy Dimension consists of: Power, Interest, and Strategy of Actor Involved, Institution and Regime Characteristic, Compliance and Responsiveness.

Based on the above point of view, the reason why this research uses a post-positivism approach is in accordance with one of the characteristics of post-positivism paradigm based on verification theory, this research also proposed theoretical reference in order to see and assess Government policy. Nevertheless, this study uses a qualitative approach as the principle of fostering. This is in line with Sugiyono's explanation which reveals that qualitative research methods are often referred to as naturalistic research methods because the research is done on natural conditions [1]

The design of this evaluation study is presented so that the focus of the research stage undertaken by the researcher can be clear exposure. The research design scheme as shown in the following figure 1.

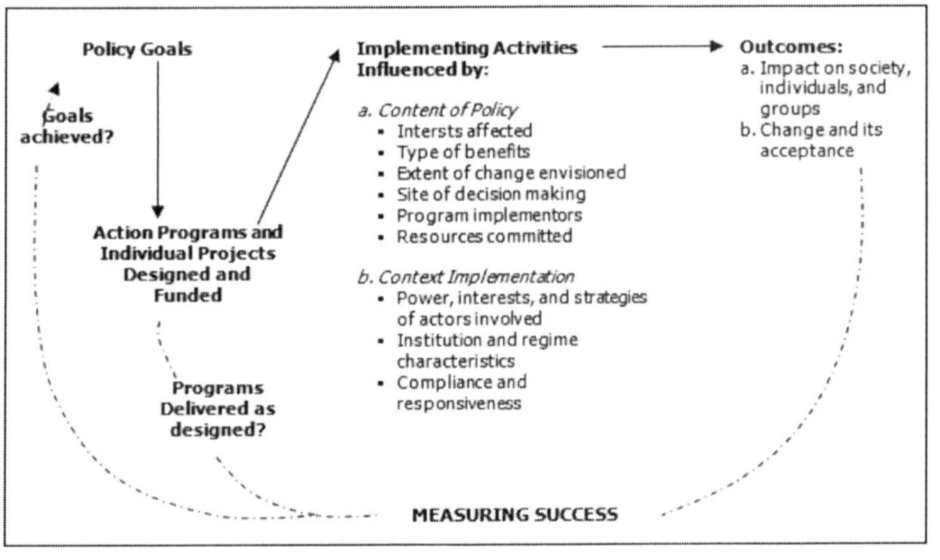

Figure 1. The research design scheme [4]

The picture above shows that a program or policy has a clear purpose as a form of policy value orientation. The objectives of the program implementation are formulated into specific programs of action and projects that are designed and financed. The implementation of policies or programs is largely influenced by the content of the policy and implementation context. Outcomes of the program are seen through their impact on the intended target both individuals and groups and communities.

Techniques of data collection in this research were observation, in-depth interview, and document study as part of research instrument [2]. The research instrument was written based on interviews or observations or questionnaires prepared to obtain information from respondents [3]. These three techniques sre common techniques used in research as processed data for research results.

3. Result and Discussion

The result of evaluation in this research is based on the evaluation model used in the form of a public policy, which is determined by the level of policy implementation itself, consisting of Content of Policy and Context of Policy. The evaluation results of components in the evaluation are as follows:

Firts, Interests Affected ; Interest affected is closely related to the various interests that affect a policy implementation. Discussion based on this indicator will argue about the many interests in connection with the implementation of the Konawe Regional Regulation number 5/2007 on public mining management when it is associated with the form of supervision and environmental management. In the context of decision-making, interests that greatly affect the implementation of this policy are the head of the Region through an extension of the Mining Service. From the interviews, it can be said that there are many interests that affect the implementation of Konawe Regional Regulation number 5/2007 on the Public Mining Management. Although the statements are general in nature, there is a clear picture that different forms of interest are concerned. This condition is reinforced by community involvement system in the form of supervision and environmental management especially post-mining.Then, based on the form of guidance and supervision and environmental management, it was found that the involvement of various elements had an influence on the interests of both the government, non-government institutions and the community itself as well as a mining business actor.

Second, Type of Benefits ; At this point the content of policy shows that regional regulation number 5/2007 is a policy that has benefits that demonstrate the positive impacts resulting from the implementation of policies on general mining management, especially in the field of environmental monitoring and management. Based on the research, the real picture behind the economic benefits that can be felt by the business community in the fulfillment of the necessities of life, of courseunfortunate impact member also for other communities if the mining process and then not accompanied by environmental conservation. From the research it can be concluded that the activity of guidance and supervision and management of the environment, especially post-mining, becomes very useful. Konawe regional regulation number 5/2007 can control guidance and supervision, especially in environmental management over general mining management activities. Then the management of the environment, especially after mining activities, of course, has a lot ofbenefits, not only limited to the observance of the rules of the shrine-law invitation but also as a form of our interpretation as social beings who must be responsible for environmental sustainability.

Third, Extent of Change Envisioned (Degree of Change to be reached); Every policy has a target to be and wants to achieve. The content of policy to be explained at this point is that the extent to which changes are wanted or want to be achieved after Konawe Regional Regulation no. 5 of 2007 on public mining management. With the Regional Regulation Konawe number 5 of 2007 on Public Mining management, the form of mining management more focused and well monitored.The content of policy to be explained at this point is that the extent to which changes can be achieved after Konawe Regional Regulation number. 5/2007 on public mining management is implied. Based on the results of research, it is clear that there is a change in the implementation of the Konawe Regional Regulation number 5/2007. Tuoy becomes the central point of mining activities on the C min scale. It can be concluded that the level of change to be achieved has a difference. However, the differencecan be a means to improve the quality of implementation policy issued [5]. Researchers assume that the rate of change that occurred has not been too proportional to what is expected for the activity of public mining management conducted.

Fourth, Site of Decision Making (Location of Decision Making);Decision-making plays an important role in the implementation of a policy. Decision-making on the local regulation must originate from the government as an executive institution through the head of the region with the approval of the DPRD as the legislative body. In this Decision Making Declaration, it can be concluded that basically the location of decision making through the policy towards general mining management is appropriate even though at the level of reality field there are various kinds of depth in its implementation. The existing constraints are only technical in nature where things can be affected by the lack of field personnel to implement the forms of supervision as mandated by the local regulation. Even so what is the purpose of the Konawe Regional Regulation number 5 of 2007 is judged appropriate.

Fifth, Implementors In implementing ; The policy on the Konawe Regional Regulation no. 5 of 2007 on public mining management should be supported by a competent and capable policy implementer for the success of a policy. It is therefore necessary to know whether the policy has specified the details of the perpetrator or the implementor in detail. In chapter two on the authority and responsibility of the affairs in the field of general mining business (Peratutan Daerah Konawe regency number 5 of 2007) concerning general mining management mentioned that the authority and responsibility of general affairs in the field of mining is done by the head of the Region. The Regent as the head of the Region then assigns the Mining and Energy Agency to plan and carry out governmental affairs in the field of general mining business. In relation to the implementer of the program stated in the policy through the regional regulation of Konawe no 5 of 2007 stated that the implementation of government affairs in the field of general mining include: (a) Arrangement; (b) Licensing; (c) Guidance and Supervision; (d) Evaluation and Reporting; While the authority and responsibilities of Regional Government in the field of general mining include: a. To organize, manage, develop and develop general mining business activities; b. Conducting survey activities, inventorying and mapping of minerals; c. Development and stipulation of licensing procedures for permit disclosure; d. Development and establishment of procedures for guidance, supervision and inspection; e. Arranging and and determining the competence of positions in the organization of the mining office; f. Arranging and stipulating the procedures for managing information on energy and mineral resources in the regions; g. Arranging and establishing procedures for evaluation and reporting; h. Resolving the issue of overlapping areas between general mining business and other business or other purposes; i. Planning and development of mining areas and areas; j. Unauthorized mining prevention and mitigation; k. Determination and supervision of environmental management and occupational health and safety (K3) in the framework of general mining business; l. Determination and supervision and development and implementation of community development programs around mining business areas; m. Granting of mining business permit; n. Conduct regional levies or state levies; o. Determination of people's mining area (WPR) Implementers of the policy will certainly be oriented to something that contains the benefits banya not only the perpetrators need to feel it, but also for society and other community groups. Therefore, it is not surprising that in this policy received a positive appreciation from many circles.

Referring to some statements submitted by the resource person, then in this research can be concluded that the implementation of government policy in the region can be said to succeed by being driven by the implementer of policy programs in the field of general mining management that has been organized as the rules apply. Although the program implementers at each level based on the interaction of researchers with each resource show different behaviors, but generally they run the policy program in accordance with the established according to the rules that apply and based on the duties and functions of each.

Sixth, Resources Committed (Resources Used);Implementation of a policy must also be supported by the resources that support implementation well. Empowerment of personnel who have competence in the management of mining of course is needed so that when doing mining activities really understand what to do. Therefore, when they finish the mining process, they do not leave a negative impact such as the excavation of excavation pits potentially causing a puddle of either the place of the nest mosquitoes and other bad possibilities bad that can affect the whole community.Referring to the results of research it can be concluded that the resources committed indicate the conditions that are naturally undertaken by the ministry of mining and energy. The resources that contribute to support in relation to general mining management activities are generally sufficient.

Seventh, Power, Interests, and Strategies of Actors Involved (Power of Interest, and Strategy of the Actors Involved) ;in a policy needs to be taken into account also the power or power, interests, and strategies used by the actors involved in order to smooth the course of implementation of a policy implemtasi. If this is not calculated carefully, it is very likely that the program to be implemented will be far away from the fire. This is intended to find out how much power, interests and strategies that are owned by the actors involved in the implementation of the policy of the regency of Konawe number 5 of 2007. In Konawe District many entrepreneurs are also made up of families of government officials so that when dealing with project management including mining it can not be denied that kinship relationships also give color in the affairs. This can be known as easily by obtaining the permission from the relevant agencies to run the business. These interests are then become a target that requires starategi pencapainnya. Many later people who when going to run the mining business handed the affairs to certain people to help make it easier in the stewardship. Likewise when there is a violation or spotlight as the existence of environmental damage, then then diupuayakan how resolved without being brought because the law. This condition then becomes a reinforcement as well as entrenched in the environment of miners in the village of Tuoy. Based on the results of research on the dimensions of Power, Interests, and Strategies of Actors Involved that power, interests and strategies run by business actors are still so great influence. That's why when found some irregularities or even violations in the management of mining so that the impact on the destruction of the environment is still strongly influenced by the existence of certain elements because it has an interest. This condition also affects the occurrence of environmental damage due to less assertive in giving sanctions due to influence of interest owners.

Eighth, Institution and Regime Characteristics (Characteristics of Regimes and Institutions);In any environment of activity in which the policy is implemented, it will affect the level of achievement because that is then what will become its benchmark. In reality, the intended conditions will certainly be influenced also by the characteristics of the regime and institutions.

The environment where the implementation of regional regulation of Konawe number 5 of 2007 on general mining management applies to the governance system in the region until now. That's why in this section, researchers need to know how the characteristics of governance are in power. In such environment where the policy implemented is very influential on its success.

In this context, the Regional Government of Konawe Regency through the Mining and Energy Agency has established a policy of contracts for processing permits for C quarry mining located in Tuoy urban village. This is intended to make it easier for business actors in the administration of administrative completeness. This policy is not in conflict with Konawe regional regulation number 5 of 2007 on public mining management.

The local government through the mining and energy agencies also involves the urban village administration in general mining management affairs managed by the dominance of the local community. It is intended that in addition to being a form of extension of the Regional Government, the village is enhanced, as well as the concern of the local government in providing additional contingent of Tuoy urban village as the point of management of sand mining.

The results of the study also obtained information that the Regional Government also provides opportunities and opportunities to non-governmental institutions that have competence in it including the community to sit together / with an opinion if there are things that need to be addressed in terms of general mining management. But on the other hand, the goodwill of the Regional Government in involving all elements in filtering aspirations in mining development, especially in the village of Tuoy municipality still less directly proportional to the level of concern of the Local Government against the impact. This condition can be known by the lack of firmness of the government in sanctioning members against the occurrence of environmental damage to the ex-mining area.

Based on the results of the analysis of the information obtained in the research in connection with the Institution and Regime Characteristics, it is known that the characteristics of the governing government are still more inclined to the utilization of available mining potential but not firmly in response to the impacts caused by mining activities, especially environmental damage the point is located in the Tuoy urban village. Therefore, the form of regional regulation no 5 of 2007 from the duties and functions of the Regional Government needs to be further developed, especially on the part of the government's concern in dealing with the environmental impact caused by mining activities.

Ninth, Compliance and Responsiveness (Compliance and Responsiveness); Another thing that is important in the implementation process of a policy is the compliance and response of the implementers, then that will be explained at this point is how far compliance and response from the implementer in response to a policy in this case the implementation of Konawe regional regulation number 5 of 2007 about the management general mining.

Administratively it can be seen that the compliance of business actors to the implementation of what is mandated in the local regulation is in accordance with what is expected. But in reality, based on the results of the research, it is still found indication of lack of compliance of business actors against the rules that have been set.

In Article 32 Paragraph (2) of Konawe regional Regulation No. 5 of 2007 mentioned that the Regional Government shall supervise and supervise in the implementation of environmental management and monitoring carried out by the license holder in accordance with the prevailing laws and regulations. The process of guidance and supervision in the management and environmental monitoring of these business actors who have become the duties and responsibilities set forth in the local regulation has been implemented but not yet maximally. When the government is obliged to deliver it is impressed because it has become imperative. Due to this condition, the response or response for the recipient of information or in this case the business actor in dealing with compliance with the rule is also lacking.

As mentioned in the above mentioned regulations, especially environmental monitoring is becoming highly visible in the insensitivity to have a sense of responsibility. The government should be firm in applying the rules so that business actors' compliance is really proven even if deemed necessary and very urgent, the government should revoke the mining permits they have when the environmental damage is getting worse.

Based on the results of research in terms of Compliance and Responsiveness note that the sensitivity of business actors to what is the point of emphasis contained in regional regulation number 5 of 2007, especially in the environmental management is considered very less. This condition is reinforced by the result of obsancation of dumping of mined excavation of mining location. Thus re-affirmed the level of compliance with what the provisions of the Local Government and the responsiveness of business actors is still very low.

4. Conclusion

Based on the results of research on Evaluation of Implementation of Regional Regulation of Konawe Regency Number 5/2007 on Public Mining Management by focusing on C class mining activities located in Tuoy Village by using model applied by Merilee S. Grindle, it can be concluded as follows:

1. The content of policy formulated by the Regional Government of Konawe Regency through the Mining and Energy Agency generally does not encourage the achievement of the management conservation of mining in observing the impacts that can be used for mining extraction.
2. Context Implementation of regulatory policies Regions on the management of public miningdoes not entirely run in accordance with the expected targets and targets.

Based on the above conclusions, for the Regional Government through the Mining and Energy Agency, non-governmental institutions, business actors and the public in connection with the implementation of Konawe regulation number 5/2007 on public mining management is recommended as follows:

1. The need to improve the supervision and management of the environment with the involvement of elements that have an influence on those interests either the government, non-government institutions or the community itself as well as the mining business actor.
2. The need for more intense efforts in raising awareness of mining business actors in Tuoy Village to be more concerned with the awareness of the importance of paying attention to aspects of the environment.
3. The need to increase the number of employees, especially those who will conduct monitoring and supervision in the mining location
4. Establish and tighten regulations on violations of regulations that have been specified, especially those concerning the issue of maintaining environmental sustainability in the mining area
5. The Regional Government needs to take a rare effort to repair the environmental damage caused by mining activities, especially by restoring the ground surface conditions that are left behind by miners, or by utilizing the excavation pits for the purpose of aquaculture tasteless

References

[1] Babbie E 2005 *The Basic Of Social Research* (Belmont: Thompson Wadsworth)
[2] Berger P L and Locman T 1976 *The Social Construction of Reality: A Treatise in The*

Sociology of Knowledge (New York: Doubleday Books)
[3] Daniel L S 200 *Evaluation Models Viewpoints on Educational and Human Service Evolution* (New York: Kluwer Academic Publishing)
[4] Fischer F 2015 *Handbook Analisis Kebijakan Publik (Teori, Politik, dan Metode)* (Bandung: Nusa Media)
[5] Grindle M S 1980 *Politics and Policy Implementation in the Third World* (New Jersey: Princeton University Press)
[6] Hasibuan P M 2016 Dampak Penambangan Bahan Galian Golongan C Terhadap Lingkungan Sekitarnya di kabupaten Deli Serdang *Jurnal Equality* **11**(1): 19-23
[7] Manheim J B, Rich R C and Willnat L 2001 Empirical Political Analysis: Research Method in Political Science, 5thEditio (New York: Longman)
[8] Sugiyono 2010 *Metode Penelitian Administrasi* (Bandung: Alfabeta)
[9] Sugiono 2011 *Metode Penelitian Kualitatif, Kuantitatif dan R&D* (Bandung : CV Alfabeta)
[10] Van M D and Van H C The Policy Implementation Process: A Conceptual Framework. Administration and Society
[11] W Gulo 2005 *Metode Penelitian* (Jakarta: PT. Gramedia Widiasrana Indonesia)

The identifications of the availability of facilities and infrastructure at the temporary evacuation of Padang

Eva Rita[1], Nasfryzal Carlo[1,*], Sugiono[1] and Yusrizal Bakar[2]

[1] Civil Department, Faculty of Civil and Planning, Bung Hatta University, Padang, Indonesia

[2] Technology Industry Department, Faculty of Technology Industry, Bung Hatta University, Padang, Indonesia

*carlovana113@ymail.com

Abstract. The building of Temporary Evacuation Sites (shelter tsunami) has been done in West Sumatra. One of them is the Padang shelter, located in North Ulak Karang Village, Padang. The location has met the requirement of PEMA 646. However, the facilities and infrastructure are still insufficient. Consequently, this study is conducted to inventory, analyze and recommended the facilities and infrastructure so that the mitigation of tsunami disaster will be optimal. The method used is descriptive evaluative, method by doing field observations. The data in the field is tabulated and analysed based on Regulation of Public Works No 19 / PRT/ M / 2001, Indonesian National Standard number 7766: 2012, and 7743: 2011. The results of the study show that the early warning facility (EWS) has functioned well, the logistics room, medical room, public kitchen, and warehouse have not got the equipment yet; clean water and electricity have not functioned. There are 22 evacuation road meeting the wide requirement of environment streets out of 71 existing streets, evacuation signs on 9 streets. To cope with tsunami disaster risk optimally, Padang shelter needs to be equipped with public facilities, additional evacuation maps, evacuation signs, and the improvement of evacuation streets to Padang shelter.

1. Introduction

The West Sumatra Province is one of the most prone earthquakes provinces in Indonesia, because it lies between the Eurasian and Indo-Australian plates [1], Semangko faults in the land and a meeting of the Australian plate and Eurasian plate at the base the ocean, making the city of Padang one of the cities with a high risk of earthquakes and tsunami [2, 3, 4].

In order to reduce tsunami disaster risk and to anticipate the victims in Padang, Disaster Management National Agency (BNPB), built the Temporary Evacuation Site (TES) in Padang started in 2013. One of the vertical evacuation sites built is Padang TES, located in North Ulak Karang Village, North Padang sub-district, Padang City, West Sumatra Province. North Ulak Karang is 1.53 km2 wide of the height of 0-2.5 m above the sea level. In North Ulakkarang e there are 6 Neighborhood units (RW) and 22 Neighborhood sub-units (RT)) with the population of 7,928 people.

Padang TES is located in RW 04 at Wisma Indah Housing Complex on Sumatera street, North Ulak Karang, Padang. The TES distance to the sea side is about ± 699 meters with capacity of 3,898 people. This distance is suitable with the PEMA 646 requirement [5]. The location is shown in Figure 1.

Figure 1. The Map location of TES Padang in North Ulakkarang Village

The main function of the test is to accommodate temporarily the neighboring community from the threat of a tsunami, so beside the suitability of the TES location, care and attention to the facilities and supporting infrastructure like early warning device, public kitchens and medical posts and minimum facilities are necessary [6]. The minimum TES facilities consist of equipment in the location of evacuation like basic medical equipment (first aid kit), sanitation, communication and emergency facilities [7].

The large number of components needed to be considered in building TES require a more detail technical plan. Hopefully, TES can reflect the priority on site, structure, capacity, and sensitivity to the environment as well as the accessibility of the user's base on the mandate of the Law No. 28 year 2002 about Building Construction [8].

The evacuation route in Padang is still limited and many evacuation roads are narrow and broken [9]. Meanwhile, the evacuation roads in North Padang is still inappropriate for evacuation route because in each tsunami evacuation route there are barrier point which are able to cause traffic jam [10]. Based on Regulation of Public Works Minister No. 19 / PRT / M / 2011-year 2011 [11], environmental roads into class III roads and can be used by the vehicles with the width of 2.1 (two point one) meters at the most, the length of maximum 9 (nine) meters, the high of maximum 3.5 meters, and the heaviest loading capacity of 8 (eight) tons. The standard of circumference roads is as shown in Figure 2.

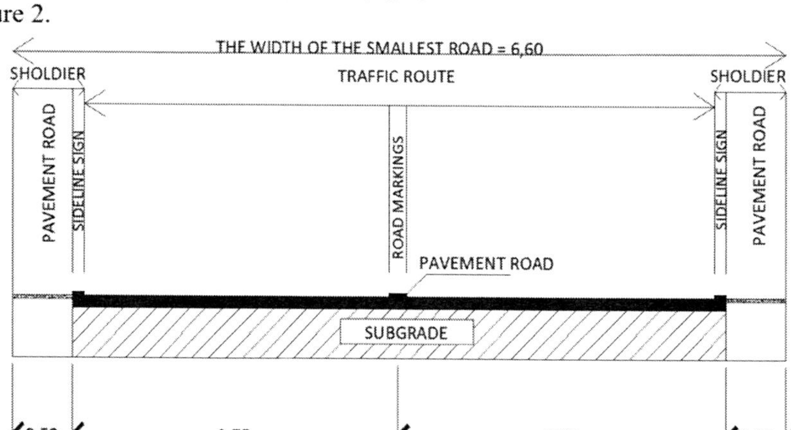

Figure 2. Standard of Circumference Road [11]

Referring to SNI 7766-2012[12], tsunami Evacuation route for the technical requirements of making the evacuation route it is necessary to determine the major route. In the prone tsunami areas, this route is a part of the escape procedures and is included in the evacuation map. Evacuation route is equipped with evacuation signs. Evacuation sign is intended to guide the people to the gathering spot away from disaster-prone areas. The standard of evacuation signs of materials, colors, shapes and sizes, symbols, sign information and the way to install according to the concepts of SNI 7743 - 2011 [13] about Tsunami Evacuation Signs. Accordingly, TES will give the sense of security for the surrounding community. If there is an emergency of tsunami disaster, TES will be ready and function appropriately. In order to function well, TES must be equipped with the appropriate the facilities and infrastructure. Therefore, this study is conducted to inventory, analyze and recommend the availability of facilities and infrastructure at the TES of Padang for the effort of tsunami disaster mitigation.

2. Research Methodology

The research methodology used is descriptive evaluative method. The Primary data are taken by direct field observation. Secondary data are obtained through documents from Disaster Management National Agency (BNPB), Disaster Management Regional Agency (BPBD), and the stake holders. The Analysis is done on the suitability of the location and the availability of facilities and infrastructure at TES of Padang. The analysis of data is done by tabulation. The result of data tabulation is analyzed based on Public the Regulations of Public Works No 19 / PRT / M / 2001 [11] and the Regulations of Indonesian National Standard No 7766: 2012, and No 7743: 2011 [13].

3. Result and Discussion

3.1. The Availability of Facilities and Infrastructures at TES of Padang

The facilities and infrastructure which are available at the Temporary Evacuation Site (TES) of Padang, subdistrict of North Ulak Karang now is as table 1:

Table 1. The Data of Facilities at TES of Padang

No	Kind of Facility	Condition Facility		
		Functional	Non-functional	Position
1.	Sirene/ EWS	√	-	4th floor
2.	Logistik Room		√	3rd floor
3.	Health Room		√	3rd and 4th floor
4.	Public Kitchen		√	3rd and 4th floor
5.	Public Warehouse		√	3rd and 4th floor
6.	Clean water Facility		√	4th floor

The availability of public facilities at TES of Padang is as follow: EWS siren is functional and logistics, medical room, public kitchen, warehouse is still empty without supporting facilities inside them, and the clean water facility is not functional, yet. With this condition, it means that this building does not have complete facilities and infrastructure appropriately based on the minimum standard for a TES. In order to make TES of Padang useful for the people during the tsunami disaster, it is necessary to provide it with the facilities stated in The Tsunami Ready Toolbox for the Indonesian Hotel Industry [7], namely:

1) Basic medical equipment (first aid kit).
2) Water and durable food must be available for 3 days.
3) Communication equipment such as loud speakers (TOA), radio using batteries, VHF Handy Talky and satellite phone.
4) Backup lighting such as flashlight, candle and matches and blankets, toilet tissue, plastic bag for garbage, rubber gloves, insect repellent, crowbars, whistle and cellophone tape

3.2. The Availability of Evacuation Route and Signs Infrastructure
The availability of evacuation route and signs to TES of Padang in the area of North Ulak Karang consist of 71 roads with 1 Provincial Road and 70 circumference Roads with 6 access to bridges and 9 functional evacuation signs for horizontal evacuation. The data of evacuation route and sign infrastructure are as shown in table 2:

Table 2. The Data of circumference Route and Sign Infrastructure

No	Infrastructure	Number	Note
1	Roads	71 roads	1 provincial road and 70 circumference roads
2	Bridges	6 access of bridges	1 construction and 5 conventional bridges
3	Evacuation Sign	9 evacuation signs	broken

3.2.1 The condition of Access Road. Based on the Regulation of Public Works Minister No. 19 / PRT / M / 2011 year 2011 about the technical requirements of roads [11], the width of the roads to be used by the community around the TES are as follow (i) there are 22 road out of 71 access roads which have met the width standard ; 18 roads are in good conditions and 4 roads are broken; (ii) there are 49 roads which have not met the width standard ; 24 roads are in good condition and 25 roads are broken. The details of road conditions are shown in appendix 1.

The main roads and the fastest alternative roads to the evacuation route of TES is Bunda Raya streets, Paus streets and Buton 1 street. However, the condition of the roads is heavily broken with the inundated water due to absence of road drainage and lack of maintenance as well as the files of garbage in some roads.

3.2.2 Evacuation Signs. There are 9 evacuation signs on the access roads. Based on SNI 7743–2011 [13], the conditions of the signs are as follow: (i) two signs are located on Paus Streets with the sign pole in faded condition, (ii) one sign is positioned on Flores streets with the sign note and sign pole in faded condition (iii) two signs are located on Sumatra street with the sign pole in faded condition; (iv) Two signs are located on Sumatra II street with the sign pole in faded condition ; (v) One sign is located on the alternative road Tim-Tim with the sign note and the sign pole in faded condition; (vi) One sign is located on the alternative road of TES of Padang with the sign note and the sign pole in faded condition. All the signs are for horizontal evacuation to direct people away from the seaside. The sign directing to or leading to the vertical evacuation of TES has not been found.

4. Conclusion
Based on the study, it can be concluded that: (i) The Public facilities at TES of Padang are still not appropriate because the only functional facility as the EWS siren, while the logistical and medical room, public kitchens, public warehouses and clean water facility have not been functional, yet. (ii) The Medical, water, communication and lighting equipment have not been available yet. (iii) There are 71 access road with 22 road have met the width requirement, while 49 roads have not met the with requirement yet, (iv) There are Evacuation signs on 9 roads out of 71 existing roads in bad conditions.

5. Recommendation
Improvement on the main access roads namely Bunda Raya Street, the alternative road of the side bank - 1, Buton – 1 street is necessary. Besides, it is urgent to add 26 evacuation signs (17 signs with notes to the TES and 9 sites with notes to the horizontal evacuation. Moreover, risk maps and evacuation maps at the TES area are essential.

Appendix A

Table 3. Condition of roads to TES Padang

No	Road	The Existing width of road (m)	The standard of widthr		The condition of road construction
			M	BM	
1.	Paus Main Street	5 - 6	√	×	Asphalted- Good
2.	Buton - 1 Street	4 - 5	×	√	Asphalted – Heavily damaged
3.	Buton 1 Alley	3 - 4,5	×	√	Asphalted – Lightly damaged
4.	Buton 2 Alley	3 - 4,5	×	√	Asphalted – Lightly damaged
5.	Buton 3 Alley	3 - 4,5	×	√	Asphalted – Lightly damaged
6.	Buton- 2 Street	4 - 5	×	√	concrete - good
7.	Flores Street	4 - 5	×	√	concrete - good
8.	Flores 1 Alley	3 - 4,5	×	√	concrete – moderate damaged
9.	Flores 2 Alley	3 - 4,5	×	√	Asphalted – moderate damaged
10.	P. Seram Street	4 - 5	×	√	concrete - good
11.	Timor Street	4 - 5	×	√	Asphalted- Good
12.	Timor-1 Street	4 - 5	×	√	Asphalted – moderate damaged
13.	Bahari Street	2,5 - 4	×	√	Asphalted – Lightly damaged
14.	S. Parman Street	7 - 12	√	×	Asphalted- Good
15.	Bunda Raya Street	5 - 7,5	√	×	Asphalted – heavily damaged
16.	Kalawi Street	3- 4,5	×	√	Soil
17.	Kalawi-1 Street	3 - 4,5	×	√	Soil
18.	Sejati - 4 Alley	3,0 - 4,5	×	√	Soil
19.	Bunda VI Street	4,0 - 6,0	√	×	Asphalted- Good
20.	Bunda V Street	4,0 - 5,0	×	√	Soil
21.	Bunda IV Street	3,5 - 5,5	√	×	Aspal - Baik
22.	Bunda III B Street	4,0 - 5,5	√	×	Asphalted – Lightly damaged
23.	Bunda III B -1 Street	3,5 - 4,5	×	√	Soil
24.	Bunda III A Street	4,0 - 5,5	×	√	concrete – Lightly damaged
25.	Bunda III A-1 Street	3,0 - 4,5	×	√	concrete – Lightly damaged
26.	Bunda III A-1 Street	4,0 - 5,5	√	×	Asphalted – Lightly damaged
27.	Bunda III Street	4,0 - 6,0	√	×	Asphalted- Good
28.	Bunda III-1 Street	4,0 - 5,5	√	×	Asphalted- Good
29.	Bunda II-1 Street	3,5 - 4,5	×	√	Asphalted- Good
30.	Bunda II Street	4,0 - 5,5	√	×	Asphalted- Good
31.	Bunda I Street	4,0 - 5,5	√	×	Asphalted- Good
32.	Riverside (BR)- 1 Street	4,0 - 5,5	√	×	Asphalted – moderate damaged
33.	Riverside (BR) - 2 Street	3,5 - 4,5	×	√	Asphalted – Lightly damaged
34.	Tim-Tim Street	4,0 - 5,5	√	×	Asphalted- Good
35.	Tim-Tim-1 Street	3,0 - 4,5	×	√	Asphalted- Good
36.	Tim-Tim-2 Street	3,0 - 4,5	×	√	Asphalted- Good
37.	Bridge - 2	3,0 - 4,5	×	√	Asphalted- Good
38.	Bridge -3	3,0 - 4,5	×	√	Asphalted- Good
39.	Sulawesi Street	3,0 - 4,5	×	√	Asphalted – moderate damaged
40.	Belitung -1 Street	4,0 - 10,5	√	×	Asphalted- Good
41.	Bangka Street	4,0 - 5,5	√	×	Asphalted- Good
42.	LPJK - 1 Alley	4,0 - 5,5	√	×	Asphalted- Good
43.	LPJK - 2 Alley	4,0 - 5,5	√	×	Asphalted- Good
44.	Belitung Street	3,0 - 4,5	×	√	Asphalted- Good
45.	Mentawai Main Street	3,0 - 4,5	×	√	Asphalted- Good

46.	Bangka Alley	3,0 - 4,5	×	√	Asphalted- Good
47.	Mentawai-1 Street	3,0 - 4,5	×	√	Asphalted- Good
48.	Sumatera Street	5,0 - 7,5	√	×	Asphalted- Good
49.	Masjid Babusalam Alley	4,0 - 5,5	×	√	Asphalted – Lightly damaged
50.	Sumbawa Street	3,5 - 4,5	×	√	Asphalted – moderate damaged
51.	Bali Street	3,5 - 4,5	×	√	Asphalted – moderate damaged
52.	Sumatera II Street	4,0 - 5,5	√	×	Asphalted- Good
53.	Sumbawa I Street	3,0 - 4,5	×	√	Asphalted- Good
54.	Sumbawa II Street	3,0 - 4,5	×	√	Asphalted – moderate damaged
55.	Bali-1 Street	3,5 - 4,5	×	√	Asphalted- Good
56.	Bali -2 Street	3,5 - 4,5	×	√	Asphalted- Good
57.	Sumbawa -1 Alley	4,0 - 5,5	√	×	concrete – moderate damaged
58.	Sumbawa - 3 Alley	4,0 - 5,5	√	×	Asphalted- Good
59.	Sumbawa - 4 Alley	3,5 - 4,5	×	√	Asphalted – moderate damaged
60.	Irian Street	4,0 - 5,5	√	×	Asphalted- Good
61.	Sulawesi Street	3,5 - 4,5	×	√	concrete - good
62.	Sulawesi - 1 Street	3,5 - 4,5	×	√	concrete - good
63.	Maluku Street	3,5 - 4,5	×	√	concrete - good
64.	Jalan Maluku - 1 Street	3,5 - 4,5	×	√	concrete - good
65.	Halmahera Street	3,5 - 4,5	×	√	concrete - good
66.	Halmahera - 1 Street	3,5 - 4,5	×	√	Asphalted – lightly damaged
67.	Kalimantan Street	3,5 - 4,5	×	√	concrete - good
68.	TES Street	4,0 - 5,5	√	×	concrete - good
69.	Riverside-1 S.Parman Street	3,0 - 4,0	×	√	Asphalted – heavily damaged
70.	Riverside-2 S.Parman Street	3,0 - 4,0	×	√	Asphalted – lightly damaged
71.	Bridge Access Jembatan	2.00	×	√	concrete – moderate damaged

References

[1] D Purbani, M P Harris, H L Salim, M Ramdhan, Y J Prihantono and L C 2015 Dewi Determination of Evacuation Line, Temple Evacuation Place (TEST) with the Capasity in the Pariaman City with Geografic Information System (GIS) *J Segara* **11**: 47-56

[2] J C Borrero, K Sieh, M Chlieh, C E Synalokis 2006 Tsunami Inundation Modeling for Western Sumatera. *Proceeding of the National Academy of Sciences of the United States of America* edited by B.A.Romanowicz **103** (52) pp 19673-19677

[3] J McCloskey, A Antonioli, A Pintanesi, K Sieh, S Stcacy, S Nalban, M Cooco, C. Giunchi, J D Huang, P Dunlop 2008 Tsunami Threat in The Indian Ocean from a Future Megathrust Earthquake West of Sumatera *J Earth and Planetary Science Letters* **265** (1): 61-81

[4] R McCaffer 2007 The Next Great Earthquake *Science* **315**: 1675-1676

[5] Rita E, Permata R, Yonne H, Carlo N 2017 Tsunami shelter in Padang city: location suitability and management issue *AIP ConfProc* **1903** (1): 110003 https://doi.org/10.1063/1.5011629

[6] Menteri Dalam Negeri 2007 Peraturan menteri Dalam Negeri Nomor 27 Tahun 2007 Tentang Pedoman Penyiapan Sarana dan Prasarana dalam Penanggulangan Bencana (Jakarta)

[7] Kesper, Alexander 2009 The Tsunami Ready Toolbox for the Indonesian Hotel Industry Bali Hotels Associates, Indonesian Ministry of Culture and Tourism.

[8] Undang-Undang Republik Indonesia 2002 Undang-Undang Republik Indonesia Nomor 28 Tahun 2002 Tentang Bangunan Gedung (Jakarta)

[9] Alhadi, Z 2014 Kesiapan Jalur dan Lokasi Evakuasi Republik Menghadapi Resiko Bencana Gempa dabn Tsunami di Kota Padang *Humanlis* **XIII** (1) 35-44

[10] Syam A 2016 Kelayakan Jalur Evakuasi Tsunami Di Kecamatan Padang Utara Kota Padang *School Leadership and Management Journal* **1**(1) 11-22

[11] Peraturan Menteri Pekerjaan Umum 2011 Peraturan Menteri Pekerjaan Umum Nomor

19/PRT/M/2011 Tentang Persyaratan Teknis Jalan dan Kriteria Perencanaan Teknis Jalan (Jakarta)
[12] Standar Nasional Indonesia 2012 Badan Standarisasi Nasional Indonesia (BSNI) SNI 7766: 2012 Tentang jalur Evakuasi Tsunami
[13] Standar Nasional Indonesia 2011 Badan Standarisasi Nasional Indonesian (BSNI) SNI 7743: 2011 Tentang Tanda Evakuasi Tsunami

AUTHOR INDEX

Institute of Physics
Dirac House, Temple Back
Bristol BS1 6BE UK

ISSN: 1755-1307
ISBN 978-1-5108-9343-6

CURRAN ASSOCIATES INC.
proceedings.com

Second International Conference on Mechanical and Aeronautical Engineering (ICMAE 2016)

IOP Conference Series: Materials Science and Engineering
Volume 187

Hong Kong, China
28 – 30 December 2016

IOP
Institute of Physics